Dairy Production & Processing

Dairy Production & Processing

The Science of Milk and Milk Products

John R. Campbell
emeritus, University of Missouri

Robert T. Marshall
emeritus, University of Missouri

WAVELAND PRESS, INC.
Long Grove, Illinois

For information about this book, contact:
Waveland Press, Inc.
4180 IL Route 83, Suite 101
Long Grove, IL 60047-9580
(847) 634-0081
info@waveland.com
www.waveland.com

10-digit ISBN 1-4786-1120-0
13-digit ISBN 978-1-4786-1120-2

Printed in the United States of America

7 6 5 4 3 2 1

Contents

Preface

Encouraged and inspired by instructors and others familiar with *The Science of Providing Milk for Man*, we decided to construct a new text that provides not only a fundamental understanding of dairy animals, dairy products, and the production aspects of each, but also a wealth of timely scientific and applied information on the scope of today's milk and milk products industry. The primary aim of *Dairy Production and Processing* is to transmit useful knowledge to those involved in the sciences, technologies, and business components of the dairy industry; we expect their application of such knowledge will lead to improved production, processing, and marketing of dairy products. The foremost sustaining purpose for having a productive dairy industry is to provide safe, high-quality milk that fulfills the nutritional needs of people of all ages globally.

Dairy Production and Processing was designed for use as a textbook for both community and four-year college students enrolled in animal science and food science courses, as well as persons seeking background, principles, concepts, and facts pertaining to the broad dairy industry. It serves as a practical reference to students and teachers of vocational agriculture, to milk producers, milk processors, and those who provide products and services to the dairy industry. Since many educational institutions provide an introductory course in dairy science, we included materials covering each aspect of production and processing of milk, as well as informative discussions regarding management and marketing as applicable to the industry and government.

Basic sciences and technologies are applied vigorously throughout the 33 chapters of this text. In order to highlight the global advancement of the dairy industry, we have included a chapter devoted to water buffalo, which rank second only to dairy cows in the total amount of milk produced worldwide, as well as chapters on sheep and goats, which also rank high in the production of both milk and meat globally for human consumption.

To guide students in learning essential principles and facts, and to enable students to test their knowledge and understanding of the materials presented, study questions are included at the conclusion of each chapter. Selected references and informative websites are provided throughout the text. An extensive glossary is provided to enable readers to expand their knowledge of selected terms.

Acknowledgments

Numerous professional colleagues have contributed to several chapters of the text. In addition, our deep gratitude is expressed to Benita Bale for her efficient and timely clerical assistance and editorial suggestions. And, we are grateful to our spouses, Eunice Campbell and Shirley Marshall, for their encouragement and support.

Finally, we thank Don Rosso, Diane Evans, and their associates at Waveland Press, who demonstrated effective leadership and initiative to revise and expand this textbook.

John R. Campbell
Robert T. Marshall

About the Authors

Dr. John R. Campbell earned three degrees and was conferred an honorary DSc from the University of Missouri, where he taught and conducted research for two decades before serving as Dean of Agriculture at the University of Illinois and as President of Oklahoma State University. Dr. Campbell has authored four and coauthored three other books and has received numerous teaching awards among other honors, such as serving as President of the American Dairy Science Association. He received the Distinguished Service Award from the University of Missouri Alumni Association in 2012. But the honor from which he derived the greatest personal pleasure was the privilege to participate in the learning experiences of more than 12,000 young people.

Robert T. Marshall, PhD is Professor Emeritus of Food Science at the University of Missouri (1960–2000). He has authored, edited, and contributed to five books. He was President of the American Dairy Science Association (1983) and the International Association for Food Protection (1984). He was Food Science Communicator for the Institute of Food Technologists (1975–1990); a committee member of the National FFA Dairy Foods Career Development Event (1962–2015) and Superintendent at the State Events (1975–2015); official judge of dairy products at the state fairs in Missouri, Illinois, and Kentucky (2000–2014); and received the Distinguished Faculty Award from the University of Missouri Alumni Association in 1996.

1

The Nutritional Contributions of Milk to Humans

The cow is the foster mother of the human race.
From the day of the ancient Hindoo to this time have the thoughts of men
turned to this kindly and beneficent creature as one of the chief sustaining forces of human life.

William Dempster Hoard (1836–1918)

1.1 Introduction

Throughout much of human history, the first food provided to newborn infants is milk. Before the development of agriculture, the sole source of milk was from women's mammary glands. If a newborn child did not have a lactating mother, the baby either suckled another woman, or died. As milk-producing animals were domesticated, their milk became available to provide essential nutrients to both infants and adults. Although humans must learn to ingest other foods, they, in common with other mammals, are born as milk drinkers. Milk has evolved through the corridors of time specifically for the nutrition and well-being of mammalian infants—to bridge the nutritional gap between the dependent intrauterine and the independent adult life.

Finding milk good, throughout the world humans domesticated various species of mammals for dairy purposes. However, the cow provides most of the *milk* in the United States. Therefore, except when noted otherwise, references to milk throughout the book will be to that of cows.

1.2 The Composition of Milk

Every person, young and old, should drink milk.
Milk contains a large variety of nutritional constituents and, considering its cost per pound, more food for the money than any other food material available.
Charles H. Mayo, MD (1865–1939)

Although milk is a liquid and often considered a drink, it contains on average 13% solids—an amount that is comparable to or greater than the solids content of many other foods (lettuce and tomatoes, for example, have a solids content of only 5 and 6%, respectively). The solids of milk contain protein, carbohydrate, fat, minerals, and vitamins. Altogether, approximately 250 chemical components have been identified in milk (about 140 individual fatty acids have been identified). These constituents vary in amount in accordance with the procedures used in the final product's preparation. Their physical and chemical properties are discussed in chapter 21.

The relationship between the composition of milks among various mammalian species and the growth rates of the respective species is an interesting phenomenon. The milk of a given species promotes the growth and development of its offspring; there is a parallel between growth rate and concentrations in milk of protein and ash, the nutrients needed to develop muscle and skeleton (table 1.1). For example, it requires about 180 days for a newborn human infant to double its birth weight, but only 47 and 8 days for the calf and puppy, respectively.

A number of minor constituents of milk—pigments, enzymes, gases, nitrogenous compounds, **cholesterol**, and others—are discussed in chapters 21 and 22.

1.3 Milk as the Most Nearly Complete Food

I have never had an outstanding successful athlete who was not a hard milk drinker.
Vincent T. Lombardi (1913–1970)

The nutritional merits of milk are indicated by the fact that daily consumption of a quart of cows' milk furnishes an average person approximately all the fat, calcium, phosphorus, and riboflavin that is needed daily, in addition to one-half the protein; one-third the vitamin A, ascorbic acid, and thiamine; one-fourth the calories; and, with the exception of iron, copper, manganese, and magnesium, all the minerals. Considerable amounts of nicotinic acid and choline are also provided.

1.3.1 Nutrients of Milk

Hippocrates recorded his principles of medical science some 400 years before the birth of Christ and is generally recognized as the father of medicine. His wisdom and his observation related to the nutritional contributions of milk to humans were conveyed in his early recorded statement that "milk is the most nearly perfect food."

To better appreciate the nutritional contributions of milk to humans, let us consider the merits of each major component, or grouping, of its nutrients.

Table 1.1 Milk Composition and Growth Rates of Selected Mammals.

	Milk Composition (%)					Time for a Newborn to Double Its Weight (Days)
Mammal	**Protein**	**Lactose**	**Milkfat**	**Ash**	**Total Solids**	
Woman	1.6	7.0	3.7	0.2	12.5	180
Mare	2.2	5.9	1.3	0.4	9.8	60
Cow	3.3	5.0	4.0	0.7	13.0	47
Goat	3.7	4.2	4.1	0.8	12.8	19
Ewe	5.9	4.8	7.4	0.9	19.0	17
Sow	4.9	5.3	5.3	0.9	16.4	18
Dog	7.1	3.7	8.3	1.3	20.4	8

Source: Adapted from Campbell et al., 2010.

Protein is the major structural component of all cells in the body. The amino acids found in protein serve as precursors for nucleic acids, hormones, vitamins, and other important molecules. The Recommended Daily Allowance (RDA) for both men and women is 0.80 g of good quality protein per kg of body weight per day. Milk provides an abundance of premium-quality protein.[1] The protein in 1 quart of milk is approximately equivalent to that in 5 ounces of meat or fish, 5 large eggs, 4 ounces of American or cheddar cheese, or 16 slices of bread.

There are two main proteins in milk—casein and lactalbumin. (Smaller amounts of other proteins, especially lactoglobulin, are also present.) Casein comprises about 80% of the total amount of milk proteins. Milk protein contains all of the essential amino acids, and in appreciable amounts. The estimated average minimal daily requirements of nearly all of the essential amino acids for adult humans can be provided by one pint of milk (table 1.2). Moreover, the relative surplus in the amino acid lysine makes milk proteins valuable in supplementing vegetable proteins, particularly those of cereals low in lysine (see section 1.3.2).

Dairy products and their milk proteins increase satiety and reduce food intake and blood glucose response when consumed alone or with a carbohydrate. Data from epidemiological and intervention studies show a modest association between milk and dairy product consumption and lowered body mass index (Dougkas et al., 2011). Milk proteins, especially derivatives of whey proteins, are more satiating than either carbohydrates or fat. Milk proteins have physiologic functions that contribute to maintenance of healthy body weight. These benefits can be achieved within the range of usual consumption of dairy products (Anderson et al., 2011). Camfield et al. (2011) concluded, based on a broad search of research reports, that lowfat dairy products, when consumed regularly as part of a balanced diet, may have a number of beneficial outcomes for neurocognitive health (e.g., dementia) during aging. Beneficial constituents are most likely bioactive peptides, anitbodies, proline-rich polypeptides, α-lactalbumin, vitamin B_{12}, and calcium.

Table 1.2 Essential Amino Acids in One Cup of Milk (3.25% Fat) versus the RDI for Adult Humans (150 lb).

Amino Acid	Amount (g)	RDI (%)
Histidine	0.17812	25
Isoleucine	0.39284	28
Leucine	0.6344	23
Lysine	0.33428	16
Methionine	0.17812	
Phenylalanine	0.35136	
Threonine	0.3416	33
Tryptophan	0.17812	64
Valine	0.45872	25

Source: World Health Organization, 2007.

As a complete food for neonates, milk provides nutrients and bioactive substances, supports development and integrity of the intestinal mucosa, establishes favorable gut microflora, and provides natural defenses against pathogens. In addition to known antimicrobial compounds (lysozyme, lactoperoxidase, and lactoferrin), there are antimicrobial peptides (AMP). Antimicrobial peptides are short chains of amino acids, which can enter cell membranes and form pores that kill bacteria. They may be produced by the organism or released from food-derived proteins (e.g., milk proteins) by digestive enzymes. Within the human gastrointestinal tract, AMP can supplement innate immunity against pathogens combined with gastric acidity, digestive enzymes, peristalsis, and the mucus layer. For neonates, who have limited intestinal microflora, protection by milk-derived AMP appears to be essential for healthy gut development, including intestinal epithelial homeostasis and mucosal healing. Several milk-derived AMP have been reported from lactoferrin, the caseins, lactalbumin, and lactoglobulin.

Caseinomacropeptide (CMP) is a glycolphosphopolypeptide fragment produced by pepsin hydrolysis of κ-casein. It is also released into whey during chymosin-induced renneting of casein micelles during cheesemaking.

Research has shown that CMP used as a food **supplement** can regulate intestinal microbial flora and prevent infections, decrease blood pressure, and act on gastric hormones. Laboratory (**in vitro**) experiments provide evidence that CMP-derived peptides have immunostimulating, anti-adhesive, antihypertensive, antithrombotic, and antioxidative properties. Other in vitro tests showed that these peptides promote growth of certain bacteria, for example *Bifidobacterium*, while destroying others, including *Escherichia coli*.

The acidic stomach environment is the first barrier against pathogens. The acidity of the human stomach is dependent on food intake (type and amount) and on time elapsed after a meal. The stomach's pH of around 2.0 can increase up to pH 5 after meal ingestion. *E. coli* are rapidly killed at pH 2, but 50% may survive after 80 minutes of exposure at pH 3.5. Since food passes the stomach relatively rapidly, many bacteria survive the acid environment and pass into the small intestine. Generation of food-derived AMP during gastric digestion could protect significantly against pathogens. However, the gastric environment may kill probiotic bacteria, which confer health benefits on the host when administered in adequate amounts.

In recent experiments of the effects of pepsin-treated CMP, bactericidal activity was observed at pH 3.5 against *E. coli*, decreasing survival by more than 90% within 15 minutes at 0.25 mg/mL (Robitaille et al., 2012). At a pH greater than 4.5, the bactericidal activity disappeared, indicating that pepsin-treated CMP was efficient at low pH only. The effectiveness of pepsin-treated CMP at pH 3.5 was not affected by the bovine or caprine origin of milk. In contrast, addition of pepsin-treated bovine or caprine CMP to the media increased the acid resistance of *Lactobacillus rhamnosus*, a **probiotic**. Viability reached 50% after 60 minutes of incubation at pH 3 compared with 5% survival in the media without added pepsin-treated CMP. This suggests that bioactive peptides released by the pepsin treatment of CMP had an antibacterial effect on *E. coli* in acidic media, but improved the resistance of *L. rhamnosus* to acid stress.

Osteopontin (OPN) is a key component of human milk, which contains approximately 140 mg/L. The content of OPN in bovine milk, from which most infant formula is made, is only 20 mg/L. Therefore, additional protein is needed in infant formula. In 2011, Arla Foods Ingredients announced the development of a patented process to extract OPN from bovine milk. Due to the tiny quantity of OPN present in bovine milk, about 70,000 gallons of whey are required to extract a kilogram of this special protein. OPN has been sold primarily to infant nutrition manufacturers in Asia. The protein is also believed to be beneficial in relation to dental health.

1.3.1.1 Lactose (Milk Sugar)

Lactose is the principal carbohydrate of milk—it is interesting that milk is the only natural source of lactose. Moreover, it is an amazing fact that the milks of all mammals contain lactose.

Most mammals stop drinking milk soon after weaning. In mammals generally, and in some human races, the gene for producing the enzyme lactase is switched off, and the ability to digest lactose is lost. In most northern Europeans, however, the gene remains active and well over 90% can digest lactose throughout their life (Sahi, 1994). Bersaglieri et al. (2004) assessed population-genetics evidence for selection of adult lactose tolerance by typing 101 single-nucleotide polymorphisms covering 3.2 Mb around the lactase gene. They estimated that strong selection occurred within the past 5,000 to 10,000 years, consistent with an advantage to lactase persistence in the setting of dairy farming; the signals of selection observed were among the strongest seen at that time for any gene in the genome.

In one study, students who were breast-fed as infants made significantly higher scores on examinations than their bottle-fed contemporaries (Campbell, 1972). Why or how could this be possible? Human milk is especially rich in lactose (7%), which coincides with the large brain size (as related to total body mass) of humans. Since lactose contains galactose, a constituent of the central nervous system (galactosides and cerebrosides form part of the nerve and brain tissue), lactose may be a brain food, a special nutrient for growth and development of the central nervous system of mammalian young.[2] Lactose constitutes 56% of the dry matter of a woman's milk, contrasted with 36 and 6% in cow's and rabbit's milk, respectively.

Another favorable feature of lactose is its hygienic value. Unlike sucrose and other simple sugars, lactose is relatively insoluble and passes the ileocecal valve to be digested and absorbed slowly from the small intestine. Its presence in the intestines stimulates growth of microorganisms that produce organic acids and synthesize many B-complex vitamins (e.g., biotin, riboflavin, nicotinic acid, and folic acid). Additionally, the high acid concentration suppresses protein **putrefaction** and also impedes growth of many pathogenic organisms because of their sensitivity to high acidity.

Lactose is readily converted to lactic acid in the intestines. Lactic acid has a favorable influence during the oral administration of antibiotics because it encourages reestablishment of desirable intestinal microflora (thus enabling synthesis of certain vitamins).[3] These organisms replace those destroyed by the oral antibiotics and thereby enhance the production of lactic acid. Physicians sometimes prescribe cultures of *Lactobacillus acidophilus* for patients who have received oral antibiotics.

The presence of lactose enhances the absorption of calcium, phosphorus, magnesium, and barium from the intestines. This unique quality of lactose also makes milk an excellent antirachitic food (i.e., preventing the disease rickets), even when milk is low in vitamin D. Vitamin D enhances the absorption of calcium and phosphorus from the small intestine and also is important in the prevention of rickets. The addition of lactose to a rickets-producing diet proved especially beneficial in the **prophylaxis** of rickets among preschool children who received minimal levels of calcium in the form of a low-milk diet supplemented with dicalcium phosphate. It is interesting that women's milk is more antirachitic than cow's milk, even though women's milk contains only one-third as much calcium and phosphorus as cows' milk, which may be explained by the higher level of lactose in women's milk (7 versus 5%).

Although lactose and sucrose have the same empirical formula, they differ in many respects. Lactose is about one-fifth as sweet as sucrose and is less soluble in water. During digestion, each molecule of lactose yields a molecule of glucose and one of galactose. The low solubility of lactose makes it less irritating to stomach and intestinal mucosa than the highly soluble sugars. This makes milk valuable in diets used in the treatment of ulcers of the stomach and duodenum (see also section 1.4). These unique nutritional contributions of lactose would not be predicted from its chemical composition.

Oligosaccharides (i.e., complex sugars) in human milk are thought to positively influence infant health by reducing the binding of pathogenic bacteria to intestinal cells, modulating immune function, and encouraging the growth of beneficial bacteria in the gastrointestinal tract. They are the third most abundant component of human milk and the only component that plays a significant role by stimulating growth of beneficial bacteria in the gastrointestinal tract. Research suggests that the biological activities of oligosaccharides in cow's milk (e.g., helping to protect infants from disease-causing organisms) may be similar to those in human milk. A recent study (Chichlowski et al., 2011) revealed that 7 of 15 oligosaccharides present in **whey permeate** had the same composition as some human milk oligosaccharides. Interestingly, cows' milk oligosaccharides from whey permeate, a by-product of cheesemaking, may be used in the future to improve the functionality of infant formulas.

1.3.1.2 Milkfat

Butterfat is found in the milk of all mammalia.
It is physically and chemically unlike any other fat in existence.
It was designed by nature for the food and sustenance of infant offspring, having the most delicate of all digestions.
William Dempster Hoard (1836–1918)

The historical aspects of the nutritional importance of milkfat[4] are generally known. Vitamin A was first discovered in milkfat, and the first severe cases of clinical **xerophthalmia** were described in children fed skim (fat-free) milk. The role of vitamin A, abundant in milkfat, in good eyesight is now well established.

Milkfat is characterized by a relatively high content (7%) of short-chain **volatile fatty acids** (mainly butyric and caproic), which are easily digested by humans and other mammals. Vegetable fats contain fewer fatty acids with less than 12 or 14 carbons than do animal fats.

Since body fat can be derived from the carbohydrates of food, one might expect that dietary fat is unimportant. On the contrary, certain fatty acids (unsaturated) are **essential fatty acids**, or dietary elements whose deficiency produces certain effects similar to vitamin deficiency. These essential fatty acids include linoleic (two double bonds), linolenic (three double bonds), and arachidonic (four double bonds) acids. Any one of these apparently can meet the body's needs for essential fatty acids. One quart of whole milk (3.25% fat) contains about 18 g of saturated, 8 g of monounsaturated, and 2 g of polyunsaturated fatty acids (PUFA) (cf. chapter 21).

Fat aids in calcium absorption, and since milk contains abundant calcium, the complementary effect of fat in whole and lowfat milks is especially important to human nutrition and health.

1.3.1.3 Atherosclerosis and Milkfat

For a number of years heart disease has been the leading cause of death in the United States. Atherosclerosis (also called hardening of the arteries) occurs when fat, cholesterol, and other substances build up in the walls of arteries and form plaque. Over time, plaque can block the arteries and cause problems throughout the body. Atherosclerosis is the major underlying cause of coronary heart disease (CHD). Furthermore, a high intake of certain saturated fatty acids (SFA) may increase concentrations of low-density cholesterol (LDL) in the blood, and LDL, the major carrier of cholesterol, is associated with increased risk of CHD.

CHD is the most common type of heart disease, killing more than 385,000 people annually in the United States. Heart disease is the leading cause of death for both men and women—about 600,000 people die of heart disease every year, which is 1 in every 4 deaths (Kochanek et al., 2011). For these reasons, many individuals advise against the consumption of milkfat, which contains cholesterol and SFA.

To determine the weight of such recommendations, let us briefly review the consumption trends of milkfat. The annual per capita consumption of milkfat decreased from 29.7 lb in 1910, to 20.6 lb in 1970, and to 15.4 lb in 1988. This reflects, in large part, the decrease in per capita butter consumption (18.3 lb in 1910, 5.2 lb in 1970, and 3.4 lb in 1988) (Raper et al., 1992). It is important to note that prior to 1990, the consumption of milk was dominated by whole milk. After 1990, the consumption of reduced fat, lowfat, and nonfat milk began to surpass that of whole milk. Skim milk is essentially cholesterol-free. Furthermore, milk is an excellent medium in which to add fats containing omega-3 fatty acids, the major deterrent being the propensity for those highly unsaturated fatty acids to oxidize. US consumers obtain about 8 to 10% of their dietary fat and 12 to 15% of their dietary cholesterol from milk and milk products (Hiza et al., 2008).

The intake of SFA is a strong predictor of concentrations of cholesterol in the blood. However, not all SFA have the same affect: palmitic (C16:0), myristic (C14:0), and lauric (C12:0) raise the cholesterol level, whereas stearic (C18:0) and medium-chain SFAs have little or no effect. Short- and medium-chain fatty acids are absorbed directly into the portal vein and not transported to the liver. Since milkfat contains relatively high amounts of stearic and short-chain fatty acids, its hypercholestrolemic effect is expected to be small compared to SFAs from meat and coconut oil (Ginsberg et al., 1992).

Considerable emphasis has been placed on the detrimental effects of trans fats in the diet. These fats contain isomers of **cis fatty acids** (cf. chapter 21). They form in vegetable fats during hydrogenation. In a recent report on effects of trans fat on cardiovascular and diabetic diseases (Tardy et al., 2011), the authors concluded that moderate intakes of trans fat from dairy foods are not a threat to cardiovascular health or blood sugar maintenance. Furthermore, the avoidance of consuming dairy products threatens the realization of a nutritionally healthy diet.

Milkfat contains conjugated linoleic acid (CLA). This is a collective term describing the geometric isomers of this 18-carbon dienoic (two double bonds) fatty acid. These double bonds, separated by only one carbon atom rather than two carbons in nonconjugated usual forms (cf. chapter 21), occur at several positions in the 18-carbon chain, but the cis-9, trans-11 isomer is predominant (75 to 90%) in milkfat. Furthermore, vaccenic acid can be converted to CLA in humans. Together these fatty acids have been found to inhibit development of tumors of various types, including those of the mammary gland and colon (Huth et al., 2006). Experiments with animals suggest 700 to 800 mg/d as an effective dose in humans, whereas human intake is estimated to vary between 140 and 420 mg/d. It is possible to enrich milkfat with CLA and vaccenic acid. Milk from cows on green pasture is high in PUFA and CLA (Ellis, 2006). However, cows fed legume **silages** (alfalfa, red clover, white clover) produced more milk with higher concentrations of PUFA, especially α-linolenic acid, than those fed grass silage (Dewhurst et al., 2003).

As a final note on this subject, Huth and Park (2012) reviewed the published research on the relationship between milkfat in dairy foods and cardiovascular health. Their findings indicated that the majority of observational studies have failed to find an association between the intake of dairy products and increased risk of coronary heart disease and stroke, regardless of milkfat levels.

1.3.1.4 Minerals

Milk is noted for its abundant supply of minerals. Moreover, they occur in milk in the right proportions or ratios for optimal absorption into blood from the digestive tract. Thus, milk is the best nutritional source of calcium, not only because of its richness in calcium, but also because of a favorable calcium to phosphorus ratio (about 1.4:1).

The mineral fraction is composed of macroelements (Ca, Mg, Na, K, P, and Cl) and microelements (Fe, Cu, Zn, and Se). Depending on their nature, the macroelements are distributed differently in the aqueous and micellar phases of milk. Ions of K, Na, and Cl are mostly in the aqueous phase, whereas Ca, P, and Mg are partly bound to the casein micelles. About two-thirds of the Ca, half of the P, and one-third of the Mg are located in the suspended (globular and micellar components) phase of milk.

The dramatic supplementary value of milk's minerals on growth of children was noted by Brody (1945) in his account of an experiment in which growing boys were given an extra pint of milk daily while receiving a diet rated as adequate. The result was that the control group (boys who received only the "adequate" diet) gained 4 lb in weight and 1.8 in in height and the milk-supplemented group gained 7 lb and 2.6 in. This growth-accelerating effect on weight and especially on height (skeletal growth) can result from skim milk (nonfat milk) as well as whole milk.[5] The richness and availability of calcium and phosphorus in both types of milk leads one to conclude that the minerals found in milk are a major factor in increased skeletal growth.[6]

Adolescence is an important period of nutritional vulnerability due to increased dietary requirements for growth and development and variable dietary habits. Calcium needs are elevated as a result of intensive bone and muscular development and thus adequate calcium intake during this period is extremely important to reach the optimum bone mass and to protect against osteoporosis in adulthood. However, most children and adolescents worldwide fail to achieve the recommended calcium intake. The hormonal changes associated with the pubertal period promote greater mineral utilization, which needs to be satisfied with adequate consumption of calcium. Diet, therefore, must contribute nutrients in sufficient quality and quantity to allow maximum bone mass development.

Ordinary human diets are more often deficient in calcium than in any other nutrient. It is especially important that children and nursing mothers receive an adequate dietary calcium intake. In the United States, about three-quarters of the population receives its dietary calcium from milk and milk products as well as one-third of its dietary phosphorus (Hiza et al., 2008). Thus, milk and milk products are practically an indispensable source of calcium.[7] In countries where milk is not available, bones are commonly consumed as such and in soups cooked in vinegar. The vinegar disintegrates the hard bone and renders its calcium available.

A curious feature having practical importance to the health of our senior citizens is the fact that the skeleton is, like other constituents of the living body, in rapid flux. Experiments using radioisotopes indicate that about one-sixth of skeletal calcium is turned over annually. Thus, the need for a consistent and reliable source of dietary calcium exists throughout one's life. Bone mass in adult life is determined primarily by an individual's peak level of bone mass, of which more than 90% is developed by age 20 and 99% by age 26. Individuals with a high bone mass at age 30 will likely be in the high bone mass group at age 70. Unfortunately, milk consumption among 20- to 25-year-olds in the United States is much too low to ensure adequate dietary calcium.

If a diet is low in calcium, calcium is taken away from the bones, which may become soft and begin to disintegrate, a condition called osteoporosis. This disease predominates in older persons, especially women. Inadequate calcium intake, probably of many years in duration, is a major contributing factor and is associated with the large number of fractured bones in older people. The surgeon general recommends three servings of milk products per day (US Department of Health and Human Services, 2004).

Findings that calcium in dairy foods decreases body weight and accelerates fat loss to a greater extent than calcium alone suggest that other bioactive components (e.g., protein) in addition to calcium in dairy foods may be involved (Zemel, 2005).

Because the health and growth of bones are of great importance to overall health and function, it is the effects of milk and dairy product consumption on chronic disease incidences that are of the greatest relevance to survival. Elwood et al. (2008) concluded that there is

> fairly clear evidence of a reduction in vascular disease and Type 2 diabetes by milk and dairy consumption.

> Taken together with the probable reduction in colon cancer and allowing for some increase in prostate cancer there is fairly convincing overall evidence that milk and dairy [foods] consumption is associated with an increase in survival in Western communities. (p. 732S)

Dietary data of more than 10,000 American adults provided an inverse relationship between dietary calcium and blood pressure levels: dietary calcium intakes of 1,000 mg/d or more were associated with a 40 to 50% reduction in the prevalence of hypertension (Huth et al., 2006). Lamarche (2008) concluded that data on the beneficial lowering of blood pressure are convincing, and the association between dairy food consumption and prevention of metabolic syndrome is promising and of great interest. Metabolic syndrome is a condition characterized by the presence of five metabolic abnormalities, including abdominal obesity, high blood pressure, and an impaired glucose and lipid metabolism, which are risk factors for cardiovascular disease. Data from a nine-year prospective study in France (Fumeron et al., 2011) found that a higher consumption of dairy products and calcium was associated with a lower incidence of metabolic syndrome, type 2 diabetes, and abnormal levels of fasting blood glucose. The study also found that cheese consumption was significantly associated with a reduced risk of metabolic syndrome.

Although milk is deficient in iron and copper, the presence of these minerals in large amounts would be destructive to certain vitamins and would catalyze oxidation, thereby producing a metallic or oxidized flavor. So, instead of having the need to consume high concentrations of these two minerals, the body provides liver stores of them at birth to carry neonates through the suckling period. The iron content of women's milk is about threefold that of cow's milk, mostly likely in relation to the longer suckling period of a human infant than a calf. Also, it is interesting that iron is recycled in the body. Therefore, it is only during growth, gestation, menstruation, significant blood loss, diarrhea, and intestinal disease that considerable dietary iron is needed.

1.3.1.5 Vitamins

Milk contains all of the known vitamins and is an especially good source of riboflavin (vitamin B_2) (US consumers obtain about one-half of their riboflavin from dairy products) and certain other B vitamins.[8] Most fluid milks are enriched with vitamins A and D so that milk and milk products provide humans with a well-balanced supply of fat-soluble vitamins (A, D, E, and K). They also provide most of the water-soluble vitamins (the B vitamins), except ascorbic acid (vitamin C). The vitamin C content of women's milk is about sixfold that of cows' milk.[9]

Can humans safely ingest an excessive amount of vitamins? With the possible exception of vitamins A and D, there is little or no danger of overconsuming vitamins. Certainly, there is no danger of excessive vitamin intake from vitamin-rich natural foods, and typically this is the form in which most people ingest vitamins. Vitamin intake should, however, increase with increasing age to compensate for reduced absorption, utilization, and storage, greater needs for detoxification, and especially because of the declining oxygen supply to the tissues. This latter need may be compensated, in part, by increasing the concentration of certain vitamins (oxidative enzymes) in the body. These facts, coupled with the abundant supply of minerals and the favorable protein to calorie ratio of milk, are important reasons why senior citizens should increase their consumption of milk and milk products.

1.3.2 Supplementary Value of Milk

It is a rational deduction from Liebig's "law of the minimum" that a deficiency of a ration in an indispensable nutrient will impair the utilization of all other nutrients and hence the net energy of the ration.

Harold H. Mitchell (1886–1966)

The above discussion of the nutritional contributions of milk to humans did not take into account the significant and unique supplementary value of milk to other dietary foods. We have already noted that milk contains numerous essential nutrients including vitamins, minerals, and amino acids. Recent attention has been given to nonessential nutrients in milk. Both human and bovine milks contain complex oligosaccharides, which, since they pass undigested into the intestines, would appear to have no function. However, they play important roles in selection of microorganisms in the intestinal ecosystem, serving as **prebiotics** (cf. chapter 23). They support growth of protective bacteria while limiting infections by pathogens. They interact with the body to heighten immunity and modify the metabolism (Newburg, 2009). Failure to acquire a beneficial microflora has been linked with allergies and chronic inflammatory diseases as well as obesity and diabetes. The oligosaccharides of bovine and human milks are closely related in structure. Many of them pass into whey during cheesemaking and through ultrafiltration membranes with permeate during concentration of whey proteins (Barile and German, 2011). The cheese industry has a significant opportunity to concentrate and market these oligosaccharides.

Milk proteins are nutritionally superior to plant proteins because they possess a better balance of amino acids. Zein (corn protein) is an example of an incomplete plant protein: it is deficient in the amino acids lysine and tryptophan. Milk protein is an excellent source of these amino acids. Therefore, the biological value of proteins in milk is 85% contrasted to that of whole corn or navy beans (cooked) at 60 and 38%, respectively.[10] The nutritional value of the proteins found in corn, potatoes, white bread, or navy beans is substantially increased when consumed with milk because milk is rich in the amino acids that are missing from these foods. This same phenomenon also may be observed in other monogastrics. For example, when growing pigs are fed only cereal, about 30% of the cereal protein is utilized for growth; but if milk is added to the diet (liquid milk equal in weight to the cereal weight), 60% of the cereal protein is utilized for growth.

The high nutritive value of milk protein makes it of special value in the treatment of **kwashiorkor**, a type of protein malnutrition commonly found among children in developing countries where people subsist on plant foods.[11] Nonfat dry milk (35.9% protein) and whey protein are used in many bakery and canned foods as protein supplements.

The supplementary value of milk is especially important to those groups (particularly the low-income group) who rely heavily on refined cereals and sugars in their diet. A high percentage of the total calories ingested by the average US consumer is furnished by sugar, other nutritive sweeteners, and flour and cereal products. In processing wheat for flour, the milling process removes about three-fourths of the copper and iron, one-half the calcium, three-fourths of the riboflavin and thiamine, and nearly all of the manganese and magnesium.[12]

1.4 The Roles of Milk and Milk Products in Daily Life

In addition to offering consumers important food nutrients at low cost, there are many other significant aspects of milk and milk products. Let us identify a few. First, we should mention the pleasing palatability or taste of milk and milk products (a mild, sweet taste). Few children dislike the mild flavor of milk, although some prefer the addition of chocolate or other flavors. Secondly, milk is essentially 100% digestible—a desirable characteristic of all foods, although few approach this degree of digestibility.[13] Additionally, there is no waste associated with milk. All of milk is edible, contrasted with the inedible bones, hides or skins, feathers, and viscera of animals from which we obtain meat. This also holds true for many vegetables, fruits, and other foods having seeds, cores, peelings, and stems. Since milk requires no mastication, it can be ingested by the young and old regardless of the status of their teeth.

Many people have observed the relaxing effects of a cup of warm milk taken just prior to going to bed at night. Although the authors have found no controlled experimental documentation of this, it is interesting that infants almost always sleep after consuming warm milk, either breast-fed or bottle-fed. It is believed that the tranquilizing effect of milk is primarily due to its high content of calcium and tryptophan. Calcium increases the force of the heartbeat and tends to dilate coronary arteries. It regulates excitability of nerve fibers and nerve centers and decreases irritability.

Milk is valuable in the control of peptic ulcers. The clinically accepted principle of no acid, no ulcer still holds. Skim milk and whole milk are equally effective since it is the protein fraction that neutralizes acids.

Of special interest is the observation that the risks of stomach cancer decrease as milk consumption increases. A study of 250,000 adult Japanese found that the stomach cancer death rate was about twice as high among persons who drank no milk as among persons who drank milk daily (Hirayma, 1968).

1.4.1 Incompatibilities of Milk and Humans

Can all persons drink milk with no ill effects? Some individuals are allergic to milk protein, especially that of cows' milk. Milks of all species contain the same types of proteins, although they differ immunologically. Thus, persons allergic to the proteins found in cows' milk may be able to ingest goats' milk without ill effects. Moreover, goats' milk has a soft curd and is *naturally* homogenized (cf. chapter 17), which makes it easier to digest than cows' milk.[14]

Certain people do not possess sufficient lactase (β-galactosidase) enzyme to digest the lactose of milk properly. This phenomenon is called **lactose intolerance** (also malabsorbance) and is discussed in chapter 21. These persons can usually digest yogurt and other cultured milk products in which bacteria have converted much of the lactose to lactic acid (hence the sour or acid taste). Moreover, most individuals, even those thought to be intolerant to lactose, can tolerate the lactose in at least a cupful of milk taken daily.

1.5 Measuring Milk Consumption and Utilization

Traditionally, per capita overall consumption of milk in all forms was computed largely on the milk equivalent basis, which means calculations were based on consumption of milkfat. As consumption of butter decreased and sales of nonfat and reduced fat fluid milks displaced sales of whole milk during the 1980s, there appeared to be a gross decrease in milk consumption. The preferred approach to accounting for utilization of milk is on the basis of valuable measurable solids, viz., milkfat, protein, lactose, and "other solids." In some markets prices paid producers for raw milk are based on this approach (cf. chapter 20).

1.6 Comparative Efficiency of Producing Milk

Although the commercial production of beef, eggs, and pork requires less human labor in proportion to food yield than does the production of milk, the dairy cow is more efficient in converting feedstuffs into edible human food than any other farm animal. It is interesting that women and goats (of the same approximate weight class) produce milk in about the same quantity and at the same gross efficiency. Lactating women, cows, and goats (high-producing) have been observed to convert about one-third to one-half of their food calories to milk calories (Brody, 1945).

The energetic efficiency of milk production is so much greater than that of the production of red meats that, in spite of the greater labor cost of milk production, the US consumer can purchase protein and food solids of milk and dairy products at a lower cost than the food solids of red meats (table 1.3). Moreover, dairy cows consume forages that poultry and swine cannot utilize, and unlike those in nonruminants, the microorganisms that reside in the rumen synthesize complete proteins and virtually all the B vitamins (cf. chapter 8). As noted in table 1.3, consumers can purchase protein and food solids (although of lower quality) at a lower cost in plant foods than in dairy products.

Many high-producing dairy cows secrete 20,000 to 25,000 lb of milk in 305 days. Moreover, they can annually provide a replacement **heifer** or a **bull**, which can be fed as a dairy steer (cf. chapter 15). A cow can provide humans with about 620 lb of high-quality protein, 1,000 lb of lactose, 740 lb of fat, and 140 lb of minerals (including 24 lb of calcium and 20 lb of phosphorus). The milk protein produced by this cow is approximately equivalent to the protein in the trimmed, edible cuts of eight 1,200 lb steers or of 28 200 lb pigs. Moreover, these 20,000 lb of milk contain enough protein to supply the needs of an adult male for about 12 years, enough calcium for 37 years, phosphorus for 30 years, riboflavin for 23 years, and energy (calories) for 6 years. Note this favorable protein to calorie ratio of milk—proportionately there is twice as much protein as energy in milk (based on human dietary needs). The protein to calorie ratio is 4:1 in skim milk, from which one-half of the calories is removed in the form of cream. Thus, milk (especially lowfat milk) is an ideal food for weight watchers.

1.7 Summary

Tell me what you eat and I will tell you what you are.
The destiny of a people depends on the nature of its diet.
J. A. Brillat-Savarin (1755–1826), *The Physiology of Taste*

The nutritional contributions of milk are significant in that it is the chief sustaining force for all mammalian young. Moreover, the components found in the milk of each species are specifically targeted for the growth and development of that species.

It is fortunate indeed for human nutrition when milk and milk products are available since they provide an excellent source of high-quality protein, an abundance of minerals in the right proportions for good absorption and utilization, and a sugar source (lactose) that is unique and available only through milk. In addition, from a food-production efficiency standpoint, the dairy cow is the most efficient converter of feed into food. Therefore, as rising populations tax the sustaining power of the land, we should turn more to milk and milk products produced by cows from plants grown on noncultivatable lands.

Milk's value as a whole is greater than just the sum of its known components. Not only do the components of milk complement each other in their digestion and assimilation, but also they supplement other foods, especially the cereals. When the total intake of food must be curtailed either to lose or to control weight, it becomes especially important to include chiefly those foods that provide generous amounts of essential dietary nutrients, yet moderate amounts of calories. Lowfat and nonfat milk are such foods. Other dairy foods (e.g., cottage

Table 1.3 Costs of Selected Foods versus Protein Content.

	Price[a] (100 g)	Protein (g)	Cost of Protein	Food Solids (g)	Cost of Food Solids
Chicken (broiler, whole)	$0.25	18.3	$1.36	35.7	$0.70
White flour (unenriched)	$0.22	12.0	$1.83	86.6	$0.25
Milk (instant, nonfat)	$0.66	35.1	$1.88	96.1	$0.69
Pinto beans (dry)	$0.44	21.4	$2.05	89.0	$0.49
Eggs (chicken, large)	$0.29	12.6	$2.30	56.0	$0.52
Pork loin (fresh, lean only)	$0.55	21.4	$2.57	27.7	$1.99
Milk (grade A lowfat)	$0.11	3.3	$3.33	10.8	$1.02
Ground beef (80% lean)	$0.75	17.2	$4.36	38.0	$1.97
Cheddar cheese	$1.10	24.9	$4.41	63.0	$1.75
Oatmeal (multigrain)	$0.55	11.3	$4.86	89.0	$0.49
Cottage cheese (lowfat)	$0.44	11.8	$8.47	11.3	$3.89
Ice cream (vanilla)	$0.44	3.5	$12.57	39.0	$1.13

[a] Based on observed retail food prices in 2012.

Source: Agricultural Research Service, 2013.

cheese and yogurt) also fit well into low-calorie diets, and their nutritional contributions are discussed in subsequent chapters.

Although growth and bone health are of great importance to health and function, it is the effect of milk and dairy product consumption on the incidence of chronic disease that is most relevant to survival. The known nutritional contributions of milk to humans prompt us to conclude this chapter by stating that the human diet throughout life should include generous amounts of milk and milk products, thereby minimizing the possibility that nutrition will be a limiting factor in the quality and length of life.

Study Questions

1. What significance did the domestication of milk-yielding animals have upon the practices of feeding infants?
2. Define fluid milk. Why should milk be considered a food rather than a drink?
3. Discuss the relationship between the composition of milk and the growth rate of a given species.
4. Compare the RDAs for protein, calcium, phosphorus, fat, riboflavin, and vitamin A and the amounts of these same nutrients found in one quart of milk.
5. What are the major proteins of milk? Describe their functions in the human diet.
6. What is lactose? Describe its functions in the human diet.
7. Why do physicians sometimes prescribe cultures of *L. acidophilus* to patients who have received oral antibiotics?
8. Why is women's milk more antirachitic than cows' milk?
9. What are the major components of milk? Describe their average percentages in women's and cows' milks.
10. Why is milk valuable in the treatment of ulcers of the stomach and duodenum?
11. Describe the importance of vitamin A to the human body. What are the best sources for vitamin A?
12. Is milkfat easy or difficult for healthy individuals to digest? Why?
13. Explain the complementary effect of milkfat in milk as related to the minerals found in milk.
14. Describe the relationships between milkfat and coronary heart disease. How do levels of intake of milk and milk products affect this disease?
15. How much additional milk should women in late pregnancy consume to assure adequate calcium for the developing fetus? For lactating women to support lactation?
16. How do people in countries not equipped with an adequate milk supply often obtain dietary calcium?
17. What is osteoporosis? Why is the consumption of milk important in preventing the disease?
18. In which minerals is milk deficient? Why is this the case? How does this affect human development?
19. What is meant by enriched milk?
20. Why are milk proteins nutritionally more desirable than plant proteins for humans?
21. What is meant by biological value?
22. Why does milk have *supplementary value* to people consuming refined cereals and sugars?
23. How valid is the suggested relationship between the consumption of milk and milk products and human life expectancy?
24. Compare the iron and vitamin C content of women's milk with that of cows' milk. What explanation can you offer for these differences?
25. What is meant by milk equivalent?
26. Why is goats' milk easier for people to digest than cows' milk?
27. Is the current relative cost of milk protein in keeping with its contributions? Why?
28. How does the dairy cow rank with other food-producing animals in converting feed into human foods?
29. What is the protein to energy ratio of milk? Of skim milk?
30. Do you believe the dairy industry should involve itself more in consumer education? Discuss.

Notes

[1] Proteins are broken down during digestion to their constituent amino acids, which are absorbed and reassembled into specific types of body proteins. Certain amino acids can be formed within the human body and apparently need not be provided preformed by foods. Some amino acids, however, are not produced in sufficient quantities, or not at all. These *essential* amino acids must be obtained from digested food proteins. A protein that contains all of these essential amino acids in appreciable amounts (e.g., milk proteins) is a high-quality protein.

[2] The human brain achieves about 70% of its adult weight by the end of the first year of life. Inadequate nutrition during infancy may prevent the brain from attaining its full potential in size. Malnutrition causes irreversible brain damage in infants and young children. Undernourished children have been found to have lower IQs than do those receiving balanced and nutritionally complete diets.

[3] It is generally accepted that biotin and vitamin K are produced by intestinal bacteria in sufficient amounts to meet human adult needs. However, sulfa drugs and certain antibiotics may destroy many intestinal microorganisms and thereby impair or inhibit their production of these vitamins.

[4] Although the term *butterfat* is often used to indicate the fat faction of milk, we consider it more appropriate to identify this component of milk as *milkfat* and have done so throughout the book.

[5] Research with rats has demonstrated that animals fed a high calcium diet gain weight about 25% faster than those fed a low calcium diet. Thus, the efficiency of dietary energy utilization (as measured by weight gains at a given dietary level) is about 25% greater on high than on low calcium dietary levels.

[6] The average mature human body contains about 6 lb of ash. The composition of this ash is approximately 40% calcium, 22% phosphorus, 5% potassium, 4% sulfur, 3% chlorine, 2% sodium, 0.7% magnesium, and other minerals in much smaller amounts. Approximately 99% of the calcium and 90% of the phosphorus in the body is in the skeleton. Of course, calcium is an important dietary essential for sound teeth as well.

[7] To provide for the calcium needs of the developing fetus during the second and third trimesters of pregnancy and for milk secretion, women should consume an additional 0.5 g of calcium daily (equivalent to the calcium in about 2.5 glasses of milk) during pregnancy and subsequent lactation.

[8] It is well accepted that food utilization is significantly depressed in a vitamin-deficient diet. This results, in part, from differences in completeness of food oxidation, and riboflavin is an essential component of perhaps a dozen flavoprotein enzymes that are vital in oxidation of foods. It is known that a decrease in dietary riboflavin is reflected in a decrease in the corresponding flavoenzymes. Perhaps this is another supplementary value of milk.

[9] This causes one to conjecture that the diets of certain species may have, through the centuries, included ample ascorbic acid, resulting in a loss (if it ever was present) of an inherent ability to synthesize vitamin C. For example, if the guinea pig evolved in the tropics, surrounded by herbage rich in ascorbic acid, survival would not depend on ascorbic acid production and, in the course of evolution, this species would lose the ability to produce it, or would not need to develop it. The same thinking may be extended to primates, who cannot produce ascorbic acid; they too probably originated in the tropics.

[10] Biological value considers the efficiency with which food nitrogen (protein) is used in the body, but fails to take into account the digestibility of that food. Milk has a digestibility of near 100%, whereas those of whole corn and cooked navy beans are 95 and 76%, respectively.

[11] Even in the United States a large number of persons are malnourished.

[12] White flour is enriched with the B vitamins thiamine, niacin, and riboflavin, and with iron. Research has shown, however, that the "enrichment" of white flour with these three vitamins and iron *does not* compensate for all losses in milling. The "enriched" bread is deficient in the amino acids lysine and valine as well as in certain vitamins and minerals present in whole wheat flour. Incidentally, oatmeal (rolled oats) is a whole grain (rather than a milled one like decorticated wheat) and, nutritionally considered, it is richer in vitamins, minerals, and protein than even whole wheat. Nutritionally fortunate, then, are children whose mothers prepare warm oatmeal and milk for breakfast rather than serving the more convenient (and more costly) sweet rolls made with white flour, or the highly advertised dry cereals.

[13] It is especially important that an infant's first food be highly digestible to safeguard his/her untried digestive system. Human milk is, of course, the best source of nourishment for the infant; its curd is exceedingly fine and flaky and easy to digest. Physicians report fewer feeding problems among breast-fed babies.

[14] When cows' milk is fed to infants, it should have the curd modified by homogenization (cf. chapter 24). Colic, distress, and vomiting may occur when infants are given nonhomogenized milk (such milk is also slower to be digested and remains in the stomach longer). Infants that suckle their mothers regurgitate less than those fed other milks.

References

Agricultural Research Service. 2013. USDA National Nutrient Database for Standard Reference, Release 26. Nutrient Data Laboratory Home Page, http://www.ars.usda.gov/ba/bhnrc/ndl

Anderson, G. H., B. Luchovy, T. Akhavan, and S. Panahi. 2011. "Milk Proteins in the Regulation of Body Weight, Satiety, Food Intake and Glycemia." In R. A. Clemens, O. Hernell, and K. F. Michaelsen (eds.), *Milk and Milk Products in Human Nutrition*. Nestlé Nutrition Institute Workshop Series Pediatric Program 67:147–159. Basel, Switzerland: Nestec Ltd. Vevey/S. Karger AG.

Barile, D., and B. German. 2011. Wheys and means: The new model for human diet and health. *Prepared Foods* 180:87–90.

Bersaglieri, T., P. C. Sabeti, N. Patterson, T. Vanderploeg, S. F. Schaffner, J. A. Drake, M. Rhodes, D. E. Reich, and J. N. Hirschhorn. 2004. Genetic signatures of strong recent positive selection at the lactase gene. *American Journal of Human Genetics* 74(6):1111–1120 [epub].

Brody, S. 1945. *Bioenergetics and Growth*. New York: Reinhold.

Camfield, D. A., L. Owen, A. B. Scholey, A. Pipingas, and C. Stough. 2011. Dairy constituents and neurocognitive health in aging. *British Journal of Nutrition* Feb. 22:1–17 [epub].

Campbell, J. R. 1972. *In Touch With Students—A Philosophy for Teachers*. Columbia, MO: Educational Affairs Publishers.

Campbell, J. R., M. D. Kenealy, and K. L. Campbell. 2010. *Animal Sciences: The Biology, Care, and Production of Domestic Animals*. Long Grove, IL: Waveland Press.

Chichlowski, M., J. B. German, C. B. Lebrilla, and D. A. Mills. 2011. The influence of milk oligosaccharides on microbiota of infants: Opportunities for formulas. *Annual Reviews in Food Science and Technology* 2:331–351.

Dewhurst, R. J., W. J. Fisher, J. K. S. Tweed, and R. J. Wilkins. 2003. Comparison of grass and legume silages for milk production. 1. Production responses with different levels of concentrate. *Journal of Dairy Science* 86:2598–2611.

Dougkas, A., C. K. Reynolds, I. D. Givens, P. C. Elwood, and A. M. Minihane. 2011. Associations between dairy consumption and body weight: A review of the evidence and underlying mechanisms. *Nutrition Research Reviews* Feb. 15:1–24 [epub].

Ellis, K. A., G. Innocent, D. Grove-White, P. Cripps, W. G. McLean, C. V. Howard, and M. Mihm. 2006. Comparing the fatty acid composition of organic and conventional milk. *Journal of Dairy Science* 89:1938–1950.

Elwood, P. C., D. I. Givens, A. D. Beswick, A. M. Fehily, J. E. Pickering, and J. Gallacher. 2008. The survival advantage of milk and dairy consumption: An overview of evidence from cohort studies of vascular diseases, diabetes and cancer. A review. *Journal of the American College of Nutrition* 27(6):723S–734S.

Fumeron, F., A. Lamri, and C. A. Khalil. 2011. Dairy consumption and the incidence of hyperglycemia and the metabolic syndrome. *Diabetes Care* 34:813–817.

Ginsberg, H. N., S. Barr, C. Johnson, W. Karmally, S. Holleran, and R. Ramakrishnan. 1992. Comparison of the effects of AHA step 1 diets enriched in dairy, meat or coconut oil products on plasma lipids and lipoproteins in normal men. *Circulation* 86(4):I–404 (abstr.).

Hirayma, T. 1968. The epidemiology of cancer of the stomach in Japan; with special reference to the role of diet. *Gann Monograph* 3:15–27.

Hiza, H. A. B., L. Bente, and T. Fungwe. 2008, March. *Nutrient Content of the US Food Supply, 2005*. Washington, DC: USDA Center for Nutrition Policy and Promotion.

Huth, P. J., D. B. DiRienzo, and G. D. Miller. 2006. Major scientific advances with dairy foods in nutrition and health. *Journal of Dairy Science* 89:1207–1221.

Huth, P. J., and K. M. Park. 2012. Influence of dairy product and milk fat consumption on cardiovascular disease risk: A review of the evidence. *Advances in Nutrition* 3:266–285.

Kochanek, K. D., J. Q. Xu, S. L. Murphy, A. M. Miniño, and H. C. Kung. 2011. Deaths: Final data for 2009. National Vital Statistics Reports 60(3) (accessed http://www.cdc.gov/heartdisease/facts.htm).

Lamarche, B. 2008. Review of the effect of dairy products on non-lipid risk factors for cardiovascular disease. *Journal of the American College of Nutrition* 27(6):741S–746S.

Newburg, D. S. 2009. Neonatal protection by an innate immune system of human milk consisting of oligosaccharides and glycans. *Journal of Animal Science* 87(13 Suppl.):26–34.

Raper, N. R., C. Zizza, and J. Rourke. 1992. *Nutrient Content of the U.S. Food Supply, 1909–1988* (Home Economics Research Report No. 50). Washington, DC: US Department of Agriculture.

Robitaille, G., C. Lapointe, D. Leclerc, and M. Britten. 2012. Effect of pepsin-treated bovine and goat caseinomacropeptide on *Escherichia coli* and *Lactobacillus rhamnosus* in acidic conditions. *Journal of Dairy Science* 95:1–8.

Sahi, T. 1994. Hypoclasia and lactase persistence. Historical review and terminology. *Scandinavian Journal of Gastroenterology* Suppl. 202:1–6.

Tardy, A. L., B. Morio, J. M. Chardigny, and C. Malpuech-Brugère. 2011. Ruminant and industrial sources of trans-fat and cardiovascular and diabetic diseases. *Nutrition Research Reviews* Feb. 15:1–7 [epub].

US Department of Health and Human Services. 2004. *Bone Health and Osteoporosis: A Report of the Surgeon General.* Rockville, MD: US Department of Health and Human Services, Office of the Surgeon General.

USDA Agricultural Research Service, Nutrient Data Laboratory. 2015, May. USDA National Nutrient Database for Standard Reference, Release 27 (http://www.ars.usda.gov/ba/bhnrc/ndl).

World Health Organization. 2007. *Protein and Amino Acid Requirements in Human Nutrition* (WHO Technical Report Series 935). Geneva, Switzerland: Author.

Zemel, M. B. 2005. The role of dairy foods in weight management. *Journal of the American College of Nutrition* Dec. 24(6 Suppl.):537S–546S.

2

The Dairy Industry

To make the most out of humans,
to raise them to the desirable level of productive capacity,
to enable them to live as they ought—they must be well fed.
The power of a person to do work depends upon nutrition.
Milk provides the surest source for human nourishment.
Wilbur O. Atwater (1844–1907)

2.1 Introduction

The foremost function of the dairy industry is to provide milk and milk products for humans. Although it is not known when people began competing with calves for cows' milk, it is believed to have been as far back as 9000 BC. There are many references in the Bible to milk being used by humans; the Promised Land was described as "a land flowing with milk and honey." Historically, the origin and naming of the Milky Way is a curiosity, but Greek myth asserts that it was formed from spilled milk.[1] Hippocrates recommended milk as a health food and medicine about 400 BC. Ever since the cow was domesticated, she has been a vital part of each progressive, developing society.

Before the invention of money, a person's wealth was measured by the number of cattle owned, so cows became an accepted medium of exchange. When coins were first made (probably about 700 BC), coin makers placed images of cows on the metal to impress skeptical people that these shiny metal objects were indeed valuable. Even today, some countries still imprint the image of a cow on their currency.

Christopher Columbus (1451–1506) brought cattle to the West Indies on his second voyage to the region. The pilgrims' failure to bring cows to the New World contributed to their high death rate, especially among the children.[2] Later, cows were required to be brought on ships at the rate of one for every six people. In 1614, the Dutch brought cows to New Netherlands, which the Indians apparently confiscated and consumed. Cows were brought to the Plymouth Colony in 1642. However, it wasn't until the eighteenth and nineteenth centuries that appreciable numbers of dairy cattle were imported into the United States.[3] It is interesting that Admiral Byrd (1888–1957) recognized the nutritional merits of milk and took cows on his second expedition to Antarctica in 1934.

In the first chapter we learned of the nutritional merits of milk. Now let us consider the cosmopolitan cow and her industry, both nationally and internationally, including factors that influence the existence, intensity, and growth of the dairy industry in a given country.

2.2 Factors Influencing the Dairying Industry Worldwide

The man who views his herd merely as dollars and cents has lost the art of husbandmanship. He who looks on his cattle merely as meat and milk has lost the art of living. There is a dignity and beauty in the scene of feeding cattle that provides a spiritual satisfaction equaling that instilled by the art of music or that of the theater. Both associate man with the meaningful things in life.

William Dempster Hoard (1836–1918)

Although not necessarily presented in order of importance, the following factors influence the development, intensity, and success of dairying within a country: climate, perseverance of the occupation within families, greater consumer demand due to urbanization, and the carbon footprint of the industry.

Dairy cows perform best in temperate climates and when given access to high-quality forages. In many areas of the world, sheep and goats are better adapted than are dairy cows to negotiating mountainous terrain in search of forage. Collectively throughout the world, cows, water buffalo, goats, and sheep harvest tremendous volumes of forage from land not suited for cultivation and thereby convert unused land into a resource that can supply nutritious, health-enhancing milk and meat.

Certain countries have been recognized as having special interests in art, some in music, others in literature, and still others in farming. People truly interested in dairying often have a special closeness to the cow and other milk-producing animals. Traditionally, the older generation has passed to the younger generation its interest in its own profession. One has only to observe dairy farming throughout western Europe to see two, three, or more generations of a family engaged in the dairy enterprise.

2.2.1 The Cosmopolitan Cow

The cosmopolitan cow and her associate mammalian mothers—dairy goats, sheep, and water buffalo—have made the dairy industry universal in scope. Of course, in one place or another, mares, camels, reindeer, and other mammals have provided milk for humans.

We can find many different forms of dairy farming in essentially all parts of the world. Methods are often primitive in developing countries compared with those of the industrialized nations. However, there is considerable variation in the intensity of dairying among countries, from one cow for every 1.1 persons in New Zealand to one for every 97.8 persons in Japan. Even allowing for differences in milk production per cow and in per capita consumption of dairy products, it is readily apparent why some countries are heavy exporters and others are heavy importers of dairy products.

Large metropolitan centers and a high degree of urbanization encourage **fluid milk** production for drinking, whereas densely populated countries often import large quantities of manufactured dairy products (cheeses, butter, dried milks). Conversely, in the more rural countries, more milk is produced and marketed as **manufacturing milk**. Examples of the latter are Australia, New Zealand, and the Scandinavian countries, all of which export substantial quantities of most major manufactured dairy products.

Livestock products produce a large "carbon footprint" compared with other foods. The major contributing factors are emissions related to feed use and manure handling. Productivity of the land required to produce the feed in question is a main variable. The farming stage is a crucial determinant of the carbon footprint of the industry because as much as 70 to 90% of the emissions in the total chain occur before products leave the farm gate. Hermansen and Kristensen (2011) suggested the following as the most important mitigation options:

- Improve feed conversion at the system level,
- Use of feeds that increase soil carbon sequestration versus carbon emission,
- Ensure that the manure produced is a substitute for synthetic fertilizer, and
- Use manure for bioenergy production.

Methane (CH_4) is produced during fermentation of wastes. Carbon dioxide (CO_2) is produced during combustion of fuel. Nitrous oxide (N_2O) is emitted from fields receiving manure and artificial fertilizers. The impact of these emissions on global warming has become a major subject of international concern and debate. Reducing herd replacement rates by maximizing herd health and reproductive efficiency significantly lowers the carbon footprint, as does increasing milk production per cow up to about 9,000 lb per year (Vellinga et al., 2011).

A significant contributor to greenhouse gas (GHG) emissions, the dairy industry is taking steps to reduce its adverse environmental impact, as it accounts for approximately 2% of all emissions in the United States. Part of these efforts includes a voluntary commitment to reduce GHG emissions of fluid milk by 25% by 2020. For example, the Clover Stornetta Farms dairy in Petaluma, California, installed a solar cogeneration system to reduce its GHG emissions, resulting in more than 50% energy savings. The system combines solar electricity generation with solar hot water production, thereby decreasing GHG emissions by 32 metric tons annually. The 50.6 kWh, 20 module solar cogeneration system provides renewable electricity and 6,000 gallons of solar hot water daily for the clean-in-place (CIP) process at the dairy plant. The system offsets about 50% of the plant's energy used for wastewater heating, displacing approximately 11,500 kWh and 2,300 therms annually.

2.3 Development of the Dairy Industry in the United States

Many factors contributed to the growth and development of the US dairy industry. Let us identify some and note subsequent chapters in which these and related topics are discussed further.

2.3.1 Early Dairy Practices

Dairying in the United States began as a small part of general frontier farming. In frontier America, cows were kept to provide milk and meat for the family, however, little extra was available for sale or trade to other settlers. Cheese and butter were usually imported from Europe. During the eighteenth century, New England farmers began to pursue commercial dairying on a limited scale. Only cheese could withstand shipment for an appreciable distance, so it was the chief dairy product available for sale. If a family did not own a cow, a member took a gallon syrup pail (or another container) to the nearest neighbor having a milk cow to obtain the family's daily milk supply.

Early American farmers had neither shelter nor grain for their cows in the winter. Only a limited amount of **hay** was available. This resulted in low production through the winter months, which meant that city dwellers had to own a cow if they wanted a reliable and regular supply of milk. Even in such towns as Boston and New York people kept cows. Of course, the populations of New York City, Boston, and Philadelphia were less than 40,000 people as late as 1790. The great and costly Chicago fire of 1871 (an estimated 250 lives were lost with $187 million in damages) allegedly resulted from Mrs. Patrick O'Leary's cow kicking a lighted lantern into straw while being milked. Can you imagine the pollution and other problems that would accrue today by having family cows in our metropolitan areas?

During the Industrial Revolution a rapidly expanding population spawned a growing consumer base to support the burgeoning dairy industry. Not only did the influx of immigrants create a need to increase production, a growing number of newborns established, in a significant way, the potential market for dairy products. For example, the introduction of condensed milk in 1856 made available dairy products that expanded markets at home and abroad and significantly enhanced growth of the US dairy industry.

2.3.2 Education and Research

Without research progress is slow. The first state agricultural colleges in the United States were founded in Michigan and Pennsylvania in 1855. Now all states have agricultural experiment stations and most of them conduct research related to the dairy industry. Through research, scientists have discovered and identified amino acids, fatty acids, vitamins, and minerals—which are all present in milk. Through consumer education programs, our growing knowledge of nutrition has helped develop the dairy industry.

Through knowledge gained in microbiology, control of microorganisms that cause disease or deterioration of milk and milk products has been enhanced (cf. chapters 23 and 24). Additionally, we have learned to utilize bacteria in the manufacture of cheeses, cultured milk products, butter, and other dairy products. Chemists have developed a wide range of tests used in determining the quality and composition of milk and milk products (cf. chapters 21 and 22).

Healthy cows are essential to profitable dairying, and improved dairy herd health has paralleled the development of veterinary medicine. Tuberculin testing of dairy cattle, for example, was introduced in 1890 (cf. chapter 13.)

The genetic improvement of dairy cattle began thousands of years ago as cattle were domesticated and selected to meet specific needs. Today's dairy breeds have been developed primarily during the past century (cf. chapter 6). Breeding high-producing dairy cattle is important, but equally essential is an adequate diet, one that provides sufficient nutrients to enable the cow's genetic milk-producing abilities to be expressed (cf. chapter 8). Improvements in breeding have also been possible with the development of artificial insemination. This highly valuable tool was made available to dairy cattle breeders in 1936. The process, coupled with production testing, young sire proving programs, and genetic testing, has provided highly significant advances in genetic improvement of dairy cattle (cf. chapter 6).

2.3.3 Technology

Developments in technology have provided many opportunities for the dairy industry and dairy farmers alike. They also expand the abilities of the dairy industry to meet consumer demand in the United States, and around the world. For example, Louis Pasteur's (1822–1895) early finding of the effectiveness of heat treatment in food preservation allowed for the transport of food products to nonlocal areas. It was adopted about 1890 by the dairy industry. In addition to extending the shelf life of milk, pasteurization destroys pathogens that might be conveyed from contaminated milk to humans (cf. chapter 23). Milk and most milk products are highly perishable. When mechanical refrigeration became available in 1861, it extended their shelf life, which has been an important contributor to the expansion of the dairy industry.

As supplies increase to fulfill consumer demands, the ability to increase production without increasing labor or costs is critical. Therefore, the invention of many types of dairy equipment for use on farms (e.g., the Thistle milking machine in 1895) and in milk processing plants (e.g., the centrifugal cream separator in 1878) has contributed substantially to the development of the dairy industry. Dr. Samuel M. Babcock of the University of Wisconsin introduced the first simple, rapid, accurate, and inexpensive test for determining the fat content of milk in 1890 (cf. chapter 22). This test provided a means of establishing fat composition and milk sale prices. The test is still used, especially in laboratories of small operators and as a reference test for instrumental testers. Homogenization, a milk processing innovation of the 1920s, breaks fat globules into minute sizes, thereby providing milk that is more easily digested by infants and others with sensitive digestive systems (cf. chapter 26).

2.3.4 Management Information

The founding of the *Journal of Dairy Science* in 1917 by the American Dairy Science Association provided a means of communicating research and scholarly writings to those in education and industry. Farm publications and scientific journals have aided the growth and development of the dairy industry. The practical aspects of dairy farming have been effectively communicated through such reliable publications as *Hoard's Dairyman* (founded in 1885).

Organizations such as the Dairy Cattle Breed Association and the Purebred Dairy Cattle Association (PDCA) have provided substantial leadership in the development of the dairy industry. Not only do their efforts further the industry, they also benefit their members as they offer support to dairy farmers and disseminate information. Production testing associations for dairy cows were first organized in Denmark in 1895 and in the United States in 1905. Production testing gives dairy farmers a solid base on which to form management decisions (cf. chapter 4). Numerous marketing organizations have been developed to help distribute and merchandise dairy products (cf. chapter 20). Dairy farms, which are overwhelmingly family owned and managed regardless of size, are generally members of producer cooperatives.

2.4 Milk Production in the United States

Milk in the United States is produced on farms varying greatly in climatic conditions, size, types of buildings, sources and types of feeds, equipment used, and overall management. Approximately 99% of milk is produced under grade A regulations described in *Grade "A" Pasteurized Milk Ordinance* promulgated by the US Food and Drug Administration. The remainder meets manufacturing grade milk standards as described in *Milk for Manufacturing Purposes and Its Production and Processing* promulgated by the US Department of Agriculture's (USDA) Agricultural Marketing Service.

Milk has a farm value of production second only to beef among livestock industries and is equal to corn in crop production. The trend since World War II has been toward fewer dairy farms with larger herds (table 2.1).

Whereas in 1960, 16.8 million cows in about 1,800,000 dairy operations were needed to produce 123.1 billion lb of milk, by 2014 only 9.3 million cows in about 50,000 operations produced 206 billion lb of milk (figures 2.1 and 2.2). Farms with more than 2,000 cows produce about 34.7% of the total milk supply (table 2.2). Cash receipts from milk marketed during 2014 totaled $49.3 billion. Producer returns for milk averaged $24.07 per hundredweight. Milk used on farms totaled 962 million lb with 90% being fed to calves.

Although confined animal operations are not new, their use in livestock farming has been increasing. During

Table 2.1 Numbers of Cows, Dairy Farm Operations, Average Number of Cows per Operation, and Total US Milk Production, 1950–2010.

Year	Number of Cows (1,000,000)	Number of Operations (1,000)	Cows per Operation	Pounds of Milk (1,000,000,000)
1950	21.3	3,681	6	116.6
1960	16.8	1,837	9	123.1
1970	12.0	648	19	117.0
1980	10.8	334	32	128.4
1990	10.0	203	52	147.7
2000	9.2	105	88	167.7
2010	9.1	65	229	188.9

Source: USDA Economic Research Service, *Dairy Yearbook*.

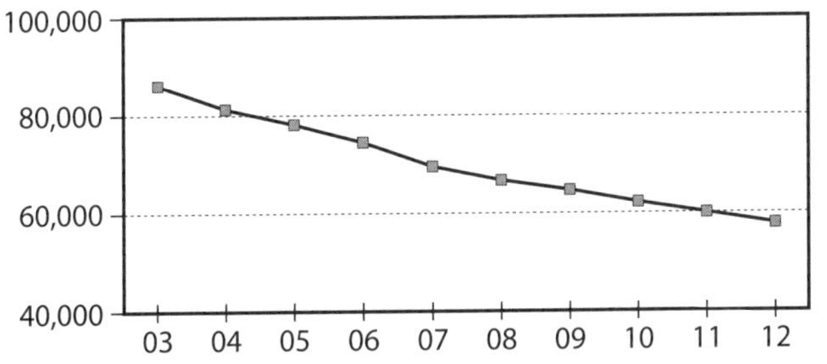

Figure 2.1 Changes in numbers of milk cow operations in the United States, 2003–2012 (USDA National Agricultural Statistics Service, 2013).

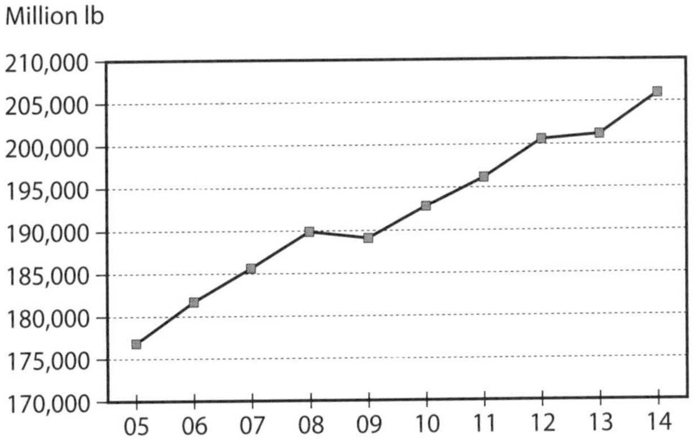

Figure 2.2 Changes in quantities of milk produced in the United States 2005–2014 (USDA National Agricultural Statistics Service, 2015).

the 20-year period 1987–2007, the mid-aggregate number of dairy cows per farm increased from 80 to 570. This manner of classification places half of the cows of smaller and larger herds equally on each side of the mid-aggregate number. The increasing dominance of confined animal feeding operations for many types of livestock and a growing reliance on production contracts have contributed to the growth of large, specialized poultry, swine, and dairy operations. Changes in the relative prices of

Table 2.2 Number of Operations, Percent of Inventory, and Percent of Milk Produced by Group Size, 2012.

Number of Head	Number of Operations[a]	Inventory (%)	Production (%)[b]
1–29	18,800	1.6	1.0
30–49	9,700	4.3	3.2
50–99	14,500	11.3	9.5
100–199	7,900	11.8	10.7
200–499	3,800	12.5	12.6
500–999	1,570	11.9	12.4
1,000–1,999	950	14.0	15.9
2,000+	780	32.6	34.7
Total	58,000	100.0	100.0

[a]An operation is any place having one or more head of milk cows on hand on December 31.
[b]Percents reflect average distributions of various probability surveys conducted during the year.
Source: USDA National Agricultural Statistics Service, 2012.

land, labor, and capital over the past three decades may have encouraged the substitution of less expensive capital (in the form of more mechanized animal housing, feeding, and manure management facilities) for more expensive land and labor.

Milk production is concentrated largely in three regions of the United States: Pacific, Lake, and Northeast. This is further reflected in production records by state, which show the leaders as California and Wisconsin (table 2.3 on the following page). Interestingly, states in the Mountain region lead in production per cow with annual averages of more than 24,000 lb. The highest average production per cow in 2014 was in the state of New Mexico (25,093 lb). The top 10 states produce about 74% of the nation's milk (table 2.3).

For many years, northeast Texas with its grass-based dairy system dominated the Texas dairy industry. As herds grew larger in confined operations, milk production began to concentrate in the High Plains where average temperatures and low humidity meant cows spent

Table 2.3 Ten Leading States in Milk Production in 2014.

State	Milk Produced (1,000,000 lb)	Number of Cows (1,000)	Milk per Cow (lb)
California	42,337	1,780	23,785
Wisconsin	27,795	1,271	21,869
Idaho	13,873	575	24,127
New York	13,733	615	22,330
Pennsylvania	10,683	530	20,157
Texas	10,310	463	22,268
Michigan	9,609	390	24,638
Minnesota	9,127	460	19,841
New Mexico	8,105	323	25,093
Washington	6,584	273	24,117
United States	206,046	9,257	22,258

Source: USDA Economic Research Service, 2015a.

Table 2.4 Changes in Number and Annual Processing Volume of Fluid Milk Plants in the United States, 1950–2010.

Year	Number of Plants	Average Volume (million gallons)
1950	8,195	4.3
1960	5,328	8.8
1970	2,216	23.6
1980	1,066	50.1
1990	605	93.9
2000	405	143.2
2010	332	189.2

Source: USDA Economic Research Service, 2011.

much of every year in their "**thermoneutral zone**," so they could expend less energy for maintenance and more for milk production. In addition, irrigated crops provide a consistent forage supply and the rail system enables economical shipping of grain from the Midwest. Although in 2000 milk production in eight High Plains counties represented only 4% of the state's production, in 2010 it accounted for 54% of the total.

For record purposes states have been divided into 10 regions by the USDA. Below is a breakdown of each region; the approximate percentage share of milk per region is shown in parentheses (USDA Economic Research Service, 2015b).

Pacific: WA, OR, CA, AK, HI (25%)
Lake States: MI, WI, MN (23%)
Mountain: MT, ID, WY, CO, NM, AZ, UT, NV (16%)
Northeast: ME, NH, VT, MA, RI, CT, NY, NJ, PA, DE, MD (15%)
Corn Belt: OH, IN, IL, IA, MO (8%)
Southern Plains: OK, TX (5%)
Northern Plains: ND, SD, NE, KS (3%)
Appalachian: VA, WV, NC, KY, TN (2%)
Southeast: SC, GA, FL, AL (2%)
Delta States: MS, AR, LA (.24%)

2.5 Main Branches of the Dairy Industry in the United States

Milk can be used to produce a very large number of products, as chapters 28–33 reveal. Here we highlight those segments of the industry that utilize most of the milk supply. It is important to note that the number of dairy foods plants has decreased greatly over the past 60 years. Of course, the volume of milk processed per plant has increased due to large increases in the numbers of consumers and fewer plants to supply that demand. For example, in 2010, only 4% of the number of existing processing plants compared to those that were in operation in 1950 processed more than 45 times as much milk (table 2.4).

2.5.1 Fluid Milk

The fluid milk (**market milk**) branch of the dairy industry processes and distributes beverage milks, creams, and cultured milk products. Large population centers and dependable transportation are essentials to its growth. Total beverage milk consumption in the United States has ranged between 51 and 55 billion lb from 1975 through 2010. However, during those years, sales of whole milk (3.25% milkfat minimum) dropped from 68 to 28% of beverage milks while sales of lower fat milks (1 and 2% milkfat) increased from 22 to 48% and sales of skim (nonfat) milk rose from 4.7 to 15%. Other classes of beverage milk products include flavored milks and cultured buttermilk whereas eggnog, yogurts, and cream products (including sour cream) are grouped among total fluid milk products. Sales of yogurt increased from 425 million lb in 1975 to 4.65 billion lb in 2013. Similarly, sales of sour cream increased from 350 million lb in 1975 to nearly 1.3 billion lb in 2013. Americans consumed a per capita average of 189 lb of fluid milk products, including eggnog and yogurt, in 2013.

2.5.2 Cheese

This growing industry thrives in the same general conditions as the butter industry except for cottage cheese, which is more perishable and follows more nearly the conditions favoring production of fluid milk. In 2014, about 47% of all cheeses produced in the United States were manufactured in Wisconsin and California, 2.9 and 2.4 billion lb, respectively. Concurrently, Idaho and New York produced 895 and 785 million lb, respectively. Worldwide more than 400 varieties of cheese are produced (cf. chapter 30). Cheeses are subdivided into natural and processed types. The latter consist of natural cheeses that have been given some kind of physical and/or chemical treatment. In 2014, 27 plants in the United States produced about 1.2 million lb of processed cheeses.

American cheeses (cheddar, colby, monterey, and jack) comprise about 40% of total cheeses produced in the United States. Cheddar makes up 72% of American cheeses. Wisconsin, Minnesota, and Idaho lead in cheddar production while producing 567, 547, and 450 million lb, respectively. Wisconsin leads all states in number of plants (60) making American cheeses. California is second with 20.

Production of mozzarella (pizza) cheese totaled 3.9 million lb in 2014. It now exceeds that of cheddar with leading states being California (37%) and Wisconsin (25%).

Per capita consumption of cheeses in the United States increased as follows from 1975 to 2013: American type, 8.1 to 13.4 lb; other cheeses except cottage, 6.1 to 20.1 lb. Thus, the average U.S. citizen consumes about 33 lb of cheeses annually. Concurrently, per capita consumption of cottage cheese fell by about 50% from 4.6 to 2.1 lb. New York and California led in production of cottage cheeses and yogurt. Consumer purchases were 57% reduced fat (lowfat and nonfat) and 43% regular creamed cottage cheese.

2.5.3 Frozen Dairy Desserts

The frozen desserts industry of the United States was relatively small until the ice cream cone was introduced at the 1904 World's Fair in St. Louis, Missouri. Improved refrigeration and transportation were vital to its growth. Ice cream is the major frozen dairy dessert and it is classified as regular, lowfat, and nonfat (cf. chapter 31). Other frozen products are sherbets, frozen yogurt, and water and juice ices. About 1.5 billion gallons of frozen dairy desserts were produced in 2014, including both the hard frozen and soft frozen forms. Most are hard frozen. Market shares of regular, lowfat, and nonfat ice creams constitute approximately 57, 42, and 1%, respectively, of the total frozen dessert market. California, Pennsylvania, and Missouri lead the states in production of regular hard ice cream, whereas Texas and California lead in production of lowfat ice cream.

2.5.4 Butter

The butter industry originally developed to the greatest extent in areas of the United States with inexpensive land, good pastures, and homegrown feeds. However, high amounts of surplus cream in California, where land costs are high and drylot management prevails, have raised it to the major producing state with 613 million lb in 2014. Pennsylvania is second in production with 94 million lb. Total production of butter in the United States was 1.9 billion lb in 2014. The fat from approximately 22 lb of milk is required to make 1 lb of butter.

2.5.5 Dry Milk Products

These products include regular and instantized nonfat dry milk, dry whole milk, dry buttermilk, dry **whey**, and dry whey products, including highly nutritious whey protein. The high solubility of instant dry milks and significant international markets create a strong demand for dry milk products. The same general conditions that favor production of butter and cheese also favor the production of dry milk products; therefore, California and Wisconsin commonly lead the United States in production of dry milk and dry whey, respectively.

2.6 Importance of the Dairy Industry to the US Economy

The dairy industry is recognized as one of the largest and most important agricultural industries in the United States—both economically and nutritionally. Economically, it accounts for about 9.4% of total agricultural commodities in the United States (table 2.5). Its nutritional contributions to consumers were noted in chapter 1.

Table 2.5 Top Five Agricultural Commodities of the United States, 2014.

Commodity	Total Receipts ($1,000)	Percent of US Total Value
Cattle and calves	81,251,415	19.3
Corn	53,679,393	12.8
Dairy products, milk	49,349,226	11.7
Soybeans	41,283,575	9.8
Broilers	32,724,667	7.8
All commodities	420,145,646	

Source: USDA Economic Research Service, 2015c.

2.6.1 Export and Import of Dairy Products

Annual exports from the United States of nonfat dry milk, dry whey products, and lactose range from 40 to 75% of amounts produced. The value of US exports of dairy products was $7.1 billion in 2014. Mexico, Canada, China, Philippines, Japan, and South Korea are the largest importers of US dairy products. Mexico is the principal importer of US cheese and skim milk powder. During the three-year period 2006–2009, prices of butter, nonfat dry milk, and cheese on the international market varied widely on a US dollar per pound basis: butter, $0.77–$2.20; nonfat dry milk, $0.80–$2.45; cheddar cheese, $1.20–$2.50. These price fluctuations reflect changes in milk production, drought in Australia and New Zealand, and a severe world economic downturn beginning in 2008. The inability of the dairy industry to quickly respond on the supply side to changes in demand resulted in wide variations in the global market for these products, which comprise the bulk of imported and exported dairy foods.

The importation of dairy products is regulated by the US government and is especially important economically to dairypersons. The major product imported to the United States is cheese, including cheese curd. Total value of cheese imports averages about four times the value of US exports and 3.2 times their weight. The United States is a major importer of **casein** and milk protein concentrates, annually about 600 and 180 million lb of each, respectively. Only small amounts of these products are produced domestically. The total value of US dairy imports in 2013 approximated \$2.65 billion; imports of cheese and related products were 43.2% of the total.

The Foreign Agriculture Service of the USDA provides data on milk production and international trade for selected countries. The Dairy Export Incentive Program (DEIP) helps exporters of US dairy products meet prevailing world prices for targeted dairy products and destinations. Under the program, the USDA pays cash bonuses to exporters, allowing them to sell certain US dairy products at prices lower than the exporter's costs of acquiring them. The major objective is to develop export markets for dairy products where US products are not competitive because of the presence of subsidized products from other countries.

The DEIP was announced by USDA on May 15, 1985, and was reauthorized by the Food, Agriculture, Conservation, and Trade Act of 1990; the Uruguay Round Agreements Act of 1995; the Federal Agriculture Improvement and Reform Act of 1996; the Farm Security and Rural Investment Act of 2002; and the Food, Conservation and Energy Act of 2008. As part of its World Trade Organization commitments resulting from the Uruguay Round Agreement on Agriculture, the United States has established annual export subsidy ceilings by commodity with respect to maximum permitted quantities and maximum budgetary expenditures. Readers should note that economies of the world depend not only on product attributes and competitive pricing, but also on agreements among nations as well as legislation within those nations.

2.6.2 Exports of Dairy Animals

World demand for dairy cows and heifers is strong with much of the strength coming from oil-rich Middle Eastern countries. During the period 2010–2014 exports from the United States of female dairy cattle averaged 55,000 with a variation of about 20,000 per year. USDA Foreign Agricultural Service records show exports to Mexico, the largest market for US dairy cattle over most of the past decade, surged from about 1,800 in 2007 to more than 12,000 in 2009. And, for one three-year period (2006–2008), exports to Saudi Arabia actually outpaced those to Mexico, averaging nearly 4,300 head per year during that period. China, also eager to grow and enhance its dairy industry, has been buying cattle from Australia and New Zealand. A big boost to US cattle exports has been an end to live cattle subsidies by the European Union. Strict environmental laws and a shrinking agricultural land base have diminished Europe's export potential. A significant obstacle to cattle exports is the small number of large transport ships for livestock in the world (only 13 in 2010). Demand for them included all types of livestock.

Exports depend not only on available purchasing power but also on political policy and disease. A resurgence of bovine spongiform encephalopathy (BSE) (cf. chapter 13) or other contagious disease would close the doors to exports, much as it did for Canada and the United States in recent years. After an outbreak of bluetongue in Europe a few years ago, several countries closed their borders to imports of cattle from European Union countries.

2.7 World Production of Milk

It seems agreed that most people are alike in needing milk; unfortunately, they are not alike in their ability to get it, and the basic problem is how to bring it within their means.

Samuel Brody (1890–1956)

The estimated annual world milk production by cows in 2009 was about 1,287 billion lb (585,234,624 metric tons), or about 190 lb (86 kg) per capita. Total milk production from all species was estimated at 703,998,075 metric tons with the following amounts (metric tons) being produced by other ruminants: buffalo, 92,140,146; goat, 15,510,411; sheep, 9,272.693; and camel, 1,840,201 (http://chartsbin.com/view/1492).

If distributed evenly, this would provide each person of the world with approximately one-half liter of milk daily (about 50% of the minimum for dietary needs). However, milk production is not uniform worldwide. Instead, it is concentrated in the temperate climate areas. An estimated 85% of the world's milk supply is produced in the regions whose peoples represent only about one-third of the world population. Thus, people of these countries are provided with some 500 lb of milk per capita annually in contrast with other more heavily populated areas in which annual per capita milk consumption averages only about 60 lb (28 qt).

The United States ranks second in total milk production to the European Union, which is composed of 28 states; however, the United States leads all countries in average production per cow (table 2.6). About one-third of US milk is consumed in fluid products, in India it is 99%.

New Zealand utilizes approximately 2% of its domestic milk production as fluid milk. This readily explains why that country is an exporter of dairy products. Even though they produce only about 18% as much

Table 2.6 Milk Production Statistics for Major Producing Nations.

Country	Number of Cows (1,000)	Production: Total (1,000,000)	Production: per Cow (metric ton/head)
EU-28	23,052	139,100	6.03
United States	9,220	91,444	9.92
India	48,250	57,500	1.19
China	8,380	34,500	4.12
Brazil	20,450	32,380	1.58
Russia	8,515	31,400	3.69
New Zealand	5,043	19,678	3.9
Argentina	2,193	11,796	5.38
Mexico	6,300	11,270	1.79
Ukraine	2,554	11,160	4.37
Australia	1,650	9,570	5.8
Canada	961	8,535	8.8

Source: USDA Foreign Agricultural Service, 2013.

milk as do farmers of the United States, they exported nearly as much nonfat dry milk during 2005–2009, 270 versus 294 metric tons. Concurrently, New Zealand exported an average of 652 metric tons of whole milk powder, whereas the United States exported only about 10 metric tons. New Zealand typically exports nearly 400 metric tons of butter, whereas the United States exports varied from 6 to 89 metric tons during 2005–2009.

In 2010, the Fonterra Cooperative Group with headquarters in Auckland, New Zealand announced (in a joint venture with Nestlé as the fresh milk processor) plans to establish in Brazil two pasture-based dairies with 3,300 cows. The climate in the area allows year-round farming on the pastoral model. The cooperative developed a third large dairy farm in China in 2012. This illustrates how the milk production industry is developing on an international scale.

In the early years of the twenty-first century, the Chinese began building their dairy cattle herd by importing dairy heifers and bovine semen. Whereas in 2006 they imported fewer than 10,000 animals, by 2011 they imported nearly 103,000. Since 2002, China increased in rank from 45 to 9 in numbers of bull semen vials imported from the United States, accepting 366,000 vials in 2011.

2.8 Sustainability of Dairy Agriculture

Humans cannot survive without food production from agricultural lands, which comprise about one-third of earth's land mass. With the world population expected to increase about 50% by 2050, global demand for food will increase greatly. However, demand for land for infrastructure, housing, and businesses will also increase. The issues of climate change and energy use are prompting consumers to ask how their food choices may affect the environment, the safety of themselves and society, nutritional quality, and a sustainable planet. Agricultural lands must be used in ways that preserve them for future generations while meeting the needs of the current population.

Sustainable agriculture has been defined as meeting the needs of the present while improving the ability of future generations to fulfill their needs by focusing on increasing productivity for nutrition while enhancing the environment and ensuring sustainability, profitability, and high-quality life for farm families.

One type of tool that can be used to evaluate the success of the sustainability of dairy agriculture is the Vital Capital Index, which describes the quality and quantity of vital capital provided by farms using the following categories (Whitman et al., 2010):

- *Sustainability stewardship:* economic performance, management, regulatory compliance, certification systems/verification/audits, and sourcing/supply chain.
- *Social stewardship*: employee conditions, worker relations/quality of life, worker health/safety, worker rights, product quality/safety, and traceability/origin.
- *Environmental stewardship:* water availability/use/protection, soil management, biodiversity, waste management/recycling, animal health/husbandry, energy, use of chemical/synthetic substances, pest management, animal welfare/humane treatment, greenhouse gas emissions/air quality, and landscape/land use.

Another factor in sustainability issues is the carbon footprint of the industry. Improvements must be made to allow for the growing issues involving the industry, the environment, and the consumer. For example, to reduce GHG emissions from dairy operations, research and production practices are emphasizing: (1) microbial functions in the rumen, (2) feed quality and digestibility, (3) milk yield per cow, and (4) lifetime production per cow.

2.9 Summary

Largely because of its nutritional merits, milk has provided nourishment in many different cultures. In addition, the cow and her products have been used for religious and cosmetic purposes. The cow has held religious significance in India for well over 4,000 years. Tibetans, for example, use butter as hair oil, and in Switzerland and Germany whey was a common medicine at the turn of the twentieth century.

Notwithstanding its religious and cosmetic significance, milk has been recognized and respected by humans for centuries as an important source of nutrients important to good health. Furthermore, it can be processed into numerous foods, and even nonfood products. This is the basis of the essential, stable dairy industry.

However, there is significant variation in the availability of milk. Millions throughout the world eat inadequately and because of increasing population, there is the risk of less availability in the future.

Variations in the amount of milk produced per person coupled with their ability or inability to purchase dairy products result in substantial differences in per capita milk consumption worldwide. Additionally, these same factors influence the import and export of dairy products in a given country.

During the balance of this century, the dairy industry will receive increased attention as developing countries expand the agricultural output of their economies and as humans seek to improve their nutritional status.

Study Questions

1. What is the foremost function of the dairy industry?
2. What early association was made between cows and coins?
3. What lesson did the first American settlers learn about their need for milk?
4. Identify and discuss factors that influence the development of dairying in a given country.
5. Cite some of the technological options available for mitigating the carbon footprint of dairying enterprises.
6. What three important greenhouse gases are produced during milk production, processing, and distribution?
7. Name the top three regions in the United States in the production of milk. Discuss some of the reasons why states such as California, Wisconsin, and Texas are milk production leaders.
8. Name two US regulatory documents under which the dairy industry operates in producing grade A and manufacturing grade milk.
9. What major factors are responsible for the consolidation of the dairy industry into large farms and processing facilities?
10. What are the main branches of the US dairy industry? In what areas and conditions does each thrive?
11. Of what significance is the dairy industry to the economy of the United States?
12. To which countries do US farmers commonly export the largest numbers of dairy cattle?
13. Name the dairy products most commonly involved in international trade. Which product is the number one import into the United States? Which countries import the most US dairy products?
14. What is the rationale for economic support of the Dairy Export Incentive Program (DEIP) by the USDA?
15. What is the significance of the variation in intensity of dairying in a country to their imports and exports of dairy products?
16. In which country is 2% of the milk produced used in fluid form? In which country is 99% of the milk produced used in fluid form?
17. In regard to sustainability of dairying:
    a. Define sustainable agriculture.
    b. List the categories in the Vital Capital Index.
    c. Name four practices that are emphasized to reduce greenhouse gas emissions.

Notes

[1] In Greek myth Heros dropped some milk while wet-nursing Maia's son Hermes, and that spilled milk allegedly formed the Milky Way.

[2] Of the 102 persons who came on the Mayflower, 42 (including almost all of the children under two years) died the first winter.

[3] American statesmen and orators Daniel Webster (1782–1852) and Henry Clay (1777–1852) were among the prominent early dairy farmers of the United States who imported dairy cattle.

References

Hermansen, J. E., and T. Kristensen. 2011. Management options to reduce the carbon footprint of livestock products. *Animal Frontiers* 1(1):33–39.

USDA Economic Research Service. 2011. Number and Size of Milk Bottling Plants Operated by Commercial Processors (http://www.ers.usda.gov/data-products/dairy-data.aspx).

USDA Economic Research Service. 2015a, May 1. Milk Cows and Production by State and Region (http://www.ers.usda.gov/data-products/dairy-data.aspx).

USDA Economic Research Service. 2015b, May 8. State Fact Sheets: United States. Top Commodities, Exports, and Counties (http://www.ers.usda.gov/data-products/state-fact-sheets/state-data.aspx#Pf330ad6e68864058beca4e8d94bf2921_3_586iT21R0x0).

USDA Economic Research Service. 2015c, September 14. Top 5 Agricultural Commodities, 2014 (http://www.ers.usda.gov/data-products/state-fact-sheets/state-data.aspx?StateFIPS=00#Pe751828fd8524f31b7f672ee6294cce5_2_586iT21R0x0).

USDA Foreign Agricultural Service. 2013, December. *Dairy: World Markets and Trade.* Washington, DC: Author.

USDA National Agricultural Statistics Service. 2012. *Agricultural Statistics.* Chapter 8, Dairy and Poultry Statistics. Washington, DC: Author.

USDA National Agricultural Statistics Service. 2013, February 19. Milk Cows: Number of Operations by Year, US (http://www.nass.usda.gov/Charts_and_Maps/Milk_Production_and_Milk_Cows/cowoper.asp).

USDA National Agricultural Statistics Service. 2015, February 20. Milk: Production by Year, US (http://www.nass.usda.gov/Charts_and_Maps/Milk_Production_and_Milk_Cows/milkprod.asp).

Vellinga, T. V., M. H. A. de Hann, R. L. M. Schils, A. Evers, and A. van den Pol–van Dasselaar. 2011. Implementation of GHG mitigation on intensive dairy farms: Farmers preferences and variation in cost effectiveness. *Livestock Science* 137:185–195.

Whitman, A., G. Clark, and J. Beane. 2010. *Measuring and Managing for Sustainability in Dairy Agriculture: The Vital Capital Index and Toolkit—Final Report* (Natural Capital Science Report NCI-2010-06). Brunswick, ME: National Capital Initiative, Manomet Center for Conservation Sciences.

Websites

DairyBusiness.com (http://dairybusiness.com)
Hoard's Dairyman (http://www.hoards.com)
Innovation Center for US Dairy (http://www.usdairy.com)
Purebred Dairy Cattle Association
(http://www.purebreddairycattle.com)
US Dairy Export Council (http://www.usdec.org)

For Further Study

1. As a further supplement, the authors recommend readers visit the website of Dairy Management, Inc. and take the Dairy Farming Today virtual tour (http://www.dairyfarmingtoday.org/Dairy-Interactive/Videos/Pages/VirtualTour.aspx).
2. The International Dairy Foods Association (IDFA) publishes annually *Dairy Facts*, a booklet that provides important information regarding the overall impact of the dairy industry (http://www.idfa.org).
3. To learn more about milk production in the United States, visit the National Agricultural Statistics Service website (http://www.nass.usda.gov). Data are provided for inventory, operations, and production by month, year, and group.

3

Breeds of Dairy Cattle

The dairy cow means more to human welfare than any other animal,
save man himself. We would be far better off as a people
to forego many of the aspects of our affluent life than
to let milk and its products become of less importance.

Glenn. S. Pound (1914–2010)

3.1 Introduction[1]

Dairy cows and their product, milk, have made important contributions to the health and standard of living of humans. Throughout the centuries the cow has fulfilled many functions for humans. In addition to providing milk and meat, she has served as a sacrificial victim, draft animal, and beast of burden. In the Stone Age, cows were hunted along with bison, reindeer, and the prehistoric wild horse. Cavemen captured aurochs (ancestors of our dairy cattle) in pits and hunted them with spears and arrows. Early artworks depicting hunting scenes of this kind have been preserved in the caves of Spain, France, Swaziland, and elsewhere.

The oldest domestic ox yet discovered (in northern Mesopotamia) is estimated to be over 9,000 years old. The aurochs were not domesticated in Europe until about 3500 BC. A relief from the eleventh dynasty in Egypt (2100 BC) depicts a cow crying because she is being deprived of milk intended for her calf. The first domesticated cattle must have resented being milked because paintings of early milkings indicate that the cow's rear legs had to be tied before milking. Of course, milk yield was low; it is estimated that in early tribes more than 10 cows were milked to obtain enough milk for one family. Through centuries of selective matings, dairy cows (*Bos taurus*) have been developed that can yield 150,000 or more liters of milk during a lifetime.

In the United States, the Purebred Dairy Cattle Association (PDCA) recognizes seven breeds of dairy cattle. These are discussed briefly in this chapter.

3.2 Origins of Dairy Cattle

The word "cattle" did not originally mean bovine animals. It was borrowed from the Old French *catel*, meaning movable personal property, especially livestock. The word is closely related to chattel (a unit of personal property) and capital in the economic sense. The word "cow" came from Anglo-Saxon *cū*, "a bovine animal." In older English cattle refers to livestock, while deer refers to wildlife.

Cattle were originally identified as three separate species: *Bos taurus*, the European or "taurine" cattle; *Bos indicus*, the zebu or Brahman cattle from South Asia (humped cattle); and the extinct *Bos primigenius*, the aurochs. Aurochs, which originally ranged throughout Europe, North Africa, and much of Asia, were ancestral to both zebu and taurine cattle. The ability of cattle to interbreed with closely related species complicates history. Hybrid individuals, and even breeds, exist between taurine cattle and zebu as well as some other members of the genus *Bos*, such as yak, banteng, and gaur. Hybrids can also occur between taurine cattle and bison. The hybrid origin of some types may not be obvious. For example, genetic testing of the Dwarf Lulu breed, the only taurine-type cattle in Nepal, found them to be a mix of taurine, zebu, and yak. Cattle cannot successfully be hybridized with the more distantly related bovines, including the water buffalo.[2]

3.3 What Is a Dairy Breed?

The term *dairy breed* refers to cattle that are especially well-equipped genetically for milk production. This includes ample body capacity to process large quantities of bulky plants as well as grains essential to sustain high milk production; large, high-quality mammary glands for secreting milk; and other genetic traits important to high milk production. Since members of this general group of cattle differ widely in size, appearance, and other traits, they have been separated into named breeds of similar characteristics. A **purebred** dairy animal is one of a recognized breed that usually meets the requirements for registration in the official herdbook of that breed.[3] A **grade** dairy animal is an offspring resulting from mating a purebred with a grade, or from mating dairy animals not purebred but having close purebred ancestors. Grade animals possess most (but usually less than 100%) of the traits of a particular breed. A **scrub** is a genetically unimproved animal that lacks dairy characteristics. Such an animal may be inferior in either breeding or individuality, or both. Selected information pertaining to dairy breeds of the United States is summarized in tables 3.1 and 3.2. Now let us discuss briefly the seven major dairy breeds of the United States in alphabetical order.

Table 3.1 Lactation Production Averages by Breed for Cows Calving in the United States.

Breed	Milk (lb)	Fat		Protein		Total Solids (%)	Number of Records
		(%)	(lb)	(%)	(lb)		
Ayrshire	18,295	3.84	703	3.14	574	12.8	5,977
Brown Swiss	21,996	3.97	874	3.31	729	13.4	18,243
Guernsey	17,349	4.41	766	3.27	567	14.5	6,076
Holstein	25,987	3.67	955	3.04	791	12.2	2,103,095
Jersey	19,004	4.70	893	3.62	688	14.6	209,844
Milking Shorthorn	18,319	3.65	669	3.11	570	12.8	3,288
Red and White	23,439	3.72	872	3.05	714	12.2	2,306

Source: Adapted from Norman et al., 2013.

Table 3.2 Selected Information about US Dairy Cattle Breed Associations.

Breed Association	Address	Year Founded	Official Breed Publication
American Guernsey Association (http://www.usguernsey.com)	7614 Slate Ridge Blvd. Reynoldsburg, OH 43068-0666	1877	*Guernsey Breeders Journal*
American Jersey Cattle Association (http://usjersey.com)	6486 E. Main St. Reynoldsburg, OH 43068-2362	1868	*Jersey Journal*
American Milking Shorthorn Society (http://www.milkingshorthorn.com)	800 Pleasant St. Beloit, WI 53511-5456	1948	*Milking Shorthorn Journal*
Brown Swiss Cattle Breeders' Association (http://www.brownswissusa.com)	P.O. Box 666 7614 Slate Ridge Blvd. Reynoldsburg, OH 43068-0666	1880	*Brown Swiss Bulletin*
Holstein Association USA, Inc. (http://holsteinusa.com)	P.O. Box 808 Brattleboro, VT 05301	1885	*Holstein Pulse*
Purebred Dairy Cattle Association (http://www.purebreddairycattle.com)	70 Main St. Peterborough, NH 03458	1940	
Red & White Dairy Cattle Association (http://www.redandwhitecattle.com)	308B Ogden Ave. Clinton, WI 53525	1964	*The Red Bloodlines*
US Ayrshire Breeders' Association (http://www.usayrshire.com)	1224 Alton Darby Creek Rd., Ste. B Columbus, OH 43228	1875	*Ayrshire Digest*

3.4 Ayrshire

A capacious and well-set udder
is the chief point of excellence in the Ayrshire cow.
Archibald Sturrock (1866)

3.4.1 History and Development

The aristocratic Ayrshire breed was developed in southwestern Scotland (county of Ayr) during the latter part of the eighteenth century. The first importation into the United States was in 1822. Although Ayrshires are scattered throughout the United States, they are located principally in the northeastern states. Ayrshires are found also in Canada, Great Britain, Finland, Sweden, Australia, New Zealand, South Africa, and throughout Central and South America.

3.4.2 Breed Characteristics

The color of Ayrshires varies from almost pure white to nearly all cherry red or brown, along with combinations of these colors.[4] Ayrshires are medium-sized cattle and commonly weigh over 1,200 lb (550 kg) at maturity. They are strong and **rugged**, adapting well to management systems that include group handling on dairy farms with freestalls and milking parlors. Ayrshire cows are noted for their symmetrical udders (figure 3.1A on the following page), which extend well forward and back with minimal tendency to be **pendant** (figure 3.1B, also on the following page). Meat of Ayrshire cattle is characterized by white fat. Production averages for Ayrshires are shown in table 3.1.

3.5 Brown Swiss

3.5.1 History and Development

Frequently called the Big Brown Cow, the Brown Swiss breed was developed in the Alpine country of Switzerland. This, the oldest of our dairy breeds, was probably developed at least by 500 BC. Bones of their ancestors have been found in the ruins of Swiss lake dwellers dating back to near 4000 BC. Brown Swiss are stabled in the valleys during winter months, but are moved up the mountains as grass becomes available in early summer. Much of the milk harvested in summer is made into butter and cheese in the mountains, although some milk is carried down the mountains in special backpack cans. Life on the mountainous slopes contributed to the selection of strong animals and the development of the breed.

The first importation of Brown Swiss into the United States was in 1869, but only 155 head were imported.[5] Brown Swiss comprise a small but important proportion of US dairy cattle. Registrations total about 11,000 annually. The world population of Brown Swiss cattle is approximately 7 million, which ranks first or second in the worldwide population of dairy cattle.

The milk of the Brown Swiss cow provides an excellent fat to protein ratio for production of most cheeses. The breed is popular for crossbreeding for production of dairy beef (cf. chapter 15), the objective being to get more milk and to produce larger calves at weaning.

Historically, the most famous Brown Swiss cow in the United States was Jane of Vernon (figure 3.2 on p. 29). As a two-year-old she was judged by many authorities as having the best udder among the Brown Swiss breed. She was named Grand Champion at the National

Dairy Cattle Congress in Waterloo, Iowa, in 1932, 1933, 1934, and 1936. Considered by many Brown Swiss breeders as the most influential female of her breed, she bore six outstanding daughters and two famous sons.

3.5.2 Breed Characteristics

Brown Swiss cattle vary in color from light to dark brown, and some are nearly gray. There is often a light stripe of gray along the back. The nose, **switch**, tongue, and horn tips are black; spotting is considered undesirable. Brown Swiss mature less rapidly than other breeds but are noted for longevity and strong legs (figures 3.3A and 3.3B). Brown Swiss cows are also noted for their persistency of lactation (i.e., their level of milk production is maintained well throughout lactation).[6] Additionally, they are hardy, rugged, thrifty cattle that quickly adapt themselves to different environmental conditions. Average weight of mature cows is 1,500 lb (682 kg). Milk production characteristics are given in table 3.1.

An advantage of using Brown Swiss bulls in breeding beef cows is the superiority of F_1 heifers as breeding stock. Typically large and fertile, they calve with ease and their additional milk adds weight to calves at weaning. This makes it possible to produce more beef per cow per acre. However, some sacrifice in marbling of the beef muscle was seen in research at Iowa State University.

Figure 3.1A Evaluating an outstanding Ayrshire cow at the Royal Show, Kenilworth, Warwickshire, England. Note the well-attached and balanced udder.

Figure 3.1B Maple-Dell Zorro Dafourth. Classified excellent (95). Produced 30,950 lb (14,068 kg) of milk, 1,387 lb (630 kg) of fat, and 939 lb (427 kg) of protein (courtesy of Ayrshire Breeder's Association).

Figure 3.2 Jane of Vernon. She was classified excellent at 15 years of age and produced a record of 23,569 lb (10,713 kg) of milk and 1,075 lb (489 kg) of fat (courtesy of Brown Swiss Association).

Figure 3.3A Examples of Old Mill E Snickerdoodle OCS during her first through eighth lactations (from right to left). Classified excellent (94). All American eight times and Grand Champion at the World Dairy Expo six times. Snickerdoodle is the only cow to win every milking class at the World Dairy Expo for all breeds. Lifetime production record: 241,114 lb (109,597 kg) of milk, 11,5[illegible] lb (5,261 kg) of fat, and 9,098 lb (4,135 kg) of protein (courtesy of Allen and Tammy Bassler, Old Mill Farm, Upperville, Virginia).

Figure 3.3B Dublin Hills Treats. Classified excellent (93). Grand Champion Brown Swiss in three shows in 2010 (courtesy of Elite Dairy, Copake, New York).

3.6 Guernsey

The sculptor works with lifeless clay that responds to his every touch, reflecting his degree of genius. The livestock breeder works with flesh and blood. His art is affected by the laws of heredity. The drag of Nature is his constant deterrent. Facts, ideas, and counsel are his guides. His rewards for creating superior seedstock are financial returns, public recognition, and personal satisfaction.

Fred S. Idtse (1896–1975)

3.6.1 History and Development

Recognized for their "golden goodness" in the form of a carotene-rich milk (carotene provides the deep golden color), the Guernsey breed was developed on the isle of Guernsey (located in the English Channel near the north coast of France). It is generally accepted that this breed was developed by crossing the large, red-and-brindle cattle of Normandy with the small, red cattle of Brittany, France (initially, development occurred during the tenth century, but more importantly during the seventeenth century). Introduction of the Guernsey to America occurred around September 1840, when Captain Belair of the Schooner Pilot brought three Alderney cows to the port of New York. Later, Captain Prince imported two heifers and a bull from the Island. These animals were the original stock of a great majority of those that make up the national Guernsey herd. Guernseys are America's second leading dairy breed; approximately 3 million have been **registered**.

3.6.2 Breed Characteristics

Guernsey cattle are usually fawn-and-white (fawn predominates), but a light cherry red is also found. The switch is usually white and the tongue light in color; their skin and body fat are yellow (figure 3.4A). Skin pigmentation is an important factor in placing Guernseys in show competition on the Island of Guernsey (especially bulls). At the Royal Guernsey Show (figure 3.4B), the

Figure 3.4A Rolling Prairie Bella Blue. Classified excellent (94) in 2012. Production in her last lactation was greater than 30,000 lb (13,636 kg) of milk and 1,246 lb (566 kg) of fat. Sired by Rolling Prairie Maxie Pistol (courtesy of *Hoard's Dairyman*).

Figure 3.4B Evaluating a class of cows at the Royal Guernsey Show, isle of Guernsey. If the two judges disagree on placings, a third party is consulted.

authors were told that this results from the relationship between skin pigmentation and the golden color desired in Guernsey milk. Mature cows and bulls weigh about 1,100 and 1,700 lb, respectively.

3.7 Holstein–Friesian

The high levels of production and prolific reproduction of Holstein–Friesian cows form a practical basis for the widespread popularity of this breed.

Maurice S. Prescott (1892–1976)

3.7.1 History and Development

Descendants of the ever-popular Holstein–Friesian breed were developed in northern Netherlands, especially in the province of Friesland, and in the neighboring provinces of northern Germany.[7] Many believe the breed has been selected for their dairy qualities since about the time of Christ. Thus, it is probably our second oldest dairy breed (figure 3.5).

In Europe the national derivative of the Dutch Friesian became very different from cattle developed by breeders in the United States, who used Holsteins primarily for milk production. Because both milk and beef were produced from dual-purpose animals in Europe, breeders imported specialized dairy Holsteins from the United States to cross with the European black and whites. For this reason, the term Holstein is used to describe North American stock. In Europe, especially in the north, Friesian denotes animals of a traditional European ancestry, which are bred for both dairy and beef use. Crosses between the two are described by the term Holstein–Friesian.

Prior to 1885 there were two associations representing the breed in the United States; one maintained a Holstein herdbook, the other a Dutch–Friesian herdbook. In 1885 the two associations combined into the Holstein–Friesian Association of America. Now named Holstein Association USA, it is currently the world's largest dairy cattle registry organization (see table 3.2).

Holsteins were first imported into the United States in 1621, but their breeding was not kept pure. Heavy importations were not made until the mid to late 1800s. Because of the prevalence of foot-and-mouth disease in Europe, relatively few Holsteins were imported after 1905. However, a limited number continued to enter the United States from Canada. More than 19 million animals are registered in the Holstein Association's herdbook.

3.7.2 Breed Characteristics

Holsteins are large, stylish animals with color patterns of black and white or red and white (figure 3.6 on the next page). At birth healthy Holstein calves weigh about 90 lb (41 kg). A mature cow weighs about 1,500 lb (680 kg) and stands 58 in (147 cm) tall at the shoulder. Holstein heifers can be bred at 13 months of age when they weigh about 800 lb (360 kg). Holstein females should calve initially between 23 and 26 months of age. Although some cows may live considerably longer, the average productive life of a Holstein is approximately four years. Average production data for slightly more than two million US Holstein cows enrolled in production-testing programs and eligible for genetic evaluations are given in table 3.1. Top producing Holsteins milked three times a day have been known to produce over 72,000 lb (32,727 kg) of milk in 365 days.

Based on records from 3,933 cows in 1,079 herds, the US Department of Agriculture (USDA) Genomic Evaluation report for the sire Ensenada (figure 3.7 on the following page) showed that he added 2,267 lb (1,030 kg) of milk, 66 lb (30 kg) of protein, and 72 lb (33 kg) of fat in productive capacities to his offspring (reliability 99%).

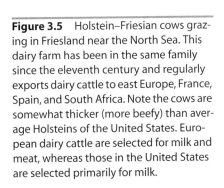

Figure 3.5 Holstein–Friesian cows grazing in Friesland near the North Sea. This dairy farm has been in the same family since the eleventh century and regularly exports dairy cattle to east Europe, France, Spain, and South Africa. Note the cows are somewhat thicker (more beefy) than average Holsteins of the United States. European dairy cattle are selected for milk and meat, whereas those in the United States are selected primarily for milk.

Merit for producing **fluid milk** and cheese, respectively: $728 and $709. Other data available on a Genomic Evaluation report assess the effects of the sire on udder, feet and legs, body capacity, and dairy character of his offspring.

3.7.3 Red-and-White Holsteins

The laws of inheritance set forth by Mendel explain the recessive character of the "red" gene. If one black-and-white Holstein parent has the red gene and the other does not, the offspring will be black and white. But if both black-and-white parents have the red gene, the probability is 25% that their progeny will be red and white. If both parents are red and white, their offspring will be red and white. If one parent is red and white and the other black and white, there is a 50% probability of the offspring being red and white only if the black-and-white parent carries the red gene, too.

Throughout western Europe the authors observed numerous red-and-white and black-and-white Holstein–Friesians side by side. European breeders of dairy cattle have held the view that coat color is not correlated with meat and milk production, and it is, after all, the latter which is most important to a profitable dairy enterprise. However, in earlier years, the Holstein Association USA

Figure 3.6 Scarlet-Summer RB Gwendelyn. Classified excellent (94); produced 58,270 lb (26,480 kg) of milk in 365 days of twice daily milking (5.2% fat and 2.9% protein). Owner: Scarlet Summer Holsteins, Muncy, Pennsylvania (courtesy of Holstein Association USA).

Figure 3.7 Ensenada Taboo Planet-et. Classified excellent (90). Born on March 3, 2003 (courtesy of David A. Bishop, breeder, Doylestown, Pennsylvania).

refused to register red-and-white Holsteins.[8] Instead, their aim was to eradicate the recessive red gene. Most red-and-white calves were sold as **vealers**. A few dairypersons kept red-and-white calves from their best cows.[9] According to Dr. Larry W. Specht:

> There are several variants from "true" red inheritance. The first one discovered was called Black/Red. The calf would be red at birth but would turn black or partially or mostly black with some reddish hairs down the backline, around the muzzle and at the poll. This happened usually by six months of age.

The earliest reference was found in the June 15, 1889, issue of *The Holstein–Friesian Register.* A letter to the editor stated "a calf may be red and white when born, and in a few weeks, at the shedding of the first hair, assume the black and white."

3.8 Jersey

3.8.1 History and Development

The stylish Jersey breed was developed on the island of Jersey[10] (figure 3.8) by selective matings of the large, brindle cattle of Normandy and the small, black cattle of Brittany (the latter predominating). In 1789 a law was passed that prohibited importation of cattle to the isle of Jersey and is in effect today. However, Jersey cattle (the only breed on the island) can be exported. The first importation of Jerseys into the United States was about 1815; the first Jerseys registered in the United States were imported in 1850. Leading countries in numbers of Jerseys are the United States, New Zealand, Australia, Canada, South Africa, Costa Rica, and Denmark. In 2011, there were almost 400,000 Jersey cows in the United States.

3.8.2 Breed Characteristics

Jersey cattle are the smallest of the dairy breeds in the United States (see table 3.1). Jersey calves weigh about 55 lb (25 kg) at birth. Mature Jersey females usually weigh about 950 lb (432 kg) and males about 1,500 lb (682 kg). They mature quickly and cows are often in milk before 24 months of age. Recent research established the optimal age for calving of Jerseys is 20 to 21 months based on relationships to milk, fat and protein yield, persistency, fertility, and lifetime production. The coat color of Jerseys is usually a shade of fawn or cream, although different shades of mouse color, gray, and brown are common; some individuals approach black. They may be solid in color or spotted with white. Their muzzles and tongues are usually black or lead colored. The animals pictured in figures 3.9A and 3.9B on the following page show excellent **breed type** (conformation). Jersey milk is rich in milk solids (highest in milkfat of all dairy breeds). Jerseys are more tolerant of heat than the larger breeds and are adapted to a wide range of climatic and geographical conditions.

The Jersey is highly efficient in milk production, yielding more pounds of milk per pound of body weight than other breeds. An average Jersey in the United States produces 16 times her body weight in milk each year, or about twice her body weight in cheddar cheese. Highest annual milk yield of a Jersey through 2011 was 49,250 lb (22,386 kg), over 40 times her body weight. Capper and Cady (2012) used population-based data from Dairy Records Management Systems to study comparative environmental impacts of Jersey versus Holstein milk production sufficient to yield 500,000 tons of cheese. Average milk yield and composition for Jerseys were 46 lb/d (20.9 kg/d), 4.8% fat, and 3.7% protein; for Holsteins, 58.2 lb/d (26.5 kg/d), 3.8% fat, and 3.1% protein. Also considered were age at first calving, **calving interval**, and culling rate. Each population contained lactating and dry cows, herd replacements, and bulls for which rations were formulated according to DairyPro (Agricultural Modeling and Training Systems) at breed-appropriate mature body weights: 1,000 lb (455 kg) for Jerseys and 1,500 lb (682 kg) for Holsteins. The study showed that to produce equivalent amounts of milkfat, protein, and other solids, Jerseys require 32% less water, 11% less land, and have a total carbon footprint that is 20% less than that of Holsteins.

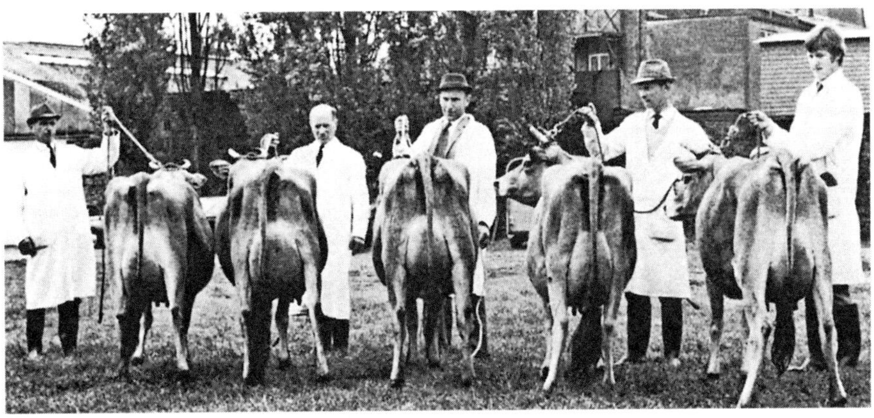

Figure 3.8 First prize get of sire of Dreamlike Pierre Royal at the Royal Jersey Show (courtesy of H. W. Maillard).

3.9 Milking/Dairy Shorthorn

Whatever our concerns—be they directed toward the improvement of our breed of cattle or general livestock improvement— there is only one limiting factor—the mind of man.

Harry Clampitt (1916–1983)

3.9.1 History and Development

This breed was developed about 250 years ago in northeastern England (at Durham in Yorkshire), where it is called Dairy Shorthorn and Durham. They are widely distributed throughout the British Isles, America, and Australia. (The Illawarra breed of Australia is closely related to the Milking Shorthorn breed.) The first importations of

Figure 3.9A Hawarden Jace Pix. Classified excellent (95). In five completed lactations she averaged 20,717 lb (9,417 kg) of milk, 1,024 lb (465 kg) of fat, 811 lb (369 kg) of protein, and 2,740 lb (1,245 kg) of cheese yield equivalent (courtesy of Frank Robinson and the American Jersey Cattle Association).

Figure 3.9B JerseySpringer-DH in a typical front view (courtesy of American Jersey Cattle Association).

Milking Shorthorns from England to America were made in 1783 when they were preferred for producing milk, meat, and power. This docile animal efficiently converts feed into milk and has a long productive life, at the end of which these large cows have a high salvage value.

Breeders began recording their Shorthorn cattle in 1846 with the first volume of the *American Herdbook*. In 1882, the American Shorthorn Breeders' Association was formed to register and promote both Milking and Scotch (beef) Shorthorns. In 1912, a group of Milking Shorthorn breeders organized the Milking Shorthorn Club. The American Milking Shorthorn Society incorporated in 1948.

3.9.2 Breed Characteristics

Shorthorns are either red, red and white, white, or **roan**. Although most Milking Shorthorns are horned, about 8% are polled. They commonly have a deep chest, well-sprung and long ribs, strong loin, level rump, capacious udder, straight legs, and weigh 1,400 to 1,500 lb (640 to 680 kg) (figure 3.10). Among the dairy breeds, they are intermediate in milk production (see table 3.1). They rank high in salvage value and in sale value of animals intended for veal and dairy beef purposes. Other attributes of the breed include ease of calving, ease of management, and economy of production, especially on home-produced **roughages** and grass. The hardy, efficient nature of the breed and its suitability to pasture/forage-based dairying are causing increases in numbers after years of decline. The breed has embarked on a program of genetic progress for dairy characteristics while concurrently counteracting a diminishing gene pool by incorporating some of the best genetics from other red dairy breeds.

3.10 Red and White

3.10.1 History and Development

In 1964, the Red & White Dairy Cattle Association (RWDCA) was established, which led to the establishment of the RWDCA "open" herdbook. It encouraged and promoted the breeding and development of a new breed of red-and-white dairy cattle, referred to as Red and Whites, that are high producers of milk and milk solids and possess good utility value (figure 3.11 on the following page).

3.10.2 Breed Characteristics

The herdbook of the association consists of the Red Registry and the Extended Registry. The Red Registry includes animals that are red and white, solid red, or solid white in color. Red and white is defined as ranging from very light yellow/rust to deep red. The Extended Registry includes animals of other colors. Animals eligible for shows must be registered in the Red Registry and must be true red and white, solid red, or solid white in color. The following breeds are accepted: Ayrshire, Brown Swiss, Guernsey, Holstein, Jersey, Milking Shorthorn, Red Dane, Angler, Swedish Red, Norwegian Red, Illawarra, Aussie Red, Rouge Flammande, Normande, Lineback, Meuse-Rhine-Issel, Dutch Belted, Gelbvieh, Red Poll, Simmental, White Park, Montbeliarde, and Canadiane.

Milk production records for 2,370 Red and Whites are given in table 3.1.

Figure 3.10 Gold Mine Poppy's OT Kay. Classified excellent (95). Produced 46,666 lb (21,212 kg) of milk, 3.3% fat, and 3.1% protein in 365 days (courtesy of American Milking Shorthorn Society).

Figure 3.11 Golden-Oaks Perk Rae-Red. Classified excellent (90). Produced 31,030 lb (14,105 kg) of milk with 3.7% fat and 3.2% protein in 365 days. An eighth generation polled cow in the red-and-white breed. Recognized in 2012 in Holstein International magazine as Red & White Impact Cow of the Year (courtesy of Red & White Dairy Cattle Association, Clinton, Wisconsin).

3.11 Why Purebreds?

Purebred breeders of dairy cattle are motivated not only by the extra income that purebreds bring them, but by their broader interest in helping create something better for others.

Harold R. Searles (1890–1972)

A purebred dairy animal is a direct descendant of foundation animals of a given breed. Purebreds may or may not be registered, but all registered dairy cattle are, or should be, purebreds. Exceptions include some animals registered by the RWDCA.

The use of purebred dairy cattle increased rapidly after Dr. S. M. Babcock of the state universities of Missouri and Wisconsin introduced the Babcock test for milkfat in 1890. Using this test, dairypersons could identify cows that yielded the most milk solids, especially fat. Thus, it provided incentive and method for improving herds and a reason for registering cattle. But why invest the time and coins required for maintaining registration papers?

3.11.1 To Sell Breeding Stock

Purebred cattle and calves command a higher average price than grades. This results, in part, from higher milk production among purebreds.[11] Registration papers, per se, do not increase production. Instead, improved genetics and management are responsible and these tend to follow purebreds. Purebred dairy cattle breeders should merchandise their cattle at home and abroad. Otherwise, there is little reason to maintain registration papers. Whereas the price of premium-quality milk is relatively inelastic, the price of a premium-quality dairy animal *is* elastic and may reach levels several times that of the average of the breed.

3.11.2 Pride of Ownership

A great deal of personal satisfaction comes to those who breed and own attractive, high-producing dairy cattle, especially purebreds. Improvements in records and management commonly accompany the personal pride and interest associated with owning registered cattle.

3.11.3 To Maintain Breed Purity

A small percentage of dairy cattle in the United States are registered. Operators of large commercial dairy farms are usually most concerned with the economics of milk production and not with the registration of cattle or breed purity. Yet, they look to dairy breed associations for leadership in providing genetically superior cattle, for milk marketing and consumer education, and for breed promotion (sale of milk and cattle).

3.11.4 Purebred Dairy Cattle Association

The destiny of any breed of registered stock lies in the hands of a dedicated and devoted governing body and in the goals of breeders with high ideals and unquestioned integrity.

Marvin L. Kruse (1919-2000)

The PDCA was founded in 1940 and includes representatives from each of the seven dairy cattle breed associations. Among its noteworthy accomplishments are: (1) the Dairy Cow Uniform Scorecard (see figure 5.2), (2) unification of type classification rules and classes for dairy cattle at breed shows and fairs, and (3) a code of ethics for private and public sale of purebred dairy cattle.

3.12 Choosing a Dairy Breed

The authors hold the view that a good cow is where one finds her, and this is irrespective of breed, location, or farm name. However, certain factors are worthy of consideration in selecting a dairy breed. Let us identify a few without regard to order of importance.

3.12.1 Personal Preference

As surely as people differ in choice of car colors and candies, so they have personal preferences for a breed of dairy cattle. The preference of some is influenced by relatives, neighbors, and friends, while others allow coat color and/or markings, temperament, longevity, and economic considerations to influence their choice. Preference, however, based on prejudice is often poorly conceived and should be exercised only when other considerations yield no definite choice.

Advantages accrue when several farmers have the same dairy breed in a community. Attracting prospective buyers, body type conferences, local breed shows, and meetings are but a few. Of course, economics is the major reason a given breed predominates.

3.12.2 Economic Issues

When the market for fluid milk is excellent, milk producers benefit from the larger breeds even though their milk contains less solids than that of smaller ones. Milk from smaller breeds is preferred for cheese production. In some markets, milk from certain breeds demands a premium price and a special label (e.g., All-Jersey or Golden Guernsey), thus making more valuable the breed that produces it.

Throughout much of the world, cattle are used for both meat and milk. In some countries, cattle are selected and mated to produce either milk or meat. Dual-purpose cattle have not been well received among US dairypersons, as noted by W. D. Hoard:

> One of the most destructive notions that has ever pervaded the ranks of dairymen; that has cut the sand from under their feet; that has prevented progress, prevented development; and that has caused the business to be unprofitable is the growth of the very old notion entitled "a general-purpose cow."

Notwithstanding the above quotation, the salvage value of cull cows, veal calves, and dairy steers makes meat prices an important consideration in selecting a dairy breed. The ratio of milk price to meat price may at times favor the dual-purpose breeds.

3.12.3 Physical Traits of Various Breeds

There are four traits that should be taken into account: size of adults, vigor of calves, and the age of maturity and longevity. A study by the USDA found that heavier cows (those averaging up to 100 lb more than herd average) are more productive. The genetic association between body weight and milk was 0.28. This indicates that selection based on milk yield and body weight would increase feed efficiency. Larger cows also have a higher salvage value when sold.

Regarding the vigor of calves, although well-controlled research pertaining to calf mortality among dairy breeds is insufficient to formulate a clear statement, unpublished data at the Missouri Agricultural Station indicate that calves of the larger dairy breeds are somewhat more hardy than those of the smaller ones (cf. chapter 7). Until recent years, many believed the larger dairy breeds performed especially well in colder climates, whereas the smaller ones were more well-fitted to areas of high temperatures. However, it is now known that all dairy breeds can perform well throughout temperate climates.

Cows of the smaller dairy breeds enter motherhood at an earlier age than those of the larger ones. Our observations of dairy cattle throughout western Europe indicate that longevity of European dairy cattle exceeds that of cattle in the United States. We believe this may be attributed, in large part, to feeding and management. In general, European cattle receive limited quantities of costly concentrates (grain-based rations) and obtain nutrients primarily from forages. Moreover, herd size is small and personal attention given to dairy cattle in Europe is much greater than that given to the larger, more commercialized US dairy herds, which are fed abundant quantities of grains.

3.13 Summary

Throughout the recorded history of the human struggle for a full and abundant food supply the people in the best fed nations of this world have benefited from the vision, the idealism and constructive work of livestock breeders.

Fred S. Idtse (1896–1975)

At various stages in human progress, cattle have served as beasts of burden, objects of worship, and sources of food. Centuries of selective matings have given us breeds of dairy cattle genetically equipped to provide an abundant supply of milk. Citizens of the Western Hemisphere benefited from progressive dairy cattle breeders in England, the Channel Islands, and on the continent of Europe. Based on foundation animals imported from abroad, programs of dairy breed societies, agricultural experiment stations, artificial insemination studs, and individual dairy cattle breeders have resulted in development of superior dairy cattle breeds in the United States, followed by their export to many other nations.

A small percentage of milk cows in the most commercial herds are purebreds. Yet improvement in dairy cattle of a given breed is accomplished principally through the use of purebred bulls. For this reason, purebred dairy cattle have and will play an important role in the dairy industry.

Study Questions

1. Name the dairy breeds recognized by the PDCA.
2. What is a purebred? A grade? A scrub? Are all purebred dairy cattle registered? Why? Are all those registered purebred?
3. Give the native home, coat color(s), approximate size of adults, and milk component percentages of each dairy breed.
4. Which dairy breeds mature the earliest? Which are slowest to mature?
5. Identify one or more dairy breeds noted for their production persistency.
6. For which dairy breed is skin pigmentation emphasized (in the native home)? Why?
7. Which dairy breed was probably the first brought to America? Which dairy breed was most recently brought to the United States?
8. How is it possible for black-and-white Holsteins to have red-and-white progeny? Can red-and-white Holstein parents have black-and-white offspring? Why? If both black-and-white parents are carriers of the red gene, what is the probability of their progeny being red and white?
9. How have certain diseases in Europe (e.g., foot-and-mouth) affected importation of dairy cattle into the United States?
10. Identify several contributions of the PDCA to dairying in the United States.
11. Give two or more reasons for maintaining purebred dairy cattle.
12. What effect has artificial insemination had on the genetic gap between grade and purebred dairy cattle?
13. Discuss the factors one should consider when selecting a dairy breed.
14. Compare the types and numbers of registered dairy cattle of today to those that existed in 1920. Which breeds have gained and which have lost in relative numbers of dairy cattle?
15. Discuss and contrast the feeding and management strategies found in the United States and those found in Europe.

Notes

[1] The authors acknowledge with appreciation the contributions to this chapter by the dairy breed associations named in table 3.2.

[2] Much of this history came from Wikipedia.

[3] An animal may be purebred and yet be ineligible for registration because of coat markings, physical defects, and other reasons established by the respective dairy breed associations.

[4] Early Ayrshires were reported to have been black-and-white. Then, about 1780, the red-and-white color became fashionable. Since the gene for red-and-white color is recessive and the mating of red-and-white cattle results in red-and-white progeny, it was possible for Ayrshire breeders to selectively breed for the color they desired.

[5] The small number of imports was due, in large part, to regulations resulting from the prevalence of foot-and-mouth disease in Europe. More than 700,000 Brown Swiss descendants have been registered from the original 130 cows and 25 bulls imported from Switzerland.

[6] Cow cost or cow depreciation in a dairy herd depends on the length of a cow's milking lifetime. Brown Swiss are noted for long life, thus giving a long period of profitable production. An example of productive longevity is Ivetta, a Brown Swiss cow that had 10 consecutive milk production records exceeding 24,000 lb (10,909 kg) milk and 1,000 lb (455 kg) milkfat. Her lifetime production totals were 308,569 lb (140,259 kg) milk and 13,607 lb (6,185 kg) milkfat.

[7] The climate and soil of the Netherlands favors the luxuriant growth of grass during summer months; this also ensures high-quality forage for winter feeding. Holsteins respond well to nutritious grasses and close contact with people. They get the latter in winter months since Dutch dairy farms are small and the terrain is flat, so owners frequently observe and associate closely with their cattle during the summer and winter.

[8] On July 1, 1971, the Holstein–Friesian Association opened its herdbook to include both red-and-white and black-and-white Holsteins.

[9] An interesting discussion of this subject is available as an article and a podcast called "Red and White Holstein History," by Dr. Larry W. Specht, professor emeritus of dairy science, Pennsylvania State University (http://dasweb.psu.edu/publications).

[10] The island of Jersey is the largest of the Channel Islands group (45 square miles). In 2009 the human population was 92,500 and the cattle count was 5,090 (down from 12,000 in 1970).

[11] The widespread use of superior sires through artificial insemination has narrowed the genetic gap between grades and purebreds. Thus, the genetic difference between registered and nonregistered cows on a *within-herd* basis is small and probably insignificant when all dairy breeds are considered (see Van Vleck and Burke, 1965).

References

Capper, J. L., and R. A. Cady. 2012. A comparison of the environmental impact of Jersey compared with Holstein milk for cheese production. *Journal of Dairy Science* 95:165–176.

Norman, H. D., T. A. Cooper, and F. A. Ross, Jr. 2013. *State and National Standardized Lactation Averages by Breed for Cows Calving in 2011.* Beltsville, MD: USDA Animal Improvement Programs Laboratory.

Van Vleck, L. D., and J. D. Burke. 1965. Production differences between registered cows and their nonregistered herdmates. *Journal of Dairy Science* 48:962–967.

Websites

American Guernsey Association (http://www.usguernsey.com)
American Jersey Cattle Association (http://usjersey.com)
American Milking Shorthorn Society (http://www.milkingshorthorn.com)
Brown Swiss Cattle Breeders' Association (http://www.brownswissusa.com)
Holstein Association USA (http://holsteinusa.com)
Purebred Dairy Cattle Association (http://www.purebreddairycattle.com)
Red & White Dairy Cattle Association (http://www.redandwhitecattle.com)
USAyrshire Breeders' Association (http://www.usayrshire.com)

4

Dairy Herd Records

The most important management tool is production records.
Almost every decision made regarding the dairy herd is based on records.
When to turn dry, how much to feed, which cows to cull,
and level of herd health are determined from production records.

Clinton E. Meadows

The authors thank Dr. Barry Steevens, Professor Animal Sciences, University of Missouri, for contributions to this chapter, especially with the dairy herd improvement section.

4.1 Introduction

Beliefs, estimates, guesses, and hunches are unsatisfactory in managing today's dairy farm enterprise. Records are essential if profits are to be maximized.

Low- and high-producing cows require the same investment in housing and equipment, and approximately the same amount of labor in breeding, feeding, and milking. The only significantly larger expenditure associated with high production is an increased feed cost. But with other costs held constant, high-producing cows return much greater profits than low-producing ones. The business goal of milk producers is not to minimize costs (as in less feed for low producers), but rather to maximize net income. Records serve as the cornerstone in building a financially successful dairy enterprise. Dairy herd improvement (DHI) records and summaries indicate that there is a steadily widening margin of returns per cow above feed costs as herd average production increases (cf. chapter 19).

Fortunately, the commercial value of dairy cows can be determined early in their productive lives, provided cattle are enrolled in a production testing program. A DHI program provides opportunities to test cows and discover if their apparent genetic ability to produce milk is real or false.[1] DHI records provide information pertaining to animal identification, milk and fat production, income over feed cost, and information related to feeding cows. Additional information relating to management includes recommendations on dates to breed cows, discontinue milking (**dry off** cows), check pregnancy, and when to expect parturition.

Of course, records only have value when used. In the dairying industry they prove useful when they supplement the manager's judgment in accurately assessing factors directly related to net profit or when they are used to prove the genetic merits of dairy bulls (cf. chapter 6). Let us now discuss various aspects of dairy herd records.

4.2 Brief History of Production Testing

When or who first advanced the idea of determining productivity of cows is lost in the night of history. The first measure of productive performance in the United States was apparently the butter churn test developed in 1853. A record was made of the pounds of butter produced from the milk of a cow in seven days. Certain dairy breed associations began production testing in 1883. The big boost in production testing came in 1890 when Dr. Samuel M. Babcock released his test for milkfat (cf. chapter 22). This accurate, simple, quick, and inexpensive test provided a means of determining how much milkfat a cow produced.

The idea of a cow-testing association is credited to a group of dairypersons in Denmark, where the first association was started in 1895. In 1905, a few progressive Michigan dairy farmers joined Danish-born Helmer Rabild in organizing the Cow-Testing Association, the first of its kind in the United States. The foremost objective was to improve herd profitability. A **tester** was hired to weigh and test each cow's milk and to keep records of feed costs and income. Information was needed for culling low-producing cows and to serve as a basis in feeding high producers. The efforts of these dairypersons were crowned with success, for within four years average profit per cow doubled. Since then, thousands of other herd owners have followed their example with equally striking results.

The US Department of Agriculture (USDA) began sponsoring cow-testing associations in 1906. The movement grew and in 1926 expanded into the National Cooperative Dairy Herd Improvement Program (NCDHIP). Currently, the NCDHIP is sponsored by the National Dairy Herd Improvement Association (NDHIA) in conjunction with the Extension Services of the USDA. Records are now used not only to appraise the productive worth of cows but also to prove the genetic merits of dairy bulls.

4.3 Benefits of Production Testing

The cow-testing association, DHIA, probably has done more, directly and indirectly, toward raising milk production and placing dairying on a sound business basis than any other single agency.

Helmer Rabild (1876–1948)

The primary purpose of dairy herd improvement associations (DHIAs) is to provide members with information useful in improving the profitability of their herds. Production records can be used in identifying profitable and unprofitable cows. The latter can be sold, which results in more uniform production throughout the year, higher overall production for the herd, and an increased profit. Average productivity of cows enrolled in DHIA testing programs is substantially above that of all dairy cows in the United States.

Milk production records are essential for evaluating milk-transmitting ability of dairy bulls. With more production-tested cows and herds there is a larger sampling of dairy sires and a resulting decrease in time required to obtain reliable sire proofs.

Records are beneficial to both the buyer and seller of dairy cattle for dairy purposes. They assist the buyer in evaluating the milking ability of animals of interest, and they benefit the seller by giving greater confidence in the quality of animals being offered for sale and in the sale price.

4.3.1 Production Testing of Registered Dairy Cows

Every herd of purebred dairy cattle should be production tested. In fact, to encourage production testing and to avoid registering dairy cattle of genetically inferior quality, it would be good if breed associations required production testing for registration. Of course, a much larger percentage of registered than of grade dairy cows is production tested. This does not mean that animals registered with a breed society are superior to their grade counterparts. However, the overall productivity and sales price of registered dairy cows generally exceeds that of grades.

4.4 The DHI System

The DHI system is composed of a combination of organizations working together to serve milk producers and other components of the dairy industry with herd records and management information systems. The system is decentralized, reflecting its information-intensive nature. The complexity involved with testing, collecting, analyzing, and sharing of data is managed by dividing the system into separate functions, which are carried out by the appropriate organizational entity within the system.

In earlier years, state dairy extension specialists were responsible for the educational aspects of the DHI system. This included training supervisors and other personnel and assisting DHI associations in developing and maintaining the abilities to certify official milk production records. However, these functions generally have been taken over by private organizations under the umbrella of the NDHIA.

4.4.1 National Dairy Herd Improvement Association

The centerpiece of the DHI system is the NDHIA. Established in 1965, the objective of the NDHIA is to coordinate the NCDHIP and serve as the umbrella organization over its affiliates (who are comprised of state and regional dairy improvement associations, producer-controlled cooperative associations, and agricultural membership organizations that provide farm management information services). It also provides producers with a national vehicle for setting policy and developing and enforcing rules associated with the program and represents DHIAs on issues involving other national and international organizations. It serves as a liaison among DHIAs and supporting institutions for delivering dairy management information systems to the industry. The NDHIA is governed by a 12-member board of directors elected from the Agricultural Research Service (ARS), Cooperative Extension System (CES), National Association of Animal Breeders (NAAB), and Purebred Dairy Cattle Association (PDCA). Member organizations conduct the business of the NCDHIP within their jurisdictions.

The NDHIA was established to promote accuracy, credibility, and uniformity of DHI records; to represent the DHIA system on major issues; and to organize activities within the industry to benefit the members of national DHIAs. Specific goals deal with communications, quality certification, coordination of educational activities, coordinating the testing and application of devices used within programs, and cooperating with the USDA, NAAB, and PDCA.

4.4.2 National Cooperative Dairy Herd Improvement Program

The DHI system includes organizations that work together to conduct the NCDHIP. The foundation of the NCDHIP is the collection of dairy cow records from farms. The NCDHIP is a voluntary cooperative effort to improve level and efficiency of milk production and to increase profitability of dairying. It involves milk producers, DHI organizations, extension services of land-grant colleges and universities, and the ARS and CES of the USDA.

Records are collected by weighing and testing each cow's milk on a monthly basis. Milk is tested for its content of fat, protein, somatic cells, and other attributes. Seven dairy records processing centers (DRPCs) process collected data and then collate it to provide comprehensive herd records and to assist in management decisions (e.g., feeding and breeding). Records are used also for research and industry-related purposes.

Dairy farmers benefit directly and indirectly from the program. Direct benefits accrue from information on production and costs, which prompt (1) culling of unprofitable cows, (2) feeding for maximal production, and (3) making precise management decisions for efficiency and profit. Producers benefit indirectly by submitting records to be pooled with those of other participants for use in sire evaluation, summary reports, analyses, and research.

4.4.3 Animal Genomics and Improvement Laboratory

The Animal Genomics and Improvement Laboratory (AGIL) of the ARS in the USDA conducts research to discover, test, and implement improved genetic evaluation techniques for economically important traits of dairy cattle and goats. Research is directed at genetic improvement of efficiency of yield traits (milk, fat, and protein) and nonyield traits that affect health and profitability (longevity, conformation, fertility, calving, and disease resistance). AGIL aids in conducting and distributing results of the sire and cow evaluation phase of the DHI program. Additionally, it assists in coordinating state record-keeping programs; furnishes necessary materials; compiles summary information; analyzes production, feed, cost, and related data; and researches various aspects of the program. It also provides records and summaries to national dairy breed registry associations and furnishes an annual report of participation and statistical data.

4.4.4 Memorandum of Understanding

A national memorandum of understanding among NCDHIP sponsoring groups outlines responsibilities of each party and how programs will be conducted:

- NDHIA—enforces the rules, policies, and quality certification standards of the NCDHIP;
- ARS—conducts the national genetic evaluation research program using NCDHIP data;
- Council on Dairy Cattle Breeding—ensures the long-term future of genetic evaluation and analysis of herd management information in the United States; and
- CES—national coordination and leadership of education programs in record collection, evaluation, and use.

The CES and a state's DHIA guide how the program is conducted and certify records within that state. Authority for program rules, policies, and quality standards is vested in the NCDHIP policy board. A memorandum of understanding exists between NDHIA and the Council on Dairy Cattle Breeding (CDCB) regarding the flow of genetic information. A memorandum of understanding also exists between the CDCB and the ARS to ensure the flow of DHIA records for industry purposes, including genetic evaluation programs.

4.5 How a Typical DHIA Operates

A typical DHIA is a cooperative organization of 25 to 30 dairy farmers who employ a supervisor to keep identification,[2] production, feed, income, breeding, and other records on cows. The supervisor visits each herd in the association monthly (a minimum of 10 times every 365 days). Milk weights for two consecutive milkings are recorded.[3] Proportional samples of milk from each milking on the test day are used to determine the production of milkfat and other milk solids.

The DHIA supervisor is an employee of the local DHI association.[4] Sometimes called a tester or field representative (called a *sampler* in the Netherlands), this person is the cornerstone of the entire DHI program. Only herd managers/owners are more basic and essential for the program to succeed. Through close contact with the producer, the supervisor can assist in fully utilizing results of the DHI record-keeping program to improve the efficiency and profitability of the dairy farm enterprise.

4.5.1 Testing Periods and Intervals

The DHIA supervisor is responsible for scheduling test days within his/her association. Once a monthly schedule is established, it is usually followed in subsequent months. Therefore, the test interval method (TIM) is used in calculating DHIA records. Production credits for the first half of the test period are calculated from the previous test-day information and those for the second half are based on current test-day information (the interval between tests can vary from 15 to 45 days). Production before the first test after calving is based on the first test, and production after the last test before drying off is based on the last test.

4.5.2 Devices for Quantifying Milk Produced

Pipeline milking systems (cf. chapter 12) have virtually replaced bucket milking, which has made the weighing of milk a simple matter. The need for accurate, durable, easily cleaned devices for obtaining milk weights and samples for use in determining milk solids has been fulfilled with approval of several different proportional sampling and metering devices (figure 4.1). Most of these devices are approved for accuracy by the International Committee for Animal Recording (ICAR).

4.5.3 DHIA Testing Plans

Adopted in 1926, the Standard DHIA is the most complete of all DHI production testing and record plans.

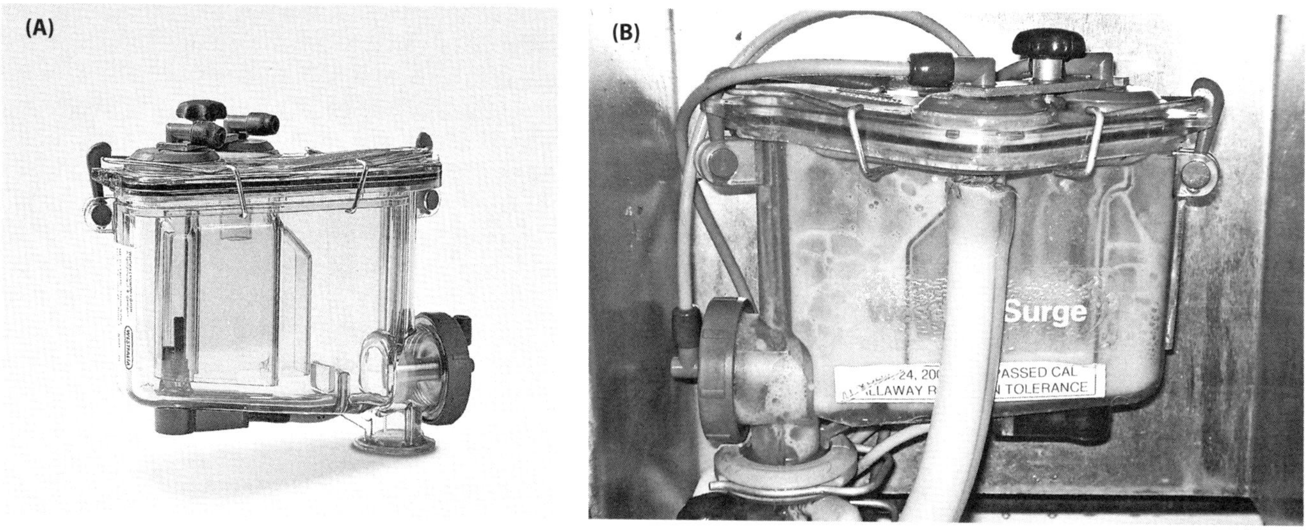

Figure 4.1 The Metatron MB milk meter measures milk quantity and conductivity for each cow: (A) new and (B) in use (courtesy of GEA Farm Technologies Australia Pty. Ltd., Tullamarine, Victoria, Australia).

Both registered and grade cows can be enrolled. Information provided on an individual cow basis by these records includes monthly and accumulative records for milk, fat, protein, amount and cost of feed, and income over feed cost.

The NDHIA has 20 test plans within the following categories: (1) supervised (conducted by an official technician), (2) supervised with adherence to breed association rules, and (3) unsupervised. Within each there are variations in numbers of weights and samples taken. A code number is assigned to each plan. For example, code DHI 00 means that milk is weighed and sampled on test day by a certified field technician/representative.

Total enrollment in DHI testing as of January 2015 was 17,875 herds in which 4,417,597 animals were on test. Among these were 3,668,546 Holsteins. Highest participation (48.3%) was in the DHI-AP plan. DHIAs also offer recording programs to meet management needs of the individual dairy. Four commonly used plans are:

- **DHI (Supervised)**: The DHIA technician weighs and samples the milk from each milking for all cows in the herd during a single 24-hr period.
- **DHI-AP (Supervised)**: The DHIA technician weighs and samples alternately at AM and PM milkings.
- **DHI-APCS (Supervised)**: The DHIA technician weighs the milk from each milking during a single 24-hr period, and collects samples for component testing at one of the weighed milkings.
- **DHI-MO** and **DHI-MO-AP (Supervised)**: The technician weighs the milk from each milking or selected milkings during a single 24-hr period. No samples are collected for component testing.

The yield of individual cows must be measured at the time of milking with a minimum of interference to the normal routine. Milk samples must be representative of all milk taken from the cow during the measured milking. All weighing and sampling devices must be used strictly according to the manufacturer's written instructions at all times. Data for each test day for each herd must be labeled to identify the degree of supervision applied.

The **Dairy Herd Improvement Registry (DHIR)** is the electronically computed Standard DHIA record with added requirements to satisfy needs of breed associations. For example, additional verification of records is accomplished by **surprise tests**.[5] These are made when milk production of certain cows exceeds breed average or another specified amount. Only registered dairy cows are eligible for DHIR records. Grade cows in herds on the DHIR have DHIA records. Production records of herds enrolled in DHIR are sent to respective national breed registry organizations for official recording.

On each test day, the Herd Summary (form DHI-202) provides a comprehensive herd analysis and management report including production, reproduction, genetics, udder health, and feed cost information. All compiled data result from input into a dairy management program by farm staff and/or by the DHI technician on test day. Data presented in the DHI-202 are used in calculations each test day. If the herd is enrolled in the "permanent string" option, a DHI-202 is also prepared for each string of cows. In multiple breed herds, all cows of each breed can be in separate "permanent strings" and their yearly averages can be considered for recognition. When all cows in the herd are the same breed (registered and/or grade), information on the entire herd is used to calculate the herd averages.

Other recording programs are available through DHIA affiliates. A list of the type of test codes and plan descriptions is available from the NDHIA (http://www.dhia.org).

4.5.4 Costs of DHI Testing

Since the DHI program exists to provide a production testing service to milk producers, who are primary beneficiaries of the program, they pay most of its costs. Such costs include: DHIA affiliate fees, lab and DRPC fees, and technician fees. The fees for most organizations that provide for the management of dairy records are based on cow numbers in a herd on the previous test day, except for DHIA operating fees and individual testing services. Some organizations also offer non-DHIA members lab services on a per sample basis.

4.5.5 Unofficial Testing

Records are accepted by the dairy industry for many purposes, including management, research, and genetic evaluation. Since unofficial testing plans do not have to conform to all the rules, they are not certified. Records from these plans are referred to as management records. In several management plans some official rules apply. Management testing plans include **owner-sampler (OS)**, weigh-a-day-a-month (WADAM), milk-only record (MOR), and alternate AM-PM. Other plans are used on a state or regional basis. Unofficial record plans are designed to aid within-herd management at minimal cost.

OS production testing was introduced in 1926. This plan is beneficial to dairypersons interested in records for private use and in minimizing the cost of production testing. The supervisor distributes sample bottles monthly. The OS member samples and weighs the milk of each cow for two consecutive milkings. Then the supervisor takes the samples to a central laboratory for testing and recording data. The main advantage of OS records is that they cost only about one-half as much as Standard DHIA records.

The accuracy and completeness of OS records depend on the herd owner/manager. Since there is no supervision of weighing and sampling of milk and no positive identification of cows by a disinterested party, these records are unofficial and are not used in sire proving programs.[6] Cows enrolled in the OS and alternate AM-PM plans include about one-fourth of all DHI production-tested cows. Although there exists the possibility of a few intentionally falsified records, most producers are honest to themselves and interested others. Therefore, breeders of dairy cattle would benefit if cows enrolled in these two record plans could be included in sire proofs.

The alternate AM-PM test requires supervision of only one milking per month rather than two—AM one month, PM the next. It permits the supervisor to enroll more herds and still continue monthly farm visits to weigh and sample milk and to collect other data essential for maintenance of the DHIA record system. However, these records may not be acceptable for genetic evaluation due to their higher variability than Standard DHI records.

MORs are based on milk weights recorded by the DHIA supervisor. No compositional tests are made. Although MORs are unofficial, they can be used in publications (e.g., in sale catalogs) since milk weights are recorded by a DHIA supervisor.

4.5.6 How DHIA Records Are Processed

After the DHIA supervisor obtains the basic information from the dairy farm during a monthly visit, the barn sheet is mailed to a state or regional processing center. There, the data are checked for completeness and accuracy then processed electronically and combined with records of previous tests. Appropriate computations are then made. Current herd and cow records are sent to cooperating milk producers. Costs of processing are paid by herd owners. Summaries and analyses of data are provided to dairy extension personnel in each state on a monthly and annual basis.

The firm Dairy Records Management Systems (DRMS) processes DHIA test-day data and maintains a database of this information. DHIA management reports can be printed and mailed monthly to producers, downloaded from DRMS, or received by e-mail. Mailed reports include over 60 predesigned lists and summaries as well as custom designed reports. Producers can choose between eight monthly reports with individual cow data, four cow pages, three somatic cell count summaries, as well as a variety of herd data summaries and high level management reports.

DRMS provides a glossary that explains codes used in reporting dairy records. For example, for code 305-2X-ME, the lactation record represents a cow with at least 50 days in milk that is extended to 305 days in milk; the data have been adjusted for location, age, and season of calving to a mature cow basis and adjusted to twice-a-day milking basis if milked more than twice a day (see section 4.6).

The exact form of the DHIA report varies among states and regions. All include calving or dry dates and number of days in production during the current lactation. Most have a system of "flagging" cows due to be bred, pregnancy checked, or dried off. Most reports of DHIA production records project the expected production of cows 60 to 75 days beyond calving, thus providing a reliable estimate of productivity for the balance of that lactation. Additionally, complete lactation records are computed and reported for every cow on test (all cows in participating herds must be tested) for each calving.

4.5.6.1 Individual Cow Report

Several forms and reports are available to participants in the DHI program through DRMS. All are important and contribute to the total dairy herd records system. Form 200 is most important; it is the monthly individual cow report and is a summary sheet projecting a complete cost-production analysis and status of each cow in the herd. Included are sire ID, milk weights (last test/this test/lactation), percentages of fat and protein, amount of concentrates fed, days in milk, projected 305-day ME, breed or **estrus** date, calving due date, persistency of lactation curve, **days dry**, days open, age at calving, and body weight. For most dairies somatic cell count (SCC) data on individual cows are vital in management of the herd.

4.5.6.2 Herd Summaries

The Standard Herd Summary (DHI-202) provides a comprehensive herd analysis and management report including production, reproduction, genetics, udder health, and feed cost information. If a herd is enrolled in the permanent string option, a herd summary is printed for each string.

4.6 Standardizing Milk Production Records

In comparing and evaluating production records of cows it is desirable that they be adjusted to a common basis of 305-day (305D) lactations, twice daily milking (2X), and **mature equivalency (ME)**. In research they are also commonly adjusted to reflect an approximately equal energy base, that is, 4% **fat-corrected milk (FCM)**.

4.6.1 DHIA 305D Projection Factors

These are used to project incomplete DHIA lactations to a 305D basis (also called 305D-2X-ME).[7] Projection factors vary significantly among dairy breeds and age groups.[8] Therefore, records on milk and fat are published by breed, above or below three years of age, and according to days in milk. For example, suppose three 4-year-old Holstein cows are injured and must be dried off at 200, 250, and 300 days. To project these three lactations to 305D, we would multiply pounds of milk produced by the appropriate factors of 1.30, 1.11, and 1.01, respectively. Note that the factors become progressively smaller as we approach 305D. Conversely, if a cow milks beyond 305D and we wish to convert the actual record to 305D, a factor of less than 1.0 is used.

A persistency of lactation curve (cf. chapter 10) that projects the cow's 305D milk equivalent (ME) production can be constructed after 80 days in milking. A persistency of 100 indicates that the lactation curve is similar to the average for animals of the same breed age and season of calving. The formula is:

$$\frac{\text{new projection}}{\text{base projection}} \times 100 = \text{persistency of lactation curve}$$

4.6.2 Predicted Transmitting Ability

The predicted transmitting ability (PTA) is the predicted difference (PD) of a parent animal's offspring from breed average due to the genes transmitted from that parent. Each PTA is given in the units used to measure the trait: milk in pounds, productive life in months. All first lactation animals in milk 40 days or more and all second and later lactation cows properly identified by sire should have USDA PTA values (cf. chapter 5).

4.6.3 Mature Equivalency

Adjusting production records for age differences is necessary to compare genetic merit of females of different ages since average milk yield increases with age to maturity and then slowly declines. Therefore, age correction factors are used to adjust production records to a common ME basis (McDaniel et al., 1967). This permits greater accuracy in comparing records and in evaluating the PD of dairy bulls. ME factors for a particular age are determined by dividing average production of mature cows within a given breed by average production of cows of that age. Thus, the ME factor for 24 months equals the average of mature cows divided by the average for cows calving at 24 months of age. Let us calculate a sample ME factor. Assume the average mature Guernsey cow yields 17,400 lb of milk at maturity (66 to 84 months of age) and 13,900 lb of milk when calving at 24 months. The age factor at 43 months would be 1.25 (17,400 / 13,900 = 1.25). Age factors decrease with age to maturity and then increase slowly. There are breed differences in ME conversions, which reflect differences in rate of maturity among dairy breeds.

4.6.4 Frequency of Milking

As noted in chapter 10, more milk will be obtained as the number of milkings per day increases. However, labor availability and cost are major factors. Automated/robotic milking can facilitate 3X milking, but investment cost must be carefully considered. The producer needs to have minimal debt before investing in a robotic system. In the United States, $6,000/cow is considered the maximal investment for a well-managed herd.

To compare and evaluate production records appropriately, data from 3X are converted to 2X. A 2-year-old cow milked 2X daily would be expected to produce only 83% as much milk as she would if milked 3X daily during the same lactation. A mature cow would be expected to produce 86% as much milk. Therefore, a 2-year-old cow yielding 18,000 lb of milk on a 3X basis would have her record adjusted to a 2X equivalent by multiplying the actual record times the conversion factor of 0.83 (18,000 × .83 = 14,940).

4.6.5 Fat-Corrected Milk

Since solids of milk vary and since milkfat contains approximately one-half the calories of milk, it is of interest, especially in research, to convert milk records to an approximately equivalent energy basis. This permits researchers more appropriately to compare milk production efficiency of cows that yield milks of varying composition.[9] Dr. W. L. Gaines, University of Illinois, developed the commonly accepted formula:

$$\text{FCM} = (0.4 \times \text{milk production}) + (15 \times \text{milkfat production})$$

The following example applies that formula. Suppose a Brown Swiss cow yielded 15,000 lb of milk testing 3.8% milkfat.

$$\begin{aligned}\text{FCM} &= (0.4 \times 15{,}000) + [15 \times (3.8\% \times 15{,}000)] \\ &= 6{,}000 + (15 \times 570) \\ &= 6{,}000 + 8{,}550 \\ &= 14{,}550 \text{ lb}\end{aligned}$$

Now let us assume a Jersey cow (A) yields 12,000 lb of milk testing 5.0% fat and a Holstein cow (B) yields 18,000 lb of milk containing 3.5% fat. Computations that follow show that the FCM of cows yielding milk with a fat percentage above 4.0% will be greater than the poundage actually produced, whereas FCM of cows yielding milk testing below 4.0% will be less than that actually harvested. Thus:

Cow A:

$$\begin{aligned}\text{FCM} &= (0.4 \times 12{,}000) + [15 \times (5.0\% \times 12{,}000)] \\ &= 4{,}800 + (15 \times 600) \\ &= 4{,}800 + 9{,}000 \\ &= 13{,}800 \text{ lb}\end{aligned}$$

Cow B:

$$\begin{aligned}\text{FCM} &= (0.4 \times 18{,}000) + [15 \times (3.5\% \times 18{,}000)] \\ &= 7{,}200 + (15 \times 630) \\ &= 7{,}200 + 9{,}450 \\ &= 16{,}650 \text{ lb}\end{aligned}$$

4.6.6 Energy-Corrected Milk Ratings

A letter rating is used to categorize cows in five production groups based on their current lactation. Breed adjustments are applied when calculating ratings for multibreed herds. This adjustment standardizes comparisons for breed differences in average energy-corrected milk (ECM). To determine the 305-2X-ME rating records for all cows are adjusted to an ECM basis. The ECM for each cow is divided by the ECM lactation average for the herd and the results, compared to the herd average, are designated as follows:

A = Top cow (more than 110%)
B = Above average (100–110%)
C = Below averagc (90 to 100%)
D = Marginal cows (80 to 90%)
E = Probable cull cows (less than 80%)

4.6.7 Linear Somatic Cell Count Score

The somatic cell score (SCS) is a linear score that is assigned based on estimated numbers of somatic cells (cf. chapter 14) in the raw milk, and it has a direct relationship to milk loss. Table 4.1 shows this relationship. The milk loss is based on second or later lactation animals. First lactation animals would have half of the loss listed.

The lactation average linear score (LALS) is the average of all linear scores accumulated over the current lactation of the cow. The LALS should be compared to the current test day linear score to monitor changes in udder health during the current lactation. Average linear scores are less influenced than average SCC scores by one or two high SCC results. Cows with chronic **mastitis** have a high LALS and should be considered for culling.

The lifetime average linear score is the average of all linear scores over the lifetime of the cow (or the time on the DHI somatic cell count option). This lifetime score should be compared with the lactation score to compare udder health from previous lactations with that of the current lactation.

4.6.8 Merit Value (Net, Cheese, or Fluid)

Merit value estimates the additional net profit that an offspring of an animal will provide during its lifetime. The measure makes adjustments for both income and expenses for a typical dairy operation to produce a measure of overall net profit. There are three formulas for merit and the primary difference between them is the emphasis placed on milk's components (cf. VanRaden, 2011). Producers should select the index that is closest to the milk payment policy in their area.[10]

Table 4.1 Relationship of Linear Somatic Cell Count Score to Milk Loss.

Score	Linear SCC Score Midpoint (1,000/mL)	Estimated Milk Loss (lb/day)	(lb/lactation)
0	12.5	0	0
1	25	0	0
2	50	0	0
3	100	1.5	400
4	200	3.0	800
5	400	4.5	1,200
6	800	6.0	1,600
7	1,600	7.5	2,000
8	3,200	9.0	2,400
9	6,400	10.5	2,800

Sources: Western Canadian Dairy Herd Improvement Services, n.d.; National Mastitis Council, n.d.

4.7 Dairy Production Records

The most valuable time any dairyman spends is in studying his cows.
Harry A. Herman (1905–1995)

4.7.1 Using Dairy Production Records

The more effectively records are used, the more valuable they become. For example, in studies at Michigan State University (Erickson and Heisner, 1973) of 37 high-producing herds (17,694 lb per cow) and 37 average-producing herds (13,397 lb per cow), a striking difference was noted in time spent studying the monthly DHI report (4.3 and 2.1 hours per month, respectively). Obviously, managers in the high group made better use of DHI records in management decisions. Recommended management practices were more frequently performed in high-producing herds, for example, vaccinating calves for IBR, BVD, and PI-3 (65 versus 27%); dry treatment for mastitis (89 versus 57%); percentage of lactating cows sired by proven bulls (68 versus 48%); and testing hay for nutrients (32 versus 14%).

When studied and evaluated properly, production records serve as the herd manager's launch pad. Armed with production and profit data, managers can make wise decisions relating to culling, feeding, management, and sire selection. Let us discuss each briefly.

4.7.1.1 Culling

Almost every herd has unprofitable cows. Records provide the means of identifying them. Factors to consider are estimated and demonstrated producing abilities, production in comparison with that of **herdmates**, profit potential (present and future), and the milk and profit potential of replacement females. Culling decisions involve an analysis of income and expenses and include such factors as feed supply, labor availability, housing and equipment space, age of cow, speed and completeness of milking, conception rate, market price of beef, and health aspects of cows. Proper timing of culling can greatly increase profits.

4.7.1.2 Feeding

The feeding program is focused on meeting the herd's nutritional needs. Feed represents the largest single cost of producing milk or meat. A good feeding program complements good breeding and efficient milking by allowing cows to produce near their genetic potential. Production records aid in determining how much grain and protein to feed and whether feed costs and returns above feed costs are optimal.

4.7.1.3 Management

Decision making is easier for some managers than others. But it is easier for all when they have facts. Dairy records provide these facts and offer accurate information. Records are useful in determining if herd production is satisfactory: How does it compare with other herds of similar size and conditions? Is it above or below breed average production (table 4.2)? What about costs and returns? Has an excessive investment been made in equipment and/or facilities for the size of the operation? Does the herd production peak with seasonally high milk prices?[11] Is the operation yielding above average earnings per employee? Are there too many days dry or are the average calving intervals too long?

4.7.1.4 Sire Selection

The proper selection of sires is one of the most important decisions a herd manager can make to ensure profits three to six years forward. Careful selection of sires with high PD for milk and income leads to higher-producing, profitable replacements for the present herd. Without production records, selecting and mating animals to produce the next generation is largely a guessing game. Records provide the basis for sire selection and for measuring the effectiveness of this selection. DHI records provide the base for sire proofs (predicted difference). Owners who use artificial insemination (AI) but who do not participate in DHI testing are sharing benefits made possible and paid for by those who participate in DHI testing.

4.7.2 DairyMetrics©

This is a web-based system from DRMS, which enables producers and consultants to benchmark a herd's performance with the performance of other herds (cohorts). Managers can compare a herd's information relative to herds in a similar "state" or to herds in a more desirable "state." Cohorts can also be chosen by identity and placed in peer-groups to compare against a target herd or against past performance. Regardless of the method, almost everyone needs the ability to identify weaknesses and acknowledge improvements. Comparison analysis can also be used to evaluate dairy manager performance.

The DairyMetrics© database is updated nightly and includes information from more than 15,000 herds that are routinely processed by DRMS. DairyMetrics© filters are used to identify cohort herds. For example, filters could include only Holsteins, 2X milking, 200 to 499 cows, summit milk on first lactation cows over 85, and SCC less than 200,000. Eighty-five filters and items can be displayed from the areas of general, production, udder health, reproduction, and genetics.

4.7.3 Identification of Dairy Cattle

There are several ways and means of assuring animal identity. An important service of DHIAs is assisting in development of reliable animal identification (ID) records. These allow every animal in the herd to be identified positively with sire, dam, and date of birth. The more managers know about individual cows, the better prepared they are to manage well.

All cows in the milking herd must be identified with an eartag or registration number. The national numbering plan prevents coded eartags from being duplicated. Eartags used by AI technicians, veterinarians, and DHIA supervisors comply with this numbering plan and are used interchangeably for animal ID. The first two digits of the coded tag are the state code, followed by three letters and four numbers. Cows are identified by both index number and barn name on most DHI reports.

Registration numbers are required in the United States for registered purebred dairy cattle. They are cross-referenced with supplemental ID, such as sketches of color markings or photographs in the broken-colored breeds (Ayrshire, Guernsey, and Holstein) or ear tattoos in other dairy breeds.

Several of the breed associations offer nationally approved animal identification tags with the prefix 840, the US identifier. This number also meets the requirement for animal traceability, a requirement adopted by the USDA in 2013. All animals crossing state lines must have official identification numbers unless they are to be delivered directly to a slaughter plant.

Table 4.2 Changes in Average Milk Production by Breed for Herds on DHIA Testing 1980–2011.

Breed	1980	1990	1995	2000	2005	2006	2007	2008	2009	2010	2011
Ayrshire	13,144	14,799	15,684	17,389	18,303	18,232	18,220	18,145	18,238	18,275	18,295
Brown Swiss	14,172	16,250	17,493	20,300	21,405	21,413	21,533	21,522	21,736	21,857	21,996
Guernsey	11,666	13,297	14,051	16,043	17,154	17,394	17,243	17,274	17,269	17,351	17,349
Holstein	17,566	20,178	21,618	24,517	25,644	25,674	25,639	25,567	25,635	25,757	25,987
Jersey	11,437	13,407	14,842	17,038	18,087	18,159	18,265	18,328	18,315	18,565	19,004
Milking Shorthorn	11,560	14,011	15,341	16,548	17,582	17,502	17,461	17,040	17,607	18,137	18,319
Red and White	16,470	18,832	20,161	22,845	23,815	23,764	23,637	23,459	23,042	23,342	23,439

Source: Norman et al., 2013.

In order to assure proper identification, options include ear, leg, and tail tags; neck chains; photographs; or brands (fire or freeze). Eartags are the most widely used means of identifying dairy animals, especially grades. They are made of steel, aluminum, nylon, or plastic. Leg and tail tags are used in some commercial herds for quick ID of cows; they are seldom permanent, however. Freeze branding is increasing in popularity. It is not as painful as fire branding and does not damage the animal's skin. In freeze branding, dry ice and alcohol or liquid nitrogen may be used to superthin a branding iron which, when applied to the animal's hide, will destroy the pigment-producing cells responsible for pigmentation of hair. Destruction of these cells results in regrowth of white hair in the area branded. This is especially helpful in identifying cows milked in herringbone milking parlors (cf. 12.2.3).

Permanent identification devices include (1) conventional plastic or metallic eartags, which may have both visual and machine readable symbology (e.g., numerals and barcodes) and (2) electronic identification devices, including transponders and corresponding transceivers. Radio-frequency identification (RFID) involves a wireless noncontact system that uses radio-frequency electromagnetic fields to transfer data from a tag attached to an object for the purposes of automatic identification and tracking. The RFID devices most widely used on animals are passive in nature. They have no battery or source of power of their own, but pick up the energy they require from the scanner. Devices of this type are called transponders, and small transponders are commonly called microchips. This method of energizing the transponder severely limits the effective operating distance between the scanner and the transponder.

The International Standards Organization (ISO) has published RFID standards for animal identification. ISO standard 11784 describes the structure and information content of the radio-frequency identification code for animals. It sets the length of the 132-bit binary message sent by the transponder to the scanner and the meaning of every bit in it. It is a national responsibility to ensure that transponders carry a unique number within that country, and ideally every country should maintain databases of information about all issued codes and the associated animals. Number uniqueness is supported by the use of a country code (based on another ISO standard) and/or a manufacturer code assigned by ICAR. The ISO 11785 standard defines the technical aspects of communication between transponder and scanner. It sets the frequencies of activation and response, the encoding format, and the precise interaction between the scanner and the transponder.

The purposes of ICAR, headquartered in Bonn, Germany, are (1) to standardize procedures and methods used in recognizing and recording identities of livestock and (2) to establish test procedures for the approval of equipment and methods of recording. For many years ICAR has had the responsibility for registration systems used in animal recording including individual animal recognition through reliable permanent identification devices.

The following variables are considered in testing the performance and reliability of identification devices: ease of application and use, efficiency of animal recognition, durability, tamperproof quality, and animal welfare.

4.7.4 Health Records

The dairy manager and veterinarian should work together in maintaining herd health records (cf. chapter 13). These should include animal name and ID; immunizations and vaccinations; dates of estrus, breeding, pregnancy diagnosis, and calving; sex, size, and health of calves; diagnosis and treatments of reproductive problems and/or diseases; injuries; mastitis and udder health information; and therapy of other diseases or disorders (e.g., ketosis, parturient paresis, ingested hardware, and grass tetany).

4.7.5 Other Aspects of Records

We have noted that records aid milk producers in making profitable decisions. They are useful in other ways. Complete herd records are vital to the reporting and computing of taxes and obtaining credit. Former students with production credit associations and other lending agencies have advised the authors that they give preference in lending money to clients with complete records.

4.8 Summary

Production testing is a powerful tool for improving the milk-producing ability of dairy cattle. Successful herd owners consider their production records as vital evidence that, when studied objectively, yields information concerning success or failure of their breeding, feeding, and management programs. A dairy herd is seldom so outstanding that it cannot be improved through effective culling, feeding, and use of production-tested breeding stock. Records hold the key to a herd improvement program. Participation in a DHI association is the best way a dairy farmer can obtain reliable records that help in management.

There are two categories of DHI production testing—official and unofficial. The former includes Standard DHIA and DHIR; the latter includes unofficial DHI testing plans such as owner-sampler, milk-only record, and alternate AM-PM records.

Central laboratories perform instrumental testing of milk for fat, protein, and somatic cells and produce records that provide milk producers with the most accurate information in the history of production testing.

In making meaningful comparisons of milk production between individual cows and their herdmates, records must first be standardized for age (ME), times milked per day (2X), and length of lactation (305D).

A reliable system of individual animal identification is a vital part of an efficient dairy herd record and management program. Breeding, feeding, and culling practices are dependent on accurate ID of individual animals. Eartags, tattoos, photographs, branding, and other means of identifying dairy cattle are employed. Use of freeze branding by dairypersons as a permanent, rapid, accurate method of ID is increasing.

DHI testing is a means of determining the true worth of a cow. Basic production records obtained in the DHI program provide information needed by managers to improve efficiency and profitability of their dairy enterprises. Additionally, they provide data by which animals to be used in producing the next generation can be identified and selected to assure genetic improvement. Hence, results obtained from the DHI testing program are the foundation for improved dairy cattle breeding in the United States.

Dairy farmers could do few things more essential in increasing the profitability of their operations than to keep reliable records and to fully exploit them as they strive to provide an abundant and reliable supply of milk.

Study Questions

1. Discuss briefly labor and cost differences between low- and high-producing cows.
2. Cite specific information provided to dairy managers through DHI records.
3. What European country pioneered cow-testing associations? When was the first one organized in the United States?
4. Describe the use of production testing programs in the United States. Is the percentage of users increasing or decreasing? How does use in the United States compare to use in Europe?
5. Describe the factors needed to standardize milk production records to 2X, 305D, ME. How is 4% FCM calculated?
6. What are the purposes of the NCDHIP? How do extension services work within this production-testing program?
7. Describe how a typical DHIA operates. What are the major responsibilities of the DHIA supervisor?
8. Discuss the four principal DHI production record plans.
9. Differentiate between DHIA and DHIR.
10. Discuss how DHIA records are processed.
11. What management information may be gleaned from the Herd Summary report? What other DHI forms and reports are available to dairy farmers?
12. Cite examples of benefits that dairy cattle managers may derive from production testing.
13. From an economic and management perspective, why is it important for dairy managers to use production records generously?
14. Name several alternatives available to dairy farm personnel for use in identifying dairy animals.
15. What information should be provided in herd health records?
16. Describe what you have learned from your study of chapter 4. How will that knowledge influence your career within the dairy farm enterprise?

Notes

[1] The term dairy herd improvement (DHI) is used to refer to the entire testing-records program in the United States. A dairy herd improvement association (DHIA) is a local organization through which testing and record-keeping are facilitated.

[2] The DHIA supervisor is responsible for eartagging nonregistered or previously untagged cows.

[3] The question naturally arises if, on test day, the cow might yield more or less on that day compared to other days of the month. A cow is more likely to yield significantly less than significantly more milk on a given test day when compared with average daily production. However, small differences spread over a lactation tend to average out. Thus, the one-day test per month has been shown experimentally to be an accurate representation of the cow's milk production.

[4] Local DHI associations are assisted and advised by personnel of the CES. Usually, an extension dairy specialist of the college of a land-grant university serves as extension adviser to the DHI program of that state.

[5] Surprise tests are also conducted on cows enrolled in DHIA testing programs.

[6] Sire proving was initiated on a national basis as part of the DHIA program in 1936 (cf. chapter 6).

[7] This procedure became effective July 1, 1965 and is used in USDA sire and cow evaluations to standardize records for days in milk. The projection factors are based on more than 162,000 DHIA lactation records and were published in B. T. McDaniel, R. H. Miller, and E. L. Corley. 1965, August. DHIA factors for projecting incomplete records to 305 days. *Dairy Herd Improvement Letter* 41(6), ARS-44-164.

[8] Because persistency of lactation changes with age, two sets of factors have been developed—one for cows less than three years of age and one for cows greater than three years of age.

[9] It has been shown experimentally, for example, that approximately 15% more dietary energy is required to secrete milk of 4.5% milkfat than that of 3.5% fat.

[10] More information about net merit may be found at http://www.ars.usda.gov.

[11] Missouri studies showed that herds enrolled in DHI testing exhibit significantly less seasonal variation in milk production than non-tested dairy herds.

References

Erickson, R. W., and B. M. Heisner. 1973. What the top herds do. *Hoard's Dairyman* 118:166–167.

McDaniel, B. T., R. H. Miller, E. L. Corley, and R. D. Plowman. 1967, February. DHIA age adjustment factors for standardizing lactations to a mature basis. *Dairy Herd Improvement Letter* 43(1), ARS 44-188.

National Mastitis Council. n.d. The Value and Use of Dairy Herd Improvement Somatic Cell Count (http://nmconline.org/dhiscc.htm).

Norman, H. D., T. A. Cooper, and F. A. Ross, Jr. 2013. *State and National Standardized Lactation Averages by Breed for Cows Calving in 2011*. Beltsville, MD: USDA Animal Improvement Programs Laboratory.

VanRaden, P. M. 2011. *Net Merit as a Measure of Lifetime Profit: 2006 Revision Updated 2008/2011* (Multi-State Project S-

1008). Beltsville, MD: USDA Animal Improvement Programs Laboratory.

Western Canadian Dairy Herd Improvement Services. n.d. Somatic Cell Counts (http://www.agromedia.ca/ADM_Articles/content/dhi_scc.pdf).

For Further Study

1. While the costs of testing will vary by region and/or state, take a look at an example provided by Mid-South Dairy Records (http://www.mydhia.org/page.php?id=2).
2. The NDHIA is the gateway to policy and policy making in DHI.
   a. Learn more about the goals and objectives of the NDHIA by visiting http://www.dhia.org/objectives.asp.
   b. Familiarize yourself with the NDHIA's *Uniform Data Collection Procedures* (http://www.dhia.org/udcp.pdf).
3. Explore and learn more about DHI reporting:
   a. Review Dairy Records Management Systems' guide, *DHI Report Options* (http://www.drms.org/pdf/materials/optman.pdf).
   b. The Herd Summary Report (form DHI-202) contains vital information that can help managers make decisions. Take a look at a sample report from the DRMS (http://www.drms.org/pdf/Reports/202-Herd-Summary.pdf).
   c. In order to decipher the DHI-202 form, DRMS provides a glossary (http://www.drms.org/PDF/materials/glossary.pdf).
4. The International Committee for Animal Recording provides guidelines for identification devices (http://www.icar.org/Documents/sc_recording_devices/Permanent%20 identification%20devices.pdf).
5. For links related to breeds and breeding in the United States as well as around the world, visit the AGIL website (http://ars.usda.gov).

5

Evaluating Dairy Cattle

Emphasis on type must be directed primarily on those traits which will assist the cow in reaching and maintaining high levels of production over long periods of time with those being: mammary system, feet and legs, size, and body capacity.

James C. Rennie (1926–)

5.1 Introduction

Mammary glands are basic building blocks of the dairy industry. Yet the **type** or **conformation** of animals to which these glands are attached is also important in producing milk because conformation affects intensity, length, and number of lactations. Desirable dairy conformation involves functional traits associated with high milk production over a long productive life.

There are two principal criteria for evaluation of dairy cows: (1) performance in the pail (cf. chapter 4 and 6) and (2) physical characteristics that affect attractiveness, milk production, and longevity. Since only a relatively small percentage of dairy cows are production tested, conformation or dairy type is often the major basis for selection. Therefore, it is important to increase our knowledge and understanding of conformation. In this chapter, we attempt to relate physical characteristics with functional aspects of dairy cattle.

5.2 Evaluating the Physical Characteristics of Dairy Cattle

There are two general means of evaluating the physical characteristics of registered dairy cattle, type classification and show-ring judging.[1]

The purpose of **type classification** is to encourage the breeding of dairy animals that are both functional and beautiful. It is sponsored by dairy breed associations who employ experienced and competent dairy judges to serve as classifiers. Each animal is scored (at the farm) in comparison with a theoretically perfect animal of the breed. Points are assigned to important physical characteristics and a total score classifies the animal as excellent (90–99), very good (85–89), good plus (80–84), good (75–79), moderate (70–74), and poor (65–69) (table 5.1). The following is a brief list of these characteristics:

- **Size**—stature, length, width, and body capacity in relation to age
- **Type**—strength and general **openness**
- **Udder**—teat size and placement, udder depth, strength of the suspensory ligament
- **Legs and feet**—set and bone quality plus functional use

The characteristics are scored in the range from 50 (or 60) to 100 depending on the breed. The average score per general characteristic of cows classified is traditionally set at 80 points. Therefore, due to the progress made in conformation over the years, a score of 85 points for a general characteristic in 2000 is significantly better than in 2010. The more the cow corresponds with the conformation standard (ideal cow), the higher the score.

Show-ring judging is practiced at local, state, regional, and national shows in which dairy judges consider all animals in a class and arrange them in order of excellence, giving due consideration to age differences and stage of lactation (in cow classes). Classes are usually based on breed, sex, and age in accordance with guidelines published by the Purebred Dairy Cattle Association (PDCA). Class winners within each breed compete for championships and grand championships within each sex.

Information pertaining to strengths and weaknesses of individual cows, cow families, and sire progeny is useful to herd owners in making adjustments in breeding programs (cf. chapter 6). The seven dairy breed registry associations in the United States are represented in the PDCA, which leads in developing uniform rules for the following: (1) official milk production testing programs, (2) governing artificial insemination of dairy cattle, (3) animal health regulations, (4) dairy cattle exporting, (5) show-ring classifications, and (5) sale ethics and practices.

5.3 Parts of the Dairy Cow

Whether student, herd owner, dairy cattle judge, or veterinarian, it is important to know the parts of a cow. The student will find such knowledge useful in many ways; for example, in giving a set of oral reasons (one measure of competency in evaluating the physical merits of dairy cattle in judging). If a cow gets injured severely, the herd owner will want to use the proper terms in communicating with the attending veterinarian. Thus, one of the commandments for those interested in dairy cattle is to learn their parts (figure 5.1).

Table 5.1 Dairy Breed Type Classification Scores and Terms.

Description	Holstein	Guernsey	Jersey	Ayrshire	Brown Swiss	Milking Shorthorn
Excellent	90–97	90+	90+	90+	90+	90+
Very good	85–89	80–89	80–89	85–89	85–89	85–89
Good plus/desirable[a]	80–84	70–79	70–79	80–84	80–84	80–84
Good	75–79			75–79	75–79	75–79
Fair/acceptable[b]	65–74	60–69	60–69	65–74	65–74	65–74
Poor	50–64	50–59	50–59	50–64	60–64	50–64

[a]Guernsey and Jersey classifiers use desirable.
[b]Jersey classifiers use acceptable.

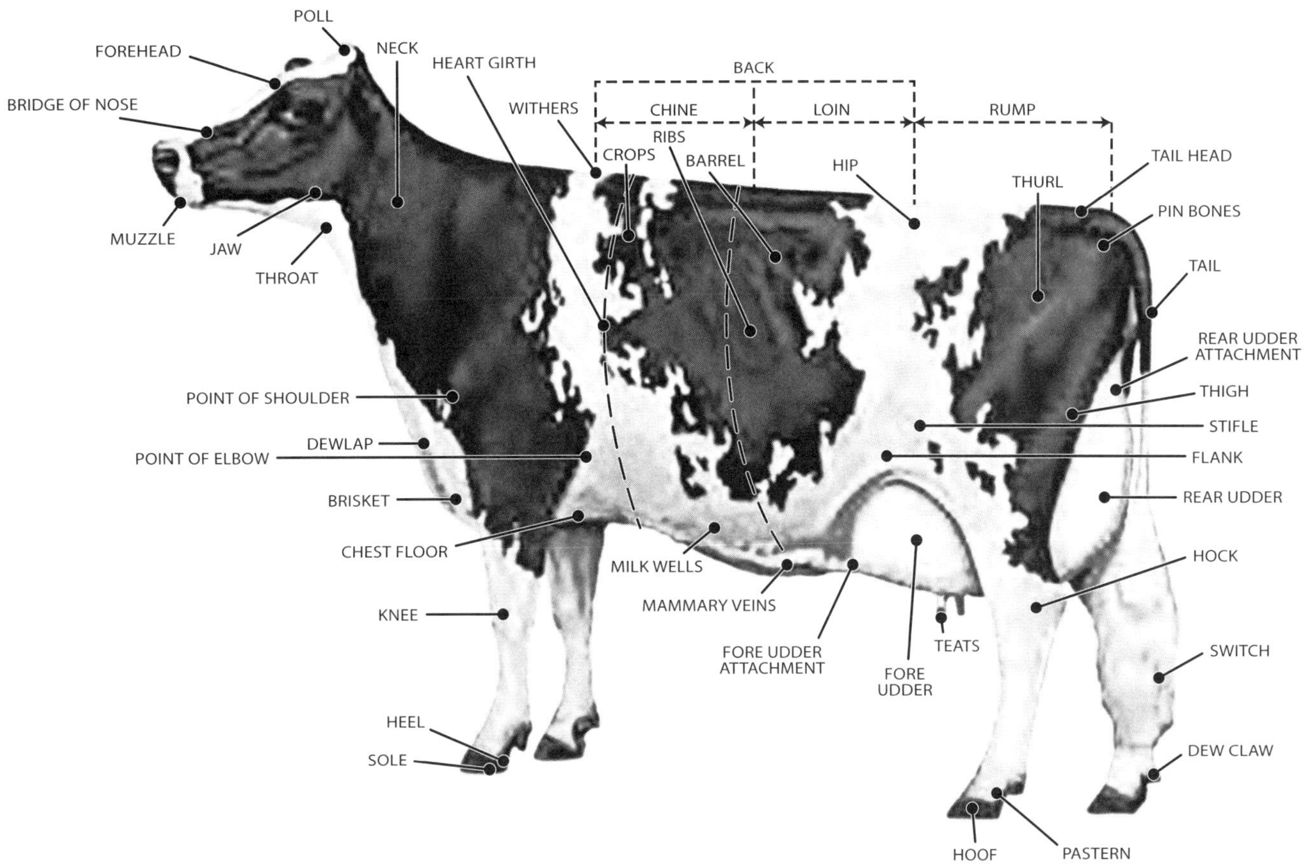

Figure 5.1 Parts of the dairy cow (courtesy of Purebred Dairy Cattle Association).

5.4 True-Type Dairy Cows

Writing in 1 BC, Varro described ideal Roman cattle as:

> . . . well made, sound, long and deep bodied with long horns, broad forehead, large black eyes, hairy ears, close-set jaws, flat noses, with the back gently sloping downwards from broad high shoulders, wide nostrils, black muzzles, thick long necks, dewlaps hanging from the throat, a wide body, well ribbed, good rump, with tail hanging down to their heels, knees wide apart and large, the feet not broad nor sounding as they go, the two parts of the hoofs not spreading far apart, the different hoofs of the same size, and small, the hide not rough nor hard to touch. In color black is preferable, the next red, third pale red, and the fourth white.

After nearly 2,000 years we still seek many of the aforementioned traits in dairy cattle. The first dairy cattle scorecard was published in 1828 by breeders on the isle of Guernsey; it set forth their ideas of type for the Guernsey breed. Of interest was the early emphasis given skin and its pigmentation. "A cow is judged by the mellow feeling of the hide, by the deep yellow circle around the eyes; the tip of the tail and the inside of the ears also should be yellow" (John Jacob, 1830).

The first American Guernsey scorecard was developed in 1887. It was revised several times, and in each revision emphasis on general appearance was reduced while that given the mammary system was increased. Patterns of development and revision of scorecards were similar among other dairy breeds. In 1943, the PDCA developed, adopted, and published a unified scorecard for all dairy breeds; it was revised five times, the last being in 2009 (figure 5.2 on the next page). This all-breed dairy cow unified scorecard serves as a guide in evaluating dairy cattle. The scorecard itself is not used, however, in the show-ring.

5.5 Studying the Scorecard

The most important thing a dairy cow does is give milk.

J. L. McKitrick

The Classification Scorecard used by Brown Swiss, Milking Shorthorn, Red and White, Dutch Belted, and Linebeck associations consists of the following five divisions: strength and substance (15%), dairy quality (15%), rump (10%), mobility (20%), and mammary system (40%). Each trait is broken down into body parts to be

DAIRY COW UNIFIED SCORECARD

Breed characteristics should be considered in the application of this scorecard.

MAJOR TRAIT DESCRIPTIONS

Perfect Score

There are four major breakdowns on which to base a cow's evaluation. Each trait is broken down into body parts to be considered and ranked.

1) Frame - 15%

15

The skeletal parts of the cow, with the exception of rear feet and legs. Listed in priority order, the descriptions of the traits to be considered are as follows:

Rump (5 points): Should be long and wide throughout. Pin bones should be slightly lower than hip bones with adequate width between the pins. Thurls should be wide apart. Vulva should be nearly vertical and the anus should not be recessed. Tail head should set slightly above and neatly between pin bones with freedom from coarseness. **Front End (5 points):** Adequate constitution with front legs straight, wide apart, and squarely placed. Shoulder blades and elbows set firmly against the chest wall. The crops should have adequate fullness blending into the shoulders. **Back/Loin (2 points):** Back should be straight and strong, with loin broad, strong, and nearly level. **Stature (2 points):** Height including length in the leg bones with a long bone pattern throughout the body structure. Height at withers and hips should be relatively proportionate. Age and breed stature recommendations are to be considered. **Breed Characteristics (1 point):** Exhibiting overall style and balance. Head should be feminine, clean-cut, slightly dished with broad muzzle, large open nostrils and strong jaw.

2) Dairy Strength - 25%

25

A combination of dairyness and strength that supports sustained production and longevity. Major consideration is given to general openness and angularity while maintaining strength, width of chest, spring of fore rib, and substance of bone without coarseness. Body condition should be appropriate for stage of lactation. Listed in priority order, the descriptions of the traits to be considered are as follows:

Ribs (8 points): Wide apart. Rib bones wide, flat, deep, and slanted towards the rear. Well sprung, expressing fullness and extending outside the point of elbows. **Chest (6 points):** Deep and wide floor showing capacity for vital organs, with well-sprung fore ribs. **Barrel (4 points):** Long, with adequate depth and width, increasing toward the rear with a deep flank. **Thighs (2 points):** Lean, incurving to flat and wide apart from the rear. **Neck (2 points):** Long, lean, and blending smoothly into shoulders; clean-cut throat, dewlap, and brisket. **Withers (2 points):** Sharp with chine prominent. **Skin (1 point):** Thin, loose, and pliable.

3) Rear Feet and Legs - 20%

20

Feet and rear legs are evaluated. Evidence of mobility is given major consideration. Listed in priority order, the descriptions of the traits to be considered are as follows:

Movement (5 points): The use of feet and rear legs, including length and direction of step. When walking naturally, the stride should be long and fluid with the rear feet nearly replacing the front feet. **Rear Legs-Side View (3 points):** Moderate set (angle) to the hock. **Rear Legs-Rear View (3 points):** Straight, wide apart with feet squarely placed. **Feet (3 points):** Steep angle and deep heel with short, well-rounded closed toes. **Thurl Position (2 points):** Near central placement between the hip and pin bones. **Hocks (2 points):** Adequate flexibility with freedom from swelling. **Bone (1 point):** Flat and clean with adequate substance. **Pasterns (1 point):** Short and strong with some flexibility, having a moderate, upright angle.

4) Udder - 40%

40

The udder traits are evaluated. Major consideration is given to the traits that contribute to high milk yield and a long productive life. Listed in priority order, the descriptions of the traits to be considered are as follows:

Udder Depth (10 points): Moderate depth relative to the hock with adequate capacity and clearance. Consideration is given to lactation number and age. **Rear Udder (9 points):** Wide and high, firmly attached with uniform width from top to bottom and slightly rounded to udder floor. **Teat Placement (5 points):** Squarely placed under each quarter, plumb and properly spaced. **Udder Cleft (5 points):** Evidence of a strong suspensory ligament indicated by clearly defined halving. **Fore Udder (5 points):** Firmly attached with moderate length and ample capacity. **Teats (3 points):** Cylindrical shape; uniform size with medium length and diameter; neither short nor long is desirable. **Udder Balance and Texture (3 points):** Udder floor level as viewed from the side. Quarters evenly balanced; soft, pliable, and well collapsed after milking. **(Note: In the Holstein breed, an equal emphasis is placed on fore and rear udder (7 points each). All other traits are the same as listed above.)**

TOTAL 100

Ayrshire

Brown Swiss

Guernsey

Red & White

Holstein

Jersey

Milking Shorthorn

evaluated and scored. The Dairy Cow Unified Scorecard (figure 5.2) includes four major sections with percentages of overall value assigned to each: frame (15%), dairy strength (25%), rear feet and legs (20%), and udder (40%). The following discussion will follow this organization.

Study the two cows pictured in figure 5.3 (on the following page). Note the pleasing appearance of cow (A). Her udder, dairy strength, frame, and rear legs and feet are each superior to those of cow (B). Make frequent reference to these two cows as we now consider each category of the scorecard, beginning with the most important, the udder or mammary system.

5.5.1 Udder

A cow without an udder is like a ship without a rudder.

H. W. Willard

The mammary glands of a cow, collectively called an **udder**, are as basic to the dairy industry as the wheel is to an automobile. Essentials of a desirable udder are attachments, size, and quality (figure 5.4 on the next page). The attachments, fore and rear, should be wide and strong as should the median suspensory ligament. Size should come from width and length and not merely from depth (the latter is often associated with a **pendulous** udder). Mammary gland quality is the ratio of secretory (glandular) tissue to other mammary gland tissues (cf. chapter 10). An udder of desirable quality is soft, pliable, and free of lumps, which indicate scar or connective tissue. A hard, meaty udder is undesirable and lacks quality.

It is desirable to have moderate cleavage between glands (halves). For many years dairy cattle breeders selected for flat udder floors. It is now known, however, that flat udders do not remain tightly attached, and soon the teats move to the sides of the udder, making machine milking difficult.

Dairy cattle breeders are now selecting for an udder that throughout life will remain above the hocks where it is more accessible for milking and where there is a low probability of injury (see figure 5.4). A relationship has been established between height of udder and injuries (such injuries may predispose an animal to mastitis). The length, width, and height of rump are related to the length, width, and levelness of the udder floor. The wider a cow is across her hooks and pins, the more room provided for a higher, wider rear udder. Extra length and width of rump, then, provides the cow with a greater capacity for udder development.

5.5.2 Dairy Strength

The small, 900 lb cow is efficient per 100 lb of cow, but so is the pickup truck efficient for a ton of trucking. A football coach would prefer a little guy who works hard to a big fellow who is lazy. But better yet, he prefers the big player who works hard.

L. O. Colebank

Let us define dairy strength as external evidence of milking ability. The inherent capabilities of a dairy cow enable her to utilize nutrients for secreting milk, even to such extent that she will, in times of negative caloric balance, withdraw from her tissues to provide the required nutrients for making milk. Our former teacher, Dr. Samuel Brody,[2] appropriately called this an animal's "lactation drive." (However, thinness resulting from improper feeding should not be confused with indications of **dairy character**.) Dairy strength (also called milkiness, dairy temperament, milkability, and dairyness) is usually indicative of long and high productivity in dairy cows.

5.5.2.1 Body Size

Body size may be expressed in three dimensions: length, breadth, and depth. The relationship is readily apparent between large body capacity and the ability to utilize the large quantities of feed essential to continued high milk production. Three components of body capacity, rib width and length, chest depth, and barrel capacity, are emphasized in the Dairy Cow Unified Scorecard (see figure 5.2).

Sizes of the barrel and chest are closely related to rib configuration. Cows (with rare exception) have the same number of ribs. Therefore, body capacity is gained by increasing depth and spring of ribs. Cows with ribs wide apart are usually longer, and those with long, well-sprung ribs are deeper. It is interesting that most short, round-bodied cows, with ribs short and close together, also lack dairy character and udder capacity. Additionally, they are often too small to provide the volume of milk sought from cows today.

"Substance" and "strength" are terms commonly used to indicate the overall size of bone and body of an individual; whereas, "**constitution**" refers to chest and barrel development and is dependent on breadth between and just back of the forelegs, since the all-important organs of respiration and circulation are located in the chest cavity. "Style" indicates physical fitness and includes symmetry, balance, and a graceful walk.

Figure 5.2 *(opposite page)* Dairy Cow Unified Scorecard developed by the PDCA and the ideal type dairy cow of seven dairy breeds (courtesy of Purebred Dairy Cattle Association).

Figure 5.3 Comparison of (A) an excellent example of Holstein breed character and quality with (B) the Holstein cow of undesirable dairy conformation (courtesy of Holstein-Friesian Association of America).

Figure 5.4 Desirable type udder of a Jersey cow (courtesy of Frank Robinson).

5.5.3 Frame

Frame refers to the skeletal parts of the cow (except for the feet and legs). The evaluator is seeking overall style and balance enhanced by femininity and straightness of lines. Bone size and strength in dairy cattle are dependent, in large part, on bone structure. Rib bones that are wide, flat, and long usually accompany large barrels and are indicative of dairy character.

5.5.3.1 Topline

One of the first physical characteristics noted in observing a cow is topline. A strong topline is usually indicative of strength throughout the cow. A strong back is essential to support the desired large barrel. A strong loin (back muscle) is important to aid in udder suspension, and a wide, long, level rump is needed to facilitate **parturition**. The rump posture is also related to udder carriage. A tilted rump is almost always accompanied by a tilted udder floor. Moreover, rump length and width are fundamental to udder length and width:

> It is pretty difficult to put a 14 ft room on a 10 ft foundation. Since the udder should not be forward of the hocks nor behind the pins, the rump should be long. Additionally, since we seek udder size from width and not depth, the rump needs to be wide. (L. O. Colebank)

5.5.3.2 The Head Is Important

Let a flock of sheep or a herd of cattle lift their heads from their grazing, or a herd of swine raise their snouts from their rooting, or a stable of horses reach their heads through their stall doors, and I'll say whether or not I am interested in selecting breeding stock from among them.

Anonymous

The head of a dairy cow is an indicator of her capabilities as a milk producer. To an experienced breeder the head of a cow reveals strength and dairy character and her eye often gives a clue to disposition. According to Melvin Scholl,

> Character in dairy cattle starts with the head. No other part of the cow or bull is so indicative of general appearance, overall conformation, quality of bone and skin, hardiness and prepotency as the head. Character is truly the hallmark of quality in cattle, and the head is where this hallmark is stamped.

5.5.4 Feet and Legs

Whether kept on concrete (**drylot** management) or allowed to roam the range in search of forage, when a cow's feet and/or legs pain her, she will not stand the "extra hour" or go the "extra mile" for feed and forage needed to sustain high milk production.

The forelegs should be straight and wide apart allowing for width in the chest floor (to accommodate a strong heart and capacious lungs). In dairy **heifers** and dry **cows**, the forelegs carry slightly over one-half of an animal's weight. In high-producing cows, however, the hind legs carry the greater proportion of a cow's weight. Thus, the hind legs should be strong and nearly perpendicular from hock to pastern when viewed from the side, and straight when viewed from the rear.

Cows have two shock absorbers—their pasterns and their hocks—which need, therefore, to be strong. The latter should not be too straight, however. The effect of **post-leggedness** in cows might be compared with the experience of riding over a bumpy road in the back of a truck standing flat-footed and with knees locked. That ride would about jar your teeth out. But if you come up on your toes and bend your knees, you have two shock-absorbing mechanisms—in your ankles (a cow's pasterns) and in your knees (a cow's hocks).

Cows need depth of heel because as they raise their heel in walking, the toe strikes the ground. As a shallow-heeled cow walks, her toe goes into the air without striking the ground. This means the toe does not wear off and must be trimmed more frequently. Toes on which there is little weight do not wear properly. As they become longer more weight is transferred to the heel, which is soft and easily injured. When toes are overgrown, tendons and joints are strained.

During hot, dry weather, the hard shell of the toe has a tendency to crack, causing the foot to be more vulnerable to infections and foot rot. Such animals are not competitive at the feed manger, and milk production and reproductive efficiency reflect this. Cows with foot and leg problems are inefficient in milk production and reproduction.

5.6 Type Classification

Beauty is bought by judgment of the eye.

G. C. Humphrey (1875–1947)

It takes little time around a sale ring in which registered dairy cattle are being auctioned to learn that type is a significant influence on sale prices. Purebred breeders know that dairy cows with higher type classification scores sell at premiums above animals of lower type classifications.

In a sale summary made of 13,127 Holsteins, nearly half the animals sold were either classified or from classified dams. They sold for an average of one-third more than unclassified animals. Females with production records sold for an average price equivalent to those classified. Thus, type classification ranks with production records in establishing sale prices of registered dairy cattle.

Dairy herd classification was first initiated in the United States by the Holstein–Friesian Association in 1928, the American Jersey Cattle Club followed in 1932, and other dairy breed associations soon thereafter. Type classification does not make dairy cows better—neither does production testing make cows secrete more milk, nor registration papers make cows consume less or more feed. However, each can be beneficial if properly used.

Table 5.2 International Standard Linear Traits for Evaluation with Descriptors to be Rated on a Scale of 1 to 9.

	Linear Score Range		
Trait	**1**	**5**	**9**
Stature	Short	Intermediate	Tall
Chest width	Narrow	Intermediate	Wide
Body depth	Shallow	Intermediate	Deep
Angularity	Lacks (close ribs, coarse bone)	Intermediate angle with open rib and intermedian bone quality	Very angular (open ribbed, flat bone)
Rump angle	High pins	Intermediate	Extreme slope
Rump width	Narrow	Intermediate	Wide
Rear legs (rear view)	Extreme toe-out	Intermediate, slight toe-out	Parallel feet
Rear legs set	Straight (160 degrees)	Intermediate	Sickle (134 degrees)
Foot angle	Very low	Intermediate	Very steep
Fore udder attachment	Weak and loose	Intermediate, acceptable	Extremely strong and tight
Front teat placement	Outside of quarter	Middle of quarter	Inside of quarter
Teat length	Short	Intermediate	Long
Udder depth	Below hock	Intermediate	Shallow
Rear udder height	Very low	Intermediate	High
Central ligament	Convex to flat floor	Slight definition	Deep definition
Rear teat position	Outside of quarter	Midpoint	Inside of quarter

Source: World Holstein Friesian Federation, 2005.

Although some studies show a low correlation between general appearance of cows and milk production, it is well known that general appearance influences final placings in show and sale rings.

Type classification programs emphasize the economically important conformational characteristics related to utility (those influencing the ability of a cow to hold up under contemporary production and management practices). Concurrently, they seek to supply important information for sire progeny evaluation (cf. chapter 6). Type classification is not an alternative to production testing, but rather a supplemental tool to assist breeders and AI organizations in developing balanced and effective breeding programs.

Information obtained through type classification, when evaluated on a sire progeny basis (and on a breed basis), can provide data needed to identify weaknesses within a herd and/or breed which, through the use of selected sires, can be corrected (cf. chapter 6).

Milk production is the most economically important trait of dairy cattle. Therefore, type classification should stress those physical characteristics important to lifetime milk yields. These include especially well-developed and attached mammary glands, strong legs and feet, large body capacity, and dairy character.

The World Holstein Friesian Federation has established recommended guidelines for classification of the black-and-white breeds, thus providing for worldwide comparisons of conformational traits. The guidelines include assessments of 16 linear descriptive traits (table 5.2) and of general characteristics (table 5.3). Linear type traits are the basis of modern type classification systems, and are the foundation of systems describing the dairy cow. Linear classification is based on measurements of individual type traits instead of opinions. It describes the degree of trait, not the desirability. Advantages of linear scoring are: traits are scored individually, scores cover a biological range, variation within traits is identifiable, and degree rather than desirability is recorded. Traits are

Table 5.3 General Characteristics Recommended for Evaluation During Classification of Dairy Cows.

Characteristic	Major Considerations
Size	Stature, length, width, and capacity of the cow in relation to age
Type	Youthfulness (heifers), strength, and general openness in relation to the conformational standard
Udder	Qualification of the total udder with emphasis on teat placement, suspensory ligament strength, and udder depth
Legs and feet	Set and bone quality of the legs, quality of feet, use of legs and feet
Final score	Total appearance of the cow with emphasis on the udder, legs, and feet

Source: World Holstein Friesian Federation.

evaluated by scoring within a range of 1 to 9, which covers the biological extremes.

This type of linear scoring of conformational traits yields a clear and semiobjective description of the appearance of the cow. Following this linear scoring, the classifier judges the animal's general characteristics according to descriptions in table 5.3.

Breed associations have standards applied to classification scores. For example, for Brown Swiss the overall maximum scores based on number of lactations are as follows: first lactation—V89, second lactation—E91, third and subsequent lactations—E93 ("V" = very good, "E" = excellent, etc.). For a cow to be scored E94 in the fourth or later lactation she must be currently scored 93 and classified by a different person than when scored E93. When a cow is reclassified "excellent" multiple times, her "E" rating is preceded by the number of times she has been awarded that rating. For example, an "E" rated cow in her third lactation with a minimum age of six years would receive a 2E rating. Cows rated 3E and 4E must have reached the ages of nine and 12 years, respectively. The Ayrshire Breeder's Association sets maximum scores for the first through fourth lactations as 89, 92, 94, and 95, respectively. Raising the score from 94 to 95 requires appraisal by a committee of three.

5.7 Heritability of Type Traits

Heritability (cf. chapter 6) is the proportion of measurable (observable) variation in a population that is due to genetic variation among individuals. Phenotypic variation among individuals may be due to genetic and/or environmental factors. The heritability of many individual type traits is relatively low (see table 6.7). Exceptions include stature (which is 40% heritable), feet and legs (30%), and dairy character (25%).

Statistics show that the Holstein breed is getting taller, causing dairy farmers to remodel stall barns to accommodate larger cows. Stature is the most heritable type trait measured by Holstein Association USA, at 0.42. Continual selection for stature in the population causes a phenotypic response that can be measured easily.

A change was made in 2011 in the way stature is evaluated in Holstein Association USA's Classification Program. Any cow falling outside of the breed's standard height of between 56 and 62 inches (as measured at the spine, between the hips) will be deducted two points from the front end/capacity category. This equates to a deduction of 0.4 points from a cow's final score. Cows evaluated at less than two years of age will have their stature score age-adjusted. For example, a cow scoring 87 for front end/capacity would only receive a score of 85 for that breakdown if she was shorter than 56 inches or taller than 62 inches as measured at the spine, between her hips.

5.7.1 Total Performance Index

The Total Performance Index (TPI)® is Holstein Association USA's sire selection index, which combines 11 traits in a formula that ranks sires based on their ability to transmit selected traits in a way that has an economic impact on the ability of their progeny to produce more milk. It is a predictor of a bull's ability to transmit based on established weightings. As a result, the TPI will encourage genetic progress in the Holstein breed. In April 2015, a revised TPI was implemented with national genetic evaluations. Major weightings of this revised TPI formula include: production (46%), health and fertility (28%), and conformation (26%) (cf. chapter 6).

5.7.2 Using Registration to Promote Genetic Progress

Registrations are the core of all pedigreed livestock associations. The Holstein Association USA offers Holstein COMPLETE℠, a program that integrates registration, pedigrees, classification, production records, genetic reports, and mating information into one package. In 2014, over 289,000 cows in 1,853 herds were enrolled in the program.

The Holstein Association USA also offers the TriStar℠ program. DHIA rules provide flexibility to users of DHI data. This flexibility places responsibility on the end-user of production data, including breed associations, to set their own standards for the use of production records. TriStar administers production records, cow and herd genetic performance reports, and recognition.

Approximately 300,000 cows are enrolled in the TriStar℠ Premier program. In 2015, four US registered Holstein herds qualified for the Herd of Excellence recognition. These herds achieved the following rigorous criteria: 25% above breed average ME for milk, fat, and protein; classified within the last year with an average score of 83 points or higher; and at least 70% of the herd homebred.

5.8 Type and Productive Longevity

Raising a replacement heifer to producing age is costly; therefore, it is important to have cows that "wear well" (stay in the herd a long time). Certain aspects of a cow's physical appearance may adversely affect productive longevity. These include a pendulous udder,[3] extrusion of teats resulting from a weak median suspensory ligament, evidence of arthritis (lameness), and other signs of illness.

As one might expect, there is an association between the nearness of udder floor to the ground and incidence of mastitis in the cow (Young et al., 1960). USDA studies indicate that young cows with deeper udders also become older cows with deeper udders.

Longevity affects the percentage of animals that must be raised to maintain or to increase herd size.[4] This, in turn, affects the degree of selection intensity that can be applied to females (cf. chapter 6). Additionally, a long-lived dairy cow of superior genetic merit will contribute more offspring than the short-timer. Other advantages that accrue from increased productive longevity include (1) lower annual cost of replacements (less cow depreciation expense), (2) increased average production per cow, and (3) an increase in the amount of **culling** possible. This latter point would allow greater culling intensity to be applied on such heritable traits as milking rates and disposition.

Although type classification is not always indicative of productive performance, it is indicative of *how long* cows perform and is usually associated with milk production. In the data of table 5.4 there were more lactations completed per cow as type classification scores increased. Similarly, average production per lactation increased (with the exception of cows classified good plus) as type classification went up. Moreover, total lifetime milk production followed classification; cows classified as excellent had lifetime yields nearly double those of animals rated good or fair.

5.9 Show-Ring Judging

The objective in dairy cattle judging is to enable one to select cows that will be not only economical milk producers, but also possess a pleasing appearance.

A. B. Nystrom

There is a thrill to a dairy cattle show that brings both exhibitors and ringside enthusiasts from miles away (figure 5.5). A number of breed programs are furthered through dairy cattle shows. Additionally, many 4-H members and students of vocational agriculture (FFA members) acquire and build interests in dairy cattle through exhibiting animals and participating in judging contests at dairy cattle shows.

Judging of dairy cattle at shows and fairs is based largely on the physical appearance and **condition** of animals as they appear on show day. They are placed in order of nearness to how well they meet the breed ideal for beauty, symmetry, mammary system, and size. Experience is the best teacher at becoming proficient in judging dairy cattle. The art of giving reasons also requires work and experience. Reasons are an explanation of one's placings in a show-ring. Several websites (see end of chapter) outline the basic fundamentals important to becoming a competent judge of dairy cattle. Additionally, dairy breed associations have judging literature and publications related to their breed.

5.9.1 To Touch or Not to Touch the Udder

To gain information about the quality of an udder and its attachments, it is often helpful to handle it. However, aside from the "best udder" class, most dairy judges refrain from this practice. One reason is that massaging the udder will frequently cause milk letdown. Then, if not milked out, tightly filled mammary glands may exert such an intense internal pressure that milk protein may enter the cow's bloodstream and cause **anaphylactic shock**. Such cows develop whelps or hives, their hair coat becomes rough, and they experience substantial pain. In such cases, a veterinarian should be called immediately to administer epinephrine in conjunction with a milkout.

5.9.2 Should Production Affect Show Placings?

Nearly a century ago, R. R. Graves (1917) introduced a plan for judging cows on the basis of combined ratings for type and milk production. Such a plan for placing cows appeals to the authors for it would give additional emphasis to the most important economic consideration of dairying—milk production—in aligning cows in order of desirability within the show-ring. A procedure in which 25 to 50% of the final placing would be based on milk production seems appropriate. Reasons why such a plan has not been adopted, however, probably include (1) nonavailability of production records on some animals entered in dairy cattle shows, which would restrict entries; (2) requirement of additional time to make the placings; and (3) negation of the thrill and satisfaction that comes from exhibiting a beautiful animal that may have had due reason for yielding less milk than she is genetically capable of producing.

5.9.3 Evaluating Dairy Bulls

We have purposely left discussion of evaluations of dairy bulls to last because we believe the major thrust in appraising the genetic merits of bulls should be through progeny testing (cf. chapter 6).

Table 5.4 Records Completed and Production Yields of Jersey Cows Scored in 2002 by Classification Category.

Classification	Score	Number of Cows	Lactations per Cow	Milk per Lactation (lb)	Total Milk per Lifetime (lb)
Excellent	> 90	624	4.8	19,715	91,846
Very good	80–89	28,387	4.0	17,541	68,921
Desirable	70–79	17,413	3.5	16,348	56,932
Acceptable	60–69	1,048	2.8	15,245	42.599
Poor	50–59	86	2.1	15,399	32,583
All cows	50–99	47,558	3.8	17,078	64,187

Source: American Jersey Cattle Association.

Figure 5.5 Showing young Holstein heifers at the Southern Dairy Show in 2014 (courtesy of Kate Geppert, Columbia, Missouri).

Bulls are evaluated on their breed characteristics, conformation, and masculinity, although, with the expanding information available on sires (cf. chapter 6), less emphasis is being given the showing of bulls. The authors favor this trend but caution breeders of dairy cattle not to consider the genetics of milk production alone in evaluating bulls, but rather their size and strength, feet and legs, and those other physical traits important to high milk production, longevity, and conformation among their progeny.

5.10 Summary

The dairy cow is a wonderful machine; but to be worthy of the name she should have, in addition to dairy breeding and good size for her breed, certain other characteristics which include: feed capacity, dairy character, good constitution and health, and well-developed milk organs—all of which are essential to profitable milk production.

G. C. Humphrey

Stressing the importance of type or conformation is not merely dressing up the package in which the milk-making machine is housed, it also places emphasis on traits affecting productive longevity, sale price, and personal satisfaction. The latter accompanies the breeding and ownership of dairy cattle that are both functional and attractive.

The primary purposes of type classification are to increase the accuracy of selecting dairy cows for increased production and to provide dairy cattle breeders with another tool to use in strengthening the physical traits associated with productive longevity.

Although being endowed with beauty is not always indicative of high milk production, we have presented data and facts indicating that cows of good conformation and dairy character are likely to yield larger amounts of milk and remain in the dairy herd longer than those of undesirable type.

We noted that type is often related to a cow's health. For example, an upstanding (high stature) cow with strong mammary gland attachments is less likely to be troubled with mastitis than a low-set, short-legged cow having a pendulous udder. Practically all physical defects are accentuated with age and milk production. This is a reason for selecting against poor type which, if done with a study of herd strengths and weaknesses, can be an important part of a progressive dairy cattle evaluation and breeding program. All of this is to provide more milk for humans.

Study Questions

1. What are the two main means of evaluating dairy cows for type?
2. Approximately what percentage of US dairy cows are production tested? Of what significance is this to selection and breeding?
3. Differentiate between type classification and show-ring judging.
4. How can information gained through type classification be useful to dairy cattle breeders?
5. What are the five most commonly used type classifications? Do all breeds use the same point breakdown to rate a cow good or fair?
6. What organization publishes the Dairy Cow Unified Scorecard? What is the percentage allotted for each major breakdown on the scorecard?
7. Identify and discuss briefly three important considerations of a desirable udder.
8. Why is it recommended that the udder of a dairy cow be above the hocks?

9. Be able to explain the association between length, width, and levelness of rump and those traits of a cow's udder.
10. Define dairy character. What other terms are used to indicate this important trait of dairy cows?
11. Describe what the following terms refer to in dairy cattle: constitution, style, substance, and strength.
12. Discuss briefly the importance of topline, head, and bone in evaluating dairy cattle.
13. Why are the feet and legs of dairy cattle receiving an increased amount of attention?
14. Of what significance is depth of heel in dairy cattle?
15. Relate barrel size and heart girth of cows with rib configuration.
16. Discuss the merits and limitations of type classification of dairy cows.
17. Of what economic significance is productive longevity of cows to their owners?
18. Does conformation affect the sale price of dairy cows?
19. Which type traits among dairy cattle are the most heritable? The least heritable?
20. Discuss briefly the merits of dairy cattle shows.
21. Of what importance are oral or written reasons in judging dairy cattle?
22. What consideration should be given the handling of a full udder in the show-ring?
23. Do you believe milk production should affect placings in the show-ring? Why?
24. Why is less emphasis being given the exhibition of dairy bulls than of cows?

NOTES

[1] Commercial milk producers usually rely on conformation of dairy cattle to aid in buying, selling, and making management decisions.

[2] Dr. Brody was a world-renowned scientist in the field of bioenergetics and growth. He was a staff member of the University of Missouri (1920–1956) and authored the monumental book, *Bioenergetics and Growth,* published by Reinhold, New York, 1945.

[3] The authors recall visiting a dairy farm in which they observed a young cow with an udder so deep and pendulous that, to be milked by machine it was necessary to train her to back up on a wooden stand about six inches high. When asked why he kept the cow, the owner smiled and said, "Because she is milking 92 lb per day." When we visited the farm two years later, the cow had been sold.

[4] Although probably not a record age-wise, the authors learned of a Jersey cow in Tennessee that was reproductively sound and lactating at 27 years of age.

REFERENCES

Graves, R. R. 1917. Production in the show ring. *Hoard's Dairyman* 53:44.

World Holstein Friesian Federation. 2005, June. International Type Evaluation of Dairy Cattle (http://www.whff.info/info/typetraits/type_en_2005-2.pdf).

Young, C. W., J. E. Legates, and J. G. Lecce. 1960. Genetic and phenotypic relationships between clinical mastitis, laboratory criteria and udder height. *Journal of Dairy Science* 43:54–62.

WEBSITES FOR JUDGING DAIRY CATTLE

Mississippi State University Extension Service (http://msucares.com/pubs/publications/p0194 pdf)

Oregon State University Extension Service (http://extension.oregonstate.edu/catalog/4h/4-h1109.pdf)

University of Arkansas, Division of Agriculture (http://www.uaex.edu/Other_Areas/publications/PDF/MP-469.pdf)

University of Georgia, College of Agricultural and Environmental Sciences (http://www.caes.uga.edu/publications/pubDetail.cfm?pk_id=7145)

University of Minnesota Extension (http://www.extension.umn.edu/youth/mn4-H/projects/docs/dairy-judging.pdf)

University of Vermont (http://www.uvm.edu/~jagilmor/judging/judging.html)

University of Wisconsin Extension (http://www.uwex.edu/ces/dairyyouth/judging.php)

6

Breeding Dairy Cattle

Creation lies at the base of the world's entire civilization . . . to breed good dairy cattle is to create and, when done in full recognition of economic requirements, is profitable. Animals of superior heredity were sought by the earliest breeders. And they will be sought by future generations.

Karl B. Musser

This chapter was revised by Roger D. Shanks, PhD. He is Editor-in-Chief, *Journal of Dairy Science*; a Dairy Genetics Consultant, Holstein Association USA; and Professor Emeritus, Dairy Cattle Breeding and Genetics, University of Illinois, Urbana–Champaign (UIUC). Shanks grew up on a registered Holstein farm outside Mundelein, Illinois. During his 31 years at UIUC, Shanks taught applied animal genetics, introduction to genetics, graduate seminar, statistics, freshman discovery, and beyond-the-classroom activities. He has about 300 publications, including 94 articles dealing with qualitative and quantitative genetics of dairy cattle, most significantly DUMPS (deficiency of uridine monophosphate synthase). Special honors include the J. L. Lush Award in Animal Breeding and Genetics, 2007; Agway Inc. Young Scientist Award in Dairy Production, 1984; and D. E. Becker Award for excellence in undergraduate teaching and counseling, 2004.

6.1 Introduction

Dairy cows serve humans by providing an abundant quantity of nutritious milk. The greater quantity of milk produced per cow, the greater service rendered people. No cow can produce milk beyond the limits imposed by the genetics of her ancestors. Therefore, breeders strive to eliminate undesirable characteristics and to accumulate and preserve those genes that enable dairy cows to demonstrate high milk production throughout a long, productive life.

Great strides have been made in breeding more useful dairy cattle during the past several decades. Average milk production per cow in the United States increased from 5,250 lb (2,386 kg) in 1943 to 10,271 lb (4,669 kg) in 1972 to over 22,000 lb (10,000 kg) in 2012. This quadruple increase represents the combined breeding and management efforts of progressive dairy farmers who are aided by personnel of dairy **breed** associations, artificial insemination (AI) organizations, colleges of agriculture and experiment stations, the US Department of Agriculture (USDA), and cooperative extension programs.

Technological advances also have facilitated genetic progress. Artificial insemination provides means to distribute semen from premium-quality dairy **bulls**. Superovulation and embryo transfer propagate outstanding female genetics and, more recently, genomics identifies superior dairy elite. There is little doubt—science is a large part of breeding dairy cattle today.

Although it is a difficult task to condense the many aspects concerned with dairy cattle breeding into one chapter, we attempt to present an overview of the subject here and trust the reader will seek additional references for further understanding of this most fundamental aspect of providing milk for humans. Breeding dairy cattle is defined by the perseverance of dairy producers, their desired goals in the selection process, and the methods available to achieve these goals.

6.2 A Brief History of Genetic Selection in Dairying

Historically, the goal of a dairy producer has been to increase milk yield per cow as a means of maximizing profit. Members of breed organizations have an additional goal to improve the conformation or appearance (type) of animals in their herds. Much debate has occurred on whether the goal should be to increase milk or to improve type.

When it comes to the genetic evaluations dairy producers use to make breeding decisions, USDA has been the source of information for more than 75 years, dating back to the first bull evaluations calculated from daughter-dam comparisons in 1936. Since the 1960s, the industry has relied on regular "**sire summaries**"—calculated and provided by the USDA's Animal Genomics and Improvement Laboratory (AGIL)—to gain production, health, fertility, and conformation information that they can use to make genetic decisions.

In 2013 the Council on Dairy Cattle Breeding (CDCB)—comprised of representatives of dairy cattle breed associations (Purebred Dairy Cattle Association), AI firms (National Association of Animal Breeders), and dairy herd improvement (DHI) organizations—established a program to provide this type of information to users.

Their mission is to maintain a high standard of integrity in data collection systems providing data for the National Dairy Genetic Evaluation Program (GEP) database; to provide data to AGIL for use in research and education; to provide data and summaries of information to research and extension personnel, and to others for educational purposes; to improve the genetic merit and production efficiency of US dairy cattle; and to enhance the world market competitiveness of the US dairy industry.

Genetic evaluations for **selection** have changed significantly over the decades. Initially, animals were evaluated based on their own performance. Then their performance was compared to that of their **dam** (mother). Obviously, if the dairy producer had improved her/his management at all, the daughter would appear better than the dam. Then the system evolved to compare an animal's performance to those of her **herdmates**. This was an improvement over the comparison between daughter and dam because herdmates performed within the same environment and under the same conditions.

National genetic evaluations then moved to modified contemporary comparisons. These comparisons were adjusted for the genetic level of herdmates, the genetic trend that had been occurring over time, and environmental effects of herd, year, and season. As mixed model procedures for genetic evaluations were developed, additional improvements in the methods were noted. Originally, the methods outpaced the computers. But computer and software advances allowed implementation of outstanding methods. Examples include simultaneous evaluation of environmental and genetic effects and the simultaneous estimation of genetic evaluations of both males and females. About the time the industry considered that it may have exhausted all opportunities for improving methods of evaluation based on **phenotypes** and mixed models, along came genomics.

Table 6.1 highlights significant advances in genetic evaluations in the United States and quantifies the percent gain from each new process, method, or trait. Additional details about genetic evaluation methods can be found in a review by Wiggans and Gengler (2011).

6.2.1 Role of Computers in Breeding

Obviously, the breeding of dairy cattle has moved from an art to a discipline that incorporates science. Computers

Table 6.1 Genetic Evaluation Advances and Increases in Genetic Progress.

Year	Advance	Gain (%)
1935	Daughter-dam comparison	100
1962	Herdmate comparison	50
1973	Records on progress	10
1974	Modified contemporary comparisons	5
1977	Protein evaluated	4
1989	Animal model	4
1994	Net merit, PL, and SCS	50
2008	Genomic selection	>50

Source: VanRaden et al., 2008.

are an integral part of genetic evaluations and, for many, the selection of animals and the mating of them. Computers have evolved from large mainframe units housed in air-conditioned buildings to even more powerful ones that can be held in one hand by classifiers when they visit a farm.

Computers are used for genetic evaluations of individual animals, to verify parentages based on genomics, to record data, to transfer data, and to assist in making mating decisions. However, before concluding that dairy producers are no longer involved, one needs to recognize that dairy producers define the goals that are used by computers to calculate values.

6.3 Genetic-Economic Indexes

Better genes in dairy cows mean
more dollars in a dairy farmer's jeans.
Anonymous

The advent of a tool called an *index* greatly reduced the debate whether the goal should be to increase milk or to improve type because dairy producers had the opportunity to practice selection for a combination of milk and conformation. Subsequently, the index included (1) milk quality in terms of percentages of fat and protein and (2) health traits. Also incorporated into the index was information on productive life. To facilitate progress, subindexes were developed for combinations of type traits of the udder and of the feet and legs.

6.3.1 Achieving Goals Using Indexes

A history of the main changes in USDA genetic-economic indexes for dairy cattle and the percentage of relative emphasis on traits are included in table 6.2. Emphasis on yield traits declined as fitness traits were introduced. For example, as protein yield became more important, milk volume became less important because of the high **correlation** of these two traits. One advantage of index selection is that many traits can be evaluated as part of the selection decision. Progress will be faster for traits with higher heritability (milk yield), but progress can still be obtained for traits with low heritabilities (health and fertility traits). A more complete history and comparisons with selection indexes used by various countries are available (cf. Cole et al., 2012; Shook, 2006; VanRaden, 2004).

Comparisons of relative value for three indexes are in table 6.3 on the following page. Indexes of net merit, cheese merit, and fluid merit contain most of the same traits, but emphases on traits differ. Two specific indexes have been widely accepted by the dairy industry: net merit and total performance index.

Table 6.2 History of Genetic-Economic Indexes and Percentage of Relative Emphasis on Traits.

	Genetic-Economic Index (Year Introduced)								
Traits Included	PD$ (1971)	TPI (1976)	MFP$ (1977)	CY$ (1984)	NM$ (1994)	NM$ (2000)	NM$ (2003)	NM$ (2006)	NM$ (2010)
Milk	52	60	27	–2	6	5	0	0	0
Fat	48		46	45	25	21	22	23	19
Protein			27	53	43	36	33	23	16
Productive life					20	14	11	17	22
Somatic cell score					6	–9	–9	–9	–10
Udder composite						7	7	6	7
Feet/legs composite						4	4	3	4
Body size composite						–4	–3	–4	–6
Daughter pregnancy rate							7	9	11
Service sire calving difficulty							–2		
Daughter calving difficulty							–2		
Calving ability								6	5

PD$ = predicted differences; TPI = type-production index; MFP$ = milkfat and protein; CY$ = cheese yield; NM$ = net merit.

Sources: VanRaden, 2002; Cole et al., 2012.

Table 6.3 Comparison of 2010 Genetic-Economic Indexes.

	Relative Value (%)		
Trait	**Net Merit**	**Cheese Merit**	**Fluid Merit**
Milk (lb)	0	−15	19
Fat (lb)	19	13	20
Protein (lb)	16	25	0
Productive life (months)	22	15	22
Somatic cell score (log_2)	−10	−9	−5
Udder composite	7	5	7
Feet/legs composite	4	3	4
Body size composite	−6	−4	−6
Daughter pregnancy rate (%)	11	8	12
Calving ability ($)	5	3	5

Source: Cole, 2012.

6.3.1.1 Net Merit

Net merit (NM$) was developed by US dairy geneticists and implemented into the genetic evaluations performed by the AGIL. The concept underlying the development of net merit was the determination of relative economic values of the many traits associated with improving dairy cattle. These traits included not only milk, but also milkfat, protein, health traits, productive life, and fertility.

6.3.1.2 Total Performance Index

Another index that has been widely developed and identified as the "gold standard" is total performance index (TPI). The abbreviation originally stood for type-production index. More recently, it has been identified as total performance index. This index was developed by the Holstein Association USA in collaboration with their members. Many components in TPI are similar to those of net merit, but it also includes contributions to TPI based on the type conformation. One of the underlying motivations for TPI was to give direction for breeding programs of Holstein cattle. Because many dairy producers worldwide accepted and respected the direction defined by TPI, the Holstein breed has made much genetic progress in the direction defined by the index.

Major weightings of the April 2015 revision of the TPI formula (figure 6.1) emphasize production at 46%, health and fertility at 28%, and conformation at 26%. Each trait has a **predicted transmitting ability (PTA)** that has been estimated by combining traditional evaluations and genomic evaluations.

The value of 2187 (found at the end of the formula) adjusts for the periodic base change, allowing TPI values to be compared across time. The value of 3.9 is a scaling factor to facilitate recognition of differences between individuals in their TPI. The denominator of each trait is an approximation of the standard deviation of each trait. The coefficient of each trait is a relative economic value.

The predicted transmitting ability value of a bull is an average number that best estimates the genetic merit of that bull. However, the **genetic value** and performance of a particular daughter contains an unpredictable component because each daughter receives a different set of genes from the bull. Chance determines the actual genetic merit of the offspring at the time of fertilization of the ova by the spermatozoon. When two animals of high genetic merit are mated, the genetic value of the offspring will likely be higher than average, but it is possible that its genetic merit will be below average.

The initial goal in dairy cattle breeding called for selection based on a single trait, milk yield. Through the implementation of indexes, significant genetic progress has been made in many traits.

6.3.2 What Does the Future Hold?

Indexes have been highly beneficial in facilitating genetic progress for breeders. Indexes are linear. The next challenge that geneticists are evaluating and addressing is something that could be described as nonlinear or multidimensional identification, also sometimes described as 2D or gene interactions. The focus on multitrait complexes is necessary because animals are comprised of

$$\left[27\left(\frac{PTAP}{19}\right) + 16\left(\frac{PTAF}{22.5}\right) + 3\left(\frac{FE}{44}\right) + 8\left(\frac{PTAT}{.73}\right) - 1\left(\frac{DF}{1}\right) + 11\left(\frac{UDC}{.8}\right) + 6\left(\frac{FLC}{.85}\right) + 7\left(\frac{PL}{1.51}\right) - 5\left(\frac{SCS}{.12}\right) + 13\left(\frac{FI}{1.25}\right) - 2\left(\frac{DCE}{1}\right) - 1\left(\frac{DSB}{.9}\right)\right] 3.9 + 2187$$

** The value 2187 adjusts for our periodic base change, allowing TPI values to be comparable across time.*
Formula last updated December 2014

PTAP	= PTA Protein	PTAF	= PTA Fat
FE	= Feed Efficiency	PTAT	= PTA Type
DF	= STA Dairy Form	UDC	= Udder Composite
FLC	= Feet & Legs Composite	PL	= PTA Productive Life
SCS	= PTA Somatic Cell Score	FI	= Fertility Index
DCE	= PTA Daughter Calving Ease	DSB	= PTA Daughter Stillbirth

Figure 6.1 The TPI formula as of April 2015 (Holstein Association USA).

complex traits that have a complex genetic architecture that defines them. An example of application of information about genetic profiles is the finding that small differences in the amino acid composition of caseins between some breeds can affect the proteolytic release of biopeptides by yogurt bacteria.

With the acceptance of computer-based software, the opportunity to make selection and mating decisions based on a matrix will become available. The significant focus of this matrix is to allow for a selection that is balanced and to maintain outstanding relationships among and between traits.

6.4 Genomics

Prior to genomics, geneticists attempted to predict **genotypes** of animals based on phenotypic information of relatives. As DNA technology and related biology improved, scientists became capable of measuring the genetic information directly (although not inexpensively). This direct measurement of genetic information became known as genomics and is sometimes labeled "the genomic revolution."

Currently, it is cost prohibitive to genomic test all animals. However, the cost of genomic testing has decreased as the technology has improved and more animals are being tested routinely. Table 6.4 indicates how many Holsteins were genome tested over a two-year period.

Unofficial genomic evaluations were available for some bulls in April 2008. The first official publishing of genetic evaluations based on genomic testing and calculation by the Animal Improvement Program Laboratory (AIPL) of the USDA occurred in January 2009. By September 2012, a total of 278,347 dairy animals had been genotyped. By June 2015, a total of more than one million dairy animals had been genotyped. Table 6.5 indicates how those genotypes occurred by breed, sex, and chip type. The 50K chips are the standard used for genome evaluations.

Table 6.4 Cumulative Numbers of Holsteins Genotyped by Selected Months (2010–2012).

	SNP Estimation[a]		Young Animals[b]		All
Date	**Bulls**	**Cows**	**Bulls**	**Heifers**	**Animals**
April 2010	9,770	7,415	16,007	8,630	41,822
Aug 2010	10,430	9,372	18,652	11,021	49,475
April 2011	12,152	11,224	25,202	36,545	85,123
Aug 2011	16,519	14,380	29,090	52,053	112,042
Feb 2012	17,710	17,679	36,597	80,845	152,831

[a]Traditional evaluation.
[b]No traditional evaluation.

Source: Cole, 2012.

Table 6.5 Genotype Counts by Chip Type, Breed Code, and Sex Code.

	AY		BS		GU		HO		JE		MS		XX		
Chip Type[a]	**F**	**M**	**F**	**M**	**F**	**M**	**F**	**M**	**F**	**M**	**F**	**M**	**F**	**M**	**Total**
3K	3	0	473	11	5	0	49,022	3,912	9,677	196	0	0	0	0	63,299
LD	957	14	529	18	0	0	154,699	4,230	10,269	197	0	1	1	0	170,915
GGP	55	5	293	288	0	4	40,340	13,420	12,513	1,511	5	1	1	0	68,624
50K V1	0	21	91	5,652	0	0	20,758	34,462	913	4,941	0	0	0	0	66,838
50K V2	129	363	135	8,168	10	152	32,130	36,340	879	3,057	2	0	5	0	81,370
HD	12	520	3	182	26	121	561	1,986	31	164	0	0	0	0	3,606
Total	3,112	1,549	3,689	16,459	1,829	610	721,141	164,188	97,717	18,832	10	12	29	0	1,029,177

Breed codes: AY = Ayrshire, BS = Brown Swiss, GU = Guernsey, HO = Holstein, JE = Jersey, MS = Milking Shorthorn, and XX = crossbred.
Sex codes: F = female and M = male.

[a]Chip types were defined by size from smallest to largest number of SNP tested: 3K (~3,000 SNP), LD (~6,000 SNP), GGP (~9,000 SNP), 50K V1 (first version chip of ~50,000 SNP), 50K V2 (second version chip with ~50,000 SNP), and HD (~777,000 SNP).

Source: Council on Dairy Cattle Breeding, 2015.

6.4.1 Genomic Evaluations

What does genomic evaluation mean? Genomic evaluations are predicted based on phenotypic information from progeny of many bulls and information on their DNA. When the genome evaluation first became official in the United States, 9,770 bulls were the basis for making the genomic predictions. These bulls had phenotypic information and traditional genetic evaluations based on a large number of progeny for each bull. Researchers were able to take segments of the genome, defined as **haplotypes**, and associate them with good or poor performance among the daughters of the bulls.

This population of almost 10,000 bulls became a predictor population. The associations that were defined from this population could then be applied to young animals with no progeny. These young animals were then considered "outstanding" if they had a large collection of favorable haplotypes. It was no longer necessary to wait to evaluate a bull until after he had many progeny. The bull could receive a genomic evaluation based on the association determined among the older population of bulls. These associations between haplotypes and phenotypic (observable) performance of the bull's daughters could also be applied to determine genomic evaluations of females. For example, DNA analyses of hair samples from two females at birth provided genomic data sufficient to reveal which animal would be expected to produce more milk two years later. Needless to say, the availability of this information has the potential to greatly alter selection programs. Many young animals, without progeny themselves, became "in favor" as the best in the breed.

6.4.2 Early Restrictions to Genomic Testing

One of the first descriptions of genomic prediction was provided by the AIPL. The report is inserted here to convey the importance and change of paradigm that has occurred in dairy cattle breeding as a result of being able to test genotypes directly.

> Genomic predictions of genetic merit were released for the first time in April 2008. Predictions are based on genotypes derived from blood samples (or other DNA) provided by animal owners through arrangements made by participating artificial-insemination (AI) organizations. The DNA is extracted from the samples in most cases by GeneSeek (Lincoln, NE). The extracted DNA then is placed on a chip developed by Illumina (San Diego, CA), USDA's Bovine Functional Genomics Laboratory (Beltsville, MD), and other research partners. That chip provides genotypes on more than 50,000 single nucleotide polymorphisms (SNPs) evenly distributed across all 30 chromosomes; of those SNPs, nearly 40,000 are informative for Holsteins. The genotypes document which genes each animal inherited, and that completely new source of information can be included in genetic evaluations. Predicted transmitting abilities (PTAs) have used pedigrees to calculate probabilities that relatives share genes. With genotypes, the actual genes that are shared can be determined.
>
> The DNA of proven Holstein bulls contributed by members of the National Association of Animal Breeders (Columbia, MO), Semex Alliance (Guelph, ON, Canada), and some other projects as well as DNA from a small number of cows is used in estimating genetic effects for each SNP. Each animal's PTA or parent average (PA) for each trait as well as its net merit index then is adjusted for the sum of the estimated genetic effects. The primary focus is to calculate genomic evaluations for young bulls and heifers nominated by the participating AI organizations. Evaluations of the nominated animals are distributed to the U.S. owners (see example mailer) and to the organizations that paid for genotyping to aid in selection decisions. The AI organizations that contributed to the research have a 5-year period of exclusive rights to obtain genomic evaluations of males. Evaluations of females are available to anyone who provides a genotype through a cooperating organization.
>
> Two different evaluations predict the merit of an animal's daughters and sons separately. The difference is the sum of genetic effects on the X chromosome. The standard deviation of the difference between daughter and son merit is about one-tenth of the genetic standard deviation for most traits. (VanRaden et al., n.d.)

6.5 Genetic Evaluation Using Sire Indexes

A good bull is half the herd; a poor one is all of it.
Anonymous

Because most dairy bulls **sire** a sizable number of progeny, the selection of bulls carries greater responsibility than that of cows. In three generations, a maximum of 87.5% of the genes in a herd could trace directly to the sires used in those three generations (figure 6.2).

Currently most sires are evaluated based on their genome as well as **pedigree** and progeny when available. Needless to say, evaluation methods have become more elaborate as more and more information contributes to the evaluations. However, the good news is that more dairy cattle breeders are embracing the science behind the genomic evaluations and using the evaluations to make decisions about how best to provide milk for humans.

Traditionally, the most important estimates of the value of dairy cows for breeding purposes were those involving milk yields. Health and longevity aspects have also become available to breeders for adding value to their cows (see table 6.1).

6.5.1 Daughter Average

An early method of sire evaluation (circa 1930) was daughter average. This sire index uses the average production of a bull's daughters as being indicative of his transmitting ability for milk production. It does not con-

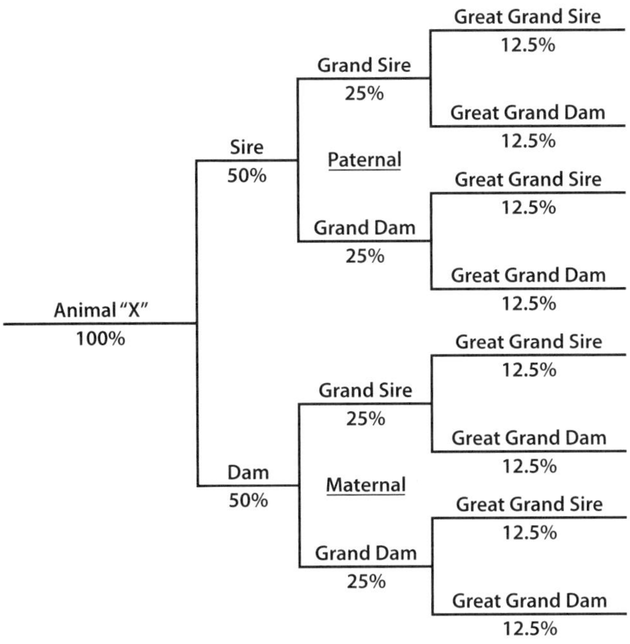

Figure 6.2 Diagram indicating the average genetic contribution made to animal "X" by each ancestor. On a herd basis, when a single bull sires many progeny, his influence is readily apparent.

sider the production of their dams, herdmates, or environmental opportunities. Although daughter average favored the practices of the best managers, it was a good first step.

6.5.2 Daughter-Dam Difference

Another early method of sire evaluation was daughter-dam deviations. This sire index considers the amount of increase or decrease in milk produced by a bull's daughters in comparison with their dams. A limitation of the daughter-dam difference in evaluating bulls is that records of dams and daughters are usually made at different times so calving dates and environmental conditions are not standardized.

The method may or may not adjust for the level of production of dams on which the sire was used. The production level of dams is important in considering the apparent genetic worth of a bull (based on his daughters' milk production).

Daughter-dam deviations were appropriate for those who wished to improve their management practices. However, early evaluation results included more than just genetics as a result of management biases. Additional biases occurred because many daughters of a single bull were raised in only a few herds, differences among herd environments were unknown, and bulls were mated to dams nonrandomly. Bias was reduced after 1942 partly because frozen semen facilitated **progeny testing** in many herds across a random sample of environments and more progeny could be obtained per bull.

6.5.3 Herdmate Comparison

Herdmate comparisons replaced daughter-dam comparisons in 1962. Records were adjusted to a standard of 305D-2X-ME. A genetic index was published for each bull.

In the USDA-DHIA sire summary records, a bull's daughters are compared with cows (herdmates) that **freshen** in the *same herd* during the *same season* of the *same year.* Thus, a sire's daughters and their herdmates are theoretically provided equal opportunity to produce.

The herdmate comparison is useful because it largely removes the effects on milk production of variations of herd, year, and season of calving. Herd-year-season yield is based on a five-month moving average, as computed by the USDA. The herdmate average for each daughter of a sire is obtained by averaging all of the records of daughters of other sires within the same breed calving in the herd in the same month as that daughter, in the previous two months, and in the succeeding two months. For example, the herdmate average for a sire's daughters that calved in March is obtained by averaging the production of all cows (except her paternal sisters) that calved in that herd from January through May.

This index was called predicted difference between 1968 through 1989. Benefits were great and included improved evaluation of genetic worth, availability to the industry, acceptance and use by the industry, and practical means of teaching concepts of population genetics. However, assumptions of herdmate comparisons became invalid. These assumptions were: no environmental trend, no genetic trend, semen distributed at random among herd-year-seasons, cows culled at random, and adjustments for age and lactation length were unbiased.

6.5.4 Predicted Difference

Predicted difference (PD) for milk and fat yields is a measurement of the ability of a bull's progeny to respond to an environmental opportunity as determined by the productivity of his daughters compared with their contemporaries; it reflected the genetic transmitting ability of a bull for milk production. PD was considered one of the most significant tools ever developed in dairy cattle breeding. It provided a reliable estimate of the expected average deviation of a bull's progeny from their herdmates in breed average herds.

6.5.5 Modified Contemporary Comparisons

Modified contemporary comparison (MCC) was the method of choice from 1974 to 1989. The MCC method adjusted for genetic level of herdmates, accounted for genetic trend, and improved adjustment for environment. Pedigree information was used to group bulls for expected genetic merit. This pedigree index (PI) combined information from the sire and maternal grandsire (MGS) via the formula:

$$PI = \frac{1}{2}\text{ sire PD} + \frac{1}{4}\text{ MGS} - \text{PD}$$

Additional improvements with MCC included better **repeatability**, incorporating records in progress, > 40 days in genetic evaluations, and updating adjustment factors.

6.5.6 Best Linear Unbiased Prediction Procedures

The animal model, or best linear unbiased prediction (BLUP) procedures, replaced the MCC method. *Best* equates to minimum variance of prediction. A *linear model* was used for analysis. The analysis was *unbiased,* reflecting a fair treatment of all individuals, and the procedure determined *predictions* of genetic merit. Advantages of BLUP were stabilized values from one sire summary to the next, accounting for both genetic competition and genetic trend, using all available information on relatives and progeny, simultaneously evaluating bulls and cows, and making environmental adjustments simultaneous with genetic prediction. Disadvantages of BLUP were complexity, expense to compute, and difficulty to explain. The reader who desires more detailed mathematical information about BLUP should see Henderson (1975).

6.6 Genetic Evaluation Using Other Tools

6.6.1 Difference from Herdmates

Commonly called *herdmate difference,* this value represents a comparison of the individual cow's record with the average of herdmates (within herd, breed, year, and season). All records are compared on the basis of 305D-2X-ME. Herdmates are all other cows within a herd of the same breed calving in the same year and season. For example, if the milk yield of cow A was 22,000 lb and that of her herdmates was 21,000 lb, the herdmate difference would be +1,000 lb. Herdmate differences are especially valuable and meaningful because they minimize environmental effects (e.g., season and management). The basis for daughter yield deviations (DYD) currently used in the calculation of genomic evaluations is analogous to herdmate comparisons previously estimated. A fundamental component of genetic evaluations is competition between individuals.

6.6.2 Estimated Average Transmitting Ability[1]

Estimated average transmitting ability (EATA) is an estimate of the portion of **breeding value (BV)** a cow will transmit to her offspring. With the animal model, simultaneous estimation of PTA for males and females occurs. Previously, male and female estimates were calculated separately and the female estimates were EATAs. The EATA calculations were based on five sources of milk production records: (1) the cow, (2) her dam, (3) her daughters, (4) her paternal **half-sibs** (other daughters of her sire), and (5) maternal half-sibs (other daughters of her dam). Information from each pedigree source was weighted in accordance with how much the information contributed to the EATA. Thus, emphasis given these sources of genetic worth was influenced by several factors. Take, for example, the degree of relationship to the cow being evaluated. A cow is twice as closely related to her dam or daughters as she is to her maternal or paternal sisters. Therefore, in calculating EATA, greater weight was given each production record of a cow's daughter than to each record of her half-sibs. Of course, EATA provides no assurance that a cow will transmit in accordance with its estimates, but for its time, it provided dairy cattle breeders with a fair, unbiased, and comparatively accurate appraisal of the breeding worth of each cow. Today, predicting dairy herd productivity depends on genomic evaluations, which are based on many of the same principles used to calculate EATA. Data from many sources are combined to predict genetic value.

6.6.3 Equal-Parent Index

The equal-parent index was based on the premise that the sire and dam contribute equally to the inherent milk-producing ability of their progeny. It must be remembered, however, that the quality of inheritance received from each parent may not be equal. The producing ability of daughters is expected to be halfway between the producing abilities transmitted by sire and dam. More recently, the equal-parent index has been replaced by the parent average.

6.6.4 Parent Average

A new tool used by breeders to interpret genetic evaluations is parent average, which is similar to equal-parent index. To review, the breeding value of an offspring is the sum of the PTA of each parent (BVoffspring = PTAsire + PTAdam). Also the PTA of an individual equals one-half the breeding value of the individual (PTAindividual = ½ BVindividual). The parent average is an estimate prior to inclusion of any genomic information on the individual and is the average of the PTAs of the parents:

$$\text{parent average} = \frac{(\text{PTAsire} + \text{PTAdam})}{2}$$

The genomic evaluation of an individual is compared with the parent average to give an indication of favorability of the segregation that produced the individual. Some individuals receive a greater proportion of desirable genes from their parents.

6.7 Selecting Dairy Females

Who would grow spirited stallions for the Olympic prizes or strong bulls for the plow, let him choose carefully the females who will be their dams.

Virgil (70–10 BC)

The opportunity to improve dairy cattle by selecting females is much smaller than by selecting sires because so few cows can be culled if herd size is to be maintained. However, especially useful tools are available for appraising the genetic worth of cows, their milk-producing ability, and their genome. To evaluate cows on a consistent basis, one should consider production records in reference to age, feeding and management conditions, length of lactation, calving interval, and number of times milked per day (cf. chapter 4). Historically, breeders have also utilized information available on milk production of full- and half-sibs (both maternal and paternal). Even without genomic information, the accuracy of PTA will increase as numbers of records per animal increase. For example, if a cow has only one record, the accuracy is only 50%, but accuracy increases to 67% and 75% with two and three records, respectively (table 6.6).

Table 6.6 Effect of Increasing the Numbers of Milk Production Records on PTA Accuracy.

Number of Records	Weighting Factor
1	0.50
2	0.67
3	0.75
4	0.80
5	0.83
6	0.86
7	0.88
8	0.89
9	0.90
10	0.91

Source: Adapted from Eastwood, 1971.

6.7.1 Selecting Dairy Heifers

Of greatest economic importance is consideration of a heifer's potential for producing milk. Her genomic evaluation will give a prediction for many traits, including milk production, conformation, health, and fertility. Once a heifer has a phenotypic record of her own, that information can be combined with the genomic evaluation to improve overall accuracy (Hayes et al., 2009; VanRaden et al., 2009).

6.7.2 Selecting for Longevity

The simplest way for increasing the longevity of farm animals is by selective mating of the long-lived individuals. Dairy cattle, for example, do not pay for themselves in milk and calves until they are about four years old, and the longer thereafter they maintain a satisfactory yield the greater the clear profit on the growth investment.

Samuel Brody (1890–1956)

Dairy farmers recognize that the cost of herd replacements, whether raised or purchased, is of major economic importance in the dairy enterprise. The average life span of a cow normally far exceeds her productive years in the herd. Therefore, we must consider important reasons why cows are culled, and then decide if the heritability of valuable traits is high enough that selection can effectively increase the length of a cow's productive life. Low production, reproductive inefficiency, and disease (especially mastitis) are the foremost reasons why cows are culled (cf. chapter 7).

In chapter 5 we discussed how type or conformation affects productive longevity among dairy cows. Fortunately, heritabilities of many type traits, and overall type, are sufficiently high for dairy cattle breeders to extend the productive lives of their cows by giving attention to type in their breeding program. Foster et al. (1989) had one of the first reports estimating the association between linear type traits and herdlife. In this study, more than 17,000 cows had both lifetime and type information.

> Cows with extreme dairy character, excellent dispositions, and moderate rump width were associated with greatest production and herdlife. Tall first lactation cows with deep udders had highest herdmate deviations for milk and milkfat, but cows with average height and udder depth had longer herdlife. (Foster et al., 1989)

6.8 Evaluating Breeding Value of Dairy Bulls

Altogether too rarely there has appeared in almost every breed of livestock a great sire whose inheritance has shaped the future of his breed.

Rex V. Conn

Although bulls secrete no milk, they transmit genes that affect the milk production of their daughters. Most early sire evaluations were based on daughter averages or herdmate comparisons. Traditionally, proving a sire meant the evaluation was based on a large number of daughters, which would represent high reliability. More recently, distinctions among bulls have been determined on whether evaluations are based on genomic information only or if performance of progeny has been included. The amount and type of information is reflected in the reliability associated with each PTA.

Progeny testing has evolved into genomic evaluations. Each process is built on phenotypic records and relationships between relatives.

6.8.1 Sire Lists

The CDCB makes available sire lists sorted by net merit ($), cheese merit ($), and fluid merit ($). Available lists include active AI bulls and foreign bulls with semen distributed in the United States.[2] These lists contain the most recent estimates of transmitting ability for production, health, and net merit traits. Genetic evaluations are based on information about the bulls' daughters in herds participating in official production testing programs (DHIA and DHIR), as well as DNA information in the form of haplotypes.

6.8.2 Predicted Transmitting Ability for Type

PTA for type estimates the genetic abilities of bulls to transmit desirable characteristics of conformation (type classification) to their daughters. They are expressed in points above or below (plus or minus) breed average. A bull with a PTA of +1.20 would be expected to sire daughters whose average scores exceed those of herdmates by 1.20 points.

6.8.3 Interpretation of Sire Summaries

The best available estimate of a bull's transmitting ability for milk production is PTA, which is one-half of an animal's breeding value. Breeding value is measured based on an animal's complete genotype, whereas PTA is a measure of the average worth of a random one-half of the genotype present in a spermatozoon or an egg. The USDA-DHIA cow index values are also expressed in terms of PTAs. The PTAs of sire summaries are computed using genomic evaluations as well as traditional evaluation methods. The format of sire summaries is presented in figure 6.3. The bulls listed here were not necessarily the best or worst, but represent the diversity in genetic evaluations for many traits.

The reliability indicates how confident we can be that the PTA for each trait reflects a bull's true transmitting ability; it is measured on a percentage scale from 1 to 100. A confidence interval can be set for each bull's PTA indicating the probability that the bull's true transmitting ability will fall within these limits. The greater the reliability (confidence factor), the greater the probability the PTA is near the true transmitting ability for each trait.

Choosing between two bulls with similar PTAs involves less chance when the bull having the higher reliability is selected. However, it is PTA, not reliability, which determines genetic progress in the long run. Therefore, it is generally preferable to use a bull with a high PTA and low reliability than one with a high reliability and low PTA. Additionally, because of genetic (biological) variation and the sampling nature of inheritance, any one daughter of a sire may deviate considerably from the performance level expected based on her sire's PTA. Therefore, a bull's transmitting ability should not be judged merely on the production performances of one or only a few daughters. All considered, PTAs provide the most useful information in determining the amount a bull is expected to raise or lower his daughters' production in comparison with their herdmates in breed average herds.

6.8.4 Sire Summary Limitations

Any one of several situations can dilute the accuracy of sire summary estimates of a dairy bull's transmitting ability. These include (1) preferential treatment of daughters over their herdmates, (2) representation by daughters of only one or a few bulls among herdmates, (3) no or few herdmates of the same age as daughters of the bull, (4) availability of only a selected sample of the bull's daughters for the sire summary, and (5) misidentification of daughters or herdmates. These situations are more likely to occur among bulls with low (less than 40%) reliability, although they may occasionally occur in bulls with moderate (40 to 70%) reliability. Therefore, dairy breeders are cautioned against breeding a large proportion of their cows to *any single* low-reliability bull. Dairy producers are encouraged to select a group of bulls for matings with females in their herds. Diversification spreads the risk of inadvertently choosing a poor bull.

Finally, because of biological variation, it is important to understand that a plus PTA bull will not always sire daughters that are plus for milk produced. However, on the average, breeders will be more successful in increasing milk production in their herd when they use bulls rated plus for predicted transmitting ability, especially when they are rated as having high reliability.

6.9 Heritability Estimates

Heritability estimates indicate the genetic progress dairy cattle breeders can expect to make in changing a particular trait. For example, assume the herd average is 20,000 lb (9,100 kg) of milk and animals selected as parents for the next offspring average 4,000 lb (1,800 kg) above herd average. We would expect only 1,280 lb (580 kg) of that superiority to be transmitted to the progeny, because the heritability estimate for milk production (ME basis) is 32% (table 6.7 on p. 74).

milk (production of parents)	24,000 lb
milk (herd average)	– 20,000 lb
	4,000 lb
heritability of ME milk yield	× 32%
	1,280 lb

6.9.1 Heritability and Genetic Progress

Heritability is a technical term used by animal breeders to describe what fraction of the difference in a trait (e.g., milk production) is due to differences in genetic value and what fraction is attributed to environmental factors. Stated another way, it is variation due to genetic effects divided by total variation.

When heritability is relatively high, selection can be effective in improving the trait (i.e., provided superior animals can be identified). If heritability is low, however, genetic progress for the trait is slowed. More information on relatives' phenotypes or genomes can partially compensate and allow some genetic progress to occur.

We have grouped heritability estimates of selected traits in dairy cows as high, intermediate, and low (see

Ayrshire Bulls – Complete Evaluation List – December 2014																						
								Milk & fat														
Identification					Index					PTA			Protein		PL		SCS		DPR	
Number	Name	NAAB code	Samp code	% US dtrs	% ile	Rel	NM$	Herds	Dtrs	Rel	Milk lb	Fat %	Fat lb	PTA %	PTA lb	Rel	PTA mo	Rel	PTA	Rel	PTA %
NR10100				100	87	40	381	2	22	63	1533	−0.04	52	0.01	50	14	0.9	35	3.07	11	−0.2
NR10285				100	93	35	399	2	10	54	1920	−0.21	36	−0.05	51	15	2.1	30	2.9	12	0.6
NR10402				100	87	39	365	2	17	60	1949	−0.06	64	0.01	64	16	−0.7	35	3.42	16	−1.7
50803818	HIDDEN-VALLEY GARTH'S BRANDON	76AY755	S	100	25	55	42	6	10	60	−276	0.05	−3	0.05	0	44	0.7	48	2.91	35	0.1
63970273	MILLER'S CONN BUCK			100	6	46	−104	2	10	56	−485	0.14	4	0.06	−5	35	−2.7	43	2.99	30	−1.1
64394151	ANMAR PETER POWER-ET	7AY87	S	100	81	82	352	43	91	89	623	0.14	49	0.05	28	69	2.3	79	2.91	65	0.7
64394181	ANMAR JO-JO	200AY4	O	0	43	73	103	48	68	80	967	−0.04	33	−0.02	26	60	−2	74	2.79	63	−1.3
64394220	ANMAR PASSELI PASSION	1AY337	S	100	75	73	336	26	44	81	2075	−0.09	63	−0.06	54	58	−0.3	67	3.08	50	0.6
65002995	PALMYRA MILL BENNETT	1AY329	S	100	25	77	40	28	52	84	−42	0.1	16	0.04	6	63	−0.2	73	2.98	58	0.1
65003025	PALMYRA BELDON	200AY334	O	0	43	73	111	40	46	78	703	−0.02	23	−0.02	18	64	−1.2	74	2.75	66	−1
65003032	PALMYRA RAVEN BILLICK	7AY86	S	100	6	86	−71	81	135	92	480	−0.19	−14	−0.11	−3	76	0.5	84	2.91	72	−1.9
65003033	PALMYRA HERMAN CORNELL	7AY85	S	100	25	73	28	37	49	84	406	−0.11	−2	−0.04	6	60	−0.7	70	2.6	51	0.2
65003035	PALMYRA CARSON	200AY335	O	0	12	74	−15	58	63	80	629	−0.1	6	−0.07	9	63	−2.2	75	2.51	65	−1.3
65003921	PALMYRA BURNSIDE	1AY335	S	100	43	72	98	22	42	80	58	0.04	9	0.02	6	57	0.9	67	2.83	49	−2.2
65003950	PALMYRA RAVEN BOSTON	1AY336	S	100	12	75	−13	42	73	86	715	−0.08	14	−0.04	15	61	−1.3	74	3.22	55	−3.2
65006996	JACKSON-HILL LANDSCAPE JOKER			100	62	48	201	1	11	58	600	−0.05	15	0.05	28	35	0.8	42	2.98	27	0.3
65459499	PALMYRA POKER RIGGINS	7AY89	S	35	25	86	48	221	394	93	158	0.11	25	0.01	6	72	−0.2	88	3.43	70	−0.3
65866439	BROW-AYR POKER NITRO	1AY339	S	100	43	70	96	26	43	79	457	−0.04	11	−0.02	11	53	0	63	2.87	46	−0.7
66596063	PALMYRA RAVEN BENEVOLA	76AY754	S	100	56	70	167	18	29	75	531	0.08	36	0.02	21	57	0	62	3.09	48	−1.6
66596095	PALMYRA BESTER	200AY345	O	0	25	72	34	85	101	81	−262	0.09	5	0.06	1	53	0.2	73	2.88	39	−0.6
66848126	ONWORD ANGELO			100	43	47	76	1	11	58	159	0	6	0.03	10	35	0.1	43	2.89	28	−0.7
69261805	PALMYRA BENDIG BERKELY	7AY93	S	100	75	56	272	10	14	63	1174	−0.07	31	−0.07	24	41	1.6	49	2.8	38	−1.8
100135646	HIDDEN-VALLEY GARTH	76AY732	S	48	56	86	128	77	217	91	−311	0.04	−5	0.05	−2	77	2.8	86	2.62	70	0.4
100145223	GRAND-VIEW B. BOWIE	76AY733	S	100	6	65	−92	15	25	74	−264	0.03	−6	0.03	−4	54	−1.4	60	3.03	48	−0.8
100147346	DALTONDALE BIG NICK	7AY73	S	100	6	80	−99	33	58	86	660	−0.1	8	−0.05	12	69	−3.1	76	3.14	62	−1.7
100156614	COVEY-FARMS RESTLESS HEART	1AY320	S	38	43	89	72	152	350	93	626	−0.1	7	−0.06	9	82	0.4	90	2.8	80	−1.9

Figure 6.3 Example of the Complete Sire List (courtesy of the Council on Dairy Cattle Breeding).

Table 6.7 Heritability Estimates of Selected Traits in Dairy Cows.

High		Intermediate		Low	
Nonfat Solids (%)	0.59	Milk Yield (ME)	0.32	Height of Thurls	0.19
Fat (%)	0.58	Feet and Legs	0.30	Levelness of Rump	0.17
Feed Efficiency	0.56	Minerals (%)	0.27	Strength of Rear Udder Attachment	0.16
Lactose (%)	0.53	Fat Yield (ME)	0.26	Tightness of Shoulders	0.16
Protein (%)	0.50	Dairy Character	0.25	Depth of Udder	0.15
Birth Weight	0.47	Milking Speed	0.23	Strength of Pasterns	0.13
Body Weight	0.40	Topline	0.23	Teat Placement	0.13
Stature	0.40	Overall Type	0.23	Strength of Front Udder Attachment	0.12
		Persistency	0.21	Strength of Head	0.12
		Sharpness	0.21	Udder Quartering	0.11
		Rump (Length and Width)	0.21	Breeding Efficiency	0.08
		Depth of Body	0.20	Mastitis Susceptibility	0.08
				Temperament	0.08
				Twinning	0.08
				Udder Quality	0.06

Source: Adapted from Norman and Van Vleck, 1972; Gaunt, 1973. Averages calculated from other selected studies.

table 6.7). Note that the heritability of major milk components is relatively high. Milkfat, protein, and lactose are positively correlated genetically with each other, but each is negatively correlated with milk yield. Thus, as we select for increases in any one of the three major milk components, the percentages of the other two will likely increase. However, as we select for greater milk yield, the percentages of milk components tend to drift downward. With current consumer demands for lowfat dairy products, the dairy industry is predicted to move toward placing greater economic emphasis on milk protein (cf. chapter 1).

Data in table 6.7 indicate that significant genetic progress can be made by selecting for the following traits among dairy cattle: milk yield and composition, stature, milking speed, feed efficiency, body depth and weight, sharpness, length and width of rump, birth weight, feet and legs, dairy character, topline, and overall type.

Genetic progress would be slower for traits having heritabilities of less than 0.20, especially if selection were directed toward more than one trait. In selecting for two or more traits, dairy cattle breeders consider (1) the economic importance of each trait, (2) standard deviation and heritability of each trait, (3) the genetic and phenotypic correlations of traits with one another, and (4) whether the trait can be measured accurately enough to identify superior animals.

6.10 Keys to Genetic Progress in Breeding Dairy Cattle

Although the authors hold no master key that will assure dairy producers of progress in breeding more useful dairy cattle, let us consider three important inputs: (1) heritability, which represents the average fraction of the total difference among individuals caused by genetic differences (this factor cannot be effectively changed); (2) selection differential; and (3) **generation interval**. Genetic progress (GP) per generation is limited by the first two inputs. By dividing the product of the first two factors by the generation interval we can estimate annual genetic progress:

$$\text{GP/yr} = \frac{\text{heritability} \times \text{selection differential}}{\text{generation interval (in years)}}$$

This formula is true when dairy producers are selecting for one trait based on phenotypic information of one trait. Heritability is replaced by a function of accuracy of selection when more than one trait or more than one phenotypic record is the basis of selection. Selection differential represents the superiority of animals selected to be parents of subsequent generations. Unless dairy managers select parents superior to the current generation, no genetic improvement can be expected. Factors comprising selection differential include intensity of selection and phenotypic variation.

6.10.1 Accuracy of Selection

Improvement of dairy cattle is achieved chiefly through selection. Accuracy of selection (AS) is the correlation between our *estimated* genetic value and the *true* genetic value of an animal. If our AS is 0%, we are guessing; whereas, if our AS is 100% we know exactly the genetic merit of an animal. The latter is improbable to achieve because AS depends on what fractions of the observed differences among cows are due to genetic and environmental effects. Even if the genotype of an individual is known precisely, that individual is transmitting only a random half of its genotype to the offspring. For example, the heritability (i.e., the portion of variation due to genetic differences) of the milk production of cows commonly represents about 32%. If all differences in milk production were genetically determined, then heritability would be 100%. We should recognize that the higher the heritability, the easier it is to predict the genetic value of a cow from her production records.

Production records provide a reliable tool for increasing AS. The more milk production records on a given cow and her close relatives, the greater the AS (table 6.8). Because the breeder influences the number and kind of records available, he/she influences accuracy of selection.

Accuracy of selecting dairy sires may be increased by a careful study of sire proofs (see section 6.8.3). The accuracy of sire proofs is increased with increasing numbers of production-tested daughters, especially when these daughters are distributed throughout many herds since this minimizes the effect of environmental factors.

6.10.2 Intensity of Selection

Intensity of selection (IS) determines whether breeders select the best 1 of 10, 5 of 10, 9 of 10, or some other fraction of dairy cows to use in producing progeny that will serve as herd replacements and/or for herd expansion. Of course, the greater the intensity applied, the faster the genetic progress expected. Factors influencing IS include reproductive inefficiency, mastitis, other diseases that may prematurely terminate the cow's productive life or an expression of her full genetic potential, the desire (or lack thereof) to increase herd size, and the practice of mating **first-calf** heifers with beef bulls.

Through AI, it is possible to exercise substantially greater IS on sires than dams.[3] USDA estimates indicate that selecting bulls out of cows listed in the Elite Cow List[4] will add about 500 lb (225 kg) of milk yield to the *expected* daughter superiority of young dairy bulls. Intensity of selection is perhaps the most important tool with which the breeder has to work.

6.10.3 Genetic Variation

Variation is the Rule; not the Exception.

Chinese proverb

Ideally, we seek to eliminate low-producing dairy cows and retain high-producing ones. This, in itself, acknowledges the existence of genetic variation (GV). Production records are especially useful in quantifying genetic variation in milk production. We know, for example, that one genetic standard deviation (includes about 68% of the cows within the population) is approximately 1,200 lb (545 kg) of milk that will be yielded for first-calf Holsteins. This means that about two-thirds of Holstein cows will yield within 1,200 lb of **breed average** during their first lactation. The smaller dairy breeds have a smaller variation in milk production (about 1,000 lb [455 kg]) from their respective breed averages. Yet the point is clear; genetic variation provides breeders with opportunities to select and mate dairy cattle. Be aware that in the formula for genetic progress that genetic standard deviation of the trait (or index) is the measure to indicate genetic variation (not genetic variance).

Studies at Cornell University and other agricultural experiment stations indicate that milk production records of daughters of dairy bulls follow a normal bell-shaped distribution in keeping with biological variation that characterizes other forms of animal life (figure 6.4). This acknowledges the fact that even the "best" bulls

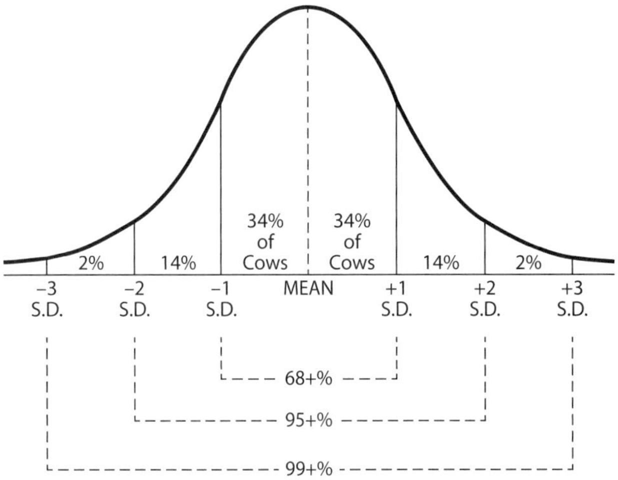

Figure 6.4 The normal bell-shaped distribution showing percentages of data within 1, 2, and 3 standard deviations above and below the mean. This can be applied to many other types of data (e.g., breed averages for milk production).

Table 6.8 Accuracy of Predicting Genetic Merit for Milk Production Based on a Cow's Records and Records of Relatives.

Number of Records on Cow	When Only the Cow's Records Are Available (%)	Cow's Records Plus Dam or Daughter Records (%)	Cow's Records Plus Dam or Daughter Records (%)	
			50 Records	200 Records
0	0	25	44	48
1	50	53	60	63
3	61	63	67	68
6	65	67	70	71

Source: Van Vleck, 1968.

have some below-average daughters while the "undesirable" sires have a few daughters above breed average for milk production.

6.10.4 Generation Interval

The generation interval (GI) is the average length of time between the birth of a cow and the birth of her replacement heifer (or the births of a sire and his replacement). Because many parents have several offspring, generation interval should be defined as the average age of the parent when all of the offspring are born. A common misconception is to estimate GI from the age of the parent when the first offspring is born; this would underestimate the GI. Traditionally for females, the GI has been four to six years. With multiple ovulation and embryo transfer schemes (see section 6.12.3) and genomic evaluations, the GI of females has been greatly reduced. Traditionally, the GI for bulls was seven to nine years for sires used in AI, and commonly five years for those used in **natural service**. Genomic evaluations are greatly reducing generation intervals of males because bulls with high genomic evaluations are used in AI much more frequently at a younger age. Genomic evaluations have increased genetic progress (see table 6.1).

GI plays an important role in GP, as noted in the following formula and examples:

$$\text{GP/year} = \frac{\text{AS} \times \text{IS} \times \text{GV}}{\text{GI}}$$

Example

	(1)	(2)	(3)
AS	0.25	0.75	0.75
IS	0.50	0.50	0.50
GV	0.50	0.50	0.50
GI	6 years	6 years	4 years

$$\text{(1) GP/year} = \frac{.25 \times .5 \times .5}{6} = 1.0\%$$

$$\text{(2) GP/year} = \frac{.75 \times .5 \times .5}{6} = 3.1\%$$

$$\text{(3) GP/year} = \frac{.75 \times .5 \times .5}{4} = 4.7\%$$

To maximize annual genetic progress, we want to increase AS (records and genomic evaluations are keys to this), IS (healthy, well-managed cows are especially important), and GV (records also help identify this), and to decrease GI. In the above examples, an increase in AS from 25% (example 1) to 75% (example 2) was reflected in a three-fold increase in GP. By reducing the GI by two years (example 2 versus 3), the GP per year was increased from 3.1% to 4.7%. The genomic revolution allowed for a decrease in GI and an increase in accuracy to increase GP.

6.10.5 Number of Traits for Which Selection Is Made

When managers select for only one trait, 100% of selectivity can be devoted to that trait. However, when selection is made for two and three traits concurrently, only 71% and 58% as much selectivity, respectively, can be expected for each trait. Therefore, the fewer the traits selected, the faster that genetic progress can be expected for the traits selected (figure 6.5).

6.11 Mating Systems for Dairy Cattle

Breeders of dairy cattle attempt to design and develop breeding plans so future herd replacements will possess the most advantageous genotypic combinations. Broad classifications of breeding systems available are inbreeding (includes line breeding and close breeding), outbreeding (includes outcrossing and crossbreeding), and grading up.

6.11.1 Inbreeding

This means mating dairy animals that are more closely related than the average of the population. It tends to make the offspring more **homozygous** than if the parents are of average (breed-wise) relationship to each other. Although inbreeding can be beneficial, its use should be confined to breeders familiar with principles of animal genetics.

The **line breeding** process is a mild form of inbreeding in which breeders attempt to concentrate the genetic merits of an outstanding individual (usually bulls because they have a greater reproductive rate than cows). An example would be mating a cow to her grandsire or great-grandsire. **Close breeding** utilizes intense inbreeding in an attempt to concentrate superior genetic worth. Examples include mating parents to progeny and/or mating full- or half-sibs and cousins.

6.11.2 Outbreeding

This is the process of mating unrelated individuals within the same breed. This is the breeding system used by most dairy producers through AI, although it may not be intentional and is limited by the genetic diversity within a breed.

6.11.3 Crossbreeding

Crossbreeding is the process of mating individuals of different breeds together. Anticipated advantages to crossbreeding dairy cattle include (1) reduced fluid carrier, (2) improved fertility, and (3) lengthened productive life. Anticipated disadvantages to crossbreeding include (1) potential loss of breed identity, (2) only short-term cure for inbreeding, and (3) difficulty in maintaining performance at the level of first generation **crossbred**.

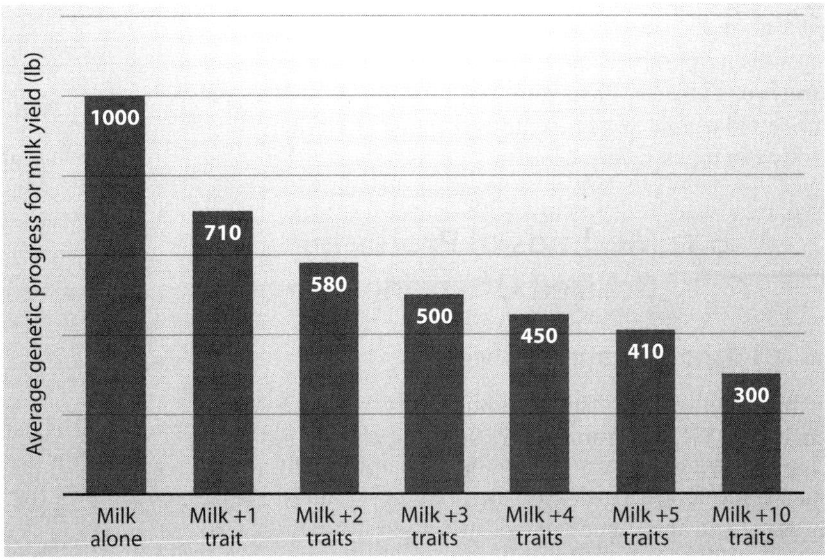

Figure 6.5 Relative expected genetic progress for milk yield if selection emphasis (pressure or selectivity) is equal for the number of traits indicated. This assumes selection for milk alone to give a genetic progress of 1,000 lb.

In general, crossbreeding can (1) take advantage of **heterosis** (hybrid vigor), (2) take advantage of good qualities of two or more breeds, and (3) cover undesirable recessives and increase heterozygosity. Heterosis is defined by how much (percentage) the average of a crossbred offspring exceeds the average of purebred parents. It equals the difference between offspring and parental averages. This is sometimes called individual heterosis (the amount that the crossbred offspring exceeds the average of purebred parents). Maternal heterosis is the difference between performances of the offspring of crossbred dams and the performance of offspring of purebred dams.

$$\text{Heterosis} = \frac{(F - P)}{P} \times 100\%$$

Where:
F = mean of offspring averages
P = mean of parent averages

In cattle, the average milk production of purebred Jersey and Holstein cows is 16,052 and 23,374 lb, respectively. Crossbred cows from the mating of Holstein bulls and Jersey females may result in cows that have an average milk production of 21,624 lb. The reciprocal cross may produce cows that average 21,744 lb. To calculate the heterosis of milk production, first we calculate the mean averages of milk production for crossbreds and their parents:

$$F = \frac{(F_1 + F_2)}{2}$$

$$= \frac{(21{,}624 + 21{,}744)}{2} = 21{,}684$$

$$P = \frac{(P_1 + P_2)}{2}$$

$$= \frac{(16{,}052 + 23{,}374)}{2} = 19{,}713$$

To calculate the heterosis of milk production:

$$\text{Heterosis} = \frac{(F - P)}{P} \times 100\%$$

$$= \frac{(21{,}684 - 19{,}713)}{19{,}713} \times 100\%$$

$$= 10\%$$

Due to heterosis and these estimates of performance of crossbreds, a 10% increase in performance is gained by matings between Holstein and Jersey. This example is plotted in figure 6.6. However, please note that even with this positive heterosis, the average Holstein is expected to outproduce the crossbreds for milk production. In figure 6.6, a heterosis of 20% or 30% illustrates examples of crossbreeding exceeding performance of both parent breeds.

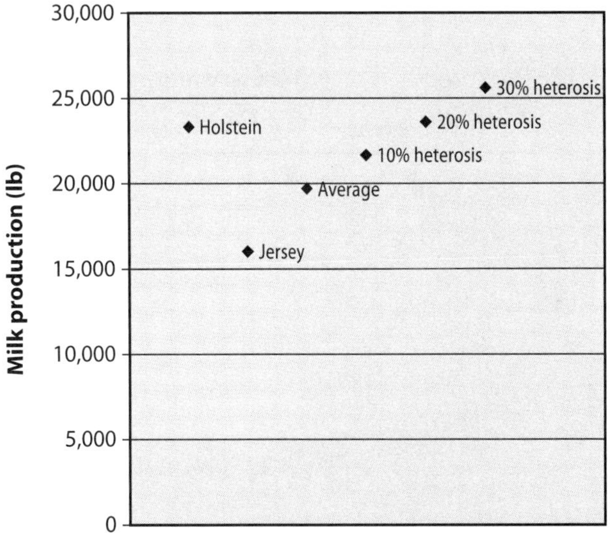

Figure 6.6 An example of performance from crossbreeding Holstein and Jersey breeds (courtesy of Roger D. Shanks).

6.11.4 Grading Up

Using purebred bulls on grade dams is called **grading up**. (This assumes the use of superior bulls.) An example of this system of improving dairy cattle is illustrated in figure 6.7.

6.12 Methods of Producing Desired Offspring

6.12.1 Genetic Mating Service

Not only do breeders of dairy cattle practice planned parenthood, they also have available computer matings. Several organizations provide a service through which dairy farmers can identify herd or cow family weaknesses, and the computer will use the information provided (e.g., weak mammary gland attachments, weak legs) to mate individual cows with what seems to be the best bull available. To undergird such decisions, it is imperative that managers have information (genome, production, type, etc.) available on their cows. Of course, knowledge of ancestors and genetic potential are important in making planned matings for the improvement of specific characteristics.

6.12.2 Sexed Semen in Dairy Cattle Breeding

Bull semen can be separated to alter the sex ratio of offspring. Dairy producers can choose to create more female than male offspring or vice versa. As more female calves are born, more selectivity is possible. Some producers can also gain added income from the sale of extra heifers. A 50:50 sex ratio causes most dairy producers to raise about 70% of their heifers merely to maintain herd size. Even raising all heifers for herd replacements does not allow much flexibility in culling. But, if 90% of the calves born were females, breeders would be able to exercise considerably greater intensity of selection.

6.12.3 Multiple Ovulation and Embryo Transfer

Accurately evaluating young animals so the best genetic candidates can be identified at a younger age is one motivation for implementing a multiple ovulation and embryo transfer (MOET) scheme. Genomic evaluations greatly enhance the opportunities to capture genetics of young females. Genetically superior females (donor dams) are super ovulated and their embryos are collected and transferred to recipient females. The surrogate female (recipient) carries the calf to term and delivers an offspring with genetics of the donor dam. All of this can occur before the donor dam would have naturally given birth to her offspring. One outcome of a MOET scheme is improved genetic change through increased accuracy of females, increased intensity of selection, and reduced generation interval.

6.12.4 Inheritance and Environment

The breeder sets the limits within which the feeder must operate.

J. M. Kays

Our discussion in this chapter has focused on the genetic aspects of increased milk production. However, good inheritance alone is not enough; the environment must be such that a cow can demonstrate her inherent ability to secrete milk. Approximately 70% of the productivity of cows is attributable to environmental factors. These factors include feeding, housing, weather, and management. Cows cannot choose their environment; they are dependent upon people to care for them properly. A champion swimmer cannot demonstrate his/her ability without water. Similarly, a champion cow cannot demonstrate her milk-producing ability without the

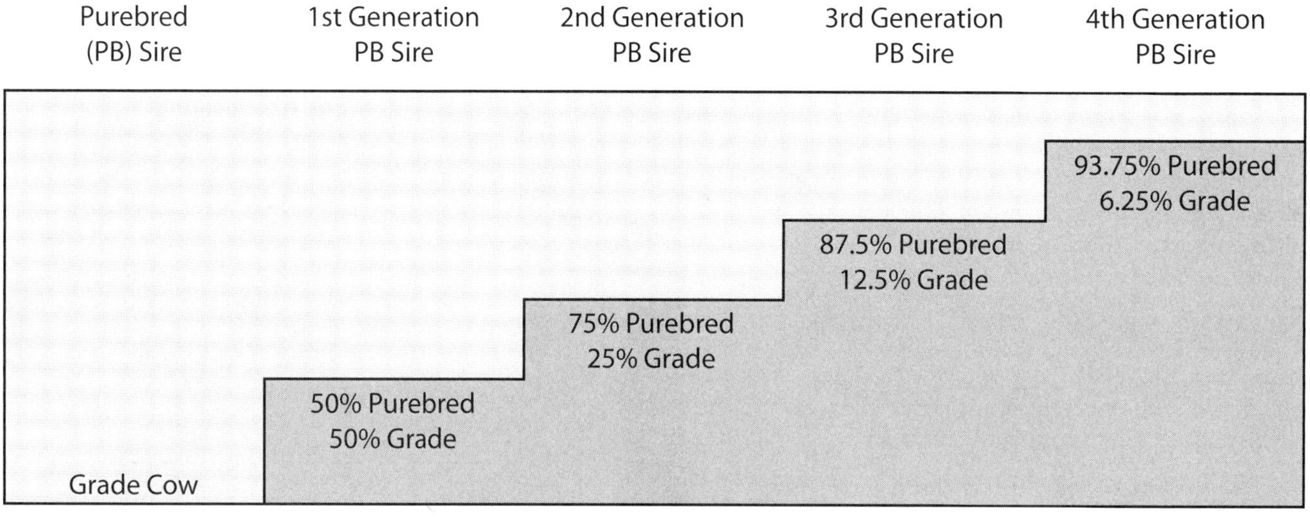

Figure 6.7 An example of how grading up increases the percentage of purebred blood.

proper environment. In general, there are more environmental than genetic differences among dairy herds.

6.13 Summary

Better cattle for better living.

National Association of Animal Breeders motto

Since humans first domesticated the cow, they have attempted to select and mate the best with the best in hopes of producing superior animals. Through use of production records, AI, embryo transfers, genomics, and the computer, progressive breeders and geneticists have greatly advanced the genetic merits of today's dairy cattle. Although developed shortly after the turn of the twenty-first century, genome evaluations are having a significant impact on genetic advancement.

A commendable goal of dairy cattle breeders is development of an outstanding herd of cows that exceeds breed average in both production and conformation. Production records, careful choosing of sires, wise selection of females, and application of sound management practices are important entries to this worthy goal. But perhaps the most basic single key to genetic progress is that of identifying genetically superior animals to use as parents of the next generation. Fundamental to identification of these *elite* parents are milk production records, not only of the animals under selection, but of those used to improve genomic evaluations.

Selection of both males and females is an important tool in improving milk production in dairy cattle. However, herd replacement needs and/or desire to expand herd size restrict intensity of selection among cows. But dairy producers can, and should, exercise intense selection pressure on the sire. Use of AI has greatly extended opportunities for sire selection. Sire summary lists are excellent resources for progressive breeders who seek assistance in identifying bulls to sire their future cows.

The foremost objective in breeding more useful dairy cattle is to determine which traits have long-range economic value, and then seek to change *those traits that can be affected genetically.* Indexes greatly facilitate productive decisions.

Heredity gives dairy cows the *ability* to milk, but environment gives them the *opportunity* to produce. In general, dairypersons have bred in more genetic ability to secrete milk than they have managed to breed out. In the final analysis, feeding and management exert a greater influence on herd production than does genetics. This is why many have said that herd averages for milk production evaluate dairy managers as much as their cows. High-production herd averages do not always indicate which herds have the best cows, but they usually indicate which cows have the best managers!

6.14 Caveats

Many abbreviations, words, or phrases are commonly confused in dairy genetics. A few of the more common are identified here.

1. SD may be selection differential or standard deviation.
   - *Selection differential*: the phenotypic difference between performance of parents and population (contemporary) average.
   - *Standard deviation*: square root of the variance. Variances may be genetic, phenotypic, or environmental.
2. ΔG (read as delta G) is genetic change or genetic progress or response to selection.
   - ΔG is often per generation but may also be per unit of time.
   - ΔGt is genetic change per unit of time.
3. If one measurement per individual is used to evaluate a trait and selection, response to selection will be heritability of trait multiplied by selection differential. For more complicated selection schemes, ΔG is a function of four components (accuracy, intensity of selection, genetic standard deviation, and generation interval). (For multiple traits genetic standard deviation becomes standard deviation of the aggregate breeding value.)
4. Lower case r = correlation. It may be used as a symbol for correlation, repeatability, or accuracy. Correlations may be genetic, phenotypic, or environmental. Repeatability and accuracy can both be defined in terms of correlations.
   - *Repeatability* is the correlation between past and future performance.
   - *Accuracy* is the correlation between phenotypic measurements and genetic value, or is the correlation between index and aggregate breeding value.

Study Questions

1. Which type (conformation) characteristic has the highest positive correlation with milk production? Why should conformation be considered in a dairy cattle breeding program?
2. Can dairy herd managers effectively select for productive longevity in dairy cattle? What should a manager consider before attempting it?
3. Of what significance is net merit to dairy cattle breeders?
4. What are herdmates? Why are they sometimes referred to as contemporaries?
5. What production records were used in calculating the EATA of dairy cows? Were all records given equal emphasis? Why?
6. Why does selection of dairy bulls carry greater responsibility than that of dairy cows?

7. What are sire indexes? Of what use are they to dairy cattle breeders?
8. Identify limitations of the (1) daughter average and (2) daughter-dam difference methods of evaluating genetic merit of dairy bulls.
9. Given the following information, and desiring to increase milk production only, which bull (A or B) would you select?

   Bull A was matcd to cows with average milk production of 18,000 lb and their daughters averaged 22,000 lb, while bull B was mated to cows with average records of 22,000 lb and their progeny averaged 23,000 lb of milk.
10. What is the basis of the parent average index? How is it determined?
11. Why is herd average used in calculating the daughter-contemporary herd difference and daughter-contemporary herd index?
12. Why is herdmate comparison useful in estimating a dairy bull's breeding value? What considerations does it include?
13. What are two multitrait indexes? How are they different?
14. What are sire summaries? When is a sire summary compiled for a dairy bull? Why are production records standardized before being used in sire summary data?
15. What is a PTA? Of what use are PTAs to breeders of dairy cattle? Why are daughter yield deviations considered in computing a PTA?
16. In appraising PTA ratings of dairy bulls, should consideration be given to the genetic level of herdmates? Why or why not?
17. Of what significance to dairy cattle breeders is the genetic principle that animals tend to drift toward breed average?
18. What is meant by reliability of predicted transmitting abilities for milk and fat yields? Identify factors influencing reliability.
19. Of what significance to dairy cattle breeders is PTA for type? What affects reliability of this genetic tool?
20. What is the basis of the formula used to calculate net merit in a sire summary? Cite an example of how net merit can be used to determine if the expected daughter's productivity will compensate for extra semen cost often associated with bulls of high plus PTAs.
21. Differentiate between transmitting ability and breeding value.
22. Why might we consider the reliability value of PTA for milk production a confidence factor?
23. Why does mating cows to bulls having higher reliabilities for milk involve less chance than breeding to those of low reliability?
24. When dairy bulls are used in herds above breed average for milk production, would their daughters be expected to exceed the PTA above their herdmates? Why?
25. Identify four factors that may affect the validity of a sire summary. Would these situations be more likely to occur among bulls with low reliabilities or bulls with high reliabilities? Why?
26. On average, do early PTAs for milk production tend to increase or decrease with time? Why?
27. What constitutes a proven sire? Does mating to proven sires guarantee superior genetic worth? Why?
28. What is a heritability estimate? What is its relationship to genetic progress?
29. Identify five or more traits in dairy cattle having high, intermediate, and low heritability estimates.
30. What is the formula for genetic progress in breeding dairy cattle? Which components of the formula can dairy cattle breeders affect most?
31. What is selection differential? Identify factors that comprise selection differential.
32. Why is genetic variation an important tool of dairy cattle breeders? What is meant by a standard deviation for milk production?
33. What is accuracy of selection? What affects it? How can it be increased?
34. What is meant by intensity of selection (IS)? How does AI affect the IS of bulls? Identify factors affecting IS of dairy females.
35. How does the number of traits for which dairy cattle breeders select concurrently affect genetic progress? Cite an example.
36. Define generation interval (GI). How does this relate to genetic progress in dairy cattle? Cite an example of how GI may be shortened by MOET.
37. Is large-scale inbreeding recommended in breeding superior dairy cattle? Why?
38. Differentiate between line breeding and close breeding.
39. Should crossbreeding of dairy cattle be practiced widely? Why or why not?
40. What is meant by grading up?
41. What is meant by computer matings of dairy cattle?
42. How could changing sex ratio (i.e., number of males and females) contribute to dairy cattle breeding?
43. Why is management important in identifying and breeding superior dairy cattle?
44. What is genomic evaluation? How did genomic evaluation impact genetic progress?
45. What is MOET? How did MOET influence genetic progress?
46. Cite step-by-step how to practice selection to achieve maximum genetic progress.

Notes

[1] Based on research conducted at Iowa State University under the direction of Dr. A. E. Freeman (see Eastwood, 1968).

[2] Dairy breed associations also publish sire summaries.

[3] Because of selection restraints applied by AI firms, less than 1% of the bulls within a breed are available to dairy farmers.

[4] The Elite Cow List is published with each genetic update. It identifies cows of outstanding genetic merit for milk production.

References

Cole, J. 2012, May. New Tools for Genomic Selection of Livestock. Department of Animal Sciences, University of Sassari; USDA Animal Improvement Programs Laboratory.

Cole, J. B., P. M. VanRaden, and Multi-State Project S-1040. 2012, September. *Net Merit as a Measure of Lifetime Profit: 2010 Revision*. Beltsville, MD: USDA Animal Improvement Programs Laboratory.

Council on Dairy Cattle Breeding. 2015, June 29. Genotype Counts by Chip Type, Breed Code, and Sex Code (https://www.cdcb.us/Genotype/cur_freq.html).

Eastwood, B. R. 1968, September. *Lactation Summary, Herd Ranking, and Preliminary Sire Summary Procedures*. Dairy Extension Mimeo (DyS-961), Iowa State University.

Eastwood, B. R. 1971, December 10. Does her record really tell you her ability? *Hoard's Dairyman* 116:1277, 1299.

Foster, W. W., A. E. Freeman, P. J. Berger, and A. Kuck. 1989. Association of type traits with production and herdlife of Holsteins. *Journal of Dairy Science* 72:2651–2664.

Gaunt, S. N. 1973. Genetic and environmental changes possible in milk composition. *Journal of Dairy Science* 56:270–278.

Hayes, B. J., P. J. Bowman, A. J. Chamberlain, and M. E. Goddard. 2009. Invited review: Genomic selection in dairy cattle: Progress and challenges. *Journal of Dairy Science* 92:433–443.

Henderson, C. R. 1975. Best linear unbiased estimation and prediction under a selection model. *Biometrics* 31:423–447.

Norman, H. D., and Van Vleck, L. D. 1972. Type appraisal: II. Variation in type traits due to sires, herds, and years. *Journal of Dairy Science* 55:1717–1725.

Shook, G. E. 2006. Major advances in determining appropriate selection goals. *Journal of Dairy Science* 89:1349–1361.

Van Vleck, L. D. 1968, May. The four keys to genetic progress. *Herd Management* 5:48–51.

VanRaden, P. M. 2002. *Selection of Dairy Cattle for Lifetime Profit*. Beltsville, MD: USDA Animal Improvement Programs Laboratory.

VanRaden, P. M. 2004. Invited review: Selection on net merit to improve lifetime profit. *Journal of Dairy Science* 87:3125–3131.

VanRaden, P. M., C. P. Van Tassell, G. R. Wiggans, T. S. Sonstegard, R. D. Schnabel, J. F. Taylor, and F. S. Schenkel. 2009. Invited review: Reliability of genomic predictions for North American Holstein bulls. *Journal of Dairy Science* 92:16–24.

VanRaden, P. M., G. Wiggans, C. Van Tassell, T. Sonstegard, J. O'Connell, B. Schnabel, J. Taylor, and F. Schenkel. 2008, September. North American Cooperation in Genomic Prediction. Canadian Dairy Cattle Improvement Industry Forum. USDA Animal Improvement Programs Laboratory.

VanRaden, P. M., G. Wiggans, C. Van Tassell, T. Sonstegard, and L. Walton. n.d. Genomic Prediction (http://aipl.arsusda.gov/reference/genomic%20prediction.html). Accessed January 16, 2013. USDA Animal Improvement Programs Laboratory.

Wiggans, G. R., and N. Gengler. 2011. "Selection: Evaluation and Methods." In J. W. Fuquay, P. F. Fox, and P. L. H. McSweeney (eds.), *Encyclopedia of Dairy Sciences*, Vol. 2 (2nd ed., pp. 649–655). San Diego, CA: Academic Press.

Websites

American Jersey Cattle Association Green Book (http://greenbook.usjersey.com/Bulls.aspx)

Council on Dairy Cattle Breeding (https://www.cdcb.us/Main/site_main.htm)

Elite Cow Lists (https://www.cdcb.us/eval/summary/elite_menu.cfm)

Genetic evaluations available from the CDCB (https://www.cdcb.us/eval.htm?)

Holstein Association USA, Holstein Genetic Evaluation Reports (http://holsteinusa.com/genetic_evaluations/GenUpdateMain.html)

National Association of Animal Breeders (http://www.naab-css.org)

Top Bull Lists (https://www.cdcb.us/dynamic/sortnew/current/index.shtml)

University of Sydney, Online Mendelian Inheritance in Animals (http://omia.angis.org.au)

USDA Agricultural Research Service, Bovine Functional Genomics Laboratory (http://www.ars.usda.gov/main/site_main.htm?modecode=12-45-29-00)

For Further Study

1. To learn more about genomic prediction and how to use it effectively, see the article by VanRaden et al., *Genomic Prediction* (http://aipl.arsusda.gov/reference/genomic%20prediction.html).
2. To learn more about the key economic values behind net merit, see the article by VanRaden and Cole, *Net Merit as a Measure of Lifetime Profit, 2014 Revision* (http://aipl.arsusda.gov/reference/nmcalc-2014.htm).

7

Dairy Herd Replacements

Raising dairy herd replacements is a vital part
of making the good herds of today better ones tomorrow.
William Dempster Hoard (1836–1918)

The authors acknowledge with sincere appreciation the timely contributions to this chapter by Dr. Alois F. Kertz and Dr. Paul L. Nicoletti. Dr. Kertz (BS and MS, University of Missouri and PhD, Cornell University) spent 28 years as a nutritional scientist with various Purina US and International companies, currently has his own consulting firm, and is a regular contributor to *Hoard's Dairyman* and other dairy industry publications. Dr. Nicoletti is Professor, Department of Pathobiology, College of Veterinary Medicine, University of Florida, Gainesville.

7.1 Introduction

Future provisions for people's milk supply are dependent, in large part, upon successful programs of raising dairy calves and heifers. The average length of time a cow stays in the milking herd in the United States has progressively decreased. In 1971, the average length of time was between 3.0 to 3.5 years (Andrus, 1971); in 2006, it was 2.0 to 2.5 years. Cassell (2006) observed that average productive life was 24.6 months for Holsteins, 29.5 months for Jerseys, and 25.5 months for Brown Swiss. This decrease is due to many factors, including economic pressure to cull older cows to make room for younger heifers with higher genetic merit and the availability of lower cost heifers while market prices for older cows are relatively high. The total cost to raise heifers varies from year to year and by location, but was $1,649 in a 2007 study of 49 operations in Wisconsin (Huibregtse et al., 2008; Zwald et al., 2007). After feed costs, replacement costs may be the second largest cost component, even before labor costs, on many dairies (Bethard, 2009).

7.2 The Need for Dairy Herd Replacements

The number of heifers a dairy producer needs annually to maintain herd size is dependent largely upon herd management, animal health, reproductive efficiency, and culling pressure applied to lower producing cows. In a survey of DHI herds in 10 states, the major reasons for culling cows were injury (27.4%), poor reproduction (19.5%), mastitis/udder (16.2%), low production (12.8%), sold for dairy purposes (8.3%), feet/legs (3.6%), and diseases (2.8%). The overall culling average has increased from 25% in 1971 to 35% (Hadley et al., 2006). This means a dairy must raise all of its female calves to maintain a 110-cow size, assuming a 50:50 sex ratio and average mortality of 18% up to calving age (National Animal Health Monitoring System, 2007).

7.2.1 Freemartins

An estimated 90% of heifers born twin to a bull are sterile due to fusion of the fetal circulatory system. Apparently, normal development of female reproductive organs is prevented by male hormones entering the female's circulation. In most other mammalian species the female twin is protected against the masculinizing effects from the male because their placentas contain an enzyme capable of converting androgens to estrogens; this enzyme is apparently absent in the bovine placenta. Diagnosis of a **freemartin** is usually based on clinical examination, but may also be determined from a sex-chromatin test[1] and from blood typing and skin grafts. Freemartins are highly tolerant of skin grafts from their male twins. Interestingly, bull calves of heifer-bull twin pairs are apparently normal.

The degree of intersexuality in freemartins varies considerably; the uterus may be rudimentary and the vagina may be only about one-third its normal length. A freemartin may have a fold of skin extending from the navel towards the udder which represents a rudimentary penis. The mammary gland is usually underdeveloped and teats are often shorter than those of normal heifers.

Aside from the frequency and infertility of the freemartin, there are other reasons to prefer single over multiple births. Multiple births lead to increased calving difficulties, more **retained placentas**, impaired breeding efficiency in the cow, reduced viability in early stages of the calf's life, and increased **stillbirths**. The gestation period for twins is usually three to 10 days shorter than for single calves. Ultrasound is used to differentiate a male from a female by looking for a fetal penis. This is not 100% accurate and is done between 65 and 85 days of gestation.[2]

7.3 Determining the Best Method of Replacement

The calf that is stunted seldom will make a cow that distinguishes herself. On the other hand, no matter how well an animal is fed, unless it has the inherited ability to grow, it never will make a large animal.

Henry La Franchi (1921–2001)

7.3.1 Raising Dairy Herd Replacements

Most dairy managers raise their own heifers for a variety of reasons. These include: (1) more direct control over a progressive breeding program; (2) usual lower cost; (3) maximal use of available feed, labor, and facilities; (4) greater biosecurity against diseases brought in from outside the dairy; and (5) potential additional income from heifer sales. Disadvantages include: (1) lack of return on investment for two years and (2) substantial feed, labor, land, housing, and managerial time diverted from an immediate source of income—the production of milk (additional lactating cows, in lieu of the heifers, would return greater immediate profits). The percentage of operations that raised heifers off the dairy was 9.3% in 2007. This accounted for 11.5% of all heifers born on an operation but raised off that dairy. Larger operations (500 or more cows) were the major users of contracts for raising calves/heifers: 35% preweaned calves, 44% weaned calves, and 23% bred heifers (National Animal Health Monitoring System, 2007).

Calves and heifers should be fed to meet their genetic potential, but without fattening. Fatter heifers have more problems calving and more metabolic problems following calving. This increases costs and reduces milk production. Calves can be fed properly to gain about 1.5 lb/day during the first two months, which will result in a doubling of their body weight from birth to two months of age. After two months, they should grow an average of

1.8 to 2.0 lb/day until they give birth at about 24 months of age. Height increase is especially critical during the first six months of age when heifers are capable of and need to realize 50% of their total height increase by first calving. Feeding program guidelines (table 7.1) illustrate key parameters, but are subject to many variables. Quantities for smaller breeds differ somewhat as related to body weight. Poor quality forages can result in poor performance, even with proper protein, mineral, and vitamin supplementation. When calves become fat, the energy level of the total ration that is fed by free choice may need to be reduced by (1) substituting lower quality forage for some of high quality and/or (2) by including **roughages** in the ration. Clean water must be readily available, as calves and heifers consume four times more water than **dry matter**.

7.3.2 Purchasing Dairy Herd Replacements

Milk producers wishing to expand herd size rapidly should purchase dairy heifers. Additionally, buying replacements provides opportunity to obtain genetically superior females. This is especially beneficial if milk production is below breed average. Lactating replacements are expected to bring a return on investment immediately.

Limitations of buying herd replacements include:

1. Milk production information is usually inadequate, especially when purchases are made from cattle brokers. Production potential of heifers is difficult to evaluate when production data on sire and dam are unavailable.
2. Many heifers available for purchase are of low or mediocre producing ability. Is it not logical that owners would retain the best heifers as replacements for their own herd? Exceptions to this are heifers sold in breed-sponsored sales, herd dispersals, and those purchased from an enterprise designed solely to raise heifers on a contract basis.
3. Costs of purchased replacements are generally greater than costs of those raised.
4. Disease is more likely to be introduced into the herd.

7.3.3 Contract Raising of Dairy Heifers

An alternative to purchasing replacements is to have them raised by someone else under contract. Two general types of contracts are employed: (1) a sell-purchase contract in which calves are sold to the raiser with an option to purchase when replacement status is reached (usually prices are agreed upon ahead of time) and (2) a direct contract in which ownership of heifers is retained and the raiser is paid an agreed amount for growing them. The latter may be based on increases in body weight, a fixed price to a certain age, monthly payments, or other means of payment.

One reason for entering into custom heifer growing is the opportunity to engage in business while utilizing available forage crops and unused livestock facilities (Wolf, 2003). The primary reasons dairy clients provided for outsourcing heifer growing were almost equally apportioned at about 20% for each of the following: more facilities to expand milking herd, insufficient facilities for heifers, labor, and management time. There were significant differences in many categories studied by regions of the country.

Important points to consider before milk producers and calf growers enter into a calf-raising contract include length of contract, how costs will be determined, age and value of the calf when delivered to grower, method of

Table 7.1 Feeding Program Guidelines[a] for Large Breed Dairy Animals.

Age (mo)	Body Weight (lb)	Wither Height (in)	Dry Matter Intake (lb)	Feed
Birth	90	29–30	—	2–4 qt of colostrum at first feeding, colostrum/transition milk up to 3 days
2	180	33.0	5.0	Milk/milk replacer, grain-based starter
4	280	37.0	7.5	Starter/grower, < 50% good quality forages
6	390	41.0	10.0	Protein/mineral/vitamin supplement, 50% forage
8	500	43.0	12.5	Protein/mineral/vitamin supplement, forage
10	615	46.0	15.0	Protein/mineral/vitamin supplement, forage
12	720	48.0	17.5	Protein/mineral/vitamin supplement, forage
14	830	50.0	19.0	Protein/mineral/vitamin supplement, forage
16	940	51.0	20.5	Protein/mineral/vitamin supplement, forage
18	1,050	51.5	22.0	Protein/mineral/vitamin supplement, forage
20	1,160	52.5	23.5	Protein/mineral/vitamin supplement, forage
22	1,280	53.5	24.5	Protein/mineral/vitamin supplement, forage
24	1,400	54.0	25.5	Protein/mineral/vitamin supplement, forage

[a]Guidelines can vary considerably with breed, genetics, health, facilities, and quality of feedstuffs.

identification, nutrition level at which calves are to be fed for optimal growth, age or size at which to breed and which sires to use, insemination charges, death losses, animal health practices and cost, age or stage of development of replacements when returned to the owner, method of payment, insurance, and how disagreements between parties will be arbitrated. A formal written contract including these points, as a minimum, is essential.

7.3.4 Leasing Dairy Cows

When capital is limited, or for those who wish to restrict the dairy farm enterprise to milking cows, the opportunity to rent or lease cows may have appeal. Older producers who wish to breed and raise herd replacements, but who do not wish to milk cows, and others who specialize in raising replacements often offer cows for lease. Such leases may be based on charges per month, per lactation, per year, or on pounds of milk produced.

7.4 Contributions of Colostrum to Calves

Colostrum is the first milk secreted at parturition. This product of mammary glands is so precious and important to farm mammals that it could be called liquid gold. Colostrum is an especially rich source of proteins[3] (more than 20%; table 7.2), minerals, and vitamins. Colostrum contains from 10 to 100 times more vitamin A and from 12 to 15 times more iron than normal milk (the deep yellow or creamy color of colostrum results largely from carotene). The "extra" iron and vitamin A is stored in the liver.

7.4.1 A Provider of Antibodies to Newborn Calves

Since neonatal calves are unable to produce antibodies for some time after birth, they must acquire preformed antibodies through colostrum, which is especially high in immunoglobulins.[4] To be effective in disease prophylaxis, colostrum must be ingested within a few hours **postpartum**, preferably within 15 to 30 minutes, because gut closure occurs within about 24 to 30 hours (the period of permeability may be less in cold weather). Subsequently, instead of the absorption of immunoglobulins into the bloodstream, the **neonate** digests these large molecular weight proteins, causing the loss of their immunization properties.

The amount of antibody absorbed depends on the time after birth the antibody is fed and the amount of antibody consumed. The calf's ability to absorb antibody is maximal at birth (about 50% efficiency). By 20 hours of age the efficiency is only 12%, and by 36 hours there is little or no further absorption of antibody (figure 7.1).

Several studies indicate that proper ingestion of colostrum reduces calf mortality by more than one-half. Should the mother die at parturition, the calf should receive colostrum from another cow or from a supply that has been frozen or otherwise preserved for such emergencies. Since the calf can absorb antibodies excreted by other farm animals, including horses and sheep, colostrum from these sources may be fed. However, certain diseases are species-specific, and colostrum from another species might not afford the desired level of protection. Colostrum replacers and extenders are also available commercially, but efficacy data should be provided in order to ensure the purchaser that antibodies are absorbed from the products.

Table 7.2 Physical Characteristics and Composition of Colostrum and Whole Milk (data mostly from Holsteins).

	Postpartum Milking					
	1	2	3	4	5	Whole Milk
Specific gravity	1.056	1.040	1.035	1.033	1.033	1.032
pH	6.32	6.32	6.33	6.34	6.33	6.50
Total solids (%)	23.9	17.9	14.1	13.9	13.6	12.9
Fat (%)	6.7	5.4	3.9	4.4	4.3	4.0
Solids-not-fat (%)	16.7	12.2	9.8	9.4	9.5	8.8
Total protein (%)	14.0	8.4	5.1	4.2	4.1	3.1
Casein (%)	4.8	4.3	3.8	3.2	2.9	2.5
Albumin (%)	0.9	1.1	0.9	0.7	0.4	0.5
Immunoglobulins (%)	6.0	4.2	2.4	—	—	0.09
IgG (%)	3.2	2.5	1.5	—	—	0.06
NPN (% of total N)	8.0	7.0	8.3	4.1	3.9	4.9
Lactose (%)	2.7	3.9	4.4	4.6	4.7	5.0
Ash (%)	1.11	0.95	0.87	0.82	0.81	0.74

Source: Foley and Otterby, 1978.

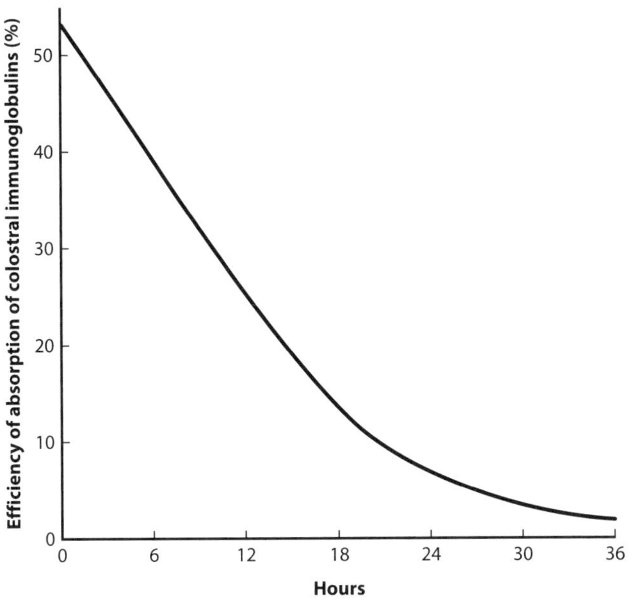

Figure 7.1 Efficiency of absorption of colostral immunoglobulins during early life of the calf.

If a cow is partially milked once or twice before calving to relieve udder congestion, the secretion upon calving differs little from normal colostrum, whereas if the udder is completely milked out for several days **prepartum**, the secretion post-calving will resemble normal milk. Therefore, to provide optimal antibody protection for the calf, it is advisable to minimize prepartum milking.

Because of the nutritional, laxative,[5] and immune contributions of colostrum to calves, excess colostrum should be saved. Importantly, care must be exercised in selecting colostrum to ensure it does not come from cows that may be positive for *Mycobacterium avium* subsp. *paratuberculosis* (Johne's disease; see chapter 13). Some dairy managers store excess colostrum so it will be fermented by lactic acid-producing bacteria and therefore can be kept several days for feeding. Since it contains about 2.2 times as much solids as normal milk, colostrum is usually diluted 1:1 with warm water when fed to older calves.

To provide calves with milk of high antibody concentration, maximal use should be made of first-milking colostrum. In addition to decreases in the calf's efficiency of absorption with time, the concentration of antibody in colostrum decreases with each successive milking. At 24 hours postpartum (third milking), colostral antibody is only about one-third the usual concentration at parturition. Therefore, to ingest an equivalent amount of antibody, a calf must consume three times as much third- as first-milking colostrum. Furthermore, to assure an equal amount of antibody is absorbed into the blood 24 hours after birth (when efficiency of absorption is only about one-fifth that at birth), the calf must ingest about 15 times as much colostrum. We emphasize, therefore, that both the presence of antibody in the digestive tract of the young calf and its absorption into the blood are highly important. Small quantities of these antibodies are present in normal milk, but they are usually absent in milk replacers. This is another reason for feeding fermented colostrum to calves for seven to 10 days.

In summary, three factors ensure that calves receive adequate colostral antibody protection: *quickness* (within an hour following birth), *quantity* (preferably 4 qt at first feeding so calves receive a minimum of 100 g of immunoglobulin G [IgG]), and *quality* (minimum of 50 mg/mL IgG concentration).

7.4.2 Pathogens Found in Colostrum

Analyses of colostrum samples from dairy farms in the United States showed 33 to 75% had total microbial counts in excess of 100,000 cfu/mL and 10,000 cfu/mL coliforms. Thus on many dairies, even if the quickly/quantity/quality parameters are met, contaminated colostrum may inoculate the calf with pathogens and reduce the intended benefits. The three major sources of bacteria in colostrums (Godden, 2007) are those shed from soiled udders; contaminated equipment, including buckets and bottles used in feeding (the primary source); and bacterial proliferation in improperly stored colostrum. Prevention requires clean udder preparation; cleaned and sanitized collection, storage, and feeding equipment; not pooling colostrum to prevent the spread of Johne's and other diseases; cooling within two hours to 40°F (4.4°C); and use within three days. Colostrum can be frozen, but this requires rapid cooling to limit bacterial growth. The best containers to use are flat resealable gallon bags (which, when laid flat on a cold surface, provide maximum surface area for cooling and rewarming); and they should not be stacked in the freezer until frozen. Household refrigerators/freezers often do not have the capacity to quickly cool or freeze colostrum. Dairies are increasingly pasteurizing colostrum, but this requires batch pasteurizers and a closely controlled time/temperature process to kill most pathogens, including those of Johne's disease (Kertz, 2010).

7.4.3 Colostrum, Gamma Globulin (IgG), Serum Protein, and Mortality

Two tests can be made of calves' blood to indicate how well colostral antibodies (IgG) are absorbed. The most conclusive is an IgG **assay**, but it is expensive and time consuming to deliver to a laboratory for assay. The target minimum, 10 mg/mL of IgG, is closely related to calf mortality. Based on data from the National Dairy Heifer Evaluation Project (1991–1992), calves with blood levels of IgG < 10 mg/mL had 3.5 times greater mortality than calves with blood levels > 10 mg/mL (figure 7.2 on the next page). Serum protein levels are more commonly used than IgG assays because they may be run quickly and easily on the farm with a **refractometer**. Target level

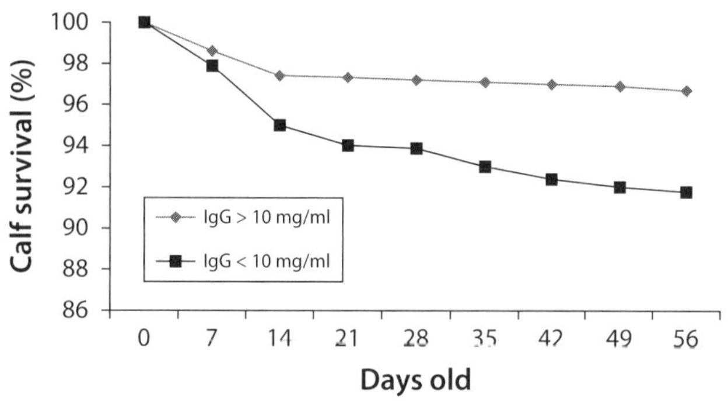

Figure 7.2 Relationship between calf serum IgG levels and mortality (1992 National Dairy Heifer Evaluation Project).

for calves is > 5.0 mg/mL serum protein. In a German study (Kaske et al., 2005), administering 4 rather than 2 liters of colostrum at the first feeding increased serum protein levels in calves by 10%, providing strong evidence of more colostral antibodies absorbed. Additionally, serum protein levels were consistent for seven days then decreased by 10% by the end of 14 days. Concentrations of serum protein continued to decrease until three to four weeks of age, after which they increased as calves began to synthesize their own antibodies. This indicates how critical colostral antibodies are to a calf since their absorption is needed to sustain the calf for weeks until it can synthesize its own. It is during the first two weeks of life that most **scours** and deaths occur in calves.

Use of disposable irrigation units designed for humans enables delivery of one quart of colostrum directly into the stomach (figure 7.3). Thus, one can be certain the calf receives colostrum soon after birth. It also permits separation of the cow and calf at birth before suckling begins. This practice minimizes transfer of mastitis-causing bacteria from infected to noninfected mammary glands (cf. chapter 14).[6] It is advisable to use colostrum from older cows that, on the average, have been exposed to greater numbers of pathogens than younger ones. Such colostrum usually has a comparatively broader and higher complement of antibodies. Subcutaneous administration of 100 cc of noncoagulated whole blood obtained from older cows also affords calves protection against many infectious diseases by activating calves' mechanisms for producing immunoglobulins.

Antibodies of IgG have multiple functions, including **opsonization** of pathogens to enhance **phagocytosis**, complement fixation for **lysis** of pathogens, prevention of adhesion of pathogens to tissues, inhibition of microbial metabolism by blocking enzymes, **agglutination** of bacteria, and neutralization of toxins and viruses. An interesting related note is that human colostrum contains approximately 100-fold higher concentrations of immunoglobulins than does bovine colostrum.

7.5 Alternatives to Fresh Milk for Calves

7.5.1 Milk Replacers

Milk is the best food for raising calves. However, economics make it impractical to feed fresh milk to dairy calves. The most accepted alternative is a substitute called milk replacer, for which costs vary but average about one-half the producer price of milk. We define a milk replacer as a formulated feed designed to replace mother's milk in providing the nutritional needs of neonatal mammals during the critical, early suckling or milk-feeding stage of life. Although milk replacers are also available for colts, kids, lambs, piglets, and other farm animals, we limit discussion here to those fed calves.

Figure 7.3 Administering colostrum with a disposable irrigation unit made for humans.

In formulating milk replacers we must appreciate that although the baby calf has four stomach compartments, only the **abomasum** (true stomach) functions appreciably during the first three weeks of age. Milk is shunted past the underdeveloped rumen of the calf into the **omasum** and abomasum through the esophageal groove (figure 7.4). Therefore, digestion in the calf is dependent upon enzymes present in the abomasum rather than on the action of ruminal microorganisms, as in older ruminants.

Since it functions as a nonruminant in early life, the calf at first requires preformed dietary nutrients. It then gradually becomes, over a period of five to eight weeks, a functional ruminant requiring fewer preformed nutrients.

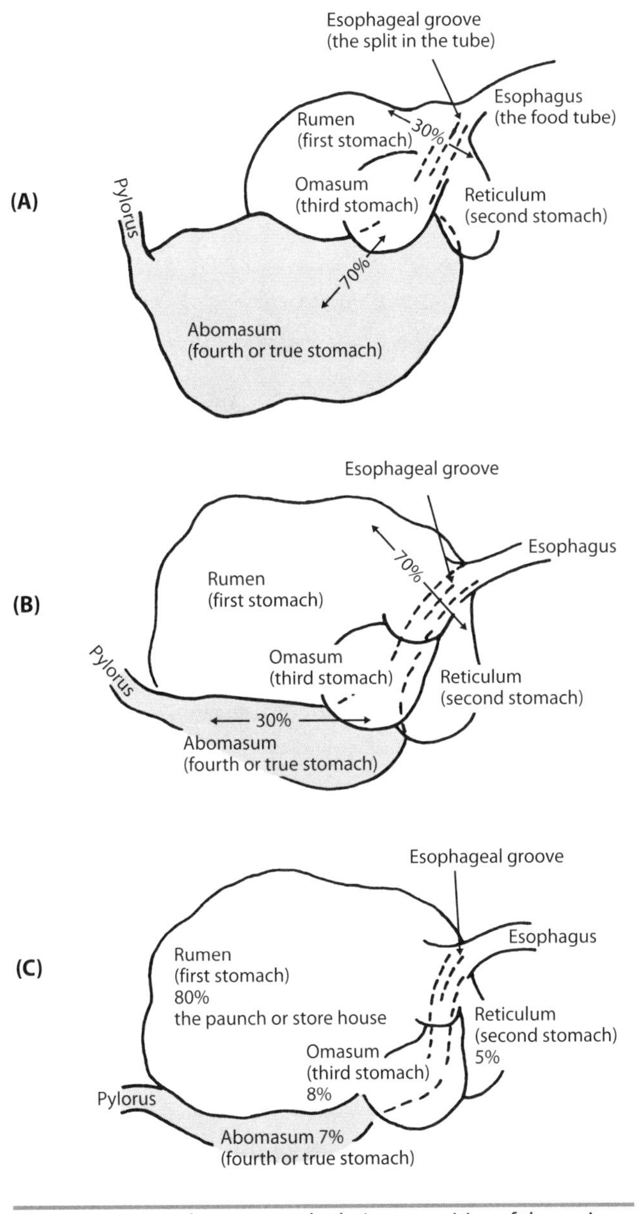

Figure 7.4 Development and relative capacities of the various stomach compartments of the (A) calf at birth, (B) calf at 2 months, and (C) mature cow (Campbell et al., 2010).

Calf starter, commonly formulated with dry grains, is fermented in the rumen, leading to functional development of rumen papillae and a functioning rumen. This is why the young calf's digestive system is unable to digest refined sugars, starch, or vegetable proteins efficiently prior to ruminal development.

Milk replacers include by-products of milk (e.g., dried skim milk, dried whey, and whey protein concentrates) as their base, as a calf's digestive system can assimilate milk's unique components of casein, lactalbumin, and lactose. About 15 to 20% homogenized fat is added to provide energy and to improve hair coat and general appearance of calves. Soy lecithin or other emulsifiers are added to improve fat digestibility and to help disperse the fat in water when it is reconstituted at feeding.

Ingredients commonly used in formulating milk replacers include:

1. *Dried skim milk.* This is an ideal ingredient; however, most of it is manufactured as human grade, making it an expensive component of milk replacers. Since skim milk contains no fat-soluble vitamins, these should be added (especially vitamins A and D) to milk replacers containing dried skim milk.[7]
2. *Dried whey.* Most research indicates that dried whey may constitute up to 50% of the milk replacer without apparent restriction of weight gain or health impairment of calves. Quantities greater than one-half, however, are reported to increase the incidence of diarrhea among calves.
3. *Dried whey proteins.* These may contain from 30 to 70% protein. Increasingly, these protein sources are going into human foods because of their physical properties and nutritious protein content.
4. *Plant proteins.* Using limited amounts of plant materials in milk replacers, especially soybean concentrates or isolates, wheat protein isolates, and certain other high protein plant sources, has resulted in acceptable calf performance. However, plant proteins have not been found nutritionally equivalent to milk proteins for young calves. Most commercially prepared plant proteins intended for use in milk replacers result in satisfactory gain among calves when they do not exceed one-half the total protein in the replacer. In general, the greater the proportion of plant carbohydrates, such as polysaccharides and starch, the lower the protein proportion and the overall digestibility of the ingredient since calves do not have enzymes in the abomasum to digest carbohydrates other than lactose or glucose.
5. *Animal proteins.* Fish protein concentrates and isolates are more digestible than products lower in fish-derived protein. But fish protein sources may have fat and other components that impart odors that are not well accepted, at least by people feeding calves, and do not dissolve or suspend well when used in liquid milk replacers. Fish protein sources may also be low in B

vitamins and magnesium and fail to support normal **tocopherol** levels in the blood of young calves. Moreover, meal made from certain fish (e.g., dogfish and sharks) is low in lysine.

6. *Blood serum proteins.* These proteins are also used at lower levels of 5% or less in milk replacers.

Calf milk replacers tend to be fed at a rate of about 500 g/day when they contain 18 to 20% fat and 22 to 24% protein. Milk replacer containing 20% fat and 24% protein will support a growth rate of up to 400 g/head/day under favorable environmental conditions. Feeding of 750 g/day of milk powder containing 20% fat and 28% protein during the first few weeks of life should support growth rates of about 700 g/day. However, for every decrease in ambient temperature of 9°F (5°C) below 60°F (15°C), calves will lose 0.2 lb body weight, resulting in the need to feed an additional 100 g/day of milk powder. For the younger calves, adjustments need to occur in the liquid feeding program. Older calves may require additional energy through both the milk replacer and calf starter.

In all cases calves should receive warm water during cold weather. How critical is warm water? A South Dakota State University study showed that when calves drank water with temperatures of 46°, 63°, or 90°F, rumen content temperatures dropped by averages of 20°, 5°, and 3°F, respectively.

7.5.2 Milk Enhancers

Adding direct-fed microbials (also called probiotics) to milk or milk replacer may support calf intestinal integrity and overall health. Most research has shown minimal effect of direct-fed microbials on animal growth or feed efficiency. Rather, improved intestinal bacterial flora may reduce the risk of diarrhea, particularly when animals are exposed to significant immunological, environmental, or other stressors.

In 2011, the Bovine Alliance on Management and Nutrition (BAMN) published the guide *Direct-Fed Microbials (Probiotics) in Calf Diets.* These products contain microorganisms, mostly bacteria, that can benefit very young calves. Beneficial bacteria (commensal) are an important part of the calf's gastrointestinal tract, which is constantly challenged by large numbers of bacteria, viruses, and protozoa found in feed, bedding, and the environment. Important functions of these bacteria are to: (1) ferment carbohydrates, thereby producing short-chain fatty acids, including acetic, lactic, and butyric acids, that reduce intestinal pH and inhibit growth by some pathogens while improving digestion and absorption; (2) compete against pathogens for nutrients; and (3) promote development and maintenance of the immune system (both structure and function) by signaling it to produce immunoglobulins and other health enhancers.

When pathogens, stress, metabolic upset, the use of antimicrobials, and other causes upset the balance of intestinal bacteria, animal susceptibility to disease increases. Providing probiotic bacteria to assist in reestablishment of the normal bacterial flora can boost animal performance. A probiotic is defined as a live microbial feed supplement that beneficially affects the host by improving its intestinal microbial balance (Heyman and Ménard, 2002).

Direct-fed microbial feed supplements may contain prebiotics along with the probiotic microorganisms. Prebiotics are compounds (e.g., yeast culture, oligosaccharides) that promote the growth of gut bacteria, but are not living organisms. Feeding both together as symbiotics may promote the growth of intestinal bacteria, thereby improving the microbial profile in the gut. Most direct-fed microbial products for calves are sold as feed additives, and are added to milk or milk replacer just prior to feeding. Other forms are administered as gels, pastes, or **boluses**.

Bacteria in direct-fed microbial products are normally listed on the product label under their scientific name (e.g., *Lactobacillus acidophilus*) followed by the numbers of viable organisms in colony-forming units (cfu) per unit weight of the product. Most direct-fed microbial products utilize one or more of the following bacteria:

1. *Lactobacillus—acidophilus, lactis, plantarum*, and *casei*
2. *Bacillus—subtilis* and *lichenformis*
3. *Bifidobacterium—bifidum, longum*, and *thermophilum*
4. *Enterococcus faecium*

In one study, feeding a mixture of *Bacillus* spores and *Enterococcus faecium* resulted in significantly higher average weight gain and feed efficiency of calves over eight weeks compared with responses of control calves.

Direct-fed microbials are regulated as feed ingredients by the American Association of Feed Control Officials (AAFCO) and the US Food and Drug Administration (FDA). The *Official Publication* of the AAFCO provides a list of microorganisms appropriate for use in animal feeds and is recognized by the FDA. Additional regulations are included in 21 CFR 573 and 579.

7.6 Early Weaning and Once-a-Day Feeding

7.6.1 Early Weaning

Milk producers (backed by scientific research) have reduced the costs of feed required to raise herd replacements by weaning at five to six weeks of age, although the national average is eight weeks (National Animal Health Monitoring System, 2007). Feeding only once daily has not gained favor because it is often impractical to feed 6 to 8 qt at one time. The amount of milk or milk replacer fed daily was traditionally about 10% of the calf's birth weight, but greater feeding rates of 15 to 20% of birth weight have enabled calves to grow more rapidly without fattening as long as milk replacer protein level is 26 to 28% with 15 to 20% fat. Successful weaning at any age from liquid feeds requires access to high-quality dry starter mixture, fresh water, but no hay until after weaning or two months of age. Feeding dry feed enhances early development of rumen fermentation by helping establish ruminal bacteria and protozoa, which produce **volatile fatty acids** in the rumen. The order of volatile fatty acid production in the rumen that most facilitates development of rumen papillae and function is butyric > propionic > acetic. This order greatly aids the calf in making the transition from liquid to dry feeds.

7.6.2 Calf Starter

Calf starter is a dry feed (table 7.3) that is fed when calves are shifted from liquid feeds. A desirable calf starter is high in energy (about 75% **total digestible nutrients**, or 1.5 Mcal/lb of **metabolizable energy**) and contains 16 to 18% protein. It should be fortified with vitamins A and D, contain an **antibiotic** (most likely against coccidiosis), and be palatable. Adding sweeteners such as molasses (4 to 6%) or dry sugars (sucrose or dextrose) improves the palatability and early acceptance of calf starters. Calf starters, which are texturized with coarse grains rather than being all-pelleted, facilitate rumen papillae development and performance (Porter et al., 2007).

7.6.3 Automated Calf Feeding

As herd size and labor costs increase, many dairy farm operators seek an automated means of feeding calves (figure 7.5). Most calf feeders have two nipples per group of about 25 calves. These feeders are programmable to allow for increasing daily feeding of milk replacer, and decreasing milk replacer feeding during the weaning transition period.

Table 7.3 Sample Calf Starter Ration.

Ingredient	%
Corn, yellow no. 2, crimped	20–40
Oats, rolled or crimped	10–25
Pelleted concentrate	40–60
Molasses	4–6

Source: Adapted from National Research Council, 2001.

7.7 First Breeding of Dairy Heifers

The question of when to breed heifers encompasses many considerations. These include age, body size and weight, seasonal milk price, milk yields, longevity, calving difficulty, reproductive efficiency, and even show-ring competition. Due to confounding influences of levels of nutrition, management practices, and genetic factors, it is difficult to assess the literature and obtain clear answers. However, let us discuss briefly the above considerations.

7.7.1 Age

Perhaps age, per se, at first **service** is less important than level of nutrition, body size and development, and physical health and well-being. When nutrition is optimal, dairy calves grow most rapidly between three and nine months of age. This is determined by body measurements such as height at withers, width at hook and pinbones, and body length.

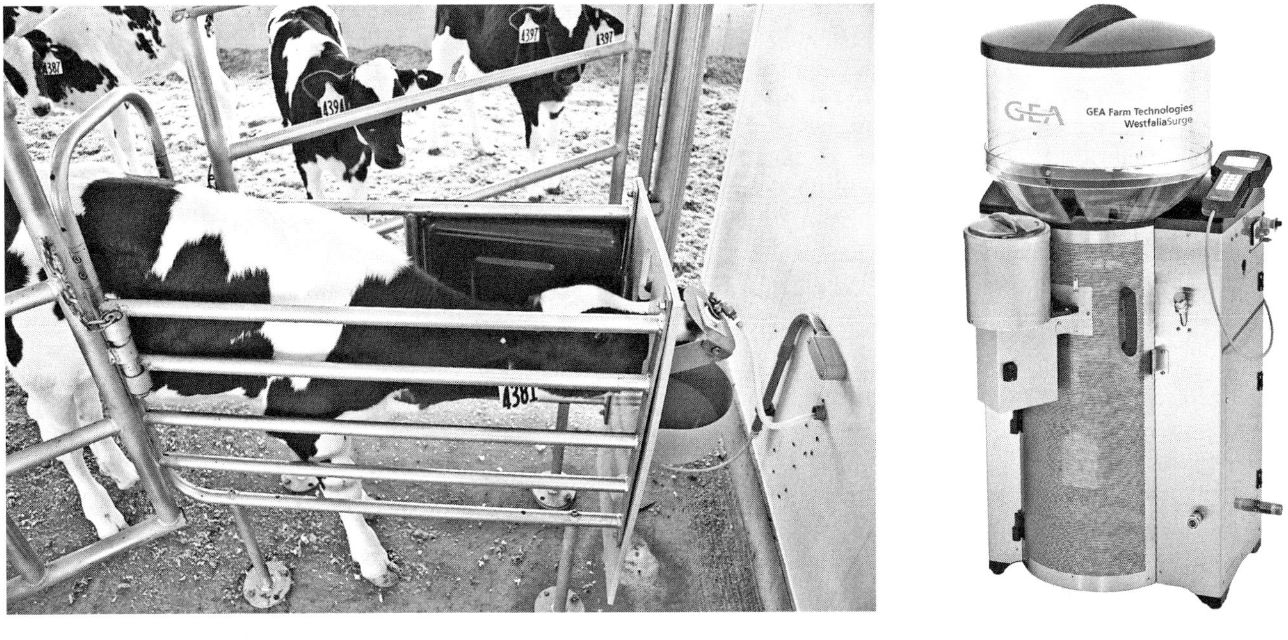

Figure 7.5 Examples of automatic calf feeders (courtesy of GEA Farm Technologies).

There has been a trend to revise recommended breeding age downward. Two important reasons for this are (1) most dairy heifers are better fed and, therefore, grow out more quickly than those of yesteryear and (2) dairypersons are aware of increasing labor and investment costs, which favor early breeding. Recommended ages and weights for breeding dairy heifers are given in table 7.4.

Table 7.4 Recommended Ages and Body Weights for Breeding Dairy Heifers.

Breed	Age (mo)	Body Weight (lb)
Ayrshire	13–15	600–700
Brown Swiss	15–18	750–900
Guernsey	13–15	550–650
Holstein	14–17	750–900
Jersey	13–15	500–600
Milking Shorthorn	14–17	750–850

7.7.2 Genetic Factors

7.7.2.1 Body Size and Weight

In general, early breeding of dairy heifers should not retard growth postbreeding because pregnancy is not nearly as much a nutrient drain as lactation will later be. Early bred high-producing animals usually attain normal size after gestation: 11% more weight and 2% more wither height during their first lactation; they will reach mature size in their third to fourth lactations (Kertz et al., 1997). If genetics and mature body weight (MBW) are known for a herd's population, targeting growth rates and breeding heifers at a percentage of this MBW (e.g., 60%) has been proposed. But there are practical limitations to measuring MBW and height of cows and then taking the same measurements in growing heifers.

Body size and weight and sexual maturity are closely related to plane of nutrition. Experiments conducted at the Rowett Research Institute in Scotland showed that, when levels of nutrition varied, dairy heifers attained sexual maturity at the same stage of physical development but at different ages (Crichton et al., 1960). Management's effect on rate of maturity has been estimated to be from two to three times as important as the effect of sires (Hillers and Freeman, 1965).

7.7.2.2 Calving Difficulty

In general, as age at first calving increases, calving difficulty decreases. This is often confounded, however, by inadequate nutrition. Moreover, underfeeding is often accompanied by increased calving difficulty.

The advisability of mating first-calf dairy heifers with Angus bulls to obtain smaller calves and thereby minimize calving difficulty is questionable since it reduces numbers of dairy herd replacements. Although it is generally accepted that the dam exerts the greatest influence on birth size and weight of calves, sire selection for small birth weight is possible. Many bulls used in AI have calving ease scores that are useful in selecting semen that can help minimize calving difficulties of first calving heifers.

7.7.3 Economic Factors

7.7.3.1 Milk Yields

When dairy heifers are bred young, milk yields trend lower during the first lactation. This is attributed, in part, to the need of smaller animals to use a relatively high percentage of their dietary intake to support body growth during the first lactation. However, total lifetime milk production is the most important consideration; average milk yield per day of herd life is perhaps a more accurate method of appraising milk yields. Research indicates that cumulative milk production is favored by early breeding. Moreover, since feed accounts for more than one-half the cost of replacements, it is sound economics to shorten the "boarding" period and bring heifers into lactation early.

7.7.3.2 Productive Longevity

There are insufficient experiments of the effect of early breeding on productive longevity to permit definite conclusions. Most such studies are confounded with other variables. Wisconsin studies indicate that there is no detrimental effect of early breeding on longevity (Schultz, 1969). Overfeeding, however, may decrease longevity (Hansen and Steensburg, 1950).

7.7.3.3 Reproductive Efficiency

Most research indicates that age at first breeding has no significant effect on reproductive efficiency. Ideally, every dairy cow should produce a healthy calf each year. We would rate this reproductive efficiency at 100 on a 0 to 100 index.

7.7.3.4 Seasonal Milk Pricing

Milk prices in the United States are usually highest in the fall. Therefore, cows calving in late summer or early fall (August to October) are commonly more profitable than those calving in spring or summer. Fall calving tends to perpetuate itself since calves born then will calve about 24 months later.[8] This assumes, of course, the ideal 12-month **calving interval**.

7.7.3.5 Exhibition of Dairy Animals

Although of little economic significance related to milk production, age of calves is a consideration in dairy show competition. Most show classes of dairy animals specify either January 1 or July 1 birthdates. Older heifers in a given class commonly have an advantage in body size and development. In certain instances, breeders of purebred dairy cattle consider this when determining the best time to breed dairy heifers.

7.8 Diseases of Calves

The number of replacements available for herd expansion and/or culling is closely related to the health of dairy calves. Many diseases affect the health and well-being of calves. We have categorized these as infectious and noninfectious diseases.

7.8.1 Infectious Diseases

7.8.1.1 Calf Scours

This acute disease of newborn calves is commonly characterized by diarrhea and rapid exhaustion. It is also called white scours, dysentery of the newborn, and calf pneumonia-enteritis. The susceptibility of the calf to scours usually decreases for the individual calf during the first two weeks of life. Yet it causes more calf deaths than any other disease.

The primary infective agent of calf scours is generally considered to be *Escherichia coli*, although one or more other bacterial and viral agents may be involved. Bacterial invasion is commonly related to a disrupted physiological equilibrium of the calf. The stresses of crowding, insanitary surroundings, substitute diets, chilling (aggravated by wind and/or dampness), malnutrition of the calf's dam, and failure of the calf to ingest sufficient colostrum increase susceptibility to the disease. It is more prevalent in cold months and when established in a group of calves. Animal passage enhances the virulence of the infecting microorganism.

The acute, fatal form of scours occurs principally in calves one to three days old and is characterized by subnormal temperature. The subacute form is characterized by a white diarrhea, normal to slightly elevated body temperature, dehydration, and emaciation. Secondary complications frequently lead to pneumonia.

Various drugs and biological agents may be used to treat calf scours and electrolytes are often administered to replace those lost because of diarrhea. Prevention is the preferred means of controlling this disease. Attention to sanitation, nutrition, and protection from inclement weather is of prime importance. Providing a clean, dry environment is one of the best **prophylactic** practices.

Ensuring ingestion of colostrum within six hours of birth is also a good prophylactic measure. Additionally, introduction of a modified (or "tamed") form of the live virus that causes scours was reported to confer considerable immunity. In a field trial cosponsored by the University of Nebraska and the Norden Laboratories of Lincoln, Nebraska, that involved 56 herds in 14 states, the incidence of all forms of calf scours in vaccinated herds was reduced from 50.0 to 16.1%, and death loss from all forms of the disease in those herds dropped from 9.0 to 1.2% (Anonymous, 1973). Prepartum vaccination of the dam will increase antibodies in the colostrum and prevent or decrease the severity of scours in the newborn calf.

7.8.1.2 Salmonellosis (Paratyphoid Infection)

This disease is common to humans and animals and, for this reason, every animal infection has potential public health significance. It is characterized in calves and adult cows by diarrhea, **septicemia**, elevated body temperature, malaise, and lung lesions.

Salmonellosis is a bacterial infection. One of the causative organisms, *Salmonella dublin*, may be viable in dried soil for up to three years. Therefore, once the infection is diagnosed in calves, there is a tendency for the disease to recur **endemically**. The disease is spread through feces also. Adult cattle infected with *S. dublin* may become fecal excretors of infection for life. Infected calves are also capable of becoming fecal excretors.

Most bovine infections occur in dairy calves between two and four weeks of age, although the age of susceptibility extends from calves one week old to adult cattle. Practices that may increase the likelihood of infection include crowding, improper sanitation, inappropriate feeding, and inadequate housing.

7.8.1.3 Hemorrhagic Enterotoxemia

This acute disease of dairy and beef calves is characterized by a sudden onset of hemorrhagic enteritis, followed by early death. It is caused by *Clostridium perfringens* type C. The disease has also been reported in neonatal piglets and lambs.

Calves affected are two to 14 days old (greatest incidence occurs at three to five days). Cold, wet, windy, unfavorable weather at parturition is commonly associated with this disease. Unfortunately, it is usually calves of high-producing cows that are affected. Some calves are found dead without having shown signs of illness. Typically, however, sick calves stop nursing for 12 to 24 hours then demonstrate signs of acute abdominal pain (uneasiness, straining, and kicking at the abdomen). The presence of hemorrhagic diarrhea is dependent upon the rapidity with which the disease acts. Body temperature usually remains normal during the two to 24 hours from onset of the disease to death.

Administration of large doses of hyperimmune horse serum (*C. perfringens* antitoxin type C or types B, C, and D combined) is of therapeutic value when injections are made early in the course of the disease. However, because a heterologous serum is used, **anaphylactic shock** is possible. (Of course, the calf must first be sensitized by an initial exposure to the foreign protein before anaphylactic shock will occur.)

Since *C. perfringens* type C is usually a part of the normal bacterial flora of intestinal tracts of healthy adult cattle, eradication of the disease is unlikely. The recommended prophylactic practice is to vaccinate cows with two injections one month apart during pregnancy. A whole-broth, type C, toxoid culture will stimulate formation of antitoxin, which will be secreted in the colostrum. A single booster dose usually suffices for subsequent pregnancies.

7.8.1.4 Viral Pneumonia (Enzootic Pneumonia of Calves, Calf Influenza)

Viral pneumonia of calves is an infectious respiratory disease. It is most common during fall and winter and affects calves from two weeks to six months of age. Calves housed in drafty and/or poorly ventilated facilities are most susceptible to this infection. About 25% of infected calves die. Recovered calves are usually immune to reinfection.

Environmental stress and a lack of sufficient colostrum precipitate calf pneumonia, which is caused by a combination of viral and bacterial infections. The principal agent of viral pneumonia of calves is *Myxovirus parainfluenza* type-3. Several mycoplasmae and bacteria have been identified as being involved with the disease. These include *Pasteurella multocida, Mannheiea haemolytica*, and *Histophilus somni.* Clinical symptoms of viral pneumonia of calves include elevated body temperature (104°F [40°C]), coughing, nasal discharge, and difficulty breathing. Diarrhea and dehydration may accompany the disease. Viral infection alone is usually mild, but it is aggravated by secondary bacterial infections, which intensify symptoms and increase mortality. Deaths usually occur three to seven days after onset.

There is no specific treatment for viral infection. Sulfonamides and **broad-spectrum antibiotics** aid in controlling secondary bacterial infections. Expectorants, atropine, and proteolytic enzymes assist in alleviating respiratory distress. Since chilling, drafts, inclement weather, starvation, fatigue, and insanitary facilities predispose calves to the disease, it is readily apparent that good management practices are the key in preventing viral pneumonia.

7.8.1.5 Navel Ill or Joint Ill (Omphalitis, Pyosepticemia)

This infection of the umbilical cord of newborn calves results from one or more microorganisms: streptococci, staphylococci, *E. coli,* salmonella, *Corynebacterium pyogenes,* and *Brucella abortus.* Morbidity and mortality rates are often high when cows calve in insanitary facilities.

Postnatal infections commonly appear at three to four weeks of age. The umbilical infection may be accompanied by septicemia with secondary sites of infection in the liver, bladder, kidneys, brain, eyes, or joints. Inflammation and pus are present at all sites of involvement.

The "joint phase" of navel ill usually involves many joints of the extremities. Pus in a joint cavity is especially damaging to articular cartilage because of digestive enzymes found in leukocytes. The result is arthritis, which produces ulcerations that can be detected in a radiograph. The ulcerations may result in permanent lameness in calves.

Systemic antibiotics aid in treatment, but the best control is prevention. A clean, dry pasture is the best provision for parturition. Clean, well-bedded maternity stalls that help minimize umbilical contamination are second choices. At birth, the umbilicus should be cut to leave a 1.5 in stump, which is then bathed in a disinfectant (iodine).

7.8.1.6 **Clostridium perfringens** *Type D Enterotoxemia*

This disease is most frequent among calves six to 10 weeks old and cattle eight to 24 months of age. Affected animals are usually being well fed, and there is a rapid course from onset to death. Clinical signs of type D intoxication of calves resemble those of heavy metal and plant poisonings and involve the central nervous system: incoordination, blindness, prostration, and convulsions. Affected calves usually die in one to two hours.

Because of the speed with which the disease acts, therapy is ineffective. However, passive immunity with *C. perfringens* type D (or types B, C, and D antitoxins combined) will give protection, and **toxoids** of the same types stimulate lasting active immunity.

7.8.1.7 Viral Diarrhea (Mucosal Disease)

Mucosal disease is caused by the bovine viral diarrhea virus (BVD). As with most viral diseases of calves, the condition does not usually affect colostrum-fed calves until they are over one month of age; it occurs principally in dairy cattle six to 24 months of age. It is commonly characterized by a *dry* cough (a cough unaccompanied by mucus or phlegm), nasal discharge, and salivation. Body temperature may rise to 105 to 108°F (40.5 to 42°C) and persist three to five days with ulcerations in the mouth and reddening of nostrils and oral mucosa, especially around the teeth. **Leukopenia** is present four to five days. Mucoid diarrhea may then develop. Rinderpest may be mistaken for viral diarrhea.

No specific treatment for viral diarrhea is known. Sulfonamides and antibiotics are only effective against potential secondary bacterial infections. Vaccines are available that stimulate immunity in calves.

7.8.1.8 Blackleg (Black Quarter)

The causative organism of blackleg, *Clostridium chauvoei,* survives as a spore in soil. It is difficult to kill and may persist, especially in swampy land, for many years. Blackleg bacilli are believed to enter puncture wounds caused by barbed wire, thorns, or other sharp objects. The organism produces a toxin that rapidly leads to death. Infected animals have a fever (105 to 107°F [40 to 41°C]), and gaseous swellings form beneath the skin, usually on hips and/or shoulders. Accumulated toxin causes stiffness, lameness, or even paralysis. The skin over the affected muscles becomes hard and stiff, and a crackling noise may be heard when pressure is applied (due to stiffness of the skin and to gas produced by bacteria beneath the skin). An infected animal loses its appetite, stops ruminating, breathes rapidly, and becomes depressed. Just before death, which commonly occurs within two to five days after exposure, body temperature decreases and convulsions often develop. Muscle tissue is blackened and streaked with dark red areas—hence the name blackleg.

Blackleg occurs in the United States and throughout most of the world. Animals 6 to 18 months of age are most susceptible. Vaccination of calves 4 to 12 weeks of age, or a few weeks prior to being turned onto pasture, is the recommended prophylactic measure. The first vaccine was developed in France in 1883.[9] Blackleg bacterins produce a high degree of immunity in 10 to 12 days, which persists for 12 months or longer. In problem areas, calves should be vaccinated at two months of age and again at four months. On most farms it is sufficient to vaccinate against blackleg only at four months. Treatment with antibiotics and serum may be effective in early stages of the disease.

Since the carcass may be a source of infection due to release of viable bacteria and spores, it should be burned or buried deeply.

7.8.1.9 Infectious Bovine Rhinotracheitis

The infectious bovine rhinotracheitis (IBR) virus of calves is related to the herpes virus of humans. It causes enteritis and necrosis of the upper respiratory tract mucosa; pneumonia may result from a secondary infection. Symptoms include an intense reddening of the **conjuctivae** and mucous membranes of the nose. Discharge from the eyes and nose is clear at first but gradually becomes thick and **purulent**. Body temperature increases and the condition may be accompanied by a moist cough.[10] The IBR virus is considered one of the inciting agents of shipping fever and a vaccine against it is available.

7.8.1.10 Screwworm Infestations

An important cause of calf losses in some locations is infestation by screwworm flies. This fly lays eggs in living tissue only. In calves, infections occur near navels of newborns and in wounds associated with castration, dehorning, earmarking, and injuries. Fortunately, a release of large numbers of male screwworms sterilized by irradiation virtually eliminated the worm, and they are no longer a problem in the United States.

7.8.2 Noninfectious Diseases

7.8.2.1 Poisoning by Pesticides

Several chlorinated hydrocarbons (e.g., lindane and toxaphene) and organic phosphorus compounds (e.g., Co-Ral and Malathion) have been used previously in controlling external parasites on cattle and insect pests on plants that may be consumed by cattle. Such products may poison young calves. Due to tissue residues and chronic toxicity, the use of chlorinated pesticides has been drastically curtailed, and many are no longer registered for use in the United States. Pyrethrins and synthetic pyrethroids are much safer to use.

7.8.2.2 Lead Poisoning

For many years, lead poisoning caused significant losses of young dairy calves. The common source of lead has been lead paints used on barns, sheds, fences, and equipment. Young calves have a tendency to lick freshly painted equipment or old paint that is flaking from painted surfaces.

Clinical symptoms of lead poisoning in calves are depression, **anorexia**, intermittent constipation and diarrhea, visual impairment, lack of coordination, difficulty in swallowing, muscular tremors, excitement, and convulsions. Clinical signs may not develop until several days following ingestion of a fatal dose of lead.

A recommended treatment is to administer magnesium sulfate or sodium sulfate. These medications convert soluble lead to insoluble lead sulfate. The disease is easily prevented by simply removing old lead paint from surfaces.

7.8.2.3 Arsenic Poisoning

This minor cause of calf loss results from calves ingesting arsenic materials such as insecticides, herbicides, and wood preservatives. Acute poisoning is characterized by evidence of pain, staggering, salivation, weakness, diarrhea, collapse, and death. If administered soon after ingestion of arsenicals, British anti-Lewisite (BAL) is an effective antidote. Not allowing calves access to arsenicals is the key in avoiding this potential cause of calf loss.

7.9 Dehorning and Removing Extra Teats

To reduce the incidence of future injuries, most dairy calves in the United States are dehorned. The application of caustic potash or use of an electric cauterizer is preferred for destroying horn buds at three to seven days of age.

Although not usually harmful, extra teats have no value and detract from the appearance of the udder. Therefore, when present they should be removed when the calf is about one month old. The teat and surrounding area of the udder should be disinfected with iodine before and after teat removal, which simply entails cutting off the unwanted teat close to the udder with a pair of sharp, *sterile* scissors. Usually there is minimal bleeding, and it is sufficient to hold cotton wool to the site momentarily.

7.10 Summary

An important aspect of a profitable dairy farm enterprise is having available a dependable supply of herd replacements of high milk-producing potential. Dairy herd improvement is possible only when culled cows are replaced with well-bred, well-fed, and well-managed replacements.

The importance of colostrum in sustaining the health of neonatal calves is well established. Colostrum is the means by which the cow transmits passive immunity (i.e., preformed antibodies) to the calf. Thus, colostral immunity significantly decreases losses of calves under excellent management practices. For maximal protection against disease, calves must receive colostrum soon after birth.

This is because decreases in concentrations of the protein fractions of bovine mammary secretions tend to follow a logarithmic decreasing curve the first four milkings postpartum, after which the decline is less rapid. Calves need immunoglobulins early in life, but their absorption efficiency decreases rapidly during the first 24 hours.

Although growth rate of suckled calves is greater than that of artificially raised ones, feeding high-quality protein and sufficient amounts of milk replacers can provide similar gains. There are, however, certain factors that should be considered in formulating milk replacers. The calf's digestive processes favor milk proteins. However, as our human population and nutritional needs expand, we find high-quality milk protein in greater demand. This means substituting, whenever possible, vegetable proteins for milk proteins in milk replacers intended for calf feeding. Present recommendations are to include up to one-half vegetable protein. Additionally, sucrose and starches are not well utilized when fed to young calves in milk replacers. It is, therefore, recommended that dried buttermilk, whey solids, and other milk by-products, which provide a good source of lactose, be used maximally in milk replacers.

Early breeding produces the advantages of lower feed cost to first calving, earlier return on investment, and higher cumulative production per day of age. This requires an optimal nutrition program to grow calves economically.

Study Questions

1. What is the average length of time dairy cows remain in the milking herd? Approximately what percentage of milk cows are replaced annually?
2. Identify the three most important reasons for disposing of dairy cows.
3. What are the advantages and disadvantages of dairy farmers raising their own replacement animals?
4. Why should dairy heifers not be overfed or fattened?
5. For what reasons would dairypersons prefer to purchase herd replacements? Identify limitations that may accrue when replacements are purchased.
6. Describe two types of heifer-raising contracts. What considerations should be included in a calf-raising contract?
7. What circumstances might prompt a dairyman to lease milk cows?
8. What is a freemartin? How can one be identified early in life?
9. Describe the contributions of colostrum to calves. How does colostrum differ in composition from normal milk?
10. Why is it important that calves receive colostrum early in life? Discuss gut permeability to antibodies and efficiency of absorption of antibodies with time.
11. How would you provide colostrum for a calf if its mother died at parturition?
12. Should cows be completely milked out prepartum? Why or why not?
13. Why save excess colostrum?
14. Why do older cows commonly have a broader complement of antibodies than younger ones?
15. Is blood concentration of gamma globulin related to early calf mortality? Why?
16. For what reasons may dairypersons choose to give colostrum directly to calves and not allow them to nurse their mothers at all?
17. What is a milk replacer? How does a calf's early digestive capabilities affect its formulation?
18. Discuss the relative merits of plant and animal proteins for use in milk replacers.
19. What considerations should be given the practices of early weaning and accelerated feeding of dairy calves?
20. What are the components of a calf starter? At what ages should it be fed to calves?
21. Identify factors influencing the decision of when to breed dairy heifers. What can be the major disadvantage of early breeding of dairy heifers?
22. Do the authors recommend mating dairy heifers with Angus bulls? Why?
23. Outline management practices that serve to minimize infectious diseases of dairy calves.
24. What are the principal noninfectious diseases of calves? How may they be eliminated?
25. Why should dairy heifers be dehorned? At what age should extra teats be removed?

Notes

[1] In singly born heifers, only cells with the XX-shaped chromosomal pair are observed, and in singly born bulls, only the XY composition is observed. However, in male-female twin pairs, both animals frequently have both types of cells. When the male cell type is observed in the heifer, she is a freemartin. This rather simple laboratory test can be performed immediately after birth to assist in identifying female twins (with a bull) that are worth raising for replacements.

[2] Professor Raymond Nebel, Virginia Polytechnic Institute and State University, personal correspondence, May 2010.

[3] Colostrum secreted 6, 12, and 24 hours postpartum contains only about 67%, 34%, and 28% as much protein, respectively, as that secreted just prior to and at calving time. Decreases in the quantity of the immunoglobulin fraction account for most of this compositional change (an exponential decrease).

[4] Gamma globulins are apparently secreted into milk directly from the cow's blood. We might consider colostrum an intermediate between blood and normal milk.

[5] The high mineral content of colostrum has a laxative effect on the calf, thus initiating the increased flow of nutrients through the digestive tract.

[6] This results, in part, from frequent suckling by the calf. With each nursing (or milking), the cow's sphincter is opened and may remain open for an hour or more.

[7] The Bovine Alliance on Management and Nutrition (BAMN) is comprised of representatives from American Association of Bovine Practitioners (AABP), American Dairy Science Association (ADSA), American Feed Industry Association (AFIA), and US Department of Agriculture (USDA). BAMN publishes the following guides in relationship to calves: *A Guide to Calf Milk Replacers: Types, Use and Quality* (2008), *A Guide to Colostrum and Colostrum Manage-*

ment for Dairy Calves (2001, Spanish version available), *A Guide to Dairy Calf Feeding and Management: Optimizing Rumen Development and Effective Weaning* (2003), *Feeding Pasteurized Milk to Dairy Calves* (2008), *Managing a Pasteurizer System for Feeding Milk to Calves* (2008), and *Heifer Growth and Economics: Target Growth* (2007).

[8] A Nebraska study revealed that fall-born calves were 7% heavier at eight months of age than those born in spring.

[9] The Navajo Indians are credited with the first immunization against blackleg in the United States. They soaked a blanket in the body fluids of an animal that died of the disease. They then allowed the blanket to air-dry while exposed to the warm sun. Pieces of the dried blanket were then stitched into the brisket of healthy animals, resulting in their developing immunity against blackleg.

[10] A moist cough is accompanied by phlegm, which is a viscid, stringy mucus secreted by mucosa of the air passages. Phlegm is one of the four humors (cardinal body fluids) of the Hippocratic formulation of general pathology.

References

Andrus, D. F. 1971. How long will she last? *Hoard's Dairyman* 116:498–499.

Anonymous. 1973. New calf scour vaccine. *Sunbelt Dairyman* 11:26.

Bethard, G. 2009. Don't let replacement costs drain you. After feed, it's your biggest hit. *Hoard's Dairyman* 154:517.

Campbell, J. R., M. D. Kenealy, and K. L. Campbell. 2010. *Animal Science: The Biology, Care, and Production of Domestic Animals.* Long Grove, IL: Waveland Press.

Cassell, B. 2006. Changes are in store for productive life. *Hoard's Dairyman* 151:299.

Crichton, J. A., J. N. Aitken, and A. W. Boyne. 1960. The effect of plane of nutrition during rearing on growth, production, reproduction, and health of dairy cattle. *Animal Production* 1:145; 2:45, 159.

Foley, J. A., and D. E. Otterby. 1978. Availability, storage, treatment, composition, and feeding value of surplus colostrum: A review. *Journal of Dairy Science* 61:1033–1060.

Godden, S. 2007, March 20. "Practical Methods of Feeding Clean Colostrum." Pre-Conference Calf Seminar, pp. 39–47. Burlington, VT: Professional Dairy Heifer Growers Association.

Hadley, G. L., C. A. Wolf, and S. B. Harsh. 2006. Dairy cattle culling patterns, explanations, and implications. *Journal of Dairy Science* 89: 2286–2296.

Hansen, K., and V. Steensburg. 1950. The endurance and production of cows reared on different rations. *Bretn. Forsgslab Kbh.* 246.

Heyman, M., and S. Ménard. 2002. Probiotic microorganisms: How they affect intestinal pathophysiology. *Cellular and Molecular Life Sciences* 59:1151–1165.

Hillers, J. K., and A. E. Freeman. 1965. Differences between sires in rate of maturity of their daughters. *Journal of Dairy Science* 48:1680–1683.

Huibregtse, A., T. Anderson, P. Hoffman, and A. Zwald. 2008. In just nine years . . . heifer-raising costs jumped over 50 percent. *Hoard's Dairyman* 153:171.

Kaske, M., A. Werner, H. J. Schuberth, J. Rehage, and W. Kehler. 2005. Colostrum management in calves: Effects of drenching vs. bottle feeding. *Journal of Animal Physiology and Animal Nutrition* 89:151–157.

Kertz, A. F. 2010, March 8. Heat-treated colostrum raises IgG absorption. *Feedstuffs* 82:12–13. This article is a synopsis of the following: Elizondo-Salazar, J. A., and A. J. Heinrichs. 2009. Feeding heat-treated colostrum or unheated colostrum with two different bacterial concentrations to neonatal dairy calves. *Journal of Dairy Science* 92:4565-4571.

Kertz, A. F., L. F. Reutzel, B. A. Barton, and R. L. Ely. 1997. Body weight, body condition score, and wither height of prepartum Holstein cows and birth weight and sex of calves by parity: A database and summary. *Journal of Dairy Science* 80:525–529.

National Animal Health Monitoring System. 2007, October. *Dairy 2007. Part I: Reference of Dairy Health and Management in the United States.* Fort Collins, CO: USDA Animal Plant and Health Inspection Service, Veterinary Services.

National Research Council. 2001. *Nutrient Requirements of Dairy Cattle* (7th Rev. Ed.). Washington, DC: The National Academies Press.

Porter, J. C., R. G. Warner, and A. F. Kertz. 2007. Effect of fiber level and physical form of starter on growth and development of dairy calves fed no forage. *Professional Animal Scientist* 23:395–400.

Schultz, L. H. 1969. Relationship of rearing rate of dairy heifers to mature performance. *Journal of Dairy Science* 52:1321–1329.

Wolf, C. A. 2003. Custom dairy heifer grower industry characteristics and contract terms. *Journal of Dairy Science* 86:3016–3022.

Zwald, A., T. L. Kohlman, S. L. Gunderson, P. C. Hoffman, and T. Kriegl. 2007. *Economic Costs and Labor Efficiencies Associated with Raising Dairy Herd Replacements on Wisconsin Dairy Farms and Custom Heifer Raising Operations.* Madison: University of Wisconsin Dairy Team Extension.

Websites

American Feed Industry Association (http://www.afia.org/afia/home.aspx)

Association of American Feed Control Officials (http://www.aafco.org/)

University of Wisconsin Extension, Pierce County (http://pierce.uwex.edu)

USDA, APHIS, Animal Health Program (http://www.aphis.usda.gov/wps/portal/aphis/ourfocus/animalhealth)

For Further Study

1. For more information about calf feeding, see this video by GEA Farm Technologies: http://www.gea-farmtechnologies.com/us//en/bu/milking_cooling/youngstock/us_calf_feeder/default.aspx.
2. The *Nutrient Requirements of Dairy Cattle* is a widely used reference that reflects the changing face of the dairy industry, allowing users to pinpoint nutrient requirements more accurately for individual animals. The authors provide guidance on how to conduct nutrient analysis of feed ingredients, insights into nutrient utilization by the animal, and formulation of diets to reduce environmental impacts. It is available from the National Academies Press (http://www.nap.edu/catalog.php?record_id=9825).

8

Feeding Dairy Cattle

The most important ingredient in any dairy ration is the dairyman.
Louis L. Rusoff

The authors are grateful to D. Wayne Kellogg, PhD, Professor of Dairy Cattle Nutrition, University of Arkansas, for his contributions to this chapter.

8.1 Introduction

A lack of food and proper nutrition is the greatest and most urgent problem facing humans. Malnourished people fail to function efficiently and productively. Similarly, inadequate nutrition limits the efficiency and productivity of dairy cows more than any other factor. A great asset of the dairy cow is her ability to utilize multiple feeds efficiently. Yet dairy farmers must not take feeding dairy cattle lightly. More low production has resulted from improper nutrition than from breeding inferior animals. Thus, if dairy farmers are to realize maximal milk production per cow, **rations** must be fitted to cows and not cows to rations.

Feeds are, for the most part, composed of the same constituents. They differ in the percentages of these constituents, and in their **digestibility**, palatability, bulk, and the physical effect they may have on animals. Ruminants can utilize straws and **stovers**, which are by-products in the processing of grains for human uses. However, such **roughages** often are low in digestibility and are deficient in protein,[1] vitamin A (carotene), and certain minerals. To be utilized efficiently these roughages must be properly supplemented to provide a **balanced ration**.

Current recommendations for providing complete daily nutrient requirements (balanced rations) for dairy cattle were produced by the National Research Council (NRC) (2001) Subcommittee on Dairy Cattle Nutrition. The nutrient requirements given in the NRC publication are, under normal environmental and animal health conditions, adequate to prevent deficiencies and to provide acceptable rates of growth, reproduction, and milk production with feeds of at least average composition, palatability, and digestibility. Variation among animals in ability to digest feed components is relatively small, but differences in appetite, body size (consequently feed capacity), growth rates, and intensity of milk secretion are indeed great.

Additional reliable sources of feed recommendations and of compositional data pertaining to feedstuffs include *Feeds and Feeding* (Morrison, 1956), a book that has been acclaimed worldwide. First published in 1898, this book has had 22 revisions and is still a widely used reference book in feeding farm animals. Requirements also have been published by the Agricultural Research Council (1980) in Great Britain, which is now part of the Biotechnology and Biological Sciences Research Council.

Since feed represents approximately one-half the cost of milk production and about three-fourths of the cost of raising dairy herd replacements, it is indeed appropriate that we devote a chapter to feeding dairy cattle. The feeding of calves is discussed in chapter 7, and that of dairy steers in chapter 15. Therefore, discussion in this chapter is devoted primarily to cows. Let us first consider the uses dairy cows make of feed nutrients.

8.2 Animal Uses of Nutrients

Nutrients are consumed to support four basic needs of dairy cattle: (1) maintenance, (2) growth, (3) pregnancy, and (4) milk production. Let us consider each briefly.

8.2.1 Maintenance

It has been estimated that the maintenance requirement is between 11 and 15 percent above the starvation minimum.

Erle E. Bartley (1923–1983)

Examples of maintenance—the nutritional price paid to keep an animal from losing weight—include use of nutrients for proper body functioning (**basal metabolism**), voluntary activity, maintaining body temperature, and circulating blood and other body fluids. Cows of the same size and breed may vary in their maintenance requirements by as much as 8 to 10%, even under controlled activity conditions.

The amount of energy required for maintenance of a lactating dairy cow has been set by the National Research Council (2001) at 0.08 Mcal of NE_L/kg of $BW^{0.75}$. Although there is no apparent difference per kilogram of body weight among breeds of cows, managers should be aware that maintenance depends, in large part, on an animal's activity and environmental conditions. Maintenance requirements of lactating cows are approximately 10% greater than those of dry, nonpregnant cows. This results, in part, from the increased secretion of thyroxine among lactating cows, which is reflected in a faster heart rate, an increased **basal metabolic rate (BMR)**, and large quantities of blood circulating through lactating mammary glands. For instance, on a daily basis a cow producing 60 lb (about 27 kg) of milk pumps approximately 12 tons (about 10,900 kg) of blood through her mammary glands (cf. chapter 10). Maintenance levels recommended are based on studies made when cows were confined and do not allow for grazing or feedlot activity, which may increase maintenance needs by 25 to 50%. If feed is procured for the cows, less energy is expended by the cow for maintenance. Approximately one-half the feed consumed by lactating cows is used to satisfy maintenance needs (figure 8.1); however, that percentage varies dramatically depending on milk yield. Thus, as the genetic capability of cows improves and more milk is produced, less of the energy fed is partitioned to maintain the lactating cow. Collectively, the dairy herd of the United States has become more efficient during recent decades because fewer high-producing cows are needed to produce milk compared to an era when less milk was produced per cow. This improved efficiency has eased pressure on the environment, even after accounting for the energy required to harvest and deliver feed to the cows (Capper et al., 2009).

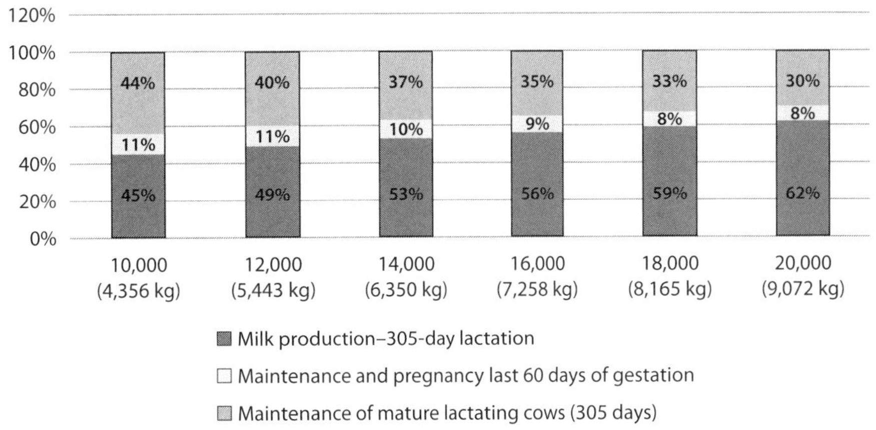

Figure 8.1 Relative proportions of feed (^{NE}lactating cows) utilized by mature 1,433 lb (650 kg) cows for maintenance, pregnancy, and milk production. Pounds of feed required for maintenance and pregnancy are quite similar irrespective of quantity of milk produced. However, more feed is required for each increment of milk produced, thus making the percentage used for maintenance and pregnancy proportionately less and less as production increases. Calculations are based on assumed levels of production from 10,000 to 20,000 lb/yr during a 305-day lactation and 60-day dry period (data obtained from National Research Council, 2001).

8.2.2 Growth

Nutrients needed for growth vary with age, breed, sex, and rate of growth. Dairy cows achieve mature weight when they are five to six years old; therefore, young cows continue growing while producing milk. A review of certain principles from chapter 7 may help understand the additional nutrients required. In relation to body weight, older animals have lower requirements for protein, energy, vitamins, and minerals than younger ones. Feeding more than the required allowance for growing animals commonly increases rate of growth, whereas feeding less slows growth. Importantly, excessive feeding of energy fattens the animal and can impair reproductive ability and shorten lifetime production of milk. Conversely, if growth is stunted in early life there is risk of permanently reducing mature size and weight. Some recently devised feeding programs achieve rapid growth by balancing rumen undigestible protein (RUP) with rumen digestible protein (RDP) while providing adequate energy with the goal of calving by 22 months of age rather than the traditional 24 months of age. Regardless of age and weight at calving, the young cow will continue to grow. Therefore, additional nutrients are required to support growth for two- and three-year-old lactating cows. Compensatory growth, i.e., gaining weight rapidly when feed becomes available, does occur after periods of slow weight gains, but there are risks associated with underfeeding.

8.2.3 Pregnancy

Proportionately, the nutrient requirements for reproduction are relatively low. However, research has showed that during the final two months of pregnancy, energy needs of pregnant heifers are approximately 50% greater than those for nonpregnant heifers of the same size.

The nutritional state of the cow is important in achieving pregnancy, especially among high-producing cows. In Louisiana State University studies, dairy cows losing weight during the month of insemination demonstrated reduced conception rates and required more inseminations per pregnancy (2.45) than did those gaining weight during the month of breeding (1.63 services/conception). Weight loss by cows is common during early lactation, and the transition period from dry (nonlactating) status to peak milk production has been the focus of much research. The goal should be to improve feed intake as quickly as possible after calving. This is achieved by preventing excessive body weight gain during the dry period and by adjusting the diet a few days before calving to avoid an abrupt change in the type of feed. Because cows reduce **dry matter** intake to cope with heat stress during hot weather, this partially explains the low reproductive efficiency of cows bred during summer. Limited availability of vitamin A, or its dietary precursor β-carotene, can affect the epithelial tissues of the body (including the reproductive tract) and lead to reproductive problems. Other nutritional factors that result in impaired energy metabolism, such as a phosphorus deficiency, will reduce reproductive efficiency of dairy cows. Yet, this is not a reason to overfeed phosphorus. It is an expensive feed ingredient, and extra phosphorus must be excreted, which will contribute to excessive levels in soil.

8.2.4 Milk Production

Dairy farmers realize greatest profits from feeding when cows convert maximal proportions of feed nutrients into milk. Nutrient requirements for production depend on amount and composition of milk secreted. Although high-producing cows require more total feed than low-producing ones, they utilize proportionately more nutrients for milk production (see figure 8.1). Suggested provisions for meeting nutrient needs of milk production are presented in subsequent discussions.

Dairy herds are made up of animals varying in age, weight, stage of lactation, stage of gestation, and genetic producing abilities. Table 8.1 illustrates the extent to which requirements for dry matter intake, dietary energy,

Table 8.1 Daily Nutrient Requirements of Large Breed Cows in Early and Mid-Lactation Based on Diets with 78% Total Digestible Nutrients.

Days in Milk	Milk (kg/d)	Milkfat (%)	DMI (kg)	LWC (kg)	NE_L (Mcal)	RDP (%)	CP (%)
11	30	3.0	14.0	−0.8	30.1	11.2	19.3
11	30	4.0	15.1	−1.0	32.8	11.1	18.3
11	40	3.0	16.0	−1.5	36.5	11.0	20.9
11	40	4.0	17.4	−1.8	40.2	10.9	19.8
90	35	3.0	22.7	1.1	33.2	10.4	15.4
90	35	4.0	24.5	0.9	36.5	10.3	14.7
90	45	3.0	25.7	0.5	39.7	10.2	16.3
90	45	4.0	28.1	0.3	43.8	10.0	15.4

DMI = dry matter intake; LWC = live weight change; NE_L = net energy for lactation; RDP = rumen digestible protein; CP = crude protein
Live weight of large breed cows = 680 kg (1,499 lb)
Early lactation = 11 days in milk
Mid-lactation = 90 days in milk

Source: Adapted from National Research Council, 2001.

and protein are affected by stage of lactation (early and mid-lactation), amount of milk produced, and milk's content of fat when true milk protein is held constant at 3%. The diet used for this table consisted of 15% immature legume silage, 33% normal corn silage, 34% ground high-moisture shelled corn, 12% soybean meal, 2.5% tallow, 1.5% fish oil, and 2% vitamin-mineral mix. In *Nutrient Requirements of Dairy Cattle*, the NRC (2001) provides extensive tabular data related to dietary requirements for large and small breed cows.

8.3 The Mysterious Vital Rumen

Chemists have their laboratories for research and investigation; mechanics have their mechanisms to aid them in the search for truth, but no human agency has been devised to enable us to discover how the dairy cow transforms the hay and grain that she eats into milk.

William Dempster Hoard (1836–1918)

Nutritionists have learned much about the rumen of cattle since Mr. Hoard made his observation. However, it has been difficult to study anaerobic fermentation in the rumen of a large animal and to follow the impact of changes on the physiology of the cow. It can be difficult to understand that much of the feed consumed by a cow is actually substrate for microorganisms, which later become "food" for the cow.

The digestive tract and its contents comprise about one-fourth of the body weight of a mature cow. But the most important anatomically different feature between **ruminant**[2] and nonruminant (monogastric) animals is the unique four-compartment stomach of the ruminant. More precisely, it is the rumen—representing about 80% of the total stomach capacity of mature cows—that performs absolute wonders. For it is this large "fermentation vat" that enables a ruminant to digest **cellulose**, upgrade low-quality feed protein, utilize nonprotein nitrogen (e.g., **urea**), and synthesize B-vitamins and vitamin K.

Cows can utilize feeds that are not digestible by humans because they serve as hosts to literally billions of bacteria and protozoa within each gram of rumen contents. It has been estimated that the number of bacteria in a single teaspoonful of rumen content is greater than the US national debt—and that is indeed significant! These microorganisms produce the enzyme cellulase, which helps convert cellulose to **volatile fatty acids**. The volatile fatty acids, primarily acetic acid, are formed by the bacterial metabolism of feed nutrients. Actually, volatile fatty acids are excreted by ruminal bacteria much as CO_2 is excreted by humans in respiration. It is both interesting and significant that **forage** remains in the **rumen** and **reticulum**, the first two stomach compartments in which fermentation occurs, approximately 48 to 60 hours,[3] whereas feed remains in the omasum and abomasum, the second pair of stomach compartments, only about 8 and 3 hours, respectively.

A microbiome is the collective genomes of the microorganisms residing in an environmental niche, otherwise known as the microbial complex of the host. Certainly the microbiome of the bovine rumen makes a vital contribution to ruminants and humans. Protozoa can compose up to 50% of ruminal biomass.

Microorganisms of the rumen generate methane in the process of digestion. Methane expulsion through belching represents a net loss of feed energy for livestock. A high-producing dairy cow typically emits 450 to 550 g per day of ruminal gas produced by fermentation. A **supplement** added to the feed of high-producing dairy cows could have ramifications for global climate change. In a study conducted at Pennsylvania State University (reported in 2015), cows that consumed a feed regimen supplemented by the novel methane inhibitor 3-nitrooxypropanol (3NOP) emitted 30% less methane and gained more bodyweight than did cows in a control group. The spared methane energy was used partially for tissue synthesis, which led to a greater bodyweight gain by the inhibitor-treated cows. Feed intake, fiber digestibility, and milk production by cows that consumed the supplement did not decrease.

8.4 Providing Energy for Dairy Cattle

Energy enables an animal to perform work and other productive processes. All forms of energy are converted into heat. Thus, energy as related to body processes is expressed in heat units (calories).[4] The *small* calorie is the amount of heat required to raise the temperature of 1 g of water 1°C (actually from 14.5°C to 15.5°C). The **kilocalorie** (kcal) is the amount of heat required to raise the temperature of 1 kg of water 1°C and is equal to 1,000 calories. A **megacalorie** (or **therm**) is 1,000 kcal. One kcal is approximately four British thermal units (BTU), or 4.184 kilojoules (kJ), which is the amount of heat required to raise the temperature of 1 lb of water 1°F.

Utilization of energy by lactating cows is influenced by many factors, including (1) composition of ration (e.g., high grain versus high forage), (2) physiological function for which it is used (e.g., maintenance versus lactation), (3) environmental conditions (e.g., warm versus cold temperatures), and (4) size of animal.

In this section, we will focus on the composition of the ration, as well as how to evaluate it. A problem in evaluating systems that estimate the feeding value of feedstuff is the interaction of forages and **concentrates** when fed together. Adding concentrate to a forage ration is often associated with a decrease in fiber digestibility. It can be especially dangerous to feed large meals of concentrates. However, the readily available energy of sugar or starch from limited quantities of concentrates can also stimulate microbial activity and improve digestibility of the diet. Thus, it is important to balance the types of carbohydrates in the diet and to arrange the feeding schedule to preserve that balance. Total mixed rations (popularly termed TMR) have become popular because, theoretically at least, each bite will contain the designed balance.

To better understand digestibility and utilization of feeds, let us briefly discuss various means of expressing their energy values.

8.4.1 Total Digestible Nutrients (TDN)

The TDN content of a feed is expressed as a percentage and can be determined only in a digestion trial in which feed and feces are analyzed for **crude protein** (nitrogen × 6.25), **crude fiber**, **nitrogen-free extract**, and **ether extract** (lipids × 2.25). These data are then used to calculate **digestion coefficients** and TDN. The factor 2.25 is used because **lipids** possess about 2.25 times more energy per unit weight than do carbohydrates. Carbohydrates have sufficient molecular oxygen present to balance the hydrogen, so that heat is released only from the **oxidation** (combination with oxygen) of carbon. In fats, there is much less oxygen present in the molecule, and oxygen outside the fat molecule unites with both the carbon and hydrogen in fat, yielding more heat. The combustion of 1 g of hydrogen yields about four times more heat than does the combustion of a similar weight of carbon. Oxidation of carbohydrates, fats, and proteins provides about 4,100, 9,450, and 5,650 calories per gram, respectively. However, deamination of protein results in an energy loss so that proteins and carbohydrates have similar usable energy values.

A limitation of TDN as a measurement of feed energy is that it does not account for losses, such as (1) urinary energy, (2) those of **gaseous products of digestion**, and (3) the **heat increment (HI)**, that occur when a feed is consumed by an animal (figure 8.2 on the next page). These losses, especially gaseous products of digestion and heat increment, are much greater for forages than for concentrates. This means that a unit of TDN from forages has considerably less value for productive purposes in an animal than does the same amount of TDN in a concentrate (grain mixture). For example, 1 lb (0.45 kg) of TDN in corn grain yields about one megacalorie of **net energy**. In the better hays (forages), 1 lb of TDN yields approximately 0.75 megacalorie, and in low-quality roughages about 0.50 megacalorie. The TDN content of feeds is commonly related inversely to fiber content. Concentrates are low in fiber and high in TDN. Hence, the TDN system tends to overrate forages in relation to concentrates. This may result in underfeeding cows, especially high-producing ones, that are fed medium to high amounts of forages.

The TDN system of evaluating feeds and balancing rations has been widely used in the United States and Canada since the early 1900s and still has some value for dairy cows that are fed forages. However, TDN values have a pair of potential weaknesses. First, there is often considerable variation in quality of feedstuff on individual farms and, second, digestibility decreases with increases in feed intake. The "true" TDN value is less at a high than at a low level of feed intake. Most TDN values are based on digestion trials conducted with restricted feeding near maintenance levels. In general, feed intake decreases as moisture and/or fiber content increase.

8.4.2 Gross Energy (GE)

The **gross energy** (heat of combustion) of a feed can be determined by burning a sample of feed completely until the oxidation products (carbon dioxide, water, and other gases) are formed. A **bomb calorimeter** is used to measure heat liberated from such combustion. A bomb calorimeter consists of a closed vessel in which the feed is burned. The bomb is enclosed in an insulated jacket con-

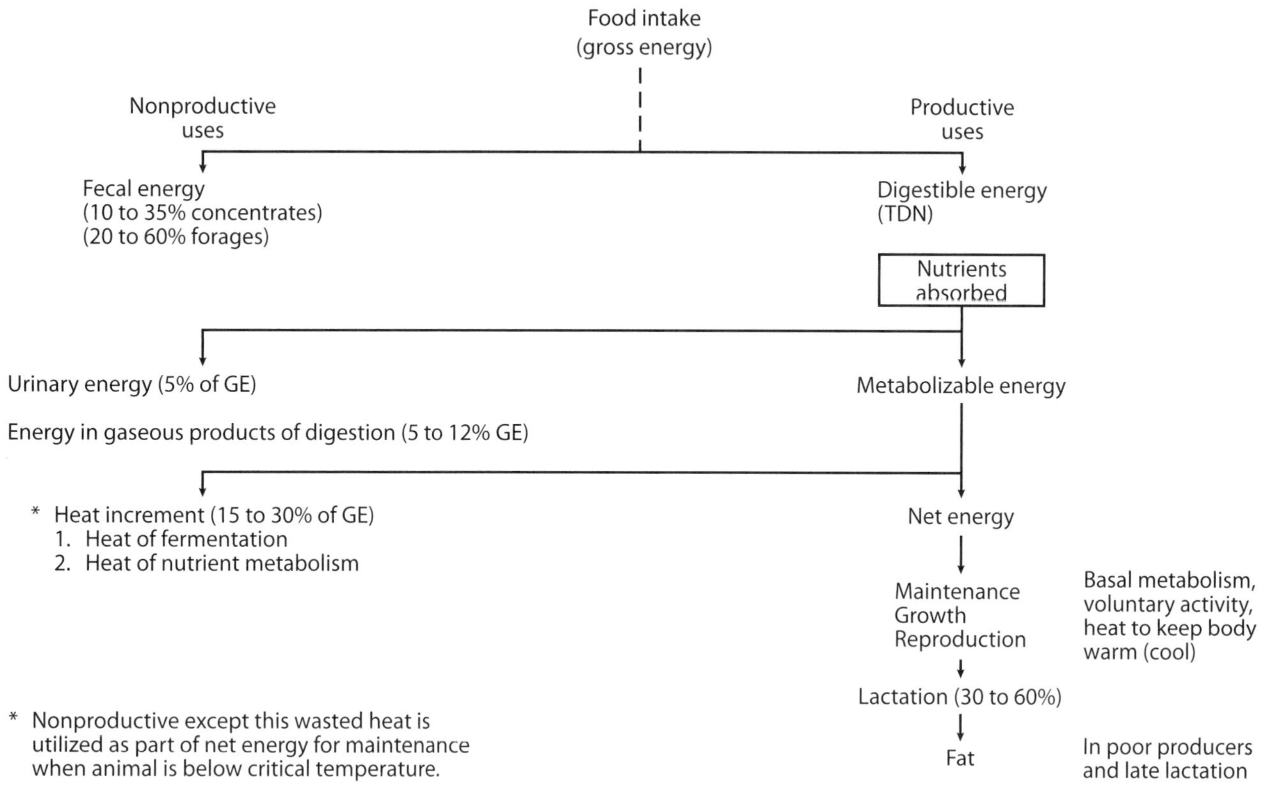

Figure 8.2 Utilization of food energy by the dairy cow (Campbell & Lasley, 1969).

taining water that absorbs the heat (calories) produced. The gross-energy determination reveals total caloric energy in the feed being tested.

8.4.3 Digestible Energy (DE)

The digestible energy of a feed is represented by that portion of feed consumed that is not excreted in feces. In the determination, the feed is fed to an animal, and the energy present in feces is determined by means of bomb calorimetry. The difference between gross energy in original feed and that in feces is called **digestible energy**. Digestible energy can be calculated from TDN values by using the factor of 2,000 kcal/lb of TDN, or from the digested nutrients by using the average gross caloric factors for proteins, carbohydrates, and fats of 5.65, 4.10, and 9.45 kcal/g, respectively.

8.4.4 Metabolizable Energy (ME)

The **metabolizable energy** of a feed is determined by subtracting energy losses in feces, urine, and gaseous products of digestion[5] (primarily methane in ruminants) from total energy in the feed. A desirable characteristic of the metabolizable energy statistic is that more of the gross energy that is unavailable to the animal for productive purposes is accounted for than by the measurement of TDN or digestible energy content.

Energy is not used with equal efficiency for all physiological functions. Studies by US Department of Agriculture researchers showed that the approximate ranges for percentage efficiency of use of ME are: cold stress, 100; maintenance, 70 to 80; lactation, 60 to 70; growth, 40 to 60; and pregnancy, 10 to 25 (Moe and Tyrrell, 1973).

8.4.5 Net Energy (NE)

Determination of the **net energy** content of a feed goes one step beyond determining metabolizable energy because the heat increment (heat resulting from digestion and assimilation of feeds) is determined and subtracted from the total energy in feed.[6] Heat increment is measured by the amount of heat given off by an animal's body. Dr. Samuel Brody appropriately called heat increment the "food utilization tax." Heat increment is approximately twice as high in lactating as in nonlactating cows, but this difference varies with lactation level and amount of food consumed by the lactating cow. During winter a portion of the heat increment is used to keep an animal warm, and this is advantageous. Heat increment is a disadvantage to animals in warm summer months when they are challenged to keep cool.

The net energy value of a feed can be determined by feeding experiments. A feed (such as corn) in which net energy is known is compared with a feed having an unknown value. The net energy can also be determined by the combined use of a bomb calorimeter and a respiration calorimeter. The former is used to determine gross energy in a feed and in feces; the latter is employed to determine the heat increment and losses of combustible gases. However, since these techniques of determining net energy are time-consuming and costly, most net energy values were estimated from TDN. Consequently, any errors in original TDN figures were introduced into resultant net energy values.

Nonetheless, most nutritionists agree that, in theory, net energy is superior to other measures since energy losses are better accounted for in net energy measurements. The net energy system is, in essence, a closed one that conforms to the first law of thermodynamics (that is, energy is neither destroyed nor created). Net energy in dairy cattle, then, is that portion of the total (or gross) energy in a feed that is retained in an animal's body for useful purposes: maintenance, growth, reproduction, milk production, and muscular work.

Net energy varies inversely with cell wall content of feed, which is an expression of quality, especially of forages. Net energy is measured in units called megacalories (therms). Thus, net energy values are expressed in metric units—megacalories per kilogram (Mcal/kg). Because net energy can be used more efficiently for maintenance (NE_M) and lactation (NE_L) than for growth (NE_G), different values have been assigned to feeds and should be matched with the purpose for which they are fed. A single value (NE_L) is appropriate for feeds offered to lactating cows as opposed to values for NE_M and NE_G, which are needed for feeds provided to a growing ruminant.

8.5 Providing Protein for Dairy Cattle

Milk yield will suffer by the degree to which the ration is deficient in protein.

J. Thomas Reid (1919–1979)

Primary concerns in meeting a cow's protein needs are providing protein for maintenance and milk production. As noted earlier, young cows require additional protein for body growth, and all cows in late pregnancy need extra protein to support the developing fetus.[7] The daily protein supplementation required by a fattening dairy steer fed a corn-based diet can be met with approximately 1 lb of **protein equivalent**, whereas a high-producing dairy cow, because of the large quantity of protein secreted in milk, may require up to 5 lb of protein equivalent daily from a supplement when fed a corn-based ration.

To appreciate better how protein requirements of dairy cattle are met, let us briefly review the role of the rumen in providing protein needs. Ruminants (e.g., cattle and sheep) are not completely dependent on their diets for the nutritionally **essential amino acids** (as is the case of nonruminants such as chickens and swine). The rumen serves as a fermentation vat in which flourishing populations of microorganisms convert simple non-amino, nitrogenous compounds into amino acids for the synthesis of body protein; these proteins in the cells of microorganisms are, in turn, used as food by their host.[8] Thus, ruminants convert proteins from roughage and nonprotein nitrogenous compounds (e.g., urea) into superior-quality proteins of milk and meat. Ruminants, especially the dairy cow and goat, thus occupy a unique position in the overall efficiency complex. These facts are of immense national and international importance, permitting the cow, water buffalo, goat, sheep, and other domesticated lactating mammals to secrete an almost perfect food from feeds that are low in protein and B vitamins (cf. chapter 1).

In high-producing cows, microbial protein is insufficient to meet maintenance and milk production needs for protein, and it is estimated that microbial protein will supply only a 12 to 13% protein diet to the lower gut. In cows fed specific diets high in rumen undegradable protein (RUP), substantial protein (estimated at 40%) escapes microbial degradation in the rumen and passes on down the GI tract. This is contrary to the old concept that **biological value** is unimportant in ruminants. Therefore, RUP, so-called bypass protein, is important in lactation when microbial synthesis is inadequate to meet protein demands of high producers and offers the opportunity to protect specific amino acids from degradation by microorganisms in the rumen.

The amount of protein required for a specified unit of milk production is largely dependent upon the protein content of milk. Since milk of high fat content is also high in protein (and total solids), it follows that cows producing high-fat milk require more protein per unit of milk secreted than those yielding low fat (and low protein) milk. Additionally, high-producing cows require slightly more dietary protein per unit of milk secreted than low producers. This is in keeping with what we learned concerning the somewhat reduced efficiency of converting NE_L into milk at high levels of production (section 8.4.1).

The NRC reviewed 82 research reports, concluding that milk yield of cows increased as dietary protein concentrations increased, but the effect diminished at higher concentrations. Increasing dietary protein from 15 to 16% of the diet increased milk yield by 0.75 kg/d, whereas an increase from 19 to 20% produced a gain of only 0.35 kg/d. However, a response surface curve was needed to show the responses when percentages of rumen degradable and undegradable proteins within the dry matter were considered (see National Research Council, 2001 [figure 5.2]). These additional measures of dietary protein are extremely important for high-producing cows. This complicates ration formulation compared to earlier NRC recommendations of 782 g/d of protein for maintenance of a mature, 650-kg (1,433-lb) cow plus 173 g/d for maintenance during the last two months of pregnancy, and 1.3 to 1.5 g/kg of protein in the milk. If the cow were 100% efficient in converting dietary protein (that is, above that protein required for maintenance) into milk protein, the figure would be equal to the grams of protein per kg of milk. Expressed another way, the dietary protein requirement for milk production was estimated as 78 g/kg of 4% **fat-corrected milk (FCM)**, and that may be accurate for low-producing cows. Studies indicated that the cow loses efficiency in converting dietary protein into milk protein at high levels of production.

8.5.1 A Sample Calculation

Computer programs are available to balance rations and formulate the diet on a least-cost basis using a combination of feeds that are available or can be purchased. However, to understand the process it is helpful to establish nutrient requirements (normally by checking the current NRC tables), select some feeds that are commonly used, and calculate a simplified ration for a lactating cow. Although it may not fit the high-producing cow, let us use an example to show how protein needs for maintenance and milk production may be satisfied.

Assume a mature, nonpregnant, 650-kg cow that yields daily 30 kg of milk testing 4.0% milkfat. Maintenance requirements for protein would be 782 g and the milk production needs for protein would be 78 g/kg of milk, or 2,340 g (78 g × 30 kg of milk). Total daily protein requirements would be 3,122 g (782 g + 2,340 g). Assume further that our forages consist of corn silage and a limited amount of alfalfa hay. If we feed 24 kg of silage (2.3% protein) and 4 kg of hay (15.3% protein), we will provide 552 g protein from silage (24,000 g × 2.3%) and 612 g protein from hay (4,000 g × 15.3%) for a total of 1,164 g from forage (552 g + 612 g). To meet protein requirements in this example (a mature, nonpregnant, 650-kg cow), the concentrate mixture must provide 1,958 g of protein (3,122 – 1,164).

Assuming our sample cow will consume 3.3 kg of dry matter per 100 kg of body weight, we would calculate her dry matter intake at 21.4 kg (650 kg × 3.3%). The dry matter of 24 kg of silage is 6.84 kg (24 kg × 28.5% dry matter) and that of hay is 3.62 kg (4 kg × 90.5% dry matter) or 10.46 kg (6.84 kg + 3.62 kg). Thus, our concentrate should provide 10.9 kg of dry matter (21.4 – 10.5). Since concentrates average approximately 88% dry matter, the cow should consume 12.4 kg (10.9 ÷ 88% dry matter) of concentrates on an *as-fed basis.*

Our next step is to determine the amount of protein needed in the concentrate mixture. Since the balance of protein needed is 1,958 g, our concentrates should contain 15.8% protein ([1.958 kg ÷ 12.4 kg] × 100]). Assume further that we have corn (8.7% protein) and soybean oil meal (45.7% protein) available. Let us use the Pearson square to determine the ratio of each needed to give a 15.8% protein-grain mixture (see figure 8.3).

In this example, percent protein desired in the grain mixture is recorded in the center of the square. The percentage of protein of grain (corn) and protein supplement (soy-

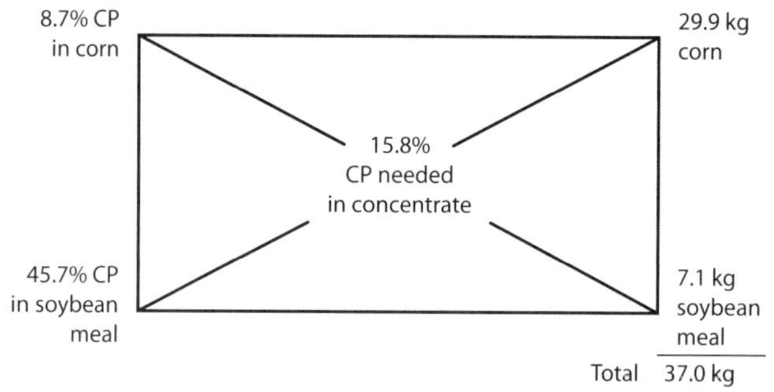

Figure 8.3 Pearson square.

bean meal) are recorded at the upper and lower left corners, respectively, of the square. By subtracting each percentage from the total percentage needed (15.8%), we determine the proportionate amounts of corn and soybean meal to use. In this example, our grain mixture would contain 29.9 parts of corn and 7.1 parts of soybean meal. For ease of determining the physical quantity of each item to use per batch of mix, the proportionate amounts may be converted to percentages. The amount of corn is 81% (29.9 ÷ 37.0) and that of soybean meal is 19% (7.1 ÷ 37.0). A ton (2,000 lb) of mix would contain 1,620 lb of corn (2,000 lb × 81%) and 380 lb of soybean meal (2,000 lb × 19%).

In determining the amount of protein needed in the concentrate mixture, it is important first to determine the amount of protein in forages to be fed by analyzing representative samples of each feed source. Specialized laboratories are available, and test costs are low compared to the expense of overfeeding or underfeeding. Grasses (e.g., brome grass, timothy, orchard grass, and Bermuda grass) usually contain less protein than legumes (e.g., alfalfa and clovers) when harvested at the same degree of maturity (protein content commonly decreases with advancing age of plants). Level of soil fertility also affects protein content. For example, high applications of nitrogen and potassium fertilizer to soil are reflected in increased **digestible protein** of grasses. Rain damage of forages at harvest may reduce protein content.

Calculations similar to those shown previously are also used in calculating NE_L (or TDN) in the ration, and a check of other nutrient requirements and ration provisions must be done carefully. For instance, fiber, mineral, and vitamin requirements must be met. Adjustments may be required after adding supplements. If available, a computer program will account for many specific nutrients, conduct the iterations, and simultaneously devise the least-cost ration.

There is merit in including several feed ingredients in the ration. In general, more components in the diet give a greater probability of having balanced nutrients. Additionally, one may expect greater animal acceptance when a variety of ingredients is included. Theoretically, every bite of a total mixed ration provides all needed nutrients and helps prevent sorting. Fortunately, cows can tolerate wide variation in proportions of dietary constituents but need time to adapt to changes of ingredients.

8.5.2 The Effects of Under- and Overfeeding Protein

Inadequate dietary protein depresses milk production. In Missouri experiments, cows fed low-protein diets yielded less milk and had significantly lower **persistency** of milk yield than those receiving adequate dietary protein. Similarly, studies at Michigan State University showed that persistency of high-producing cows fed an 8% protein ration decreased more rapidly than those of cows receiving higher levels of protein.

Ingesting more protein than is needed is not harmful, but it is uneconomical since protein commonly costs more than energy on a weight basis. If the energy requirement is not met, then part of the protein is used for energy. When more protein is fed than is needed, the extra nitrogen[9] will be excreted as urea via urine and the remainder of the protein will be used for energy purposes.

8.5.3 Feeding Urea

Urea, an organic compound, is made synthetically from ammonia (NH_3) and carbon dioxide (CO_2). It is not protein and is, therefore, called **nonprotein nitrogen (NPN)**. Feed grade urea commonly contains 42% nitrogen, and is potentially equivalent to 262% protein (42% × the protein factor of 6.25).[10] Thus, 1 kg is equivalent to 2.62 kg of protein. On a nitrogen basis, 1 kg of urea and 6 kg of corn equal approximately 7 kg of soybean meal.

From both economic and humanitarian viewpoints, it is important that ruminants can use NPN rather than plant proteins and that they can derive an increasing portion of dietary carbohydrate from sources inedible by humans. Minnesota studies showed that urea can be used to supply all the supplementary nitrogen in high-energy diets. Wisconsin research indicates that the best utilization of urea is in rations containing less than 13% protein in the total ration on a dry matter basis. Moreover, NPN is utilized most efficiently in low-protein, high-energy rations and is poorly utilized in high-protein, low-energy rations. Preformed plant proteins contain nitrogen, energy, minerals, and vitamins.[11] These latter nutrients are not present in urea and must, therefore, be added to the diet if urea is to be utilized effectively as a source of nitrogen. The most important prerequisite for efficient use of urea is a balanced ration.

Cattle can utilize urea because the enzyme urease is produced by rumen microorganisms. Urease hydrolyzes urea to CO_2 and NH_3 in the rumen. The NH_3 is then combined with alpha-keto acids to form amino acids, most of which are incorporated into microbial protein.[12] Microbial protein is digested in the intestinal tract of the host and absorbed into the body as amino acids (figure 8.4).

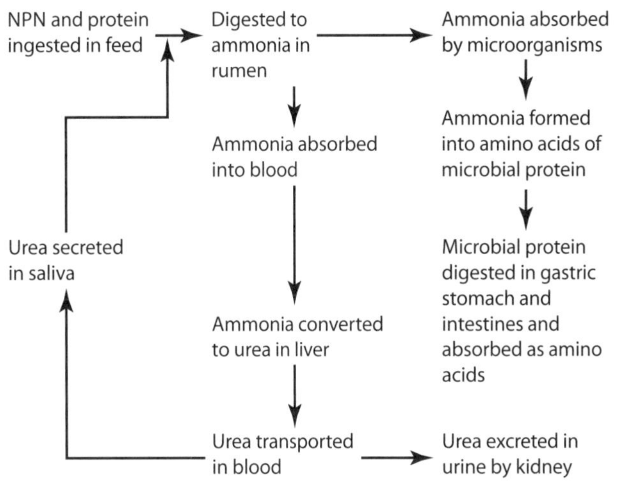

Figure 8.4 Metabolic pathways of protein and NPN in ruminants.

Rumen bacteria can obtain and use ammonia from three sources: (1) plant proteins that are degraded in the rumen to form ammonia (representing more than half the true proteins ingested); (2) NPN, such as urea, which is broken down into ammonia in the rumen; and (3) saliva, which contains urea that is broken down to ammonia. The quantity of ammonia that can be utilized by rumen microorganisms depends on the number of bacteria present and how rapidly they multiply. The latter depends on the amount of energy available for the bacteria that, in turn, depend on the amount of fermentable feed consumed.[13] Feeds high in energy are more fermentable than those low in energy. Therefore, more NPN can be utilized when high-energy feeds are used. No value of NPN is realized if ammonia concentration in the rumen exceeds the ability of the bacteria to use it. In cattle, urea constitutes approximately 70% of the total urinary nitrogen: its concentration in urine is related to urine excretion and blood urea concentration. Urinary nitrogen losses are commonly greater among cows fed urea-treated corn silage than among those ingesting an equivalent amount of nitrogen from feeds.

Because corn silage is low in protein but contains a reliable source of energy (which is needed by microbes to "fix" urea nitrogen) (see table 8.2 on the next page), dairy cattle nutritionists theorized that the protein equivalent of corn silage could be improved with the addition of urea during **ensiling**. The use of urea in corn silage was first recommended by dairy scientists at Michigan State University in 1963. A commonly recommended level of urea in corn silage varies; e.g., 0.5% and 0.75% for silages containing 30 or 40% dry matter, respectively. Added urea will increase crude protein of silage on a dry matter basis from approximately 8.5 to 13.5%. Therefore, many dairy farmers add urea to corn silage at the rate of 10 lb of urea (45% nitrogen) per ton of corn silage (0.5% urea). Urea can be spread over the top of chopped corn in

Table 8.2 Average Composition of Selected Feeds Commonly Used in Dairy Rations.

Feed	Dry Matter (%)	Protein Total (%)	Protein Digestible (%)	Energy ENE/cwt (therm)	Energy TDN (%)	Fat (%)	Fiber (%)	Calcium (%)	Phosphorus (%)	Carotene (mg/lb)
Dry Forages										
Alfalfa hay, all anal.	90.5	15.3	10.9	40.6	50.7	1.9	28.6	1.47	0.24	8.2
Grass hay, mixed, good quality	89.0	7.0	3.5	39.3	51.7	2.5	30.9	0.48	0.21	6.2
Timothy hay, all anal.	89.0	6.6	3.0	37.3	49.1	2.3	30.3	0.35	0.14	4.4
Silages										
Alfalfa, wilted	36.2	6.3	4.3	17.6	21.5	1.4	11.4	0.51	0.12	11.4
Corn, dent, well-eared	28.5	2.3	1.3	15.2[a]	19.8	0.9	6.3	0.09	0.07	5.8
Green Forages										
Alfalfa, green, all anal.	24.4	4.6	3.5	12.9	14.8	0.9	6.7	0.40	0.06	28.3
Grasses, mixed pasture	28.1	4.9	3.7	17.1	19.4	1.4	6.1	—	0.09	36.0
Sudan grass, pasture	21.6	3.3	2.4	12.2	14.3	0.6	5.6	0.12	0.10	21.5
Energy Concentrates										
Barley, common, non-Pacific states	89.4	12.7	10.0	80.1[a]	77.7	1.9	5.4	0.06	0.40	0.2
Corn, dent, no. 2	85.0	8.7	6.7	80.1	80.1	3.9	2.0	0.02	0.27	1.3
Corn and cob meal	86.1	7.4	5.4	72.1	73.2	3.2	8.0	0.04	0.22	—
Milo grain	89.0	10.9	8.5	77.8	79.4	3.0	2.3	0.03	0.28	0.1
Molasses, cane	73.4	3.0	—	71.3[a]	53.7	—	—	0.66	0.08	—
Oats, non-Pacific states	90.2	12.0	9.4	72.1[a]	70.1	4.6	11.0	0.09	0.33	—
Wheat, avg.	89.5	13.2	11.1	80.0	80.0	1.9	2.6	0.04	0.39	—
Wheat bran, all anal.	90.1	16.4	13.3	66.9[a]	66.9	4.5	10.0	0.13	1.29	1.2
Protein Concentrates										
Corn gluten meal, all anal.	91.6	43.2	36.7	80.2	79.7	2.2	3.8	0.14	0.41	7.4
Cottonseed meal, 41%	92.9	41.6	33.3	71.7	71.7	6.0	10.7	0.20	1.11	—
Linseed meal, sol., 36%	91.0	36.6	30.7	71.7	70.3	1.0	9.3	—	—	—
Soybean oil meal, sol., 44%	90.3	45.7	42.0	79.6	78.0	1.3	5.8	0.26	0.63	0.1
Urea, 45% N	—	281.2	—	—	—	—	—	—	—	—

ENE = estimated net energy; TDN = total digestible nutrients.
[a] For dairy cows.

Source: Morrison, 1956.

self-unloading wagons at time of ensiling, but it is essential that thorough mixing occur before feeding time.

About 10% of the urea nitrogen may be lost when urea is added to corn silage. Losses increase as the dry matter of the silage increases. Wisconsin studies showed that less urea is lost in corn silage when dry matter content is between 30 and 40%. In Michigan studies, nitrogen losses from urea-treated (0.5%) silage harvested in late maturity (44% dry matter) represented 69% of the added urea, but only about 16% was lost from less mature forage (30% dry matter). Furthermore, urea-treated corn silages lost more dry matter than did control silages (12.8% versus 6.1%).

Use of urea in corn silage rations tends to reduce mineral intake from natural ingredients, because plant proteins commonly used are relatively rich in minerals compared with the cereal grains that would otherwise be fed (see table 8.2). For example, supplemental sulfur should be added to rations of cows fed urea-treated corn silage. Sulfur is needed by bacteria in making important amino acids that contain sulfur, cystine, and methionine. The nitrogen:sulfur ratio of corn silage is about 13:1; an addition of urea equal to 0.5% of the silage would widen the ratio to about 18:1.

Although urea is the most common source of NPN added to corn silage, ammonium phosphate and diammonium phosphate—each of which provides both phosphorus and nitrogen—are useful sources. Convenience in handling coupled with a cost advantage may make advisable the addition of liquefied anhydrous ammonia to grass, small grain, or sorghum silages.

8.5.3.1 Toxicity of Urea

If ingested in large quantities during a short period of time, urea can be toxic to ruminants. Symptoms of toxicity include dullness, muscle and skin tremors, excessive salivation, frequent urination and defecation, rapid respiration, lack of coordination, stiffening of front legs, prostration, **tetany**, and death. Since excessive urea is toxic to cattle, it is especially important that urea be mixed well into grain rations. Guidelines provide for including urea for up to one-third of the nitrogen of the grain ration in low-producing cows and up to one-fifth that of high producers. The reason for suggesting a lower level (percentage-wise) for high-producing cows is that they normally consume substantially more grain than do low producers.

Since urea hydrolysis occurs faster in the rumen than does protein synthesis, and since urea releases ammonia more rapidly than natural protein, some urea may be lost. This explains, in part, why feeding urea in small quantities at frequent intervals results in improved animal performance (Campbell et al., 1963).

In a recent review, Dr. A. F. Kertz (2010) noted that some nutritionists have recommended excessive amounts of urea in feed for dairy cows. He recommends less than 1% in the concentrate (about 135 g/d/cow) with a maximum of 20% of the total dietary protein as added NPN from urea. He noted that after urea was added to ensiled forages, the final nitrogen content was improved and protein degradation of the silage was reduced. When urea was also added to the concentrate, no negative effects were observed if the total supplemental NPN was less than 20% of the total dietary nitrogen. Hydrolysis of urea elevates rumen pH and allows more rapid absorption of rumen ammonia into the blood, often leading to classical ammonia toxicity.

Scientists at Kansas State University developed and researched the product Starea—an intimate mixture of gelatinized starch and urea—which they reported as less toxic and more palatable than urea mixed with ground grain. It is necessary to keep all grain mixtures dry to prevent growth of molds, which may produce urease, an enzyme that splits urea into ammonia and water. If water or wet feeds are added to a total mixed ration, it should be fed soon after mixing. This is especially important if the temperature is warm.

8.5.4 Liquid Feed Supplements

These supplements are commonly formulated to contain a combination of molasses, NPN (urea and/or ammonium phosphate and diammonium phosphate[14]), phosphorus (phosphoric acid, ammonium polyphosphate, diammonium phosphate), trace minerals, and vitamins. Other (natural) sources of protein, minerals, and fish solubles are also sometimes added.

The problem of controlling intake of liquid **protein supplements** must be resolved. Missouri studies showed that cows are unable to balance their own rations with respect to protein when liquid supplements are fed **ad libitum**. Consumption of liquid supplements is often regulated by varying the amounts of phosphoric acid, urea, and other unpalatable ingredients.

8.5.5 Supplementation of Specific Amino Acids

Until recent years it was generally agreed that little benefit could be realized by feeding supplemental amino acids to ruminants. Research at Pennsylvania and Virginia Agricultural Experiment Stations indicated that feeding supplemental methionine as the synthetic hydroxyl-analog of methionine, at the rate of 25 to 40 g/d, to high-producing cows (those producing 30 kg/d or more of milk) results in increased milk production. The analog is nontoxic and has been used successfully in diets of poultry and swine. It can be used by the cow to produce part of the methionine of milk proteins. Economic benefits include reducing total protein intake and permitting use of more low-cost products that limit lysine in the diet. Those savings can more than offset the cost of supplements with the potential for improving milk components, breeding efficiency, and health of high-producing cows.

Since methionine is the most limiting amino acid of soybean meal, and since soybean meal is a common supplemental plant protein fed dairy cows, it logically follows that methionine would be the first-limiting amino acid of high-producing cows (figure 8.5). However, corn products are also commonly fed, and corn is especially low in lysine. This has been a difficult topic of research with high-producing cows, and specific experiments have determined several amino acids to be first-limiting. Usually studies find methionine or lysine as first limiting. The complicating factors are establishing the requirements for each amino acid at various milk production levels, measuring synthesis of protein by rumen microorganisms, and measuring intestinal digestion of proteins and absorption of amino acids. The goal is improving the balance of amino acids available in blood flowing to the udder, and that can be complicated if the body tissues use amino acids to supply energy. It is important to develop accurate models to define the precise need for supplementation of specific amino acids.

If we accept methionine as the first-limiting amino acid, then a form of methionine that is protected from rumen microorganisms should increase milk yield of high-producing cows. An alternative approach would increase methionine concentration in the diet, but it must be in the RUP of dietary protein. Both approaches have been successful, but universal application is not likely given the variable conditions on dairy farms. The next step is to consider lysine as the second-limiting amino acid. It has been recommended to design a ratio of 3:1 (lysine to methionine in metabolizable protein) to supply amino acids for synthesis of proteins in body tissues and milk. However, much is not yet known about efficiencies of converting metabolizable protein into specific purposes within the body.

Figure 8.5 A cow's total production is restricted by the lowest stave in the barrel of nutrients needed. In this example, protein is the limiting factor in production.

8.6 Providing Lipids for Dairy Cattle

Since fat can be formed from other dietary nutrients (and perhaps because ruminants can ingest and digest large quantities of low-fat forages), the dietary fat needs of a ruminant are nominal. Most milkfat is synthesized from volatile fatty acids (cf. chapter 10) excreted during microbial metabolism of plant nutrients. Forages commonly contain 1 to 4% ether extract (lipids),[15] which is apparently sufficient to satisfy lipid needs of ruminants.

Because most soybean meal and other oil seed meals used in dairy rations are now made using the solvent process,[16] and because other components of the concentrate mix are relatively low in lipids, rations of dairy cows usually contain less than 3% lipids. Vegetable oils are not recommended as supplements because the unsaturated fatty acids are toxic to rumen bacteria and reduce the digestion of fiber. However, small amounts of fatty acids are saturated in the rumen, and this has been suggested as a strategy to reduce methane losses. The saturation of fatty acids by the rumen has been noted historically because animal fats, such as tallow from ruminants, are more saturated than fats, such as lard, produced by the monogastric hog. It is possible to feed **saturated fats**. Illinois research suggested that liquid fat should be mixed with grain rather than with the final total mixed ration, and that ration changes should be made slowly because products may change palatability and to allow adaptation of rumenal microorganisms.

Protected fats, such as calcium salts of long-chain fatty acids or saturated free fatty acids, can be added to diets of high-producing cows. The method of providing a concentrated energy source to the cow assumes that the products are palatable, are protected or stable in the rumen, and are hydrolyzed and absorbed in the intestines. Once the protected fatty acids pass the rumen, they are used as an energy source or placed in milkfat. Whole cottonseed is a popular feed that provides fat, protein, and fiber. It can extend limited forage supplies but must be kept dry during storage. **Gossypol**, a polyphenolic pigment, is toxic in the free form and therefore limits the amount of whole cottonseed that cows should consume. Recommendations are to feed up to 150 g/kg of diet (Arieli, 1998). Roasted or extruded soybean seed would provide more protein, but less fiber. Because additional fat can form soaps with calcium and magnesium in the small intestine, those minerals should be increased about 0.2 to 0.3% above normal levels in the ration dry matter.

8.7 Energy and Protein Deficits in High Production

Being versed in the science of feeding is not enough. For highest production and greatest profits not only must one feed cows economical and well-balanced rations, but also he/she must have the watchful eye and good judgment of a true dairyman who must feed frequently and generously.

Frank B. Morrison (1887–1958)

Dairy farmers have long known that mature, high-producing cows commonly lose 200 to 400 lb of body weight in early lactation when feed intake is insufficient to support high yields of milk. It has been estimated by researchers of Cornell University that mobilization of 100 lb of body fat provides energy for the secretion of 880 lb of 4% FCM. There is, however, an increased incidence of ketosis during such energy-deficient periods (cf. chapter 13). The cow is efficient in converting stored energy (primarily fat or protein) to milk. Lactating cows convert body tissue to milk with an efficiency of 82%. Thus, when cows cannot consume sufficient feed energy to support high milk production, they can draw on body reserves for additional energy.

Lactating cows convert extra dietary feed energy (above maintenance and milk production needs) to body weight with greater efficiency (75% conversion) than nonlactating cows or steers (60% conversion). This suggests that dairy farmers should feed high-producing cows enough during late lactation that they regain weight lost during early, peak production. USDA scientists concluded that when body energy used in early lactation is replaced in late lactation, a net feed efficiency of 62% was realized (82% × 75%). This is only slightly less than the 64% efficiency they observed when cows converted metabolizable energy of feed directly into milk. Of economic significance is the fact that if the cow converted only 60% of her energy to body tissue (as does the dry cow or steer), then feed conversion efficiency would be only 49% (82% × 60%) rather than 62%.

Through **nitrogen balance** studies, in which dietary nitrogen consumed is balanced against nitrogen secreted in milk and excreted in feces and urine, there are periods in which high-producing cows are in a negative nitrogen balance. During such periods, body protein is being utilized. It is estimated that a 1,300-lb cow has about 180 lb of protein within her body (exclusive of digestive tract contents). Furthermore, it is estimated that not more than 5 to 7% of the body protein is in a form that can be mobilized in times of dietary shortage. Using a value of 6%, it can be calculated that a 1,300-lb cow could obtain enough protein from body stores to secrete only about 327 lb of milk containing 3.3% protein (180 × 6% = 10.8 lb protein—enough to produce 327 lb of milk containing 3.3% protein). This assumes 100% efficiency of mobilizing body protein for milk secretion. Actually an efficiency of 90% is more realistic, reducing the yield to 294 lb (327 × 0.9). It is readily apparent, then, that the amount of body protein that can be mobilized for milk production is indeed small compared with the amount of body fat that can be stored and mobilized. Thus, it is vitally important to provide adequate dietary protein to high-producing cows. Otherwise, protein becomes the limiting factor in milk production.

8.7.1 Appetite Is Important

High-producing cows are noted for hearty appetites; they are often the first to the feed manger and the last to leave. Watch the cows that continue grazing on into the heat of day—they are usually the ones providing the greatest quantities of milk. Cornell research showed a close association between intensity of milk production and appetite for forage. An increase of 100 lb of FCM was associated with an increase of 26 lb/d in forage dry matter intake. Studies at Kansas State University showed that dairy cows can consume up to 350 lb/d of grass.

High-producing, healthy cows will consume approximately 3.0 to 3.5 lb of dry matter per hundredweight daily. Thus, a 1,500-lb cow may ingest 45 to 52 lb/d of dry matter. But what prevents cows from eating even more? Appetite is apparently influenced by ruminal pressure, as demonstrated by English researchers when they placed a balloon containing 100 lb of water in the rumen and observed a decrease in daily feed intake from 21.9 to 16.5 lb. When they poured 100 lb of water into the rumen without its being held in the balloon, daily feed intake was unaffected. Moreover, removing hay through the rumen **fistula** increased feed intake. Many other factors affect appetite, the discussion of which is beyond the scope of this book (Allen, 2000; Baile and Della-Fera, 1981).

Lactating cows eat more frequently than nonlactating ones. Is this related to the frequent intervals at which suckling calves tug at their mothers' teats? Under natural conditions, ruminants **graze** intermittently throughout the day and thereby keep a rather uniform rumen fill. Additionally, products of rumen fermentation (e.g., volatile fatty acids) are in more constant or uniform supply than when animals are fed once or twice daily. Products of rumen digestion increase rapidly within a few hours postfeeding and then gradually decrease until the next feeding. In Missouri studies, the authors observed cows to yield more milk when fed four or seven times daily than when fed twice daily. Most of the additional milk produced could be accounted for by an increased feed intake at the more frequent feeding intervals. Dairy farmers who strive for high milk production watch their cows carefully and attempt to feed them according to appetite.

A fattening dairy steer is eating well when he consumes feed at a level of 2.5 times his maintenance requirement, whereas a high-producing cow should consume 5 to 6 times her maintenance requirement in early lactation.[17] When cows ingest large quantities of feed daily, they digest feed at a slightly reduced efficiency and lose more energy in their feces. However, they lose proportionately less energy as urine and methane and, therefore, still utilize almost as much energy as cows producing less milk and consuming less feed.

Cows consume dry concentrate at a rate of 0.4 to 0.7 lb/min. The rate may be increased to 0.7 to 1.0 lb if pelleted concentrate is fed. In studies at Washington State University, cows adjusted to milking parlor feeding consumed 0.76 lb of ground feed and 0.94 lb of pellets per minute.

Cows can be trained to eat grain faster. For example, Nebraska researchers reported that Holstein cows in a stanchion barn, with unlimited time for eating, consumed 0.46 lb/min of concentrate, whereas those in a milking parlor, with limited time for eating, consumed 0.65 lb/min. Nebraska and Michigan studies showed that cows eat concentrates more slowly when pastured (0.46 lb/min) than when fed hay or silage (0.75 lb/min). Most dairy farmers know that when cows have free access to lush, green pasture, they consume concentrates less vigorously than when subjected to **drylot** feeding conditions. Could this be related to the high-moisture content of pasture, resulting in high rumen fill? Or do cows simply prefer green grass?

8.8 Feeding High Levels of Concentrates

Even if we continue to milk cows only twice daily and they remain about the same size, which directly affects feed intake, we expect them to yield larger quantities of milk than those of yesteryear. To do this, dairy farmers must provide feeds of higher energy concentrations.[18] Most grains are high-energy feeds and are called *concentrates* since they contain a concentrated source of energy. However, simply feeding more grain does not assure more milk production. Cows must have the genetic ability to convert feed to milk (cf. chapter 6), and management practices must be favorable (cf. chapter 19).

For cows genetically geared to yield large quantities of milk, it is especially important they be provided liberal amounts of grain during their first five to six months of lactation. However, after about six months, most research indicates that 1 lb of grain for each 4 lb of milk produced is adequate when good quality forage is available ad libitum. Electronic systems permit controlled consumption of concentrates up to a preset amount and can record grain intake of cows. However, most commercial dairy farms devise systems to feed grain mixtures according to production, often by grouping cows according to production levels.

As a ruminant, the cow needs a certain amount of forage to keep her digestive system functioning properly. The genetic ability of cows to respond to grain frequently exceeds their capacity to consume grain and still maintain adequate forage intake. Dr. J. T. Reid of Cornell University observed that feeding approximately 10 lb of grain does not significantly decrease forage intake. However, above that level, each pound of grain ingested decreases forage intake by about 0.5 lb. For the long-term health and well-being of high-producing cows, many dairy cattle nutritionists, physiologists, and veterinarians agree that the amount of grain fed daily to dairy cows should not exceed 2.0 to 2.5% of body weight. Thus a 1,500-lb cow should not be fed more than 30 to 38 lb/d of grain. Otherwise, forage intake (and fiber) decreases and ruminal activities and digestive functions may be impaired.

What are the effects of high-grain, low-forage diets on dairy cows? Do they precipitate diseases, disorders, and shorten productive lives of cows? It is well known that such diets result in diminished rumination, changes in rumen microbial populations, a decreased acetate/propionate ratio, and a reduced rumen pH. A reduced ruminal pH results, in part, from reduced saliva secretion among cows fed high-grain rations. Saliva serves as a potent buffering agent in the rumen. Copious quantities of saliva (100 to 200 l/d in mature cows) are secreted by cows fed adequate levels of forages and other nutrients.

In Michigan studies conducted over a four-year period in which 170 cows were fed concentrates ranging from 464 to 4,790 kg per lactation, there were no **statistically significant** differences in the incidence of mastitis, udder edema, metritis, or conception rate. In chapter 13, we note that the incidence of abomasal displacement is greater among cows fed high levels of grain. Ketosis can increase the risk of abomasal displacement through an unknown mechanism that may be associated with decreased rumen fill. It is well accepted, however, that the incidence of ketosis is reduced when cows are fed adequate amounts of concentrates needed to support high milk production. Research at the Texas and Arizona Agricultural Experiment Stations showed that high-producing dairy cows fed low-fiber, high-energy rations were under less stress and produced more milk during hot weather than when fed high-fiber diets. This relates to the heat increment (see section 8.4.5).

8.8.1 Costs Associated with Feeding High Levels of Concentrates

Profitability of feeding high levels of grain is determined by the value of extra milk produced and forage saved, subtracting for the cost of extra grain. High energy rations are most profitable when fed to high-producing cows and when milk prices are high relative to feed price.

It is economically sound to feed cows up to the point where the additional feed just pays for itself by the value of increased milk produced. Although this concept is relatively simple in principle, it is difficult to achieve under practical dairy farm operations since herd size and labor considerations preclude the feeding of cows on a completely individualized basis. An entire corral of cows can be fed at a manger in less time than is needed for some individual cows to finish their grain after being milked in the milking parlor. Additionally, it may be difficult to calculate, especially on an individual cow basis, the point at which marginal return (value of last unit produced) equals marginal cost (cost of last unit produced).

Milk production per cow has improved significantly with the increased feeding of concentrates. However, the worldwide food shortage and the fact that much more food energy can be provided for humans when grains are consumed directly, rather than when processed through animals to animal products, raises a curious question: How long can producers afford the luxury of feeding high amounts of grains to dairy cattle and other farm animals? At times, US farmers have witnessed a substantial increase in grain prices, therefore less concentrate feeds were offered per cow. The authors suggest the trend toward feeding less grain will continue because of worldwide demand for food grains for human consumption.

Ethanol production from corn has increased dramatically and made a huge impact on feed costs. Distiller's grains, wet or dry, have been fed to cows for many years and can supply about one-fifth of the dry matter in dairy rations. The various by-products of the distilling process commonly replace concentrates. Starch is removed during fermentation, concentrating other ingredients, so care is needed in formulating rations. When soluble components are included, the phosphorus composition will be elevated.

8.8.2 Effect of Concentrate/Forage Ratio on Milkfat Production

A high ratio of concentrate to forage depresses milkfat production. Factors commonly associated with the phenomenon of reduced milkfat among cows fed high-concentrate, low-forage rations include (1) decreased proportions of acetate and increased propionate in the rumen, (2) decreased ruminal pH, (3) decreased acetate in blood and acetate uptake by mammary glands, and (4) decreased ketones in blood. Interestingly, diets that induce secretion of lowfat milk are those that cause greatest deposition of adipose tissue.

Feeding low-forage, high-grain rations results in decreased total saliva secretion and, as a consequence, the total buffering capacity of ruminal contents is reduced, thus allowing a reduced ruminal pH. This reduced pH favors microorganisms that produce above-normal amounts of propionic acid, thus tending to inhibit milkfat synthesis. This explains, in part, why low-forage, high-concentrate rations result in the secretion of lower-fat milk.

Most research indicates that a minimum of 15 lb of *long* forage (unchopped and/or unground—hay equivalent of 1.5 lb/cwt of body weight) daily is needed to prevent reductions in milkfat. Coarser forages are more effective in maintaining milkfat percentages than finer ones; however, low-quality forages reduce intake by slowing the rates of digestion and passage. Dr. P. J. Van Soest (1973) of Cornell University concluded that when particle size is adequate, about 20% **acid detergent fiber (ADF)** and 30% **neutral detergent fiber (NDF)** are needed for maintenance of milkfat production. Adding sodium and/or potassium bicarbonate (0.8 to 1.0 lb/cow/d) or magnesium oxide (0.3 to 0.4 lb/cow/d) helps maintain milkfat synthesis by cows fed low-fiber, high-grain rations.

Certain methods of grain processing, such as flaking or pelleting, have resulted in milkfat depression when grains so processed were fed in large amounts to high-producing cows. Additionally, finely ground feeds promote milkfat depression (and less efficiency of lactation) in dairy cows. Moreover, fine grinding of feeds increases **rate of passage** and thereby decreases digestibility of forages.

8.9 Providing Minerals for Dairy Cattle

Mineral elements known or believed to be required by dairy cattle are calcium, phosphorus, magnesium, potassium, sodium chloride, sulfur, iodine, iron, copper, cobalt, manganese, zinc, selenium, and chromium. Other trace elements may be needed but are available without supplementation. These are needed for (1) bone and teeth formation; (2) constituents of proteins and lipids that comprise muscles, organs, blood cells, and other soft tissues; (3) enzyme systems; (4) maintenance of osmotic relationships and acid-base equilibriums; and (5) functioning and irritability of muscles and nerves.

All dairy cattle require minerals on a routine basis. Growing animals need certain minerals for skeletal development, pregnant animals require extra minerals for the developing fetus, but the greatest need for minerals is to support secretion of mineral-rich milk. All animals require certain minerals for body maintenance. Now let us discuss briefly the mineral needs of dairy cattle (giving special emphasis to lactating cows) within the groupings of macroelements and microelements.

8.9.1 Macroelements

It seems a remarkable manifestation of instinct or intelligence that physiological deficiency in cattle should be reflected in a specific craving for bones, the only accidentally available source of phosphorus capable of relieving the craving.

Carl E. Coppock

These mineral elements are needed in relatively large amounts, especially calcium, phosphorus, and salt (sodium chloride). Magnesium, potassium, and sulfur are other especially important macroelements that we will consider.

8.9.1.1 Calcium (Ca)

The main demand for calcium in lactating cows is for milk secretion. However, calcium is also important in bone formation, blood coagulation, muscle contraction, and other body processes.

The concentration of calcium in milk is approximately 1.23 g/kg and is independent of dietary or skeletal calcium. When dietary calcium is deficient, the cow may withdraw calcium from her bones,[19] but the calcium content of milk remains the same as that of milk produced on diets adequate in calcium. However, when the amount of calcium available is inadequate for synthesis of milk, total output of milk is reduced to compensate for lack of the element. Each kilogram of 4% FCM contains an average of 1.23 g of calcium. Since it varies slightly with milkfat content, the values are 1.22, 1.45, and 1.37 g of absorbed calcium for Holsteins, Jerseys, and other breeds, respectively. Approximately 45% of the calcium in feeds is absorbed by cows. Therefore, a dietary intake of 2.7 g calcium/kg of milk produced is recommended to maintain calcium equilibrium.

The National Research Council's recommended allowance of absorbed calcium is 0.031 g/kg of body weight, or 20 g/d, for maintenance of a 1,433-lb (650-kg) mature cow and, in addition, 2.7 g/kg of milk produced. Thus the dietary requirement is about 0.61% calcium. An additional 36 g/d is recommended during the last two months of pregnancy.

Even when high-producing cows are fed rations rich in calcium, phosphorus, and vitamin D, they often secrete more calcium into milk at peak production than they can assimilate from their diets.[20] They do this by drawing on skeletal calcium stores. Calcium withdrawn in early lactation is normally replaced later during lactation and the dry period.

8.9.1.2 Phosphorus (P)

Phosphorus was discovered in 1669 by Hennig Brand when he evaporated human urine to dryness and heated the residue in sand. The material glowed in the dark, caught fire in the presence of air, and burned with great vigor. The word phosphorus is derived from *phos*, "a light," and *pherein*, "to bear."

Phosphorus has important functions in (1) structural formation of bones and teeth; (2) efficient feed utilization, growth, and reproduction; (3) buffer systems of the body; (4) energy metabolism (e.g., **adenosine triphosphate** [**ATP**] is important in the storage, transfer, and utilization of energy within the body); (5) nucleic acid structure (e.g., DNA and RNA); and, of course, (6) making milk. Recent Arkansas studies showed that increased dietary phosphorus decreased the incidence of parturient paresis. Phosphorus is also important in the efficient utilization of urea.

A phosphorus deficiency is first demonstrated by subnormal concentrations of inorganic phosphorus in blood plasma (normal levels are 4 to 6 mg/100 mL for cows and 6 to 8 mg/100 mL for calves). With phosphorus deficiency, bones become fragile, appetite declines, and growth rate is greatly reduced. Phosphorus deficiency in cattle is also associated with delayed puberty and reduced conception rate.

Since forage crops in general, and legumes in particular, are comparatively high in calcium and low in phosphorus, there is greater likelihood of cows being deficient in phosphorus than calcium. Phosphorus-deficient cattle have depressed appetites, often chew on bones or wood, and may become stiff in their joints. Young animals may develop rickets, which involves poor calcification of bones, arched backs, and crooked legs.

University of Wisconsin research showed that the phosphorus requirement can be met if the diet contains 0.32 to 0.38% phosphorus. Calves digest approximately 94% of the phosphorus in milk, whereas older animals and lactating cows digest only about 55% of their dietary phosphorus. The percentage may be considerably greater for the phosphorus of alfalfa hay. A dietary phosphorus deficit is believed to be the most common mineral deficiency affecting reproduction in dairy cattle of the United States, but other concerns—negative impact on metabolic diseases, mastitis, and foot problems—are seen as barriers to precision feeding of phosphorus. In a recent national survey in the United States, nearly all nutritionists indicated that they were feeding less phosphorus (self-reported averages of 0.36% of the diet). It is an important issue because about 85% of phosphorus imported to the farm comes in feed, and overfeeding of phosphorus adds excessive phosphorus to soil. In 2000, depending upon the watershed, as much as 50% of the total phosphorus loads to surface water came from concentrated animal agriculture.

Calcium/Phosphorus Ratio. Large skeletons and high yields of milk create a need for much more phosphorus in dairy cows than in beef cows. Body ash contains about 17% phosphorus. Approximately 83% and 87% of the total body phosphorus is present in skeletons of young and adult cattle, respectively. Since the ratio of calcium to phosphorus in bone is about 2:1 in older animals (about 1.1 to 1 in young animals), and since the calcium/phosphorus (Ca/P) ratio is approximately 1.3:1 in milk, it would seem logical that the Ca/P ratio of the total dairy ration should be between 1:1 and 2:1, especially for high-producing cows. The importance of maintaining proper Ca/P ratios in the prophylaxis of parturient paresis is discussed in chapter 13.

Providing Calcium and Phosphorus. In general, forages are high in calcium and concentrates are low. The reverse is true of the phosphorus content of feeds. This fact has special application and significance as dairy farmers change from high- to low-forage rations, and vice versa. Corn silage, a forage commonly fed to dairy cattle, is somewhat low in both calcium and phosphorus. Therefore, both calcium and phosphorus supplementation are needed for rations containing large amounts of corn silage. Lactating cows fed liberal amounts of legume hays usually ingest adequate calcium. Protein-rich feeds such as soybean, linseed, and cottonseed meals and wheat **bran** are reliable plant sources of phosphorus. As noted earlier, dried distiller's grains with solubles are high in phosphorus.

Mineral supplements for most dairy rations should be high in phosphorus and contain no more than twice as much calcium as phosphorus. Calcium and phosphorus supplements include **bone meal** and mono-, di-, and tricalcium phosphates (table 8.3). Sources of calcium alone include ground limestone or oyster shell flour; those of phosphorus include diammonium phosphate, mono- and disodium phosphates, and sodium tripolyphosphate. Of these, the most palatable calcium and phosphorus sources, respectively, are bone meal and sodium tripolyphosphate. In high-producing cows it is advisable to add 0.5 to 1.0% dicalcium phosphate or a defluorinated phosphate supplement to grain rations.

8.9.1.3 Salt (Sodium Chloride)

Since time immemorial, animals have forged trails to natural salt licks to satisfy their physical needs for salt. American pioneers followed these animal trails as they moved westward. The story of salt is filled with romance and adventure. At least 27 centuries BC—4,700 years ago—the earliest known treatise on pharmacology was published in China, and a major portion of it was a discussion of more than 40 types and forms of salt. Ancient Greece's widely practiced trade of salt for slaves gave rise to the expression "not worth his salt." Application of the Latin phrase *salarium argentum* to special salt rations given early Roman soldiers was the genesis of the English word *salary.* There are more than 30 references to salt in the Bible. The deaths of thousands of Napoleon's troops during their retreat from Moscow are believed to have been caused by lack of salt in their scanty diets. Thousands of Britons languished in prison for smuggling salt into their country in violation of laws promulgated by ancient British monarchs establishing salt monopolies; oppressive salt monopolies were a major contributing factor in the rise of the French revolution. On our own continent, the British commander Lord Howe was reported jubilant and scenting victory when he captured the salt supply of General George Washington in 1777.

Table 8.3 Calcium and Phosphorus Content of Selected Mineral Supplements.

Mineral Supplement	Calcium (%)	Phosphorus (%)	Ca/P Ratio
Bone meal, steamed	30	14	2.1/1.0
Dicalcium phosphate	26	20	1.3/1.0
Limestone, ground	34	—	—
Monocalcium phosphate	15	25	1.0/1.7
Monosodium phosphate	0	22	—
Diammonium phosphate[a]	0	22	—
Sodium tripolyphosphate	0	25	—
Ammonium polyphosphate[b]	0	16	—
Defluorinated phosphate[c]	33	18	1.8/1.0

[a] Contains 18 to 21% nitrogen.
[b] Contains 10 to 12% nitrogen.
[c] Must contain 1 part or less of fluorine to 100 parts of phosphorus, or 0.18% or less.

Dr. S. M. Babcock first studied salt deficiency in dairy cattle at the University of Wisconsin in 1900. He noted that salt-deficient cows stopped to lick the streetcar tracks, which were salted to melt snow and ice, as the cows were driven to pasture. In the early 1950s, Cornell University scientists observed that within two weeks after cows were fed a low-salt ration they licked barn stanchions, clothes of farm employees, and ate the soil of the barnyard. Moreover, salt-deficient cows attempted to drink urine of cows receiving salt-adequate rations. Babcock reported that daily salt needs of lactating cows were about 30 g for maintenance and 1 g for each pound of milk secreted. He concluded that natural feeds usually provide sufficient salt for maintenance but not for milk production.

Salt-deficient cows conserve salt. Cornell University researchers observed that urine of salt-depleted cows was low in sodium. However, since milk contains about 0.1% salt, one way salt-deficient cows conserve salt is by secreting less milk. In the Cornell studies, cows receiving 60 g of supplemental salt daily yielded approximately 59% more milk than unsupplemented controls.

Salt deficiency is characterized by intense craving for salt, loss of appetite, haggard appearance, lusterless eyes, and a rough hair coat. Additionally, lactating cows lose body weight and produce less milk. Terminal symptoms include shivering, lack of coordination, weakness, **cardiac arrhythmia**, and death. Cows recover quickly when rations are supplemented with salt.

Since sodium chloride (common salt) is not stored to any appreciable extent in the animal's body, and since it is required for normal body functioning, and especially for milk production, it is needed routinely in the diet of dairy cattle.[21] Most feeds of plant origin are low in sodium. The salt content of feeds varies but, in general, forages contain more sodium and chlorine than grains. For example, average alfalfa contains 0.15% sodium and 0.28% chlorine, whereas average corn contains only 0.01% sodium and 0.04% chlorine. Therefore, cows receiving high-grain diets need more supplemental salt than those fed high-forage rations. Also, corn silage contains less salt than alfalfa or grass silages.

According to the NRC, the dietary salt requirement for cows is 0.34% sodium in the ration dry matter during early lactation and 0.10% sodium during the dry period. Adding 0.5% salt to concentrate feeds is usually sufficient. In addition, salt should be available to dairy animals on an ad libitum basis. Cows commonly consume more granular than block salt when both are offered free-choice, though either is satisfactory because of variable intakes. Excessive intakes of salt are apparently not harmful to cows as long as fresh water is available, and except at high levels when they may cause diarrhea.

8.9.1.4 Magnesium

About 0.05% of an animal's body is magnesium. Retention of dietary magnesium by dairy cows is related to body needs and any excess is excreted via urine. Although most body magnesium is stored in bones, this reserve is not readily mobilized. Therefore, when dairy animals are switched from normal to magnesium-deficient diets, hypomagnesemia may result within a few days. Perhaps this explains, in part, why cows turned on lush, green pasture in which the average availability of magnesium to dairy cows is only about 17% may soon demonstrate grass tetany (cf. chapter 13).

Symptoms of magnesium deficiency include **anorexia**, irritability, **hyperemia**, frothing at the mouth and profuse salivation, greatly increased excitability, and tetany. Hypomagnesemic tetany apparently arises from low blood magnesium, which reflects (1) a dietary deficiency, (2) low availability of magnesium in feed (especially pasture forage), and (3) inability to mobilize skeletal reserves effectively. The dietary magnesium recommendation of the NRC is about 0.2% in diets of lactating cows.

8.9.1.5 Potassium

Although most diets of ruminants provide adequate potassium, deficiencies may occur. Maryland studies showed that when lactating cows are fed potassium-deficient diets, they demonstrate deficiency symptoms in three to four weeks. These include **pica**, licking hair of other animals, licking floors, and chewing wooden pen partitions. Hair coats of the studied animals were rough and the hides less pliable. Milk contains 0.15% potassium. Most forages provide adequate levels of potassium, but amounts in concentrates are often low. Therefore, dairy farmers should check high-concentrate rations to be certain the diet is providing 1.0 to 1.1% potassium. The level should be increased to 1.5% potassium in the diet during hot weather.

8.9.1.6 Sulfur

Sulfur comprises about 0.15% of a cow's body tissue and approximately 0.20% of milk. Much of this is in the form of the amino acids methionine and cystine. Sulfur administered orally as radioactive sodium sulfate was incorporated into cystine and methionine of milk protein. Although specific sulfur requirements for dairy cattle have not been established, the NRC recommends a dietary sulfur level of 0.2% of ration dry matter supplied as sodium sulfate. Dairy scientists at the Ohio State University found that the dietary sulfur level of lactating cows producing between 8 and 37 kg of milk daily should be about 0.18%. They noted that use of urea as a nonprotein nitrogen supplement in ruminant rations has increased the need for sulfur supplementation because protein-rich feeds replaced by urea are the usual source of sulfur. Additionally, corn silage, fed extensively to dairy cattle, is often low in sulfur (0.05 to 0.10%). A sulfur deficiency adversely affects cellulose digestion and affects the proportions of volatile fatty acids produced by rumen microorganisms. For efficient utilization of urea by ruminants, a nitrogen/sulfur ratio of 10:1 is recommended.

8.9.2 Microelements (Trace)

Because mineral elements within this category are needed in relatively small quantities, they are often called trace minerals. The term *trace mineral* is probably not a good one, but it was coined when analytical methods were not sufficiently sensitive to measure small amounts of certain minerals in plant and animal tissues. The term *micronutrient element* is now used more frequently.

Minerals needed by dairy cattle in minute amounts include chromium, cobalt, copper, fluorine, iodine, iron, manganese, molybdenum, selenium, and zinc. These are often fed as trace-mineralized salt. Microelements can also be provided in protein balancers (e.g., soybean meal), complete grain feeds, and other mineral mixtures.

8.9.2.1 Chromium

Chromium is needed for the metabolism of carbohydrates. The trivalent form (Cr+3) is needed to activate the glucose tolerance factor that is needed for insulin function. Although studies suggest that dietary chromium is required by dairy cattle, the NRC did not find enough data available in 2001 to establish a requirement. The recommended dose for adult humans is from 50 to 200 micrograms (μg) of chromium per day, and supplements have been taken for many years. Chromium that is complexed with organic compounds and chromium from brewer's yeast are more available than inorganic forms. While the hexavalent forms are toxic, the trivalent forms are generally accepted as nontoxic and did not contaminate milk or meat when fed at excessive levels (Lloyd et al., 2010). The extremely minute concentrations are difficult to measure in cattle diets. But, studies with cattle have shown a positive response to chromium supplementation, especially when animals were under stress before and after parturition and during times of heat stress. Energy status and immune function, dry matter intake, and milk yield were improved with supplementation.

8.9.2.2 Cobalt

Cobalt is definitely needed by microorganisms in the rumen for synthesis of vitamin B_{12}. Deficiency symptoms include loss of appetite and body weight, slowed growth, listlessness, development of anemia, pale mucous membranes, muscular incoordination, a stumbling gait, rough hair coat, decreased milk production, and high mortality rate of calves.

The NRC set the requirement at 0.11 mg/kg of dry matter in feed. Mixing 60 g of cobalt sulfate or 40 to 50 g of cobalt carbonate with 100 kg of salt would provide the requirement. Supplementing pregnant animals with cobalt significantly increases cobalt content of newborn calves and of **colostrum**. Growing dairy animals can safely ingest 40 to 50 mg/d of cobalt per 100 kg of body weight.

8.9.2.3 Copper

A copper deficiency in ruminants was first identified in 1931. Deficiency symptoms include diarrhea, loss of appetite and body weight, cessation of growth, rough hair coat, and anemia. Copper-depleted cows may fail to conceive, have retained placentas, and give birth to calves with congenital rickets. California studies showed that excessive copper intake may increase the copper content of milk and its susceptibility to oxidized flavor. Dairy cows require 11 mg/kg in dry matter. Cattle are not as sensitive as sheep, but 40 mg/kg of diet is the maximum safe level, except where soils are high in molybdenum.

Molybdenum is an antagonist to copper, so copper supplements will need adjustment in some areas. Organic supplements (copper-lysine and copper-proteinate) have been used successfully in some studies. Additionally, if zinc is supplemented, additional copper may be needed. Calcium, iron, and sulfur have also been identified as antagonists to copper. Adding 0.5% copper sulfate to salt is a recommended prophylactic measure in copper-deficient areas, but to ensure regular intake the copper supplement should be in the grain mixture. Copper is more likely to be toxic if over-supplemented compared to other minerals. Excesses are stored in the liver and released suddenly during a stressful period, causing a hemolytic crisis.

8.9.2.4 Fluorine

The essentiality of fluorine for dairy cattle has not been established. However, it is important nutritionally because of its toxic effects. Ingestion of more than 1.4 mg/d/kg body weight produces toxicity, which is characterized by reduced feed intake and milk production, stiffness of legs, enlarged bones, and in severe cases, death. Growing heifers can tolerate 30 mg of fluorine per kg of feed and lactating cows 40 mg/kg of feed. Fattening cattle can tolerate higher levels of fluorine because of the relatively short feeding period. Fluorine accumulates in an animal's skeleton throughout life. Excessive amounts induce bone change and mottled teeth.

8.9.2.5 Iodine

Iodine was discovered while a French chemist was seeking a way to make synthetic sodium nitrate for the manufacture of gunpowder for Napoleon's armies. Even though it was discovered by accident, we now know this element is essential to animal life. Not long after the discovery of iodine a Swiss doctor, familiar with the use of burnt sponge as an age-old treatment for goiter, experimented with small doses of pure iodine with his patients. The large and unsightly neck swelling of goiter was reduced, and his patients improved greatly.

As knowledge of iodine's role in the human body increased, it was discovered that the human thyroid gland provides the hormone thyroxine, which is about two-thirds iodine. Thyroxine influences metabolic activity of body cells. When there is not enough iodine in the body the thyroid gland attempts desperately to gain the vital mineral from the blood and it enlarges in the process, causing goiter.

Lactation and gestation increase the cow's need for iodine. Iodine is present in all milk, but is highest in colostrum. Adding iodine to feed usually increases the iodine content of milk. The requirement of 0.6 mg/kg of body weight meets the needs of dairy cows. Except in unusual conditions (for instance, when relatively large amounts of **goitrogenic** substances are present in feeds), feeding iodized salt satisfies minimal iodine requirements of dairy cattle. Calves whose mothers were fed corn silage as the only forage and large quantities of soybean meal during pregnancy had enlarged thyroid glands. Apparently the soybean meal contained goitrogenic substances. Iodine supplementation is especially important in an all-corn silage forage feeding program.

8.9.2.6 Iron

Iron deficiency is uncommon among dairy cattle, except in calves fed only milk and in cases in which injury or parasites result in large losses of blood. The iron requirement of dairy cows is about 15 mg/kg of diet, but milk-fed calves need 30 to 60 mg/d of iron to grow fast and efficiently. Young, dairy calves fed milk to eight weeks of age made faster and more efficient weight gains than untreated control calves when injected with iron dextran shortly after birth.

8.9.2.7 Manganese

The specific requirements of manganese for dairy cattle have not been established. However, heifers fed low manganese diets (30 mg/kg of ration) were slower to exhibit **estrus** and to conceive than those fed higher levels (60 mg/kg of ration). Except in poultry, manganese deficiencies are difficult to demonstrate experimentally among farm animals. Deficiencies have been reported in dairy cows consuming 16 to 17 mg/kg of diet. Although the dietary need for manganese by dairy cattle is low, the requirement is increased by high intakes of calcium and phosphorus. Feeds are quite variable in manganese concentration. The NRC recommendation of manganese for dairy cattle is 20 mg/kg of feed.

8.9.2.8 Molybdenum

Quantitative requirements for molybdenum have not been established. However, levels in forages below 3 mg/kg have been considered normal, whereas those above 20 mg/kg are commonly associated with molybdenum toxicity. Supplementing rations of high molybdenum content with copper is helpful since molybdenum and copper are antagonistic to each other in the animal body. Thus, when copper content of feed is low, a relatively small quantity of molybdenum is poisonous, but as copper increases, tolerance to molybdenum also increases. Symptoms of dietary excesses of molybdenum (molybdenosis) are diarrhea, **unthriftiness**, rough hair coat, loss of hair color, dehydration, arching of back, listlessness and weakness, brittle bones, and **emaciation**.

8.9.2.9 Selenium

Biological functions of selenium have not been clearly identified but are apparently closely connected with those of vitamin E. Toxic properties of selenium have been recognized for many years, but its usefulness as a dietary component was not recognized until about 1960, when it was discovered that selenium deficiency in farm animals resulted in nutritional muscular degeneration, or white muscle disease. Supplementing rations of cows during gestation and lactation with 0.3 mg/kg (as sodium selenite or sodium selenate) affords protection against white muscle disease. The safe range of dietary selenium is narrow. Indeed, approximately 5 mg/kg in the ration may be toxic to animals. Alkali disease and blind staggers result from selenium toxicity.

8.9.2.10 Zinc

Many metalloenzymes containing zinc are involved in metabolism of carbohydrates, proteins, lipids, and nucleic acids. **Parakeratosis** in cattle grazing forage containing zinc at 18 to 42 mg/kg (dry basis) was alleviated with oral or injected zinc. Deficiency symptoms in calves include retarded growth and reduced feed efficiency, listlessness, swollen feet with open scaly lesions, **alopecia** and scanty hair, and rough skin and general dermatitis (especially of legs, neck, head, and around the nostrils). Additionally, wounds of zinc-deficient animals fail to heal normally. Zinc-deficient lactating cows have reduced fertility and milk production. Production responses to supplemental zinc may be related to improved mammary health, immune function, and mobility (fewer hoof disorders). Some amino acid and peptide complexes have improved the efficiency of intestinal absorption of zinc. As mentioned previously, copper and zinc are antagonistic to each other. The more likely problem exists when zinc is supplemented with marginal copper levels. Cadmium and lead interfere with zinc metabolism. The estimated zinc requirement of dairy cattle according to the NRC is about 55 mg/kg of ration. Supplemental zinc will likely be needed for dairy cattle.

8.10 Providing Vitamins for Dairy Cattle

Rumen microorganisms synthesize the B-complex vitamins, although cobalt must be fed to achieve synthesis of vitamin B_{12}. Supplementation of biotin in the diet has successfully strengthened hooves of cattle. Although there is no question that niacin and other B vitamins are necessary for metabolism, attempts to alleviate related metabolic disorders have not been uniformly successful. Vitamin C and vitamin K are synthesized in tissues of cattle. Thus, dairy farmers must be concerned with supplying vitamins A, D, and E. These fat-soluble vitamins, or their precursors, are present in natural feeds in varying amounts. Dietary supplementation of vitamin A is required, and additional vitamin D and vitamin E are needed in some circumstances.

8.10.1 Vitamin A

The need for vitamin A in dairy cattle has been known for many decades. Its deficiency is characterized by degeneration of the mucosa of the respiratory tract, mouth, salivary glands, eyes, tear glands, intestinal and reproductive tracts, kidneys, and outer skin. An inadequate vitamin A intake reduces an animal's resistance to infection, and colds and pneumonia are more likely to occur. Other symptoms include diarrhea, loss of appetite, emaciation, and reduced rate and efficiency of gain. In advanced stages, excessive keratitis, softening of the cornea, **xerophthalmia**, and blindness may occur. A deficiency in pregnant cows is commonly associated with shortened gestation periods, high incidence of retained placentas, and birth of dead, uncoordinated, or blind calves. Some research trials indicated that intake of β-carotene may boost reproductive efficiency, but the effect has not been observed consistently.

Vitamin A activity is defined as retinol equivalents. Retinol is not in plants, but many feeds contain β-carotene, the precursor of vitamin A. Research indicates that for dairy cattle 1 mg of β-carotene is equivalent to approximately 400 IU of vitamin A, or 120 mg of retinol. Connecticut studies indicated that β-carotene at 10.6 mg/100 kg of body weight is the minimum requirement for growing dairy calves. The requirement for growing dairy cattle has been set at 80 IU/kg of body weight, and the requirement for adult dairy cattle has been increased to 110 IU/kg of body weight. Since levels of dietary β-carotene and/or vitamin A are reflected in the vitamin A potency of milk, and since β-carotene and/or vitamin A are important for normal reproduction and other bodily functions, it is wise to provide supplemental dietary vitamin A, especially when poor-quality or low quantities of forage are fed. Moreover, ruminants need a generous supply of vitamin A because much of it is destroyed in the rumen. If rations are not properly supplemented, vitamin A deficiency may occur among cows receiving low-forage, high-concentrate rations.

Colostrum milk contains 10 to 100 times as much vitamin A activity as normal milk, thus providing this nutrient vital for **neonatal** calves. This has special significance because dietary excesses of vitamin A can be stored in the liver. The vitamin A potency of milk reflects a cow's dietary vitamin A intake and/or body stores of the vitamin. Hence, unlike almost all other components of milk, which vary little from day to day, the vitamin A content varies considerably. During the pasture season, milk may contain 3,000 IU of vitamin A per liter, but it may contain less than one-half that amount in winter months.[22]

Reliable dietary sources of vitamin A include green forages (freshly chopped crops) and, especially, carotene-rich pastures and green leafy hays.[23] Fish liver oils and synthetic crystalline vitamin A compounds are also excellent sources of the vitamin.

8.10.2 Vitamin D

This vitamin is vital to proper deposition of calcium and phosphorus in bones and for calcification of growing bone. A deficiency leads to a weakening of bones and the disease rickets; concentrations of calcium or inorganic phosphorus, or both, in blood decrease and serum phosphatase concentrations increase. The parathyroid hormone and vitamin D are regulators of calcium and phosphorus metabolism.[24] The parathyroid functions to maintain a relatively constant concentration of calcium in the blood, whereas vitamin D is involved in the assimilation of calcium into bone and in calcium absorption from the gut.

Vitamin D requirements are difficult to define. The tendency to house dairy cattle reduces the photochemical conversion of 7-dehydrocholesterol to vitamin D in the skin. Dairy cattle exposed to sunlight may not require supplemental vitamin D. Sun-cured hay may also supply enough vitamin D. The daily requirement for calves from birth to seven months has been 660 IU/d/100 kg of body weight. Although quantitative requirements of vitamin D for maintenance, reproduction, and lactation of mature cows are not well defined, 5,000 IU/cow/d will prevent deficiency symptoms and 10,000 IU/cow/d should be adequate during gestation. The requirement for lactating cows has been set at 21,000 IU/day. Again, the probability of vitamin D deficiency is small when dairy animals are fed sun-cured forage or are exposed to sunlight or ultraviolet light. Even green forage, barn-cured hay, and silage have considerable vitamin D activity, apparently from irradiation of tissues on growing plants. However, animals housed indoors for extended periods, and especially those fed high levels of corn silage, may be deficient in the vitamin unless rations are properly supplemented (see chapter 13 for a discussion of the involvement of vitamin D in parturient paresis).

Reliable sources of vitamin D include sun-cured hays, irradiated yeast,[25] fish liver oils, and synthetic vitamin D compounds. When hay is artificially dried, its exposure to sunlight after cutting is reduced, which, in turn, results in low vitamin D content.

8.10.3 Vitamin E

A deficiency of vitamin E (tocopherol) is uncommon among dairy cattle. Pasture and fresh **green chopped** feed are excellent natural sources of vitamin E activity for dairy cattle. However, concentrations in feeds are highly variable. Silage and cured hay contain 20 to 80% less vitamin E than do fresh forages.

This fat-soluble vitamin is a natural cellular antioxidant. It maintains cellular membranes and is involved with immune function. Many of the functions overlap with those of selenium.

Supplemental vitamin E fed during the dry period reduced the prevalence of retained placenta (even when selenium was adequate) and reduced incidence of intramammary infections at calving. Experimental vitamin E deficiencies have been produced within calves, who demonstrated the same symptoms as calves with white muscle disease developed under field conditions due to a deficiency of selenium. Therefore, vitamin E is commonly supplemented with selenium in diets in the periparturient period when immunosuppression is a problem. The suggested requirement is 545 IU/cow/d.

Adequate dietary vitamin E is important in preventing an oxidized flavor in milk. Since green forages are rich in vitamin E, this may explain, in part, why the incidence of oxidized flavor is less in summer than in winter. California studies indicated that supplementation of an alfalfa hay concentrate diet with *dl* alpha-tocopherol acetate was a practical method for delaying development of oxidized flavor in milk (cf. chapter 25).

8.10.4 Vitamin K

Since this vitamin is synthesized in the rumen, it presents minimal nutritional problems for dairy cattle. The primary practical significance of vitamin K to dairy farmers is its use and effectiveness in treating sweet clover poisoning. This disease is caused by the high **dicoumarol** content of moldy sweet clover hay, which delays clotting time of blood and may result in generalized hemorrhaging.

8.10.5 B Vitamins

Since B vitamins are synthesized by rumen microorganisms and are relatively abundant in most feeds, they have been of minimal concern in dairy cattle nutrition. However, until the rumens of young calves become functional at about six weeks of age, the calves need dietary B vitamins. The preferred source of B vitamins for young calves is nonfat milk solids, which may be provided in dry or liquid form (cf. chapter 7).

8.10.6 Biotin

Biotin is produced in the rumen, and like most B vitamins it is a cofactor for many enzymes. However, increased supplementation resulted in elevated concentrations of biotin in serum and milk. Biotin is not metabolized extensively in the rumen. Hoof health and milk production have been improved by feeding supplemental biotin at 20 mg/d. A summary of 11 studies indicated that with the addition of biotin to lactating dairy cattle, dry matter intake increased by 0.87 kg/d and milk production improved by 1.66 kg/d (Chen et al., 2011).

8.10.7 Vitamin C (Ascorbic Acid)

Normally, ruminants do not need dietary ascorbic acid, for it is synthesized in body tissues. However, the rate of its synthesis apparently depends, in part, on vitamin A intake, age, and general condition of animals.

8.11 Providing Water

Although water is usually the least expensive and most readily available nutrient for dairy cattle, its importance should not be overlooked. Animals suffer more quickly from a lack of water than from a shortage of any dietary nutrient. Body size, levels of feed intake and milk production, and **ambient** temperature (cows drink substantially more water in hot than in cold weather) affect water consumption of dairy cows.

The body of a mature cow contains approximately 70% water; that of a younger animal contains even more. Water needs of lactating cows are influenced largely by level of milk production; milk averages about 87% water. Therefore, normal body functions and milk secretion require large amounts of water. A cow yielding 30 kg/d (66 lb, 8.3 gal) of milk would require about 26 kg (7 gal) of water to satisfy milk secretion needs. In general, cows drink approximately 4 kg of water per kilogram of milk produced as well as per kilogram of dry matter consumed. All feeds contain water; most grains and hays average about 10%, high-moisture silage is approximately 66% water, and pasture varies from 50 to 80% water depending on season, state of maturity, and soil moisture.

Several researchers have observed that cows drink water about 10 times daily when it is available constantly. Studies conducted at Clemson and Iowa State University showed that cows having access to automatic waterers drank 18% more water and yielded 3.5% more milk than control cows watered only twice daily and 6 to 11% more milk than those watered only once daily. High-producing cows commonly drink 25 to 50 gallons of water a day.

Although cows apparently cannot detect nitrate in water, large quantities can be detrimental. Therefore, when nitrate problems are suspected, water should be analyzed for nitrate content. Missouri studies showed that cows will drink more water if it is warmed slightly during cold weather. No differences in milk production or water consumption were observed among cows provided hard or soft water.

8.12 Guidelines for Profitable Feeding

Selection of feeds is of prime importance in the profitable management of dairy cattle and, next to the selection of cows of proper type and breeding, is the factor of greatest importance in profitable dairying.

Edwin B. Hart (1874–1953)

Several guiding principles allow for better decision making when it comes to providing adequate nutrition for dairy cattle at minimal cost. These are treated in the following discussion.

8.12.1 Testing and Program Development

It is just as economically unwise to overfeed low-producing cows as to underfeed high producers. But how do dairy farmers identify high and low producers? It is possible only through production records and participation in a production-testing program (cf. chapter 4). The quantity of milk produced by a cow is largely determined by her ability to convert feed into milk and by the quality and quantity of forages and concentrates fed. Feeding levels should precede—not follow—high daily milk production.

When weighing the quality of a forage against its cost, the following should be considered: high-quality forage cannot make an inherently low producer profitable, but low-quality forage can make an inherently high-producing cow unprofitable. A cow's appetite can be satisfied with low-quality forages but, while such feed fills the rumen, it fails to provide the nutrition adequate for high milk production. Forage, whether homegrown or purchased, serves as the base on which to build a balanced ration. Table 8.2 provides compositional data representative of feed ingredients commonly used in dairy rations.

Once a decision is made regarding the level of quality desired, it then must be obtained. Because of soil fertility, stage of maturity, and weather effects, there is considerable variation in composition and digestibility of feeds, especially forages, from year to year and farm to farm. This is especially important for silages. Dairy farmers should obtain an analysis of the composition of forage to be fed, and an essential priority is to determine moisture content. Information about composition of byproducts is also important. Before a builder bids on a building contract, the quality, quantity, and price of each material is determined. This enables him/her to make a more accurate cost analysis. Similarly, dairy farmers should determine the quality, quantity, and price of feeds available if milk production is to be maximized and feed costs minimized.

In many areas a good **pasture** program affords dairy farmers a means of providing nutrients for dairy animals at the least cost. With pasture-based systems, the dairy farmer is concerned with maximizing kilograms of milk sold per hectare. Excellent fencing equipment is available to reduce labor requirements and there are novel ideas for economical watering systems in small paddocks. However, management skills are required to devise and operate intensive grazing of milking cows. Flexibility is also required to accommodate unexpected weather conditions that impact pasture availability. However, we should note that very high-producing cows cannot consume enough pasture alone to sustain high milk production, and supplements need to be designed to take advantage of the high nutritive quality of pasture.

8.12.2 The Economics of Profitable Feeding

8.12.2.1 Least-Cost Ration Formulation

It is relatively simple to balance a dairy cow's ration when cost can be disregarded and variables are few. However, when a combination of several ingredients is to be used to provide a palatable, balanced ration at minimal cost, calculation of the least-cost and profit-maximizing ration presents a greater challenge. This is best accomplished with the aid of a computer, and several feed manufacturers and agricultural colleges provide such a service using linear programming. There are computer programs available for purchase, but some expertise will be required in their use. Dairy farmers who wish to have a ration formulated need to provide the following information: (1) results of feed analyses, (2) milk production data, (3) local (or delivered) feed prices, and (4) any constraints they wish to specify (e.g., minimum and maximum energy, protein, urea, calcium, phosphorus, fiber, carotene, and forage level). Since feed is the largest expenditure in producing milk, its cost holds the greatest potential for affecting profit. Prices of alternative feeds and supplements should be included to maximize the cost-saving aspect.

Milk price and production potential are not necessarily considered when formulating least-cost rations. Instead, the objective of a least-cost ration is to provide adequate nutrient levels in a specific quantity of feed at the lowest cost consistent with certain constraints previously mentioned. Therefore, the constraints should receive careful consideration, and changes may be needed if the formulation calls for dramatic changes in the constituents of the ration. Major changes may require gradual adjustment. The goal should be rations that maintain high milk production at the lowest cost.

The value of a feed to dairy farmers is best estimated based on the quantity of required nutrients it provides (i.e., assuming palatability and digestibility are high). Of course, dairy feeds should be purchased on a nutrient basis and not merely on a weight basis, as is commonly done. Dairy farmers who must purchase large quantities of feed can benefit most from least-cost ration formulation.

Fortunately, there are no required feedstuffs for dairy cattle. Instead, only certain nutrients are required. Therefore, no single ration can be cited as ideal. Nutritional contributions and cost of ingredients are more important than number of ingredients used. Dairy rations vary depending upon availability of feedstuffs and nutrient cost in a given locality. Thus, each dairy farmer has a different set of prices on feeds available for use in formulating rations. With volatile feed prices, it may be necessary to re-evaluate the ration regularly.

8.12.2.2 Milk-to-Feed Ratio (M/FR)

Feeding for maximum production is subject to the economic law of increasing costs, while increasing the potential ability for high production is subject to the law of decreasing costs.

Carl F. Huffman

An efficiency measure of what the price realized from milk will purchase becomes the preferred economic estimate for dairy farmers. When the price received from 100 lb of milk will buy 100 lb of feed, we have a M/FR of 1:1; when it buys 200 lb of feed, the M/FR is 1:2. This is significant because the cost of feed is the dominant cost in producing milk.

The M/FR may refer to the milk/grain price ratio, which would be more meaningful to the dairy farmer who raises forage and purchases only grain and supplements. There is considerable seasonal variation in milk/grain price ratios. Similarly, there have been long-term variations and trends.

The milk/grain price ratio does not consider (1) diminishing milk-producing value of the ration as feed intake increases, (2) diminishing expense of maintenance (percentage-wise) as daily milk yield increases, or (3) labor economy of high production per cow. The most economical level to feed concentrate to individual cows is associated with milk production responses to variable quantities fed, and to milk-feed and grain-forage price relationships. However, this may be tempered by availability of homegrown feeds, which may or may not be of such quality as to sustain high milk production.

Some sacrifice in milk production may prove profitable by saving out-of-pocket costs for feed when poorer quality homegrown feed is available. Similarly, group feeding or additional mechanization may be profitable, even if it results in less than maximal production in certain cows.

8.12.3 Forage as the Foundation of Dairy Rations

Feed represents the greatest single cost of producing milk or meat. A good feeding program complements good breeding and efficient milking by allowing cows to produce nearer their genetic potential. Forages serve as the foundation in formulating profitable dairy rations.

Forages consist of cell wall constituents and cell contents. The latter are about 98% digestible, whereas cell walls contain considerable **lignin**, which is nondigestible and reduces digestibility of nonlignin plant constituents, including celluloses and hemicelluloses. Chemical bonding of lignin with other structural nutrients reduces digestibility, thus interfering with the action of ruminal bacteria. As forage plants mature, the concentration of acid detergent fiber and lignin increases, thus reducing digestibility.

Emphasis given to forage feeding in future years will depend, in large part, on costs of grain relative to costs of forage and value of milk. The extent to which (1) grains are used to produce fuels, (2) governments commit grain supplies to feeding a hungry world, and (3) exports of feed grains grow will greatly influence the price of grain. With the prospect of returning to a situation in which dairy farmers rely more heavily on feeding greater quantities of forage, an increasingly important aspect of feeding in the future will be forage *quality*. Early-cut forages are superior to late-cut ones for dairy cows, because they are consumed more readily, contain more nutrients, and are more digestible.

As forage quality decreases, rumination time increases. Poor-quality roughage remains in the rumen longer and, thereby, reduces daily feed intake. Of course, as cows increase forage intake, the consumption of concentrates decreases, and vice versa. Studies at Cornell University showed that great differences exist among individual cows in their choice between excellent- and poor-quality forage.

8.12.3.1 Forage Feeding and Productive Longevity

The cow is a ruminant with a digestive system designed to process forages. Because the cow is about the same size as that of a few decades past, and because we expect her to produce four to five times as much milk as her ancestors of only a century ago, we must maximize the concentration of energy inputs to sustain high milk production. So we cram concentrates into a system originally designed to process forages. Are we thereby shortening the cow's longevity? In this case, it is interesting to observe longer productive lives among the dairy cows of Europe than those of the United States. Europeans feed minimal amounts of concentrates. Instead, they depend heavily on forages to maintain milk production. Therefore, although the authors found no long-term studies designed to determine the productive longevity of dairy cows fed either high-forage, low-grain rations or high-grain, low-forage rations, we hold the view that high-forage feeding favors longer productive lives. Admittedly, the reduced milk production and the exercise that cows receive during grazing are factors that affect longevity.

8.13 Feeding Silage

Since plants contain carbohydrates, it is possible to utilize the process of fermentation to preserve forages in various types of **silos** (figure 8.6). A Frenchman, Auguste Goffart, first preserved forage using a silo in 1865. The discovery was so great the French government awarded him the Cross of the Legion of Honor. Silos were introduced into the United States from France in the early 1870s.

Corn is often called the king of crops, and corn silage is a basic forage component of many dairy cattle feeding programs, comprising approximately three-fourths of all silage fed in the United States. Advantages of corn silage over dry forages include (1) greater yield of dry matter per acre, (2) less loss of food nutrients in harvesting, (3) greater palatability, (4) less waste in feeding, (5) greater flexibility in dealing with the weather in harvesting, (6) greater uniformity of quality, and (7) greater adaptability to mechanization of handling.

Corn silage production has especially benefited from improved technology and crop management practices. Minimum tillage, dense plant population per acre, increased rates of fertilizer, substitution of chemicals for cultivation, improved harvesting and handling machinery and equipment, and lesser costs of storage have increased yields and reduced per unit cost of producing corn silage.

Quantities of silages are commonly converted to hay equivalent by assuming that 3 kg of silage equals 1 kg of hay. The moisture content of silage is the foremost factor in estimating the value of silage relative to hay. Approximately 2.0, 2.3, 2.6, 3.0, 3.6, and 4.5 kg of silage are required to equal 1 kg of hay when silages contain 55, 60, 65, 70, 75, and 80% moisture, respectively.

Recommended moisture for good fermentation, digestibility, palatability, and **feed efficiency** are: corn silage, 65%; hay silage, 55%; high-moisture shelled corn, 28%; milo, 30%; and cereal silages (barley, oat, wheat), 45 to 60%.

High-moisture silage commonly reduces ration dry matter intake by one-fourth compared with good-quality hay or **low-moisture silage**. Therefore, high-producing cows fed high-moisture silage ad libitum are almost surely underfed energy-wise. In high-producing cows, silage should be limited to about 3.5% of body weight. Thus, a 1,500-lb (680 kg) cow would receive a maximum of 50 to 55 lb (25 kg) of silage daily.

Corn silage is usually cut while in the hard dough or dent stage. Research indicates that the optimum maturity of corn silage is from 30 to 35% dry matter when measured by sampling the chopped whole corn plant. Because of its grain content, corn silage contains more energy per unit of dry matter than most hays.

When feeding corn silage (which contains only 2.3% crude protein) to high-producing cows, dairy farmers must be especially careful to provide sufficient supplemental protein to meet protein needs of maintenance and high production.[26] Corn silage is also low in calcium, phosphorus, sulfur, manganese, and salt. Therefore, proper mineral supplementation of high-corn silage diets is especially important.

8.13.1 Feeding Hay and Silage

When high-quality hay is the sole source of forage, cows may consume hay up to 3 kg/100 kg of body weight. Each kilogram of hay is equivalent in dry matter to approximately 3 kg of either corn or grass silage. It is apparent, then, why a high intake of silage restricts potential nutrient intake to a point below that required for high milk production. In general, high-moisture silage should not constitute more than 50% of the nutrient intake. Otherwise, its bulk and high-moisture content restrict dry matter intake to a point of restricting milk production.

Since the nutrients are in the dry matter, feed intake and economic value should be calculated on a dry matter basis. For example, a sample of high-quality alfalfa hay might contain 16% protein and 90% dry matter, whereas a sample of **haylage** from the same field might contain 8% protein and 45% dry matter. When both nutrient content and intake are considered, the feeding value of these two forages would be about equal. However, the haylage is only half as valuable as the hay on a weight basis, and even less if its moisture content restricts nutrient intake to the point of restricting milk production.

The dominant hay fed to dairy cows in the United States is either alfalfa or mixed alfalfa. The Arabic meaning of alfalfa (a native of Asia) is "the best **fodder**." This is an appropriate meaning since alfalfa is the "Cadillac" of forages for dairy cattle.

8.13.2 Newer Methods of Handling Hay

Costs for cubing exceed those for baling, but the additional costs may be compensated for by automated handling, reduced shipping and storage space, less waste, and increased animal intake and production. Cubing and

Figure 8.6 (A) The art and science of silage making. Heavy plastic is used to ensile forage in temporary silos (riekephotos/Shutterstock.com). (B) The wrinkled appearance of the plastic of the mini-silos results when vacuum is applied to remove air and thus maintain silage quality (Taina Sohlman/Shutterstock.com).

artificial drying of alfalfa and other hays minimize interference by weather in preserving hay quality. Use of 1-ton square bales has also increased, but many of the newer methods require more field drying before baling and fit best in semiarid climates where alfalfa is irrigated. Obviously, specialized equipment is needed to handle and feed those large, heavy bales. In more humid areas, large round bales are most popular and can be chopped or sliced before adding to a total mixed ration. Weather effects on hay preservation and quality are also significant internationally (figure 8.7).

8.13.3 Feeding Haylage

This low-moisture silage is formed by first wilting grasses and/or legumes to about 40 to 50% moisture and then **ensiling**. It is important to silage quality that air be excluded during ensiling and storage. Otherwise, butyric acid is formed, which greatly reduces palatability and intake of silage. Haylage is a palatable, high-quality feed that reduces losses from leaf shatter and rain damage that occur during hay processing. Additionally, cows consume more dry matter in the form of haylage than of silage.

Research indicates that wilted silage and haylage harvested at between 40 and 70% moisture have the lowest harvest and storage dry matter losses of the legume-grass forages (figure 8.8).

8.13.4 Complete Rations

Complete rations are comprised of combined forages, grains, supplements of proteins and minerals, and additives such as vitamins A and D. Automation for incorporation into a complete feeding system is easier for silage than for hay.

8.14 Agricultural Chemicals and Feed

Great public interest has been generated concerning the merits of using various agricultural chemicals designed

Figure 8.7 Because of the distribution of rainfall and relatively mild temperatures, haymaking is indeed a problem throughout much of Western Europe. (A) Freshly cut forage is draped over wooden racks for drying. Because of frequent showers, several days are required for drying, and forage quality deteriorates (Ann Louise Hagevi/Shutterstock.com). (B) Large round bales can be covered with plastic and removed from fields using heavy equipment (Pablo Debat/Shutterstock.com).

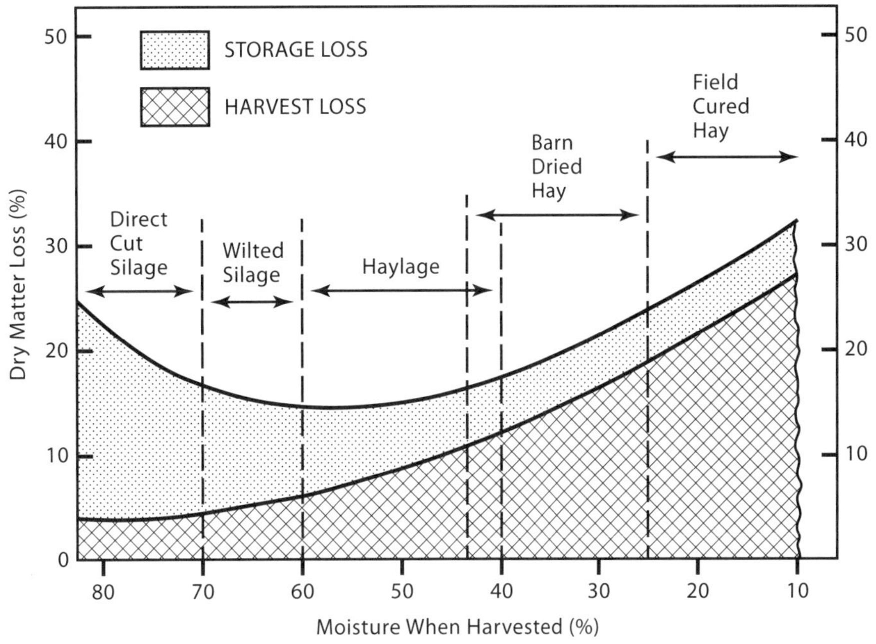

Figure 8.8 Estimated harvest and storage losses when legume-grass forages are harvested at varying moisture levels and by alternative methods (adapted from Hoglund, 1964).

to aid in controlling weeds, insects, and other pests of plants. Although in-depth discussions are beyond the scope of this book, we hold the view that many agricultural chemicals are needed to provide a reliable, high-quality supply of food. Obviously, chemicals and medications should be stored safely away from feed or feed handling equipment. Chemicals must be used with caution, and some not at all, especially if the feed is for lactating cows. Dairy cows must not be fed crops that have been sprayed with chemicals until the crop is safe to use. Restrictions on labels should be read carefully, and farmers must insist on strict adherence to approved clearance times. In some instances the cow may not be harmed, but many chemicals, or a metabolite of the chemical, can appear in milk, thus causing risk to consumers. Rapid detection by milk handlers and public health officials safeguards the public, but a quarantine of the offending farm is difficult to manage and, of course, is very costly—especially if detoxification or disposal of cows is involved. Serious mistakes rarely happen, but the appropriate authorities should be alerted immediately if feed becomes contaminated.

A more common occurrence is the inadvertent natural contamination of feed by mycotoxins (mold-produced toxins). The most typical scenario is that the dairy farmer is notified by the marketing cooperative that the latest shipment of milk was rejected because of contamination with M1, a metabolite of aflatoxin. Milk samples are screened routinely to avoid contamination of food with this carcinogen. The most likely source is moldy corn grain (especially drought-stressed corn), peanut meal, or whole cottonseed (that became wet during storage), and it should be removed from the cow's diet immediately.

It is good practice to sample and freeze labeled containers of each batch of feed. Laboratory tests can detect mycotoxins. It is often possible that only a small area of stored feed has become wet enough to support growth of molds. During warm weather molds grow rapidly, leaving the toxins as residues. Remaining feed may be proven safe, and shipments of milk can resume as soon as health officials lift the quarantine. In more complicated circumstances, expert assistance will be needed to overcome the problems. For instance, cottonseed and cottonseed meal can be treated by ammoniation to eliminate the aflatoxins, thus providing a product that is safe to feed to cattle.

8.15 Summary

Although good nutrition, per se, does not guarantee high milk production, poor nutrition will prevent attainment of a cow's full genetic potential to secrete milk, just as surely as will poor management or other environmental conditions. Therefore, the feeding of dairy cattle should receive high priority within dairy farm enterprises.

The improved genetic ability of dairy cows to secrete milk has challenged the formulator of rations and the manager of feeding systems. Fortunately, most high-producing cows possess the ability to store and mobilize large quantities of energy for milk secretion. However, their ability to mobilize tissue protein for milk synthesis is small relative to their ability to mobilize energy. Methods developed during the past decades help address the needs of the dairy cow during late gestation and early lactation. Dairy farmers should be especially careful to incorporate modern management techniques and to provide ample dietary energy, protein, and other nutrients during periods of high milk production.

Maintenance, gestation, and lactation are the three major factors to consider in calculating rations for dairy cows. Feeding standards and nutrient requirement tables provide recommended amounts of protein, net energy, minerals, and vitamins for cows producing varying amounts of milk.

Least-cost rations for dairy cattle can provide the nutrients needed for maintenance and productive purposes (growth and milk) at the least amount of cost consistent with changing grain and forage prices. Of course, when weighing costs versus quality, consideration should be given to the following factors: the availability of homegrown feeds, palatability and physical form of rations, and effects of various feed ingredients on animal health and productivity.

Three important factors influence the amount of energy available for milk production: (1) the concentration of energy per pound of feed consumed, (2) total pounds of feed consumed, and (3) body stores of reserve energy available. Selection of dairy cows on the basis of milk yield alone increases the genetic potential for feed efficiency. More importantly, feed efficiency in dairy cows is the amount of milk produced per unit of feed consumed. It is an important factor in determining cost of milk production.

The use of corn silage as a forage in dairy rations continues to increase because of high nutrient yield per acre and adaptability to mechanization and group feeding. However, the relatively low crude protein content of corn silage presents practical problems with respect to balancing cows' protein and energy requirements over a range of milk yields. Furthermore, provisions for certain minerals and vitamins should be made when dairy cows are fed diets high in corn silage.

Dry matter intake, expressed either as percent of body weight or in grams per kilogram of **metabolic weight**, decreases as the percent concentrate in the ration increases. Also, ruminal concentration of acetate decreases, whereas concentrations of propionate and butyrate increase. We discussed the effects of this on milkfat content, and practices helpful in alleviating the situation were noted.

Factors favoring the feeding of large quantities of concentrates include (1) low price of concentrates, (2) high price of milk, (3) high inherent milk-producing capacity, and (4) low displacement of forage per unit of grain added. Although dairy farmers of the United States enjoyed an increasingly more favorable milk/feed ratio until recently, we noted trends and reasons that suggest that grain may become an increasingly greater luxury for feeding dairy cattle.

Study Questions

1. How do you interpret the following statement: "Rations must be fitted to cows and not cows to rations"?
2. Define a balanced ration.
3. Identify and discuss briefly the four basic needs for nutrients in dairy cattle.
4. Describe the process of rumination, and how it effects the ability of the cow to produce milk.
5. Describe the different ways of expressing energy value of feeds. Cite merits and limitations of each.
6. Differentiate among: crude protein, digestible protein, rumen digestible protein, and rumen undigestible protein.
7. What is TDN? How is it calculated?
8. Why does digestibility of feed change with levels of feed intake? How is this related to the efficiency of milk production?
9. How does ambient temperature affect the usefulness of heat increment in cattle?
10. Do protein and energy needs change with animal age? Why? Do they change with level of milk production? Why?
11. In evaluating NPN, what is meant by *protein equivalent*?
12. Given age, body weight, status of pregnancy, and level and composition of milk production, how are daily NE_L and protein requirements for dairy cows calculated?
13. What are the consequences of underfeeding protein and/or energy to high-producing cows?
14. How can ruminants use urea? Why is the amount and type of energy important when urea is fed? Does level of dietary protein affect the usefulness of urea? Why?
15. Discuss the merits of adding urea to corn silage. Discuss any limitations.
16. What are the metabolic pathways of protein and NPN in ruminants?
17. Why is dietary sulfur a consideration when urea is fed?
18. What are liquid feed supplements? How may their intake be controlled?
19. What is the theoretical basis for methionine supplementation in high-producing cows?
20. What are probable consequences of feeding energy- and protein-deficient rations to high-producing cows? Describe the affects of having a lack of protein and energy on the persistency of lactation in cows.
21. Cite and discuss the factors affecting the appetite of dairy cows. Is total dry matter intake related to digestibility of feeds? Why?
22. Discuss merits and limitations of feeding greater levels of grain to lactating cows. What is the basis of calling high-grain rations "cool" ones for feeding ruminants in summer months?
23. Cite examples of feeding regimes that affect the fat content of milk. Discuss briefly the role of fiber content and physical forms of feed in this regard.
24. List the minerals required in the diets of dairy cows, along with reliable sources of each mineral. Describe the circumstances and/or conditions under which deficiencies may occur.
25. Does the relative quantity of forage and concentrate appreciably affect mineral supplementation of dairy cows? Cite and discuss examples. What special consid-

erations of minerals would one give a high-corn silage diet for lactating cows?

26. Cite two examples of interrelationships between vitamins and minerals.
27. Identify one or more minerals that may be toxic, even when ingested in relatively small quantities.
28. Which vitamins are of major concern in feeding dairy cattle? Are dietary vitamin needs of neonatal calves the same as those of mature cows? Why or why not?
29. Under what conditions are vitamin D and vitamin E needed in supplements?
30. Why are water needs of cows affected by dry matter content of feeds? Cite examples of feedstuffs having substantially different water contents.
31. Discuss the basis for and merits of least-cost ration formulation.
32. Of what significance is the milk/feed ratio to dairy farmers? Does it affect the amount of concentrate feeds that are fed in today's economic climate? Explain.
33. Discuss the use of forages as the base in providing feed for dairy cattle. Why is forage quality of great importance?
34. Why has the feeding of corn silage to dairy cows increased in recent years? Discuss protein and mineral inadequacies of corn silage.
35. What is haylage? In what ways does it differ from silage?

NOTES

[1] Except when noted otherwise, we use the term *protein* to mean total or crude protein in feed as opposed to rumen digestible protein (RDP).

[2] Rumination is a characteristic of ruminants, "cud-chewing" animals, in which food, in the form of a **bolus** weighing about 90 g when made of forage and 115 g when made of grain, is regurgitated, remasticated, and reswallowed in preparation for digestion. Cows normally spend approximately 8 hours per day chewing their cuds. However, those fed high grain and ground or pelleted rations may ruminate only 1 to 2 hours per day.

[3] Bacterial degradation of fibrous carbohydrate materials is slower than for starches.

[4] Certain scientific journals also use *joule* as an energy term. One *international joule* is a unit of work or energy equal to 10,000,000 ergs, or the amount of energy expended by a current of 1 ampere flowing for 1 second against a resistance of 1 ohm. One calorie is the equivalent of 4.185×10^7 ergs or 4.185 joules.

[5] Approximately 400 liters of methane and CO_2 gases are produced in the rumen of an average-sized, mature dairy cow daily. The greater the quantity of forage ingested, the larger the amount of gases produced. However, it is difficult to measure **gaseous products of digestion (GPD)**; values are often estimated by equations. Energy losses in feces and urine are determined by bomb calorimetry.

[6] In ruminants, heat increment is comprised of (1) heat of fermentation (microbial activity in the rumen) and (2) heat of nutrient metabolism, which is higher for proteins than carbohydrates and higher for carbohydrates than fat.

[7] Interestingly, a mature cow loses approximately 15% of her body weight at parturition. The calf represents less than 10% and the balance is comprised of placental tissues and body fluids.

[8] Microbial cells contain up to 65% protein (dry weight basis) and approximately three-fourths of it is digested and absorbed by the cow. There is, however, only limited absorption of amino acids from the rumen. Instead they are absorbed further down the GI tract.

[9] Nitrogen comprises only about 16% of protein; the other 84% is largely carbohydrates. In chemically analyzing a feed for protein, the amount of nitrogen is multiplied by 6.25 (100 ÷ 16) to calculate crude protein.

[10] The only resemblance between urea and protein is that both contain nitrogen, and since a NPN such as urea provides only nitrogen, the term *protein equivalent* is used to indicate the compound's potential protein value.

[11] When dairy farmers have an abundant supply of high-quality legume hay, they will need to purchase only a limited quantity of protein supplement, which restricts the potential economic advantage of feeding urea.

[12] It is estimated that rumen microorganisms can produce 3 to 4 lb of microbial protein daily. In doing this, they use approximately 0.5 lb of nitrogen (equivalent to about 1 lb of urea).

[13] Since urea contains no energy, and since energy is vital for protein synthesis, it is important that approximately 1 kg of readily fermentable carbohydrate (e.g., corn) per 100 g of urea be added for maximal utilization of urea in adapted cows. Cattle commonly need about two weeks to become adapted for efficient utilization of urea.

[14] Diammonium phosphate contains 18 to 21% nitrogen—a protein equivalent of 112 to 133%.

[15] Ether extract of forage is commonly less than 50% fat; it also contains essential oils, waxes, resins, chloroplasts, and other lipid materials that are poorly utilized by ruminants.

[16] Wisconsin studies demonstrated the feasibility of adding whole soybeans (high in fat) to dairy rations at levels up to 20% of the concentrate when economic conditions are favorable and provided the ration contains adequate forage. Undesirable effects on rumen fermentation, milk production, and milkfat percentage were observed with the addition of high levels of whole soybeans to rations containing less than one-third forage.

[17] Dry matter intake of nonlactating cows is usually about 2% of body weight. Properly feeding dry cows commonly increases milk production in the next lactation.

[18] Appreciate, however, that the stimulus to secrete milk comes from the endocrine system and not from feed, per se (cf. chapter 10).

[19] It is estimated that by mobilizing 10% of her skeletal calcium reserves, a 1,500-lb (680-kg) cow would provide enough calcium for 2,000 lb (907 kg) of milk. This explains, in part, the ability of cows, especially in early lactation, to secrete large quantities of milk even though they are in a negative calcium balance.

[20] Studies at Kansas State University (Ward et al., 1972) showed that positive calcium balances can be maintained in early lactation provided cows are fed adequate calcium and vitamin D (300,000 IU/week). Consumption of 5 g of calcium/kg milk, plus maintenance allowance, was adequate to maintain calcium equilibrium during early lactation. The phosphorus requirement for lactation was satisfied with 2.3 g/kg milk.

[21] There is no experimental evidence indicating a specific lack of chlorine in dairy rations. Rather, it is sodium that is vital to many body processes; for example, sodium is the most abundant **cation** in saliva.

[22] The higher level of vitamin A is prudent, since high temperatures increase the vitamin A requirements of animals. Moreover, it coincides with normal intense suckling by many mammals in the "natural" state.

[23] Hays lose about 75% of their carotene after six months of storage. Missouri studies showed that corn loses from one-fourth to one-half its vitamin A activity during the first year of storage.

[24] Calcitonin (thyrocalcitonin) secreted by the thyroid gland is also involved in calcium and phosphorus metabolism. In hypercalcemia, the hormone rapidly lowers blood calcium by inhibiting bone resorption; it also increases urinary excretion of phosphate.

[25] The vitamin D content of milk may be increased significantly by feeding irradiated yeast to lactating cows. Of course, vitamin D is routinely added to processed milk (cf. chapter 28).

[26] Regular (traditional) corn silage is well known for its low protein content. In South Dakota studies, high-sugar corn silage contained significantly more crude protein than regular corn silage (8.59% versus 6.34% of dry matter). When fed to Holstein steers, apparent digestibilities of protein and fiber were greater for the high-sugar corn silage (63.5% versus 52.2% and 61.9% versus 59.3%, respectively).

References

Agricultural Research Council. 1980. *The Nutrient Requirements of Ruminant Livestock.* Slough, UK: Commonwealth Agricultural Bureaux, Farnham Royal.

Allen, M. S. 2000. Effects of diet on short-term regulation of feed intake by lactating dairy cattle. *Journal of Dairy Science* 83:1598–1624.

Arieli, A. 1998. Whole cottonseed in dairy cattle feeding: A review. *Animal Feed Science and Technology* 72:97–110.

Baile, C. A., and M. A. Della-Fera. 1981. Nature of hunger and satiety control systems in ruminants. *Journal of Dairy Science* 64:1140–1152.

Campbell, J. R., W. M. Howe, F. A. Martz, and C. P. Merilan. 1963. Effects of frequency of feeding on urea utilization and growth characteristics in dairy heifers. *Journal of Dairy Science* 46:131–134.

Campbell, J. R., and J. F. Lasley. 1969. *The Science of Animals That Serve Mankind.* New York: McGraw-Hill.

Capper, J. L., R. A. Cady, and D. E. Bauman. 2009. The environmental impact of dairy production: 1944 compared with 2007. *Journal of Animal Science* 87:2160–2169.

Chen, B., C. Wang, Y. M. Wang, and J. X. Liu. 2011. Effect of biotin on milk performance of dairy cattle: A meta-analysis. *Journal of Dairy Science* 94:3537–3546.

Hoglund, C. R. 1964. *Comparative Storage Losses and Feeding Values of Alfalfa and Corn Silage Crops when Harvested at Different Moisture Levels and Stored in Gas-Tight and Conventional Tower Silos: An Appraisal of Research Results* (Agricultural Economics Mimeo, Report 947). East Lansing: Michigan State University.

Kertz, A. F. 2010. Urea feeding to dairy cattle: A historical perspective and review. *Journal of Animal Science* 26:257–272.

Lloyd, K. E., V. Fellner, S. J. McLoed, R. S. Fry, K. Krafka, A. Lamptey, and J. W. Spears. 2010. Effects of supplementing dairy cows with chromium propionate on milk and tissue chromium concentrations. *Journal of Dairy Science* 93:4774–4780.

Moe, P. W., and H. F. Tyrrell. 1973. The rationale of various energy systems for ruminants. *Journal of Animal Science* 37:183–189.

Morrison, F. B. 1956. *Feeds and Feeding* (22nd ed.). Claremont, Ontario, Canada: Morrison Publishing.

National Research Council. 2001. *Nutrient Requirements of Dairy Cattle* (7th Rev. Ed.). Washington, DC: National Academy of Sciences.

Van Soest, P. J. 1973. "Revised Estimates of the Net Energy Values of Feeds." Proceedings of the Cornell Nutrition Conference.

Ward, G., R. C. Dobson, and J. R. Dunham. 1972. Influences of calcium and phosphorus intakes, vitamin D supplement, and lactation on calcium and phosphorus balances. *Journal of Dairy Science* 55:768–776.

For Further Study

1. To explore further the nutritional requirements of dairy cows, see the following publications:
   - National Research Council, *Nutrient Requirements of Dairy Cattle* (http://www.nap.edu/openbook.php?record_id=9825)
   - The Merck Veterinary Manual (http://www.merckmanuals.com/vet/management_and_nutrition/nutrition_cattle/nutritional_requirements_of_beef_cattle.html)
2. Several universities provide software to calculate dairy rations:
   - Michigan State University, Spartan Dairy Ration Evaluator/Balancer (http://spartandairy.msu.edu)
   - University of California, Davis, Department of Animal Science (http://animalscience.ucdavis.edu/extension/software)
   - University of Minnesota, Dairy Extension (http://www.extension.umn.edu/agriculture/dairy/feed-and-nutrition/feeding-the-dairy-herd/ration-formulation.html)

9

Physiology of Reproduction in Dairy Cattle

Life is short and the Art long; the occasion fleeting;
experiment dangerous, and judgment difficult.
Hippocrates (ca 460–377 BC)

The authors gratefully acknowledge the significant contributions of Matthew C. Lucy, PhD, Professor of Animal Sciences, University of Missouri, Columbia, Missouri.

9.1 Introduction

It is interesting that lactation is made available only because of the reproductive process. Although scientists have initiated lactation artificially through hormonal treatments, these applications of endocrinology have not been developed on a practical basis. Instead, it is only through reproduction that the cow's secretly formulated milk has been made available for the calf and humans.

The cow not only contributes the female sex cell (egg), essential to initiating development of the young, but also provides the environment in which the egg is fertilized and the new individual is nourished during fetal life. In this chapter we discuss briefly the process of bovine reproduction. This subject should not be taken lightly since reproductive efficiency is especially important to dairy farmers. Low fertility results in high costs through losses in production of milk and calves, purchases of herd replacements, and payments for additional veterinary services, medications, and breeding fees. Additionally, there is a lowered rate of herd improvement through culling and genetic advances.

9.2 Parts of the Male Reproductive Tract

The reproductive system of the dairy bull is comprised of several important organs, the most basic of which are the testicles (testes). Bulls have two testicles and several accessory glands and structures. The testes are suspended outside the body cavity in a scrotum (figure 9.1), which provides a suitable environment (about 4 to 5°C below body temperature) for spermatogenesis. Billions of spermatozoa are in continuous production within the seminiferous tubules that comprise most of the testicular mass. The seminiferous tubules join in the *rete testes,* which in turn connect with the head of the epididymis. Maturing spermatozoa accumulate within the epididymides (structures composed of tightly coiled tubes attached to the top, bottom, and sides of testes). Larger conducting tubes (called the *vas deferens*) lead from each epididymis, between muscle layers of the abdominal wall, to the area adjacent to and above the neck of the bladder. There the tubes expand and unite with the urethra, which provides a common passage from the body via the penis for both semen and urine.

Near the location where urine and sperm enter the urethra, several other glands—paired seminal vesicles, prostate, and paired bulbourethral or Cowper's glands—empty their secretions. These provide a fluid environment that serves as a suspending and activating medium and supplies energy essential for spermatozoa. Collectively, the suspending fluid and the spermatozoa are called *semen* (spermatozoa constitute only about 20% of the total volume of semen). Another important function of the testes involves secretion of the male hormone testosterone by the interstitial cells, called cells of Leydig, which lie between the seminiferous tubules. Testosterone is important in maintaining **libido** and the secondary sex characteristics of males.

The penis serves as a mechanical means of depositing semen within the female vagina in natural mating. When a bull is sexually excited, blood accumulates within the penis, which increases its diameter and length and renders the organ less flexible. Additional length essential

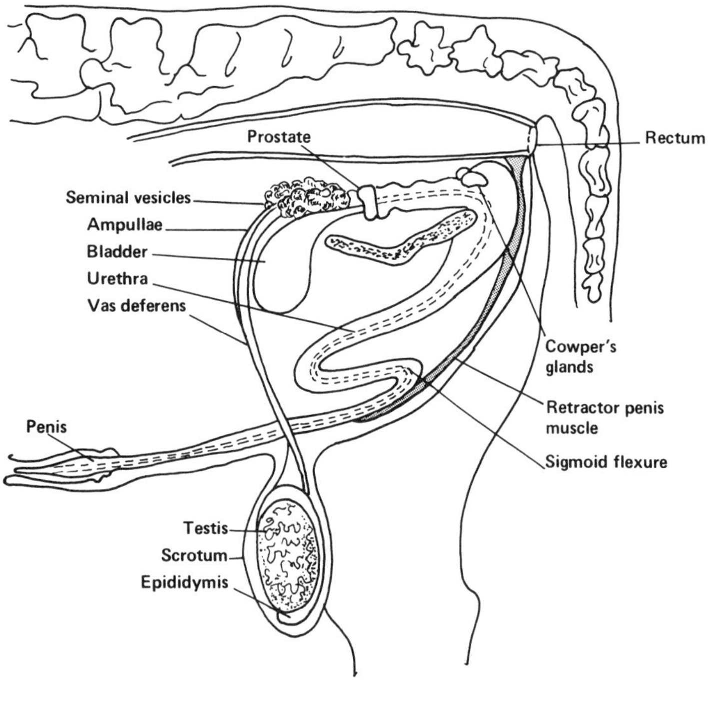

Figure 9.1 Reproductive organs of the bull (Pelissier, 1971).

for penetrating the female reproductive tract is achieved in erection by temporarily straightening the S-shaped curve (sigmoid flexure) in the penis, located at about the level of the scrotum. Ejaculation of semen occurs at the peak of sexual excitement when the mounted bull makes a single deep thrust of the penis into the cow's vagina, or into an artificial vagina used to collect semen for use in artificial insemination (AI).

Ejaculate volume varies with age, frequency of service, and preparation (teasing) time. These factors, along with environmental temperatures and season, also affect the concentration of spermatozoa. The latter is important both to fertility and to the amount of dilution possible for semen intended for use in AI. Normal bull ejaculates range from 1 to 12 mL, averaging about 4 to 5 mL. Good-quality semen commonly contains 1.0 to 1.5 billion spermatozoa/mL (about 100 million per drop), of which 60 to 80% are vigorous and actively motile.[1] To prolong the life of collected spermatozoa, semen must receive careful handling (see section 9.13). After dilution, semen may be stored indefinitely in straws and/or ampules at –196°C (–320°F) in liquid nitrogen.

9.3 Parts of the Female Reproductive Tract

The vulva is the only externally visible part of the cow's reproductive tract. The vestibule (figure 9.2) extends inward about 4 in (10 cm) from the vulva to the urethra, the opening of the bladder. The vagina is about 10 to 12 in (25 to 30 cm) long and extends from the urethra to the cervix. In **natural service**, the vagina receives the male sex cells (sperm) and also serves as part of the birth canal for the calf at **parturition**.

The cervix, about 3 to 4 in (9 cm) in length and 0.75 to 2 in (2 to 5 cm) in diameter, is a tough muscular organ with a small center opening leading to the uterus. In natural mating the bull deposits spermatozoa near the entrance of the cervix, whereas in artificial insemination a small catheter or pipette is passed through the cervix, allowing deposition of semen into the body of the uterus (figure 9.3 on the following page). During pregnancy the cervix secretes a thick wax-like substance that serves as a plug at its opening to prevent entrance of foreign materials. Between estrous periods, the mucus secreted by the cervical and vaginal linings becomes thick and sticky and, instead of flowing from the vulva, forms a cervical plug that helps prevent bacteria from entering the uterus. At **estrus**, the mucus is plentiful and less viscous, which allows it to spill from the vulva when cows **mount** other females.

The uterus is located anterior to (in front of) the cervix and serves to nourish the developing calf. It consists of a short body (about 1.5 in [4 cm]) and two horns (8 to 15 in [20 to 38 cm] in length) resembling a ram's horns. The uterus serves as an incubator for the developing calf. On the lining of the nonpregnant uterus are about 100 small elevations called *caruncles*. During pregnancy these, together with their related parts from the fetal membranes (cotyledons), become the so-called buttons through which a developing calf is nourished. These may become 1 × 3 in (2.5 × 7.5 cm) or larger in size at parturition. The fetal membranes, frequently called **afterbirth** following calving, must separate from the uterus and be discharged following parturition. Otherwise, the veterinarian should be called to unfasten the membranes and clear the uterus.[2]

The two oviducts, or Fallopian tubes, extend from the right and left uterine horns to the ovaries and provide a passageway for the sperm and egg. Fertilization, the union of sperm and egg, usually occurs in one oviduct and the fertilized egg then descends to the uterus for the remainder of pregnancy.

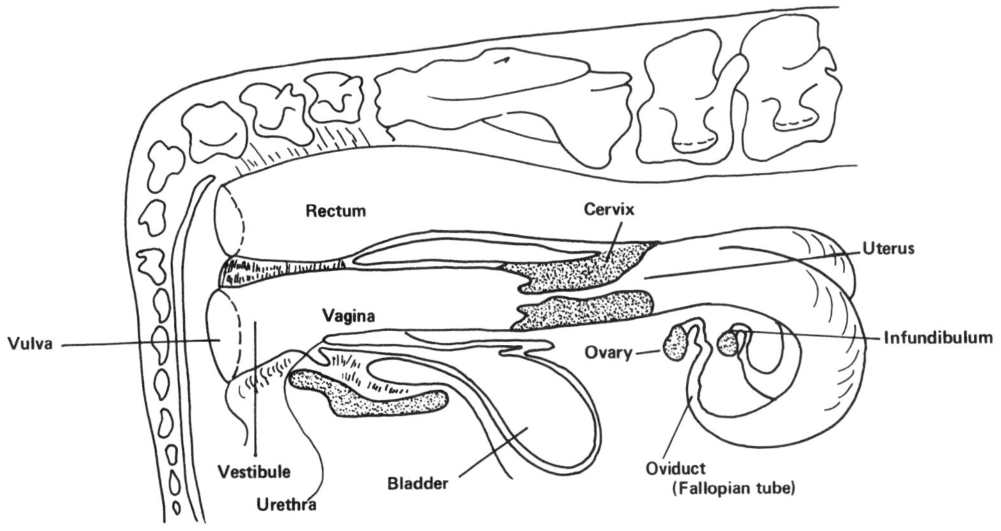

Figure 9.2 Selected reproductive organs of the cow (Pelissier, 1971).

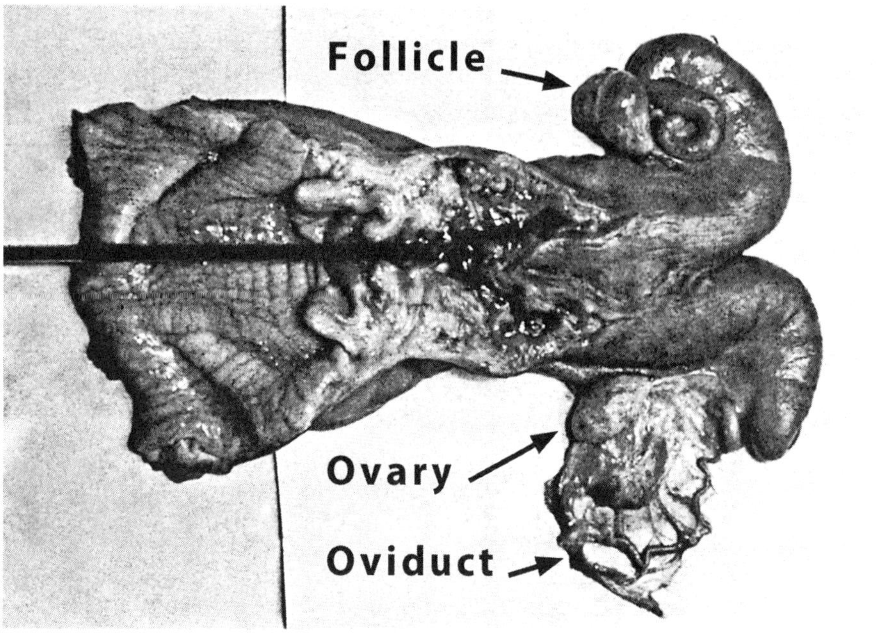

Figure 9.3 Photograph of the cow's uterus. During AI, semen should be deposited in the body of the uterus, the dark area at the end of the pipette (courtesy of Dr. J. W. Morse, Turlock, California).

9.4 Forming an Egg

The primary reproductive organs in the cow are two *ovaries*, right and left, each measuring about 1.5 × 1.0 × 0.5 in (3.8 × 2.5 × 1.3 cm) in size. Eggs (ova) and hormones are produced in these organs. Reproductive activity of the nonpregnant cow occurs in cycles averaging 21 days. During each cycle, a Graafian follicle develops and an egg is formed within it. It resembles a small blister and at maturity is about the size of a small grape. Additionally, the developing follicle secretes a hormone called *estrogen*. Secretion of estrogen peaks when the follicle is about ready to rupture and the egg is fully developed. When the egg is ready to be fertilized, estrogen causes the cow to exhibit estrus, or physical signs commonly called heat. An ovarectomized cow (one with her ovaries removed) can be brought into estrus by administering estrogen.

When the ovarian follicle ruptures and releases the egg, the follicular cavity is filled by a yellow body called the *corpus luteum*, which secretes a hormone called *progesterone*. This hormone helps maintain pregnancy. If the egg is fertilized, the corpus luteum secretes progesterone throughout pregnancy. If, however, the egg is not fertilized, the yellow body regresses, a new follicle is formed, and the 21-day cycle is repeated.

9.5 Fertilization and Gestation

The oviduct serves as the site of fertilization. The fertilized egg descends the oviduct and enters the uterus, where it becomes attached. It is in this setting that a water sac or placenta forms around the developing embryo and the young calf develops its own blood supply. Hence there is no mixing of the calf's blood with that of its mother. Maintenance of pregnancy is dependent upon the secretion of progesterone by the corpus luteum. Near the end of gestation the corpus luteum regresses and progesterone secretion is greatly reduced. Pregnancy can be prolonged and parturition delayed by injecting progesterone.

By approximately the forty-fifth day of pregnancy, all body parts of the developing calf are formed. From then until the end of pregnancy, there is a gradual increase in fetal size. Average birth weights of dairy calves range from about 60 lb (27 kg) in the smaller breeds to 95 lb (43 kg) in the larger ones.[3] Average lengths of gestation (pregnancy) in days for dairy breeds are: Ayrshire, 279; Brown Swiss, 290; Guernsey, 282; Holstein, 279; Jersey, 279; and Milking Shorthorn, 282.

9.6 Parturition

Gestation terminates in dairy cows with parturition—the process of giving birth to young—and the initiation of lactation. Parturition, or calving, is induced by hormones. The hormone relaxin (secreted by the ovaries in cows) causes relaxation of the cervix and softening of the connective tissues of the pelvic region, which allows expansion to provide a passageway for the calf at parturition. As progesterone concentrations decrease, estrogen activity increases[4] and uterine muscles become sensitive to the hormone oxytocin, which causes them to contract. A series of uterine muscular contractions, aided by action of the abdominal muscles, results in expulsion of the fetus through the cervix and on through the birth canal. Occasionally not enough oxytocin is produced, and the veterinarian should administer the hormone to aid in parturition. Cows should be undisturbed at parturition. Otherwise, excitement or nervousness may result in an incomplete release of oxytocin, or in the release of another hormone, epinephrine, which counteracts the favorable effect of oxytocin (cf. chapter 10).

9.7 Postpartum Mating

The question of how long to wait after calving before a cow is rebred is an important one. The conception rate increases significantly at 60 to 90 days postpartum for two important reasons: First, the uterus undergoes great postpartum changes. At calving time, the enlarged, thin-walled uterus weighs up to 20 lb (9 kg) and is large enough to contain a 60- to 100-lb (27- to 45-kg) calf, 5 to 6 gal (19 to 23 L) of placental fluids, and the placental membranes, which commonly weigh more than 10 lb (4.5 kg). Within about 60 days it involutes (shrinks) to a small, thick-walled organ weighing only 1 to 2 lb (0.5 to 1 kg), with a diameter of about 5 to 6 in (12 to 15 cm).

Secondly, in pregnancy, the placenta is attached to the uterine wall at 80 to 120 places (called caruncles), which must separate from the cotyledons of the placenta before the placenta can be passed as afterbirth. Then, the caruncles of the uterus must involute from a size at parturition of about 2.5 in (6.4 cm) long and 1.0 in (2.5 cm) wide to their nonpregnant diameter of less than 0.3 in (7.5 mm). It is rather miraculous that the cow's uterus and its caruncles can regress and repair to the normal, resting, nonpregnant state in as short a period as 60 days. Another important postpartum change is the reestablishment of estrous cycles. Of interest are several studies indicating that nursing a calf will delay estrus in dairy cows. French scientists showed that the suckling stimulus—and not lactation per se—is the reason lactating ewes do not come into estrus for considerable periods of time.

9.8 Detecting Estrus

Owners "miss" more heats than do bovines.

Harry A. Herman

Estrus cycles begin as the female reaches puberty—the time at which an individual gains the capacity to reproduce. Onset of puberty results from a series of developmental events that occur within the reproductive endocrine system. Most heifers attain puberty by about 11 months of age, but poor diets can delay puberty well into the second year of life. Nutritional status influences the timing of the increase in the frequency of the pulse-like release of luteinizing hormone (LH), which is the vital trigger for the onset of puberty. Until there is sufficient development of the reproductive system, a negative feedback of estradiol (a steroid produced in the ovaries) maintains a low rate of secretion of LH.

Schillo (2011) pointed to the importance of the central nervous system in his review of the reproductive endocrinology of the female. Various blood-borne signals reflecting the nutritional status are detected by the nervous system and, in turn, trigger neural pathways that converge on the hypothalamus. Certain endocrine organs (hypothalamus, anterior pituitary gland, ovaries, and uterus) produce hormones that enable communications among the various reproductive tissues. Gonadotrophin releasing hormone (GnRH) is secreted into the portal blood system in a pulsatile manner by the neurons of the hypothalamus. It is transported to the anterior pituitary where it controls release of follicle stimulating hormone (FSH) and luteinizing hormone (LH), collectively known as gonadotropins. These hormones regulate ovarian function, which involves production of ovarian steroids. These, in turn, regulate release of gonadotropin, uterine function, sexual behavior, and expression of secondary sex traits.

Seasonal breeders exhibit anestrous during periods of long day-length but resume periods of estrus as day-length shortens. Thus, a long photoperiod reduces the intensity of the pulsatile (flow with periodic variations) release of LH. Cattle are not seasonal breeders. Nonpregnant heifers will exhibit regular estrous cycles throughout the year.

Failure to achieve the desired reproductive efficiency in dairy cattle is often related to failure to detect estrus. Estrus (heat) represents the period of sexual receptivity of nonpregnant cows. The duration of heat is commonly from 4 to 12 hr, averaging about 8 hr (it averages less time in subtropical climates[5]). Estrus detection is essential when AI is employed or when cows are not bred by natural mating.

Cows in estrus commonly exhibit similar behavioral traits. They usually attempt to mount or **ride** cows not in heat.[6] Only cows in estrus will remain standing when mounted by other cows (figure 9.4 on the next page), and this sign is the best indication of when to expect ovulation. Cows in estrus follow, stand beside, and often put their heads on the backs and/or rumps of other females. Moreover, cows in heat may throw their heads as if to mount, smell other cows, wrinkle their noses, and work their rumps and tails as a bull. Many cows bawl considerably during estrus; others exhibit restlessness, pace along fences, or even try to break through in search of a bull. Frequently, clear mucus resembling the white of a raw egg may be observed to spill from the vulva when they mount other animals. This mucus is often smeared on the buttocks or as strings on the tail. Hair over the tail head is usually roughened or even rubbed off in spots. The vulva of a cow in heat commonly becomes swollen and wet so the fine wrinkles disappear and the hairs are moistened and matted. A drop in milk production and decreased feed intake on the day of estrus is common.

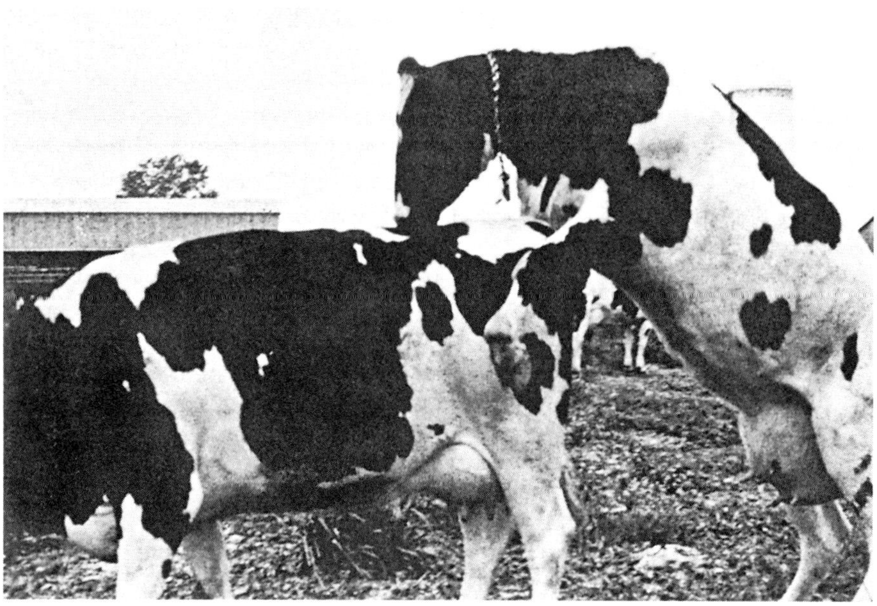

Figure 9.4 Typical mounting behavior exhibited by cows in estrus (courtesy of Missouri Agricultural Experiment Station).

Blood often passes from the vulva on the second or third day following estrus regardless of whether the cow becomes pregnant. This postovulatory bleeding is called *metrorrhagia*. Such observations may be helpful in determining that a heat period was missed. In such cases, a notation on the calendar to be especially watchful for signs of estrus in the cow about 17 to 18 days later is essential to improved heat detection and increased reproductive efficiency. Such bleeding suggests that the cow was in estrus (that is, she ovulated) two or three days previously. The bleeding results, in part, from the engorgement of blood vessels of the uterine lining during estrus, which bleed into the uterine cavity and are shed two to three days following heat.

9.8.1 Heat Detection Aids

Numerous aids have been used to improve heat detection. A vasectomized bull or **yearling** dairy **steer** to which testosterone has been administered makes a good heat detector.[7] An important consideration in using such males to detect heat is that since sexual contact is made, venereal diseases affecting reproduction can be spread. It is possible for a veterinarian to perform a surgical deviation of the penis so sexual contact cannot be completed.

Other heat detection aids involve devices containing dyes that are cemented over the tail head of females. Upon pressure from another animal riding, the device will break and release a colored dye or stain. Some devices have a scratchable surface that has a bright color undulation. The Kamar heat detector (figure 9.5) detects heat due to pressure from a mounting animal.

A California study indicated that visual observation will detect more than 90% of cows in estrus providing there are at least four observations of cows made daily by a first-class detective. Records that indicate when particular animals are due to enter estrus are especially helpful in accurately detecting heat.

9.8.2 Estrus Synchronization

Most dairy cows are inseminated after a predetermined postpartum interval (60 to 90 days). Under most systems, cows are inseminated after spontaneous estrus for a predetermined period and then anestrous and cystic cows are intensively managed. More intensive approaches to reproductive management involve programmed breeding for all inseminations after a series of hormonal treatments for the purpose of controlling the time of insemination. These approaches are referred to as timed AI (TAI) programs. Timed AI has also resulted in the development of synchronization programs that do not require estrus detection.

The hormones used to control the estrous cycle are the same as those within the cow. The earliest estrous synchronization method (developed in the 1960s) blocked ovulation by administering exogenous progestogens. Although the method provided acceptable synchrony, conception rates were depressed. In the 1970s, the discovery of prostaglandin $F_{2\alpha}$ ($PGF_{2\alpha}$) as the hormone causing luteal regression led to the development of new synchronization protocols: modified target breeding, presynchronization for timed AI, presynchronization with "cherry picking" of cows in estrus, and synchronization for timed AI (figure 9.6).

In presynchronization (presynch), the objective is to have cows at a similar stage of the estrus cycle prior to synchronization for AI. Two injections of $PGF_{2\alpha}$ are administered 14 days apart, with the second injection administered between 11 and 14 days before initiating a timed AI program. Injection of $PGF_{2\alpha}$ to regress the corpus luteum and bring cows into estrus is effective beginning 5 to 7 days after estrus. Two treatments of $PGF_{2\alpha}$ 11

Figure 9.5 Kamar heatmount detectors assist in identifying cows ready for artificial breeding. The pressure sensitive detector has a timing mechanism that is activated when the cow is mounted. Glued onto the sacrum (tail head), pressure from the brisket of a mounting animal requires approximately 3 seconds to turn the detector from white to red. This timing mechanism helps distinguish between true standing heat and false mounting activity. A shorter Kiwi-type cow will need a forward placement, whereas a Holstein will need her detector rather closer to the tail head (courtesy of Kamar Products, Inc., Zionsville, Indiana).

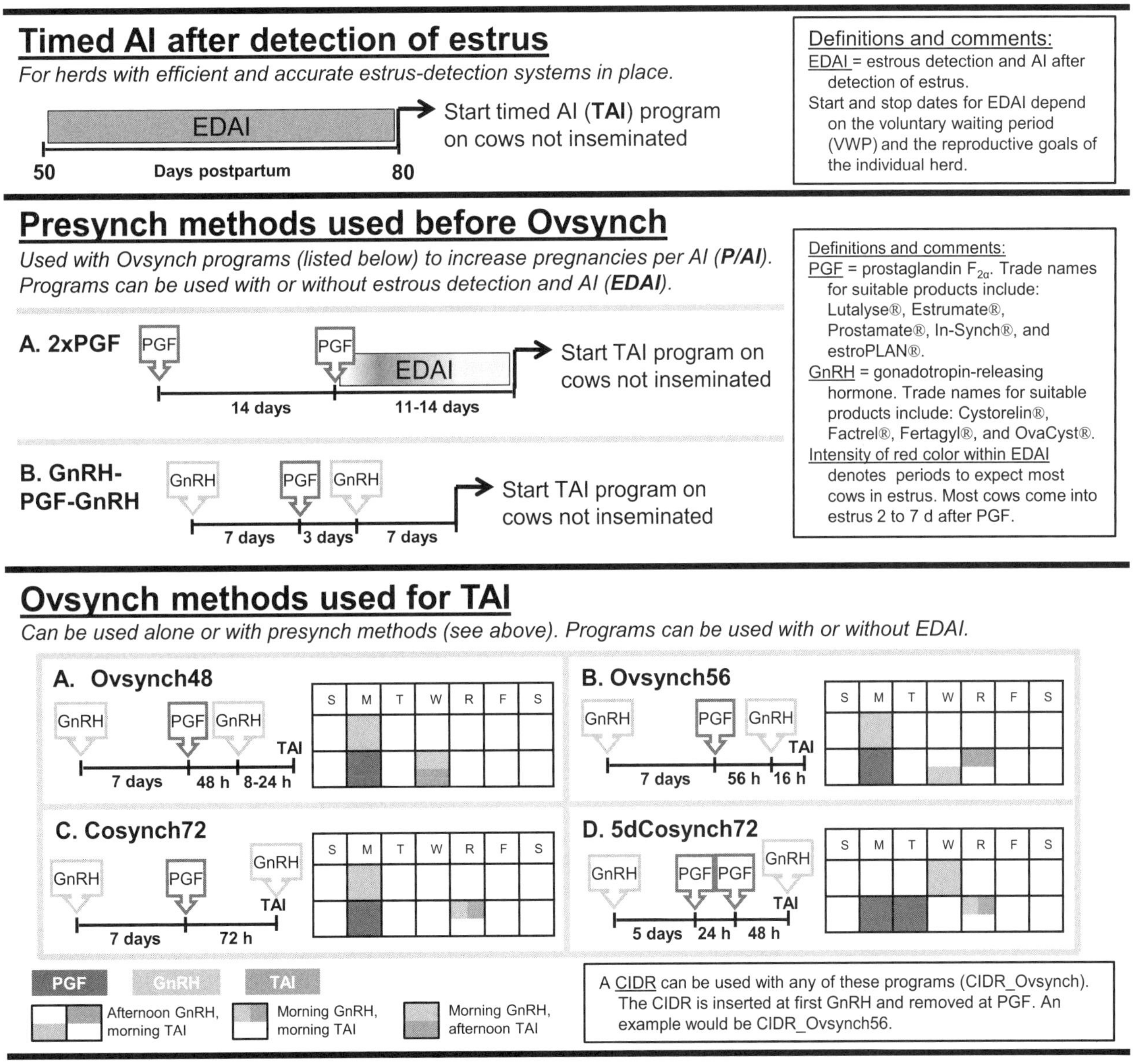
Timed AI after detection of estrus
For herds with efficient and accurate estrus-detection systems in place.
EDAI
Start timed AI (TAI) program on cows not inseminated
50
Days postpartum
80
Definitions and comments:
EDAI = estrous detection and AI after detection of estrus.
Start and stop dates for EDAI depend on the voluntary waiting period (VWP) and the reproductive goals of the individual herd.
Presynch methods used before Ovsynch
Used with Ovsynch programs (listed below) to increase pregnancies per AI (P/AI). Programs can be used with or without estrous detection and AI (EDAI).
A. 2xPGF
PGF
PGF
EDAI
Start TAI program on cows not inseminated
14 days
11-14 days
B. GnRH-PGF-GnRH
GnRH
PGF
GnRH
Start TAI program on cows not inseminated
7 days
3 days
7 days
Definitions and comments:
PGF = prostaglandin F2α. Trade names for suitable products include: Lutalyse®, Estrumate®, Prostamate®, In-Synch®, and estroPLAN®.
GnRH = gonadotropin-releasing hormone. Trade names for suitable products include: Cystorelin®, Factrel®, Fertagyl®, and OvaCyst®.
Intensity of red color within EDAI denotes periods to expect most cows in estrus. Most cows come into estrus 2 to 7 d after PGF.
Ovsynch methods used for TAI
Can be used alone or with presynch methods (see above). Programs can be used with or without EDAI.
A. Ovsynch48
GnRH
PGF
GnRH
TAI
7 days
48 h
8-24 h
S M T W R F S
B. Ovsynch56
GnRH
PGF
GnRH
TAI
7 days
56 h
16 h
S M T W R F S
C. Cosynch72
GnRH
PGF
GnRH
TAI
7 days
72 h
S M T W R F S
D. 5dCosynch72
GnRH
PGF
PGF
GnRH
TAI
5 days
24 h
48 h
S M T W R F S
PGF
GnRH
TAI
Afternoon GnRH, morning TAI
Morning GnRH, morning TAI
Morning GnRH, afternoon TAI
A CIDR can be used with any of these programs (CIDR_Ovsynch). The CIDR is inserted at first GnRH and removed at PGF. An example would be CIDR_Ovsynch56.

Figure 9.6 Dairy cow synchronization protocols (Department of Animal Sciences, University of Wisconsin–Madison).

to 14 days apart can effectively synchronize estrus in cyclic cows. The interval to estrus after the second $PGF_{2\alpha}$ injection, however, is too variable for acceptable conception rates after timed AI. Cows treated with $PGF_{2\alpha}$ in this manner, therefore, must undergo a period of estrous detection after $PGF_{2\alpha}$ injection.

Cyclic cows can be treated with progesterone for 7 to 9 days and given a luteolytic dose of $PGF_{2\alpha}$ 0 to 2 days before progesterone withdrawal. The timing of the $PGF_{2\alpha}$ treatment at the end of progesterone treatment partially determines the precision of estrous synchronization and also the fertility after insemination. Injection of $PGF_{2\alpha}$ at the time of progesterone withdrawal produces a less synchronous estrus. Cows treated with $PGF_{2\alpha}$ 1 or 2 days before progesterone withdrawal have better estrous synchrony but may have lower fertility.

To eliminate the need for estrus detection, a GnRH, $PGF_{2\alpha}$, GnRH program is used. The best-known program is called Ovsynch. The protocol is based around two injections of GnRH and $PGF_{2\alpha}$. Synchronization of both a follicular wave and subsequent estrus can be achieved if GnRH treatment is followed seven days later by $PGF_{2\alpha}$ treatment, which regresses luteal tissue. Estrus and ovulation of the newly formed dominant follicle occurs 2 to 3 days after $PGF_{2\alpha}$ injection. A second GnRH injection follows the $PGF_{2\alpha}$ injection by either 48 or 72 hours (traditional Ovsynch) or 56 hours (Ovsynch 56). With traditional Ovsynch, cows are inseminated either at 48 hours (Cosynch 48) or 72 hours (Cosynch 72). Cows treated with Ovsynch 56 are inseminated 16 to 24 hours after the second GnRH injection.

The Ovsynch protocol uses timed AI. This protocol has distinct advantages over other estrous synchronization procedures because every cow is inseminated at the end of treatment (100% submission rate). Timed AI protocols have somewhat lower conception rates when compared to protocols that employ insemination at estrus. Lower conception rates are generally offset by higher submission rates for timed AI. Thus, the total number of pregnant cows may increase when timed AI is used. The economic performance of Ovsynch relative to $PGF_{2\alpha}$ systems is highly dependent on estrous detection efficiency. Herds with low estrous detection efficiency will achieve the greatest return from the 100% submission rate in Ovsynch systems.

Progesterone blocks estrus and ovulation; thus, supplementing progesterone between the first injections of GnRH and $PGF_{2\alpha}$ improves overall synchrony with an Ovsynch program. The outcome of estrous synchronization is also improved when progesterone is used because anovulatory cows as well as cystic cows benefit from progesterone supplementation. A system is: GnRH injection and CIDR insertion (controlled internal drug release of progesterone intravaginally); wait seven days; $PGF_{2\alpha}$ and CIDR removal; wait three days; GnRH and timed AI. The most common time interval to breeding for this system is 66 to 72 hours.

Pregnancy rates for anestrous cows treated with Ovsynch were lower than pregnancy rates for cyclic cows treated with Ovsynch. When the CIDR device was applied within the Ovsynch protocol, the pregnancy rates for CIDR-treated anestrous cows were about equal to those for cyclic cows. Thus, the CIDR treatment partially corrected the conception rate failure in anestrous cows treated with Ovsynch.

9.9 Importance of Timing in Conception

It must be appreciated that a cow is fertile only when an egg is present that could be fertilized. Similarly, it should be understood that an egg lives only a short time after being shed from the ovary—about 5 to 10 hours—unless it is fertilized. The egg is not usually shed until after the close of visible heat (figure 9.7). This may be from 6 to 18 hours after the end of estrus; the average is about 12 to 14 hours. Therefore, to help maximize conception, cows should be inseminated in the latter half of standing heat or within a few hours (preferably within 6 hours) after having gone out of estrus. Sperm cells commonly retain ability to fertilize for 24 to 30 hours after being deposited in the bovine female.

Wild cattle, other wild animals (e.g., deer and buffalo), and certain domesticated animals (e.g., sheep) have distinct breeding seasons. Generally, the peak of sexual activity (mating) and gestation in females coincide with the time of year best suited for the young to be born. However, through selection and modern management of dairy cattle, much of this characteristic has been lost.

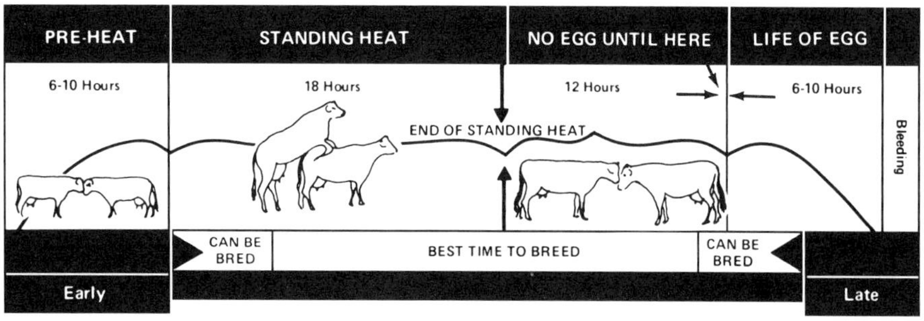

Figure 9.7 Common hours of estrus and best time to breed the cow (adapted from Wright and Oleskie, 1962; American Breeders Service, Inc., 1972).

Therefore, today's dairy cattle breed throughout the year, although there is noticeable seasonal change in their sexual activity. During spring and fall, heats tend to be longer and more intense, thus more easily detected. During the cold months, heats tend to be shorter, less intense, and require greater effort for detection. Extended periods of hot weather often reduce the intensity of estrus and conception rates.

9.10 Reproductive Efficiency

Ideally, a 12- to 13-month calving interval is optimal. To achieve this goal, cows must be mated 60 to 90 days postpartum, conceive on first service, and have no interruptions or complications in the 280-day pregnancy. This allows for a 305-day lactation and an 8-week dry period. It is obvious that high reproductive efficiency is important in the economics of producing milk and raising herd replacements. Dairy cows that fail to **freshen** regularly have long dry periods that reduce profits through wasted feed, housing, labor, and management. Reproductive inefficiency is second only to low production in reasons why dairy cows are culled (cf. chapter 7). Standards for measuring reproductive efficiency are given in table 9.1.

There are three especially important inputs that contribute to fertility in dairy cattle: (1) the male, (2) the female, and (3) the manager. The *bull* represents a factory, warehouse, and delivery system for the manufacture, storage, and deposition of semen. In addition to possessing superior genetic value (cf. chapter 6), the dairy bull must be fertile and possess sufficient libido to perform his part of the sex act. In natural insemination he may be expected to mate with 30 to 50 females annually, whereas in artificial insemination his matings may exceed 50,000/year depending upon his recognized genetic worth. Through AI he is in competition with his peer group and the computer serves as recorder and referee.

Low fertility among bulls is common[8] whereas sterility is infrequent. Several factors influence fertility. Certain venereally transmissible diseases such as bovine venereal trichomoniasis and bovine vibriosis usually do not cause sterility in the male, but when transmitted to females at coitus, after a time, they interrupt pregnancies that have been initiated (cf. chapter 13). These diseases also can be transmitted through semen in AI. The veterinarian should make routine examinations of a bull's semen and of his health. Artificial insemination provides a means of minimizing the spread of disease from bulls to heifers and cows.

Table 9.1 Methods of Measuring Reproductive Efficiency in Dairy Cows with Values Indicating Average and Excellent Management.

Method	Average	Excellent
First service conception rate (%)	35	50
Services/conception	2 to 3	2
Calving interval (months)	13.5 to 15	13.5
Calf crop (%)	70 to 80	> 90
Cows in milk (%)[a]	80 to 83	> 85
Days open (%)[b]	150 to 200	< 120

[a] Percentage of cows in the herd lactating throughout the year.
[b] Interval between calving and subsequent conception.

The female role in reproductive efficiency is great. Hormones must be in balance both for conception and maintenance of pregnancy. Embryonic mortality is a major contributing cause of low reproductive efficiency. Prenatal mortality often exceeds 20% of all conceptions. Certain factors that cause embryonic death may be expected to increase with advancing age of the cow. This may explain, in part, why reproductive efficiency usually decreases in older cows.

It cannot be assumed that cows are pregnant when they fail to demonstrate estrus. Routine pregnancy checks are recommended. An experienced veterinarian usually can determine pregnancy in heifers 30 to 40 days and in mature cows 35 to 40 days after conception. **Metritis** and/or **pyometra** often confuse the inexperienced examiner.

The *manager* influences reproductive efficiency in many ways. Proper nutrition in heifers and cows is essential to reproduction. The first prerequisite of nutrition is survival, then comes reproduction. If nutrition is inadequate, the latter may wait. In artificial and in natural insemination, when the bull is kept from the herd, the manager selects the time for mating. Close observation of cows coupled with knowledge of estrus and ovulation patterns contribute to high reproductive efficiency by providing females the opportunity to become pregnant. Additionally, it is the manager's responsibility to maintain good records and work closely with the veterinarian in checking pregnancies and in monitoring reproductive problems among dairy cattle.

It is important that managers do not look down upon mammals that are less complex socially and intellectually and consider them to be less complex physiologically. Instead, the manager must appreciate the complexity of bovine reproduction, especially among high-producing dairy cows.

The *inseminator* also contributes to reproductive efficiency. Her/his insemination techniques, sanitation, and handling of semen are important adjuncts to conception. When AI is accomplished by an off-the-farm technician, the manager must be relied upon for heat detection and timing.

9.10.1 Retained Placentas and Breeding Efficiency

Retained placentas are significantly and inversely related to breeding efficiency. Hence, more services per conception are required among females having retained placentas at parturition.[9] The incidence of retained placentas is also higher among cows having milk fever at parturition (about twice as great) than among those that do not incur **parturient paresis**. **Dystocia** is also much more prevalent among cows having milk fever than among those that do not. Could it be that low blood-calcium concentration associated with milk fever may restrict maximal uterine contractions essential for normal expulsion of the calf and fetal membrane? If this is true, the prevention of milk fever may improve breeding efficiency.

9.10.2 Health and Nutrition of Cows and Breeding Efficiency

It is well known that healthy, well-fed cows conceive more often than malnourished, diseased ones.[10] Moreover, improperly fed cows are slower to initiate estrus than those receiving adequate amounts of a balanced diet. A lack of energy is probably the most common form of nutritional infertility, whereas an inadequate intake of phosphorus is the most prevalent mineral deficiency affecting reproduction. A wide calcium/phosphorus (Ca/P) ratio has also been reported to reduce reproduction among dairy cows (cf. chapter 8). Contrary to some belief, current research does not indicate that **urea** used at recommended levels in the ration reduces reproductive efficiency.

9.10.3 Milk Production and Breeding Efficiency

Most dairy managers agree that, in general, their highest-producing cows are the most difficult ones to get and keep pregnant. Limited research indicates that milking cows four times daily (4X) instead of twice daily (2X) will delay estrus, presumably because of the additional milk yielded with the 4X milking.

The initiation of postpartum ovarian activity is related to body condition at calving and to negative energy balance after calving. Cows should be in an optimum good body condition at calving (not too fat and not too thin). Cows that are overly fat have a decreased appetite following parturition compared with more ideally conditioned cows, and thus, are in a greater negative energy balance and have a much larger change (decrease) in body condition score (BCS) during lactation. Cows in very poor body condition probably do not have energy stores available to mobilize for maintenance and lactation requirements.

Fat stored in body tissue at calving is used during the early postpartum period to meet the energy demands of lactation. There is an immediate shift in metabolism at calving where nutrients are mobilized from energy and protein stores to meet the demands of lactation. Feed intake does not reach peak levels in lactating dairy cows for a few weeks following parturition. It is during this time that negative energy balance (difference between energy intake and requirements) reaches its maximum. The resumption of postpartum estrous cycles is dependent on the severity of the negative energy balance and body condition of the cow. Cows that experience the greatest loss in body condition after parturition require more time to achieve first ovulation than those that have less change in body condition. Conception rates of cows with a greater BCS loss are lower compared with cows with less BCS loss postpartum. Similarly, cows in a lesser negative energy balance or in a positive energy balance have greater conception rates than those with a more negative energy balance.

The negative energy balance increases when milk production increases faster than feed intake. When the extent of negative energy balance decreases, events leading to the first ovulation will begin. Ovulation of a dominant follicle postpartum is dependent upon stimulation through increasing mean concentration and pulsatile release of LH. Pulsatile release of LH is greater in cows that ovulate a follicle in the early postpartum period than those that remain anestrus. In most cases, follicles that fail to ovulate do not reach full ovulatory size. Low energy availability decreases LH pulsatility, and thus, follicular stimulation and estradiol synthesis are less. Also the responsiveness of follicles to LH may be decreased because of lower circulating concentrations of insulin-like growth factor (IGF-I; a hormone produced by the liver of the cow). Circulating IGF-I concentrations are directly correlated with energy status. Cows in a greater negative energy balance have lower circulating IGF-I concentrations. The IGF-I concentrations are greater in cows that are ovulating.

9.10.4 Cow Density and Reproductive Efficiency

Social stress, or other factors, associated with high cow or heifer density adversely affect reproductive efficiency. Researchers in New Zealand concluded that short estrous cycles and reduced breeding efficiency resulted from social stress experienced by young cows in large herds. Several species of wild animals exhibit sharp reductions in reproductive efficiency when crowded with their peer group.

9.11 Reproductive Irregularities

Many biological variations and irregularities exist that relate to the reproductive efficiency of dairy cattle. Among "normal" cows, 5 to 10% show signs of estrus during pregnancy. Passing the insemination tube through the cervix and into the uterus of a pregnant cow often causes abortion.[11] Good records and pregnancy examinations help reduce the incidence of such practices.

9.11.1 Short Estrous Cycles

In some cows, there is an interval of only 8 to 10 days between estrous periods rather than the usual 21 days. This likely results from an abnormally early degeneration of the corpus luteum, with insufficient progesterone being produced to inhibit production of gonadotrophins by the anterior pituitary gland. An early degeneration of the corpus luteum allows growth of a new group of follicles sooner than normal, and estrus occurs within a shorter period of time. Cystic ovaries may also cause short and/or irregular estrous cycles.

9.11.2 Anestrus

Most so-called no-heats are actually missed heats.[12] Some cows display only mild signs of estrus and require especially close observation to detect heat. A small percentage of cows, particularly very high-producing ones, either fail to cycle regularly or fail to exhibit visible signs of estrus. No-heats or quiet heats can result from a retained corpus luteum, a condition requiring treatment by a veterinarian.

9.11.3 Nymphomania

This organic disorder, characterized by abnormally frequent heats or almost constant estrus, often reflects cystic ovaries (the cystic follicles fail to rupture, and estrogens are more or less continuously produced by these follicles).[13] Many such cows can be treated successfully by use of GnRH by an experienced veterinarian.

9.12 Breeding Problems and Genetic Progress

Culling low-producing cows is an important adjunct to dairy herd improvement. Yet, breeding problems reduce culling potential because there are fewer herd replacements; each year, numerous cows are sold because of reproductive failures, which also can impede this process. Replacements are especially limited in dairy herds using beef bulls on "problem" cows and/or virgin heifers. Breeding problems tend to discourage use of bulls having a high **predicted difference (PD)** since their semen usually commands a premium price.

9.13 Reproductive Biotechnologies

By applying selected reproductive biotechnologies, superior genetics, higher milk yield, improved milk quality, and reduced environmental impact can be achieved in much less time than by traditional breeding methods. For example, artificial insemination and embryo transplant have already provided significant advancement. Additional biotechnologies being applied in certain conditions include superovulation, in vitro fertilization, sexed semen, cloning, and gene transfer (transgenic modification of an animal's genome).

Artificial insemination (AI) is the introduction of male reproductive cells into the female reproductive tract by an artificial (mechanical) means. The foremost value of AI in dairy cattle lies in its use as a tool for genetic improvement. This is made possible by identifying genetically superior bulls through progeny testing large numbers of offspring among many herds distributed throughout various climates and environmental conditions (cf. chapter 6). When such information is coupled with the advantages of being able to dilute semen 100 or more times and storing it indefinitely, the potential benefits of AI in breeding dairy cattle become readily apparent (Campbell et al., 2010).

Proper insemination techniques are important adjuncts to increased conception when AI is employed. Two common errors of AI technicians are (1) expelling semen from the pipette too rapidly, which tends to rupture sperm cells (at least 5 seconds should be allowed) and (2) failing to deposit semen in the body of the uterus (see figure 9.3). In one study in which dyes were used, it was found that approximately two-thirds of the inseminators were depositing semen in the right uterine horn.

Superovulation with ova transplant provides a method of obtaining large numbers of progeny from mating genetically superior cows and bulls. By administering a superovulatory dosage of FSH, dairy females commonly ovulate five or more eggs. Using artificial insemination, the eggs can be fertilized and later transferred to recipient cows (ova transplant) of lesser genetic worth.[14] Estrous cycles of recipient cows must be synchronized with those of donor cows. It is also possible to freeze and store fertilized bovine eggs. If coupled with artificial insemination, this provides even greater flexibility in selecting and mating dairy cattle.

In vitro fertilization (IVF) involves collecting eggs from ovaries, exposing them to spermatozoa outside the body, generating an embryo, and planting it into a foster female. Embryo transfer (ET) following IVF provides the opportunity to produce several offspring from a single female, increasing the opportunity to obtain up to 10 offspring rather than one from a single cow. When combined with sexed semen technology, IVF and ET enable rapid advancement of certain genetic traits.

Each spermatozoon contains a single X or Y chromosome, making possible production of either a female or male offspring. The higher content of DNA in the female chromosome enables separation of X- and Y-bearing spermatozoa via flow cytometry. Sexed semen is semen that has been sorted by its X and Y chromosomes, so gender can be preselected for use in artificial insemination.

Cloned animals can be produced by splitting an embryo or by somatic cell nuclear transfer (SCNT). The latter, which is most commonly applied, involves introduction of DNA from the nucleus of an adult somatic (body) cell into an **enucleated oocyte**. The two fuse and the oocyte reprograms the SCN. The egg develops into an early stage embryo and is transferred into a recipient female to be carried to **term**. The technique is inefficient and expensive but has advantages over other breeding alternatives.

In genetic modification, a microinjection of DNA into a recently fertilized ovum can lead to expression of a desirable genetic trait. However, the gene may be inserted at random locations in the host genome, leading to various expressions of the gene and potential undesirable side effects. Alternative methods have produced desirable outcomes by placing genes in specific sites and permitting the target gene to be turned on or off. Modifying milk composition through transgenesis has the potential to increase milk's value and potential uses. For example, it may be possible to create "hypoallergenic" milk by blocking the gene for β-lactoglobulin, the major allergen in cow's milk (Gifford and Gifford, 2013).

9.14 Summary

Reproduction in dairy cattle is of prime importance in the bovine life cycle. Lactation is, in a sense, a by-product of reproduction. Therefore, a profitable dairy enterprise is dependent upon the ability of the cow to reproduce. And her ability is often restricted by several factors, for example, (1) the failure of humans to effectively assume the bull's role in heat detection, and by other human errors and management deficiencies; (2) inadequate dietary energy (underfeeding) has been shown experimentally to delay first estrus in growing heifers and to delay estrus and significantly reduce conception in lactating cows; and (3) anestrus is a more important breeding problem under hot than under cold climatic conditions. Moreover, the duration of estrus in dairy heifers is shorter under hot than under cold climatic conditions.

Reproductive efficiency is a complex and economically important aspect of dairy herd management. Moreover, fertility is a major influence on productive longevity of dairy cattle. Notwithstanding the volume of reproduction-related research conducted during the last five decades, breeding problems among dairy cattle are prevalent and have great economic significance. (The economic consequence of low reproductive efficiency approximates the net dairy farm profit.) Reproduction can be a complex problem in which both management and animal physiology overlap and interact. Reproductive physiology will be a fertile field for research investigations in the years ahead as humans continue to exercise artificial control of reproduction and to provide more milk for humans.

Study Questions

1. Why may milk be considered a by-product of reproduction?
2. Of what economic significance is reproductive efficiency to dairy farmers?
3. What are the main reproductive organs of the bovine male? What are the functions of each?
4. Name the main reproductive organs of the bovine female. What are the foremost functions of each?
5. Why is a cervical plug formed during pregnancy in the cow? Of what significance is this in rebreeding cows artificially?
6. What are caruncles? Cotyledons?
7. What hormones are secreted by the ovarian follicle and corpus luteum? What are the roles of these hormones in bovine reproduction?
8. What is the normal length of estrous cycles in cows?
9. Where is the egg fertilized? Does this have any importance to the site of semen deposition in AI?
10. Discuss briefly the hormonal control of parturition.
11. How soon after parturition should dairy cows be mated? What is the purpose for this time interval?
12. What are the typical behavioral traits of cows in heat (estrus)?
13. Why is heat detection so important to dairy herd owners/managers?
14. Identify and describe three detection methods of estrus.
15. When do cows normally ovulate? How does this relate to optimal times of insemination?
16. What is the ideal calving interval for dairy cows?
17. Discuss briefly the role of (a) the male, (b) the female, (c) the manager, and (d) the AI technician in the reproductive efficiency of dairy cattle.
18. What are some methods of evaluating reproductive efficiency in dairy cows?
19. How may AI minimize spread of venereal diseases among dairy cattle?
20. What contributions can the veterinarian make to reproductive efficiency in dairy cattle?
21. What is the relationship between the following and breeding efficiency: (a) estrous cycle length, (b) retained placentas, (c) animal health and nutrition, (d) level of milk production, (e) cow density, and (f) estrus synchronization?
22. What physiological abnormality may be involved with (a) short estrous cycles, (b) anestrus, and (c) nymphomania?
23. How may breeding inefficiency impede genetic progress?
24. What is the primary value of AI in dairy cattle?
25. Identify two common errors of AI technicians when inseminating dairy cattle.

Notes

[1] It has been estimated that if the sperm produced by *one* bull in *one* week could be placed end to end, they would extend from Chicago to Honolulu—a distance of approximately 4,300 miles.

[2] The current recommendation of some veterinarians is to allow these membranes to be sloughed naturally; antibiotics may be administered to minimize infections.

[3] A Holstein calf weighing 176 lb at birth was reported in *Hoard's Dairyman* (116:799) in 1971.

[4] Estrogen secretion increases with advancing pregnancy and, when progesterone secretion decreases, dominates the uterus.

[5] High **ambient** temperatures, as found in the tropics and subtropics, not only shorten the average length of estrus, but also have a depressing effect on the expression (apparent intensity) of estrus in cattle.

[6] This assumes that conditions are provided that encourage the mounting behavior. Slick concrete lots in which cows frequently fall when attempting to mount another cow discourage this behavior and make estrus detection more difficult.

[7] Vasectomized bulls need not receive exogenous testosterone since removing the section of the vas deferens for the vasectomy does not interfere with the cells that secrete testosterone.

[8] The two main reasons dairy bulls leave AI service are low fertility and **crampiness**.

[9] Uterine infections delay conception and often accompany retained placentas.

[10] In a Louisiana study, 61% of animals gaining weight during the month of service had first-service conceptions and 1.65 services per conception compared with 41% and 2.45, respectively, for animals losing weight during the month of conception. This suggests another possible reason why very high-producing cows have lower conception rates (i.e., because of the enormous caloric input into milk, they are frequently in a negative caloric balance and are thereby losing weight).

[11] This presents no problem in natural insemination since the bull's penis does not penetrate the cervix.

[12] Research indicates that approximately 90% of cows considered anestrous have unobserved (subestrual) heats.

[13] Many cows diagnosed by veterinarians as having cystic ovaries exhibit a laxness or sinking of the muscles around the tailhead, producing a noticeable depression.

[14] Fertilized eggs can be flushed from the donor's uterus and deposited into estrous-synchronized recipient cows.

References

American Breeders Service, Inc. 1972. A.I. Management Manual. DeForest, WI: Author.

Campbell, J. R., M. D. Kenealy, and K. L. Campbell. 2010. *Animal Sciences: The Biology, Care, and Production of Domestic Animals.* Long Grove, IL: Waveland Press.

Gifford, J. A. H., and C. A. Gifford. 2013. Role of reproductive biotechnologies in enhancing food security and sustainability. *Animal Frontiers* 3(3):14–19.

Pelissier, C. L. 1971, June 22. Herd Breeding Problems and Their Consequences. Presented at the 66th Annual Meeting of the American Dairy Science Association, Michigan State University, East Lansing.

Schillo, K. K. 2011. "Estrus Cycles: Puberty." In J. W. Fuquay, P. F. Fox, and P. L. H. McSweeney (eds.), *Encyclopedia of Dairy Sciences* (2nd ed., pp. 421–427). San Diego, CA: Academic Press.

Wright, F. A., and E. T. Oleskie 1962. *Breeding Efficiency in Your Dairy Herd* (Leaflet 321). New Jersey Agricultural Experiment Station.

Websites

American Dairy Science Association (http://www.adsa.org/)

Dairy Cattle Reproduction Council (http://www.dcrcouncil.org)

Iowa State University, Office of Biotechnology (http://www.biotech.iastate.edu)

Michigan State University, Laboratory of Reproductive Physiology and Reproductive Management of Dairy Cattle (http://dairycattlereproduction.com)

University of Arizona/New Mexico State University, Cooperative Extension (http://cals.arizona.edu/extension/dairy/az_nm_newsletter/2007/may.pdf)

University of Minnesota, Reproductive Biotechnology Center (http://ncroc.cfans.umn.edu/Research/AnimalScience/ReproductiveBiotechnologyCenter)

University of Wisconsin, Department of Animal Sciences (http://www.ansci.wisc.edu/jjp1/ansci_repro/lab/cow_project/Dairy_Protocols2011.pdf

West Virginia University, Interdisciplinary Program of Reproductive Physiology (http://reprophys.wvu.edu)

10

Physiology of Lactation

Normally each year the mammary glands of dairy cows come to life functionally and, in many respects, structurally, then decline.

Samuel Brody (1890–1956)

10.1 Introduction

The primary usefulness of a dairy cow depends upon her mammary glands. Although dairy cows are of similar size and external conformation, they vary greatly in ability to secrete milk. This is explained by variations in inheritance of genes that enable them to produce hormones that stimulate growth of mammary glands, secrete milk, and release it from the alveoli (milk letdown). The amount of milk a cow produces is determined, fundamentally, by her genetic makeup, but is greatly influenced by environmental factors (e.g., feeding, milking practices, and weather), which are discussed in this and in the subsequent two chapters.

In the cow there are four mammary glands which, collectively, are called the **udder**. The right and left halves of the udder are separated by thick connective tissue membranes, whereas the fore and rear quarters are separated by thin membranes. Milk of each mammary gland can be removed only from the teat of that gland.

Structurally, a mammary gland may be compared with a bunch of grapes, the alveoli representing grapes and the ductules branching stems. A group of alveoli is called a lobule. Several lobuli are called a lobe (figure 10.1). The lobes constitute a quarter and, finally, four quarters comprise the udder of a cow. These various divisions of an udder are bound and separated by connective tissues and muscles. Mammary glands are supplied with a rich bed of blood vessels and nerves.

Size of mammary glands is determined by the total tissue present (i.e., connective, fatty, and secretory tissues). Quality is dependent upon the amount of secretory tissue in relation to connective and fatty tissues. A high-quality mammary gland has a high proportion of secretory (glandular) tissue. Moreover, it has strong ligaments and attachments. A young cow may have an old udder, appearance-wise, if it is **pendulous** (cf. chapter 5).

Mammary glands of high-producing cows are abnormally developed, the result of decades of selective breeding. Through commercial dairying, they are subjected to great physical strain, with little opportunity for rest or repair. The entire dairy industry depends on exploitation of the evolutionary development of mammary glands.

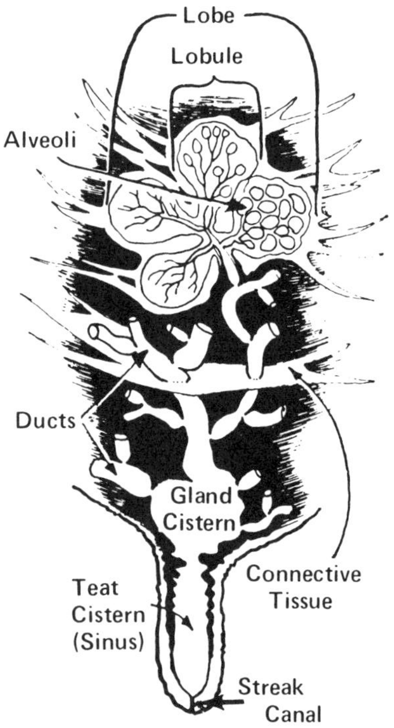

Figure 10.1 Each mammary gland in the cow is comprised of many lobes, each of which consists of smaller lobules, which in turn contain many alveoli. The teat is connected to millions of alveoli by a duct system supported throughout by connective tissue (adapted from Turner, 1952).

10.2 Growth and Development of Mammary Glands

Development of the mammary gland, the apparatus for nourishing the young, is synchronized with that of the sex organs, the apparatus for producing the young.
Samuel Brody (1890–1956)

During the female's life, structural and functional development of the mammary gland (lactogenesis) occurs in multiple cycles. The mammary gland is one of few tissues that undergo cycles of development, functional differentiation, and regression. Changes in cellular metabolism and physiology result in cell growth and death. Molecular events required for mammary development are under the control of genes that are encoded by DNA molecules. Hormonal and growth factor stimulation of mammary development during **gestation** facilitates the generation of abundant glandular alveoli. Subsequent differentiation of epithelial cells of these alveoli leads to the onset of milk synthesis and secretion in conjunction with **parturition** (Akers and Capuco, 2011).

Initial development of mammary glands occurs **in utero**. A calf's udder at birth consists of teats, teat and gland cisterns, and a small amount of connective and fatty tissues. As the young calf grows, the udder increases in size, proportional with other body parts, with little growth of ducts. At puberty[1] the hormone estrogen, which is produced by ovarian follicles, stimulates growth and development of the duct system with each recurring estrous period. To obtain maximal development of udder tissues, i.e., adequate expression of bovine mammary RNA, there must be adequate nutrition prior to the calf reaching puberty.

If estrogen secretion is low, there is an adverse effect upon the growth of mammary glands as well as secretion of **prolactin**—if the **endocrine system** will not support intense secretion of milk by large mammary glands, there is no reason to grow large glands.

During pregnancy, growth of the duct system is continuous rather than cyclic. Soon the corpus luteum secretes progesterone, which stimulates growth and development of the fine ducts and of the lobule-alveolar system. By about the seventh month of pregnancy in first-calf heifers, mammary glands are developed sufficiently to secrete milk. Limited mammary gland growth occurs the last few weeks of pregnancy. Instead, the alveoli begin secreting colostrum, and lymphatic fluids accumulate.

Occasionally, a growth develops that prohibits removal of milk from a quarter, which is then called a **blind quarter**. Such a gland will regress and soon become nonfunctional. It is interesting, however, that in such a three-quartered cow additional development of the adjacent mammary glands and use of milk precursors and hormones occurs in the three functioning glands that would have originally been used in the blind quarter. Thus, most three-quartered cows yield less than they would if the fourth gland were functional, but more than three-fourths of normal. This same phenomenon may occur in animals in which one or more mammary glands are rendered nonfunctional by mastitis.

About 2% of dairy heifers have more than four teats. The extra, or supernumerary, teats are associated with small glands and are usually located behind the rear teats. They should be surgically removed prior to puberty.

10.3 Circulatory Aspects of Milk Secretion

Mammary glands have a rich blood supply. And indeed they should, since an estimated 400 volumes of blood flow through the cow's udder for each volume of milk secreted. Two external pudic arteries—one serving each half of the udder—provide a source of milk precursors and oxygen-rich blood. These arteries enter the center of the udder from the top, via the inguinal canal. They branch into two mammary arteries—the cranial or anterior supplying the fore, and the caudal or posterior supplying the rear quarters—then they rebranch extensively throughout the mammary glands.

After arterial blood has replenished the oxygen and nutrient needs of udder tissue, it is collected by an intricate capillary-venous blood system and exits in three ways: (1) by the external pudic veins (these parallel the external pudic arteries), (2) by the subcutaneous abdominal veins, commonly called milk veins, that run forward from the fore glands beneath the skin and enter the thoracic cavity through openings called **milk wells**, and (3) by the perineal veins, which pass backward and upward from the udder through the pelvic arch. The venous blood system of the two udder halves is joined by cross connecting branches. By means of this extensive **vascular** network, blood flows from the udder irrespective of the position in which a cow lies.

Missouri studies showed that only 1.0% of the total blood passes through the mammary glands of nonlactating **(dry)** goats, whereas 3.7% circulates through a lactating udder. Each heartbeat of a cow pumps approximately one liter of blood. The heart rate of a lactating cow is about 50 to 60 beats/min and a complete cycle of blood occurs approximately every 50 seconds. Other Missouri studies show that blood volumes of lactating and nonlactating cows are approximately 8.1 and 6.4% of body weight, respectively. **Blood plasma** is also higher in lactating cows than in nonlactating cows (4.9 and 3.8% of body weight, respectively). This is important since blood constituents that contribute to milk secretion are found in blood plasma.

10.3.1 Lymphatic System

Lymphatic vessels convey lymph upward and toward the rear of each half of the udder (a one-way flow) where they pass through the supramammary lymph glands. Lymph from various body tissues eventually returns to the jugular vein where it is mixed with blood. Lymph removes bacteria and, in certain respects, serves as a backup defensive mechanism to the blood system. (Lymph resembles blood plasma in many respects although it has no red blood cells.) The lymphatic vascular system includes lymph glands (nodes), which add lymphocytes or white blood cells to lymph. The system is especially busy a few days before and after parturition when lymph commonly accumulates in the udder, and even forward of it (figure 10.2 on the next page). This condition of udder congestion and swelling is called **edema**.

10.4 How Milk Is Made

Two important processes account for the secretion or making of milk (galactopoiesis) by mammary glands: (1) filtration of certain milk constituents from the bloodstream and (2) the synthesis of other milk components by cellular metabolism within the alveoli.[2] Considerable research effort has been expended to define these processes (Mather, 2011). The use of radioactive isotopes has been especially helpful in identifying milk precursors. Using this technique, a substance believed to be a milk precursor is labeled (tagged) with a radioactive isotope and injected into an experimental animal. If the compound appears in milk, it may be assumed to be a milk precursor.

10.4.1 Protein

The mammary gland serves as a **bioreactor**. Milk proteins result from both filtration and synthesis. Since the **casein**, lactalbumin, and lactoglobulin found in milk are not present in blood, they must be synthesized from amino acid precursors derived from the blood. This occurs within the secretory epithelium in the alveoli. These proteins represent approximately 94% of the protein nitrogen in cow's milk. Arteriovenous difference studies have shown the remarkable capacity of the mammary gland to extract amino acids from the blood. Hormones essential for protein production are prolactin, insulin, and cortisol (hydrocortisone).

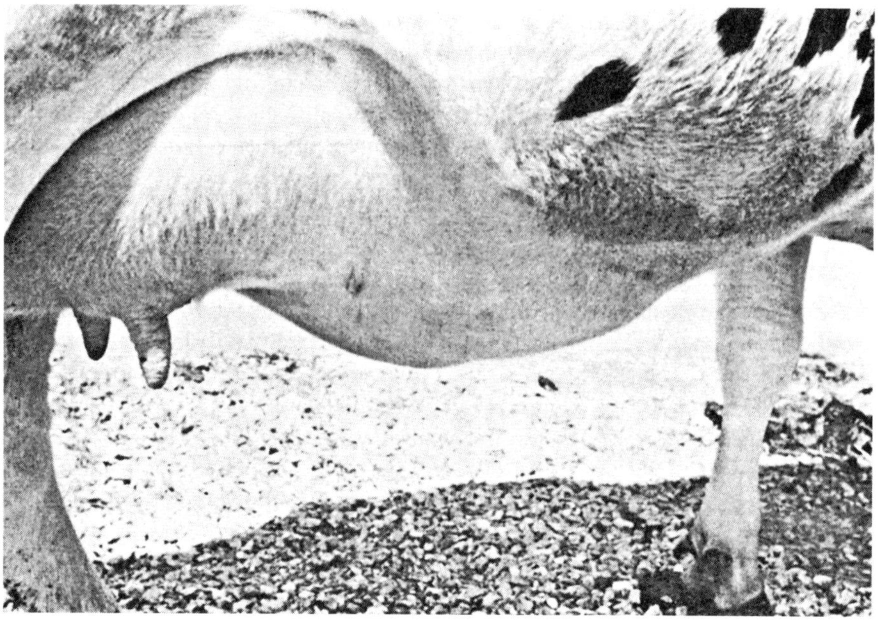

Figure 10.2 An udder showing edema at parturition. Note the result of accumulation of lymph (apparent swelling) forward of the udder (courtesy of Missouri Agricultural Experiment Station).

Immunoglobulins and serum albumin appear to be identical in blood and milk and, therefore, apparently diffuse into milk, unchanged in form from what is found in blood. Unlike humans, the bovine is unable to transfer immunoglobulins to the fetus *in utero*; therefore, calves must receive colostrum soon after birth.

Genetic modification of mammary gland cells has been shown to produce new functional proteins, some for use as pharmaceuticals, including anticlotting agents and antibodies; in New Zealand, genetic modification of these cells has increased milk's casein content. Reduction of the **lactose** content of milk, and consequently the water content, would be achieved by genetically removing the cow's α-lactalbumin gene. Genetically engineered goats have been used to increase milk's lysozyme concentration, thus decreasing renneting time and increasing curd strength of cheese (cf. chapter 30). Increased lysozyme in milk may have other benefits, including antimicrobial action in both the mammary gland (mastitis) and in raw milk, by increasing its **shelf life**. This arena of research presents numerous important opportunities for dairy scientists who have extensive backgrounds in biology with special emphases on genetics, enzymology, and biochemistry.

Milk proteins synthesized in the rough endoplasmic reticulum are processed through the Golgi complex, packaged into secretory vesicles, and secreted from the apical surface by exocytosis (meaning out of the cell).

10.4.2 Lactose

The principal carbohydrate of milk is lactose, which consists of one molecule of glucose and one of galactose. Two molecules of glucose enter the secretory cell for each molecule of lactose formed. One of the glucose units is converted to galactose. Condensation of the second glucose molecule with galactose is catalyzed by an enzyme, lactose synthetase. This enzyme is composed of two protein subunits, one of which is α-lactalbumin, a protein component of milk, and the other is galactosyltransferase.

Glucose is a normal blood component, lactose is not. Determinations of arteriovenous differences have shown glucose to be taken up by mammary tissues (arterial blood loses about 25% of its glucose content while passing through a lactating mammary gland). There are at least two active glucose transport systems in mammary gland tissues. About 70 to 80% of the carbon in lactose is derived from plasma glucose. Lactose may be derived also from short-chain fatty acids.

Since lactose creates a **hypertonic** solution within the secreting cells (Golgi), water is drawn in. Thus, lactose concentration in large part determines the volume of milk secreted and the amount of water contained by milk (dissolved salts influence this also).

In high-producing cows, especially in early lactation, the significant increase in mammary glucose uptake from the blood to be used in synthesis of lactose may cause the animal to become **hypoglycemic** and develop ketosis (cf. chapter 13).

10.4.3 Fat

More than half of bovine **milkfat** is synthesized in the mammary gland. Triacylglycerol molecules, composed of three fatty acids attached to a backbone of glycerol, are synthesized in the rough endoplasmic reticulum (figure 10.3) by membrane-bound fatty acid acyltransferases. These molecules aggregate forming microlipid droplets (< 0.5 μm diameter) and are released into the cytoplasm of the cell where they are coated with proteins and polar lipids from the reticular membrane. These tiny droplets tend to fuse, forming larger droplets as they are transported to the apex of the cell (pathway A in figure

10.3). A budding process results in expulsion of droplets from the cell. During the expulsion they are coated with more proteins and with phospholipids, forming an outer membrane (fat globule membrane, or FGM). On a weight basis milkfat globules are approximately 4 μm in diameter; on a total numbers basis more than 80% have a diameter of < 1 μm, although some can exceed 10 μm.

Synthesis of milkfat requires a major supply of energy to mammary glands; therefore, adequate nutrition is vital to production of both fat and the other components of milkfat globules. In fact, inadequate nutrition can cause the lowfat milk syndrome, commonly known as milkfat depression (MFD). However, low concentrations of milkfat can be caused also by feeding diets supplemented with fish or plant oils, diets high in **feed grains** but low in fiber, or fiber that has been ground and/or pelleted. In cases of MFD the concentration of all fatty acids is decreased, but the shortage is greatest among those of short- and medium-chain length since they are synthesized in the mammary gland. Long-chain fatty acids are largely absorbed from the circulatory system. For MFD to materialize, the diet must contain a significant amount of unsaturated fatty acids, which affect microbial metabolism in the rumen.

Increases in content of trans 18:1 fatty acids has been associated with MFD. Additionally, a potent inhibitor of milkfat synthesis is trans-10, cis-12 conjugated linoleic acid (CLA) (cf. chapter 21). Fatty acids of trans configuration are formed as intermediates during rumenal biohydrogenation of unsaturated fatty acids, especially linoleic acid. Since numerous intermediates are formed, including more than 20 different isomers of CLA, other inhibitors may be present. Since experiments in animal models have shown cis-9, trans-11 CLA (rumenic acid [RA]) to be both anticarcinogenic and antiatherogenic, the presence of this isomer in milk provides potential for application in disease prevention (Bauman et al., 2006). The typical RA content of milk can be increased significantly by manipulating a cow's diet, including grazing on new growth pasture and addition of vegetable and fish oils to dairy rations. However, RA content in milk of individual cows can vary two- to threefold, even when individuals receive the same ration.

Two additional isomers of CLA (trans-9, cis-11 and cis-10, trans-12) can depress milkfat synthesis. Note, however, that cis-9, trans-11, the predominant isomer in milkfat (75 to 80%), does not depress synthesis of milkfat. Although the double bonds in these isomers are in the same locations, their orientation is reversed (trans/cis versus cis/trans).

In ruminants, acetate is the principal precursor of fatty acids of carbon chain lengths up to 16, that of palmitic acid. Cows fed high-grain/low-forage rations commonly secrete lowfat milk because this diet reduces production of acetate in the rumen. Additionally, cows commonly secrete milk low in fat during hot weather. Missouri studies showed that ruminal acetate production is reduced in hot weather. Glucose of blood is utilized in synthesis of the glycerol portion of fat molecules.

It is interesting that human milkfat contains a higher concentration of unsaturated fatty acids than does cows'

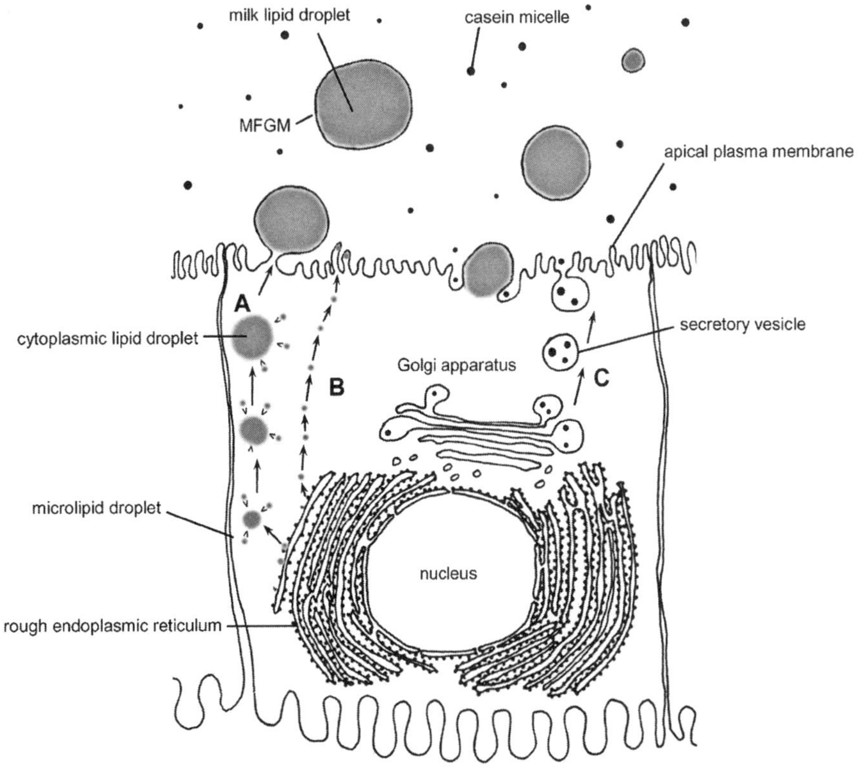

Figure 10.3 Major secretory pathways in mammary epithelial cells during lactation. A = cytoplasmic lipid droplet pathway; B = microlipid droplet pathway; C = secretory pathway for skim milk components (reprinted from Bauman et al. 2006. Major advances associated with the biosynthesis of milk. *Journal of Dairy Science* 89:1235–1243, with permission from Elsevier).

milk. Australian and USDA researchers increased levels of polyunsaturated fatty acids of milk by feeding cows a diet containing a formaldehyde-treated safflower oil–casein particle. This treatment protected the linoleic acid of safflower oil from usual biohydrogenation in the rumen. The linoleic acid content of milk increased from 3 to 35% of the total fatty acids. Total milk production and cholesterol content were unaffected.

10.4.4 Minerals

The minerals of milk are derived from the blood and reach milk through filtration. It is interesting how mammary glands can concentrate certain minerals. For example, milk contains at least 10 times more calcium and phosphorus than blood. Mastitis can change the permeability of the cells that make milk and thereby permit certain minerals—chlorides, for example—to pass into milk in above normal concentrations (cf. chapter 14). For any decrease in lactose content there is a compensatory rise in the concentrations of sodium and potassium salts. This explains why mastitic milk, with its lower concentration of lactose, may taste salty.

10.4.5 Vitamins

Vitamins filter directly from blood into milk. Quantities of B vitamins synthesized by ruminal microorganisms are rather constant in milk, whereas concentrations of fat soluble vitamins, especially vitamins A and D, are dependent upon quantities in the ration and body stores (cf. chapter 21).

10.4.6 Water

The greatest portion of milk is water, averaging 87% by weight. Water is filtered from blood and, of course, varies inversely in content with milk solids.

10.5 How Milk Is Discharged (Excreted)

Milk secretion involves the synthesis of milk during the interval between milkings. Secreting cells lengthen as they synthesize milk. After filling with milk, the cells rupture and pour their contents into the lumen of an alveolus (figure 10.4), causing a gradual expansion or ballooning out of the alveolus. Gradually mammary glands become engorged with milk, much as a sponge fills with water. As milk continues to be secreted, it slowly flows out through ducts and gradually fills the entire duct system and the gland and teat cisterns. Cows have a well-developed cistern and duct system in which large amounts of milk may be stored. This storage system is lined by elastic tissue elements, which permit great enlargement and expansion during the interval between milkings, and an elastic recoil upon removal of milk. Although this part of the udder stores considerable amounts of milk, still a large portion of the total milk present at milking is in smaller storage

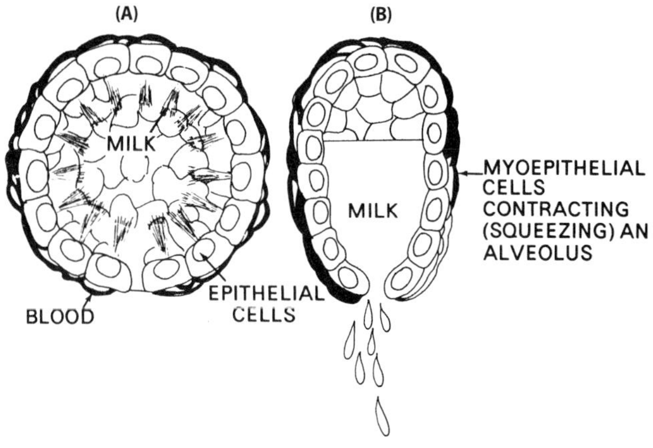

Figure 10.4 This diagram, showing cross sections of an alveolus, depicts (A) milk being secreted into the lumen and (B) the contracting of myoepithelial cells, which results in milk letdown. This process causes milk to be squeezed into the duct system and finally into gland and teat cisterns.

spaces—capillary ducts, lumina of the alveoli, and epithelial secreting cells.

Milk secretion is rapid following milking. As secretion progresses and milk is discharged, there is a gradual rise in intramammary pressure. This tends to slow further milk secretion and discharge. Research indicates that milk secretion each hour following milking is approximately 90 to 95% of that of the preceding hour. In high-producing cows, milk secretion will essentially stop within 18 to 24 hours postmilking unless milk is withdrawn.

When intra-alveolar pressure reaches 30 to 40 mm Hg (4.7 kPa), about that of capillary pressure and about one-fourth that of systemic blood pressure, milk secretion is appreciably reduced, if not stopped entirely, and reabsorption begins. This is possible because the increasing pressure of milk on blood capillaries reduces blood flow through the udder. It also makes it difficult for secreting cells to rupture and discharge their contents of milk into the lumina. Increasing intramammary pressure also changes the mode of milk discharge from the cell. With increased back pressure from accumulated milk stored in the lumen, epithelial cells are unable to rupture. The result is that milk is discharged from secreting cells only as it is possible for it to pass through the semipermeable cell membrane. A putative feedback inhibitory protein that decreases synthesis of lactose has been isolated from goats milked only once daily.

Milkfat is in a suspended emulsion in milk serum and cannot diffuse from the cell. Therefore, fat accumulates within secreting cells, and milk low in fat is discharged. Thus, milk secreted when intramammary pressure is low is comparably high in milkfat, but as the interval between milking lengthens, this milk is diluted by the discharge of lowfat milk. This explains why total yield of milkfat is reduced as the interval between milking lengthens.

If a cow is milked first at 10- and then at 14-hour intervals, it will be observed that more total milk and fat is obtained after 14 hours; however, if one divides this higher total by 14, the milk yield per hour, as well as percent of fat, will be less than the yield per hour obtained after a 10-hour period. This also explains why milking at shorter intervals, thus relieving udder pressure more frequently, permits the production of milk to continue at a higher rate and the consequent realization of greater daily production. Moreover, we can now see why the last milk evacuated at milking time (strippings) is higher in fat. Cells rich in milkfat discharge their accumulation of fat globules near the end of milking when intramammary pressures have been substantially reduced.

On the average, each fore quarter yields about 20% and each rear quarter about 30% of the total milk secreted. Each right and left half of the udder yields, on the average, approximately 50%. Of course, there is considerable cow-to-cow variation.

10.6 Milk Letdown

Contented cows will not "hold up" their milk but rather upon proper stimulation will "force" it down so that it may be removed easily by the milker or milking machine.

Charles W. Turner (1897–1975)

Milk stored in the cistern system and larger ducts can be removed from a teat by a cannula. However, milk stored in the smaller storage spaces (fine ducts, lumina of alveoli, and epithelial cells) requires active participation of the neuroendocrine system for its removal. This latter process is called *letdown*. What causes milk letdown?

The warm, moist mouth of a hungry calf stimulates milk letdown. Letdown is also stimulated by washing the mammary glands, hand milking, high internal udder pressure, and sounds commonly associated with milking. Stimulation of the reproductive tract will stimulate letdown also. Milk letdown is a conditioned reflex; it is involuntary. A conditioned reflex is caused by what the cow associates with milking.

Properly stimulated, a cow's nervous system conveys a message to her brain that it is "milking time." This causes a hormone, oxytocin, to be released from the posterior pituitary gland into the bloodstream, which conveys it to the udder. Vasopressin is also released from the posterior pituitary and may play a role in milk letdown. The effect of oxytocin in mammary glands is to cause the myoepithelial cells that surround the alveoli to contract, squeezing the alveoli, causing milk letdown (figure 10.5). Contraction of myoepithelial cells is similar to the action of a hand in squeezing the bulb of a syringe. Oxytocin, therefore, causes an increase in intramammary pressure and also an increase in blood flow to the udder. In the average cow each teat has a storage capacity of 40 to 50 mL of milk, and each gland cistern has a storage capacity of 400 to 500 mL.

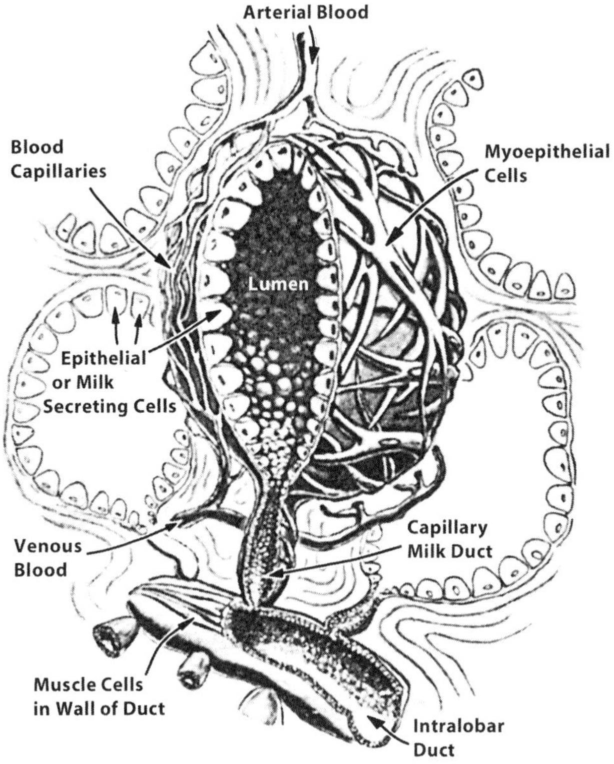

Figure 10.5 An alveolus in which milk is secreted. There are literally billions of these microscopic-sized milk-making units in a single, well-developed bovine mammary gland. Each alveolus is lined with thousands of epithelial or milk-secreting cells. Milk is stored in the lumen. Note the blood vessels (at left) and myoepithelial cells (at right) surrounding this alveolus. The myoepithelial cells surrounding each alveolus contract, more or less at once, and squeeze the alveolus causing milk to be ejected with letdown (adapted from Turner, 1952).

10.6.1 Effect of Adrenalin on Milking

If cows at milking time become excited for any reason, there is great difficulty in obtaining the cow's cooperation in this important matter.

Charles W. Turner (1897–1975)

A cow's response to excitement, fright, pain, or mistreatment is release of a hormone called *adrenalin,* which acts to inhibit the letdown effects of oxytocin. How is this possible? Although the area is not clearly defined, research supports two explanations: (1) Adrenalin leads to a marked reduction in mammary blood flow (although it apparently has little effect on intramammary pressure). By restricting mammary blood flow, oxytocin delivery to the myoepithelial cells may be impaired. (2) Adrenalin causes contraction of the blood vessels of the mammary gland, thus preventing oxytocin from reaching the myoepithelial cells in sufficient concentration to cause contractions essential to milk letdown.

A hypothesis holds that there are two types of receptors in myoepithelial cells: *alpha receptors*, which respond to adrenalin, and *beta receptors*, which are affected by oxytocin. Apparently, myoepithelial cells respond preferentially to adrenalin. Thus, if a cow is frightened adrenalin is released, and upon reaching the myoepithelial cells it causes them to relax, thereby preventing milk letdown; also if the cells had been stimulated by oxytocin prior to the cow being frightened, it inhibits further letdown. Adrenalin has been labeled appropriately the "fight or flight" hormone. Fortunately, adrenalin has a short half-life ($t\frac{1}{2}$), although it is longer than that of oxytocin, so that when its effective blood concentration has dissipated, usually in 20 to 30 minutes, the cow will again respond to stimuli of milk letdown.[3]

10.7 Residual (Complementary) Milk

The first report of residual milk was made by the Missouri Agricultural Experiment Station. This milk remaining in the udder at the end of milking commonly represents 15 to 20% of the total milk yield and 25 to 30% of the fat. By injecting 10 IU of oxytocin **intravenously** or 20 IU **subcutaneously** (which is less disturbing to a cow) and re-milking, one can obtain residual milk. This technique is useful in the therapy of mastitis to gain maximum effectiveness from **antibiotics**, which will thereby be diluted less by milk. Relative amounts of residual milk are determined using the formula:

$$\text{Residual milk (\%)} = \frac{\text{Residual milk}}{\text{Machine milk} + \text{Residual milk}} \times 100$$

Factors influencing the amount of residual milk include (1) stage of lactation (there is less residual milk with advancing lactation), (2) age (younger cows have less residual milk than older ones), (3) milk yield (high-producing cows have proportionately less residual milk than low-producing ones), and (4) letdown (anything that inhibits complete letdown, such as excitement, fright, and pain). University of Minnesota studies showed that faster-milking cows have less residual milk, are more persistent, and have higher milk yields per lactation.

How can the amount of residual milk be reduced? The most practical way may be through selection and breeding of cows that milk out rapidly and those with longer periods of intramammary contraction, which is made possible by discharge of large quantities of oxytocin, a longer $t\frac{1}{2}$, or both. The latter could be measured by tagging oxytocin with radioisotopes and measuring its $t\frac{1}{2}$ time. Cows with a short contractile period, whether because they secrete a small amount of oxytocin and/or have a fast $t\frac{1}{2}$, have higher percentages of residual milk.

10.8 Hormonal Control of Lactation

Hormones are chemical mediators released from endocrine (*endo* = inside, *crine* = to secrete) tissue into the bloodstream where they travel to target tissue and generate a response. Hormones regulate various functions including metabolism, growth and development, tissue function, and mood.

The endocrine system plays a central role in all aspects of mammary development (mammogenesis), onset of lactation (lactogenesis), and maintenance of lactation (galactopoiesis). A myriad of growth factors target both mammary epithelial and **stromal cells** of the udder. Complex networks of extracellular signals interact in controlling the size of the mature mammary gland and the capacity of the alveolar secretory cells to synthesize and secrete milk. The inherited capacity of a cow for high milk production depends largely upon her ability to secrete optimal amounts of numerous hormones and cofactors that influence productivity. Hormones, which have been called the "messenger girls," not only initiate and sustain milk secretion, but also greatly affect its intensity.

Steroid hormones, including estrogens and progesterone, appear in milk in both free and conjugated forms (those attached to other molecules). Only free forms can bind to receptors and elicit biological responses. Although epidemiological studies have suggested that steroid hormones may have adverse effects on human health, no cause and effect of these in milk has been proven scientifically.

If an animal's secretion rate of a single hormone is deficient, then her milk-secreting capacity will be limited to the level permitted by that secretion rate. Hence, some potentially high-producing cows may be low producers due to a single hormonal deficiency, whereas others may be low producers because of multiple deficiencies.

10.8.1 Prolactin

Increasing amounts of estrogen are secreted in late pregnancy by the placenta. One effect of estrogen is to stimulate secretion of prolactin, the hormone that initiates and maintains milk secretion. But what is responsible for continued secretion of prolactin after parturition? Following parturition it is the milking or nursing stimulus that maintains prolactin secretion. Prolactin is secreted and stored in the anterior pituitary gland. Milking stimulates the nervous system, causing discharge of prolactin into the bloodstream, which conveys it to the mammary glands. There prolactin stimulates milk secretion for the subsequent milking.

Prolactin stimulates increased enzymatic activity of epithelial cells. This increased activity enables the cells that secrete milk to convert various constituents of blood into milk's components. Could the effect of milking on release of prolactin explain, in part, why cows yield more milk when nursed by calves than when milked by machines? Calves nurse more frequently than the usual twice daily mechanical milking, and, thus, likely initiate release of greater amounts of prolactin.

10.8.2 Thyroxine

This hormone is secreted by the thyroid glands and is especially important in lactation because it increases appetite, heart rate, flow of blood to mammary glands, and rate of milk secretion. It is a major regulator of **basal metabolic rate (BMR)**,[4] which is the speed with which oxygen is used by cells and with which carbon dioxide is produced. Thyroxine has been compared with the draft in a furnace. When the draft is opened the fire burns more rapidly and more heat is generated. Similarly, when more thyroxine is available, body cells consume more energy and function at a higher level. Moreover, the higher metabolic plane, when coupled with optimal levels of other hormones important to milk secretion, enhances secretion of milk. For a detailed discussion, see Capuco and Akers (2011).

Missouri studies indicated that average daily secretion of thyroxine by beef cows was only one-half as much as by dairy cows. Also, dairy cows secreted only one-third as much thyroxine in summer as in winter. This partially explains why milk secretion is relatively slow in hot weather. A thyroidectomy procedure (removal of the thyroid glands) on a cow causes a marked decrease in milk secretion (up to 75%), which can be restored by administering thyroxine or desiccated thyroid. Underfeeding also depresses thyroxine secretion rate.

10.8.3 Growth Hormone (Somatotropin)

Regulatory peptides are found in milk with the highest concentration in the first colostrums. These include IgG, growth hormone (GH), prolactin, IGF-I, insulin, and glucagon. IGF survives stomach acid and passes into the small intestine, especially in **neonates**. The effects of GH and its receptors (GHRs) on milk production has been the subject of much research, resulting in development and use of recombinant somatotrophin (rbST), or Sometribove® (trade name Posilac). Although somatotrophin is primarily involved in the growth rate of young animals, it also has a marked effect on milk secretion. This secretion of the anterior pituitary influences the precursors of milk, increasing the availability of blood glucose, amino acids, and fatty acids in cells of the mammary gland needed to make milk.

Administration of bovine somatotropin (bST) to lactating cows increases milk yield within a day and reaches its peak within a week (Akers, 2006). Elevated milk yield is maintained as long as treatment is continued but drops quickly to control levels when bST is discontinued. No chemical test is available to distinguish bST in milks from treated and untreated cows. Currently approved for injection in the United States is n-methionyl-bST, a slow release formulation.

Barbano et al. (1992) observed 9.6% higher milk yield and slightly higher concentrations of protein in milks from bST-treated (prolonged-release, biweekly administration) cows compared to nontreated counterparts. No differences were observed in concentrations of other major components of the milks.

The US Food and Drug Administration (2009) reported that long-term studies were unnecessary for assessing the safety of rbGH/rbST based on studies which showed that: bovine growth hormone (bGH) is biologically inactive in humans even if injected, rbGH is orally inactive, and bGH and rbGH are biologically indistinguishable.

10.8.4 Insulin-Like Growth Factors (IGF)

Molecules similar to insulin, produced primarily in the liver, are important in many aspects of mammary development and function. The major function of IGF-I is mammogenesis. Additional forms include IGF-II, six IGF-I binding proteins and nine related proteins. The binding proteins have several functions related to IGF-I: prolonging its half-life, transporting it from circulation, and localizing it on functional target cells.

10.8.5 Epidermal Growth Factors (EGF)

The general mechanisms of EGF include ligand binding, receptor activation, and molecular phosphorylation. This family is composed of at least 10 members. They are important in normal mammary gland development. Production is likely induced by estrogen. Receptors for them are found in stromal cells of rapidly growing mammary ducts. Whether their roles are direct or permissive remains in question (Akers, 2006).

10.8.6 Parathyroid Hormone

This hormone controls calcium and phosphorus concentrations in blood. Since these minerals are abundant in milk, this hormone is vital in milk secretion. Dihydrotachysterol (AT10) has been shown to have physiological properties similar to parathyroid hormone. Feeding this preparation orally in optimal amounts may stimulate increased milk secretion in cows that are deficient in parathyroid hormone.

10.8.7 Adrenal Hormones

These are important for normal body functioning, but high levels depress milk secretion. Secretion of adrenal hormones increases in times of stress. These hormones may be administered in the therapy of certain diseases and/or metabolic disorders, for example, in the treatment of **ketosis** (cf. chapter 13).

10.8.8 Other Related Growth Factors

Modern biochemical methods and genetic tools have been used to identify numerous hormonal factors. Fibroblast growth factors (FGF) constitute a family of 20 small peptides that share a high homologous core of 140 amino acids and strong affinity for heparin and heparin-like glycosaminoglycans of the extracellular matrix (ECM). Some play a significant role in regulating epithelial cell proliferation, particularly of mammary ducts. The transforming growth factor-β (TGF-β) group is comprised of at least five proteins that have functions ranging from modification of the ECM to stimulation of cell proliferation. Most of the mammary-associated effects of TGF-β are inhibitory so that they control hyperplasia (increase in cell numbers).

10.8.9 Hormones Found in Milk

Among the components of milk that occur in trace amounts are hormones, growth factors, and **cytokines**. Study of these components has been made possible with the development of sensitive radioimmunoassay and enzyme-linked immunosorbent assays (ELISAs). Some concerns have been raised about their potential effects on human health when bovine milk is consumed. However, no adverse effects have been proven.

Five potential reasons why hormones may occur in milk were listed by Baumrucker and Magliaro-Marcina (2011) in their review of the subject: (1) it is a means of disposal via the mammary gland, (2) hormones serve **autocrine/paracrine** roles in local regulation of mammary gland cells, (3) they affect functions within the gastrointestinal tract of the newborn, (4) they are absorbed into the gastrointestinal tract and participate in extramammary functions, and (5) hormones have no function for appearing in milk other than they were involved in the secretion of the milk.

10.9 Merits and Limitations of Rapid Milking

Large herds, group housing, and assembly-line milking operations necessitate a milking routine that can be standardized and that will result in greater milk flow rates. Fast milking means minimal milking time. As milking time increases, the cost in labor of milking increases proportionally. For example, when average time devoted to a single milking is 3 versus 5 minutes and total labor cost is $20/hour, total cost per cow per year (2X/d, 305D lactation) is $610 versus $1,017.

Milk yield is usually greater with fast milking because maximal evacuation of the udder is realized before the contractile effects of oxytocin are lost. Moreover, fat content of milk is higher since more of the last milk, which is rich in fat, is obtained. If a cow is selected that has equal production in each of her four mammary glands, and they are milked one at a time, both milk and fat yield will be highest in the output of the first gland and lowest in the last gland. In Missouri studies there was a difference of 10 and 40% in milk and fat yields, respectively, in favor of the first gland milked. This emphasizes the importance of removing milk when intramammary pressure is at a maximum and the cow is "forcing down" preformed milk in the udder.

Cows that milk rapidly are more persistent, which means a longer and more productive **lactation period**. Narrow teat orifice is the common cause of slow milking rate. (This is often reflected in a lower incidence of mastitis, however.) Overmilking is less likely to occur in slower milking cows.

10.9.1 Factors Affecting Milking Rates

We define *milking rate* (average flow rate) as the total pounds of milk harvested at a given milking divided by the number of minutes the machine is attached. Average flow rate is approximately 60 to 70% as high as *peak flow,* which is the largest amount of milk harvested during any one minute of milking. Since rate of maximum milk flow is directly related to quantity of milk produced, selection for yield will increase rate of milk flow.

Mammary gland characteristics that favor fast milking include short teats,[5] and teat ends that are cone-shaped or flat. Size of teat opening has been shown experimentally to be an important factor in determining maximum milking rate. Research using radiographic techniques has shown that streak canals with larger diameters tend to milk out faster and are more susceptible to new intramammary infections. If the lower end of the teat canal is wide (more than 1.0 mm in diameter), infection is more probable than if it is narrow (less than 0.5 mm). Of course, cows' teats become more dilated as lactation progresses and with successive lactations, and this affects resistance to intramammary infection.

It is probable that the amount of oxytocin discharged and the resulting intramammary gland pressure generated affect milking rate. Studies in New Zealand utilizing 12 sets of identical bovine twins to determine the effect of premilking stimulus on milk production showed that when cows were properly stimulated before milking began, they yielded about one-third more milk than nonstimulated cows.

Design and operation of the milking machine also affect milking rate. Applicable factors include (1) vacuum level, (2) pulsation rate and ratio (cf. chapter 11), (3) fit of the teat cups, and (4) weight of teat cup assembly.

10.10 Intensity and Persistency of Lactation

Milk yield commonly increases for 30 to 90 days following parturition and gradually decreases throughout the balance of lactation. Plotting monthly milk yields gives a cow's *lactation curve.* An average dairy cow will yield approximately one-half her total 305D production during the first 120 days. Yield of low-producing cows peaks earlier in lactation and declines more slowly than that of high-producing ones. This brings us to persistency of lactation. We define **persistency** as the degree with which the rate of peak milk production is maintained as lactation advances or, inversely, the degree to which production decreases as lactation progresses (based on peak production). The average decline after peak production is reached is approximately linear and is expressed as a percentage. It depends, in large part, upon secretion rates of three hormones—prolactin, thyroxine, and growth hormone. When one or more of these hormones is secreted at less than an optimal rate, the cow's persistency is governed by the secretion rate of the most deficient hormone. Persistency curves are depicted in figure 10.6.

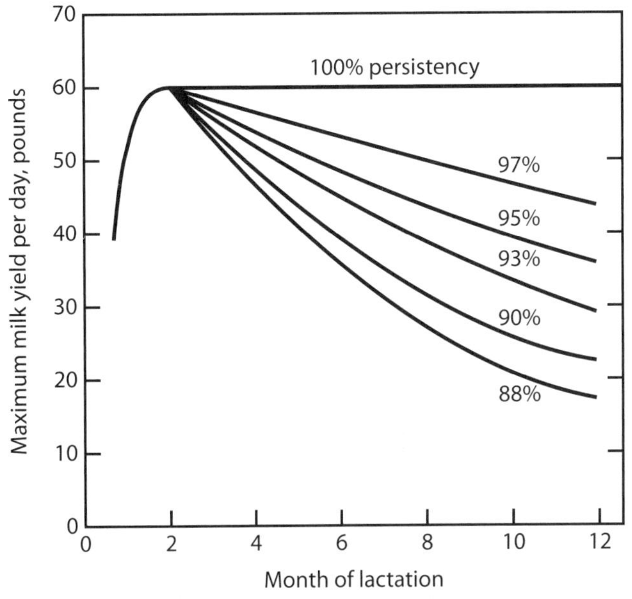

Figure 10.6 Curves showing persistency of milk secretion. Maximum daily production is 60 lb, and persistency varies from 88 to 100%. Note that daily milk production reflects changes in persistency with advancing lactation (Missouri Agricultural Experiment Station Bulletin 836).

High persistency and a flatter than average lactation curve are desirable. This would decrease the need for catabolism of depot (body) fat (occurs when energy expended for body maintenance and milk secretion exceeds caloric intake) early in lactation (cf. chapter 8). Milk producers can select for increased persistency since the heritability estimate of persistency of daily milk yield is 0.21. It follows, of course, that selection for high production is selection for high persistency. Now let us consider factors that affect persistency.

10.10.1 Age

Dairy cows commonly demonstrate greater persistency during their first lactation than in any other. This may be explained, in part, by their having smaller mammary glands as they enter motherhood than at maturity. Additionally, the amount of prolactin secreted may be sufficient to maintain high persistency of smaller glands, but insufficient to sustain high yields of larger mammary glands at maturity.

10.10.2 Season

A seasonal effect on persistency is evidenced by the usual association of high persistency indexes with fall **freshening**. Seasonal changes in milk production are least among two-year-olds and increase with age to maturity. Michigan studies showed that cows fed in **drylot** have greater seasonal variation in milk production than those on pasture. Milk prices are often highest in fall months. This is often a reason to plan pregnancies to provide maximal milk production during these months.

10.10.3 Pregnancy

Missouri studies indicated that pregnant and nonpregnant cows yield approximately the same amount of milk for five months. After the fifth month of lactation, pregnant cows declined in milk production more rapidly than their nonpregnant **herdmates**. Apparently the lower milk yield among pregnant cows resulted from the increasing divergence of nutrients from the mammary gland to the uterus for growth and maintenance of the fetus.

10.10.4 Residual Milk

There is a notable association between persistency and the percentage of residual or complementary milk. The higher the percentage of milk ejected from the mammary gland the less the intramammary pressure, thus enhancing milk secretion.

10.10.5 Nursing

It was shown in New Zealand that when a cow suckles two or three calves during the first part of lactation and is then returned to the dairy herd, she produces more total milk in her lactation than a comparable (control) animal that produces milk in the traditional commercial manner only. Calves nursing the cow provide a greater stimulus for milk production than can be generated by milking by hand or with a milking machine. Of course, the favorable effect of suckling upon milk production and lactational persistency may be due, in part, to more frequent milking by the calf. It is known that milking more than the usual twice (2X) daily increases persistency in most cows.

10.10.6 Disease

Any disease or disorder that affects milk production unfavorably will also affect persistency adversely.

10.11 Marathon Milking

Dairy managers plan for a 12-month calving interval, but what about the cow with a reproductive problem that prevents pregnancy? How long can she milk? Many cows have been observed to secrete milk continuously for several years without the benefit of a pregnancy. One cow in Nebraska produced 78 lb of milk daily after having been milked for 45 months. A Guernsey cow at the Missouri Experiment Station milked continuously for nearly 8 years. This "nonstop" milker was yielding about 25 lb daily when she was sold. But the record apparently belongs to a New Zealand cow that milked continuously and without a pregnancy for 12 years. Since lactation is under hormonal control, such cows must possess an unusually fine balance of the hormones that maintain milk secretion.

10.12 Regression (Involution) of Mammary Glands

Failure to evacuate milk from mammary glands causes **involution**, even when adjacent glands are milked. Moreover, following peak production there is a gradual reduction in mammary gland activity, evidenced by less milk being secreted, until an animal eventually goes dry (becomes nonlactating). The gradual reduction in milk secretion may result, in part, from a slow decline in the secretion of one or more hormones, but there is also a gradual net loss of cells secreting milk. Following the cessation of milking, there is a rapid shrinking of the mammary glands due to disappearance of alveoli (desquamation).

When milk withdrawal is incomplete, the drying off process is hastened. When alveoli are not emptied, pressure quickly increases, thus reducing the total amount of milk secreted between milkings. Pressure sufficient to cause constituents of blood to remain in the bloodstream impedes the secretory process and hastens involution. The process of programmed cell death, **apoptosis**, produces cell fragments called apoptotic bodies that phagocytic cells are able to engulf and quickly remove before the contents of the cell can spill out onto surrounding cells and cause damage.

10.12.1 Drying Off Cows

Opinions differ concerning the best method of drying off cows. A natural process for drying off lactating cows occurs when, for example, a cow's calf is born dead or dies while still suckling. Intramammary gland pressure increases sufficiently to inhibit further milk synthesis. Most managers agree that one can abruptly stop milking nonmastitic cows yielding up to 20 lb/day without physical complications within the mammary gland. However, persistently high-producing cows[6] should be dried off using (1) intermittent milking (i.e., skipping milkings) or (2) incomplete milking (i.e., leaving about half the milk in the udder at milking), or both. Additionally, feed, especially grains, may be reduced during the drying off process.

10.13 Merits of a Dry Period

US Department of Agriculture dairy herd improvement data show a continuous increase of average milk production per cow as her days in milk increase. This indicates that the **dry period** should be of minimal duration. But how long is long enough? Most research designed to determine the merits of a milk break (dry period) indicates that dry periods of less than six weeks have a depressing effect on milk yields during the next lactation. Research at Ohio State University showed that cows dry 45 days produced as well during the next lactation as those dry 75 days.

In an experiment using bovine twins at the University of Tennessee, cows having a 60-day dry period peaked at 10 and 15 lb (4.5 and 7 kg) more milk daily in the second and third lactations, respectively, than in the first. In comparison, those milked continuously yielded only 75 and 62% as much milk in the second and third lactations, respectively, as they did in the first. Similar results were obtained in a University of West Virginia study. Cows having no dry period yielded only 60% as much milk as in their previous lactation. Cows that abort during the seventh month of pregnancy usually yield much less milk during the lactation that follows than they should be capable of producing. This low milk yield is caused, in part, by the low rate of secretion of estrogen and prolactin. Hormonal-dry period interactions, however, remain largely a mystery.

Many believe the major benefit derived from a dry period is an improvement in the cow's nutritional status for the forthcoming lactation. That the favorable effect of a dry period on subsequent milk yield cannot be due totally to nutritional nor hormonal factors was shown in a simple, beautifully designed experiment conducted by British researchers. Ten weeks before the expected date of parturition, they dried off (by suspension of milking) the right fore and left rear (control) quarters of pregnant, first-lactation cows. The left fore and right rear (experimental) quarters continued to be machine-milked twice daily throughout the entire pregnancy. Before the quarters were dried off in the first lactation, the combined milk yields of control and experimental quarters were equal. Rates of decline in yield of the experimental quarters were not affected by drying off adjacent control quarters. During the first 90 days of their second lactation, experimental quarters yielded only 59% as much as control quarters. There was no marked difference in milk composition. The scientists theorized that continuous removal of milk from experimental quarters maintains activity of secretory tissue and prevents a renewal of epithelial cells that occurs in dried off mammary glands just prior to parturition. Other researchers have observed that the frequency of **mitosis** is considerably greater in tissue sections taken from mammary glands of nonlactating cows (8 to 9 months pregnant) than in those of lactating cows. This indicates that a dry period is important in order to allow for rapid regeneration of the secretory epithelium before the next lactation.

10.14 Immune Milk

Studies conducted at the University of Minnesota indicate that cows can be immunized with certain **toxoids**, which induce antitoxins in the cows' milk and blood. In one study, tetanus toxoid was used to immunize lactating cows four weeks prior to parturition (Marx and Cole, 1966). Injecting this **antigen** intramuscularly or infusing it intramammarily at three weekly intervals gave similar results. To **assay** the antitoxin strength of milk and blood, mice were treated with varying dilutions of milk or blood serum, after being challenged with a standard toxin dose of 10 MLD (minimal lethal dose, i.e., the minimal amount of tetanus toxin that killed the mice within four days following injection). Colostrum taken at parturition contained more than 10 times as much protective antitoxin as blood serum. However, milk and blood serum samples taken one week postpartum had similar concentrations of antitoxin. During the balance of lactation, antitoxin concentration of milk was less than that of blood, although protective dose values of milk and blood serum remained rather constant throughout lactation. Whereas mouse protection was achieved with either milk or blood serum of immunized cows, preimmunization control samples of milk and blood serum gave no protection to mice.

10.15 Summary

Growth of the cells that secrete milk is stimulated during recurring estrous cycles, pregnancy, and early lactation. Size and quality of mammary glands are fundamental to the ability of cows to produce abundant milk. The number of cells (mammary gland size) is determined primarily by the hormones estrogen and progesterone, which stimulate mammary gland growth. Of course, heredity affects the secretion of these hormones.

Milk secretion occurs most rapidly immediately after milk removal. During the interval between milkings, milk is secreted by millions of tiny epithelial cells that line each alveolus, the basic unit of milk secretion in the mammary gland. As each cell manufactures its capacity of milk, it discharges its contents into the lumen, where milk accumulates and causes the alveolus to balloon out. Accumulation of milk in mammary glands can be seen externally as an enlargement of the udder, and is proven experimentally by monitoring the gradual increase in intramammary pressure. As internal pressure mounts, milk secretion slows and finally stops. This is why high-producing cows will yield more milk daily if udder pressure is kept down by frequent milkings.

A well-fed and well-managed cow may be endowed with large, high-quality mammary glands and a rich, balanced supply of the hormones that govern milk secretion, but the "payoff" for dairypersons comes in a *complete* harvest of milk. Disturbances at milking time reduce the amount of milk harvested. Failure to remove milk at each milking reduces storage capacity of the udder and, consequently, persistency. Milk production then declines more rapidly than when fast and complete milk-out is achieved.

Dairy farmers have long recognized that certain cows become dry more quickly than others. This is reflected in a quality we call persistency, which is the linear regression of milk production from its peak to termination of lactation. Persistency commonly decreases with an increase in age. Most cows yielding large quantities of milk annually have great persistency.

We noted the favorable merits of rapid milking and high milk flow rates both on labor efficiency and on milk production. A high milk flow rate is positively correlated with milk yield. We concluded that a six- to eight-week interruption of milking between lactations is usually reflected in greater milk yields the next lactation on a global basis.

Study Questions

1. Trace the growth and development of mammary glands from birth to parturition. Include hormonal influences.
2. What is the general anatomical makeup of a cow's udder? Of what significance is the lobule-alveolar system?
3. Why is it essential that mammary glands possess such a rich vascular bed? Compare the percentages of blood flowing through lactating and nonlactating mammary glands. Do blood volume and its plasma content differ in lactating and nonlactating cows?
4. What role does the lymphatic system play in lactating cows? What is edema? When is it most commonly observed in dairy cows?
5. What are two important means by which milk is secreted? Which of these two processes can be used to explain the secretion of milk protein, lactose, fat, vitamins, and minerals?
6. How can we explain the secretion of milk low in fat during hot weather?
7. After milk is secreted, how is it discharged and stored until milking? When is milk secretion the fastest? Why? Why is the last milk obtained at milking the richest in milkfat?
8. What is meant by milk letdown? How may it be stimulated and inhibited? Identify the hormones involved.
9. Differentiate between and explain the role of the epithelial and myoepithelial cells of an alveolus.
10. What is residual milk? How may it be minimized? Of what significance is residual milk in antibiotic therapy of mastitis?

11. Discuss the role of prolactin and thyroxine in lactation. Include other hormonal relationships.
12. Name the major factors affecting the intensity of lactation and/or milk composition.
13. Why is fast milking important to high production? How does it relate to persistency?
14. Of what economic significance is milking rate? Differentiate between average flow rate and peak flow rate. Identify factors affecting milking rate.
15. During which months of the lactation period does peak milk production commonly occur?
16. What is meant by persistency of lactation? Discuss briefly several factors affecting persistency.
17. What is involution of mammary glands? How may it be enhanced or slowed?
18. Discuss briefly the basis and merits of a dry period. How long should it be? How would you convert a high-producing cow into a nonlactator?

Notes

[1] Puberty is characterized by a cascade or long series of inductive events that commonly culminate in the ability to reproduce.

[2] There is no mixing of blood and milk in the mammary gland. Instead, nutrients of blood pass through the cell wall to be used in synthesizing milk. Only when mammary glands are injured or diseased do we find blood in milk.

[3] Half-life is determined by the time required for half the injected dose of a hormone to leave the blood. For example, the favorable effects of oxytocin in milk letdown dissipate in 6 to 7 minutes. This explains why rapid milking results in more milk obtained.

[4] Adrenal cortical hormones (glucocorticoids) also stimulate basal metabolic rate.

[5] Rear teats of cows are usually shorter than fore teats.

[6] It is not uncommon for high-producing cows to complete their lactation period producing more milk daily than low producers yielded at the peak of their lactation.

References

Akers, R. M. 2006. Major advances associated with hormone and growth factor regulation of mammary gland growth and lactation in dairy cows. *Journal of Dairy Science* 89:1222–1234.

Akers, R. M., and A. V. Capuco. 2011. "Lactogenesis." In J. W. Fuquay, P. F. Fox, and P. L. H. McSweeney (eds.), *Encyclopedia of Dairy Sciences*, Vol. 3 (2nd ed., pp. 15–19). San Diego, CA: Academic Press.

Barbano, D. M., J. M. Lynch, and D. E. Bauman. 1992. Effect of a prolonged-release formulation of n-methionyl bovine somatotropin (Sometribove) on milk composition. *Journal of Dairy Science* 75:1775–1793.

Bauman, D. E., I. H. Mather, R. J. Wall, and A. L. Lock. 2006. Major advances associated with the biosynthesis of milk. *Journal of Dairy Science* 89:1235–1243.

Baumrucker, C. R., and A. L. Magliaro-Marcina. 2011. "Hormones in Milk." In J. W. Fuquay, P. F. Fox, and P. L. H. McSweeney (eds.), *Encyclopedia of Dairy Sciences*, Vol. 4 (2nd ed., pp. 765–771). San Diego, CA: Academic Press.

Capuco, A. V., and R. M. Akers. 2011. "Galactopoiesis, Effects of Hormones and Growth Factors." In J. W. Fuquay, P. F. Fox, and P. L. H. McSweeney (eds.), *Encyclopedia of Dairy Sciences*, Vol. 3 (2nd ed., pp. 26–31). San Diego, CA: Academic Press.

Marx, G. D., and C. L. Cole. 1966. Protective ability of bovine milk and blood serum antitoxins in mice. *Journal of Dairy Science* 49:718–719.

Mather, I. H. 2011. "Secretion of Milk Constituents." In J. W. Fuquay, P. F. Fox, and P. L. H. McSweeney (eds.), *Encyclopedia of Dairy Sciences*, Vol. 3 (2nd ed., pp. 373–380). San Diego, CA: Academic Press.

Turner, C. W. 1952. *The Mammary Gland*. Columbia, MO: Lucas Brothers.

US Food and Drug Administration. 2009, April 23. Report on the Food and Drug Administration's Review of the Safety of Recombinant Bovine Somatotropin (http://www.fda.gov/AnimalVeterinary/SafetyHealth/ProductSafetyInformation/ucm130321.htm).

11

Principles of Milking and Milking Equipment

Good milking depends first and foremost on the person doing the milking.
L. A. Brooks

11.1 Introduction

Although many crops are harvested only once annually, milk is commonly collected twice daily. It is the most important "harvest" dairy farmers make for two reasons. First, production of milk is the reason selected animals are bred, fed, and managed; second, milking, especially mechanical milking, is a critical process for obtaining maximal yield and maintaining healthy mammary glands.

For centuries humans attempted to invent a mechanical means of extracting milk from mammals, but none of them were entirely successful. The Egyptians, as early as 380 BC, are believed to have inserted straws (probably short stems from wheat or quill feathers) into cows' teats to remove milk. Ancient paintings in caves also depicted removal of milk from individual mammary glands by inserting straws and other hollow devices into the teat orifice. In 1852 the *New England Farmer* reported that dairyman Benjamin French was using leg bones of robins for milk tubes. Such devices would permit milk to flow rather freely from cows that had been properly stimulated for milk letdown (cf. chapter 10). However, injury and infection to the mammary glands prompted realization there must be a better way.

As early as 1819, attempts were made to design mechanical milkers in the United States. The first milking machines were an extension of the traditional milking pail. The earliest milking devices fit on top of a regular milk pail and sat on the floor under the cow. After each cow was milked, the bucket would be dumped into a milk can or holding tank. The first suction machine was produced by L. O. Colvin in 1860. To create vacuum, pistons were moved back and forth in cylinders. However, the continuous suction caused congestion of blood in the teat and was unacceptable.

Substantial improvements in means and methods of mechanical milking occurred between 1872 and 1905. During this period, the US Patent Office issued 127 patents covering milking machines and other dairy farm equipment. The first patent on a suction-type milking machine was issued to Anna Baldwin in 1878. Her hand-operated device was named the Hygienic Glove Milker. She used a board through which four holes had been drilled at proper intervals. After the cow's teats were inserted in the holes, the board was placed on top of a shallow pan, and suction was applied from an ordinary water pump. Baldwin's idea of milking a cow was to pump milk from the udder as water is pumped from a well. This was a giant step forward in mechanical milking; however, a cow's complex mammary glands and their delicate linings could not be emptied as simply as water from a well.

Introduced in 1895, Modestus Cushman of Iowa patented the first pulsating device. The Thistle milking machine is usually considered the first to use pulsation, or intermittent suction, a concept basic to today's mechanical milkers. The next important contribution was made in 1902 by Lawrence and Kennedy of Scotland, who developed a milking machine with a pulsator located on its lid. It was introduced to the United States in 1905. This provided a more reliable control of pulsation rates because of the proximity of the pulsator to inflations (teat cup liners). This contribution was made even greater by coupling it with double-chambered teat cups, which were patented by Struthers and Weir of Scotland in 1892. Their device appears to have had no commercial development, but soon thereafter Harnett and Robinson of Australia were granted a patent for a similar invention.

By 1915, still less than 1% of US dairy farmers were using mechanical milkers and only about 5% by 1925. Thus, machine milking is a relatively new innovation in harvesting milk. Depending on the type of system used, labor for milking can be the largest cost item in the care and management of dairy cows. Developments in milking machines and related milking equipment have reduced labor costs markedly while increasing the investment per cow.

In chapter 10 we learned how milk is secreted, the principles involved in milk letdown, and why it is important to harvest milk regularly, quickly, and with extreme care. Now let us discuss the principles involved in milking.

11.2 Principles of Milk Withdrawal

The three primary means of harvesting milk from cows are (1) the calf, (2) by hand, and (3) with mechanical milkers.

11.2.1 Nursing

To extract milk, a calf presses its tongue around the teat and against the palate of its mouth, creating a vacuum by separating its jaws and retracting its tongue. Approximately 100 to 120 alternating suck-and-swallow cycles may be observed per minute. A calf's sucking is the most rapid means of evacuating milk from a mammary gland.[1] It is also least damaging to the delicate tissues.

The act of nursing by the calf is cyclic and can be divided into *active* and *resting* phases. During the active phase, two events occur simultaneously to facilitate milk removal: (1) a vacuum is created at the teat end within the oral cavity, and (2) pressure is produced upon the teat sinus (cistern). The base of the teat is occluded by the tip of the tongue, dental pad, and hard palate during the active phase of suckling, which results in a pressure increase within the teat sinus. Pressure on the outside of the teat becomes sufficiently less than that on the inside of the teat (assuming a cow has been properly stimulated for efficient milking), resulting in the **teat meatus** (streak canal) being forced open and milk to flow out. During the resting phase, the vacuum at the teat end is relieved as the mouth relaxes. As a result of tissue rebound, a slight vacuum of approximately 2.7 kPa (20 mmHg) is created within the teat sinus at the end of each cycle. Pressure on the base of the teat is relieved at the end of the resting phase, and the teat sinus refills with milk. The cycle is then repeated.

Milk removal during nursing is also dependent on the differential pressure across the teat canal. A differential pressure of 73.3 kPa (535 mmHg) across the teat canal during suckling is considerably higher than with hand milking. Additionally, the cyclic rate during nursing is nearly twice as fast as during hand or mechanical milking (see section 11.6).

11.2.2 Hand Milking

There are several ways to remove milk by hand. Let us discuss briefly the preferred technique. Upon applying hand pressure to remove milk, the thumb and forefinger pinch off the upper part of the teat, which prevents milk from returning upward to the gland cistern, and then by squeezing downward, milk forces the streak canal open and passes through it (figure 11.1).

In hand milking, no vacuum is applied to the teat end. Milk is forced through the teat canal by the high intracisternal pressure that develops when the teat is compressed by hand. The amount of pressure developed in the teat sinus is variable and depends upon the force applied by the hand and the patency of the teat canal (see section 11.4). The average frequency of hand milking is 65 cycles/minute.

11.2.3 Mechanical Milking

Machine milking is accomplished through use of a double-chambered teat cup consisting of a metal or rigid plastic shell and a rubber or flexible plastic inflation (figure 11.2 on the next page). The chamber or space inside the inflation and below the teat end is subjected to a continuous vacuum, which dilates the teat canal and causes milk to flow freely. Upon collapse of the rubber inflation inside the teat cup shell, milk stops flowing because the vacuum to the teat end is virtually closed off. The pulsation chamber formed between the rubber liner and the metal shell is alternately subjected to vacuum and air atmospheric pressure by pulsator action. When the pulsator admits air into this chamber, the rubber liner collapses against the teat with a massaging action, allowing blood to circulate. This is the *resting* or massage phase of the pulsator cycle. When air in the pulsation chamber is again evacuated, the liner expands back into its original form. This is the *milking* or expansion phase. The time relationship between milk and rest phases is called the *pulsation ratio.* A pulsator should massage the teat approximately 50 to 60 times/min; this is called the *pulsation rate.* Applying continuous vacuum to the teat for 20 to 30 seconds (without the massage or resting phase) results in an accumulation of blood and hemorrhage of blood vessels within the teat.

The machine (figure 11.3 on the following page) is comprised of four teat cups (shell and inflation) that contact the cow's teats and remove the milk, a claw into which milk flows on removal from the teats, vacuum tubes that provide vacuum to the teat cups, a milk tube through which milk flows from the claw to the milk pipeline, a source of vacuum, and a pulsator that regulates the on-off cycle of the vacuum. Many milking machines have an automatic take-off (ATO or detacher) device that removes the machine from the cow as milking is completed. Some milking machines are linked to a computerized system that controls the machine and generates data about the cow and its milk during milking.

Shells for teat cups are usually constructed of stainless steel in cylindrical form. Liners, the flexible tubes that contact the teats, are made of natural, synthetic, or silicone rubber. Liner designs vary, but each is under some tension within the shell. Slippage of liners during milking causes variation in vacuum and increases risks of mammary gland infections. Tendency to slip is associ-

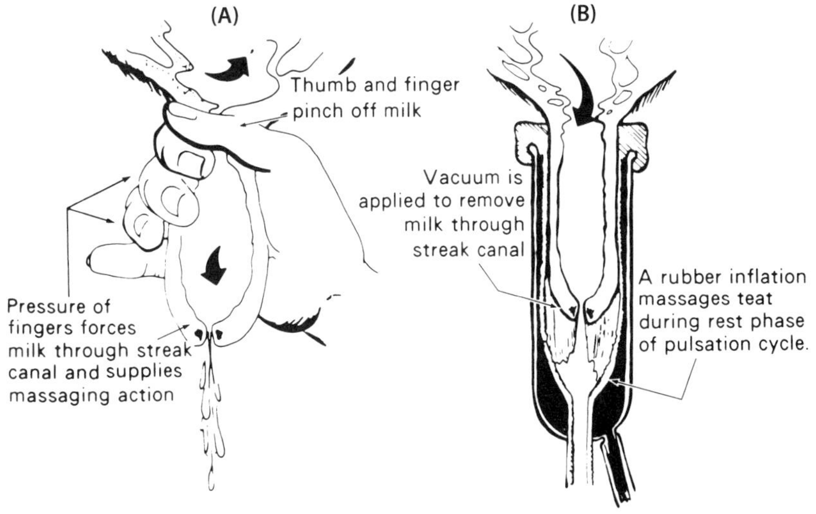

Figure 11.1 Milking a cow by hand (A) and by machine (B) employs two different principles. In hand milking the teat is concurrently squeezed for milk and massaged. Both a calf and a milking machine utilize a vacuum to withdraw milk in intermittent strokes. In machine milking, allowing air periodically into the space between the teat liner (inflation) and the shell permits the inflation to collapse around the teat and to supply a massaging action that is essential to prevent congestion of blood in teat tissues.

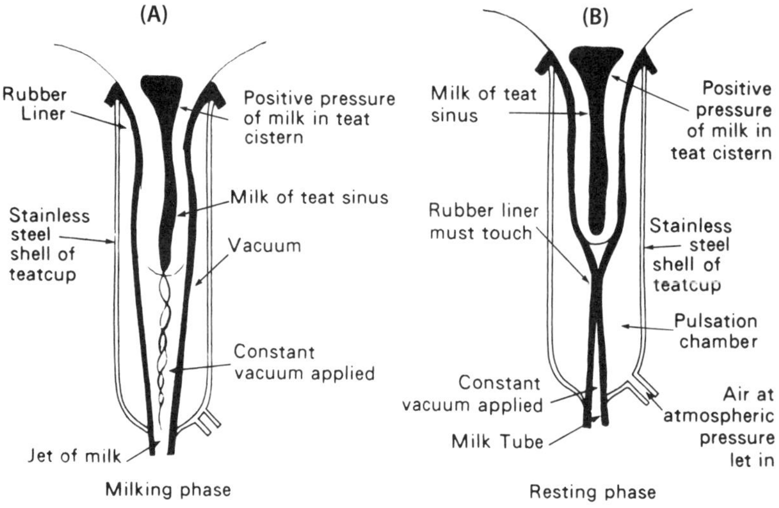

Figure 11.2 Cross section of a teat cup showing (A) the milking phase and (B) the resting phase of the pulsator cycle.

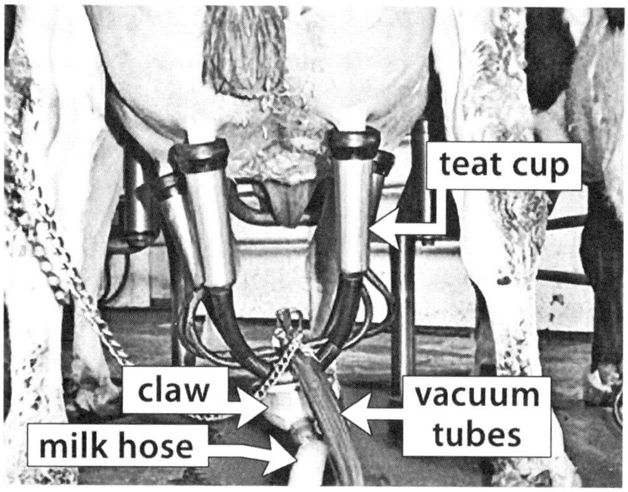

Figure 11.3 Milking machine in use with its major component parts identified. The chain leading from the claw to the left side of the image is part of the ATO of this system (courtesy of University of Illinois).

ated with udder shape, teat location, cluster position, and pressure within the mammary glands (Spencer, 2011).

11.3 Preparing the Cow for Milking

To stimulate milk letdown and to provide cleaner milk, the udder and teats should be washed with warm water containing sanitizer about one minute before milking is to begin.[2] After washing, single-service paper towels should be used to dry teats and adjacent surfaces. This further stimulates milk letdown and removes any excess water that might otherwise be drawn into the milk. Additionally, two or three streams of milk may be removed from each mammary gland into a strip cup, the top of which contains a fine mesh wire or is a shallow black dish. This procedure gives a final stimulus for milk letdown, opens the streak canal, removes milk that is commonly high in bacteria from teat cisterns, and affords the operator an opportunity to detect the presence of abnormal milk.

11.4 Milking Rate

The person that operates milking machines is just as important as the machine itself. He/she must recognize the individuality of each cow and treat her accordingly.

M. E. Senger

The rate at which milk can be removed depends, in large part, on the ease with which the resistance of the rubber band-like muscles (**sphincter** muscles) surrounding the streak canal is overcome, allowing the canal to open to its widest capacity. The term **patent** is used to indicate weak, relaxed, or incompetent sphincter muscles that fail to effectively maintain complete closure of the streak canal. The so-called hard milker has tight sphincter muscles, which are difficult to open and often cause milk to be emitted in a fine spray. Conversely, weak sphincter muscles often allow milk to drip from teats prior to milking. One function of the streak canal is to prevent bacteria from gaining entrance into the teat. For this reason easy-milking cows are more susceptible to mastitis than hard-milking ones (cf. chapter 14).

The rate at which a cow milks is dependent, then, upon size of teat opening, strength of sphincter muscles surrounding the teat meatus, pressure within the mammary glands (highest during peak production), and milking techniques. The latter includes vacuum level, pulsation ratio, stimulation of cow, and operational maintenance of the milking machine.

Research at Missouri and other agricultural experiment stations showed that rate of milking is appreciably inherited. Differences exist in maximum and average rates of milking among daughters of different bulls. Among dairy breeds of the United States, Holsteins average highest in amount and rate of milk yielded per minute of milking. Maximum peak flow of milk is about 15 lb/min. The hard-milking cow is commonly incompletely milked, and incomplete milking is associated with an increased rate of drying off of the cow. Moreover, time spent in milking is influenced greatly by milking rate.

11.5 Importance of Fast Milking

Milk is squeezed from cavities of the alveoli by millions of tiny myoepithelial cells that contract and squeeze the alveoli (cf. chapter 10). Once released it flows down through the ducts, gland cisterns, and, finally, the teat cisterns. The act of stimulating mammary glands properly at milking causes the pituitary gland to release the hormone oxytocin, which is conveyed to the udder via the bloodstream. This hormone causes muscle cells surrounding each alveolus to contract and eject milk, a process called *milk letdown*. But the important thing to remember is that this hormone is short-lived in the bloodstream; its blood concentration dissipates to ineffective levels after approximately 6 to 7 minutes. When this occurs, milk remaining in the alveoli is not let down but stays in the udder as *residual* milk. This milk is not only lost to the cow's daily production, but also tends to reduce further milk secretion. It is readily apparent, then, why fast milking is important to high yields of milk (cf. chapter 10).

11.6 The Function of Vacuum in Milking

The vacuum pump is the foundation of every milking system, although vacuum, per se, does not milk a cow. Instead, vacuum makes it possible to move milk. When properly stimulated for efficient milking, internal mammary gland pressure is greater than atmospheric pressure outside the teat. Thus, as milk is ejected, it is moved away by vacuum. Milking vacuum at the teat end is approximately 50 kPa when atmospheric pressure is about 100 kPa.

Constant vacuum is applied inside the teat cup liner (inflation), and thus to the teat, but when the pulsator allows air to enter the space between liner and shell, the liner collapses and, momentarily, separates the teat end from the vacuum being applied through the system. The liner will fail to collapse properly if (1) the pulsator does not permit enough air to enter between liner and shell (due to air port obstruction, wear, or too fast a pulsation rate), or (2) there is insufficient vacuum within the teat cup to permit air to close the liner. If the liner does not collapse completely during the resting phase, or for a sufficient duration, congestion with blood and damage to the teat end will occur because vacuum is constantly being applied to the teat.

Before continuing the discussion of milk removal using differential pressures, let us explain some technical terms. Vacuum is measured as gauge pressure and is expressed as kilopascals (kPa). Atmospheric pressure is typically about 100 kPa at sea level, but is variable with altitude and weather. Gauge pressure is referenced at zero against atmospheric pressure. Vacuum is expressed as a negative value compared to atmospheric pressure and is expressed in kPa. The standard unit of force is the newton (N), and the standard unit of area is the square meter (m^2). Thus, pressure (P) is force per unit of area (Nm^{-2}), and the derived unit is the pascal (Pa). Because pressures may be expressed in other terms, we have included millimeters of mercury (mmHg) and pounds per square inch gauge (psig) in certain expressions of pressure values.

Milk removal is dependent, in large part, upon the differential pressure across the teat canal. For example, when intracisternal pressure is 5 kPa (37.5 mmHg) and the milking vacuum at the teat end is 40 kPa (300 mmHg), the differential pressure across the teat canal is 45 kPa (337.5 mmHg). This differential pressure during mechanical milking results in expulsion of milk from the teat. Increasing vacuum at the teat end results in greater differential pressure. Although an optimal vacuum level has not been established, the usual range is 40 to 50 kPa. Machines operating with high pipelines, to which milk must be raised, need higher vacuum than those with low pipelines. Of course, the higher the vacuum at the teat end the higher the milk flow rate.

11.7 Overmilking

Overmilking disturbs the host-parasite relationship in favor of the parasite.

J. Malley

Humans and the calf wisely stop milking when milk flow ceases; however, machine operators often do not remove milkers at the proper time. Thus, the greatest cause of tissue damage during mechanical milking is leaving machines on too long. So long as it remains in the gland and teat cisterns, the milk itself serves as a cushion to the delicate teat linings. But when milk stops flowing teat cups creep upward, and with each pulsation friction irritates teat walls. The effects of unrelieved vacuum and the bruising action (mechanical injury) of pulsations under these conditions may increase susceptibility of glands to the organisms that cause mastitis. Research has shown that when cows are milked by machine their milk has substantially higher somatic cell counts than when milked by hand. This indicates that the milking machine does, in fact, exert greater stress on the delicate membranes lining the teat than does hand milking.

Intramammary pressure immediately prior to milk letdown is less than 0.7 kPa (5 mmHg). Washing the udder with warm water stimulates letdown within 30 to 60 seconds and results in intramammary pressures of 4.7 to 7.3 kPa (35 to 55 mmHg). If milking machines are not removed when milk stops flowing, residual vacuum develops within the teat sinus. Pressure changes within the teat sinus are apparently due to action of the rubber inflation upon the teat and the extension of milk line vacuum through the teat canal. This may be a factor, too, in entrance of mastitis-causing bacteria into mammary glands.

11.8 Pulsation

In mechanical milking, the purpose of pulsation is to compress the teat end periodically, thus assisting the return of blood from the teat and briefly interrupting milk flow (see figure 11.2). Pulsation rate should be in accordance with recommendations of manufacturers. Excessively high pulsation rates tend to be associated with increased udder health problems. Recommended speed of most pulsators ranges from 50 to 60 times/min.

The pulsator is a device that, by the production of an intermittent vacuum, alternately allows air into the chamber between the teat cup inflations and the shell, then withdraws this air by opening a port into the vacuum system. The conventional pulsator functions by means of a slide moving back and forth over two ports in a valve face. The slide alternately connects the port leading to the pulsation chamber of the teat cup with the vacuum line of the machine and with the atmospheric air.

Pulsation ratio refers to time the inflation is in the milking or open phase compared with the resting or collapse phase. It is expressed in a percentage ratio. Most early milking machines operated at 50:50 ratio; research indicates, however, that a pulsation ratio of 60:40 results in faster milk withdrawal than one of 50:50. This is especially true during the first 2.5 minutes of milking when peak milk flow occurs. (Average peak milk flow occurs at 1.5 minutes into milking.) When the pulsator ratio is 60:40, the unit is milking 60% of the time and resting 40%. Some milking machines have pulsation ratios as wide as 70:30. Wide milk/rest ratios shorten the length of the resting phase, and this may cause injury to teats. A short resting phase is often associated with an insufficient collapse of the liner, especially if air movement is restricted by improper design or faulty maintenance.

Pulsation can be applied to teats in different sequences: simultaneously when all four liners close concurrently, or alternately when two liners are closed and the two opposite liners are open. Alternating pulsation, which requires a dual pulse tube, can be applied side to side, front to rear, or diagonally (the latter being seldom used).

11.9 Narrow-Bore versus Wide-Bore Teat Cup Liners

A narrow-bore liner (diameter of 0.75 in [1.9 cm]) confines vacuum to the teat end throughout milking. A wide-bore liner (1 in [2.5 cm] or more) of greater diameter than the teat does this only during milk flow when the positive pressure of milk within the teat and the negative pressure outside the teat expand the teat to meet liner walls. As milk flow diminishes, there is a decrease in intramammary pressure, the teat contracts, and vacuum operates over an increased area of the teat. To help minimize the incidence of teat erosion and mastitis, the teat cup liner should have a soft mouthpiece, high-tension qualities, and a narrow bore.

11.10 Machine Stripping

Machine stripping is the application of downward pressure on the claw piece of the milking machine, accompanied by simultaneous massage of individual quarters, to force milk down into the gland and teat cisterns. As a cow becomes older, there is an increase in "machine stripping time" and "machine-on time," independent of milk yield. Researchers from Great Britain and Cornell University concluded that machine stripping of cows is not an economically sound practice. Their studies indicate that 0.5 to 1.0 minute devoted to machine stripping yielded less than 0.5 lb of milk per milking, which was not enough milk to pay for the extra labor required; persistency was unaffected.

11.11 Teat Cup Flooding

It is generally assumed that vacuum and gravity move milk from teat cup to claw. Easy-milking cows, wide milk/rest ratios, and high vacuum contribute to teat cup flooding. Additionally, teat cup liner volume, sluggish closing of liners, and short milk tubes influence the degree of flooding. A simple means of correcting the problem of teat cup flooding is to vent each teat cup at the base of the shell. Vents can be installed in the short milk tube. However, the question arises whether teat cup vents induce rancidity in milk by increasing churning. Field studies in California indicate this is not the case.

Milk lock results from insufficient removal of milk from the milk collection systems and, as a result, milk surges back and forth in the milking machine tubes. The term *flooded* is commonly used to describe a condition in which slugs of milk interspersed with air traverse the pipeline, creating disturbances that interfere with efficient movement of milk.

11.12 Pipeline Milking Systems

The milk pipeline system uses a large diameter pipe to carry milk from the cow to the farm **bulk milk tank**. A separate vacuum pipe, with quick-seal entry ports above each cow, encircles the milking facility. Milk is pulled into the milk pipe by the vacuum system and flows through a vacuum-breaker into the storage tank. Some installations include a plate cooler ahead of the storage tank. Invention of the pipeline system greatly reduced the physical labor of milking since the farmer is no longer needed to carry around heavy buckets of milk from each cow (figure 11.4).

Milk pipelines are constructed of smoothly polished stainless steel or, less frequently, of Pyrex glass. The steel pipes have welded joints in most installations. This method of conveying milk from cow to cooler ensures prompt cooling and maximal protection of milk quality. It also virtually eliminates milk's absorption of undesirable barn odors. Pipelines fit well into a mechanized cleaning system. Pipeline milking systems have been in use in the United States since 1928 and for many years longer in certain other countries, especially Australia and New Zealand.

11.13 The Four-Way Milking Cluster

A recent development is the "four-way teat cluster" known as the IQ™ Milking Unit (figure 11.5 on the next page). The device is claimed to maximize milk quality, quantity, and udder health. The unit, which has no conventional claw, is composed of four separate milk chambers and provides the advantages of more stable vacuum at the teats, no slippage on individual teats, and no teat-to-teat cross-contamination. The milk guide piece is subdivided into four chambers. Milk flows directly from individual teats to the outlet. The weight of the cluster sits 80% on the teat and 20% on the milk guide piece claw. Quick and nonstressful milking can be realized without air ingressions and with high throughput.

Stabilization of the milking cluster can provide more even milking by keeping the cluster positioned directly beneath the mammary glands. One such device, known as PosiMax by GEA Farm Technologies, Inc. (figure 11.6 on the following page), bears the weight of the long milk and vacuum tubes. As it moderates twisting and drag forces, the cluster is kept in optimal position. Height is adjustable and the cluster follows cow movements readily.

11.14 Robotic (Automatic) Milking Systems

Beginning in the 1980s, robotic milking systems were developed and were marketed commercially in the 1990s, principally in the EU. These systems permit the cow options within predefined windows for time of milking.

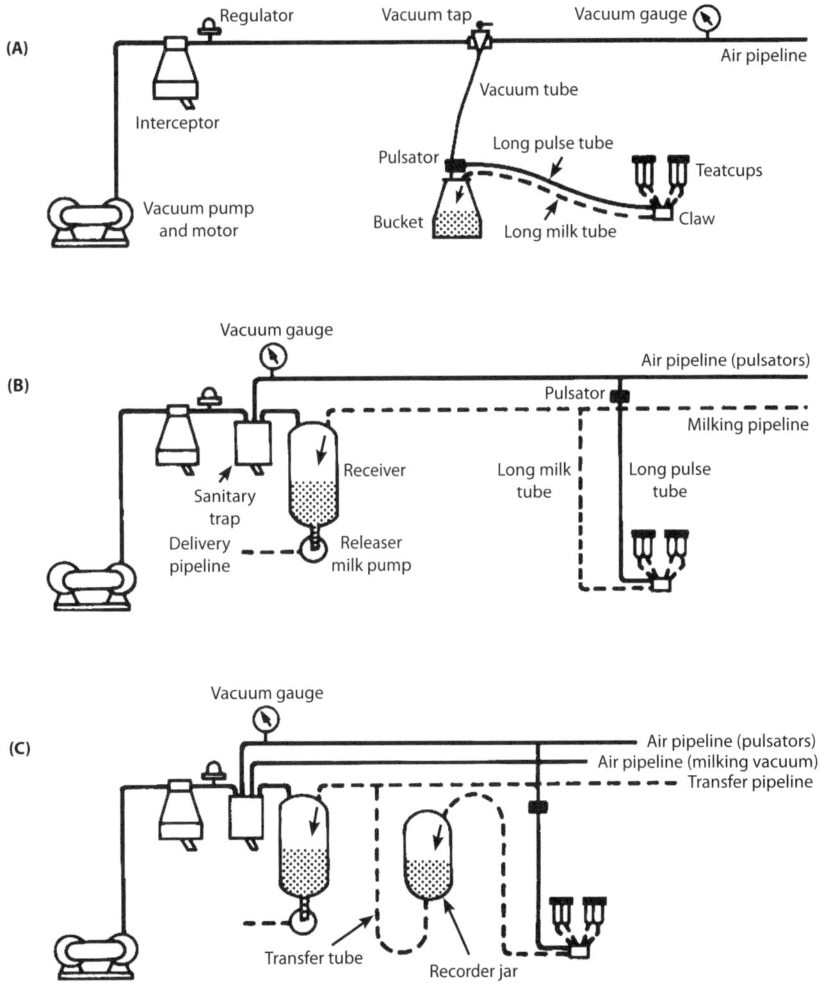

Figure 11.4 Examples of milking systems: (A) bucket milking, (B) pipeline system, and (C) pipeline system with weigh jar (often replaced with appliances that quantify milk produced) (Food and Agricultural Organization of the United Nations, 1989).

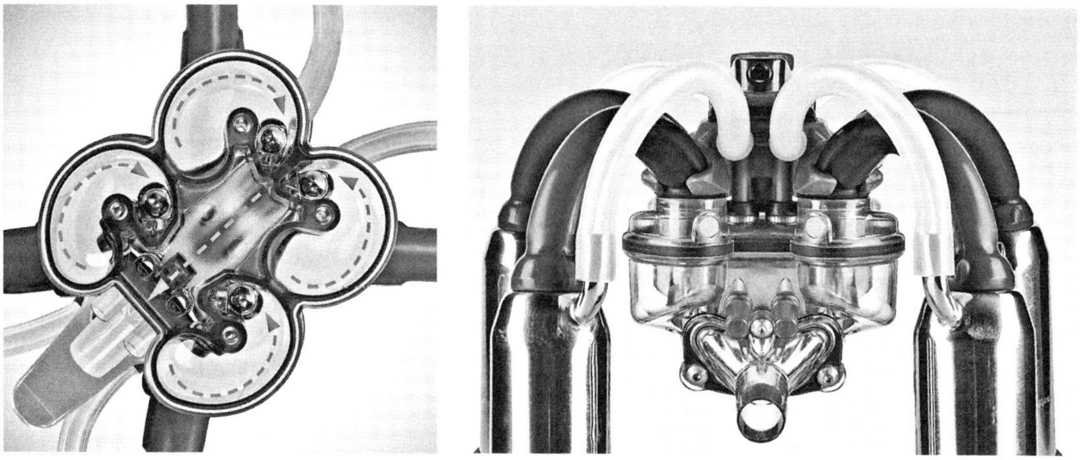

Figure 11.5 Examples of four-way milking cluster with individual milk chambers into which single teats deliver milk (courtesy of GEA Farm Technologies, Inc.).

Reduced labor, a better social life for dairy farm families, and increased milk yields resulting from more frequent milkings are important benefits of automated milking (AM) systems (Jacobs and Siegford, 2012). Average reductions in labor for milking-related tasks are 20 to 30% when robotic systems are used.

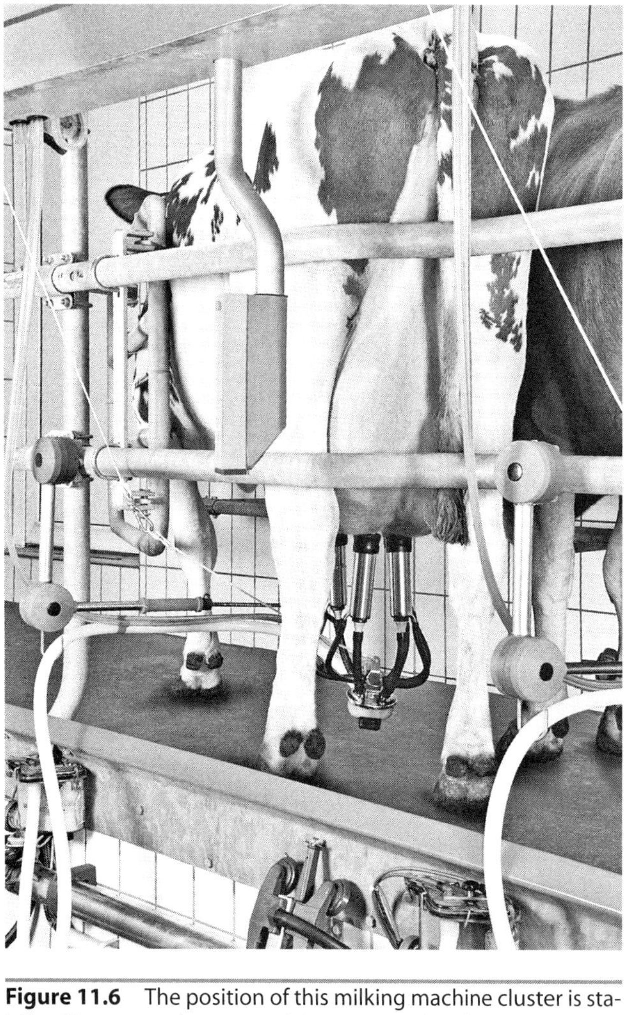

Figure 11.6 The position of this milking machine cluster is stabilized by an arm (courtesy of GEA Farm Technologies, Inc.).

Single stall systems with integrated robotic and milking functions accommodate smaller herds (figure 11.7). Multistall systems with a transportable robot device, combined with milking and detachment devices at each stall, serve larger herds. Cow-to-cow variations in number of voluntary milkings per day range from 2.5 to 3, but large differences in milking intervals are reported from commercial farms. Some cows may need to be fetched by the herdsman when they fail to visit the AM system within the maximal milking interval. It is easy to prevent cows from being milked that have inappropriate yields or during intervals that are too short. Numbers of milkings possible per day within a single AM unit depend on flow rate and amount of milk. On a practical basis, single stall systems can milk 55 to 65 cows several times a day, whereas multistall systems having two to four stalls can milk 80 to 150 cows up to three times per day.

A potential benefit of automatic milking is increased milk yield from more frequent milking. An increase from 6 to 25% during full lactations has been observed when milking frequency increases from two to three times per day. Some cows are milked less than twice per day in AM systems and may, therefore, produce less milk. European experiences indicate a production increase of 5 to 10% for herds milked automatically. In many large herds with highly automated conventional parlors, milking three times daily is commonplace.

Automation of milking started with introduction of equipment for automatic cluster removal followed by equipment for attachment. Much more is required for a completely automatic system: viz., cow identification, udder cleaning and sanitizing, plus computer-controlled sensors to detect milking progress and abnormalities in both animals and milk. De Koning (2011) lists six main modules of an AM system: (1) milking stall with electronic identifier and feed concentrate dispenser; (2) teat cleaning system; (3) teat detection system based on application of ultrasound, lasers, or camera systems; (4) robotic arm for attaching teat cups; (5) control system (including sensors and software); and (6) milking machine.

Figure 11.7 Teat cups have been attached and the cow is consuming grain while being milked in the Astronaut robotic milking system (from Lely North America).

When a cow decides to be milked, she enters an entry lock where she is identified and is allowed to pass if sufficient time has elapsed since her last milking. A robotic arm then extends beneath the cow and, guided by laser and photo sensors, cleans and dries each teat before attaching teat cups. Milk flow, quantity, and milking time are monitored for each teat and compared with previous values. Significant differences may indicate illness or injury. After milking, teat cups are retracted, and the robotic arm directs disinfectant spray to the cow's teats. The gate then opens, releasing the cow to the feeding area, and the system cleans itself before allowing the next cow to enter.

Controls and sensors perform the following functions: (1) detect and divert abnormal milk based on conductivity or somatic cell count; (2) measure and record yield, feed intake, body weight, and animal movements; (3) store data for analysis in a management program; and (4) warn managers of serious problems.

The teat detection system creates a three-dimensional pattern used to direct the robotic arm with the teat cups to the correct site of the teats. Unlike a conventional machine, there is no cluster; instead, separate teat cups are positioned on four long milk tubes, each having a pulse tube and shutoff valve (figure 11.8). Since milk fills the milk tubes, vacuum drops considerably below the teat end compared to the drop when milk from each inflation drops directly into the open cluster bowl.

Cows may be permanently housed in a barn and spend most of their time resting or feeding in the freestall area. Voluntary entry to be milked requires easy access of cows to milking stalls followed by entry into a feeding area. Thus, a well laid-out freestall barn is required. Although grazing systems are possible, long walking distances negatively affect visiting behavior of cows to the AM system. AM systems depend on cows being motivated to enter the unit for milking. Culling of cows for lack of such motivation averages about 5%. If cows are to be grazed, a selection gate is required to allow only those that have been milked to the pastures.

Milk quality is expected to be lower with AM versus conventional systems, primarily because cleanliness of teats is not visually monitored. Furthermore, time between cleaning and sanitizing of equipment may vary, and long periods between visits of cows can allow warm milk to remain in lines as well as drying of films on surfaces. Both factors can contribute to increased bacteria counts. It is reported that concentrations of free fatty acids are comparatively higher in milk from AM than from conventional milking systems.

Finally, the key factors for successful implementation of AM systems are: realistic expectations, support by skilled consultants, flexibility and discipline to control the system and the cows, ability to work with computers, appropriate barn layout and cow traffic flow, technical functioning of the AM system and regular maintenance, and healthy cows with good feet and an aggressive eating behavior.

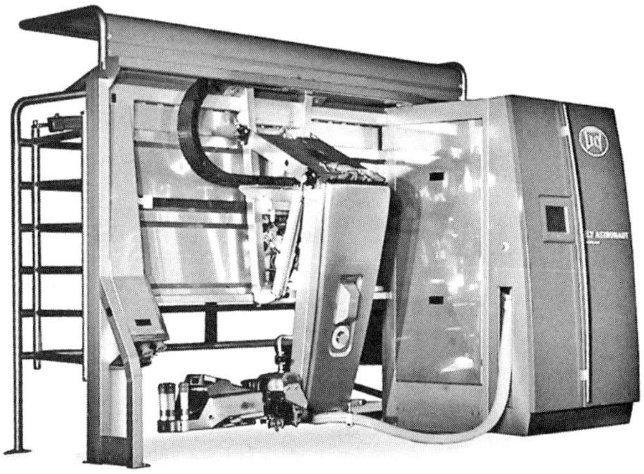

Figure 11.8 The Lely Astronaut milking robot uses a 3-D camera to direct teat cups precisely into place (from Lely North America).

11.15 Bulk Milk Cooling and Storage

The first commercial **bulk milk** handling system of the United States was in the Oakland, California, milkshed in 1936. But bulk milk tanks did not gain widespread acceptance on US dairy farms until the 1950s. In 1937, G. R. Duncan Sr. of Washington, Missouri, invented the first bulk milk cooler ever installed on a US dairy farm.

Adequate cooling and storage of raw milk is essential for the production and maintenance of high-quality dairy products. Instant precooling equipment and milk temperature recorders are widely used to preserve milk's quality. Many producers have installed in-line cooling units capable of lowering temperatures to 35°F (2°C) as milk is conveyed from milking parlor to milk room. In-line cooling is usually accomplished using a centrifugal pump to circulate ice water through a plate-type cooler in a direction opposite that of milk flow. This assures that the coldest water enters the system where milk exits. The *Grade "A" Pasteurized Milk Ordinance* (*PMO*) (US Food and Drug Administration, 2013) requires recording thermometers on farm bulk milk tanks to provide temperature histories of the milk.

Most milk is cooled and stored on dairy farms in refrigerated bulk milk coolers (figure 11.9). To meet 3-A Sanitary Standards in the United States (cf. chapter 24), which are required for grade "A" milk, the milk cooling tank should be provided with sufficient cooling capacity (figure 11.10) to lower milk's temperature to 50°F (10°C) or less within four hours of commencement of the first milking and to 45°F (7°C) or less within two hours after completion of milking. Additionally, the blend temperature after the first and subsequent milkings must not exceed 50°F (10°C). For every day or every other day pickup, the tank's cooling capacity should be rated at 50 or 25%, respectively, of the tank volume.

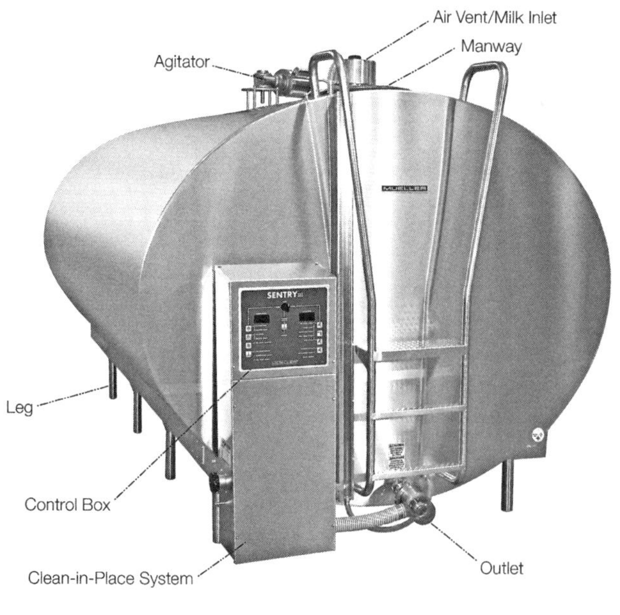

Figure 11.9 Bulk milk storage tank with manway view port and agitator motor mounted through the top; the milk outlet is located in the lower front wall (courtesy of Paul Mueller Company, Springfield, Missouri).

Partial removal of milk from bulk storage tanks is permitted only when tanks are equipped with 7-day temperature recording devices and the equipment is emptied, cleaned, and sanitized at least every 72 hours.

Bulkheading is the process of placing the access cover and outlet of the bulk milk tank in the milk room area and extending the remainder of the tank through a wall to the outside of the building or into another room within the building (figure 11.11 on p. 168). Bulkheading allows for a smaller milk room, which affords greater ease in cleaning the milk room. Bulkhead tanks should be **cleaned-in-place (CIP)**. In new installations, bulkheading can cut construction costs by reducing the total need for expensive building materials frequently used in milk rooms. Additional insulation, however, is often required to prevent freezing of milk in winter and to minimize exposure to hot weather and sun in summer. During an expansion of the operation it may be unnecessary to build a new milk room to accommodate a larger tank.

11.16 Clean and Sanitary Milking Equipment

Once outside the cow, milk is one of a few natural foods that have no protection from contamination (e.g., no skin, shell, husk, or peel).[3] Therefore, the instant it is withdrawn from the cow, milk is exposed to bacterial action. With modern milking practices, the first possible place of contamination outside the cow is with the milking equipment. This is why we mention the importance of clean and sanitary equipment here. Principles of cleaning are discussed in chapter 27, and detailed requirements can be found in Items 9r and 10r of the *PMO*.

Proper cleaning and sanitization of milking machines, pipelines, and bulk storage tanks is fundamental in assuring flavorful, high-quality milk. The *PMO* requires that product-contact surfaces of all multi-use containers, equipment, and utensils used in the handling, storage, or transportation of milk be cleaned after each milking or once every 24 hours for continuous operations. Fortunately, most washing of milking equipment can be accomplished by cleaning in place (figure 11.12 on p. 168). CIP was first exploited in the early 1950s and has contributed to both a reduction in labor and improved sanitary conditions of dairy equipment. Three important factors affecting CIP are time of washing, temperature of detergent and water, and turbulence of the wash water. The latter is accomplished by conveying cleaning solutions through pipelines at speeds of at least

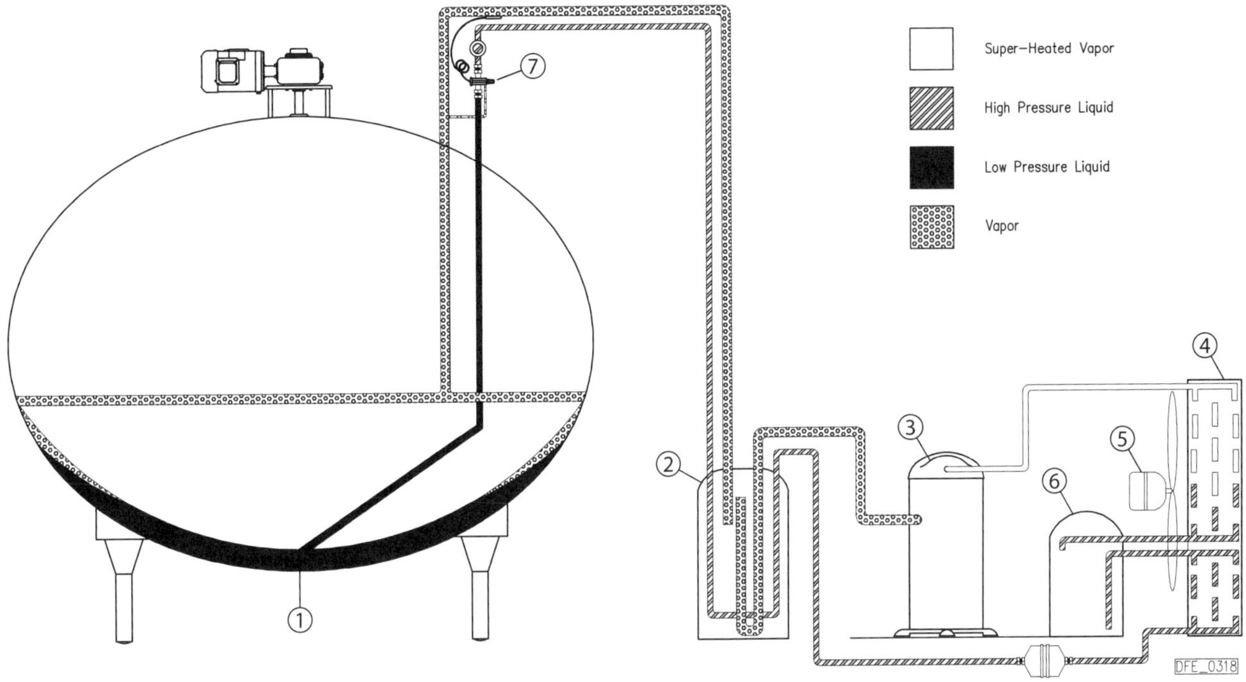

Figure 11.10 Schematic diagram of a farm bulk milk tank with a basic refrigeration system. Warm milk is cooled by exchanging heat with cold, low-pressure liquid refrigerant as it passes through the double-walled evaporator's heat-transfer surface (1) in the bottom of the milk cooler. This changes the refrigerant's state to a low-pressure vapor that passes to an accumulator (2) en route to the suction port of the motor driven compressor (3). There it is compressed into a high-temperature, high-pressure vapor and is discharged into the top of the condenser coil (4). A second heat exchange occurs as the superheated, high-temperature vapor passes through this condenser coil where it is cooled by air (5) and condensed to a high-pressure liquid that is collected in the liquid receiver (6). As the high-pressure liquid refrigerant passes through a metering device (7) pressure drops and the refrigerant becomes a low-pressure liquid ready to repeat the cycle (courtesy of Paul Mueller Company, Springfield, Missouri).

5 ft/sec and by injecting air periodically to increase agitation. Commonly a high-pressure spray ball is used to create cleaning turbulence in bulk milk tanks (cf. chapters 27 and 28).

We discuss principles of cleaning dairy equipment, including reference to **milkstone**, in chapter 27. However, we point out here that approximately half the milkstone accumulation on dairy equipment comes from milk solids and half from minerals of water (table 11.1).

Table 11.1 Composition of Milkstone and the Types of Cleaner Required to Remove It.

Component	Amount in Milkstone (%)	Cleaner Required
Water	5.0	—
Milkfat	10.0	Alkali
Protein	35.0	Alkali
Minerals:	50.0	Acid
Calcium oxide	25.0	Acid
Phosphorus pentoxide	20.0	Alkali
Magnesium oxide	2.5	Acid
Iron oxide	0.1	Acid
Sodium oxide	2.4	Acid

This emphasizes the importance of proper water conditioning and selection of proper cleaning compounds.

11.17 Maintenance of Milking Equipment

The milking system is the most important equipment on a dairy farm. Yet in many cases it is the least methodically maintained. This results partly from lack of understanding of how the equipment works and partly from failure to appreciate the need for routine care and maintenance. Rubber inflations, for example, readily absorb milkfat. Therefore, they must be cleaned adequately with an alkaline detergent. Fat-laden inflations soon lose the physical qualities essential for good milking and udder health. Therefore, even well-maintained rubber inflations must be replaced periodically, and when this is done they should be replaced in sets of four.

Finally, no dairy equipment is designed and manufactured so well that careless operators cannot use it incorrectly. Following manufacturers' recommendations usually leads to more efficient performance of equipment and a longer life.

Figure 11.11 Bulkheading of milk cooling tanks allows placement of the bulk milk tanks through the wall to the outside or in another room and removes the need to enlarge the milk room (courtesy of Paul Mueller Company, Springfield, Missouri).

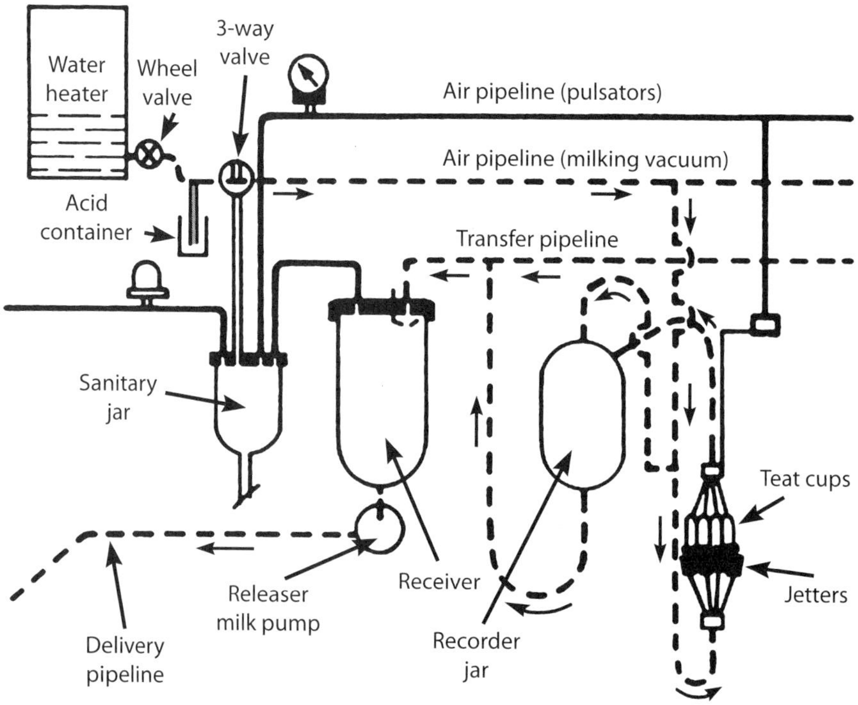

Figure 11.12 Components of a CIP system for cleaning pipeline milking equipment. Dashed line indicates flow of detergent and rinse solutions. Note the "jetters" at the teat cups to provide high turbidity there.

11.18 Summary

Dairy farmers can select, breed, feed, and manage cows correctly, but if milk is not harvested properly the potential payoff is never realized. Therefore, harvesting the milk crop is the most important activity of the dairy farm enterprise. Development of milking machines from the crude devices of yesteryear into the highly automated and engineered milking systems of today did not come about by chance. Hard work and extensive scientific research marked the progress that made possible this aspect of producing more milk.

Milking machines are no better than their operators. The human element is of major importance in fast, efficient milking, which means better udder health and more milk. A mechanical milker is rather unique in that it operates in connection with a living animal of highly developed nervous temperament. Moreover, it is used on a delicate milk-secreting organism, which produces a highly perishable product that is easily contaminated. Overmilking is a problem in milking machine design and usage. Vacuum fluctuations, flooding of inflations, and variable and/or low vacuum are major contributors to udder problems and mastitis.

Most milk is stored on dairy farms in insulated, stainless steel, refrigerated bulk milk cooling tanks. These tanks have been calibrated to facilitate ease in determining the amount of milk present (cf. chapter 26). Proper sanitation and maintenance of milking equipment are especially important adjuncts to healthy cows, efficient production of milk, and to providing tasty, high-quality milk for humans.

Study Questions

1. What is the most important "harvest" dairy farmers make? Why?
2. Discuss briefly the history and development of mechanical milking.
3. What are the three primary means of harvesting milk? Upon what principle does each depend?
4. Discuss briefly the principles involved in mechanical milking. Differentiate between the milking and the resting phases of machine milking. Why is each necessary?
5. What is the pulsation ratio? How does it affect speed of milk withdrawal?
6. Why is it important to prepare a cow for milking? What procedure should be followed?
7. What is milking rate? Of what significance is it to milk producers? To cows?
8. Define the term *patent*. How does it affect milking rate?
9. Does total milking time affect milk yield? Why?
10. Of what significance is overmilking?
11. What role(s) does vacuum play in mechanical milking?
12. How does a pulsator function? What is its purpose?
13. Identify one or more limitation(s) of wide-bore teat cup liners.
14. What is teat cup flooding? Identify the factors affecting it.
15. What benefits are derived from use of pipeline milking systems?
16. Identify the benefits and costs of bulk milk cooling and storage.
17. What is meant by bulkheading? Identify its merits and limitations.
18. Why is in-line cooling receiving favorable attention on dairy farms?
19. Discuss briefly the importance of cleaning and sanitizing milking equipment. What is CIP? Give three factors that affect CIP.
20. What is milkstone? Name its major components.
21. Discuss briefly the importance of proper care and maintenance of milking equipment.
22. What are the advantages and disadvantages of automated (robotic) milking?

Notes

[1] This is attributed to calves' rapid sucking cycle and to their exertion of a greater differential pressure across the teat canal. USDA studies found that the average differential pressures generated across the teat canal for hand milking, mechanical milking, and calf sucking were 310, 352, and 535 mmHg, respectively (25.4 mmHg equals 1 inHg).

[2] Cows in early lactation commonly require a shorter stimulation period to achieve full letdown than those of late lactation because of higher intramammary pressure associated with higher production in early than in late lactation.

[3] Of course, mammary glands secrete *and* store milk, which allows milk to be extracted directly from mammary glands by the hungry recipient.

References

De Koning, C. J. A. M. 2011. "Milking Machines; Robotic Milking." In J. W. Fuquay, P. F. Fox, and P. L. H. McSweeney (eds.), *Encyclopedia of Dairy Sciences*, Vol. 3 (2nd ed., pp. 952–958). San Diego, CA: Academic Press.

Food and Agricultural Organization of the United Nations. 1989. *Milking, Milk Production Hygiene and Udder Health* (FAO Animal Production and Health Paper 78). Rome: Author.

Jacobs, J. A., and J. M. Siegford. 2012. Invited review: The impact of automatic milking systems on dairy cow management, behavior, health, and welfare. *Journal of Dairy Science* 95:2227–2247.

Spencer, S. B. 2011. "Milking Machines; Principles and Design." In J. W. Fuquay, P. F. Fox, and P. L. H. McSweeney (eds.), *Encyclopedia of Dairy Sciences*, Vol. 3 (2nd ed., pp. 941–951). San Diego, CA: Academic Press.

US Food and Drug Administration. 2013. *Grade "A" Pasteurized Milk Ordinance*. Washington, DC: Author.

Websites

Food and Agricultural Organization of the United Nations (http://www.fao.org/home/en)
GEA Farm Technologies (http://www.westfalia.com)
Lely North America (http://www.lely.com/en/home)
Paul Mueller Company (http://www.muel.com)
University of Illinois, Department of Animal Sciences (http://ansci.illinois.edu)

12

Milking Facilities, Housing, and Equipment

Remember that a cow is a mother, and her calf is a baby.
Care of one affects the welfare and performance of the other.
William Dempster Hoard (1836–1918)

12.1 Introduction

In Ur of the Chaldees, a frieze completed in 3500 BC depicts a dairy scene showing milk containers and strainers. Throughout human history, dairy equipment has provided the means to obtain a valuable resource of nutrition. Although the intensity of labor needed to collect and process milk steadily declined during the twentieth century, in today's global marketplace, labor costs and efficiency are of prime consideration. Procedures that reduce labor costs have a marked effect upon annual production costs per cow.

Dairying in developed countries has become highly specialized. Many herds have grown larger as facilities have become more mechanized. Milking often constitutes over 50% of labor expended on a dairy farm. Depending on equipment and parlor design, an individual may milk from 20 to more than 100 cows/hr. Milking facilities may comprise 20 to 60% of the capital investment in a dairy. Together the annual cost of typical milking facilities and the labor to operate them commonly ranges from 5 to 30% of the cost of producing milk (Reinemann, 2002). These data show high variability depending on numerous factors associated with facilities, equipment, labor, and management. The desired milking routine and capabilities of the labor should be determined before selecting parlor type and size of the operation. The entire milking routine should be performed in the time required to milk a cow in order for the operator to "keep up" with the parlor.

Although annual costs of housing and equipment account for only about 10% of the total costs of producing milk, the impact is greater than is readily apparent. Type of housing and degree of mechanization affect efficiency and cost of labor and feeding, human and cow comfort, cow health, sanitary conditions, and milk yield (figure 12.1). Dairying is a risky business. The rate of return is typically low compared with the level of capital investment required. Profitable dairying requires highly specialized knowledge in a large number of different subject areas. These are some of the reasons dairy farms have continued to be family-owned businesses rather than investor-owned or corporate enterprises. Increasing the debt load increases the risks associated with a dairy enterprise. Reducing debt and financial risk will give the dairy owner/operator more management options for dealing with other risks, including volatile milk price, bad weather, poor feed quality, and disease. In this chapter we discuss milking facilities, housing, and related equipment with emphasis on trends and innovations for increasing the efficiency of producing milk.

12.2 Milking Parlors

Through the years the word *parlor* has come to mean a room for receiving guests and VIPs. And this is an appropriate name for the milking parlor since that is where cows are received to harvest milk. Adoption of the parlor concept enabled increased specialization and mechanization of dairying, especially with the introduction of loose housing arrangements. The term *milking parlor* is sometimes used interchangeably with *milking barn* or *milking shed.*

Figure 12.1 Cutaway view of a typical milking facility on a modern dairy farm (Canada Plan Service).

The type of parlor has evolved over time from the high labor, slow throughput flat station (tiestall) barn to the modern rotary (or carousel) design (figure 12.2). The goal of minimizing labor costs precipitated introduction of automated equipment and prompted dairy farmers to consider new and custom-tailored designs.

The milking parlor allows a concentration of money into a small area so that more technical monitoring and measuring equipment can be devoted to each milking station in the parlor. Automatic prepping devices, crowd gates, automatic takeoff milkers, power gates, and sequencing feed box covers are among some of the developments used to increase outputs of milk per worker hour. Rather than simply milking into a common pipeline, for example, the parlor can be equipped with fixed measurement systems that monitor milk volume and record milking statistics for each animal. Tags on animals allow the system to automatically identify each animal as it enters the parlor. There are many types and arrangements of milking parlors.

Smith et al. (1997) reported how parlor size and type affect premilking hygiene, milking routine, labor efficiency, and operator walking distance. The average times to perform premilking routines were as follows: stripping, 4 to 6 seconds; predipping, 4 to 5 seconds; wiping the udder, 6 to 8 seconds; and attaching the milking unit, 8 to 10 seconds. Based on these times, minimal premilking hygiene would require 14 sec/cow and a full premilking hygiene 25 sec/cow.

An important adjunct to the milking parlor is a well-arranged facility in which cows are held in a room or area just prior to entering the parlor for milking. In some dairies this area is used to wash cows. Others use prep stalls in washing and stimulating cows as they enter milking stalls (figure 12.3 on the next page). Crowd gates are used to move cows from the **holding area** into the parlor. Holding areas should have a rough finish to minimize slipping and sufficient slope to facilitate daily cleaning.

Now let us briefly discuss selected parlors and milking stall arrangements.

12.2.1 Stanchion Barns

In stanchion (tiestall or flat) barns, cows are milked with either bucket milkers or pipeline milking systems. Cows are moved into and through the barn via a center aisle and then positioned side by side. Most stanchion barns have feed managers and a feed alley in front of the cows and barn cleaners behind them to convey manure to an outside loader. Milking cows in this type of barn is extremely labor intensive and requires much stooping and bending to properly clean the udder and attach the milking equipment (figure 12.4 on the following page).

Common to the southern United States, Latin America, and other warm climates, a flat barn, when combined with a pipeline milking unit that is **cleaned-in-place**, is a relatively inexpensive means of milking cows in gang stanchions. In these areas flat barns have no sides, but a roof provides shelter for cows and operators during milking. Cows are handled individually, and the operator can walk around, observe, and work with a cow as needed. Slow milkers and eaters can be pampered, but at the expense of the operator.

12.2.2 Walk-Through or Step-Up Parlors

The desire to reduce the intense labor of a stanchion parlor has led to several types of milking parlor designs in

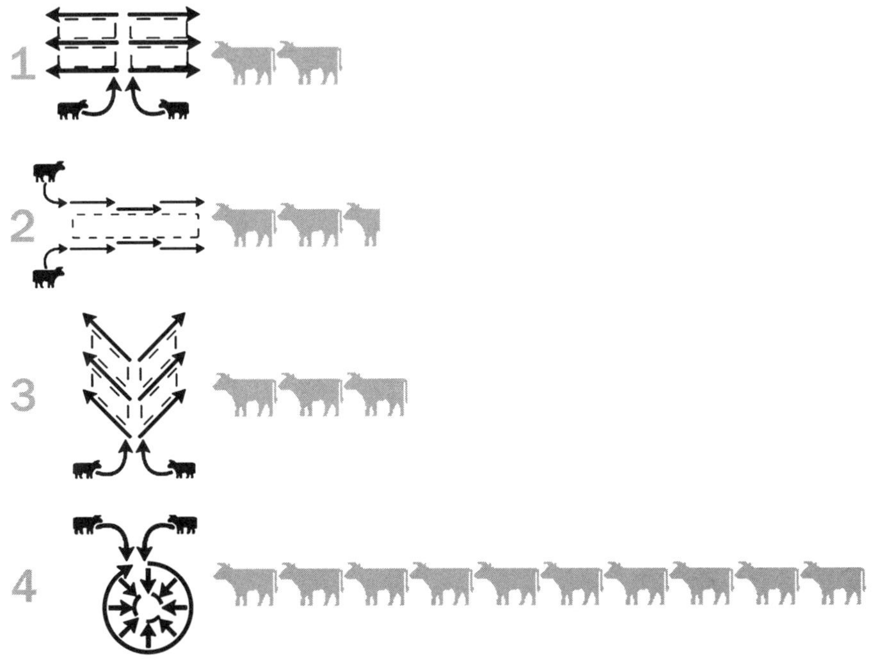

Figure 12.2 Examples of estimated throughputs (cows/hr) of four general styles of milking parlors. (1) Bali = 50; (2) swing-over = 60; (3) herringbone = 75; (4) rotary = 250 (Kleinpeter graphic, by Csand12, http://en.wikipedia.org/wiki/Dairy_farming) (https://creativecommons.org/licenses/by-sa/3.0/deed.en).

which the milker need not bend to be at the level of the cow's udder. Instead, the milker stands positioned so his or her arms are at the level of the cow's udder. Walk-through or step-up parlors are often installed or retrofitted into existing tiestall barns as a cost-effective way of alleviating the demands of the milking chore. In these parlors cows enter from the rear, step up onto an elevated platform for milking, and then exit forward through a headgate. Walk-through parlors are inexpensive, but labor demands are still relatively high.

12.2.3 Herringbone Stalls

Herringbone milking stalls permit group handling of cows, thus eliminating opening and closing of gates to individual milking stalls. Although the group milking principle and herringbone parlor originated in Australia about 1916, its introduction in the United States was not until 1957. The name *herringbone* was coined because the angle in which cows are positioned in the parlor resembles the herringbone weave of cloth.

In herringbone parlors, cows are moved into the parlor and milked one row at a time. Once milking machines have been removed from the milked row, those cows are released. A new group is then admitted into the vacant side and the process repeated. Depending on the size of the parlor, these rows of cows can range from 4 to 60 at a time.

The herringbone parlor is designed to place cows close together at about a 30° angle to the operator's pit (figure 12.5). This arrangement places the cows' udders about 3.0 to 3.5 ft apart, which is less than half the distance between udders in some elevated stall milking systems (figure 12.6). This "angle parking" of cows reduces

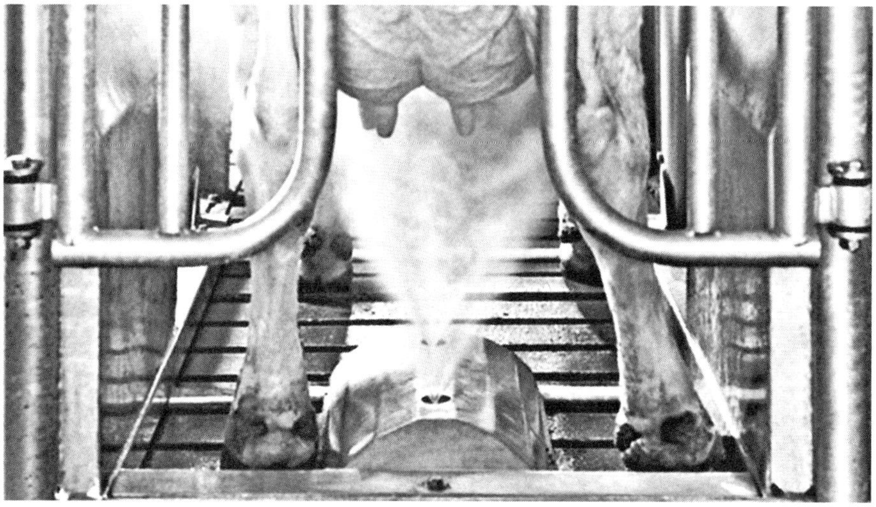

Figure 12.3 Automated prep stall washes cow's udder with warm water and detergent/sanitizer just prior to milking. This stimulates letdown and reduces cow time in milking stall (photo courtesy of GEA Farm Technologies, Inc.).

Figure 12.4 A stanchion barn with an around-the-barn-pipeline and an adjacent milk room (jack_photo/Shutterstock.com).

milker walking mileage and, coupled with the lower cost of stall and parlor construction (assuming the same number of side-opening or walk-through stalls), provides incentive for their installation. Although this parlor arrangement makes udders readily accessible, it reduces visibility and access to the remainder of the cow.[1] Additionally, cows must conform to the group in milking time and eating speed (if fed in the parlor), or the labor efficiency gained in group handling is reduced by waiting on slow cows. (Note the slow cow problem is minimal when herringbone stalls are used in the polygon parlor—see section 12.2.8.)

12.2.4 Parallel Parlors

Parallel parlors are similar to herringbone parlors except that cows stand perpendicular to the operator pit and are milked from the rear, between the cow's hind legs. Advantages are that the cows stand closer together, so the worker walks less of a distance between cows being milked. Disadvantages are that the cow's tail is often in the way and it may be a long reach to the front teats for some workers.

To accommodate the length of the cow, the platform is wider than that of the herringbone parlor. Stall fronts use small gates to position each cow. To assure that each position is filled in order, a series of interlocking fronts prevent a position from being used until the one ahead of it has been filled. The stall spacing in this parlor is too narrow to permit the use of support arms for clusters of teat cups.

12.2.5 Swing-Over Parlors

This type of parlor is designed to accommodate use of milking machines between two lines of cows positioned at a 70° angle to the operator pit. This angle does not allow sufficient exposure of the cow to permit milking from the side, so procedures and equipment are the same as those used with the parallel parlor. Milk lines (pipes) typically are mounted above the head of the operator, which requires lifting the milk a distance of about 3 ft (1 m) from the udder to the lines. Pit width ranges from

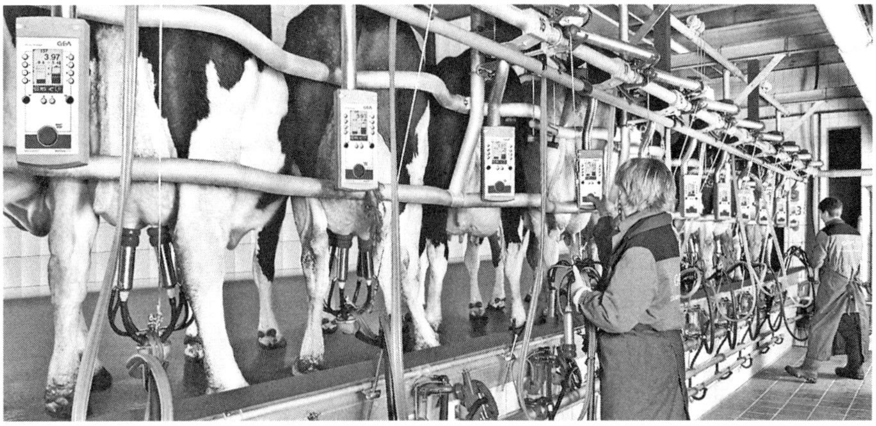

Figure 12.5 High-capacity herringbone milking parlor (from GEA Farm Technologies, Inc.).

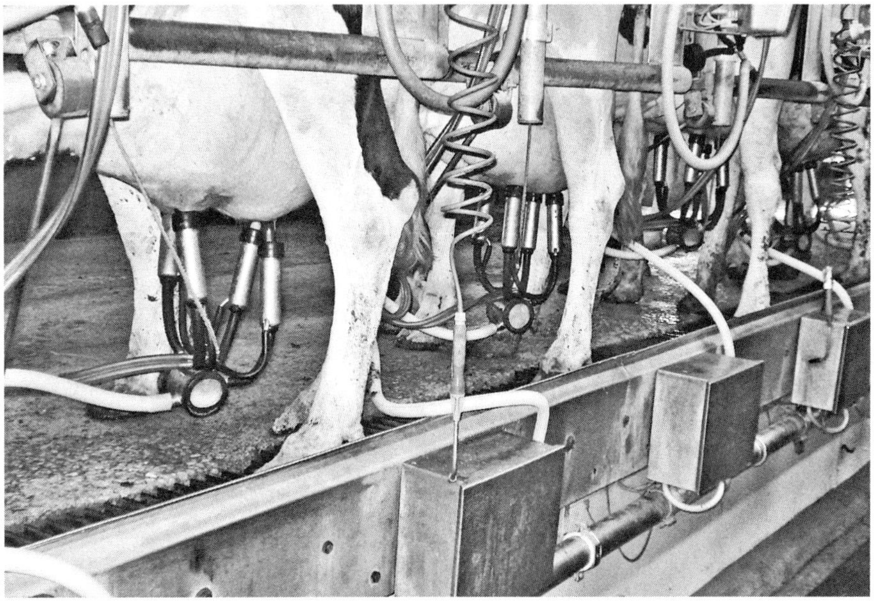

Figure 12.6 Cows being milked in a herringbone parlor with auto take offs and in-line milk yield measurement. Milk flows from the claw to the milk meter (found within the stainless steel box), thence to the stainless steel milk line, which leads to the milk receiver where the milk is pumped into the bulk tank. Milk yield data and cow identification are electronically sent to a computer at each milking (courtesy of Foremost Dairy Research Center, College of Agriculture, Food and Natural Resources, University of Missouri).

5 to 8 ft (1.5 to 2.5 m) depending on the number of operators and parlor length.

The primary advantage of the swing parlor is that fewer milking units are needed and stalls are more simply constructed, reducing initial costs. The main disadvantages are that milker unit hoses hang in the area of operator movement and switching of units from side to side results in inefficiencies due to differences in rates of milking. This parlor is not suited for use of support arms.

12.2.6 Side-Opening Stalls

Side-opening (diagonal) stalls represent a compromise between the flat barn and herringbone. Cows can be handled and cared for individually, more of the cow is visible, the operator is in a pit to minimize stooping, and slow milkers and eaters can be given additional time without disrupting a group of cows (figure 12.7). Side-opening parlors have been considered slow parlors. However, development of reliable prewashing and prefeeding systems has reduced this problem.

Side-opening parlors usually are located at the proximal end of a holding area with two entrance lanes similar to herringbone and parallel parlors. The entrance gate to the milking parlor holds the cow until an empty stall is ready. The parlor may be organized to allow cows to exit in return lanes on either side of the operator area or to move to a single return lane on one side. The use of a single return lane reduces the cost of the parlor and makcs easier catching and/or sorting of cows leaving the parlor. A single return lane does not slow cow flow in this type of parlor because cows are released individually.

Side-opening, tandem, and in-line parlors handle cows one at a time, so a slow-milking cow does not delay the completion of milking and release of other cows in the parlor. These parlors are well suited to farms that practice individual cow care in the parlor.

The number of stalls in a side-opening parlor usually is limited to 4 to 8 for one operator and 8 to 12 for two operators. Long stall lengths may result in excessive walking and difficulty in monitoring distant milking machines. These parlors have increased in popularity because of computer controlled automation. If automatic detachers are used, the detacher can signal removal of a unit, the cow is automatically let out, the gate is closed, and another cow is allowed to enter the stall.

This parlor type has a high stall use rate (7 to 8 cows/stall/hour), which makes it an economical choice for farms using a high level of automation and technology; it is well suited to farms with up to about 400 cows and practice a high level of management. This parlor type is not easily expandable, but if designed properly can be converted into a herringbone or parallel parlor with more milking stalls in the future.

12.2.7 Rotary Parlors

The circular parlor was apparently first conceived, designed, and built on the Walker–Gordon dairy farm in New Jersey during the 1930s. They appropriately named it a *rotalactor.* A rotary milking parlor involves a revolving circular platform with individual stalls that carry each cow for a complete revolution (figure 12.8). A milking unit is attached as cows enter the turntable. Upon completing the circular journey, the milking unit is removed. Parlor movement and the opening and closing of gates are remotely controlled by an operator. The platform can be stopped at any point, and problem or slow-milking cows can be allowed to circle twice.

In such parlors, speed of rotation is set by management. Theoretically, such a milking arrangement will give a full table of properly milked cows with each cycle. A 20-cow rotary parlor with a 6 to 7 min rotation cycle provides a theoretical capacity of 180 to 200 cows/hr. Three opera-

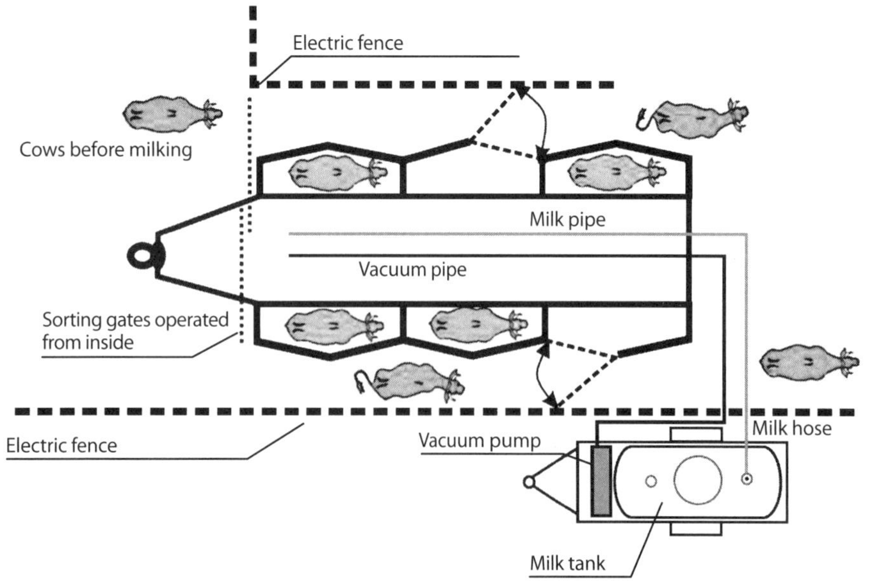

Figure 12.7 Typical diagonal (side-opening) stall parlor arrangement (from Mototecha).

tors are recommended. Duties assigned operators 1, 2, and 3, respectively, are: (1) clean udders, examine quarters, and attach machines; (2) remove machines, dip teats, and separate cows for treatment; and (3) move cow groups into and out of holding pen and move stubborn individual cows onto the table. Rotary parlors have considerable merit for large dairies. Each rotary parlor commonly has the capacity of three or four conventional parlors. Therefore, buildings for rotary parlors and adjacent holding pens can be smaller and cow transfer lanes and traffic patterns more simple.

Figure 12.8 Examples of two types of rotary parlors. (A) Operators service the rotary herringbone parlor from outside and behind the cows, whereas in (B) they work from the inside and beside the cows (courtesy of DeLaval US).

12.2.8 Polygon Milking Parlors

The polygon parlor was conceived and designed by agricultural engineers and dairy scientists at Michigan State University. The system features automated cow traffic control (optional crowd gates move and hold cows), automatic udder stimulation, milking machine detachment, and a unique parlor arrangement. Parlors commonly have four sides with 6 to 10 cows per side (figure 12.9).

Utilizing a basic herringbone stall arrangement, the system includes automatic power gates and feed bowl covers to improve cow traffic in and out of the parlor. When the entrance gate opens, all feed bowls except the one farthest from the entrance are covered. Thus, the leading cow proceeds to the bowl with available feed. Upon reaching that bowl, she actuates a switch that opens the second feed bowl. The second cow activates a switch that uncovers the third bowl, and so on down the line to the last cow, which triggers a switch that closes the rear gate. Nozzles automatically wash and stimulate with warm water the udders of cows to be milked. The operator then checks and dries the udder and attaches the milking machine. From this point forward, the system is automated. When all cows in the row are milked, the feed bowls are covered and the exit gate opens. After the last cow clears, the exit gate closes, the entrance gate opens, and a new group enters.

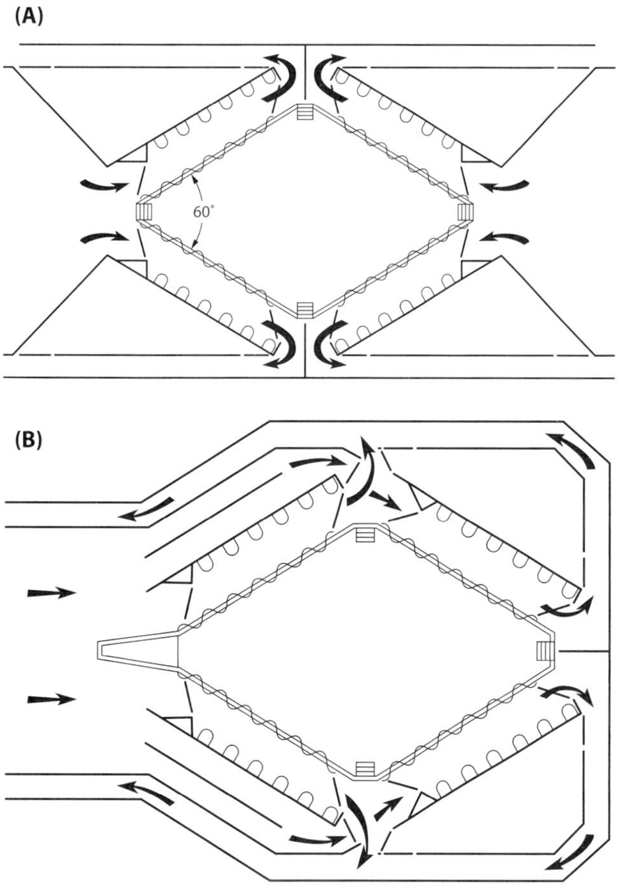

Figure 12.9 (A) A 24-stall polygon parlor with exits to two holding pens. (B) A 24-stall polygon parlor with ramps for exit to a single holding pen (Bickert and Armstrong, 1976).

Automatic detachment of machines is an important aspect of the system, permitting the parlor to be operated by one person. However, most managers use two operators in the 24-stall polygon setup. Each milking unit is fitted with a device that monitors milk flow. When milk ceases to flow, the monitor signals that milking is completed and the machine automatically detaches. During the milking phase, milk flow is monitored as it passes through a closed glass loop. Below the closed loop is a constriction calibrated to keep milk in the loop when the flow is above a predetermined rate. A large bypass tube conveys milk in excess of that rate past the constriction.

12.2.9 Unilactor

This automated means of milking cows was developed at Alfa Laval in Sweden and consists of a U-shaped belt of milking stalls that move on rails. As a cow enters a stall, the operator attaches a milking unit. The stall then moves ahead and the next cow in line enters the next stall. When the slowly moving stalls reach the exit point, milking is complete. The milking unit is removed and the milk jar is emptied. Milk jars and units are then automatically disinfected.

Milking is initiated at a low vacuum, but vacuum increases as milk flow begins. When the udder is nearly empty and the stripping point (< 0.5 lb/min) is reached, vacuum level again decreases, reducing the risk of udder injury from overmilking.

12.2.10 Complying with Sanitary Standards

In building, remodeling, or repairing milking facilities it is wise to first contact a representative of the local health department to be certain of compliance with applicable sanitary regulations. Others from whom one may obtain advice and assistance include service representatives of electric companies, dairy plant field representatives, agricultural extension specialists, dairy breed field personnel, and other dairy farmers.

12.3 Dairy Cattle Housing and Shelter

Types of dairy housing vary greatly among regions of the world. Choice of housing type and degree of mechanization are dictated by each of the following: size of operation, economic status of the owner, environment, labor costs, governmental regulations, and personal preferences. It is obvious that there is no one ideal type of housing for dairy cattle. Instead, each owner/manager must decide after weighing the merits and limitations of each housing and milking management system. Design specifications and plans for various types of dairy cattle housing units are available through experiment stations and universities, as well as from publications such as *Hoard's Dairyman* (http://www.hoards.com).

Needs for housing or shelter vary depending on climate and herd management. A dairy housing system should function to (1) provide a healthy, comfortable environment for maximal milk production; (2) comply with sanitary codes; (3) provide safe, desirable working conditions for operators; and (4) integrate housing with feeding, milking, and waste-handling systems.

In areas in which environmental conditions require housing dairy cattle, two systems are commonly utilized: In the first, cows are confined by stanchions or tiestalls on a hard-surface floor. This type of barn, within which cows are housed, fed, and milked, is prevalent in colder climates.

The second system generally employs a loose housing arrangement that permits freedom of cow movement except during milking. This system may provide a loose housing barn where bedding is allowed to accumulate for several months (figure 12.10) or a **free-stall** arrangement in which cows have access to individual cubicles. Normally, manure is scraped away daily in the free-stall system.

As early as 1913, S. S. Buckley of the University of Maryland experimented with loose housing of cattle. Cattle were not tied in barn stalls but rather allowed to come and go as they pleased. Loose housing lends itself to greater mechanization and labor efficiency than the stanchion barn arrangements of housing and feeding. Other advantages of loose housing include less investment per cow, greater cow comfort (as evidenced by hocks that are less swollen and stiff and fewer injured teats), fewer ventilation problems, and greater flexibility in expansion of facilities. Loose housing requires more bedding than does an operation with free stalls.

12.3.1 Free Stalls

In the early 1920s, Henry Ford saw the possibility of improving efficiency in producing automobiles by having the parts move to the men rather than men moving to the parts. The free-stall housing-milking parlor system brings similar advantages to dairy farming. The cow walks to her feed, to her housing, and to be milked. People do minimal walking. The concept of free stalls originated in 1959 in Washington when a dairy farmer noticed that a certain cow always rested in a nook between a hay wagon and the barn wall. She not only liked the spot, but also stayed clean. He theorized that other cows would like their own stall. So he installed a row of narrow stalls down each side of his loafing barn and the concept of free stall was thus spawned.

Free stalls (commonly called *cubicles* in Great Britain) permit uncontrolled entry and exit by cows into small individual stalls (approximately 4 × 8 ft). They conserve space and bedding; cows are cleaner than those housed in stanchion barns and loose housing (this reduces washing time in the milking parlor and aids in producing clean milk); and liquid manure handling is more feasible than with conventional loose housing. A disadvantage of free stalls is that more injuries occur to feet, legs, and pelvises. Free-stall housing is used routinely for lactating cows, dry cows, and replacement heifers.

Large dairies commonly provide free stalls adjacent to feed troughs (figure 12.11 on the next page).

12.3.2 Bedding

Several substances are used for bedding. Clay is stable and absorbent. Sand is sometimes satisfactory but is easily pushed around and requires frequent leveling. Sand is unsatisfactory with liquid manure systems because it settles to the bottom of storage tanks. Additionally, it is abrasive to manure pumps and lines. Concrete or asphalt may be used when covered with mats or shavings. Peanut hulls make satisfactory bedding material and have acceptable flow characteristics in a liquid manure handling system. Hard surfaces are sloped so urine will drain into alleys.

Figure 12.10 A loose housing barn for dairy cattle. Such an arrangement provides maximal flexibility in feeding and milking systems. In cold weather, heat generated from the accumulating manure pack helps keep cows warm. The system requires more bedding than most others, however (RHIMAGE/Shutterstock.com).

Figure 12.11 A large barn containing free stalls and a continuous feed trough alongside a driving lane for delivery of feedstuffs (courtesy of DeLaval US).

12.3.3 Heat Stress

Heat stress in cows will cause decreased production and fertility. Although some heat stress is unavoidable, its effects can be minimized if certain management practices are followed, such as (1) modify diet to maintain feed intake; (2) increase the amount of cool, fresh water available; (3) provide shade so cows can get out of direct sunlight during peak hours; and (4) ensure good air exchange and increased air flow in the housing environment.

Arizona scientists sought an answer to heat stress by using an evaporative cooling pad. When water was sprayed on the upper edge and fans were used to blow air back into the shaded area, peak air temperatures were 10 to 12°F (5.6 to 6.7°C) lower than those under conventional shades. Cows shaded under evaporative coolers produced 1,200 lb more milk per year than their uncooled **herdmates** (figure 12.12). Moreover, breeding efficiency was greater among cooled cows (Stott and Wiersma, 1973).

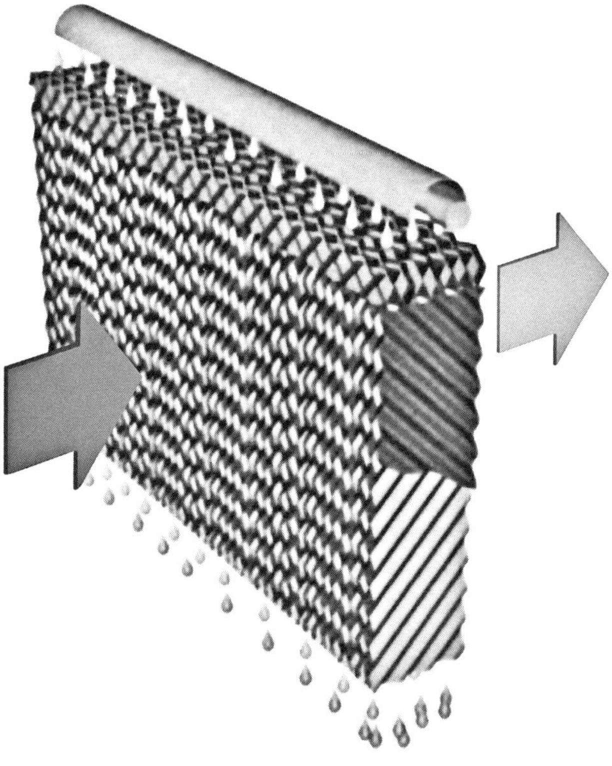

Figure 12.12 Evaporative cooling unit. Fans blow air through such heat exchangers into the shaded area where cows are cooled (courtesy of Rotor Source).

12.3.4 Additional Dairy Farm Facilities

Depending on environmental conditions and degree of specialization, other needed dairy buildings and facilities may include hay storage, silos, a bull barn, and calf, heifer, and dry cow housing. A maternity pen per 25 cows is also recommended. Additionally, space provisions should be made for confining animals for artificial insemination, postpartum examination, pregnancy diagnosis, examining and treating healthy and sick cows, and trimming hoofs and dehorning. In all housing, feeding, and milking systems, maximal use of pavement of lots and cow traffic patterns is recommended. Paved lots provide for ease of cleaning and for the production of high-quality milk.

12.4 Feeding Systems

The foremost objective of a feeding system is to provide dairy animals with optimal feed at the right time and place with minimal labor. Current trends in feeding dairy cattle in the United States are to provide more silages, especially corn silage and haylage.

High-producing cows do not have sufficient time to eat grain in milking parlors. Therefore, supplemental grain should be fed in forage bunks. By dividing cows into two, or preferably three or four, production groups, it is possible to feed additional grain to high-producing cows. Indeed, current trends are toward increased group feeding of cows according to level of milk production. This allows **challenge feeding** of cows that possess the inherent genetic ability to lactate at high levels. For example, group 1 would include cows yielding 70+ lb/d of milk (plus cows in the first 30 days of the test period); group 2 would include cows producing 40 to 70 lb/d; group 3 would be all other lactating cows; and group 4 would include dry cows and bred heifers nearing parturition.

Another exciting potential for feeding dairy cows is the possibility of individualized feeding, even when housing is by groups. For example, Dr. Peter Broadbent of the University of North Scotland developed an automatic animal recognition device that enables animals to be fed individually although they are housed in groups. Each animal carries a recognition key coil suspended from its neck by a nylon chain. Access to feeding boxes is prevented by doors. Each door can be opened only by the animal that carries the appropriate key. When the door is approached by the animal bearing the correct key, the internal door sensor mechanism recognizes the key and permits access to the feed.

12.5 Disposing of Dairy Wastes

Few people go to visit a farm because of the favorable odors of Bossy's barn. Instead, many view farm animals as being major contributors to pollution.[2] If the cow had evolved with a 100% efficiency of **feed conversion**, there would be no problem with dairy farm waste disposal. But only about 50% of the cow's food is digested and assimilated. The balance is excreted via feces and urine. The average quantity of feces and urine excreted daily by a lactating cow is approximately 7 to 8% of her body weight. For a 1,500-lb (680-kg) cow this means disposing of about 110 lb (50 kg) of feces and urine daily.

12.5.1 Methods of Manure Disposal

Basically, there are three means of handling manure: (1) as a solid, (2) as a liquid, and (3) as a processed or dehydrated product (figure 12.13). In areas in which land is abundant, this has many advantages. It can be used as a soil conditioner, a fertilizer, or as bedding for dairy cattle.

As a soil conditioner, manure furnishes food for soil microorganisms which, in turn, improves soil structure by leaving certain structure forming by-products in the soil. Dairy wastes contain, on a dry matter basis, an average of 7% fats and oils, 11% proteins, and 57% carbohydrates. These nutrients, together with many minerals present in raw manure, provide an excellent growth medium for microorganisms. Moreover, the digestible organic constituents of manure eventually form humus and improve the physical properties of soil. This, in turn, reduces soil loss from erosion.

Composting usually refers to decomposition of organic matter in a compost heap made from cut grass, leaves, vegetable debris, or manure. Substantial bacterial activity is present in organic debris—whether of animal or plant origin. Such naturally occurring microorganisms are **facultative** (that is, they can develop in either the presence or absence of oxygen).

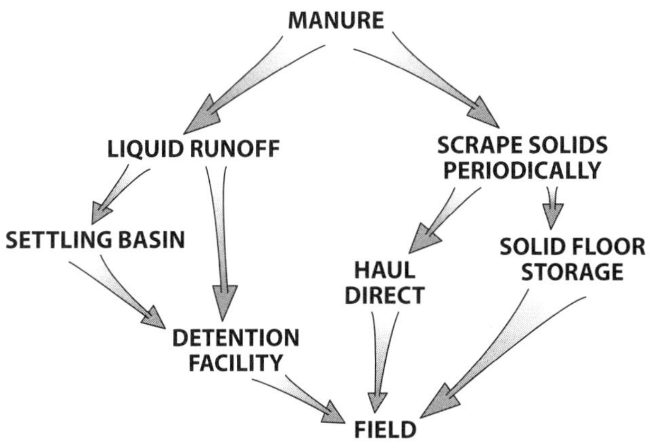

Figure 12.13 Manure disposal options (D. Jones, Purdue University).

An interesting development in waste control is the separation of fiber from the slurry for reuse as bedding or nursery mulch. This practice enables dairypersons to gain a reliable supply of bedding and to greatly improve the pumpability and biodegradability of remaining liquids. Washed fiber is clean, odor free, stable, highly absorbent, and easy to handle. With most fibrous material removed from the slurry, buildup of solids and surface crusting by the remaining liquid in lagoons is minimized.

12.5.2 Solid Manure Handling

Whenever possible, it is usually most economical to scrape alleys, lots, and barns by machine daily, and to convey the manure into a spreader for immediate spreading. Daily spreading has the advantages of no storage costs and minimal problems of odors and flies. However, there are times when the ground may be too wet or frozen. Although manure can be spread on frozen ground, there is a significant loss of nutrients unless it is plowed under within three or four days. Moreover, surface water from rain and/or snow carries much of the manure into streams or bodies of water, contributing to pollution. When daily spreading is impractical, manure can often be handled by accumulating it in reservoirs or composts (figure 12.14 on the next page). This may, however, present a fly problem in warm climates and seasons. Some farmers have confined chickens to manure storage areas to control fly breeding during the summer. Most studies indicate that fly larvae will not develop in manure more than three to four weeks old. Thus, it is fresh manure that attracts and enhances fly growth. Untreated manure can also be dehydrated and sold as organic fertilizer.

Figure 12.14 Straw and manure are added daily until crops are harvested from the nearby field so the decomposed manure can be spread (sima/Shutterstock.com).

12.5.3 Liquid Manure Handling

Ivan Bigalow of the US Steel Corporation brought the liquid manure story to the United States from Europe in 1964. He related the longtime experiences of Scandinavian and German dairy farmers in storing, pumping, and hauling liquid manure. There are several means of handling manure in the liquid state. They all involve a significant investment.

For larger dairy farm operations, a flush system holds the advantage of saving labor. Lots, barn floors, and holding areas having a 3 to 5% slope are easily flushed. A high-pressure system can be used to pump the slurry by sprinklers onto cropland. Such specialized pumping units are being used widely in dairies, beef lots, poultry and swine operations, rendering plants, wineries, and canneries.

An alternative is to inject liquid manure into suitable soil (figure 12.15). Liquid manure can be applied to agricultural cropland through an irrigation system that can benefit the producer and protect the environment. However, the following guidelines should be followed: (1) The annual application depth (in/yr) of manure must be limited to a depth where nutrients can be taken up by the crop grown. (2) The manure must have a solids content and particle size that can easily pass through application nozzles. (3) The application must be uniformly applied to the field at a rate that allows all of the liquid to soak into the soil and not flow away from the point of application into drainage ways, ditches, or bodies of water (Jarrett and Graves, 2014).

The lagoon system is an inexpensive and efficient means for handling large volumes of waste water in order to return it to the land. It requires minimal maintenance and labor input. Slotted floors, which allow animals to deposit their wastes directly into a tank or reservoir, can be used in conjunction with either a flush or lagoon system (a flush system can be linked with a lagoon).

Proper function of a lagoon system depends on bacterial and algal activity. There are two types of lagoons, aerobic and anaerobic. The primary function of an anaerobic lagoon is to liquefy and store waste material until it can be irrigated back onto land. The aerobic lagoon is intended to be self-sustaining (that is, the organic matter is essentially all decomposed). Aerobic dissimilation releases approximately 20 times as much heat per unit of waste decomposed as is released by anaerobic dissimilation. Small water plants (algae) produce oxygen needed for aerobic bacteria to digest manure. Aerobic bacterial action is limited by the quantity of oxygen produced by algae and depends upon sunlight, temperature, and other factors. This means an aerobic lagoon requires a large surface area per animal serviced. Maximal depth of a lagoon is 3 ft.

12.5.4 Environmental Impact

There is great need for communication and a better understanding of the dairy waste problem by all. It is important, for example, that dairy farmers and others be knowledgeable of the necessity of preventing pollution of potable water supplies as well as lakes and streams. For large dairy operations, guidelines need to be developed that are both effective and practical.

Some states and regions have regulations that require manure management plans (MMP). For example, the state of Iowa requires an MMP be submitted to its Department of Natural Resources for confinement feeding operations (totally roofed) that will have a capacity of more than 500 animal units. Deep pits used to store liquid manure in enclosed buildings may present a hazard due to production of high concentrations of hydrogen sulfide (H_2S). Danger to humans is highest when the water supply contains greater than 250 mg/l of sulfates, pH of the manure is below neutral, temperatures are high, and the contents are being agitated. H_2S is especially toxic to brain cells (Donham, 2008).

Figure 12.15 Liquid manure injectors maintain no till benefits and make efficient use of nutrients through proper placement. (A) Ruud Morijn/Shutterstock.com. (B) RPW de Jong/Shutterstock.com.

12.6 Summary

Rising labor and operating costs have caused dairy farmers to increase herd size, making mechanization a virtual necessity. Labor efficiency of milking systems is most appropriately determined by calculating the pounds of milk harvested per hour of labor. Of course, consideration must be given to investment costs and the effect of the system on cow health, lactational persistency, and longevity. Factors affecting cow flow (throughput) and overall milking efficiency include: holding area design, crowd gates, preparation stalls, power gates and doors, and design of cow entrances and exits. Large capital investments in dairy equipment and buildings required to achieve labor efficiency, to meet health requirements of cows, and to satisfy the sanitary requirements for producing grade A milk have increased specialization in dairy farming.

Dairy farmers have a variety of alternative housing, milking, feeding, and waste disposal systems from which to select. The most practical and economical systems or combinations of facilities and equipment for individual milk producers vary with differences in herd size, availability and cost of labor, crops produced and fed, and future plans. In making decisions, owners commonly consider the impact of different housing systems on herd health, culling rates, feed intake and efficiency, human and cow comfort, milk production, economics, and environment.

The most practical use for manure is as a fertilizer of cropland, thus ecologically completing the cycle of soil to feed to cow to soil. Emphases on ecology and environmental quality will continue to affect methods of waste disposal and systems of dairy cattle housing and management. Therefore, waste management is of prime concern among dairy farmers. Liquid storage in underground tanks is widespread but requires large investments in construction, equipment (holding tanks, pumps, and spreading equipment), and labor for distribution. Solid storage is less expensive but may present odor and fly problems. Anaerobic lagoons plus storage ponds or sprinkler systems for lagoon effluent can be used successfully. Equipment to separate fibrous material, for reuse as bedding, is especially useful in warm climates. The effluent material, after separation, can be processed to return the fluids to potable water.

As sanitary codes are changed to comply with federal, state, and local directives, it is especially important that dairy farmers confer with milk sanitarians, dairy field personnel, and extension specialists prior to building or remodeling dairy farm facilities. Also to be considered are cow comfort and health, durability, maintenance costs, mechanization, and future expansion plans.

Study Questions

1. Cite an example of specialization in dairy farming.
2. What is meant by the term *milking parlor*? What types of equipment affect its efficiency?
3. Discuss briefly the merits and limitations of herringbone, side-opening, walk-through, rotary, and polygon milking parlor arrangements.
4. What sanitary codes should be considered during the planning of dairy facilities? Where can dairy farmers go for more information on sanitary codes?
5. Where may dairy farmers obtain plans for dairy cattle housing units?
6. Describe the additional functions dairy cattle housing and shelter systems can provide.
7. Differentiate between stanchion and loose housing barns. What are some merits and limitations of each?
8. What are free stalls? Of what significance is this system of housing dairy cattle?
9. Cite an example of the value of shade to dairy cattle in warm climates.
10. When considering feeding arrangements, why may it be advisable to divide cows into groups on the basis of production?
11. What are three basic means of handling manure on the dairy farm? Discuss briefly the merits and limitations of each.
12. Why is it important that communications concerning agricultural pollution be improved at all levels?
13. Which system of handling wastes on dairy farms appeals most to you? Why?

Notes

[1] Some operators have tattooed the rear udder with 1-in numbers to aid in identifying cows milked in herringbone parlors.

[2] We define pollution as the deposit of any solid, liquid, gas, or odor into air, soil, or water that interferes with their use or potential use.

References

Bickert, W. G., and D. V. Armstrong. 1976, November. *Polygon Milking Parlors* (Extension Bulletin E-1035). East Lansing: Cooperative Extension Service, Michigan State University.

Donham, K. J. 2008, March. Work safely around liquid manure. *Iowa Pork Producer* 45.

Jarrett, A. R., and R. E. Graves. 2014, August. *Irrigation of Liquid Manures* (F-254). Penn State Cooperative Extension.

Reinemann, D. J. 2002. Milking Center Options. University of Wisconsin, Department of Biological Systems Engineering. Madison, Wisconsin (http://milkquality.wisc.edu/wp-content/uploads/2011/09/milking-center-options.pdf).

Smith, J. F., D. V. Armstrong, and M. Gamroth. 1997, March 13–15. Labor Management Considerations in Selecting Milking Parlor Type & Size. Western Dairy Management Conference, Las Vegas, Nevada (http://www.wdmc.org/1997/LaborMilkingParlo.pdf).

Stott, G. H., and F. Wiersma. 1973. Now, cooling cows can be profitable. *Hoards Dairyman* 118:719, 741.

Websites

Milking Parlors

Dairymaster (http://www.dairymaster.ie/milking-parlours/advanced-technologies)

GEA Farm Technologies (http://www.gea.com/global/en/productgroups/milking-systems/milking-parlors)

University of Wisconsin (http://milkquality.wisc.edu/milking-research-and-instruction/milking-parlors/)

Milking Parlor Designs

Chapman Dairy (http://www.chapmandairy.com/milking-parlour-design-cowflow.php)

DeLaval (http://www.milkproduction.com/Library/Scientific-articles/Management/Planning-a-dairy-expansion/)

MidWest Plan Service, Iowa State University (http://www.public.iastate.edu/~mwps_dis/mwps_web/d_plans.html)

Pennsylvania State University Extension (http://extension.psu.edu/animals/dairy/courses/technology-tuesday-series/webinars/dairy-systems-planning-and-building-milking-center-layout-and-components/milking-center-design-1)

University of Wisconsin Cooperative Extension (http://www.uwex.edu/ces/dairymod/milkingcenter)

13

Dairy Herd Health

Pure milk from healthy animals is a luxury of the rich,
whereas it ought to be the common food of the poor.
Mahatma Gandhi (1869–1948)

The authors gratefully acknowledge the contributions of Dr. P. L. Nicoletti, Professor, Department of Pathobiology, College of Veterinary Medicine, University of Florida, Gainesville.

13.1 Introduction

A herd health program is an important aspect of managing a profitable dairy farm enterprise. Without disease control, progress in dairy cattle breeding is seriously impeded and attempts toward effective and efficient feeding and management are not productive. Healthy cows are not necessarily productive ones, but productive cows are necessarily healthy ones.

Dairy herd health is the responsibility of all engaged in the production aspects of providing milk for humans. This includes milk producers, farm employees, and veterinarians. A daily examination of the health of animals is important to early identification and treatment of diseases. An effective herd health program is designed so animals are born healthy and remain that way throughout their growing and producing lives. It enables cows to more nearly demonstrate their full genetic potential to produce milk.

Sanitation is an important aspect of reducing the risk of disease among dairy cattle and is also vital in the production of clean milk. The originator of the ancient, often-quoted Chinese proverb "an ounce of prevention is worth a pound of cure" must have included sanitation in this rationale. It is interesting that by the year 3000 BC, the Chinese had compiled a list of more than 1,000 compounds used in treating diseases—many of which remain in use today—yet they stressed "prevention." We emphasize the importance of sanitation and disease prevention in dairy herd health in this chapter.

As we breed, feed, and manage cows to yield more milk, we can expect to observe more problems associated with high milk production. These include metabolic diseases, cows going **off feed**, reproductive inefficiency, and mastitis. Therefore, a comprehensive herd health program will aid in minimizing disease problems and maximizing milk production and profits.

Some diseases discussed here were not mentioned in earlier dairy science textbooks. Other diseases and/or metabolic disorders will likely be identified in the future. Over the years the trend in dairy farming has been toward larger herds, often in **drylot** conditions. Importantly, concentrating dairy cattle in small areas increases the potential for diseases.

Concerns about possible transmission of antibiotic resistance from bacteria of antibiotic-treated animals to bacteria affecting humans have prompted requests that federal authorities ban **antibiotic** use in animals, particularly in animal feeds. The president of the American Veterinary Medical Association (AVMA) stated in 2011 that

> AVMA strongly supports the judicious use of antibiotics in food producing animals. This includes the administration of antibiotics for treatment, control and prevention of animal diseases to promote food safety, as well as assure the health and well-being of food producing animals. . . .
>
> To restrict certain uses of antibiotics without careful consideration of the risks and benefits to both humans and animals removes a valuable tool in the veterinarian's medical bag for preventing and minimizing animal disease and suffering while also ensuring a safe and wholesome food supply.

Since the United States is free of the two major infectious diseases of cattle—foot-and-mouth disease[1] and rinderpest (commonly called cattle plague)—we have not discussed them in this chapter. However, foot-and-mouth disease is of great importance throughout much of the world. Fortunately, rinderpest has been virtually eradicated worldwide. Important diseases of dairy calves were discussed in chapter 7 and, because mastitis is the most costly disease of dairy cows in the United States, we devote chapter 14 to bovine mastitis. Additionally, public health aspects are discussed in chapter 24.

13.1.1 Health Organizations

The National Animal Health Surveillance System (NAHSS) is a network of government and private partners working together through surveillance to protect animal health. A key partnership component is the NAHSS steering committee, which represents the livestock, poultry, and aquaculture industries; state animal health agencies; diagnostic laboratory organizations; academic institutions; private practitioner organizations; and relevant federal agencies. The purposes of surveillance are rapid detection of introduced diseases and emerging issues, monitoring and providing information for responses against endemic diseases, and measuring regional prevalence of trade-significant diseases. To do so, NAHSS uses ongoing systematic collection, collation, analysis, and interpretation of data and dissemination of information to those who need to assure appropriate action can be taken.

The Center for Veterinary Medicine (CVM), a unit of the US Food and Drug Administration, regulates the marketing of animal drugs, food additives, feed ingredients, and devices for animals. It also performs timely research and monitors product safety and efficacy.

The World Organization for Animal Health, also known by its historical acronym of OIE, is the standard-setting organization for international trade in animals and animal products. The organization's purpose is to promote and facilitate harmonization of sanitary and phytosanitary measures on a worldwide basis with member countries using international standards, guidelines, and recommendations. As of 2015, the organization had 180 member countries and territories and almost 300 reference laboratories and collaborating centers. OIE maintains formal agreements with 35 international and regional organizations, including the Food and Agriculture Organization of the United Nations, World Health Organization, and Codex Alimentarius Commission. The scope of the organization's functions includes animal disease control and prevention, safety of animal products, animal welfare, and food safety.

13.2 The Veterinarian and Herd Health

An important partner in dairy herd health programs is the veterinarian. When health problems arise, the veterinarian should be consulted early—not merely as a last resort! Early diagnosis and treatment by veterinarians are important factors in the rapid recovery of sick animals. Administration of timely veterinary services yields large dividends to dairypersons; services administered late are often too late and result in unwarranted decreased production and/or loss of animals.

Home remedies are frequently expensive in the long run. Dairy farmers should provide good management (sanitation and husbandry practices) but leave medical aspects of herd health to a trained and experienced professional. Many owners have contractual agreements with veterinarians for routine visits to perform specific tests or procedures, for example, the veterinarian will perform rectal palpations on cows that calved 30 to 45 days previously to determine the degree of uterine involution and condition of ovaries; examine animals inseminated 45 to 60 days previously to confirm pregnancy; and attempt diagnosis of possible causes of irregular estrous cycles. Such regular visits permit the veterinarian to plan and/or administer vaccines, to **dehorn** and remove extra teats of calves, and to perform other services associated with dairy herd health and management. The veterinary profession has made great strides in developing methods and techniques for diagnosing and treating diseases and disorders of dairy cattle.

13.3 Vaccination and Herd Health

Vaccination is much like buying casualty insurance.
It is an investment to prevent great potential financial loss.
Paul L. Nicoletti (b. 1932)

The most reliable way to control certain diseases of dairy cattle is through vaccination. When an animal is immunized, it is protected against a particular disease by substances called **antibodies** that aid the white cells (phagocytes) in the blood in destroying disease-producing germs. **Immunity** can be stimulated by injections of microbial suspensions (*vaccines*). These may be killed cells (cells that have been killed in any one of several ways, called **bacterins**) or *attenuated* cells (cells that have been treated to reduce their pathogenicity without affecting their abilities to stimulate antibody production in the host). Immunizing substances can also be obtained by extracting antigens chemically from cells of microorganisms. Although 10 to 14 days are normally required after immunization for animals to develop sufficient immunity or to build enough antibodies to ensure disease resistance, this type of immunity often affords disease protection for a long period. Conversely, antiserum (blood serum taken from animals that have been immunized previously or animals that have recovered from the disease) contains antibodies that provide protection against disease, but since the animal is not stimulated to develop its own resistance, administration of antiserum provides protection of a short duration.

Toxoids are modified toxins of microorganisms (detoxified with heat or formaldehyde) that elicit production of antitoxins. Administration of toxoids provides immunity to the toxin but not the microorganism that produces it. Vaccination against certain diseases is required in exporting dairy cattle.

13.4 Diseases Affecting Reproduction

Illinois studies showed that more than a third of all visits of veterinarians to dairy farms were related to reproductive problems. More than half of all infertility in cattle results from infectious diseases of the reproductive tract. The four major infectious diseases affecting reproductive efficiency of dairy animals worldwide are brucellosis, leptospirosis, trichomoniasis, and vibriosis (table 13.1 on the following page). Let us discuss each briefly.

13.4.1 Brucellosis

This disease is caused by the bacterium *Brucella abortus* in cows and by *Brucella melitensis* in goats and sheep. The latter causes serious economic loss and threatens human health in many countries. Brucellosis is contracted mostly via the digestive tract (e.g., licking or chewing contaminated placentas and/or aborted fetuses). The disease in humans is sometimes called undulant fever and people can become infected by drinking unpasteurized contaminated milk, by handling infected tissues, or by direct contact with infected animals, especially through contact with vaginal discharges (cf. chapter 24).

The causative organisms localize in the mammary glands and in the placenta where cotyledons of the fetal chorion join the caruncles of the uterus. When pregnant cows are infected with *B. abortus*, the chorion may become inflamed and edematous. This decreases circulation of blood and exchange of nutrients to the fetus, which often results in abortion. Although abortion may occur any time after the second month of pregnancy, most abortions caused by *B. abortus* occur after the fifth month of gestation. Retained placentas often accompany brucellosis.

Infected cows develop antibodies to *Brucella* organisms. To test for brucellosis, blood serum is diluted and incubated with a *B. abortus* antigen (killed cells). If a precipitate forms (**agglutination**), the test for the disease is positive. Conversely, if no agglutination occurs the test is negative. However, the blood of previously vaccinated animals may contain antibodies, which can confuse serologic diagnosis. A herd surveillance test for brucellosis is the **ring test** of milk. This test is useful in screening for *B. abortus* antibodies in milk and is usually performed two to four times annually on dairy herds of the United States.

Table 13.1 Symptoms and Prophylactic Measures of Four Major Diseases Affecting Reproductive Efficiency of Dairy Cattle.

Disease	Symptoms	Prophylactic Measures[a]
Brucellosis	Abortion in last trimester, retained placenta, uterine infections	Vaccinate calves at 3 to 6 months of age, use AI, disinfect calving pens
Leptospirosis	Abortion any time (most commonly last trimester), depressed appetite, high fever, bloody urine, anemia, difficult breathing, ropy milk, decreased milk production	Periodic vaccination, sanitation, keep cattle separated from other farm animals, keep cows out of bodies of water, eradicate rats and other rodents
Trichomoniasis	Abortion in first trimester, uterine infections and discharge, irregular estrous cycles, low conception rate	Use AI, periodic examination of bulls, microscopic examination of semen and of swabs from the preputial cavity for causative organism
Vibriosis	Abortion in middle trimester, low conception rate, irregular estrous cycles, a cloudy discharge at time of estrus	Use AI, periodic examination of bulls, microscopic examination of semen and culturing of swabs from the preputial cavity for causative organism

[a] As in controlling most diseases, sanitation of facilities, isolation of infected animals, and a 30-day isolation of purchased animals entering the dairy herd are important prophylactic measures.

There is no practical therapy for brucellosis in cows. Most infected cows abort only once, although they may continue to shed *Brucella* organisms. Aborted placentas and uterine exudates of infected cows are commonly heavily contaminated with *Brucella* and, along with contaminated bedding, should be burned. Calving pens and shelters should be cleaned and disinfected as a prophylactic measure. The incidence of brucellosis among dairy herds is virtually nonexistent in the United States.[2] This is because of cooperative efforts of dairy farmers, veterinarians, and governmental agencies, which have sponsored test and slaughter programs and calf vaccination programs. Calves are vaccinated at three to six months of age with an attenuated vaccine (Strain 19 or RB51).[3] Since nearly all states are essentially free of brucellosis in cattle, there is considerable sentiment to abandon calf vaccination. Strain RB51 does not cause production of antibodies that may confuse diagnosis, as with Strain 19.

13.4.2 Leptospirosis

There are several species of the genus *Leptospira* that cause leptospirosis, but *L. pomona* and *L. hardjo* are the most common in cattle in North America. Leptospirosis was first reported in 1934. The organisms may survive indefinitely in ponds, lakes, or other bodies of water. They accumulate in kidneys of infected cattle and are shed through the urine. The organism is transmitted by direct contact of nasal or oral mucosa or conjunctiva with contaminated urine or water. Animals need not demonstrate symptoms of the disease to be carriers; **asymptomatic** swine are important in the transmission of leptospirosis.

Symptoms vary with the intensity of the disease. Elevated body temperatures (103 to 107°F [40 to 41°C]) are common. Cows usually show signs of depression, loss of appetite, and breathing difficulty. Milk production decreases significantly and milk from infected cows is often off-color, ropy, and occasionally blood tinged, although the milk itself is leptospirocidal (kills the disease bacteria) and the disease is not transmitted through it. Anemia and bloody urine are other symptoms. Abortion may occur at any stage but commonly is observed during the last trimester of pregnancy (usually 1 to 5 weeks following initial infection). Bred heifers and first-calf heifers are especially susceptible to leptospirosis. The aborted fetus, fetal membranes, and uterine caruncles are not commonly infected.

The agglutination test, including several **serovars**, is used for diagnosis and is quite accurate. Isolation of the organism is difficult to achieve since the **pathogen** dies quickly outside the host, except in water. Sanitation is the most dependable means of control. Cows and swine should be kept separated and cows should be kept out of ponds, lakes, and other bodies of water. Immunization every 6 to 12 months is an effective prophylactic measure. Cattle to be sold at auction or exhibited in breed shows should be vaccinated 10 to 15 days before transport to allow time for production of sufficient antibodies. Moreover, cows should be vaccinated for leptospirosis two to three weeks prior to being turned onto pasture in the spring. Antibiotics may be used in the therapy of leptospirosis.

13.4.3 Trichomoniasis

This venereal disease of cattle is caused by the protozoan *Tritrichomonas foetus*. The primary mode of transmission is natural mating of infected bulls to noninfected females. Contaminated semen used in artificial insemination (AI) also can transmit the disease, although this mode of transmission is rare. The causative organism may persist indefinitely in the **prepuce** (foreskin) of infected bulls without symptoms. Pregnancy may result even though the semen is contaminated but is usually terminated within three months after conception. The fetus dies and decomposes in utero. Fluids and pus often accumulate in the uterus (pyometra) prior to being expelled with the decomposed fetus. Discharge of uterine exudates through the vagina during the first half of pregnancy commonly characterizes trichomoniasis.

Diagnosis of trichomoniasis requires microscopic identification of the protozoan in the aborted fetus, the uterine discharges, the preputial fluids of the bull, or in the semen. When diagnosed in cows, sexual rest should be given during the period in which there are uterine discharges and for one to two additional months. During sexual rest the uterus becomes essentially clear of the organism. Infected cows commonly develop immunity to subsequent exposures to trichomoniasis.

13.4.4 Vibriosis

The bacterium *Campylobacter fetus venerealis* causes vibriosis in cows, a disease commonly transmitted through semen of infected bulls during natural mating. It is the most widespread venereal disease of cattle in the United States. Vibriosis is rarely transmitted through artificial insemination because semen used in commercial AI is treated with antibiotics, which destroy the causative organisms. Infected bulls show no outward symptoms of vibriosis although the organism may persist indefinitely in the genital tract, especially in the prepuce. Abortion between the fourth and seventh months of pregnancy commonly characterizes vibriosis in cows, although not all infected cows abort. As with brucellosis, it is the interference of the disease with the exchange of nutrients in the placenta that causes abortion. Retained placentas, irregular estrous cycles, especially prolonged intervals between estrous periods (indicative of embryonic deaths), and low conception are other symptoms of vibriosis (see table 13.1).

Diagnosis is based on (1) detection of causative organism by fluorescent antibody procedure, (2) mating a suspected bull to virgin heifers and then subjecting the heifers to vaginal mucus cultures 4 to 10 days postbreeding, or (3) isolation of the organism from the abomasum of an aborted fetus. **Prophylaxis** is best assured by using AI and semen from vibriosis-negative bulls. (AI studs have veterinarians who examine bulls routinely for vibriosis.) Infected cows should be given sexual rest for 90 days after abortion. During this time cows commonly develop immunity and return to normal fertility. Intrauterine infusion of antibiotics by a veterinarian often eliminates the bacteria from the reproductive tract of infected cows. Vaccines have also been developed and may be used when the incidence of vibriosis is high.

13.4.5 Q Fever

Q fever is a worldwide disease with acute and chronic stages caused by the bacterium *Coxiella burnetii*. Cattle, sheep, and goats are the primary reservoirs although a variety of species may be infected. The bacteria are excreted in milk, urine, and feces of infected animals. Vaginal secretions and uterine fluids released from infected does during birthing carry the highest number of *C. burnetii*. Shedding of the cells in vaginal secretions can continue for weeks after birthing. The disease is enzootic in most areas where cattle, sheep, and goats are kept. In the United States seroprevalence studies have shown antibodies to *C. burnetii* in 41.6% of sheep, 16.5% of goats, and 3.4% of cattle.

The organism is resistant to heat, drying, and some common disinfectants, enabling cells to survive for long periods in the environment. Infection of humans usually occurs by inhalation of these organisms from air containing barnyard dust contaminated by dried placental material, birth fluids, and excreta of infected animals. The disease is global in distribution, with cases reported sporadically or occasionally as outbreaks. Abortions caused by Q fever typically do not recur in sheep, but goats may abort again in subsequent pregnancies if infection recurs.

Diagnosis of Q fever abortion requires testing of the fetuses and placentas from aborted animals. Veterinary diagnosticians typically identify the organism by using special stains applied to microscopic sections of these tissues, and/or through the use of tests to detect the presence of the organism's DNA.

No licensed vaccine against Q fever in livestock is available currently in the United States. Once Q fever is confirmed as a cause of abortion in a herd or flock, treatment of pregnant animals with tetracycline may reduce the risk of further abortions, but is unlikely to eliminate the problem entirely.

Prevention of Q fever in animals is difficult, since infected animals may show no signs of infection with the organism. Newly purchased animals should be isolated from pregnant ewes or does until all pregnant animals have birthed. Animals that abort should be isolated. Tissues and fluids from an abortion should be buried or burned. Contaminated equipment and facility surfaces should be cleaned with a detergent solution and disinfected with a phenolic disinfectant.

13.4.6 Metritis and Pyometra

Calving difficulties and/or retained placentas are often associated with metritis (an inflammation of the uterus). The disease is characterized by a depressed appetite, loss of weight, and a discharge accompanied by an unpleasant odor. Treatment requires the services of a veterinarian in flushing out the uterus and administering capsules containing antibiotics.

Similarly, retained placentas and/or calving difficulties are commonly associated with pyometra (an accumulation of pus in the uterus). Pyometra is aggravated by closure of the cervix and by the inability of the uterus to contract and expel pus.

13.4.7 Retained Placentas

Placental membranes are normally expelled soon after parturition. This allows the cervix to close and lowers the probability of uterine infections. However, the cervix cannot close if membranes protrude. Therefore, if the placenta is not shed within 24 hr, it should be considered retained, and a veterinarian should be called. If the veterinarian finds that the reproductive tract is normal, antibiotics may be deposited in the uterus and within a few days the placenta should be expelled naturally.[4] However, if conditions are abnormal and/or the cow is off feed, the veterinarian may elect to remove the placenta manually and then infuse the uterus with antibiotics. Some veterinarians prefer to give Nature "a second chance." This is because manual removal of the placenta frequently results in pieces of tissue being retained within the uterus, which increases the probability of metritis. Estrogen or gonadotrophins may be administered to cause expulsion of retained placental membranes.

Nutrition may affect the incidence of retained placentas.[5] For example, vitamin A deficient cows have an increased likelihood of retaining placentas. There is a greater incidence of retained placentas in older cows than in young ones.

13.5 Metabolic Diseases of Dairy Cows

Through selection and mating programs, cows have been developed with the ability to secrete large quantities of milk. Yet the stress of high milk production predisposes cows to certain metabolic disorders. We define a metabolic disorder as a pathological condition resulting from a disturbance in a normal metabolic pathway. It usually results from a combination of an inherited weakness with the addition of an abnormal load of a feedstuff (nutrient) or environmental stress. Three important metabolic diseases of dairy cows are considered here.

13.5.1 Parturient Paresis (Milk Fever)

This metabolic disease is caused by **hypocalcemia**, a sharp decrease in blood calcium concentration from the normal 10 mg% to levels of 3 to 7 mg%.[6] It commonly occurs within 72 hr postpartum, and treated cows may have a relapse. The tendency to develop milk fever is inherited, occurring more frequently among certain cow families, certain breeds (Jerseys are the most susceptible, Brown Swiss the least), and among older cows. It is rare at first calving and uncommon in low-producing cows. Approximately 5 to 10% of the dairy cows in the United States demonstrate symptoms of milk fever one or more times during their productive lives. Although the metabolic basis of the disease has not been clearly defined, it is believed to be related to great demands for blood calcium, which are associated with the secretion of large quantities of calcium-rich milk.

Early symptoms of an affected cow include loss of appetite, muscle spasms, an unsteady gait, and/or subnormal temperature.[7] If untreated, cows go down and commonly lie with their heads toward their flanks (figure 13.1). Affected cows are unable to rise because of partial muscle paralysis. Approximately 70% of the untreated cows die within a few hours. However, mortality of properly treated cows is low (3 to 4%). Cows failing to respond to two or more treatments are often called *downer cows*. Such cows often incur injuries while attempting to rise during periods of low blood calcium. Of course, other diseases can also cause downer cows.

The usual treatment of parturient paresis is an intravenous injection of 250 to 500 mL of a solution of calcium gluconate or another suitable calcium salt (figure 13.2). Many solutions contain phosphorus and magnesium in addition to calcium. Treated cows will commonly rise within a few minutes.

Massive doses of vitamin D_2 (20 million IU/d), fed three to seven days prepartum, greatly reduce the incidence of parturient paresis. Because prolonged administration of high levels of vitamin D can have detrimental effects, treatment at this level should not exceed seven days. Studies at the University of Wisconsin indicated that oral administration of 1 mg of 25-hydroxycholecal-

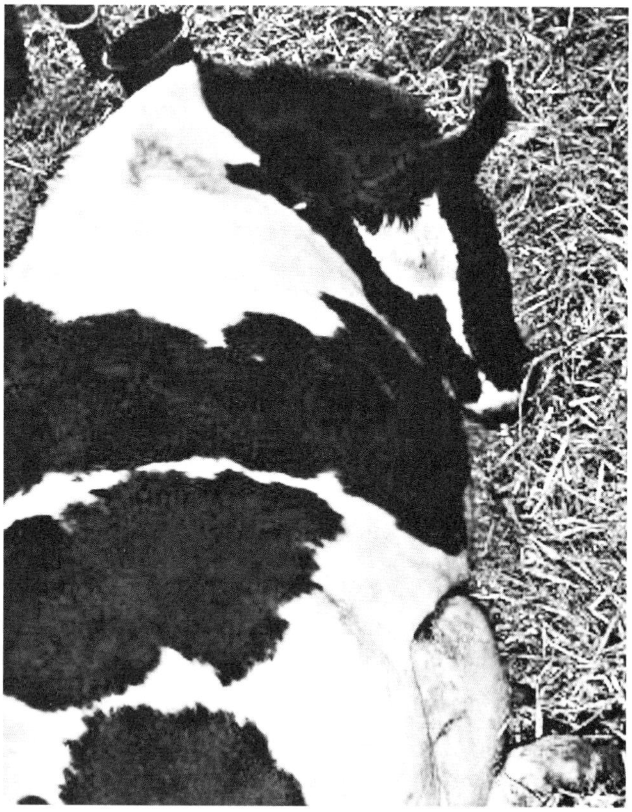

Figure 13.1 Typical position of a downer cow with milk fever (National Animal Disease Information Service, 2015) (http://www.nadis.org.uk).

Figure 13.2 Administering calcium gluconate to a cow with parturient paresis. Although affected cows are essentially immobilized, they respond quickly to an intravenous infusion of calcium salts and rise within minutes (National Animal Disease Information Service, 2015) (http://www.nadis.org.uk).

ciferol (25-OH-D_3) 72 hr before predicted calving and repeated every 48 hr until parturition reduced the incidence of parturient paresis in parturient paretic-prone cows. Vitamin D_3, either ingested in the diet or produced beneath the skin by irradiation of 7-dehydrocholesterol, is converted by an enzyme in liver mitochondria to 25-OH-D_3. Both vitamin D_2 and 25-OH-D_3 are involved in calcium absorption from the intestines and in the mobilization of calcium from bone, which is important in the prophylaxis of parturient paresis.

Another prophylactic measure is feeding a narrow calcium/phosphorus (Ca/P) ratio (1:1 to 1.5:1). One way to narrow the Ca/P ratio is by including 5% monosodium phosphate in the grain mixture, especially to the grain fed during the dry period. A high calcium intake during the dry period increases the probability of milk fever at calving. Therefore, feeds high in calcium (e.g., alfalfa) should be fed sparingly to parturient paretic-prone cows during their dry period. Moreover, several research studies have shown that excessively fat cows are more likely to have parturient paresis than those in average body condition.

Since demands of milk secretion for calcium are great, and since incomplete milking reduces total milk secreted, the practice of leaving about a third to half of the milk in the udder at milking during the first three days postpartum is often helpful in preventing the disease. The same principle of increasing intramammary pressure to decrease milk secretion was used in the early treatment of parturient paresis.

13.5.2 Ketosis (Acetonemia)

First described in the United States in 1929, this metabolic disease usually occurs in high-producing cows 10 to 60 days following calving. It is more common among older cows. It is estimated that 5 to 10% of all dairy cows in the United States demonstrate symptoms of ketosis one or more times during their productive lives. It is typically characterized by inappetance, nervous disfunction, incoordination, and bellowing. Diagnosis is possible by testing milk or urine for ketones.

Researchers have progressed in, but have not completed, their analyses of precisely what causes ketosis. Most scientists agree that carbohydrate metabolism is involved, especially since the concentration of blood sugar is lowered in ketosis. Propionate, one of the principal **volatile fatty acids** absorbed from the rumen, is converted to glucose in the liver and stored as glycogen, which, in turn, is used to maintain blood concentrations of glucose.[8] Since glucose is used to synthesize lactose (milk sugar) and is the major source of energy in lactating cows, demands for glucose become great in high-producing cows. In the absence of sufficient propionate and/or blood glucose, the cow mobilizes body fat to produce milk. However, when carbohydrate catabolism is upset, fat catabolism is also affected and three compounds (acetoacetic acid, 3-hydroxybutyric acid, and acetone—collectively called *ketone bodies*) accumulate in the blood, thus causing ketosis.[9] More than one-third of all cases of ketosis are complicated with other diseases or disorders, such as retained placentas or hardware disease (see section 13.13). Subclinical ketosis (SCK) is defined as concentrations of β-hydroxybutyrate (BHBA) greater than 1.2 to 1.4 mmol/L of blood, and it is considered a gateway condition for other metabolic and infectious disorders such as metritis, mastitis, clinical ketosis, and displaced abomasum.

Accepted treatments for ketosis include (1) intravenous glucose injections, (2) administration of adrenocorticotrophic hormone (ACTH) and glucocorticoids, and (3) feeding molasses or other materials that form glucose. The latter include propylene glycol and sodium propionate.

Thus far, research has not revealed complete prophylactic measures. Increasing caloric intake two to three weeks prior to parturition and during early lactation usually reduces the incidence of ketosis. Limited success has been achieved by daily feeding of 0.5 lb of either sodium propionate or propylene glycol in the concentrate rations of high-producing cows. Low-producing cows are rarely affected with ketosis. Since sodium propionate is somewhat unpalatable and may cause decreased feed intake, it can be administered as a drench. Research at Pennsylvania State University indicated that certain methionine-like compounds are useful in preventing and treating ketosis. Feeding high-quality roughage and a balanced ration, avoiding abrupt changes in the ration during ketosis-susceptible periods, and not allowing dry cows to become excessively fat are recommended prophylactic measures.

13.5.3 Grass Tetany (Hypomagnesemia)

This disease is accompanied by a low magnesium concentration in blood. It was first reported in the United States during the Civil War. Female ruminants are affected mainly, and cattle are more susceptible than goats and sheep. Adult lactating cattle are most susceptible because of loss of magnesium (Mg) in milk. Animals with **grass tetany** commonly walk stiffly, with little flexibility in their hind legs. This presents the impression that the animal is staggering, which gives the disease another label—grass staggers. At first, appetite decreases and the animal appears listless. Later, affected animals become nervous or wild, present an awkward gait followed by muscle tremors and rapid breathing, urinate frequently, and may have diarrhea. The head may be thrown back, and there are intense contractions of limbs and tail. Finally, animals fall, have convulsions, and if not treated, die.

Lactating, pregnant cows require large amounts of magnesium and are especially susceptible. Grass tetany often occurs in cool, wet weather (e.g., it is especially prevalent in the Netherlands). The disease is most commonly observed in spring when cattle are turned onto lush, green pasture.[10] It is also more common among older cows.

The mechanisms involved and the causes of grass tetany are not yet fully understood. Fertilization of crops with magnesium (Mg) and supplements containing magnesium will likely become more widespread. Studies in West Virginia, for example, indicated that feeding cattle blocks containing molasses and magnesium increases magnesium levels in blood plasma and reduces the incidence of grass tetany. Daily consumption of 50 g of magnesium oxide in concentrate feeds affords considerable protection and the cost is minimal. It also can be sprinkled on silage at feeding time or mixed with silage at the time of ensiling (2 to 3 lb/ton of green forage). Response to intravenous calcium and magnesium preparations is usually rapid if treatment is initiated within four hours of the onset of symptoms.

13.5.4 Indigestion

Indigestion or off feed is a condition that occurs most commonly among heavily fed dairy cows. The first symptoms are a depressed appetite and a dull appearance of the eyes. Indigestion is frequently accompanied by slight to severe **impaction** of the rumen. Impaction is characterized by a heavy mass of material in the digestive tract, a deficiency of rumen fluids, and an absence of normal movement of muscles of the rumen wall. Feeding a liter of blackstrap molasses is usually helpful in treating impaction, although stronger laxatives, water infusion into the rumen, and stimulants to reinitiate ruminal activity may be indicated. Prophylactic measures include avoiding overfeeding, making ration composition changes gradually, providing an adequate and convenient supply of fresh water, and feeding mildly laxative feeds.

13.6 Nutritional Diseases

There are several diseases of nutritional origin. These include anemia (iron deficiency), rickets (inadequate intake or balance of calcium, phosphorus, and/or vitamin D), goiter (dietary iodine deficiency), and night blindness (vitamin A deficiency). Other less common nutritional deficiencies and/or mineral imbalances also exist (cf. chapter 8). A pair of diseases—**bloat** and fescue foot—are related to feeding and management. Let us consider them briefly.

13.6.1 Bloat

Bloating is characterized by distension of the reticulo-rumen resulting from an accumulation of an excessive amount of gases, especially carbon dioxide and methane. Bloat results when gases are not expelled as rapidly as they are formed during fermentation. Bloat was first described by a Roman writer in 60 AD. *Feedlot bloat* results from a high-grain, low-forage ration and is usually less acute than is *pasture bloat*. Feedlot bloat and bloat resulting from consumption of **legumes** are similar since excessive foaming of ruminal ingests is involved in both. When cattle graze bloat-provoking legumes, viscosity of rumen contents increases, causing gases to become trapped in a foam that impairs **eructation** and distends the abdomen (figure 13.3). Plant proteins are important in producing rumen foam. This is one reason protein-rich, fast-growing legumes are more likely to cause bloat than older legumes having less protein.

Dr. E. E. Bartley and coworkers at Kansas State University tested a feed additive, poloxalene,[11] which inhibits foaming in the rumen, thus controlling bloat in cows fed fresh legumes. Poloxalene is fed at the rate of 10 to 20 g once daily before cattle are fed green legumes. The compound has no apparent effect on reproductive efficiency, milk production, or milk's composition or flavor. Providing dry forage and/or molasses salt blocks containing polox-

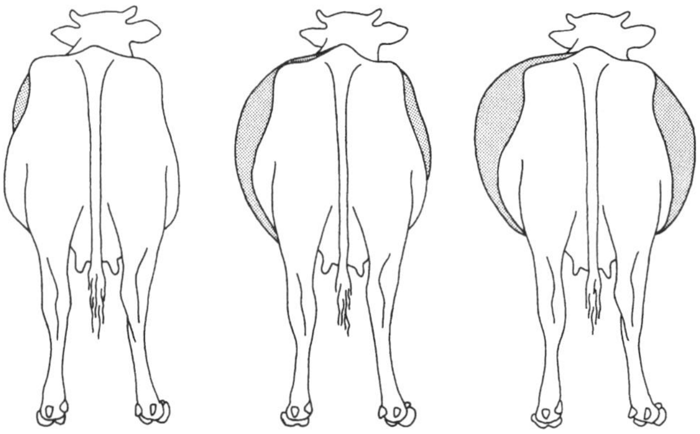

Figure 13.3 Left to right: mild to moderate to severe swelling of the rumen due to bloat. Symptoms are more readily visible on the left than the right side of the animal (Majak et al., 2003).

alene before pasturing cattle on alfalfa or clover reduces the incidence of bloat.[12] In New Zealand, bloat has been prevented by spraying certain oils (3 oz/animal/d) on pasture shortly before grazing. Another prophylactic measure used in New Zealand is the administration of 15 g/d/cow of detergent for 30 to 45 days during the spring. Some antibiotics also reduce the incidence of bloat, probably because of their inhibitory effect on certain rumen microorganisms involved in bloat. However, most antibiotics are not approved by the US Food and Drug Administration for routine use in lactating cows since antibiotics may be secreted into milk.

13.6.2 Fescue Foot

Fescue toxicity results from cattle grazing endophyte infected fescue in which a large amount of ergovaline has been produced. Normally the offending fescue will be tall, season long or lush fall growth that came along after heavy nitrogen application. If sensitive cattle eat enough of the toxin-bearing fescue, constriction of blood vessels occurs limiting blood flow to the extremities of the animal's body.

Symptoms of the disease include loss of appetite and weight, lameness, and stiffness, and in advanced stages the tail switch, ear tips, and feet may slough, and hoofs may split. The disease occurs more frequently during droughts and in cold weather. It is more common in malnourished cattle. Symptoms may appear three to four days after the first consumption of fescue but usually require one to two weeks to appear. No medication is effective in treating the disease. Cattle usually recover if an early diagnosis is made and if they are removed from fescue pasture and given other feeds. Rotating to nonfescue pastures every 10 days is a recommended prophylactic measure.

The endophyte is totally contained within the fescue plant, and can be transmitted only through the seed. Endophyte-free varieties are available, but reportedly are not quite as productive in some climates. Ironically, the endophyte gives the plant insect and disease resistance and may be one reason why fescue is so productive.

13.7 Respiratory Infections

Infectious respiratory diseases of dairy cattle cause increased body temperature, decreased appetite, and decreased production. Let us review selected infections.

13.7.1 Bovine Respiratory Synctial Virus (BRSV)

BRSV is an important virus in the bovine respiratory complex of dairy cattle because of its high frequency of occurrence, serious effects on the lower respiratory tract, and ability to lower host resistance to bacterial infections.

Clinical findings are fever, depression, decreased feed intake, cough, and nasal discharge. Since the symptoms and tissue lesions are similar to those of other respiratory infections, confirmed diagnosis requires laboratory tests.

Although there is no specific treatment, inactivated (killed) and live vaccines are available that may reduce symptoms of BRSV.

13.7.2 Infectious Bovine Rhinotracheitis (IBR)

This acute, contagious viral disease causes inflammation of the upper respiratory tract. The disease, first reported in 1950 and also called *red nose*, causes nasal discharges, loss of appetite, rapid breathing, decreased milk production, elevated body temperature (104 to 108°F [40 to 42°C]), and white pustules on the nasal mucosa of infected animals. Most animals recover within two weeks. Antibiotics and sulfa drugs are used in the therapy of IBR. Clinical diagnosis of respiratory IBR is difficult since the symptoms are also observed in other diseases including shipping fever, parainfluenza infection (PI-3), and bovine virus diarrhea-mucosal disease (BVD-MD).

The IBR virus can also cause infectious pustular vulvovaginitis (IPV) in the reproductive tract of a cow. The virus may be present in the genitalia of bulls, where it is known as infectious pustular balanoposthitis (IPB), and infection can be transmitted to females in natural mating. The virus causes abortion in about 25% of pregnant cattle two to three months after initial infection of the respiratory tract. Artificial insemination of cows using semen of bulls proven free of IPB (by testing blood twice annually) is a good prophylactic measure.

Cattle develop immunity to IBR following primary infection or vaccination. Heifers can be vaccinated from four months of age[13] to one month prior to breeding. Since the modified live virus IBR vaccine can produce abortion, its use should be restricted to nonpregnant cattle. Two IBR vaccines have USDA label clearance for use in pregnant cows. One is a live vaccine administered in the nasal passage. The second is a killed vaccine—two doses are administered two weeks apart. Vaccination usually affords lifetime protection.

13.7.3 Bovine Virus Diarrhea-Mucosal Disease Complex (BVD-MD)

The BVD-MD complex was first described in New York in 1946 and is clinically similar to IBR. MD primarily affects dairy animals 6 to 18 months of age, and when **gastroenteritis** is severe, mortality is often high. The mortality rate of BVD is low, usually less than 5% (although body temperature increases to 104 to 106°F [40 to 41°C]), and there are nasal discharges. As with other viral infections, effective treatment is lacking. Antibiotics are administered to combat secondary infections. Prophylactic vaccines are available through veterinarians but should be restricted to nonpregnant animals.

13.7.4 Parainfluenza-3 (PI-3)

The PI-3 virus was isolated in 1959 and is related but not identical to the human strain of PI-3 virus. This virus causes a mild respiratory disease in cattle, usually characterized by elevated body temperature, depressed appetite, mild coughing, and nasal discharge. Although usually mild in cattle, PI-3 injures respiratory tissues and predisposes animals to more severe infections. Cattle should be vaccinated against PI-3 at four to six months of age.

13.7.5 Hemorrhagic Septicemia (Shipping Fever)

Stress associated with shipment commonly lowers resistance to infection and thus predisposes cattle to this disease. Hence, the term *shipping fever.* This acute infection of cattle was first described in Germany in 1878 and is characterized by an elevated body temperature (104 to 107°F [40 to 41°C]), increased pulse rate, rapid breathing, **lachrymation**, decreased appetite, coughing, rough hair coat, and nasal discharges. Extensive hemorrhaging of internal organs (caused by weakened blood vessel walls), or gastroenteritis may occur. Severely affected animals may develop pneumonia and/or septicemia and die.

Shipping fever (also called pasteurellosis) of domestic animals is caused by *Pasteurella multocida* or *Pasteurella hemolytica*, possibly secondary to virus infection. Veterinarians may administer hemorrhagic septicemia bacterins, hemorrhagic septicemia antiserum, or combinations of hemorrhagic septicemia bacterins with PI-3 attenuated vaccines as prophylactic measures. The dairy herd should be vaccinated for hemorrhagic septicemia in the fall, especially if cows are entering or leaving the herd. The vaccine should contain both PI-3 virus and *Pasteurella* organisms.

13.8 Anaplasmosis (Rickettsemia)

This is an infectious disease of blood among cattle and deer. It is caused by a tiny parasite (obligate intraerythrocytic bacteria of the order Rickettsiales, family Anaplasmataceae, genus *Anaplasma*, species *phagocytophilium*) that destroys red blood cells, resulting in anemia and sometimes death. Early symptoms include depressed appetite, weight loss, and lowered milk production. Advanced symptoms include anemia, yellow skin and mucous membranes (due to accumulation of bile pigments), difficult breathing, and constipation. Difference in virulence between *Anaplasma* strains and the level and duration of the rickettsemia also play a role in the severity of clinical manifestations. Horseflies, mosquitoes, stable flies, and horn flies may transmit anaplasmosis in a herd when they bite infected animals and then bite healthy ones. The disease is also spread by certain species of ticks in which the parasites survive for days or months, and by humans through careless use of unclean needles, dehorning equipment, tattoo instruments, and/or surgical instruments. A commonly used treatment of diseased cattle consists of a single IM injection of long-acting oxytetracycline at a dosage of 20 mg/kg.

13.9 Blackleg (Black Quarter)

The causative organism of blackleg, *Clostridium chauvoei*, survives as a spore in soil (see section 7.8). It is difficult to kill and may persist, especially in swampy land, for many years. Blackleg bacilli are believed to enter puncture wounds caused by barbed wire, thorns, or other sharp objects. Infected animals have a fever (105 to 107°F [40 to 41°C]), and gaseous swellings form beneath the skin, usually on hips and/or shoulders. Accumulated toxin causes stiffness, lameness, or even paralysis. An infected animal loses its appetite, stops ruminating, breathes rapidly, and becomes depressed. Just before death, which commonly occurs within two to five days after exposure, body temperature decreases and convulsions often develop. Muscle tissue is blackened and streaked with dark red areas—hence the name blackleg.

Blackleg occurs in the United States and throughout most of the world. Animals 6 to 18 months of age are most susceptible. Vaccination is the only effective and reliable means of protecting animals against blackleg. The first vaccine was developed in France in 1883.[14] Blackleg bacterins produce a high degree of immunity in 10 to 12 days, which persists for 12 months or longer. In problem areas, calves should be vaccinated at two months of age and again at four months. On most farms it is sufficient to vaccinate against blackleg only at four months of age.

13.10 Cowpox (Vaccinia)

This infectious disease is of minor importance to dairypersons. It is caused by a virus.[15] It is commonly restricted to the skin of the udder and teats and is spread to other cows by machine or hand milking. Following a four- to six-day incubation period, small, dark-red papules appear on teats and udder. Vesicles (blister-like elevations) follow in one to two days and, if not broken, collapse in a few days leaving a brown scab, which dissipates in approximately two to three weeks. Cowpox may be a predisposing cause of mastitis, especially if a pox lesion develops at or near the teat opening. A mixture of equal parts of tincture of iodine and glycerin and/or sulfa and zinc oxide ointments are effective treatments. Of course, milkers should disinfect their hands and/or milking units between the milking of each cow and infected cows should be milked last.

13.11 Displaced Abomasum

Displacement of the abomasum (DA) has become a serious problem in dairy cattle management. It is a condition in which the abomasum becomes distended with fluid and/or gas with a concurrent migration either to the left or right (about 85% are displaced to the left) and upward within the abdominal cavity (figure 13.4). Normally, the abomasum rests on the floor of the abdominal cavity approximately over the midline of the abdomen. It is interesting that DA is minimal when cows are permitted to harvest their own forage by grazing. DA occurs more frequently in older cows and during winter months. An estimated 85% of all cases of DA occur within 20 days following calving. High-grain, low-forage rations lack bulk and are often associated with DA.

A Danish veterinarian observed that feeding high levels of concentrates increased the amount of volatile fatty acids within the abomasum resulting in fewer abomasal contractions. A similar reduction in abomasal motility was observed when hay-fed animals received abomasal injections of rumen fluid obtained from an animal fed a high-concentrate diet. Although diet seems related to DA, additional research is needed to clearly define the responsible factors. Most veterinarians are familiar with the diagnosis and treatment of DA and surgical methods are successful in correcting the problem. There is, however, significant loss of milk production, and the medical expenses incurred are often significant.

13.12 Foot Rot (Necrotic Pododermatitis)

Fusobacterium necrophorum, the major causative bacterium of foot rot, is prevalent in manure and mud. The organism may gain entrance into the hoof through a cut or break in the horny integument or the skin. Predisposing conditions include muddy pastures and barnyards (figure 13.5 on the following page) or standing in unclean pens or gutters of stanchion barns. The disease is most frequent during winter months.

As the infectious organism multiplies, it causes redness, inflammation, and swelling of the tissue and skin between the toes and dewclaws. Inflammation may extend to the pastern and fetlock. The onset of the disease is rapid, and the extreme pain leads to increasing lameness, a decreased appetite, and lessened milk production in lactating cows. As bacteria destroy the tissues, an objectionable odor develops. Foot rot is responsible for more than half of all foot disease problems of dairy cattle in the United States. The first sign is swelling and **erythema** of the soft tissues of the interdigital space and the adjacent coronary band. Typically, the claws are markedly separated, and the inflammatory edema is uniformly distributed between the two digits. In severe cases, the animal is reluctant to bear weight on the affected foot.

Foot rot should be treated by a veterinarian. Most treated animals recover in a few days. Good results are obtained by intramuscular injection with penicillin G for three days. Infected hoofs should be cleaned and kept dry. In mild cases, the caretaker may treat with creolin (a disinfectant) or a copper sulfate paste.

A preventive measure for foot rot is a 4-in deep mixture of lime and 5 to 10% copper sulfate (mix 5 lb of copper sulfate with 100 lb of chlorinated lime) so placed that cows leaving the milking parlor will step in the mixture (figure 13.6 on the next page). Both lime and copper sulfate have an inhibitory effect on the causative bacteria; in liquid form copper sulfate has greater penetrating properties than lime. On many farms, animal research units, bull studs, and other animal production facilities, visitors are required to step through a disinfectant upon entering the premises or certain buildings. Another prophylactic measure is an organic iodine compound mixed with the grain ration. The recommended dosage of organic iodine is 400 mg/d/cow for 10 to 14 days, then 100 mg/d/cow.

13.13 Hardware Disease

Foreign objects such as wire, nails, staples, or other metals may be ingested by dairy cattle. These objects may be forced into or through the stomach wall, accompanied by bacteria. The metal object sometimes penetrates the diaphragm and punctures the heart. Inflammation occurs in affected areas causing great pain to the animal. Cows with hardware disease go off feed, lose body weight, and yield less milk. If diagnosed early, veterinarians can remove the object surgically and most animals will recover.

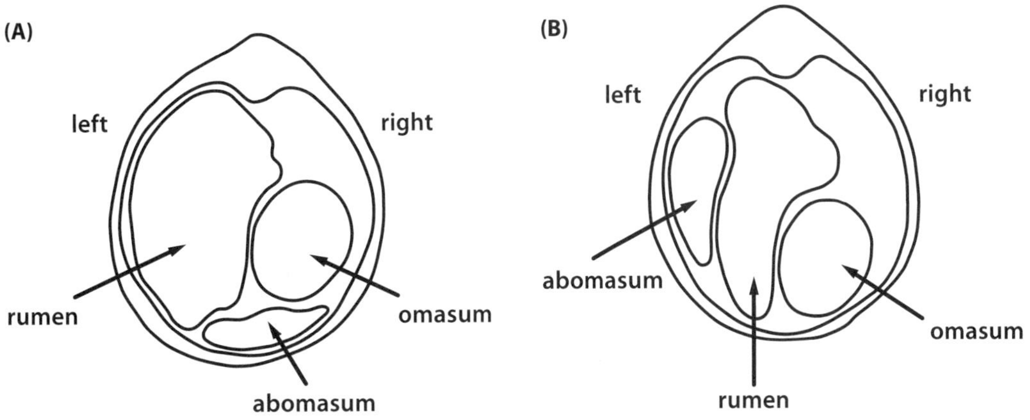

Figure 13.4 (A) Normal position of the abomasum on the right side. (B) The most common location of a displaced abomasum (courtesy of University of Missouri Extension).

To reduce the incidence of this disease, feeds, pastures, lots, and other facilities should be kept free of loose wire or nails. Magnets (figure 13.7) that attract metal objects and prevent their penetration of the stomach wall are available for oral administration with a balling gun. These magnets are made of aluminum, nickel, and cobalt and may hold up to 175 g of metal. They are about 0.5 in × 3 in with smooth, rounded ends. The magnet is installed prior to first calving and usually remains on the floor of the reticulum, where most metal objects are collected, for the life of an animal. Some magnets are designed to be retrieved, cleaned, and reinserted every 6 to 8 months.

Figure 13.5 Poorly drained and/or inadequately cleaned lots serve as excellent mediums for the accumulation and growth of many disease-producing organisms (Nordroden/Shutterstock.com).

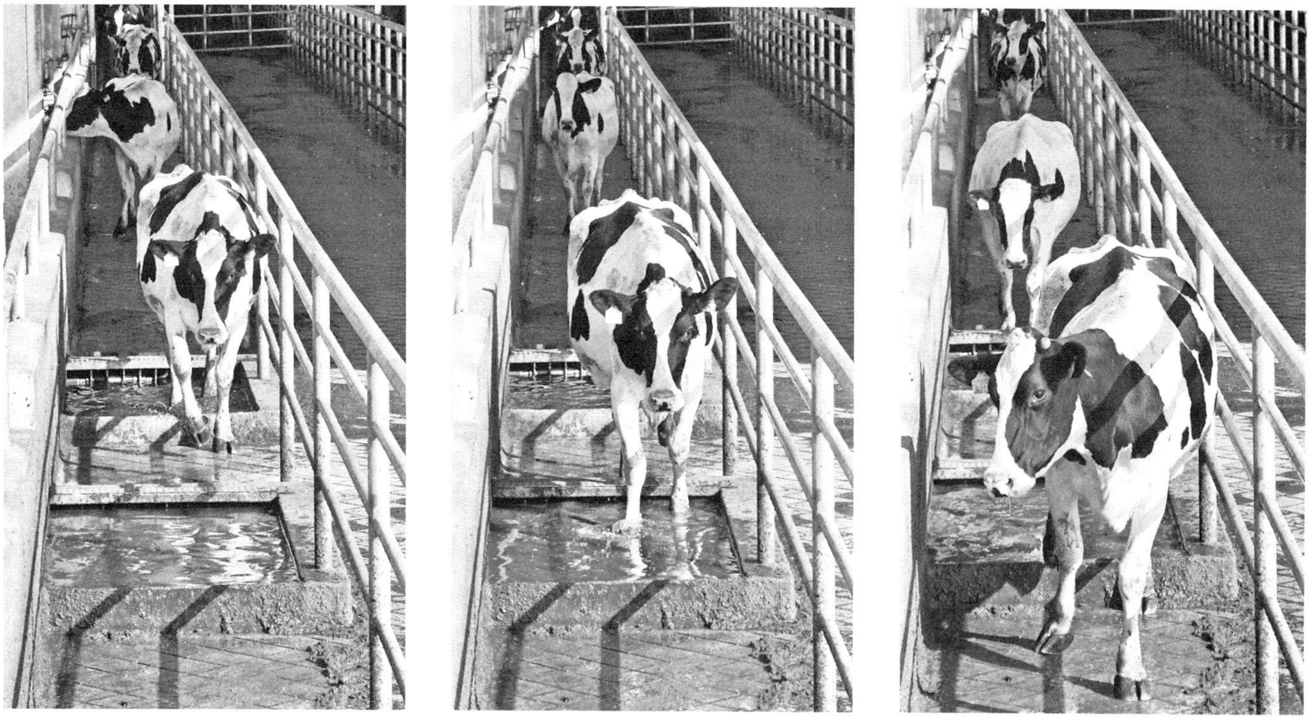

Figure 13.6 Cows walking through a footbath before entering the milking parlor (photos courtesy of *Hoard's Dairyman* magazine's western editor, Dennis Halladay; images found in Cook, 2007).

Figure 13.7 Typical magnet used in controlling metals that gain entrance into a cow's stomach.

13.14 Lumpy Jaw (Actinomycosis)

The causative organism of this disease is *Actinomyces bovis*. Penetrating wounds of the oral mucosa from wire, coarse hay, or sticks allow the organism to enter mucous membranes of the mouth from contaminated grasses and grains. Involvement of adjacent bone frequently results in facial distortion and loose teeth (making chewing difficult). *A. bovis* may attack the bony tissue of the jaw or skull and adjoining soft tissues.[16] When the tongue is involved, it is called *wooden tongue*.

Usually the condition can be arrested with proper treatment, but in advanced stages it may interfere with eating, as teeth often become loose, and may constitute a functional reason for removing affected animals from the dairy herd. Veterinarians often use surgical treatment of lumpy jaw and apply gauze packs with tincture of iodine. Treatment is rarely successful in chronic cases in which bone is extensively involved due to poor penetration of antibacterial agents into the site of infection. In less advanced cases, penicillin may be effective.

13.15 Pinkeye (Infectious Conjunctivitis)

Dr. Frank Billings of the University of Nebraska first reported pinkeye in dairy cattle in 1889. Also called *keratitis*, this contagious disease of cattle is caused by the bacterium *Moraxella bovis* and certain viral agents, including the virus of IBR. Isolation of affected animals is an important aid in controlling the disease. Pinkeye is frequently introduced into a herd by newly purchased animals that were exposed to infection in transit. This is another reason why, when facilities permit, it is a good prophylactic practice to isolate from the herd for two to four weeks animals that have been purchased or exhibited off-farm.

The first symptom of pinkeye is lacrimation (tearing) and mild conjunctivitis. The eyeball soon becomes red, eyelids swell, and the affected animal develops severe sensitivity to sunlight; loss of appetite and body weight and decreased milk production follow. Untreated animals often develop ulcers in the cornea, which result in permanent eye damage and possible loss of sight. Although the disease may occur in any season, it is most prevalent in summer when irritation from sunlight, dust, wind, flies,[17] grasses, weeds, and other irritants render eyes more susceptible.

Infected eyes can be washed with a saturated solution of boric acid followed with a 10% solution of Argyrol (silver nitrate). Ampicillin, penicillin, gentamicin, and kanamycin can be injected subconjunctivally; best results are obtained with injection into the bulbar conjunctiva. Oxytetracycline is considered the drug of choice for systemic therapy because it is concentrated in corneal tissue.

13.16 Tuberculosis and Paratuberculosis

Bovine tuberculosis (TB) is caused by the organism *Mycobacterium bovis*, which may be transmitted to humans through the milk of infected cows. Dairypersons of the United States are required by law to maintain tuberculosis-free herds.[18]

Probably no disease affecting humans and livestock worldwide has been studied as much as tuberculosis. The test-and-slaughter method has been used in the United States with indemnities for diseased cattle to virtually eliminate the disease. The practice has been used widely also in Canada, the United Kingdom, New Zealand, and Australia. The test-and-slaughter policy is the only one assured of eradicating TB and relies on the slaughter of reactors to the tuberculin test. The tuberculin test for detecting tuberculosis was developed in 1890 by Robert Koch. As recently as 1930, approximately 50% of the cattle of certain areas of the United States were infected with tuberculosis.

The USDA Animal and Plant Health Inspection Service (APHIS) (2004) provides *Bovine Tuberculosis Eradication: Uniform Methods and Rules*. In the United States any herd supplying milk to the marketplace shall be within an area declared tuberculosis-free or have passed an annual tuberculosis test. All sexually mature cattle are examined for possible lesions of tuberculosis upon slaughter. Suspected lesions are examined in special USDA laboratories and, if positive, the herd of origin is tested.

Paratuberculosis, known also as Johne's (YO-knees) disease, is common in cattle internationally. However, it is largely unnoticed in dairy herds. In 1996, a study of dairy producers by the National Animal Health Monitoring System (1997) revealed that 45% of producers were either unaware of Johne's disease or only recognized the name. In contrast, by 2007, an updated study revealed that 94% of producers were fairly knowledgeable of the disease or knew some basics about it, and only 1.5% had not heard of the disease (USDA APHIS, 2008). Although symptoms are not well known, the disease can be economically significant in heavily infected dairy herds.

Infection with the causative bacterium, *Mycobacterium avium* subspecies *paratuberculosis* (MAP), occurs most frequently in calves less than six months of age, although all ages of cattle are susceptible. The most common route of infection is by ingestion of bacteria shed in the feces of infected animals. Some calves are infected in the uterus while being carried by infected cows. Separating calves from cows and their manure immediately after calving reduces the potential of newborn calves ingesting MAP. The bacterium can be secreted in colostrum, milk, and semen. Embryo transfer is not a likely route of transmission to recipient cows; however, an infected recipient cow may transmit infection to the embryo/fetus she has received.

A major concern is that clinical signs may not appear for up to five years after infection during calfhood, even though the animal may have been shedding the organism. During that asymptomatic time, there may have been decreased milk production, increased culling rates, decreased fertility, and increased mastitis. One study revealed that in a herd in which 10% of cull cows had clinical Johne's disease, cost on a herd-wide basis was about $200/cow/yr. The majority of these costs resulted from decreased milk production.

An estimated 68% of US dairy herds are infected with *M. paratuberculosis* with large herds most likely containing infected animals. Clinical symptoms include profuse long-term watery diarrhea, marked weight loss, and sometimes, intermittent fever. Infected cows continue to eat, even with severe diarrhea, the contents of which contain no blood or mucous. So the clinical symptoms of Johne's disease are largely nonspecific and may be caused by several other agents. Often, even in severe clinical cases, Johne's disease is not recognized by producers and the animals are simply sent to slaughter without concern for the underlying cause of the disease. Within any infected herd, few infected cows will have diarrhea at any one time.

The two most common ways to confirm a diagnosis of Johne's disease are by culturing the causal bacteria from the feces of an animal or by detecting antibodies to the organism in the blood of an infected cow. Unfortunately, the causative bacterium needs 4 to 16 weeks to grow adequately on diagnostic media. Blood tests yield results more quickly, but both culture and blood tests commonly are positive only in about 50% of animals known to be infected due to variation in incubation periods, intermittent fecal shedding, and the varied immune response of individual animals to infection. Although a negative test does not rule out infection, a positive result is 99% certain to indicate an infected animal.

Infected animals without clinical symptoms shed low numbers of *M. paratuberculosis* in their feces, and serum antibodies are seldom present in adequate numbers for detection. As clinical signs appear and worsen, an animal is more likely to be positive to either a culture or blood test.

13.17 Warts (Verrucae)

Most warts are gray, keratinized, nonmalignant skin tumors. They vary in size and shape and have little economic or apparent health significance in dairy cattle. They are, however, unsightly and cause skin damage. In calves, they usually appear around ears, eyes, and muzzle or on the neck and shoulders. On cows, they occur more frequently on the skin of the udder and teats.

Warts are caused by a contagious virus, which is somewhat species-specific. The virus gains entrance through injuries or breaks in the skin. A vaccine aids in treatment, especially of lactating cows. The vaccine is prepared from virus grown in chicken embryos (bovine-origin vaccine is prepared from virus taken from the warts of cattle). Small warts can be clipped from young animals with sterile, curved scissors. The exposed wound should be treated with tincture of iodine. Since most warts have a rich blood supply they bleed easily when injured and often become infected with bacteria or infested with parasites.

13.18 Winter Dysentery (Winter Diarrhea)

This acute, infectious disease of cattle occurs seasonally in many areas of the world during winter months. A bovine coronavirus (BCV), closely related to the virus that causes diarrhea in neonatal calves, has been implicated as the **etiologic** agent. Coronaviruses survive best at low temperatures and at low ultraviolet light intensities, which can lead to a buildup of virus in the environment during the colder months. Concurrent risk factors, such as changes in diet and presence of other microorganisms, may be required before BCV will cause clinical disease in adult cattle. The disease is characterized by feces that are dark-colored, watery, and sometimes bloody. Body temperature usually increases 1 to 2°F, appetite is depressed, milk production is markedly reduced, and extreme diarrhea leads to dehydration and loss of weight.

BCV is transmitted via the fecal-oral route through ingestion of feed or water contaminated with feces from clinical cases or clinically normal carrier animals. Therefore, infected cows should be isolated to minimize spread of the disease to **herdmates**. Strict sanitation is the best prophylactic measure. Transmission of disease is promoted by close confinement of the animals. Most cows overcome the disease in three to seven days. A veterinarian should be called to treat severe cases.

13.19 External Parasites

Although not normally a major problem in most parts of the United States, external parasites can decrease production if not properly controlled. The three principal ones affecting dairy cattle are lice, mites, and the fungus that causes ringworm.

13.19.1 Lice

The tiny cattle-biting louse *Damalinia bovis* irritates cattle with its sharp claws as it feeds on particles of hair, scabs, and excretions from skin at the withers, along the backbone, and around the tailhead. Infestations are most prevalent in the northeastern parts of the United States. Outbreaks usually occur during late winter and early spring months.

Lice are wingless, flattened insects, usually 2 to 4 mm long. The claws of the legs are adapted for clinging to hairs or feathers. Louse eggs or nits are glued to hairs of mammalian hosts near the skin surface and are pale, translucent, and suboval.

Two species of blood-sucking lice are especially bothersome among young dairy animals. The short-nosed cattle louse, *Haematopinus eurysternus*, and the long-nosed cattle louse, *Linognathus vituli*, may be found attached to the skin at the poll, along the top of the neck, and on the brisket, dewlap, shoulders, and inner surfaces of thighs. These lice feed by piercing the animal's skin and extracting blood. This weakens the animal and may cause anemia, thereby predisposing the animal to other diseases. Lice cause animals to be **unthrifty** and to have rough coats of hair.

Symptoms of lice are biting, rubbing, and scratching of infested areas by animals, resulting in skin irritation and loss of hair. In general, lice are more prevalent on poorly fed and inadequately housed animals. Finding either lice or their eggs attached to hairs confirms diagnosis and calls for treatment with a pesticide, which can be applied as a dust, spray, or dip. Dips containing recommended pesticides are especially effective in killing lice and certain other external parasites (figure 13.8). The veterinarian or extension entomologist will recommend a pesticide (**pediculicide**) to use, but always *read and follow directions on the label*! Treated lice die within a few days but their eggs continue to hatch for several weeks. Therefore, treatment is repeated at least twice at three-week intervals. The life cycle of cattle lice requires about four weeks.

13.19.2 Mange (Barn Itch)

At least four kinds of mites can cause mange in dairy cattle. The most common type, chorioptic mange, is caused by *Chorioptes bovis*. It commonly appears around the pastern areas of the legs and the tailhead and, if not treated, spreads to other areas of the body, including the udder. It does not affect humans. Sarcoptic mange, caused by *Sarcoptes scabiei* var. *bovis*, is most common in horses but may affect cattle also. Lesions start on the head, neck, and shoulders and can spread to other parts of the body. It is also contagious to humans. Psoroptic mange, caused by *Psoroptes ovis*, is more common in

Figure 13.8 A walk-through dip tank used to control external parasites of dairy cattle in South Africa (De Deken et al., n.d.) (creativecommons.org/licenses/by/3.0/).

sheep than in cattle. Demodectic mange, caused by *Demodex* spp., may occur as nodules on the neck and shoulders of cattle, although it is not common.

Symptoms of mange include rubbing, scratching, and loss of hair. After sarcoptic mites burrow into the epidermis and lay eggs, hairless spots containing small papules (small elevations of skin) appear on the neck and jaw. As infestation proceeds, the skin becomes thickened, wrinkled, and covered with crusty scabs. Diagnosis is confirmed by a microscopic examination of skin scrapings. Mange can be controlled in dairy cattle by treating with an acaricide or miticide recommended by extension entomologists or veterinarians. Again, *read the label and follow directions*!

13.19.3 Ringworm

The fungus *Trichophyton verrucosum* is the most common cause of ringworm in cattle. It occurs most frequently in young calves on which round, gray, crusty lesions form, especially around the head and neck (figure 13.9 on the next page). Diagnosis is confirmed by microscopic examination of scrapings from lesions.

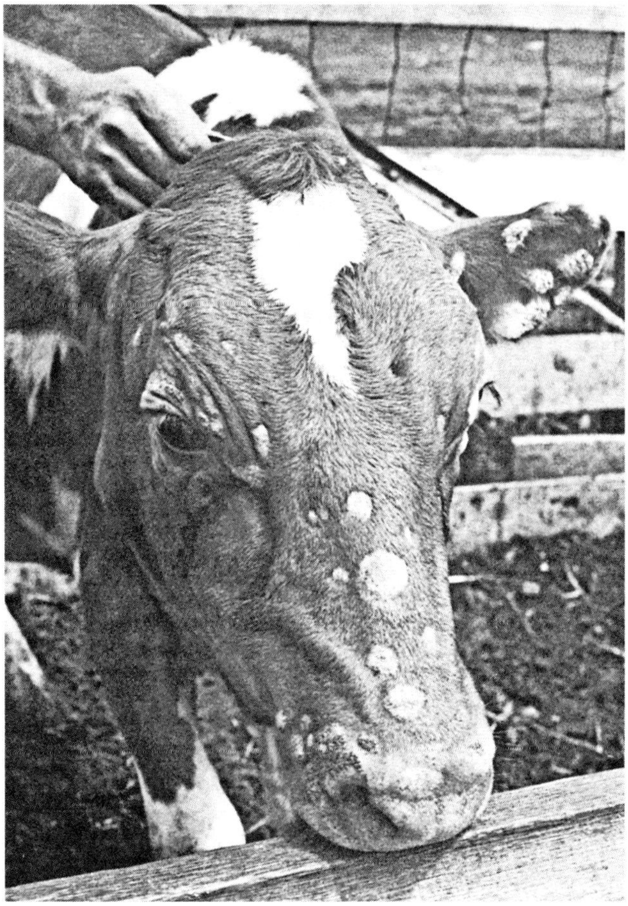

Figure 13.9 Crusty ringworm lesions in a Guernsey heifer. The incidence of ringworm is much higher in winter than in summer months (courtesy of Missouri Agricultural Experiment Station).

Infected animals should be treated by removing the crusty surface of lesions with a stiff brush and a mild soap, then applying tincture of iodine, Lugol's solution, or sulfur iodide on alternate days. The antibiotic griseofulvin is also effective systemically. The authors found that injecting high levels of vitamin A (300,000 IU) had both prophylactic and therapeutic value in Guernsey calves. Since ringworm is transmissible to humans, it is important to wash hands and arms thoroughly after handling and/or treating infected animals.

An attenuated fungal vaccine, in use in some European countries, prevents development of severe clinical lesions and also reduced the incidence of zoonotic disease in animal care workers. Unfortunately, vaccinated animals shed fungal spores for a time after vaccination. No live vaccine has been available in North America.

13.20 Internal Parasites

Internal parasites are more prevalent in large than in small herds. The most common internal parasites affecting dairy cattle are grubs and worms. Grubs may be observed throughout the United States; worms are more prevalent in the South. Infestations of either grubs or worms decrease milk production by an estimated 15 to 25%.

13.20.1 Cattle Grubs (Heel Flies)

Types of grubs and problems with them in dairy cattle vary throughout the United States. In northern states and Canada, adult flies of the northern cattle grub *Hypoderma bovis* deposit eggs at hair roots of the animal's thighs and rear legs in summer months. Larvae live in the animal until spring of the following year. In southern states, heel fly activity varies. In Texas, heel flies, or adult common cattle grubs, *Hypoderma lineatum*, are active in February and larvae emerge in the fall. Cattle have a great fear of heel flies, which explains their running on pasture to shaded areas. Stampeding is often reflected in injuries and cattle going off feed.

In Central and South America, larvae (tropical warbles) of *Dermatobia hominis* are important pests of cattle.

Eggs of cattle grubs hatch in about a week, and maggots burrow through the hide by enzymatic dissolution and move throughout the body in connective tissue, migrating finally to subdermal tissue along the animal's back from withers to tailhead. Encysted larvae elevate the skin, which, upon examination, reveals a breathing pore above each cyst. Grub holes affect the quality of cattle hides exported and sold in the United States. Heaviest infestations usually occur in younger animals under two years of age.

Since adult flies live for about a week and have a limited flying range, destruction of larvae beneath the skin is the preferred method of control. Several systemic insecticides are available for use in salt blocks, mineral mixtures, concentrate rations, and as a drench or spray. These kill grubs in the migrating stage before they damage the hide. However, some should be used only on nonlactating animals. The local veterinarian or extension entomologist can best advise on current FDA-approved control measures.

13.20.2 Tapeworms and Roundworms

With good sanitation and management, worms are of minimal concern. However, many species of worms may infest dairy cattle, decreasing the growth of young animals and the production levels of cows. Tapeworms, *Moniezia expansa*, can be large (5 to 25 ft long [1.5 to 7.5 m]) and may attach to intestinal walls.

Nematodes (small roundworms) threaten herd health if not controlled. They are especially serious in young dairy animals where they cause diarrhea and loss of appetite and weight. Certain species attach themselves to tissues and extract blood, causing anemia in calves.

Roundworms and tapeworms are spread in similar ways. Eggs of adult worms are excreted via feces onto pasture, or in manure that may be spread on pastures. After hatching from eggs, the larvae of roundworms attach to blades of grass and are ingested by grazing animals. Tapeworm eggs are ingested by mites that are later ingested by grazing cattle. Treatment and control of intestinal worms depends on the species of parasite. A veterinarian should be consulted to identify the species and to recommend an **anthelmintic**. Parasitized animals should be isolated.

13.20.3 Lungworms

The parasite *Dictyocaulus viviparius* is the most common lungworm of cattle. It is yellowish-white in color and 2 to 3 in long. The female produces large numbers of eggs, which hatch in the air passages of cattle. Larvae are coughed up, swallowed, and expelled in feces. The larvae can become infective in feces after one to three weeks in warm, moist conditions. Once infective, the larvae can be further dispersed from fecal pats mechanically or by the sporangia of the fungus *Pilobolus*. A proportion of infective larvae will survive on pasture throughout the winter until the following year but, in very cold conditions, most will become nonviable. Therefore, come spring infected carrier animals are the main source of new infections.

Inflammation and irritation of the lungs and bronchi, caused by lungworms, result in coughing. Lungworms are found throughout the United States but do not thrive in dry climates. Cattle usually become parasitized by eating contaminated wet grass or recently cut wet forage. Drinking contaminated water can also be a source of lungworms. Veterinarians make a diagnosis by isolating eggs or larvae from feces or by locating larvae in nasal secretions. The veterinarian will administer an appropriate therapeutic drug after a positive diagnosis. Pasture rotation is a recommended prophylactic measure, since the parasites die when animals are not available for continuation of their life cycle.

13.21 Poisoning

The toxic effects of certain plants have been known for many centuries. Dairy cattle may ingest lethal doses of a number of poisonous plants (e.g., bracken); chemicals (e.g., nitrates, nitrites, herbicides, and insecticides); or heavy metals (e.g., lead). Since cases are often too advanced for treatment by the time they are diagnosed, prevention is especially important.

13.21.1 Lead

Heavy metals such as lead are highly toxic to dairy cattle. Some doses may be accumulative. Most lead poisoning of cattle has resulted from ingesting paint containing lead, but since leaded paint was banned by law in the United States, the incidence of lead poisoning among dairy animals has decreased significantly. However, the danger remains on farms where lead paint was used in the past.

13.21.2 Hydrocyanic Acid (Prussic Acid)

Many plants are capable of causing cyanide poisoning when ingested by cattle because they contain prussic acid (hydrocyanic acid). (Cyanide is the lethal factor.) Such plants include arrow grass, bracken, wild chokecherry, wild black cherry, elderberry, Johnson grass, larkspur, lupine, milkweed, nightshade, sorghum, Sudan grass, poison hemlock, and water hemlock.

Overgrazing and/or lack of adequate pasture increase the probability that dairy cattle will ingest poisonous plants. New shoots from young, rapidly growing plants often contain high concentrations of prussic acid glycosides. Grazing stunted plants during drought is the most common cause of poisoning of livestock by plants that produce prussic acid. The cyanogenic glycoside potential of plants can be increased by heavy nitrate fertilization, especially in phosphorus-deficient soils.

Prussic acid poisoning causes rapid breathing, weakness, twitching, and staggering. The blood of poisoned cows is usually cherry red. When milk producers suspect prussic acid poisoning, they should promptly call upon the services of their veterinarian for therapeutic action.

13.21.3 Nitrates and Nitrites

Increased use of nitrogen fertilizer has increased the incidences of nitrate accumulations in forages. Plants accumulate nitrates more readily under certain conditions such as hot, dry weather. Oat and Sudan grass hays are especially prone to contain nitrates. Factors known to increase the nitrate contents of plants are shade (as in cool, cloudy weather), high available soil nitrogen and phosphorus, drought conditions, deficiency of certain micronutrient elements, and application of the herbicide 2,4-D.

Nitrate poisoning may be transmitted to livestock through water. Some water supplies are naturally high in nitrates, but more often they contain nitrates due to pollution by drainage from lots in which manure is concentrated or in some instances by runoff from frozen fields. Symptoms of nitrate poisoning include a staggering gait, muscular tremors, and a rapid pulse.

Nitrates are reduced to nitrites in the rumen of a cow. When the nitrite ion enters blood, it causes hemoglobin to be converted to methemoglobin (giving the blood a chocolate-brown color), reducing the capacity of red blood cells to carry oxygen. Affected animals may die from anoxia (lack of oxygen). If detected early, an intravenous injection of methylene blue will convert methemoglobin back to hemoglobin.

13.21.4 Moldy Feeds

Moldy corn and other moldy grains and hay or silage containing toxin-producing molds (e.g., *Aspergillus flavus*, *Penicillium cyclopium*, *Penicillium islandicum*, and *Penicillium palitans*) may cause toxicosis and death of dairy cattle.

13.21.5 Chemical Sprays

When properly used and stored, herbicides, insecticides, and pesticides are important adjuncts in providing food for humans and feed for animals. However, improper use or storage can result in cattle poisoning. Additionally, certain chemicals can accumulate in an animal's body fat and be secreted in milk or found in meat. As stated previously, *read and heed the label*! Wise managers learn from their mistakes *and* from those of others. Mistakes include careless handling and/or use of agricultural chemicals so that cows ingest them and then secrete contaminated milk. When this occurs, milk producers lose their permits to sell milk.

13.22 Summary

Wanted!! Milk from healthy cows. It seems such a simple thing when you run down to the store and buy a carton of milk—but there is more to it than that! In fact, it is only through the cooperative efforts of veterinarians, dairy farmers, public health officials, and others that we are able to fulfill this "want" of humans for milk from healthy cows.

Deckle McLean

A comprehensive herd health program designed to prevent disease and injury and to isolate and treat sick animals is an important part of a profitable dairy farm enterprise. Dairy managers should be prompt in calling for a veterinarian. Veterinary service at the *right time* returns large dividends, but at the *wrong time* (too late) costs money—and the animal may be lost.

Veterinarians have the responsibility of providing an effective disease prevention and control program for cows that meets the needs and requirements of dairy producers. The veterinarian is valuable in preventing, diagnosing, and treating dairy herd health problems. Agreements calling for herd visits on a regular basis are recommended, especially for large dairy farm operations.

Three important metabolic diseases of dairy cows are parturient paresis, ketosis, and grass tetany. Nutritional diseases and those affecting reproductive efficiency can render a herd unprofitable. We noted the importance of rotating pastures as well as other means of controlling internal parasites of dairy cattle. Also discussed were external parasites, poisoning of dairy animals, respiratory infections, and other selected diseases and disorders affecting dairy herd health, productivity, and profitability.

Parturition and the sudden demands of high milk production place great stress on dairy cows. A cow is probably more susceptible to infections such as mastitis and to certain metabolic disorders, such as parturient paresis and ketosis, during parturition and early lactation than at any other period. A herd health program developed jointly by the milk producer and the veterinarian will return monetary dividends as measured by increased milk production, decreased labor costs, herd value, and the satisfaction that accompanies owning and managing animals that produce milk.

Study Questions

1. Who is responsible for dairy herd health?
2. Who are the partners in the National Animal Surveillance System? What are the purposes of surveillance by this organization?
3. What is the veterinarian's role in dairy herd health programs?
4. Differentiate between antibiotics and antibodies, and between live and dead vaccines.
5. What is a *toxoid*? What is its function?
6. Be able to name and discuss briefly the major diseases affecting reproductive efficiency of dairy cattle. Include modes of transmission, symptoms, and prophylactic measures.
7. Define metabolic disorder. Describe the causes and symptoms of parturient paresis, ketosis, and grass tetany?
8. What are the prophylactic and therapeutic indications of parturient paresis, ketosis, and grass tetany?
9. Why is the term *milk fever* a misnomer?
10. What federal agency in the United States regulates the marketing of animal drugs, feed ingredients, and devices for animals?
11. What is meant by an animal being *off feed*? What is impaction and how may it be treated?
12. What causes bloat? How may it be prevented?
13. Under what conditions would one expect to observe fescue foot? Cite prophylactic measures.
14. Name and discuss briefly the major respiratory infections of dairy cattle. Include symptoms and prophylactic measures.
15. What causes anaplasmosis? Cite control measures.
16. What are the symptoms and prophylactic measures of blackleg? How did the Navajo immunize animals for blackleg? What principle did their technique involve?
17. Why is a displaced abomasum a more recent phenomenon? How can this be explained?
18. Briefly discuss ways of controlling foot rot of dairy cattle.
19. What is hardware disease of cattle? Cite prophylactic and therapeutic measures.
20. Of what significance is lumpy jaw in animal health?
21. Cite ways and means of controlling pinkeye in dairy herds.
22. Why has tuberculosis of cows received so much attention worldwide through the years? Describe the reasons why eradication efforts in the United States were successful.
23. Which department of the USDA supplies regulations regarding the eradication of bovine tuberculosis?
24. How can milk producers rid dairy animals of warts?
25. Describe symptoms of winter dysentery of dairy cattle.

26. Identify and briefly discuss important external and internal parasites of dairy cattle. Cite ways and means of controlling them.
27. Discuss the effects of poisoning of dairy animals by heavy metals (e.g., lead), hydrocyanic acid, nitrates, moldy feeds, and chemicals.

Notes

[1] Although foot-and-mouth disease does not presently exist in the United States, it has invaded the country nine times, the first in 1870, the last in 1929.

[2] Prior to World War II, brucellosis probably represented the greatest cause of abortion and reproductive inefficiency of cattle in the United States.

[3] Strain 19 produces a transient infection in most cattle, regardless of their age at time of vaccination. But fewer animals vaccinated at three to six months of age demonstrate a positive test as mature animals than animals vaccinated at an older age.

[4] Cows with retained placentas should receive intramuscular injections of broad-spectrum antibiotics to prevent blood poisoning, or septicemia.

[5] Underfed cows have a higher percentage of retained placentas than those receiving adequate nutrition. Moreover, the incidence of retained placentas is substantially higher following twin births than single births (40 versus 7%, respectively).

[6] Calcium ions have two basic neurophysiologic functions: (1) to stabilize the excitability of nerves, and (2) to control the amount of muscle-activating substance that is released from nerve endings. Hence, the depressant effect of calcium deficiency inactivates the cow.

[7] The term "milk fever" is actually a misnomer since there is nothing wrong with the milk of affected cows and they often have below normal body temperatures (no fever). The cow's muzzle becomes dry and her ears cold.

[8] Ruminants absorb only a small proportion of their glucose needs directly from the gut. The balance is produced from propionic acid, lactic acid, glycerol, and certain amino acids.

[9] An increase in blood ketone bodies is expected when energy needs of milk and maintenance exceed energy intake and body fat reserves are mobilized. This occurs in **fasting** or underfeeding, as well as in high-producing cows that are unable to ingest sufficient energy to sustain high milk production, especially in early lactation.

[10] The low dry-matter (high-moisture) content of lush, fast-growing pasture dilutes the magnesium concentration.

[11] Poloxalene is the generic name of product No. 18667, polyoxypropylene polyoxyethylene block polymer (POE) developed by Smith, Kline and French Laboratories, Philadelphia. It is marketed under the trade name Bloat Guard.

[12] The incidence of bloat is reduced when cattle ingest hay prior to being turned onto legume pasture. When suspicion arises concerning a particular feed, it should first be tried on one or more of the less valuable animals of the herd.

[13] Passive immunity from colostrum usually prevails for four to six months.

[14] The Navajo are credited with the first immunization against blackleg in the United States. They soaked a blanket in the body fluids of an animal that died of the disease, then allowed the blanket to air-dry while exposed to the warm sun. Pieces of the dried blanket were then stitched into the brisket of healthy animals, resulting in their development of immunity against blackleg.

[15] Cowpox prevailed in Europe and America in the early 1800s. Its incidence was directly related to epidemics of smallpox in humans, the smallpox virus being transmitted to cows through the act of milking.

[16] A related species, *Actinobacillus*, affects soft tissues.

[17] Control of flies is important in preventing continued eye irritation and spread of the disease.

[18] Of course, a law is only as strong as the public sentiment behind it. In this case, people of the United States believe firmly in a safe milk supply from healthy, tuberculosis-free animals.

References

American Veterinary Medical Association. 2011, December 28. Letter to Honorable Tom Vilsack, Secretary of Agriculture. Schaumburg, IL: Author.

Cook, N. B. 2007, April. Footbath alternatives. *Hoard's West* W-44.

De Deken, R., I. Horak, M. Madder, and H. Stoltsz. n.d. *Ticks: Tick Control*. Pretoria, South Africa: University of Pretoria.

Majak, W., T. A. McAllister, D. McCartney, K. Stanford, and K-J Cheng. 2003. *Bloat in Cattle*. Edmonton, Alberta, Canada: Alberta Agriculture and Rural Development.

National Animal Disease Information Service. 2015. Hyprocalcaemia and Hypomagnesaemia (http://www.nadis.org.uk/bulletins/hypocalcaemia-and-hypomagnesaemia.aspx).

National Animal Health Monitoring System. 1997. *Johne's Disease on U.S. Dairy Operations* (N245.1097). Fort Collins, CO: Author, USDA Animal and Plant Health Inspection Service.

USDA Animal and Plant Health Inspection Service. 2004, December. *Bovine Tuberculosis Eradication: Uniform Methods and Rules, Effective January 1, 2005* (APHIS 91-45-011). Washington, DC: Author.

USDA Animal and Plant Health Inspection Service, Veterinary Services, Centers for Epidemiology and Animal Health. 2008, April. *Johne's Disease on U.S. Dairies, 1991–2007*. Fort Collins, CO: Author.

Websites

US Food and Drug Administration, Animal Drug Products (http://www.fda.gov/AnimalVeterinary/Products/ApprovedAnimalDrugProducts/default.htm)

USDA Animal and Plant Health Inspection Service, Animal Disease Information (http://www.aphis.usda.gov/wps/portal/aphis/ourfocus/animalhealth/sa_animal_disease_information)

USDA Animal and Plant Health Inspection Service, Animal Health Program (http://www.aphis.usda.gov/animal_health/index.shtml)

USDA Animal and Plant Health Inspection Service, Monitoring and Surveillance (http://www.aphis.usda.gov/wps/portal/banner/help?1dmy&urile=wcm%3Apath%3A/APHIS_Content_Library/SA_Our_Focus/SA_Animal_Health/SA_Monitoring_And_Surveillance)

14

Bovine Mastitis

Of all the diseases of the dairy cow, mastitis has the most profound economic impact because it is the greatest threat to the functionality of mammary glands.

W. E. Peterson (1892–1971)

14.1 Introduction

Healthy mammary glands are essential to the secretion of milk that is wholesome to drink and in sufficient quantity to be profitable. The most important disease affecting udder health is mastitis. The word mastitis comes from the Greek words *mastos*, meaning "breast," and *itis*, meaning "inflammation of." Bovine mastitis, then, is inflammation of the mammary gland of the cow.

Some have likened mastitis to the proverbial iceberg, since most mastitis is hidden from sight, occurring in **subclinical** forms for which there are no visible signs of the disease. Unwarily, the producer may continue dangerous practices until clinical symptoms are seen. Approximately 50% of the dairy cows in the United States are infected in an average of two quarters, but only a small percentage of infected glands show visible signs of infection on a given day.

Mastitis is the most costly of diseases affecting dairy cows. Costs can be attributed to reduced milk production, treatments, extra labor, veterinary services, discarded milk, death and premature culling, decreased genetic advancement, and reduced milk quality (table 14.1). Mastitis is associated with increased activity of bovine proteases of milk, such as plasmin and enzymes from somatic cells, which cause proteolysis of the caseins and may reduce yield and quality of cheese.

High **bacterial counts** often accompany mammary gland infections requiring remedial actions to produce milk acceptable to public health agencies. Acute mastitis can cause complete loss of secretory capacity (blind quarters) in the affected gland. If the infection becomes **systemic**, the animal may die. This is especially true of infections with coliform bacteria.

Mastitis not only affects milk production. Researchers in Israel found that approximately 30% of cows with subclinical chronic mastitis exhibited delayed ovulation associated with low plasma concentrations of estradiol and a low or delayed preovulatory surge in luteinizing hormone (Lavon et al., 2010). Researchers in the United Kingdom found a decreased probability of pregnancy among dairy cows during periods of clinical and subclinical mastitis, especially among animals whose somatic cell counts (SCC) exceeded 400,000/mL. The research demonstrated that both clinical and subclinical mastitis are associated with a reduction in reproductive performance, however, this influence varies in magnitude but can be exerted over a prolonged period (Hudson et al., 2012).

Table 14.1 Annual Losses Due to Mastitis.

	Loss per Cow ($)	Total (%)
Decreased milk	140	70
Discarded milk	20	10
Replacement cost	16	8
Decreased sale value	10	5
Drug therapy	8	4
Veterinary services	4	2
Extra labor	2	1
Total	$200	100

Source: Schroeder, 2012.

Lifetime milk yield is also affected. Archer et al. (2013) studied the association between SCC at 5 to 30 d in milk and lifetime milk yield for cows in Irish dairy herds. The data set included records from 53,652 cows in 5,922 Irish herds. They observed large effects on lifetime milk yield when the SCC was high early in the first lactation.

14.2 Causes of Mastitis

Bacterial infections are the major cause of mastitis, but, unlike many other diseases, mastitis has several **etiologic** agents. The most important are the contagious pathogens *Streptococcus agalactiae*, *Staphylococcus aureus*, and *Mycoplasma* spp. Contagious bacteria can live on the skin and multiply inside infected teats and quarters. The bacteria can be spread from cow to cow through contaminated hands and equipment. With the exception of some mycoplasmal infections that may spread through the bloodstream, these organisms gain entrance to the mammary gland through the teat (streak) canal.

Of secondary importance are the environmental pathogens, which include *Streptococcus uberis*, *Streptococcus dysgalactiae*, *Pseudomonas aeruginosa*, and the coliform bacteria (*Escherichia*, *Enterobacter*, and *Klebsiella*). Environmental bacteria are commonly found in the cow's living environment, making them difficult to eradicate. They are also opportunistic in that they can take advantage of circumstances that can cause weakened defenses within the cow, such as before calving or after a drying off period.

Both streptococci and staphylococci are nearly spherical in shape and stain Gram-positive. However, streptococci divide in a single plane as they reproduce, whereas staphylococci divide in two or more planes, forming tetrads or clusters of cells.

Corynebacteria (environmental bacteria) have been found in milk drawn aseptically (without contamination). The authors and their graduate students discovered that *Corynebacterium bovis* organisms, when present, reside almost exclusively within the teat canal. These bacteria have special nutritional requirements that appear to make them well suited to inhabit this small duct. Their physiological importance is debatable, but infected quarters tended to shed more somatic cells than noninfected quarters. In a study of 250 cows, quarters infected with *C. bovis* were not as susceptible to invasion by the usual mastitis-producing bacteria as were noninfected quarters of the same cows. Whether this indicates increased resistance in quarters containing *C. bovis* or decreased ability of the invaders to pass through the teat canal remains a mystery.

14.2.1 Development of Bacterial Infection

The mechanism of entry into the teat canal can be by bacterial growth that spreads through the 10 to 15 mm tubular space from the end of the teat to Furstenburg's rosette and the teat cistern (figure 14.1).

The teat canal is the primary physical barrier that prevents the invasion of pathogens into the udder. Between milkings, the smooth muscles that surround the teat canal contract and the teat canal is tightly closed. During milking, lipid-rich keratin cells line the teat canal. The keratin traps mastitis pathogens as they attempt to pass through the canal; these pathogens are then removed from the canal during the milking process, as the effect of milk flowing through the canal removes the outer layer of keratin (figure 14.2 on the following page).

However, once inside the teat barrier, microorganisms can multiply on tissues in the lower portion of the gland. They may spread slowly, invading only the lower portions of the udder during the first few months. The more **virulent** the organism, the more rapidly will it spread. One method of preventing the spread of microorganisms is **antibiotic** infusion. However, in chronic cases of mastitis, repeated antibiotic infusions can destroy the natural sealant properties of keratin. In addition, formation of the seal can be compromised in cows that produce large amounts of milk at drying off. Use of an artificial teat sealant at the time of dry cow treatment is therefore recommended. It can be used as well on teats that are not chosen for antibiotic injections. With such use it is imperative that strict hygienic practices be observed. Both internal and external sealants exist. The latter have a relatively short effective life and need to be reapplied approximately weekly.

The entry by bacteria into the teat canal can be achieved by a pressure differential between the teat end and the interior of the gland, which may occur as the flow of milk ceases near the end of milking. This permits back flow into the teat cistern as the inflation opens after collapsing onto the teat in the teat massage phase of milking (cf. chapter 11). After overmilking, the low pressure (partial vacuum) in the milking machine may be extended into the mammary glands. Then, when the machine is removed, higher pressures of the atmosphere may assist in entry of microorganisms as the pressure differential is equalized by air entering the teat.

14.3 Physical Response to Bacterial Invasion

As stated previously, bovine mastitis is an inflammation of the mammary gland of a cow. To counteract and reduce this inflammation, large numbers of somatic cells are delivered to the gland. A somatic cell is any biological cell that forms the body of a multicellular organism, other than a germ cell or undifferentiated stem cell. The term is used in the dairy industry to describe the cellular

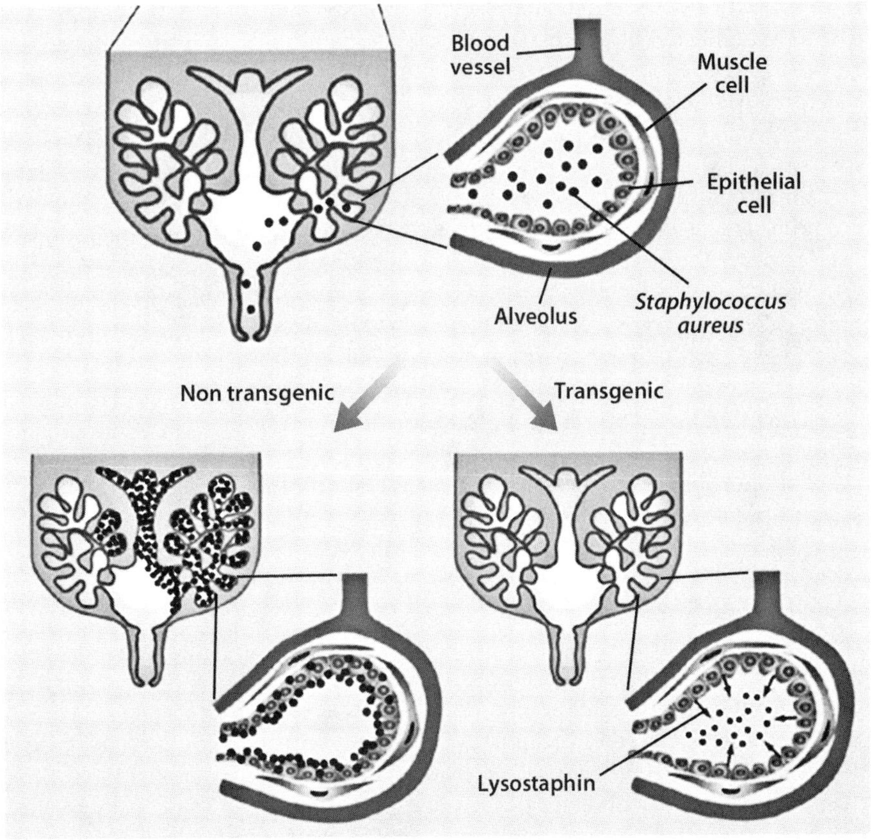

Figure 14.1 Process of microbial infection of bovine mammary glands (© Bob Crimi).

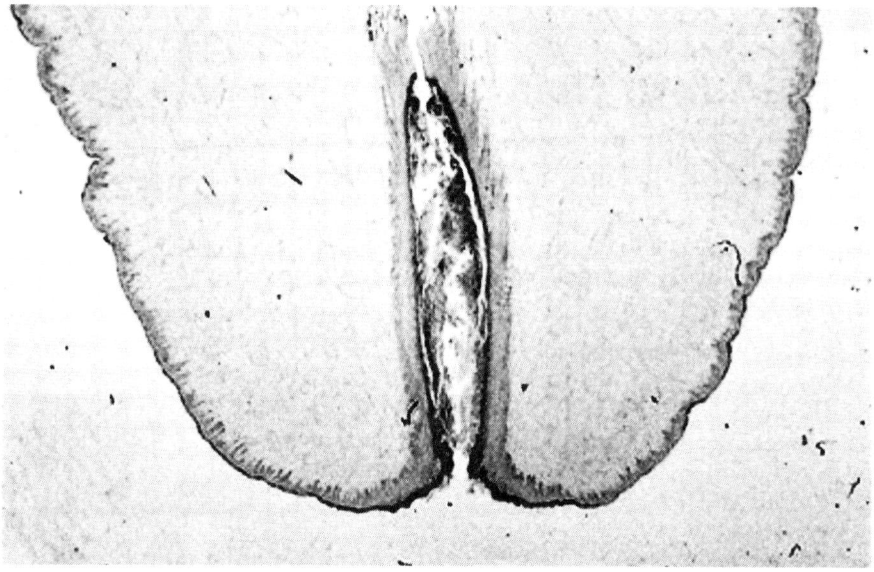

Figure 14.2 Image of the teat end of a lactating bovine cow showing the layer of keratinaceous tissue lining the streak canal (photo by Utrecht University).

components of raw milk because analytical methods do not differentiate among the cell types detected. Most somatic cells are white blood cells, or leukocytes (from the Greek word *leuko* meaning "white"), which defend mammary tissues from invading microorganisms. The five diverse functional types of leukocytes are produced in the bone marrow. They live 6 to 48 hours. **Phagocytes** are the white blood cells that ingest (phagocytose) harmful foreign particles, bacteria, and dead or dying cells. Phagocytes of animals are called "professional" or "non-professional" depending on how effective they are at **phagocytosis** (i.e., the process of digesting or engulfing). Professional phagocytes include neutrophils, monocytes, macrophages, dendritic, and mast cells. They have surface molecules called receptors, which can detect harmful bacteria.

Phagocytosis occurs after a bacterial cell has been bound to receptors on the leukocyte as the cell stretches itself around the bacterium and engulfs it. Once inside the phagocyte, the bacterium is trapped in a compartment called a phagosome. Within minutes the phagosome merges with either a lysosome or a granule to form a phagolysosome. The bacterium is then subjected to an overwhelming array of killing mechanisms and is dead within minutes.

Polymorphonuclear (containing many nuclei) **leukocytes** predominate in mastitic milk and are the principle phagocytes, or body-defending white blood cells, that infiltrate the mammary gland in response to the presence of toxic or antigenic substances, including microorganisms and their toxins.

According to current definitions of udder health in Germany, the cross over of normal cellular defense in the mammary gland into an inflammatory reaction begins at a level of > 100,000 cells/mL (Schwarz et al., 2011). These counts vary with stage of lactation, age, stress of the animals, time and frequency of milking, and season, but primarily in response to udder infection. Somatic cell count is a robust quantitative measurement, but it does not divide the cells present in milk into different cell types.

In the mammary gland, number and distribution of leukocytes are important for the success of udder defenses against invading pathogens. Lymphocytes, macrophages, and polymorphonuclear leukocytes (PMNL) perform an important role in inflammatory responses within the mammary gland. Induction and suppression of immune responses are regulated by lymphocytes, which recognize antigens through membrane receptors specific for invading pathogens. Macrophages are active phagocytic cells capable of ingesting bacteria and cellular debris. They recognize invading pathogens and release chemoattractants inducing the rapid recruitment of PMNL into the mammary gland. These PMNL cells defend against invading bacteria at the beginning of an acute inflammatory process. The distribution of leukocyte types varies in normal milk without any symptoms of mastitis, but in cases of clinical mastitis the proportion of PMNL cells can reach 95%.

Leukocytosis, infiltration from the blood into the mammary gland, can result from physical injury or chemical agents, even when no microorganisms are present. Such inflammation, however, is of short duration, and the cow is expected to exhibit only mild signs of infection as long as the affected tissues remain sterile.

14.4 Characteristics of Major Types of Mastitis

Mastitis can be caused by several different types of microorganisms, each with their own levels of contagion, intensity of effects, and **virulence**.

14.4.1 Infections by *Streptococcus agalactiae*

Streptococcus agalactiae uses the mammary gland as its primary habitat, and eradication of it is both practical and cost effective. Infections usually involve the cisterns and ductal system of the udder, where irritants that cause inflammation are produced. Secreting tissues are destroyed as waste products of the infection accumulate, making it important to remove milk frequently and completely from infected glands. Extensive scar tissues develop when infections continue over several months. Thus, mammary tissues become swollen and hard.

This **obligate parasite** of the udder survives outside the cow for a limited time so that disinfection of equipment and the immediate environment favors elimination of the organism. Most infected glands show clinical symptoms of infection infrequently, but milk usually contains high concentrations of somatic cells. When a large percentage of a herd's mammary glands are infected, somatic cell counts often exceed 1,000,000/mL.

Eradication of *S. agalactiae* infections is favored by the low resistance of the organism to beta-lactam-type antibiotics (penicillin and its semisynthetic forms) and to the fact that the organism seldom invades mammary tissues. This means that the organism can be eliminated from a herd by identifying and treating infected quarters with an effective antibiotic. Samples should be taken from all animals of the herd and cultured to identify infected glands. To limit infections of other cows, culling of those that do not respond to two regimens of treatment is recommended.

Dry cows, heifers, and new entrants into the herd must be included in the eradication program. Calves fed infected milk can spread the organism by suckling themselves or penmates. Once established in an immature gland the infection can persist into the first lactation of the animal.

14.4.2 Infections by *Staphylococcus aureus*

In mastitis staphylococci differ from streptococci in important ways: (1) they invade tissues, (2) some live and grow outside the animal, and (3) they often produce penicillinase, an enzyme that hydrolyzes penicillin, rendering them resistant to penicillin. Furthermore, they tend to develop resistance to other antibiotics.

Staphylococci produce toxins that injure epithelial cells. These substances and the bacteria themselves attract leukocytes and **antibodies** from blood into areas of the injury. Blood plasma may also infiltrate the infected area. Clots often form from plasma components (especially fibrin), leukocytes, and bacteria. These may block ducts through which milk normally flows. The affected alveoli (cf. chapter 10) **involute**. If staphylococci remain **viable**, the inflammatory reaction spreads, and an abscess may form with fibrous tissues surrounding the affected area. This is the body's way of walling off infection. Staphylococci can spread to other portions of the gland from this point. Only when the infection is eliminated before such damage occurs will a gland return to its normal capacity for milk production. Although leukocytes engulf and destroy staphylococci, some strains of this bacterium elaborate substances that destroy the leukocyte, freeing the viable cell. Coagulase-producing staphylococci are likely to resist the degradative enzymes of leukocytes.

Staphylococci vary in their abilities to cause disease. The alpha-toxin produced by some is a powerful **vasoconstrictor** and is the agent responsible for the gangrenous form of mastitis. This toxin produces a clear zone of **hemolysis** around positive colonies growing on cow or sheep **blood agar**. The most frequently produced **hemolysin** of staphylococci is beta-toxin, and it is less irritating to mammary tissues than is alpha-toxin. It causes red blood cells to darken but not to lyse. Some strains of staphylococci produce gamma-lysin or epsilon-lysin, which modify red blood cells. Most cases of clinical mastitis are caused by coagulase-positive strains. These organisms are able to coagulate rabbit or human blood serum. Milder inflammation results from infections by coagulase-negative strains. These organisms, mostly *Staphylococcus epidermidis*, are prevalent on the skin of cows (and humans) and are often found in milk during the first lactation of cows.

Localization of staphylococci within tissues and/or resistance to beta-lactam antibiotics by some strains results in failure to eliminate many infections by therapy. Experiments at the North Louisiana Hill Farm Experiment Station showed that the **efficacy** of treatment among quarters infected with staphylococci was 68%, 67%, 28%, and 13% when California Mastitis Test reactions were slight, weak, distinct, and strong, respectively.

The previous discussion makes it apparent that control and elimination of mastitis caused by *Staphylococcus aureus* is more difficult than that caused by streptococci. Nevertheless, the principles of control are similar. The chain of pathogen transfer between infected and noninfected cows must be broken.

14.4.3 Infections by Nonagalactiae Streptococci

Streptococcus uberis and *S. dysgalactiae* are the major nonagalactiae streptococci. *S. uberis* is distributed in the environment of the cow and on her udder surfaces, especially when intramammary infections by the organisms exist in the herd. It appears they become established by taking advantage of weaknesses in the udder. Though infections are not considered highly contagious, they have caused serious clinical mastitis in some herds. Within one infected herd, one of the authors observed counts of several million per milliliter, and more than 30% of the quarters of 25 cows were producing visibly abnormal milk due to infections by *S. uberis.*

S. dysgalactiae is similar to *S. uberis* in sources and in production of relatively mild inflammation under normal conditions.

14.4.4 Infections by *Mycoplasma* Species

Although antibiotic treatment for mycoplasmal mastitis is ineffective, the disease can be controlled by eliminating infected cows from the herd or milking them separately from the uninfected ones. To do so requires sampling and culturing the samples. Cross-infections with these resistant organisms can occur when antibiotics are administered to multiple quarters from multidose vials, so single-use products are essential. As with other infections, animals added to the herd should be tested before being milked with the existing herd.

Mycoplasma spp. are common residents of the bovine respiratory tract and can be transferred from the lungs to the mammary gland. Poorly ventilated barns increase the risk of mammary gland infection. Milk from cows containing *Mycoplasma* cells can cause respiratory infections in calves fed the milk.

14.4.5 Other Types of Mastitic Infections

The mammary gland can provide a suitable environment for growth of other bacteria, fungi, rickettsia, and viruses. A few of these microorganisms are found frequently enough to deserve noting here.

The coliform bacteria, *Escherichia*, *Enterobacter*, and *Klebsiella*, may cause (1) mild clinical mastitis of a few day's duration from which the cow recovers spontaneously, (2) chronic mastitis, or (3) acute mastitis, including systemic infections that may result in death. Severity depends on the virulence of the organism and resistance of the cow. Intestinal tracts of humans and animals are sources of *Escherichia coli*, whereas species of *Enterobacter* are common on vegetation. Infections by *Klebsiella pneumoniae* have been associated with the use of sawdust bedding. Animals are thus in frequent contact with these bacteria and may maintain significant levels of antibody to them. Many strains of coliform bacteria are not pathogenic.

Pseudomonas aeruginosa is often called the "bacillus of blue-green pus" because it produces a water-soluble blue-green pigment that causes intramammary infiltration of leukocytes, a high concentration of which becomes pus. Infections often disappear spontaneously. However, acute infections, as with some coliforms, are characterized by quick onset, high temperature, and a severe response of the cow. Commonly used antibiotics are often ineffective.

Other organisms identified from mastitis cases include *Bacillus cereus*, *Brucella abortus*, *Corynebacterium pyogenes*, *Leptospira pomona*, *Listeria monocytogenes*, *Mycobacterium bovis*, *Nocardia asteroids*, *Pasteurella multocida*, and *Cryptococcus neoformans* (a yeast).

14.5 Susceptibility to Infection

Cows vary in susceptibility to invasion and in their response to invading organisms—the infection process. Even quarters within the same udder may differ in response. Cows with **pendulous udders** have a relatively high incidence of mastitis. This is especially true if the udder floor extends below the level of the hocks because of the greater likelihood of physical injury.

The teat canal is a barrier to infection, so in the first lactation, when the canal is more tightly closed than in succeeding lactations, fewer organisms are able to invade. Researchers at the National Animal Disease Center in Ames, Iowa, have correlated high tendencies toward infection with larger diameter of the teat canal.

Injury at the teat end often predisposes the gland to infection, probably because the duct will not close completely and/or damaged tissues provide nutrients for bacteria. The keratinaceous material inside the teat canal is a sebum-like mass of degenerating cells and lipid materials (cf. chapter 10). The sloughed cells seal the teat, and the long chains of fatty acids within the lipids are **bacteriostatic** to *S. agalactiae*. Overmilking can erode this important protective barrier of the teat canal.

Higher rates of milking have been associated with higher rates of infection. This tends to emphasize the importance of a tightly closed teat duct. A study of first lactation animals revealed infection rates of 10% versus 67% when peak milk flow rates were 2.4 and 6.8 lb/min, respectively (Dodd and Neave, 1951).

Many factors that determine susceptibility are affected by genetic makeup of the cow, but breeding for mastitis resistance, other than for desirable udder conformation, is not yet recommended. Recent research at The Ohio State University (Gott, 2012) disclosed a relationship between ruminal acidosis and lowered immunity to endotoxin-producing bacteria. The prolonged exposure of cows with acidosis to the release of endotoxin caused by the death of rumen microbes results in the cow becoming tolerant or nonresponsive to endotoxin produced in the mammary gland by coliform bacteria. Therefore, resistance to such an infection is lowered. Ruminal acidosis can occur when cows are fed high energy diets for a prolonged period or by sudden shifts from low to high energy diets (see section 8.8).

14.6 Environmental Factors That Influence the Incidence of Mastitis

It is readily evident that a cow is exposed to infectious mastitis bacteria in direct proportion to the incidence of mastitis in the herd and pathogens in the environment. Probability of infection is increased when the environment introduces stresses including poor nutrition, physical injury, and extreme cold. Limiting physical injury by providing individual stalls (**free stalls**) is recommended. In cold climates, housing should be constructed to prevent cold drafts and the freezing of teats. The use of sand for bedding instead of straw or sawdust reduces the risk of growth of pathogens because sand is a nonorganic substrate. In any case, dirty bedding should be replaced promptly.

The milking machine and the milking act have been highly associated with the dissemination of pathogens as well as setting the stage for bacterial invasion. Bacteria are transmitted from infected glands to exteriors of other glands via milking machines. Extreme fluctuations in vacuum and ballooning of teats inside large inflations increase inflammation and lead to infection, especially by *Staphylococcus aureus*.

Infected quarters respond to irritation by milking machines by becoming more highly inflamed. British researchers milked one-half of each udder of 84 cows with molded large-bore inflations and the other half with narrow-bore inflations (cf. chapter 11). The incidence of clinical mastitis was 12.7 and 5.5% in the two groups, respectively. This difference was **statistically significant**. There were no differences in rates of new infection, occurrences of teat erosions, or increased somatic cell counts in noninfected glands milked by the two types of inflations.

14.7 Detecting Mastitis

The oldest methods of detecting mastitis are visual observation of the condition of the milk and physical examination of the mammary gland. Clinical mastitis usually causes milk to become discolored, watery, clotty, or stringy. These symptoms can be detected in the foremilk by use of a strip cup (figure 14.3). This is an effective method of detecting clinical mastitis so highly abnormal milk can be discarded and cows can be treated promptly. However, it is not a sensitive test for inflammation that results in infiltration of large numbers of somatic cells into the gland. Much more sensitive tests focus on numbers of somatic cells found in the milk.

Figure 14.3 A strip cup fitted with a screen-type lid is used to detect visible abnormalities in milk prior to attaching the milking machine (Ambic Equipment, Ltd.).

14.7.1 California Mastitis Test

The California Mastitis Test (CMT) is highly sensitive to the presence of somatic cells in milk. The CMT reagent (3% alkyl aryl sulfonate, an anionic detergent) with a pH indicator (bromocresol purple, 1:10,000) lyses the cell wall and reacts with the nuclear material (**deoxyribonucleic acid—DNA**) of somatic cells. The alpha helix of the DNA unwinds allowing its molecules to bind to each other, forming a gel. In milk from highly inflamed glands, pH is increased by infiltration of blood serum. This causes the purple color of the CMT reagent to change to maroon.

The simple procedure of the CMT requires that approximately 2 mL each of milk and reagent be mixed in a four-cup white plastic paddle (figure 14.4). The amount of gel and/or precipitate that forms is directly proportional to the concentration of somatic cells in the milk. Milk always contains these cells. However, numbers do not exceed 200,000/mL in milk from the nonmastitic gland (table 14.2 on the following page). Milk of first lactation cows usually contains less than 50,000/mL. Identifying quarters with higher CMT scores increases the probability of getting a positive culture. For example, a quarter with a CMT of 3 is three times more likely to yield a positive culture than a CMT of 1. Conversely, if the result is trace (200,000 to 400,000 cells/mL), the quarter is likely to be infected, but may be difficult to detect. Studies have suggested that a single CMT

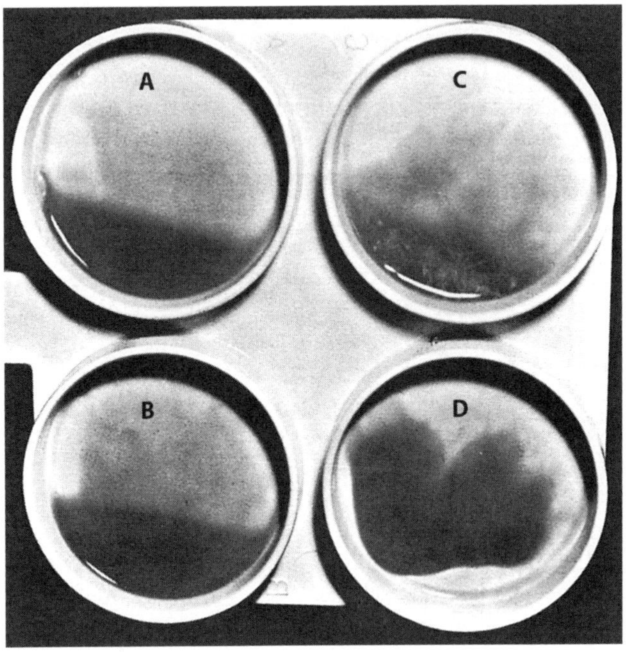

Figure 14.4 Milk from a normal quarter flows freely (cup A). The gel is fragile and forms small clumps in a moderate reaction (cups B and C). The gelatinous mass clings together in a strong reaction (cup D) (Missouri Agricultural Experiment Station).

Table 14.2 How to Read a CMT Test.

CMT Score	Somatic Cell Range	Interpretation
N (negative)	0–200,000	Healthy
T (trace)	200,000–400,000	Subclinical mastitis
1	400,000–1,200,000	Subclinical mastitis
2	1,200,000–5,000,000	Serious mastitis infection
3	Over 5,000,000	Serious mastitis infection

test may detect only 60 to 80% of infected quarters. Multiple tests increase the sensitivity of detecting infections, and may be most accurate several days after calving.

The CMT should be regarded as a screening test for identifying inflamed glands and for assessing, over a period of several months, the inflammatory status of a herd. There is only a general correlation between reactivity to the test and the ability to culture disease-producing bacteria from the milk. The CMT reaction becomes weaker as sample storage time increases, particularly if samples are not kept refrigerated. It is probable that deoxyribonuclease, an enzyme produced by some bacteria, degrades the DNA, the reactive component of the somatic cells.

It is interesting that goat's milk typically contains comparably higher numbers of somatic cells than does cow's milk. The apocrine secretory system of the goat differs from the merocrine secretory system of the cow. The allowable somatic cell count for producer herd goat milk is set at 1,500,000/mL. Much higher numbers of somatic cells are needed in goat's milk to produce the same intensity of reaction to the CMT than are needed with cow's milk.

14.7.2 Bulk Tank Somatic Cell Counts

Regulatory agencies perform electronic somatic cell counts on samples collected from farm bulk tanks. Such tests can provide a general indication of a herd's mastitis status. Results are used to control the acceptance of abnormal milk into the general milk supply (cf. chapter 25). Losses in milk production associated with elevated bulk tank somatic cell counts (BTSCC) are estimated as follows: 6% at 200,000/mL, 18% at 500,000/mL, and 29% at 1,500,000/mL.

Herds with a BTSCC of greater than 500,000 cells/mL have large losses in milk yield even though counts are below the maximum US regulatory limit of 750,000/mL for grade "A" milk. The regulatory limit in the European Union is 400,000/mL. Most BTSCC reported to producers are determined by milk plants and regulatory agencies according to requirements of Section 7, Standards for Grade "A" Milk and for Milk Products, of the *Grade "A" Pasteurized Milk Ordinance* (US Food and Drug Administration, 2013).

In 2014, results of somatic cell counts of bulk farm milk within the Central, Mideast, Southwest, and Upper Midwest federal milk marketing orders were monitored by the Center for Epidemiology and Animal Health of the USDA Animal and Plant Health Inspection Service (2015). Samples from 26,424 producers in 32 states accounted for 99.1 billion lb or 48.1% of the 206 billion lb of milk produced in the United States in 2014. Approximate percentages of shipments that fell within selected count ranges are found in table 14.3.

Table 14.3 Percentage of Milk, Shipments, and Producers by BTSCC Level in 2014.

BTSCC (× 1,000 cells/mL)	Milk (99.1 billion lb)	Shipments (284,528)	Producers (26,424)
Less than 100	5.6	5.7	1.1
Less than 200	53.9	38.3	16.1
Less than 400	95.5	86.7	64.9
Less than 650	99.6	98.2	92.1
Less than 750	99.9	99.2	95.6

The European standard of 400,000/mL was met in 95.5% of the shipments and the US standard of 750,000/mL by over 99% of the shipments. There is a seasonal effect on somatic cell numbers with monthly "milk-weighted" counts being highest in August and lowest in November.

While bulk tank somatic cell counts may be a good indicator of a herd's general udder health status, their use will not identify problem cows nor identify causes of the high counts.

14.7.3 Culture Testing

Evidence of inflammation does not disclose its cause. To do so, milk samples from individual quarters should be subjected to microbiological culture tests. Samples must be taken aseptically (without contamination) and examined within a few hours of collection. Typically, a wire loop delivering 0.01 mL of milk is used to spread the sample on one-fourth the surface of a blood agar plate. After incubation at 99°F (37°C) for 24 to 48 hr, plates are examined for bacterial colonies typical of mastitis pathogens, most of which are hemolytic. Hemolysis, the lysis of red blood cells, is evidenced by cloudiness or clearing of areas around the colonies (figure 14.5). The National Mastitis Council (1999a) has published procedures for diagnosis of mastitis in the *Laboratory Handbook on Bovine Mastitis*.

It is not unusual to observe "sterile" mastitis, that is, clinical symptoms exist but culture tests for viable bacteria are negative. Two factors are most likely the cause: (1) the infection is caused by *Mycoplasma* spp., which grows poorly on blood agar, and (2) bacteria have been engulfed by phagocytes, lowering the counts of viable cells below

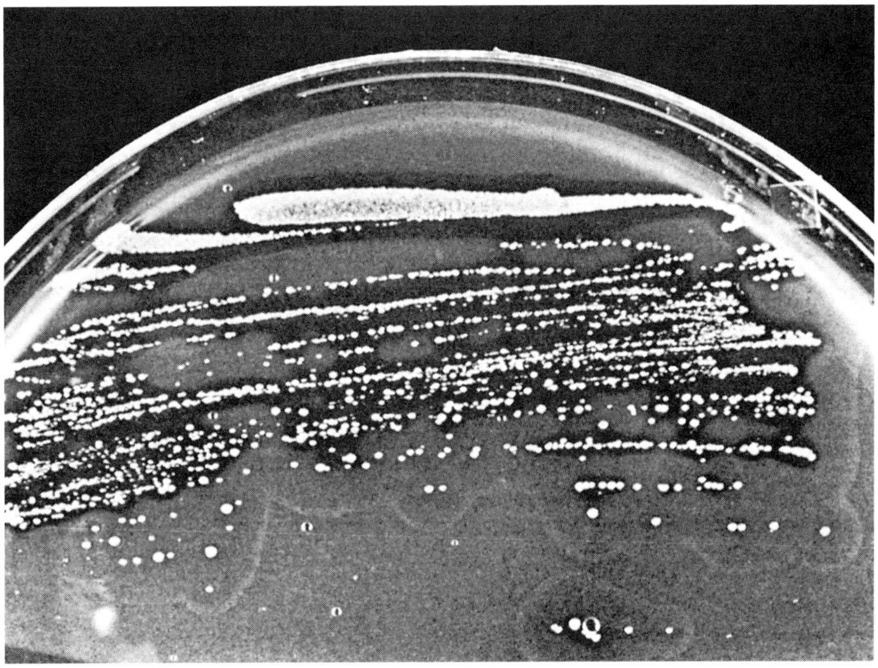

Figure 14.5 Blood agar plate showing colonies of typical mastitis-producing *Staphylococcus aureus*. Colonies are surrounded by zones of alpha- and beta-hemolysis (Missouri Agricultural Experiment Station).

the threshold of the test (100 to 1,000 cells/mL). Leukocytes remain viable in milk for several hours after milking. A concentration of 4 million leukocytes per mL caused a 50% drop in the bacteria count of milk held at 39°F (4°C) for 12 hr (Singh, 1965). Thus, phagocytic action in both the udder and in samples between collection and analysis was a probable cause of failure to culture the infecting organism. Other possible causes are infection by nonculturable microorganisms, inadequate growth conditions for the pathogen, and growth inhibitor present in the sample.

14.7.4 Electronic Counters in a Laboratory Setting

Whereas the CMT provides subjective information on the extent of inflammation within a single gland, the test is best performed on a multi-cow basis at milking time when udders are full of milk. This requires time that is not affordable in most dairy herds. Instead, an alternative method of testing involves the collection of samples from the combined milk of each cow's four quarters, followed by testing in a laboratory using electronic counters. These samples can also be tested for milkfat and protein.

Dairy Herd Improvement (DHI) reports to producers vary somewhat among the DHI processing centers. In addition to the current somatic cell counts per cow, many centers report SCC history for individual cows and herd summary data such as SCC by lactation number and stage of lactation.

14.8 Mastitis Therapy

When a cow exhibits signs of severe udder inflammation, accompanied by fever and loss of appetite, immediate treatment should be administered. Such animals are usually given both intramammary and intramuscular or intravenous injections of broad-spectrum antibiotics.

Chronic mastitis is best treated during the dry period for the following reasons: (1) infused antibiotics remain in high concentration for a longer period of time as they are not diluted by milk or removed during milking; (2) a much higher dose can be used safely; (3) milk is not likely to be contaminated with therapeutic drugs if the withholding time after calving is observed; (4) damaged tissues may be repaired before freshening; and (5) numbers of dry period infections are reduced significantly.

The rate of infection is almost seven times higher during the first three weeks of the dry period than in any comparable period of lactation, and it is more than 35 times higher than in the balance of the dry period. On average, about 25% of cows develop at least one new infection during the dry period.

Treatment of all quarters of each cow at drying off ensures that all infected glands will receive therapy without the need for expensive culture and sensitivity tests. Additionally, all quarters will have an increased level of protection from new infections during the dry period.

14.8.1 Antibiotic Treatments

Therapeutic drugs should be administered after the last milking of lactation. Such drugs require approval of regulatory authorities such as the US Food and Drug Administration and the European Medicines Agency. To prevent cross-infections only single-dose containers should be used.

Products for dry cow therapy are usually formulated to kill **Gram-positive** bacteria, especially *Streptococcus aga-*

lactiae and *Staphylococcus aureus*. They are effective against some environmental streptococci, especially *Streptococcus uberis*, but are ineffective against the Gram-negative coliform and pseudomonas bacteria. In Europe and Australia, products for dry cow therapy are reported to be effective for up to 54 days (National Mastitis Council, 2006).

Though nonlactating infected glands are treated more effectively, *Streptococcus agalactiae* are so highly sensitive to penicillin G that a high rate of cure can be obtained in lactating glands when two infusions of 100,000 units each are administered at 24-hr intervals. However, infusions after each milking (12-hr intervals) are recommended for high-producing animals (25 lb/milking or more).

Parenteral therapy is generally unnecessary unless there is systemic involvement accompanied by udder **edema** causing blockage of ducts within the udder. High doses of drugs are required for intramuscular or intravenous injections because they are distributed throughout the animal's body. Infused drugs remain largely within the confines of the mammary gland.

Antibiotics commonly used in mastitis therapy and the organisms they usually inhibit include the following:

- Penicillin G, cephalosporins, and their derivatives inhibit Gram-positive bacteria, especially staphylococci and streptococci. These drugs inhibit cell wall synthesis.
- Streptomycin and dihydrostreptomycin are especially effective against Gram-negative bacteria, e.g., coliforms and pseudomonads. These are among the aminoglycoside antibiotics that inhibit protein synthesis.
- Tetracyclines (aureomycin, terramycin, tetracycline) exhibit a broad spectrum of activity against bacteria.
- Sulfa drugs (sulfadiazine, sulfamerazine, sulfathiazole) inhibit coliforms and *Pseudomonas aeruginosa*.

Effectiveness of penicillin G is generally enhanced when used in combination with streptomycin or sulfa drugs. Some Gram-negative bacilli produce an enzyme, beta-lactamase, which inactivates penicillin G and its semisynthetic derivatives, including the cephalosporins, by breaking their beta-lactam ring structure.

Therapy usually should be based on knowledge of the infecting microorganism and the antibiotic drug effective against it. Testing for sensitivity to such drugs, however, is expensive and may require up to three days. Therefore, the common practice in treating clinical infections is to use preparations that have been effective in the past and ones that have a relatively broad spectrum of activity. Different bacteria respond differently to various antibiotics. Stated succinctly, "There is no perfect single or combined antibiotic, and there is no such thing as complete coverage of possible pathogens" (Karzan, 1970).

14.8.2 Antibiotic Sensitivity Testing

Antibiotic sensitivity testing involves isolation of the infecting bacterium, suspension of millions of cells in a solution, seeding of a blood agar plate with the suspension, implantation on the plate of paper discs containing known quantities of applicable antibiotics, incubation of the plate to produce growth of the bacterium, and examination of the plate after incubation for zones of inhibition of growth surrounding the discs (figure 14.6). Zone size depends on the sensitivity of the organism to the antibiotic, antibiotic concentration and rate of diffusion, agar depth, and the number of cells in the medium and their growth rate. Therefore, only when these conditions are standardized are zone sizes indicative of the sensitivity of the specific organism to the antibiotics tested.

The Clinical and Laboratory Standards Institute has established guidelines to be used by lab technicians when analyzing and interpreting results of antibiotic susceptibility tests as well as to predict antibiotic effectiveness against mastitis organisms when the antibiotics are given in the infected quarter and then have to travel through milk, milk components, and blood to the site of infection.

Even though sensitivity tests suggest the effectiveness of a given drug, failure may result for the following reasons: (1) antibiotic fails to reach the site of the infection, probably because the pathogen is protected by scar tissue caused by multiple or chronic infections; (2) bacteria are engulfed by leukocytes and cannot receive a lethal dose, therefore bacteria survive and become reestablished; (3) inflammatory response inhibits the antibiotic; (4) antibiotic binds to protein of milk or serum; (5) insufficient initial dose of drug; and (6) incompatible combination of drugs were used.

Multidrug resistant strains of bacteria can result from uncontrolled exposure to several classes of antibiotics. The spread of antibiotic resistance among bacteria has been associated with mobile genetic elements including **plasmids**, **transposons**, and **integrons**. The latter have the ability to distribute genes encoding resistance to

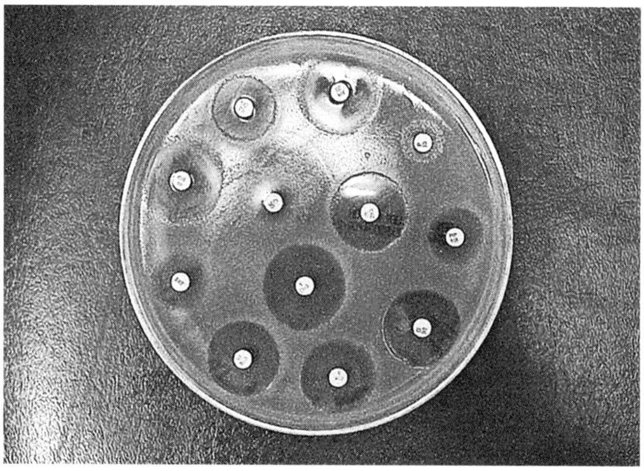

Figure 14.6 Sensitivity to antimicrobial drugs is indicated by zones of inhibition (US National Oceanic and Atmospheric Administration).

several antimicrobial drugs. Integrons capture genes by a recombination system and have been found in both **chromosomal** and extra-chromosomal DNA.

14.8.3 Storage and Labeling of Antibiotics

Rules applicable in the United States can be found in Section 15r of the *Grade "A" Pasteurized Milk Ordinance* of the US Food and Drug Administration (2013). It contains the following provisions:

1. Equipment used to administer drugs is not cleaned in the wash vats and is stored so as not to contaminate the milk or milk-contact surfaces of equipment.
2. Drugs intended for the treatment of nonlactating dairy animals are segregated from those drugs used for lactating dairy animals. Separate shelves in cabinets, refrigerators, or other storage facilities satisfy this Item.
3. Drugs shall be properly labeled to include the name and address of the manufacturer or distributor for over-the-counter (OTC) drugs, or veterinary practitioner dispensing the product for prescription (Rx) and extra label use drugs. If the drug is dispensed by a pharmacy on the order of a veterinarian, the labeling shall include the name of the prescribing veterinarian and the name and address of the dispensing pharmacy, and may include the address of the prescribing veterinarian.
4. Drug labels shall also include: (a) directions for use, and prescribed withholding times; (b) cautionary statements, if needed; and (c) active ingredient(s) in the drug product.
5. Unapproved and/or improperly labeled drugs are not used to treat dairy animals and are not stored in the milkhouse, milking barn, stable, or parlor.
6. Drugs are stored in such a manner that they cannot contaminate the milk or milk product-contact surfaces of the containers, utensils or equipment.

To avoid drug residues in market milk label directions must be followed carefully. When dry periods are short or multiple treatments are given, testing of the milk should be done using test kits available through most firms that purchase the milk, the local health department, or a veterinarian.

14.9 Prevention of Mastitis

The cost of an effective mastitis control program is normally far less than the costs of the disease, including the large losses in milk production. Ideally, a complete program involves a team composed of the dairy farmer, herd veterinarian, sanitarian, dairy field personnel, milking machine service personnel, agricultural engineer, and dairy extension specialist. Seldom is it practical to involve all of these resources. The herd veterinarian is the specialist who likely can contribute the most, provided that person has mastered aspects of control such as: (1) utilization of screening tests; (2) influence over environment, sanitation, and milking machine management; (3) cleanliness of the milking system and equipment; (4) identification of infected mammary glands; (5) bacteriology of mastitis; (6) analysis of the farm's bulk milk samples; and (7) mastitis therapy.

Practices that prevent mastitis can be placed in three categories: (1) limiting exposure to infectious microorganisms, (2) enhancing the immune system of the host, and (3) reducing physical injury.

14.9.1 Reducing the Incidence of Exposure

A highly effective and economical means of reducing exposure to infectious microorganisms is by applying a strong disinfectant to the teats after milking. Iodophors (0.1 to 0.25%), chlorhexidine (0.3 to 0.5%), and hypochlorite (0.6 to 0.9%) have been used successfully. Research has shown favorable results from applications of iodine (0.3 to 1.0%), linear dodecyl benzene sulfonic acid (1.9%), quaternary ammonium (0.5%), and lactic acid (2.9%) plus sodium chlorite (0.7%). Products used should be listed with the US Food and Drug Administration and shown to be effective by controlled research. Proof of efficacy is not required in the United States to permit labeling. Iodophor products made for sanitizing equipment, though effective in killing target organisms, can cause skin irritation because they contain high concentrations of phosphoric acid.

The National Mastitis Council (1999b) reports that more than 50% of new udder infections can be prevented by dipping or spraying teats with an effective product before milking and immediately after milking while the teat orifice is still somewhat open. Postmilking disinfection removes milk from the teat ends, thus removing moisture and food for bacterial growth. Following such treatment, it is necessary to provide the cow with a clean and dry environment.

Postmilking disinfection is especially effective against the pathogens that are spread cow to cow—*Streptococcus agalactiae* and *Staphylococcus aureus*. Control of environmental pathogens, including the nonagalactiae streptococci and coliform bacteria, requires that cows be kept in a clean, dry environment, subjected to premilking teat disinfection, and milked, after milk letdown, with properly performing machines.

Teat skin must be maintained in a healthy condition. To this end some disinfectants contain one or more of the emollients glycerin, lanoline, propylvinyl pyridine, allantoin, and aloe. During cold weather, when temperatures drop below 20°F (7°C), teats should be dry before cows are turned out. Allow 30 sec before blotting excess disinfectant, and protect cows from winds. Swollen udders and teats of fresh cows predispose them to chapping and freezing.

Other practices that reduce the incidence of cross infections include (1) milking clean cows first and infected cows last, (2) isolation of cows with clinical mastitis, (3) disinfection of milking machine inflations between cows, and (4) wearing plastic or rubber gloves during milking and disinfecting the gloves between cows.

14.9.2 Promoting the Effectiveness of the Immune System

It is the healthy animal that is able to ward off infections by her natural immune system. So, good nutrition and prevention of other diseases are important. Negative energy balance or vitamin (e.g., A, D, E) or trace mineral (e.g., selenium, copper, zinc) deficiencies can reduce immunity.

Vaccination has failed to provide a significant tool for preventing mastitis. Whereas it appears to reduce severity of mastitis due to the homologous organism (the one from which the vaccine was produced), the incidence of new infections is affected little. Fundamental problems are (1) production of specific antibody for each potential type of infecting organism (antibodies usually have limited activity against species of organisms other than the one that elicited their production) and (2) transport of sufficient concentration of antibody from the blood to milk. Antibody titers in milk rarely exceed 1:64 (1 part milk to 63 parts diluent), regardless of the nature of the antigen, method of administration of the antigen, previous history of the host, or titers in the blood (Schalm et al., 1971). Only when the mammary gland is inflamed does it appear that antibodies are transported into the milk in relatively high concentrations. Inflamed tissues are more permeable to constituents of blood than are normal tissues.

Streptococci are poor antigens; consequently, limited antigen is produced in cows vaccinated with them. Since staphylococci vary widely in type, vaccine produced for one strain may not induce antibodies effective against another strain. However, staphylococci produce several antigenic toxins against which antibodies can be produced. These antitoxins are not strain specific and can reduce severity of staphylococcal infections. However, they will not eliminate existing infections or prevent new ones.

Autogenous vaccines may be prepared from microorganisms isolated repeatedly from a herd, thus providing antigens specific for that herd. However, costs may be high and the vaccine may not be effective against other organisms in that herd.

Vaccines are composed of bacterins, toxoids, or a combination of the two. Bacterins are essentially suspensions of killed microorganisms. Toxoids are prepared from toxic extracts of cells or toxic products excreted by the cells. They are detoxified with formaldehyde or betapropiolactone. A multivalent vaccine is one that contains materials from multiple organisms.

14.9.3 Reducing Physical Injury

The most important source of physical stress on mammary glands appears to be the milking machine. Thus it is critical that the machine be properly engineered, installed, and maintained. Excessive fluctuations of vacuum at the teat end are particularly undesirable. Excessively high vacuum produces edema at teat ends and **extroversion** of the teat canals. Both probably increase the rate of infection. Overmilking tends to produce the same effect and causes injury to the base of the teat when inflations creep upward. Inflation creep may close the lumen of the teat, precluding milk flow from the gland cistern into the teat cistern. This reduces milk withdrawal and leaves a medium for growth of invading bacteria.

Milking machines should be attached after the cow begins milk letdown, usually about 30 sec after oxytocin (milk letdown hormone) is released during routine cow preparation. Stressing of cows impedes normal milk letdown, posing a risk of physical injury to gland tissues as the machine is attached.

Physical injury near the teat end provides a site for bacterial growth close to the teat canal. Bruises to the gland may release staphylococci that have been walled off, thus starting a new infection process. An environment must be provided that minimizes cuts and bruises to mammary glands. Animals should be moved carefully without running, and animals in **heat** should be segregated.

14.10 Bacteria Counts Associated with Mastitis

Bacterial numbers in milk increase significantly with the onset of mastitis. However, this does not mean that mastitis is the reason for all high bacterial counts. The number of bacteria in milk from infected cows depends on the following variables: (1) growth rate of the organism invading the gland, (2) the cow's resistance (i.e., phagocytes and antibodies produced), and (3) the amount of involvement of secreting tissues.

Staphylococci commonly do not produce counts as high as do streptococci in cases of chronic infections. Philpot (1971) never observed staphylococci in concentrations in excess of 15,000/mL in grade "A" bulk tank milk samples. However, he observed that milk from one cow infected with *Streptococcus agalactiae* contained enough bacteria to increase the bacterial count of bulk herd milk from about 10,000/mL to more than 100,000/mL. In acute mastitis, bacteria grow rapidly until the host is able to marshal her body defenses, and counts of several million per mL are not uncommon. However, when leukocytes migrate into the mammary gland, phagocytosis begins and counts are expected to decrease significantly. This accounts for cyclic variations in bacteria counts of milk from the same infected gland. The authors observed that numbers of viable bacteria continued to decrease for several hours postmilking in milk containing high numbers of somatic cells.

14.11 Decision Tree for Treatment of Mastitis

Pinzón-Sánchez et al. (2011) developed a decision tree to evaluate the economic impact of durations of intramammary treatment for the first case of mild or moderate clinical mastitis occurring in early lactation. The analysis included probabilities for the distribution of etiologies (Gram-positive, Gram-negative, or no growth), bacteriological cure, and recurrence. The economic consequences of mastitis included costs of diagnosis and initial treatment, additional treatments, labor, discarded milk, milk production losses due to clinical and subclinical mastitis, culling, and transmission of infection to other cows (only for contagious mastitis caused by *Staphylococcus aureus*). Pathogen-specific estimates for bacteriological cure and milk losses were used. The economically optimal path for several scenarios was determined by comparison of expected monetary values. For most scenarios, the optimal economic strategy was to treat clinical mastitis cases caused by Gram-positive pathogens for two days but not to treat cases caused by Gram-negative pathogens or when no pathogen was recovered. Use of extended intramammary antimicrobial therapy (5 to 8 days) resulted in the least expected economic values.

14.12 Summary

There is a close relationship between udder health and profitable dairy farming. Mastitis costs dairy herd owners more than does any other bovine disease. Milk producers are often unaware that subclinical mastitis exists, and this lack of information affects management practices. According to the National Mastitis Council, reduced milk production accounts for about 70% of total losses associated with mastitis.

Many microorganisms can cause mastitis, thus complicating efforts to prevent and cure the disease. Two genera, *Staphylococcus* and *Streptococcus*, cause the majority of cases. Yet there are several strains and species of mastitis producers within these genera.

Mastitis is the microbial invasion of the mammary gland and the consequent inflammation of its tissues. Products of inflammation, primarily infiltrating leukocytes, are the principle reactant in the screening tests for abnormal milk. Somatic cells, including leukocytes, are counted in confirmatory tests for abnormal milk.

The teat canal is the major barrier to entrance of the udder by bacteria. Antibodies and phagocytes infiltrate the udder in a cow's efforts to kill invading microorganisms. Vaccines are designed to improve the animal's ability to produce specific antibodies against the microorganism and its products. However, evidence of successful application of vaccines in commercial dairy herds is lacking.

Prevention is the key to mastitis control, and prevention depends on minimizing exposure of mammary glands to infectious bacteria, maximizing the immune response, and preventing physical injury.

Therapy is usually administered in the form of antibiotics by intramammary infusion through the teat canal. Drugs of choice include penicillin G, semisynthetic penicillins that kill Gram-positive cocci, plus streptomycin and other drugs effective against Gram-negative rods such as *Escherichia coli*.

Bacterial counts of milk are increased during mastitis, but phagocytosis by leukocytes tends to reduce counts even after milk is removed from the cow and cooled.

Treatment during the dry period has the advantages of higher chances of a cure, no loss of milk, and reduced risk of contaminating the milk supply. Considered on a herd-wise basis, treatment during early lactation has been judged economical only for cases of clinical mastitis caused by Gram-positive bacteria.

Study Questions

1. Describe mastitis. Differentiate between subclinical and clinical mastitis. What is acute mastitis? What is sterile mastitis?
2. Describe the economic losses that occur due to mastitis. How do these losses rank compared to economic losses among other diseases affecting dairy cows?
3. What are the major causes of mastitis?
4. How do most microorganisms gain entrance into the mammary gland?
5. What factors favor elimination of *Streptococcus agalactiae* from a dairy herd?
6. In what important ways do staphylococci differ from streptococci?
7. In what ways do staphylococci vary in their abilities to cause mastitis?
8. Why is it more difficult to control staphylococcal mastitis than it is to control *S. agalactiae*?
9. Name some microorganisms other than staphylococci and streptococci that can cause mastitis.
10. What physical characteristics of the mammary gland prevent invasion by bacteria and prevent establishment of the infection once invasion has occurred?
11. Identify management practices influencing the severity of bovine mastitis.
12. How may milking equipment contribute to the development of mastitis?
13. What are somatic cells? How are they related to mastitis?
14. In order to be classified as grade "A" milk, what is the maximum BTSCC score allowed in the United States? In Europe?
15. What is phagocytosis? How is it related to the "natural" control of mastitis?

16. Describe using a strip cup to detect mastitis. Is it effective? Why or why not?
17. Describe the California Mastitis Test and its procedures. Describe its strengths and weaknesses.
18. What is a white blood cell? A leukocyte? A macrophage?
19. How can one explain the apparent absence of typical mastitis pathogens in mammary glands that yielded milk having strong CMT reactions?
20. Does evidence of inflammation within the mammary gland provide information on the cause of mastitis? Why?
21. Why is the treatment of chronic mastitis most effective during the dry period?
22. Name six reasons why antibiotic treatments may fail to eliminate an infection even though drug sensitivity tests indicated effectiveness.
23. In mastitis therapy, why are larger doses of drugs necessary when administered intramuscularly or intravenously than when infused into the mammary glands?
24. Describe the purposes for sensitivity testing. What are some limitations of such tests?
25. What are broad-spectrum antibiotics?
26. What areas of expertise related to mastitis should a herd veterinarian have?
27. Describe the process of teat dipping. Why is it recommended? What precautions should be exercised in the use of teat disinfectants during periods of cold weather?
28. Identify practices that reduce cross infections among dairy cows.
29. What principle is involved in vaccinating cows against mastitic organisms? Has this means of controlling mastitis been effective? Why or why not?
30. Identify limitations of vaccination for staphylococcal mastitis.
31. In what ways can physical injury be associated with mastitis?
32. Is there always a close relationship between bacterial counts of milk and CMT scores? Why or why not?
33. Describe the decision tree process in evaluating the economic impact of the duration of intramammary treatment for mastitis in early lactation.

References

Archer, S. C., F. Mc Coy, W. Wapenaar, and M. J. Green. 2013. Association between somatic cell count early in the first lactation and the lifetime milk yield of cows in Irish dairy herds. *Journal of Dairy Science* 96:2951–2959.

Dodd, H., and F. K. Neave. 1951. Machine milking rate and mastitis. *Journal of Dairy Research* 18:240.

Gott, P. 2012. Endotoxin tolerance in lactating dairy cows. *IDF Animal Health Newsletter* 6:12.

Hudson, C. D., A. J. Bradley, J. E. Breen, and M. J. Green. 2012. Associations between udder health and reproductive performance in United Kingdom dairy cows. *Journal of Dairy Science* 95:3683–3697.

Karzan, B. N. 1970. *Antimicrobial Therapy.* Philadelphia: W. B. Saunders.

Lavon, Y., G. Leitner, H. Voet, and D. Wolfenson. 2010. Naturally occurring mastitis effects on timing of ovulation, steroid and gonadotrophic hormone concentrations, and follicular and luteal growth in cows. *Journal of Dairy Science* 93:911–921.

National Mastitis Council. 1999a. *Laboratory Handbook on Bovine Mastitis* (Revised Edition). Verona, WI: Author.

National Mastitis Council. 1999b. *Teat Disinfection Facts* (Revised, NMC Factsheet). Verona, WI: Author.

National Mastitis Council. 2006. *Dry Cow Therapy* (Revised, NMC Factsheet). Verona, WI: Author.

Philpot, N. 1971. What Research Tells Dairy Farmers about Mastitis. Proceedings, Mastitis Seminar, Missouri Mastitis Council, Jefferson City, Missouri.

Pinzón-Sánchez, C., V. E. Cabrera, and P. L. Ruegg. 2011. Decision tree analysis of treatment strategies for mild and moderate cases of clinical mastitis occurring in early lactation. *Journal of Dairy Science* 94:1873–1892.

Schalm, O. W., E. J. Carroll, and N. C. Jain. 1971. *Bovine Mastitis.* Philadelphia: Lea and Febiger.

Schroeder, J. W. 2012, July. *Bovine Mastitis and Milking Management* (AS1129 Revised). Fargo: North Dakota State University Extension Service.

Schwarz, D., U. S. Diesterbeck, S. König, K. Brügemann, K. Schlez, M. Zschöck, W. Wolter, and C. P. Czerny. 2011. Flow cytometric differential cell counts in milk for the evaluation of inflammatory reactions in clinically healthy and subclinically infected bovine mammary glands. *Journal of Dairy Science* 94:5033–5044.

Singh, B. 1965. Factors Affecting Activity of Leukocytes in Milk. PhD Thesis, University of Missouri, Columbia.

US Food and Drug Administration. 2013. *Grade "A" Pasteurized Milk Ordinance.* Washington, DC: Author.

USDA Animal and Plant Health Inspection Service. 2015, June. *Determining U.S. Milk Quality Using Bulk-Tank Somatic Cell Counts, 2014.* Washington, DC: Author.

Websites

Center for Epidemiology and Animal Health (http://www.aphis.usda.gov/wps/portal/aphis/ourfocus/animalhealth/sa_program_overview/sa_ceah)

Clinical and Laboratory Standards Institute (http://www.clsi.org)

National Animal Health Laboratory Network (http://www.aphis.usda.gov/nahln)

National Mastitis Council (http://nmconline.org)

National Veterinary Services Laboratory (http://www.aphis.usda.gov/nvsl)

15

Dairy Beef

All is not butter that comes from the cow.
Proverb

This chapter was contributed by Bryon Wiegand, PhD, Professor of Animal Sciences, University of Missouri, Columbia. Wiegand's 15-year career as a college faculty member has brought outstanding teaching awards from six organizations at the university, state, and regional levels. He is a member of the editorial board for the *Journal of Animal Sciences*; an associate editor for the *Journal of Natural Resources and Life Science Education*; an official judge for 4-H, FFA, and state fair judging events; and contributor to a chapter in *Advances in Conjugated Linoleic Acid Research* (2003), Volume 2, CRC Press.

15.1 Introduction

About 20% of the beef produced in the United States comes from dairy animals. Surplus dairy cows account for about 6% and approximately 14% comes from dairy steers, bulls, and vealers that have been fed for meat production. In view of the worldwide shortage of grains needed to feed humans, it is becoming increasingly clear that dairy farmers and livestock producers must, in the future, rely more heavily on the ability of ruminants to utilize large quantities of **forage** and **roughage** if humans are to have a reliable supply of meat and milk. Dairy steers convert pasture, hay, silage, and other roughages that are often wasted into salable, nutritious products.

Although dairy steers provide no milk for humans, they provide income from meat to owners. Thus, a two-way profit—meat from steers and milk from cows—has been realized. This has permitted some dairy farm enterprises to remain profitable in the face of increasing labor and fixed costs.

15.2 Nutrient Profile of Lean Beef

Regarding value, few dietary components are as nutrient dense as lean red meat. Beef, from bovine animals, is a good source of dietary protein, fatty acids, and certain essential vitamins and minerals. A 4-oz serving of lean

Table 15.1 Nutrient Content of Selected Cuts of Raw Beef.

		Beef, Ribeye "0" fat trim, choice, raw			Beef, Ground 90% lean, 10% fat, raw		
Nutrient	Unit	Value per 100 g	4 oz (113.3 g)	1 Steak (249 g)	Value per 100 g	4 oz (113.3 g)	1 lb (453.6 g)
Proximates							
Water	g	66.5	75.37	165.58	69.42	78.44	314.89
Energy	kcal	187	212	466	175.0	198	794
Protein	g	19.46	22.05	48.46	19.97	22.57	90.58
Total lipid (fat)	g	11.4	12.92	28.39	10.0	11.30	45.36
Carbohydrate, by difference	g	1.75	1.98	4.36	0.0	0.0	0.0
Fiber, total dietary	g	0.0	0.0	0	0.0	0.0	0.0
Sugars, total	g	0.0	0.0	0	0.0	0.0	0.0
Minerals							
Calcium, Ca	mg	6	7	15	12.0	14	54
Iron, Fe	mg	2.64	2.99	6.57	2.23	2.52	10.12
Magnesium, Mg	mg	24	27	60	20.0	23	91
Phosphorus, P	mg	210	238	523	185.0	209	839
Potassium, K	mg	357	405	889	321.0	363	1456
Sodium, Na	mg	88	100	219	66.0	75	299
Zinc, Zn	mg	7.8	8.84	19.42	4.79	5.41	21.73
Vitamins							
Vitamin C, total ascorbic acid	mg	0.0	0.0	0	0.0	0.0	0.0
Thiamin	mg	0.082	0.093	0.204	0.042	0.047	0.191
Riboflavin	mg	0.219	0.248	0.545	0.151	0.171	0.685
Niacin	mg	3.509	3.977	8.737	5.075	5.735	23.020
Vitamin B-6	mg	0.39	0.442	0.971	0.369	0.417	1.674
Folate, DFE	µg	3	3	7	6.0	7	27
Vitamin B-12	µg	3.21	3.64	7.99	2.21	2.50	10.02
Vitamin A, RAE	µg	2	2	5	4.0	5	18
Vitamin A, IU	IU	7	8	17	14.0	16	64
Vitamin D (D2 + D3)	µg	0.1	0.1	0.2	0.06	0.07	0.27
Vitamin D	IU	4	5	10	3.0	3	14
Vitamin K (phylloquinone)	µg	1.5	1.7	3.7	0.8	0.9	3.6
Lipids							
Fatty acids, total saturated	g	4.359	4.94	10.854	3.927	4.438	17.813
Fatty acids, total monounsaturated	g	5.171	5.86	12.876	4.194	4.739	19.024
Fatty acids, total polyunsaturated	g	0.453	0.513	1.128	0.345	0.390	1.565
Fatty acids, total trans	g	0.48	0.544	1.195	0.54	0.610	2.449
Cholesterol	mg	64	73	159	65.0	73	295

Source: USDA National Nutrient Database for Standard Reference, Release 27 (http://ndb.nal.usda.gov/).

whole-muscle or ground beef, roughly the size of a deck of playing cards, provides a significant contribution to the daily nutrient requirements for an adult human of average size and weight (table 15.1). Most notably, beef is a significant source for iron, zinc, niacin, and vitamin B_{12}. These micronutrients combat anemia, enhance brain function, and support a healthy immune system.

15.3 Breeding and Crossbreeding for Dual Purposes

The use of dual-purpose (milk and meat producing) animals is widely accepted in Europe and Asia where animals have been selected to optimize these traits. However, the trend for divergent selection of milk versus meat production has defined the US cattle industry. A good example of the influence of genetic selection is the Simmental cattle from Switzerland (figure 15.1). Derived and used for meat and milk production in the Simme Valley of Switzerland, Simmentals were exported to the United States and introduced as a beef breed with the genetic potential to increase yield and size of common British-derived beef breeds through crossbreeding. Thus, the Simmental cattle of Switzerland are vastly different from those of the US beef industry.

While the use of dual-purpose animals is not unique to the livestock industry (i.e., sheep for meat and fiber or chickens for eggs and meat), dairy cows, bulls, steers, heifers, and veal calves constitute a significant portion of the red meat supplied to the US and global economies. Until recently, there has been a statistical probability that one-half of the offspring produced in a dairy herd will be male. A high percentage of dairy farms utilize artificial insemination (AI) programs, requiring fewer high-quality bulls compared with the beef cattle industry. With the more recent adoption of sexed semen, the ability to choose calf gender is possible, but comes with increased costs and somewhat lower conception rates when used in AI systems. Therefore, the number of male calves born in the dairy industry is still near 50%. These animals should not be viewed as a by-product of the dairy business, but as an opportunity for meat production and added income to the dairy operation.

With Holstein being the predominant US dairy breed, it is no surprise that they also constitute the most cattle slaughtered for dairy beef production (figure 15.2 on the next page). However, some might argue that other dairy breeds, e.g., Brown Swiss and Milking Shorthorn, might be better suited as dual-purpose breeds in the United States in as much as they are, compared to other dairy breeds, better suited for production of both milk and meat. One should also consider the method by which dairy producers replace females in their herd. For some, purchasing replacement heifers for milk production is common, so mating their milking herd to beef breed sires is a viable option to produce **first-generation (F_1)** progeny that are better suited to feedlot performance and ultimately meat production. The beef sire options in this scenario would be many, but some logical considerations might be breeds that contribute optimal traits for carcass trimness, muscling, and marbling (key factors in USDA quality grading).

15.4 Slaughter and Consumption Trends

In 2014 in the United States, 30.2 million head of cattle were slaughtered for meat production. Of these, 2.8 million head, 9.5% of the total, were surplus dairy cows presented for slaughter under US federal inspection (USDA National Agricultural Statistics Service, 2015). Many of the surplus dairy cows and bulls are used to produce ground beef. Ground beef is an economical meat product that is utilized in institutional food service, quick-serve and traditional restaurants, and in homes. In 2014,

Figure 15.1 This Simmental cow (of Swiss origin) epitomizes the dual-purpose animals that originated in Western Europe to provide both milk and meat for human consumption (photo by Irmgard, https://en.wikipedia.org/wiki/Simmental_cattle) (creativecommons.org/licenses/by-sa/2.5/deed.en).

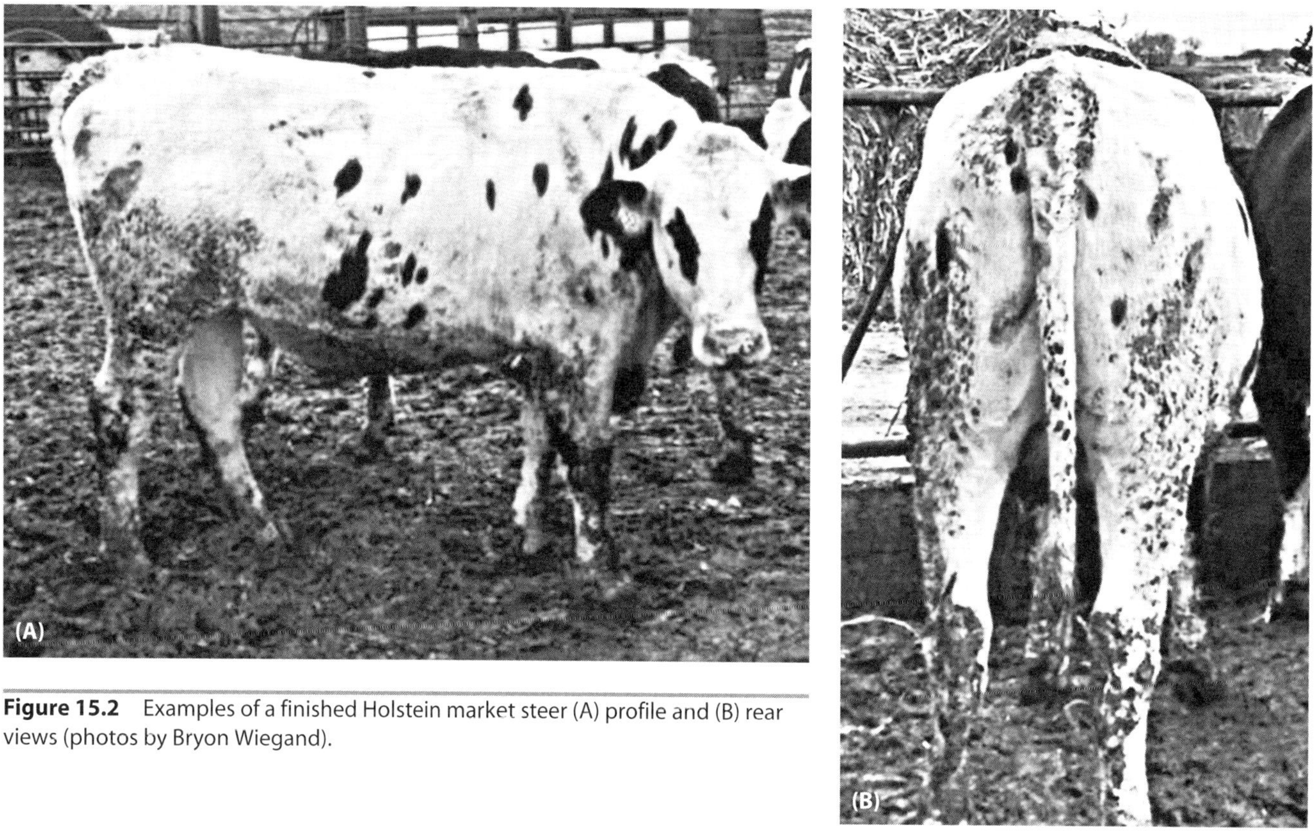

Figure 15.2 Examples of a finished Holstein market steer (A) profile and (B) rear views (photos by Bryon Wiegand).

per capita consumption of all retail beef was 54.2 lb (24.6 kg), the lowest consumption since 1955 (USDA Economic Research Service, 2015). This changed consumption pattern might be reflective of increased retail prices for beef causing consumers to choose alternative meats.

15.5 Veal Production

Veal is a term referring to beef animals less than one year of age raised for meat production. The age range for these animals at slaughter can be a few days (bob veal) up to a year. Production of veal in the United States was 327 million lb in 1990 and declined to about 100 million lb in 2014 (USDA Economic Research Service, 2015). Presumably the decline in production is related to a decrease in consumer demand and the possibility of greater profit margin in feeding traditional veal calves through as finished cattle.

Traditionally, veal calves have been housed individually in crates or huts with access to individual milk or feed and water systems. Although these systems decrease the spread of disease between individual animals, this type of housing has been abandoned in favor of group housing scenarios in some countries, especially with older calves where immune systems are more competent. Research with special attention to lighting, sanitation, and bedding has led to significant changes in housing systems for veal production in Europe and the United States.

Although any breed of cattle can be used for veal production, the largest percentage is of dairy breed origin, especially Holsteins in the United States. Consumers of veal desire retail meat cuts that are light in color, tender, and contain a very low percentage of connective tissue. Younger veal animals most likely meet consumer demand better than animals closer to one year of age because as animals age their musculoskeletal system exhibits higher concentrations of myoglobin (pigment protein) and collagen (connective tissue).

15.6 Finishing Systems

Traditionally, a large percentage of dairy calves were utilized in veal production systems and the commercial cattle feeding industry did not have large numbers of dairy animals in their feed yards. However, with more predictable carcass merit (discussed in the next section) and more uniform genetics (enhanced by AI use of fewer sires per breed), finishing of dairy cattle for beef, especially steers, is a common practice for many feedlots. Dairy breeds, compared to beef breeds, tend to produce less muscle per carcass and require more days on feed to reach common slaughter weights. Because Holsteins represent the greatest number of dairy animals finished for beef, we will consider their expected performance in a feedlot scenario (figure 15.3). It should be noted however, that other dairy breeds of similar size and stature would perform similarly to Holsteins.

Figure 15.3 Holstein steers nearing market weight in a typical feedlot scenario. Modern dairy steer feeding utilizes animals that are genetically and phenotypically uniform (photo by Bryon Wiegand).

Holstein steers are typically fed in one of three systems: (1) continuous grain feeding from approximately 150 to 1,200 lb (70 to 550 kg), (2) two-phase feeding with an early phase of forage followed by grain feeding, and (3) continuous corn silage (a blend of grain and forage). Producers commonly choose a system based on available feedstuffs, cost of feed ingredients, and compatibility with existing facilities. Dairy beef animals tend to be less efficient in converting feed to body weight compared with beef breeds, thus requiring more days on feed to achieve similar finished weights. Dairy beef animals may achieve feed efficiencies of 7:1 (lb feed/lb gain) up to 1,000 lb (450 kg) live weight and then exhibit diminished efficiency as they reach physiological maturity where rate of muscle **accretion** declines and carcass fat deposition increases.

Many compounds are approved for increasing growth rate, increasing muscle accretion, and decreasing subcutaneous fat deposition during finishing. These products can be especially useful in dairy beef production to improve carcass yields, especially in high value cuts of beef, including the rib and loin. These compounds are often delivered via subcutaneous ear implant or as feed additives. They include classes of metabolically active molecules that allow hormone-derived (naturally or synthetically) growth. These implants could include estradiol, progesterone, testosterone, zeranol, or trembolone acetate as active ingredients to induce enhanced growth. More recently, certain β-agonists have been approved for finishing cattle and include active ingredients of zilpaterol or ractopamine. These compounds work at the cell receptor level to repartition nutrients toward muscle accretion at the expense of fat deposition. Care should be taken to recognize and adhere to all label and use restrictions for these growth promoting compounds.

15.7 Carcass Yield

Dressing percentage (DP) is defined as the difference between live animal weight (LW) and hot carcass weight (HCW) divided by LW and expressed as a percentage. The average DP of beef cattle is 62%. Dairy type cattle used for beef tend to have 6 to 7% lower dressing percentages compared with beef cattle because they have been selected for large organ capacity to accommodate high milk yield. Therefore, the combination of lesser muscle mass and a greater percentage of body weight in organ mass results in lower DP.

Carcass yield is directly related to total lean, saleable cuts from a meat animal carcass. Yield grade, as defined by the USDA, considers HCW, ribeye muscle area (REA), twelfth rib subcutaneous fat depth (FT), and percentage of kidney, pelvic, and heart (KPH) fat. Factors positively influencing yield are HCW and REA, whereas FT and KPH negatively affect yield. As a result, dairy beef fed to similar endpoint weights as beef cattle will tend to produce favorable USDA yield grades that partially compensate for the differences in DP discussed previously. One noticeable characteristic of many dairy beef carcasses is the shape of the ribeye muscle and a thinner carcass conformation. Ribeye muscles in dairy beef carcasses generally have less depth and are more angular in shape (figure 15.4) compared to the deeper and more ovular shape of ribeyes from beef breed carcasses. In some instances, packing companies may apply price discounts for dairy-type carcasses.

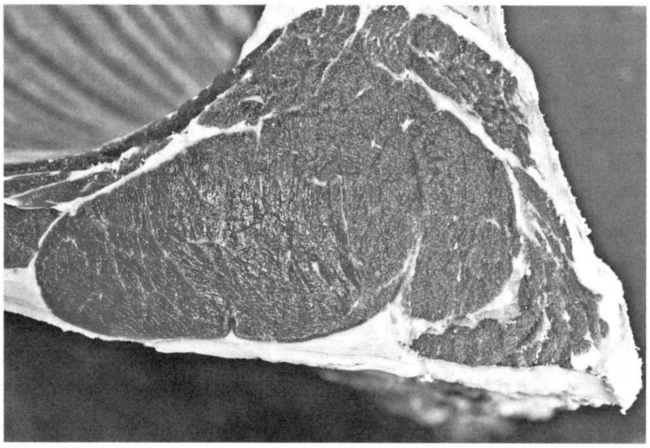

Figure 15.4 Example of ribeye shape in dairy-type carcasses (photo by Bryon Wiegand).

15.8 Meat Quality

Meat quality typically refers to those factors that influence the meat-eating experience. These include flavor, juiciness, color, and tenderness. Quality grade as defined by the USDA accounts for physiological maturity (skeletal ossification) and degree of marbling (intramuscular fat of the ribeye muscle). Dairy beef fed to similar market weights as their beef breed counterparts can reasonably be expected to grade a high percentage of USDA Choice, a desired quality endpoint (figure 15.5). In fact, given the genetic similarity in many Holstein steers, the uniformity and predictability for growth performance, yield grade, and quality is quite high.

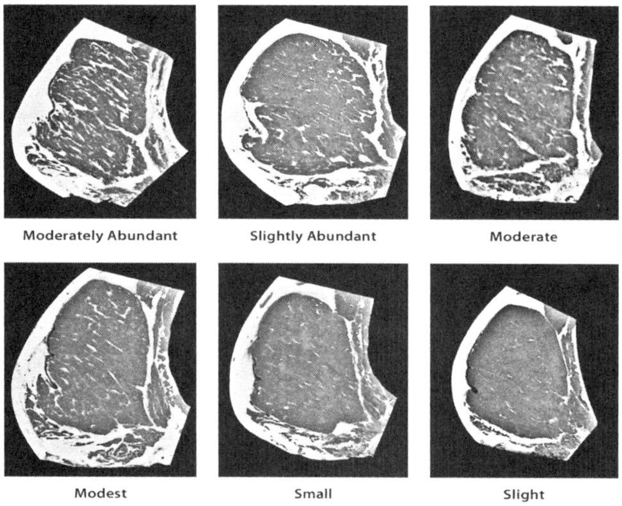

Figure 15.5 USDA marbling standards for quality grading.

15.9 Summary

Dairy beef contributes a significant tonnage of red meat to the global beef market. Meat production is an added revenue stream for milk producers and provides an outlet for male cattle (especially steers), surplus females, or animals that may not express their full genetic potential in milk production. Given the strong demand for beef globally, dairy beef should continue to experience strong consumer demand as a high quality, dietary protein source.

Study Questions

1. Approximately what percentage of the US beef supply is provided by the dairy industry?
2. Discuss the concept of dual-purpose cattle.
3. How might beef production improve the economic situation of dairy producers?
4. Discuss the significance of the Holstein breed regarding dairy beef production in the United States.
5. Why do dairy steers often have favorable USDA yield grades, but lower carcass dressing percentages compared with beef breed carcasses?
6. What trends have occurred in production systems for veal?
7. Why are younger veal animals preferred by consumers?
8. How are dairy steers currently viewed by the US feedlot industry?
9. What expectation might there be for the meat quality of dairy beef steers finished to the same approximate live weight (1,200 lb [550 kg]) as beef-type steers?

References

USDA Economic Research Service. 2015, June 30. Livestock & Meat Domestic Data (http://www.ers.usda.gov/data-products/livestock-meat-domestic-data.aspx#26091).

USDA National Agricultural Statistics Service. 2015, April. *Livestock Slaughter: 2014 Summary.* Washington, DC: Author.

Websites

Iowa State University Extension and Outreach (http://www.extension.iastate.edu/dairyteam/beef)
Pennsylvania State University Extension (http://extension.psu.edu/business/ag-alternatives/livestock/beef-and-dairy-cattle/dairy-beef-production)
University of Minnesota Dairy Extension (http://www.extension.umn.edu/agriculture/dairy/beef)
University of Wisconsin Extension, Beef Information Center (http://fyi.uwex.edu/wbic/dairybeef)
USDA Economic Research Service (http://www.ers.usda.gov)
USDA Foreign Agricultural Service (http://www.fas.usda.gov)
USDA National Agricultural Statistics Service (http://www.nass.usda.gov)
USDA National Nutrient Database for Standard Reference (http://ndb.nal.usda.gov)

16

Water Buffalo

The authors are grateful to Mr. T. J. Olson, Owner of The Turkey Creek Company, Texarkana, Arkansas, for his significant contributions to this chapter.

16.1 Introduction

In this chapter we discuss water buffalo—animals that were domesticated about 5,000 years ago and have become increasingly important nutritionally and economically as sources of milk and meat. Common domestic Asian water buffalo are the descendants of the now designated endangered species wild Asian water buffalo, bred predominantly in Asia for pulling power (draft/draught). Water buffalo provide more than 5% of the world's milk supply and 20 to 30% of the farming power (draft) in Southeast Asia. Milk from these animals is used by numerous human populations.

Buffalo meat, sometimes called "carabeef," is frequently offered as beef in certain regions and provides considerable export revenue for India, which has the world's largest population of water buffalo (table 16.1). Buffalo meat is rejected in many regions of Asia because of its toughness. However, slow cooking and addition of spices can make the meat quite palatable while extending its shelf-life, an important factor in hot climates where refrigeration is frequently not available.

Table 16.1 Countries with the Largest Populations of Water Buffalo.

Country	Number ($\times 10^6$)	Head (per km^2)	Head (per 100 persons)
India	98.70	33.1	8.4
Pakistan	28.17	36.5	16.2
China	13.63	2.4	Unavailable
Nepal	4.37	30.4	15.4
Egypt	4.10	4.1	5.1
Philippines	3.38	11.3	10.8
Vietnam	3.00	9.6	3.4
Indonesia	2.08	1.1	0.9
Myanmar	2.04	4.3	5.7
Thailand	1.74	3.4	2.6
Bangladesh	1.21	9.2	0.7
Laos	1.12	4.8	18.4

Source: Food and Agriculture Organization of the United Nations, Global Livestock Production and Health Atlas (http://kids.fao.org/glipha).

Numbers of dairy buffalo have been increasing whereas the population of swamp buffalo is decreasing due to increasing mechanization. The river buffalo represent about 75% of the buffalo population in Asia.

16.2 Major Breeds and Countries

There are many breeds of domestic water buffalo (*Bubalus bubalis*). The name differentiates the breed from bison (*Bison bison*), which has long been called buffalo in the United States. Two main types exist, river buffalo and swamp buffalo. These types differ in chromosome number with the river buffalo having 50 and the swamp buffalo 48.

The chief domestic dairy-type river buffalo is the Murrah breed, which originated in the Punjab and Haryana states of India and the Punjab province of Pakistan. Individuals of the breed have been used to improve milk production of dairy buffalo in other countries, including Italy, Bulgaria, and Egypt. Other dairy breeds include Nili-Ravi, Surti, and Kundi. The animals are mostly black but vary toward gray and brown (figure 16.1). White markings may appear on the forehead, the lower portion of the legs, and the tip of the tail, especially in the Nili-Ravi breed. Horns are commonly short and may curl tightly forward, downward, or backward. However, in some breeds horns are sickle-shaped and curl upward. There are no **polled** breeds, and farmers do not practice dehorning. Adult females usually weigh 900 to 1,000 lb (450 to 500 kg). The head and neck are comparatively small, shapely, and clean-cut in females but coarse and heavy in males. Dewlap is nonexistent. Ears are small, thin, and pendulous. The front of the barrel of females is light and narrow while the rear portion is large, giving a wedge-shaped appearance. In males the rear portion is light and the front portion heavy. Ribs are well-rounded, and there is no hump. The downward sloping rump differentiates the topline from that of most dairy cattle breeds.

The swamp buffalo, which is rarely milked, is the chief work animal of the rice-growing countries of Southeast Asia, particularly China and Thailand. Comfortable in water, they plow, harrow, and puddle rice fields before planting. They are stockily built and may weigh up to 2,200 lb (1,000 kg).

Figure 16.1 Dairy-type water buffalo (Mudgal, 1999) (http://creativecommons.org/licenses/by-nc-sa/3.01).

16.3 Buffalo Milk and Its Products

Water buffalo are the second largest source of the world's milk supply, annually producing more than 187 billion lb (85 million metric tons). This level of production is comparable to the production of milk by dairy cows in the United States. India and Pakistan are the major producers (table 16.2). Whereas the world cattle population over the last two decades has increased by less than 1% per year, the buffalo population has increased by 2% per year, with increases in India (3.5%), Pakistan (5.4%), China (3.7%), Vietnam (4.8%), and Nepal (2.0%). The higher population growth rate of buffalo in these countries can be attributed to better feed conversion efficiency than that of local cattle. In India, buffalo account for 33% of the milk animal population and 45% of the milk.

Yield of milk ranges from 3,307 to 3,968 lb (1,500 to 1,800 kg) during the first lactation and increases steadily to peak in the fourth lactation. Whereas a lactational yield of 6,614 lb (3,000 kg) has been considered outstanding, many animals now yield 8,818 lb (4,000 kg) and some yield 11,023 lb (5,000 kg).

Table 16.2 Top 10 Buffalo Milk-Producing Countries (June 2008).

Country	Production (metric tons)
India	56,960,000[a]
Pakistan	21,500,000[b]
People's Republic of China	2,900,000[b]
Egypt	2,300,000[b]
Nepal	930,000[b]
Iran	241,500[b]
Myanmar	205,000[b]
Italy	200,000[b]
Turkey	35,100[b]
Vietnam	31,000[b]
World	85,396,902[c]

Metric ton = 2,000 kg = 2,204.62 lb

[a] Unofficial/semiofficial/mirror data

[b] Official figure

[c] Aggregate data (may include official, semiofficial, or estimates)

Source: Food and Agriculture Organization of United Nations, Economic and Social Development Department, Statistics Division.

16.3.1 Composition of Buffalo Milk

Total solids content of buffalo milk is high (15.5 to 17%) compared with that of the cow (12.4 to 13%). Concentrations of fat, protein, and lactose approximate 7.5%, 4.5%, and 5.0%, respectively (table 16.3). Fat globules average 20 to 30% wider in diameter than those of most dairy cattle breeds. However, considerable variation occurs among breeds of buffalo, e.g., average globule volume of the Egyptian buffalo was found to be twice the volume of the Indian buffalo and six times that of the Egyptian cow. Compared with cow's milkfat, buffalo milkfat contains less of the lower weight fatty acids (those containing 5 to 10 carbons), but is much higher in palmitic acid (C16). Content of unsaturated fatty acids is lower than in cow's milk. The pH of milk, about 6.7, varies little between the buffalo and cow. Freezing point (31°F [–0.56°C]) is about the same also. Freezing point is kept virtually constant because it is controlled by the osmotic pressure of blood. Immediately after parturition colostrum can contain as much as 32% total solids comprised of 15% fat, 13.5% protein, and 3.1% lactose plus minerals and growth factors.

Both **casein** and **whey** proteins are in higher concentrations in buffalo milk than in cow milk. Almost all the casein in buffalo milk is in the micellar form; diameters of most **micelles** fall within 110 to 160 nm compared with 70 to 110 nm in bovine milk. Whereas the amino

Table 16.3 Milk Composition Analysis (per 100 g).

Constituents[a]	Unit	Cow	Doe (Goat)	Ewe (Sheep)	Water Buffalo
Water	g	87.8	88.9	83.0	81.1
Protein	g	3.2	3.1	5.4	4.5
Fat	g	3.9	3.5	6.0	8.0
Carbohydrate	g	4.8	4.4	5.1	4.9
Energy	kcal	66.0	60.0	95.0	110.0
Energy	kJ	275.0	253.0	396.0	463.0
Sugars (lactose)	g	4.8	4.4	5.1	4.9
Cholesterol	mg	14.0	10.0	11.0	8.0
Calcium	IU	120.0	100.0	170.0	195.0
Saturated fatty acids	g	2.4	2.3	3.8	4.2
Monounsaturated fatty acids	g	1.1	0.8	1.5	1.7
Polyunsaturated fatty acids	g	0.1	0.1	0.3	0.2

[a] These compositions vary by breed, animal, and point in the lactation period.

Source: McCane, Widdowson, Scherz, and Kloos, International Laboratory Services (http://web.archive.org/web/20070929071651/http://www.northwalesbuffalo.co.uk/milk_analysis.htm).

acid compositions of α_{s1}-casein in buffalo and bovine milk are similar, that of the buffalo has no disulfide bridges (for detailed compositional data see Sindhu and Arora, 2011).

16.3.2 Buffalo Milk Products

A recently organized system of cooperative milk marketing has led to rapid expansion of the dairy industry in India and Pakistan. The system assures the producer a stable yearly price and eliminates intermediate marketing entities, thus providing more profit to producers. Cooperatives assist in obtaining loans to purchase superior animals and market feeds, and provide veterinary and AI services.

A major fermented product of buffalo milk in Asia is dahi (dadhi). Resembling yogurt, it is made from boiled or pasteurized milk. **Starter culture** is composed, preferably, of acid-producing mesophilic lactococci combined with *Leuconostoc* to provide the usual buttery (diacetyl) flavor.

In Italy, mozzarella di buffala campana, a traditional **pasta filata** type cheese with a **protected designation of origin (PDO)**, is made only from raw buffalo milk in southern Italy. The optimal ratio of fat to protein in the milk is 7.0:4.5. By Italian law, mozzarella cheese must be made solely from buffalo milk. It is preferred for its especially high butterfat content.

Buffalo milk containing 8.3% fat and 4.3% protein yields an average of 54.2 lb (24.6 kg) of mozzarella cheese per 220 lb (100 kg) of milk compared with 28.7 lb (13 kg) yield per 220 lb (100 kg) from cow's milk.

Buffalo milk supplies a high concentration of fat for the production of ghee and butter. Production of ghee accounts for usage of fat from about 45% of India's buffalo milk. Ghee is prepared by melting butter and then simmering while stirring occasionally until the froth disappears. The residue solids settle to the bottom and the ghee is then filtered. It solidifies when completely cool. Ghee can be stored for extended periods without refrigeration, provided it is kept in an airtight container to prevent oxidation and remains moisture-free.

16.4 Milk Production

A major difference between the bodies of cattle and buffalo is the ability to cool through evaporation of sweat. The buffalo has less than one-tenth the density of sweat glands common in cattle. Moreover, their dark color promotes heat absorption from direct sun rays, and their thick epidermal layer slows heat dissipation through conduction and radiation. Thermal stresses reflect themselves in lower milk yields, slower growth, depressed signs of estrus, and high calf mortality. This explains, in part, why buffalo desire to spend time cooling in pools of water. Milk producers can reduce heat stress on the animals by providing shade, wallows, and sprinkling water during especially warm periods. In the deserts of Saudi Arabia, open-sided sheds feature misting sprays and fans to cool the animals during hot weather.

During the heat of the day, researchers in India's Uttar Pradesh exposed females to showers for five minutes every two hours and installed wet screens to minimize hot dry wind. The treatments reduced maximal temperatures to 91°F (33°C) compared to 108°F (42°C) for a nonshowered control group. In response, feed consumption by the control group was 42.5 lb/d (19.3 kg/d) compared with 29.5 lb/d (13.4 kg/d) by animals in the unprotected shed. Milk yield was 12.8 lb/d (5.8 kg/d) versus 9.9 lb/d (4.5 kg/d) for the two groups, respectively.

As feed intake rises to keep pace with milk output, metabolic heat production increases. Thus, a high-yielding, adequately fed buffalo in full lactation in a hot climate will have greater difficulty dissipating surplus heat than a lower-yielding lactator.

16.4.1 Lactation and Milk Yield

Body weight at first lactation affects milk yield. The Murrah heifer should weigh at least 1,100 lb (500 kg) at first calving to reach maximum milk yield. The highest milk yield is most common in the fourth lactation, after which it declines. The shape of the lactation curve depends primarily on feed intake and quality, diseases, managerial quality, and milking practices. The optimal lactation length in the Murrah ranges from 262 to 295 days.

Lactation and milk yield depend on both genetic and nongenetic factors. The genetic influence reflects species, breed, and individual characteristics. Further, it is affected by the ability to reproduce, e.g., fertility and thereby calving interval. Improvement on these is often the result of breeding and selection. Nongenetic factors are management, amount and quality of feed, and skill of the farmer to detect heat and illnesses. Factors outside the farmer's control include climate, temperature, and humidity.

Feeding is commonly the most important factor for increasing and sustaining milk yield. To gain maximal yield sufficient energy, protein, minerals, and water must be provided. Calving interval is closely related to lactation length and milk yield. The longer the calving interval, the longer the lactation and the higher the lactation yield. However, total lifetime yield will be substantially less than when calving intervals are shorter. As with the bovine, milking frequency of water buffalo affects yields of both milk and fat. Murrah buffaloes produced 31% more milk and 26% more milkfat when milked three versus two times per day.

16.4.2 Anatomy and Physiology of the Buffalo Udder and Teat

Gross anatomy of the buffalo udder is similar to that of cattle. The four teats vary in shape and generally are larger than those of cattle. The Murrah breed has cylindrical teats. Average teat sizes are: front = 2.3 to 2.5 in long and 0.98 to 1.0 in wide (5.8 to 6.4 cm × 2.5 to 2.6 cm); rear = 2.7 to 3.1 in long and 1.0 to 1.1 in wide (6.9 to 7.8 cm × 2.6 to 2.8 cm). The hind quarters of the udder are slightly larger than the front ones and contain more milk. As with cattle, the approximate hind to front ratio is 60:40. The anatomy of buffalo teats differs slightly from that of cattle. The epithelium of the streak canal and the sphincter muscle surrounding it are thicker and more compact in buffaloes than in cattle. Therefore, more force is required to open the streak canal, causing them to be "hard milkers." Selection for improved conformation of the buffalo udder is an important goal of successful managers.

In the bovine, the milk is synthesized in the alveoli and is periodically transferred to the large ducts and cisterns of the mammary gland and teats. In the buffalo, milk is held within the alveoli and small ducts in the upper glandular part of the udder. No milk is stored in the cistern between milkings; instead milk is expelled to the cistern only during milk ejection. The same condition is found in Chinese yellow cows and yaks. Since milk is virtually absent from teat cisterns between milkings, the teats are collapsed and soft before milk letdown. In comparison, teats of the bovine cow can be quite firm because milk is held within the teat cistern.

16.4.3 Physiology of Milking

Because of their low rate of milk ejection and hard teat muscle sphincter, buffaloes are slower milkers than the bovine cow. In buffaloes, time to begin the letdown process averages two minutes but may be as long as 10 minutes. Because the udder cistern is absent or has a very small volume, there is little or no cisternal milk available. Furthermore, little intramammary pressure exists in the cistern to cause milk flow. In cattle, the milk is already stored in the large cistern, and milk is available for extraction following stimulation of milk letdown.

The intramammary pressure is higher in buffaloes during milking than in cattle. After increasing at the onset of the letdown process, it peaks during peak flow then decreases to zero at the end of milk flow. Intramammary pressure varies among individuals and milkings, its level being poorly correlated with high milk production. Letdown time is shorter in early and middle stages of lactation than in late lactation. A faster flow of milk is observed when the yield is higher.

16.4.4 Induction of Milk Letdown

Although the female buffalo has a fondness for her calf that causes the calf's presence to induce milk letdown, other conditions can trigger the process. Physical stimulation of the teats excites receptors from which nerve impulses are sent to the posterior pituitary gland causing secretion of the hormone oxytocin (cf. chapter 10). The hormone is transported via the blood to the mammary gland. Because both hormones and nerve impulses are involved in the milk ejection reflex, it is called a neurohormonal reflex. Oxytocin stimulates contractions of the alveoli and the small ducts thereby emptying milk into the larger ducts and cistern. Thereafter milk can be evacuated from the udder.

Contraction of the alveoli may be enhanced by tactile stimuli of the udder (massaging, squeezing). When calves suckle, they butt at the udder to increase milk secretion. Massage of the udder during milking imitates this reflex.

Like cattle, buffaloes become accustomed to specific stimuli. Oxytocin release can be activated by visual or audible stimuli, such as sight of the milker, noise of the vacuum pump, or entry to the milking parlor. This is called a conditioned reflex. (An unconditioned reflex is the suckling of the calf.) However, during milking, animals are sensitive to changes in the environment. When they are stressed, the hormone adrenaline is secreted, causing constriction of blood vessels, thereby limiting the amount of oxytocin available (cf. chapter 10).

Milking should begin after letdown starts and should be done quickly without causing stress or pain. Milk removal should be as complete as possible without excessive stripping, because residual milk in the secretory part of the udder decreases milk secretion, thereby decreasing milk yield.

16.4.5 Machines for Milking Buffaloes

Since the udder and teats of buffaloes differ from those of cattle, milking machines (figure 16.2) used on them need a higher vacuum, a faster pulsation rate, and a heavier clus-

Figure 16.2 Milking machine typical of those used in milking small herds of water buffalo (from DeLaval).

ter to counter the tendency for teat cups to climb due to the higher vacuum. Highest milk yields within an acceptable time were found when using 56 kPa vacuum, 65 pulsation cycles/min, and a pulsation ratio of 50:50. Studies in Pakistan indicated that for Nili-Ravi buffaloes the pulsation rate and ratio should be 70 cycles/min and 65:35, respectively. Milking rate is much slower than for cattle.

16.5 Reproductive Aspects

Buffaloes are slower than cattle to reach puberty, so the age at first calving is 25 to 35 months with conception weights of 550 to 600 lb (250 to 275 kg). Gestation time is about 310 days compared with 290 for *Bos taurus*. Calving interval is 14 to 18 months compared with 12 to 14 months for cattle. Estrus symptoms are marginally evident, and homosexual behavior among females is practically nonexistent. Therefore, determination of the optimal time for artificial insemination can be difficult.

Buffalo tend to be seasonal breeders. The peak of the calving season in India is from August through October. This and the long gestation period profoundly affect the production of milk for commercial use. Little evidence of estrus may be seen during hot months of the year. The average inter-calving period in hot climates is about 18 months. This differs markedly from the yearly freshening that is possible with the bovine cow.

Freezing of semen and embryos is practiced successfully. However, genetic improvement is slow because females are located primarily in developing nations. Furthermore, females of breeding age are often retained in the herd for nine lactations (16 years of age). However, buffalo vary widely in genetic characteristics, providing important opportunities for selection of desired traits. Since there is a high repeatability of yield, selection of the dam should be based on first lactations. The heritability of second-lactation yield is lower than that of the first. Furthermore, to delay selection slows the rate of genetic gain per year. Additionally, heritability is high for yield during the first 250 days and its use can minimize the effect of ensuing pregnancy. Successful genetic improvement can be accomplished through the use of AI provided several conditions are met: (1) proven sires are selected, (2) trained persons do the insemination, (3) semen is highly viable, (4) females are in good nutritional condition, (5) estrus periods are effectively recognized, and (6) records are kept and used for selection and breeding.

Males also are affected unfavorably by high temperatures. During summer, Murrah bulls were observed to lose libido while semen quantity decreased and quality became poor (Misra and Sengupta, 1965).

The buffalo should be dried off approximately 2 to 3 months before expected calving to provide rest and replenishment of udder tissue. In a high-yielding machine-milked herd (above 22 lb/d [10 kg/d]) the buffalo should be dried off when the daily yield falls below 5.5 lb (2.5 kg), even if it is still more than three months to expected calving. As an alternative to drying off, the animal can be used as a foster mother to newly born calves.

When river and swamp buffalo are interbred, the F_1 hybrid has a diploid chromosome number of 49. The unbalanced **karyotypes** of crossbreeds may, however, affect the process of **gametogenesis**, resulting in infertility in males.

A team of researchers in Canada, India, and Italy investigated the possibility of producing hybrid embryos of domestic cattle (*Bos taurus*) and water buffalo (*Bubalus bubalis*) by exposing cattle oocytes to buffalo sperm and buffalo oocytes to cattle sperm. Cleavage rates and the post-fertilization features of hybrid embryos up to the blastocyst stage were compared with those of buffalo and cattle embryos. The cleavage rate in buffalo oocytes exposed to cattle sperm was low (40.8%), with only 8.8% of these hybrid embryos reaching the blastocyst stage. Cattle oocytes exposed to buffalo sperm showed 86.3% cleavage, while 25.9% of them attained the blastocyst stage. The speed of development of both types of hybrids was intermediate between that of cattle and buffalo embryos, with hatching occurring on day 7.5 in hybrids, day 8 or 9 in cattle, and day 7 in buffalo. The authors concluded that cattle–water buffalo hybrid embryos produced using interspecies gametes are capable of developing to advanced blastocyst stages and that their in vitro fate, and developmental potential, are influenced by the origin of the oocyte (Kochhar et al., 2002).

16.6 Feeding Water Buffalo

The Asian buffalo is a true herbivore that tends to eat a wider variety of plants than do cattle. They are able to efficiently digest and assimilate cellulosic components of poor quality feeds. Most of India's milk is produced from diets based on crop residues including wheat straw, paddy straw, maize, and millet **stovers** supplemented with concentrates containing oil cakes, rice polish, urea, molasses, and minerals. Since the swamp buffalo spends most of its time in marshes and swamps, aquatic plants comprise most of its diet. Factors that enable the water buffalo, compared with cattle, to utilize poor quality forages include: larger rumen volume, higher rate of salivation, slower rate of passage of digesta through the reticulorumen, slower rumen motility, and higher cellular activity.

Buffaloes used primarily for family and neighborhood purposes are commonly fed cut fodder and crop residues plus about 1.1 lb (0.5 kg) of concentrate mixture per liter of milk produced. Those located near large cities and used for commercial production are frequently fed dry fodder plus large quantities of discarded bread and other products containing flour.

Under practical conditions of feeding buffaloes in most countries, a range of temperatures is experienced over each 24-hr period. Seldom do temperatures remain over 93°F (35°C) for an extended time and commonly drop overnight. In hot climates the following factors in animal management should be considered before nutritional improvement is made. Stress from direct sunlight must be avoided, drinking water must be available, and the buffalo should be sprayed to prevent distress. Feeding cycles should be arranged to avoid exposure to high heat. If these conditions can be provided, feeding a highly nutritious diet can be expected to enhance productivity.

16.7 Diseases of Special Significance

Diseases of buffaloes and dairy cows are quite similar as are their treatments and controls. The most important infectious diseases are anthrax, foot-and-mouth disease, hemorrhagic septicemia, tuberculosis, and, previously, rinderpest.

Anthrax is caused by the rod-shaped endospore-forming bacterium *Bacillus anthracis*. Environmental presence of the organism is favored by failure to recognize and report the disease, slaughter of infected animals for use as human food, salvaging of hides of animals that died of the disease, and improper disposal of infected carcasses or offal. Wallows are particularly important sources of the organism, the spores of which remain viable and infective for long periods. Most forms of the disease are lethal, and it affects both humans and other animals. There are effective vaccines against anthrax, and some forms of the disease respond well to **antibiotic** treatment. Thanks to over a century of animal vaccination programs, sterilization of raw animal waste materials, and anthrax eradication programs in North America, Australia, New Zealand, Russia, Europe, and parts of Africa and Asia, anthrax infection is now relatively rare in domestic animals, with only a few dozen cases reported annually.

Tuberculosis-infected water buffalo respond readily to the tuberculin test and retain evidence of the skin reaction longer than do cattle (cf. chapter 13).

Leptospirosis (cf. chapter 13) is an especially important disease in water buffalo that have access to water in paddies, wallows, or muddy soils where the causative organism, *Leptospira pomona,* may reside.

Hemorrhagic septicemia is a highly fatal bacterial disease seen mainly in cattle and water buffalo (cf. chapter 13). Hemorrhagic septicemia results from infection by *Pasteurella multocida* subsp. *multocida*, a Gram-negative coccobacillus in the family Pasteurellaceae. In susceptible animals, the clinical signs often progress rapidly from dullness and fever to death within hours. Because the disease develops so quickly, few animals can be treated in time and recovery is rare. Young animals are mainly affected in endemic regions, and outbreaks are particularly common during rainy weather when the organism can spread readily. In areas where cattle have no immunity, severe disease is expected to occur in all ages. Where extensive use of vaccination for hemorrhagic septicemia has been practiced, the disease has been extensively reduced.

Rinderpest (also cattle plague), now eradicated, was an infectious viral disease of cattle, domestic buffalo, and some other species of even-toed ungulates. The disease was characterized by fever, oral erosions, diarrhea, lymphoid necrosis, and high mortality. Death rates during outbreaks often approached 100% in immunologically naïve populations. The disease was mainly spread by direct contact and by drinking contaminated water.

After a global eradication campaign, the last confirmed case of rinderpest was diagnosed in 2001. In August 2011 the United Nations declared the disease eradicated, making rinderpest only the second disease in history to be fully wiped out, following smallpox.

As a *Morbillivirus*, the rinderpest virus (RPV) is closely related to those causing measles and canine distemper. Despite its extreme lethality, the virus is fragile and quickly inactivated by heat, desiccation, and sunlight. The measles virus evolved from the then-widespread rinderpest virus, most probably between the eleventh and twelfth centuries.

16.8 Parasites

Buffaloes and cattle share a large number of parasites. A few are exclusive to one or the other host. Environment is a major variable determining presence of infective parasites. A high incidence of calf mortality is caused by *Toxocara vitulorum*, the larvae of which can be transferred from the dam to the calf during the first month of life. The following parasites are of particular significance to water buffalo.

1. *Sucking louse*: Its presence is easily detected because of the large size of both the louse and its eggs.
2. *Buffalo fly*: A sucking-type insect.
3. *Black fly* (*buffalo gnat*): Can tear the skin with its rasping teeth.
4. Sarcoptes scabiei *var.* bubalus: Induces mange that can develop into small papules that become scabbed. Particularly prevalent during drought.
5. Psoroptes: Mite that causes psoropic mange, primarily on the head and tail.
6. *Trypanosomiasis*: A parasitic disease, it is common in the Indian subcontinent and Southeast Asia and can cause hemolytic anemia. Systemic disease and reproductive failure are common and cattle appear to waste away. Some species of African buffalo appear trypanotolerant and do not suffer from clinical disease. Calves are more resistant than adults.
7. *Filaroid nematodes, shistosomes*: Parasites of the circulatory system.
8. *Amphistomes* (*flukes*): Parasites of the gastrointestinal tract.

9. *Nematodes*: Parasites of the abomasum and large intestine.
10. *Nematodes* (*helminths*): Parasites of the small intestine. At least five genera have been identified. Worm burdens, commonly heavier in buffalo than in cattle, may pack the small intestine to near occlusion. Cestodes cause coccidiosis.
11. *Trematodes and flukes*: Parasites of the liver, pancreas, and peritoneal cavity.
12. *Trichomonas and leeches*: Parasites of the genital tract.
13. *Sarcocystis*: Parasites of the muscles. Cysts 0.4 to 0.8 in (1 to 2 cm) in diameter occur in the esophagus and tongue.

16.9 Supporting Organizations

The International Buffalo Federation (IBF) is the most prominent organization that supports the science and practice of production of milk and meat by water buffalo. The IBF was founded in Cairo, Egypt, in 1985 and is an independent, nonpolitical, nonreligious, and nonprofit organization with permanent headquarters in Rome. It promotes internationally planned research, enhances contacts among involved scientists and extension personnel, and strengthens linkages among related organizations internationally.

Many countries, especially in Southeast Asia and Europe, have organizations that focus on advancement of *Bubalus bubalis* as a contributor to the welfare of humans. These organizations support international congresses that bring together authorities and producers to advance our knowledge related to the water buffalo, such as the Asian Buffalo Congress, World Buffalo Congress, Buffalo International Conference, Congress for Development of Mesopotamian Buffaloes, and Buffalo Symposium of the Americas and Europe. The Italian Water Buffalo Association is prominent among organizations supporting advancement of dairy-type water buffalo.

16.10 Summary

More than 160 million water buffalo, concentrated in countries of Southeast Asia, annually produce about 85 million metric tons of milk that contains 15 to 18% total milk solids compared to 12 to 14% in the milk of the dairy cow. The quantity of milk produced during a 305d lactation averages about one-third that of Holstein cows. However, buffalo can digest a wider variety of plants than can the cow. The anatomy of the buffalo mammary gland differs from that of the cow, and they tend to be hard and slow to milk. They reach puberty at an older age than do bovine females and show symptoms of ester much less openly. In addition to providing milk and meat, water buffalo are also used as draft animals. Genetic improvement through the use of artificial insemination is being practiced.

Study Questions

1. How does the composition of buffalo milk differ from that of the cow?
2. Name three countries that have the largest population of water buffalo.
3. List four breeds of dairy buffalo.
4. Discuss ways reproduction differs between the buffalo and the cow.
5. What are some anatomical characteristics of the water buffalo that affect milk letdown and removal?
6. Name some practices that have been recommended for genetic improvement among dairy buffalo.
7. What management practices should be followed when feeding dairy buffalo in hot climates?
8. What two factors make it difficult for producers to provide a year-round production of buffalo milk for commercial use, especially in hot climates?
9. Name five diseases especially common among water buffalo.
10. What is the major determinant of infections by parasites among water buffalo?

References

Kochhar, H. P. S., K. B. C. Appa Rao, A. M. Luciano, S. M. Totey, F. Gandolfi, P. K. Basrur, and W. A. King. 2002. *In vitro* production of cattle–water buffalo (*Bos taurus–Bubalus bubalis*) hybrid embryos. *Zygote* 10(2):155–162.

Misra, M. S., and B. P. Sengupta. 1965. Climatic environment and reproductive behavior of buffaloes. III. Observations on semen quality of buffalo bulls maintained under two different housing conditions. *Indian Journal of Dairy Science* 18:130–133.

Mudgal, V. D. 1999. "Milking Buffalo." In L. Falvey and C. Chantalakhana (eds.), *Smallholder Dairying in the Tropics* (chapter 6). Nairobi, Kenya: International Livestock Research Institute.

Sindhu, J. S., and S. Arora. 2011. "Buffalo Milk." In J. W. Fuquay, P. F. Fox, and P. McSweeney (eds.), *Encyclopedia of Dairy Sciences*, Vol. 3 (2nd ed., pp. 503–511). San Diego, CA: Academic Press.

Websites

American Water Buffalo Association
(http://americanwaterbuffaloassociation.com/index.htm)

European System of Cooperative Research Networks in Agriculture (http://www.escorena.net)

Food and Agriculture Organization of the United Nations, Global Livestock Production and Health Atlas
(http://kids.fao.org/glipha)

International Buffalo Federation
(http://internationalbuffalofed.org)

The Turkey Creek Company
(http://home.valornet.com/pcwdb/index.htm)

17

Dairy Goats

Blessed are we goat owners who really *know* the value of goat milk.
Mary A. Shanks

The authors are grateful to David Kiesling, Lincoln University, Jefferson City, Missouri, for his significant contributions to this chapter.

17.1 Introduction

Goats (*Capra aegagrus hircus*) were among the first domesticated animals. They are believed to have descended from a species of wild goat found in Asia Minor, Persia, and nearby countries. The goat is a member of the family Bovidae and in the goat-antelope subfamily Caprinae. Female goats are referred to as **does**, intact males as **bucks**. *Nanny goat* (for females) originated in the eighteenth century and *billy goat* (for males) in the nineteenth. Their offspring are called *bucklings* or *doelings*, depending on the sex (a unisex name is **kid**). Castrated males are *wethers*.

Goats are **ruminants**. They have a four-chambered stomach consisting of the rumen, the reticulum, the omasum, and the abomasum. As with other mammal ruminants, they are even-toed ungulates. Females have an udder consisting of two milk-secreting glands with one teat each, in contrast to cattle, which have four glands with teats. There are more than 300 distinct breeds of goats.

Hardy and disease-resistant, goats are natural browsers and thrive on rough, sparse pastures where a dairy cow would virtually starve. This characteristic makes the dairy goat a commonly underrated animal that is uniquely valuable as a provider of milk, meat, and more for humans. The dairy goat is frequently called the cow of the poor. The goat eats little, occupies a small area, and produces enough milk for the average unitary family. There are approximately 862 million goats in the world, up from about 590 million in 1990 (table 17.1). Of approximately 360,000 dairy goats in the United States, Wisconsin, California, Texas, Iowa, and Pennsylvania have the highest numbers.

Table 17.1 Numbers of Goats in Different Parts of the World.

Location	Number (million)	Percentage of World Total
Asia	514	59.7
Africa	291	33.8
Central America	9	1.0
Caribbean	4	0.5
North America	3	0.4
South America	2.1	0.25
Europe	1.8	0.21
Oceania	1	0.1
World	862	100

Source: Food and Agriculture Organization of the United Nations, Statistics Division (http://faostat.fao.org).

17.2 Major Dairy Goat Breeds of the United States

The first permanent importation of goats into the United States was in the early 1900s. The first breeds imported, Toggenburg and Saanen, came from Switzerland. Later Alpine and Anglo-Nubian breeds were imported from Switzerland and England, respectively. The American Dairy Goat Association (ADGA) maintains separate official herd books for purebred herds and American breed herds. A purebred goat is born of a purebred sire and a purebred dam of the same breed and conforms to breed standards. An American breed goat is one born to a sire and dam of the same breed, going back a minimum of three generations for does and four generations for bucks.

Currently there are eight registered ADGA breeds for which the organization issues certificates of registration and maintains herd books and production records; there are seven standard size breeds—Alpine, LaMancha, Nubian, Oberhasli, Saanen, Sable, and Toggenburg—and one miniature breed—the Nigerian Dwarf. Milk production data for each breed are given in table 17.2.

The standard size breeds can be registered in the Purebred and American herd books for their breed, and

Table 17.2 Top Individual Producers of Milk, Fat, and Protein among the Major Dairy Goat Breeds of the United States (2011).

	Milk[a]		Fat		Protein	
Breed	lb	kg	lb	kg	lb	kg
Saanen	6,080	2,763	255	115	193	88
Alpine	5,670	2,575	149	68	152	69
Toggenburg	4,270	1,941	145	66	110	55
LaMancha	4,140	1,882	151	69	133	60
Nubian	3,920	1,782	178	81	152	66
Oberhasli[b]	3,450	1,614	121	56	121	47
Sable	3,320	1,509	124	56	104	47
Nigerian Dwarf	1,500	681	103	47	61	28

[a] 305 days of production

[b] Based on data for 2001 from the Oberhasli Breeders of America (http://www.oberhasli.net).

Source: American Dairy Goat Association (http://www.adga.org).

crosses between these seven may be recorded in Recorded Grade Herd or Experimental Herd books.

17.2.1 Alpine (French Alpine)

This breed originated in the French Alps and in 1904 was the first breed of dairy goats registered in the United States. Alpine goats are heavy milkers and are widely used for commercial milking. The Alpine is also referred to as the French Alpine, and registration papers for it use both designations synonymously. These are hardy, adaptable animals that thrive in virtually any climate while maintaining good health and excellent production. They have cone-shaped, perky ears and an alert appearance (figure 17.1). Their color varies from pure white to pure black, but they usually have white spots or mixtures of white and darker hair. Wither height and body weight of mature does are 30 in (76 cm) and 130 lb (59 kg), while their mature mates measure 34 in (86 cm) and weigh 170 lb (77 kg). The face is straight. For registration purposes a Roman nose, Toggenburg color and markings, or all-white are discriminated against. Both sexes may have wattles and beards.

17.2.2 Nubian

Nubian dairy goats were developed by the British, who crossed their native females with bucks imported from India and Africa. Reflecting their Middle Eastern heritage, these goats can live in very hot climates and have a longer breeding season than other dairy goats. They are considered a dairy, or dual-purpose, breed. They have short hair and are various colors of brown, black, white, cream, gray, or combinations thereof. They may be solid in color or have a spotted and/or dappled pattern.

Nubians are characterized by large pendulous ears and convex Roman noses (figure 17.2). The males have beards. Most does weigh at least 135 lb (61 kg) and bucks 175 lb (79 kg). The minimum height of the breed, measured at the withers, is 30 in (76 cm) for does and 35 in (89 cm) for bucks. Nubians are known for a high milkfat yield, averaging about 5%, but average milk yield is typically lower than that of other dairy breeds.

Figure 17.1 A mature French Alpine doe. This breed of dairy goats was imported into the United States from France (www.eXtension.org).

17.2.3 Saanen

Animals of this breed are probably the largest of Swiss dairy goats. The wither height and body weight of mature females are 30 in (76 cm) and 145 lb (66 kg), respectively, whereas those of males are 35 in (89 cm) and 185 lb (84 kg), respectively. Saanens are white or creamy colored with a flesh-colored muzzle. Their hair is short, and most of them are hornless but have a beard (figure 17.3).

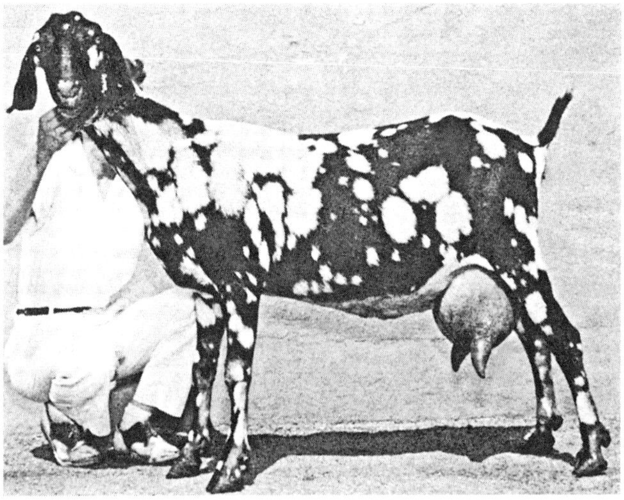

Figure 17.2 A lactating Nubian doe. Nubians are of Oriental and English origin. They are commonly large and of many colors. Their milk is rich in milkfat (courtesy of *Dairy Goat Journal*).

Figure 17.3 A lactating Saanen doe. Originally from the Saanen valley of Switzerland, the Saanen is large, white, and high-producing (courtesy of *Dairy Goat Journal*).

It is believed that the Saanen goat has remained true to type for at least 35 generations and is, therefore, a highly pure breed. This is interesting in view of the fact that goats have been continuously mixed for many centuries in Europe and Asia—dating back even to the fifth century AD.

17.2.4 Toggenburg

This popular breed of dairy goats was the first imported into the United States from the Toggenburg valley of Switzerland, from which it takes its name. Colors vary from light fawn to dark chocolate, and there is a white stripe down each side of the face and a white triangle around the tail. The legs are white from the knees and hocks to hoofs (figure 17.4). Toggenburgs are commonly hornless, and their hair may be short, long, or curly. They may have wattles and often a beard. The ears are erect and carried forward. Facial lines may be dished or straight, never Roman.

Wither height and body weight of the doe average 29 in (74 cm) and 130 lb (59 kg), respectively. Those of the buck average 33 in (84 cm) and 160 lb (73 kg), respectively. Toggenburgs are known for their excellent udder development and high milk production with an average fat content of 3.7%.

17.2.5 LaMancha

The LaMancha breed was developed in its early stages in the late 1930s by Mrs. Eula Frey in Oregon, but it is thought to have descended from Spanish goats brought by early settlers to California. This breed is characterized by small ears and excellent productivity (figure 17.5). The breed has excellent dairy temperament and is a sturdy animal that can withstand hardship and still produce quality milk with high fat and protein over a long period of time. One advantage of the breed is they can be milked for two years without being **freshened**. The

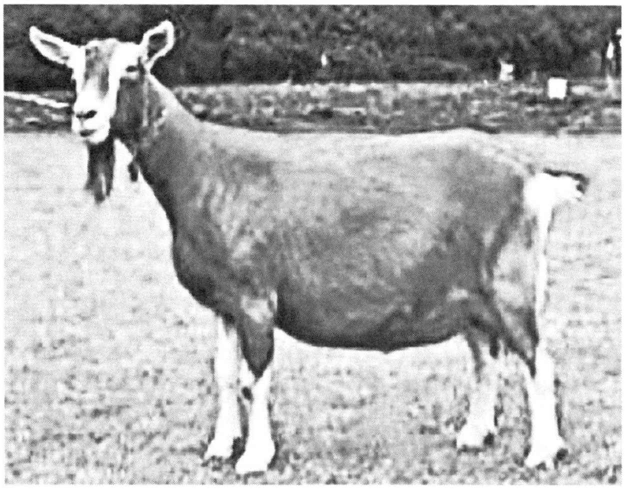

Figure 17.4 Mature Toggenburg doe (www.eXtension.org/).

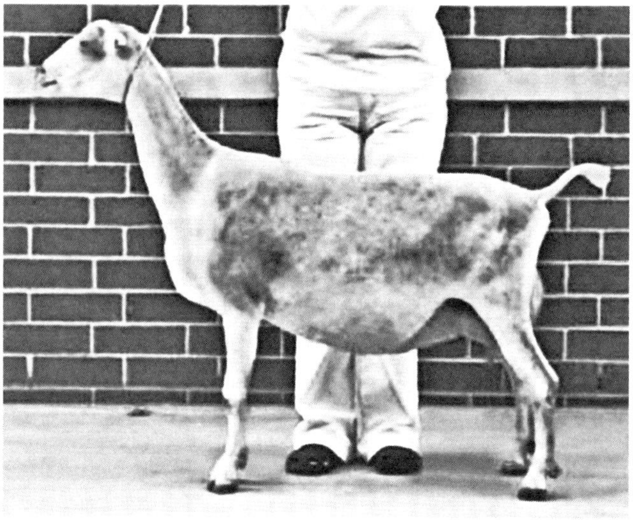

Figure 17.5 Young LaMancha doe (courtesy of Forrest Pride Dairy Goats, Lebanon, Missouri, and Lincoln University of Missouri, Jefferson City, Missouri).

LaMancha face is straight, and there are two types of ears: the gopher ear and the elf ear. The maximum length of the gopher ear is approximately 1 in (2.54 cm), or there may be very little ear, with little or no cartilage. The end of the ear must be turned up or down. This is the only type of ear that will qualify a buck for registration. The elf ear has a maximum length of about 2 in (5 cm). The end of the ear must be turned up or down and cartilage shaping the small ear is allowed. They have a straight nose. This small breed must measure at least 28 in (71 cm) at the withers; they weigh about 130 lb (59 kg). LaManchas are usually more calm and docile than goats of other breeds.

17.2.6 Sable

Sable goats are derived from the Saanen breed that originated in Switzerland. They may be any color or combination of colors except white or light cream. Their color patterns result from the interaction of recessive genes from the sire and the dam. If the offspring has a dominant gene for white color, the animal is white and thus a Saanen, not a Sable. If the animal has two recessive genes, one from each parent, a colored coat is the result (figure 17.6). Because the colored genetics were brought in with the original Saanen imports from Switzerland, Sables will continue being born as long as people are breeding Saanens.

Sable does are feminine and heavy producers of milk usually containing 3 to 4% fat. The Sable breed, like the Saanen, is medium to large, weighing approximately 145 lb (65 kg), with rugged bones and high vigor. Minimal heights at the withers are 30 in (75 cm) and 32 in (80 cm) for does and bucks, respectively. The body is deep, wide, and long with well-sprung ribs pointing to the rear. Sables have a straight back with very little slope to the rump.

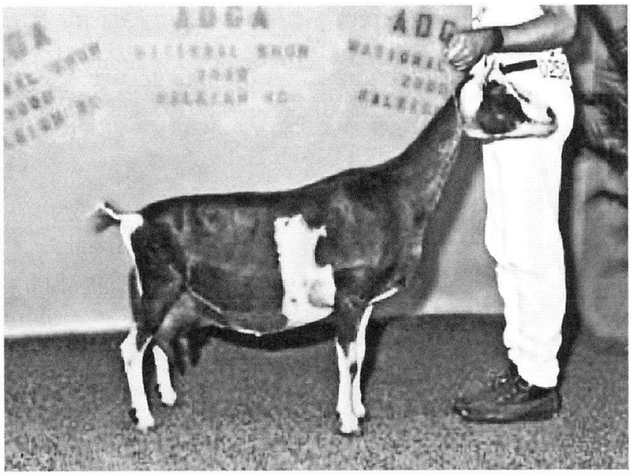

Figure 17.6 Mature Sable doe (www.eXtension.org/).

The legs are long, squarely set, strong, straight, and wide apart. The ears are large and upright, sometimes rounding at the tip and pointing forward. The hair is short and fine. The head is wide and long with a straight or dished face. They have a deep muzzle and a strong jaw.

One reason for breeding Sables is due to their color. White Saanens do not fare as well in hot, sunny climates because their light skin color makes them prone to skin cancers. Buyers from tropical countries know this and are often reluctant to buy Saanens despite their favorable size, productivity, and temperament. Sables do not have this problem.

17.2.7 Nigerian Dwarf

The Nigerian Dwarf is a miniature goat of West African origin. Its conformation is similar to that of the larger dairy goat breeds. It is characterized by a straight nose, upright ears, and a soft coat with short to medium hair. Its balanced proportions give it the appearance of the larger breeds, but in height does stand no more than 22.5 in (57 cm) and bucks no more than 23.5 in (60 cm). Ideal weight is about 75 lb (34 kg). The main colors are black, chocolate, and gold. Random white markings are common, as are spots and other color combinations, including red, white, gold, and black (figure 17.7).

Dwarf goats are gentle and loveable. They breed year-round and can be bred at seven to eight months of age, although one year is preferred. Many owners breed does three times in two years, with a six-month break between pregnancies. Dwarf does can give birth to several kids at one **kidding**. Three or four are common. Bucks are able to breed does when as young as three months of age and easily by their eighth month.

Nigerian Dwarf milk is higher in fat and protein than milk from most dairy goat breeds. They can produce 3 to 4 lb/d containing 6 to 10% milkfat.

Figure 17.7 A lactating Nigerian Dwarf doe. This doe is angular with excellent dairy character and a capacious and well supported mammary system (Jmkarohl/wikimedia.org) (creativecommons/org/licenses/by-sa/3.0/deed.en).

17.2.8 Oberhasli

The Oberhasli, also known until 1987 as the Swiss Alpine, is a very old breed from Switzerland. As a "color breed," they must be colored correctly to be registered with the ADGA. Does may be pure black or chamoisee, whereas bucks (males) must display correct chamoisee coloring defined as bay (light to dark reddish brown) with the following black points: (1) two black stripes from above the eyes down the face and merging to a black muzzle, (2) a black dorsal stripe, (3) black lower legs and belly, and (4) a gray to black udder. Individuals are typically about two inches shorter than other standard-sized (nonminiature) breeds.

In 1936, Dr. H. O. Pence imported five purebred Oberhaslis to the United States from which all purebred Oberhasli in North America are descended (Dohner, 2001). In 1978, Oberhaslis were accepted as a breed by the ADGA. Because of its rarity and close similarity in type to the multicolored French Alpine goat, the Oberhasli registrations were lumped with the American Alpines. A small number of them were kept pure with records intact, and the breed was kept alive in America almost singlehandedly by Esther Oman, a breeder in California.

The Oberhasli breed is growing in popularity at goat shows based on their attractive general appearance, body capacity, dairy character, and mammary system (figure 17.8 on the next page). Many have been added to dairy herds, thus increasing their availability. Their docile, loving nature makes them ideal in the show ring.

This breed is a good choice for those who want to produce milk from hardy and thrifty dairy goats possessing a vivid rust-red coloration and the Swiss-type head (upright ears).

17.2.9 Evaluating Dairy Goat Conformation

A prerequisite to evaluating body conformation is learning the parts of a goat (figure 17.9). The female dairy goat should be angular, have prominent hip bones and

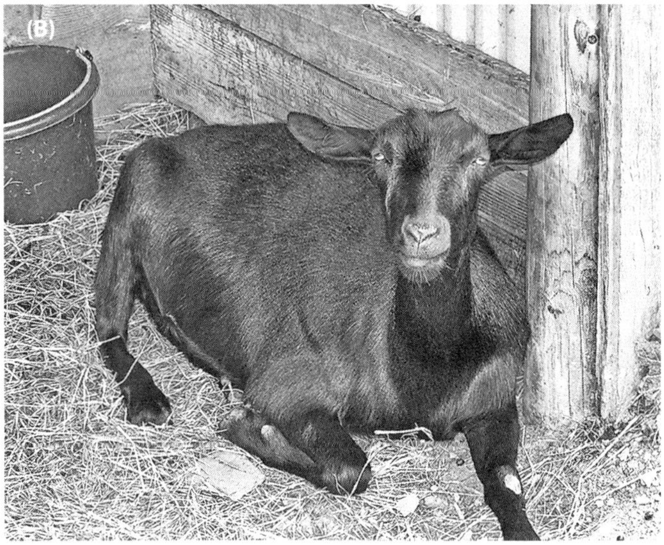

Figure 17.8 (A) Oberhasli doe in full lactation (www.eXtension.org). (B) A variation of the breed, the black Oberhasli (Thricebaked/wikipedia.org) (creativecommons/org/licenses/by-sa/3.0/deed.en).

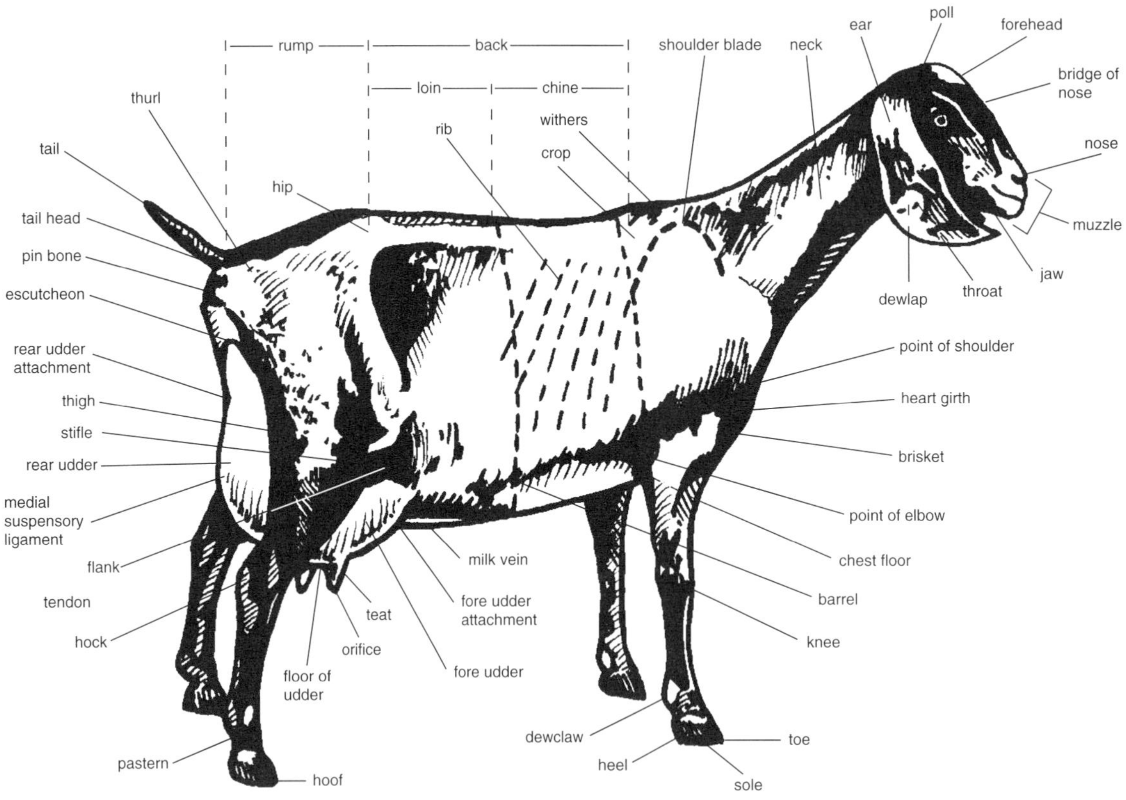

Figure 17.9 Parts of the dairy goat (McNulty et al., 1997; National Nubian Club).

thin thighs, and possess a long body and neck. She should have a strong topline. The ribs should be long, well sprung, and far apart. This openness of rib denotes dairy temperament in the goat as well as in the dairy cow. The chest floor should be wide, denoting large lung capacity and a strong constitution. Legs should be straight and hoofs kept trimmed. The skin of healthy, lactating dairy goats is smooth, thin, and pliable. A good udder is capacious, strongly attached, and of high quality (pliable and soft). After milking, the udder should be collapsed and pliable, much as a soft leather glove.

17.3 Reproductive Aspects of Goats

Doelings reach puberty by 6 to 8 months of age and are usually bred at 7 to 10 months of age, depending on breed and nutritional status. It is preferable to breed after the doe has reached 70% of adult weight, usually at 8 to 10 months of age. However, this is rarely possible in open-range herds. In temperate climates and among the Swiss breeds, breeding season commences as day length shortens and ends in early spring. In equatorial regions, goats are able to breed at any time of the year. Successful breeding in these regions depends more on available forage than on day length.

Goats are seasonal breeders. On the North American continent most dairy goats will mate from September to March. The average estrous cycle duration of a doe is 18 to 22 days. Signs of estrus include nervousness, twitching the tail, frequent urination, **bleating**, and swelling of the vulva. There is commonly a mucous discharge. Goats in estrus ride each other, as do cows (cf. chapter 9). Insemination should be made the day following observed estrus. Estrus ranges in length from a few hours to three days. If a doe is known to remain in estrus for three days, insemination should be accomplished on the third day. Artificial insemination, where available, allows access to a wide variety of bloodlines and minimizes the need of owners of small goat herds to keep a buck. Dr. C. A. V. Barker of the University of Guelph, Ontario, Canada, perfected the techniques involved in using frozen semen in goats.

The doe should be bred at 45 to 60 days in lactation and normally will produce milk for 305 days. The kidding interval should be about 12 months. The normal gestation period of goats is 150 ± 5 days. Twins are the usual result, with single and triplet births also common. Larger litters occur infrequently. Birth weight of kids is approximately 5 to 7 lb, with males weighing slightly more than females. The average male/female ratio is approximately 1.1:1.0.

A **growthy** buck kid at approximately six to eight months of age may be mated with six or eight does during his first season. During his second season (18 to 20 months of age) he may breed 25 to 30 does, and when mature he may breed 50 to 60 or more does during the breeding season. Goats commonly have high conception rates; allowing the buck to breed the doe only once per mating (estrus) is sufficient. Unlike does, bucks do not have a dormant or sterile breeding season. Bucks (intact males) of Swiss and northern breeds come into rut in the fall during the does' heat cycles. Bucks of equatorial breeds may show seasonally reduced fertility, but as with the does, are capable of breeding at all times.

17.3.1 Odor of the Buck

Some persons associate goats with unpleasant odors and unsanitary conditions. Although the male goat does emanate a disagreeable odor at certain seasons, especially during the breeding season, the female does not, and of course it is she who secretes milk. To prevent potential contamination of milk with the odor, bucks and lactating does must be kept apart, except for the act of mating. Sebaceous scent glands at the base of the horns add to the male goat's odor, which makes him attractive to the female. Some does hesitate to mate with a buck that has been de-scented.

The buck's odor results from two musk glands located behind and to the side of the horns. In a mature buck they are 1 to 2 cm^2 in size. During breeding season, they enlarge into three pleats. Both male and female goats have musk patches. In females and castrated males they are commonly small and inactive, so the odor is unnoticeable. It is believed that they are activated by the male sex hormone. Goat breeders can avoid this odor by using artificial insemination. Moreover, bucks can be virtually de-scented by applying a hot iron to the musk patches behind the horn bud areas. Or the musk patches can be removed surgically.

17.4 Characteristics of Goat Milk

The *Grade "A" Pasteurized Milk Ordinance* (*PMO*) defines goat milk as

> the normal lacteal secretion, practically free of colostrums, obtained by the complete milking of one (1) or more healthy goats. Goat milk sold in retail packages shall contain not less than 2.5% milkfat and not less than 7.5% milk solids not fat. Goat milk shall be produced according to the sanitary standards of this *Ordinance*. The word "milk" shall be interpreted to include goat milk. (US Food and Drug Administration, 2013)

This means that any goat milk product labeled "grade A" shall be produced under the same standards the *PMO* sets for cow milk (cf. chapter 24).

The goat is essentially 100% efficient in converting carotene into vitamin A; that is why goat milk is white in color.[1] It follows, then, that butter made from goat milk is essentially white. Of course, vegetable coloring can be added, as is done in the manufacture of other spreads.

Goat milk resembles that of cows in composition (see table 16.3) although it is notably lower in lactose,

4.4% versus 5.0%. Goat milk is widely used by persons afflicted with stomach ulcers. Additionally, persons allergic to certain foods often can drink goat milk without ill effects.[2] However, its fat globules are smaller (diameter averages about 2 μm compared to 3 μm in cow milk) and by nature tend to remain suspended (little cream rises to the top), so mechanical homogenization is unnecessary. Although cream does not rise readily, it can be obtained with a mechanical separator. Studies at the University of Minnesota indicate that the low creaming ability of goat milk, especially at low temperatures, results from insufficient agglutinating **euglobulins** and not from small fat globules, as often stated. Agglutinating euglobulins cause fat globules to clump and therefore to rise at an accelerated rate. Milks reconstituted from goats' cream and cows' skim milk creamed readily, whereas those made by combining cows' cream and goats' skim milk creamed poorly.

Although the protein content of cow and goat milk is similar, the curds have different characteristics. The typical major alpha$_{s1}$-casein (α_{s1}; cf. chapter 21) in cow milk is absent in goat milk, and the formation of casein curd under **rennin** action is different. The casein-based curd of goat milk forms small, light, friable flakes that are easily dissolved, much like the curd of women's milk. The softer the curd, the more easily digestible it is. Softness is reflected in the curd tension, a measure of hardness or softness of the curd formed on addition of strong acid to milk. Curd tension is largely a breed/species characteristic. Holsteins generally have the softest curd in the bovine family. As measured in grams of resistance, curds from cow milk range from 15 to 200 g (average = 70 g), whereas curds from goat milk range from 10 to 70 g (average = 36 g). Curd formed in goat's milk with acid-pepsin treatment is softer than similarly formed curd of cow's milk. However, curd formed with rennet is stronger in goat's milk than in cow's milk. Curd strength varies among individual animals and stage of lactation. Curd strength decreases to a minimum in mid-lactation and then increases to the end of lactation.

17.4.1 Flavor

Milk of goats produced under sanitary conditions is sweet, bland, and free of off-flavors, except that consumption of odorous plant materials, especially garlic and onion, can impart "feed and weed flavors" to the milk. The same factors that adversely affect the flavor of cows' milk also affect goats' milk (cf. chapter 25). Additionally, producers of goat milk must be certain the buck is kept at least 164 ft (50 m) away from lactating does to prevent the milk from absorbing the buck's odor.

17.4.2 Cooling

Milk should be cooled promptly after milking to prevent spoilage by bacteria that inevitably are present in milk from any lactating animal (cf. chapter 23). For the average small dairyperson, placing bottles of goat milk in containers of ice water is satisfactory. The container should have a capacity of three quarts of ice and water for each quart of milk. After the milk has been cooled to 50°F (10°C) or below, it can be stored in home refrigerators held at 45°F (7°C) or below. Warm goat milk should not be placed in a refrigerator without precooling because heat exchange proceeds more slowly in air than in water.

17.5 Milking Dairy Goats

Currying, petting, and gentle handling makes better stock and increases production. The more you whip a goat the less cream you will have to whip.

Kent Leach

The principles of milking cows, discussed in chapter 11, also apply to goats, except that goats have only two mammary glands. Gentleness at milking time cannot be overstressed.

Goats may be milked from either the side or the rear: most persons prefer the side approach. Does are usually milked on a stand 12 to 18 in high or higher. The support should be long enough for the doe to stand comfortably and about 18 to 20 in wide. Rotary milk parlors are used on some commercial farms (figure 17.10).

Use of the pressure method is the preferred means of drying off a doe. When milking is abruptly stopped, intramammary gland pressure increases and milk secretion stops in accordance with the principles discussed in chapter 10.

17.5.1 Milk Production of Dairy Goats

The gross energetic efficiency of milk production in dairy goats is of the same order as that of dairy cattle.

Samuel Brody (1890–1956)

Dairy goats commonly produce 6 to 8 lb (2.7 to 3.6 kg) of milk daily. Some yield 20 lb/d (9 kg/d) at the peak of their lactation. Goats are commonly milked 10 months before being given a two-month dry period prior to the next lactation. The dry period allows the mammary system time to repair and regenerate for the next lactation. As with cows, peak milk production of goats is commonly achieved at 6 to 7 years of age. Dairy goats in their prime (commonly the third or fourth lactation cycle) average 6 to 10 lb (2.4 to 3.9 kg) of milk production daily during 10 months of lactation with a production cycle similar to that of the bovine (cf. chapter 10). The average productive life of a doe is 10 to 12 years, although many are still producing at 15 to 18 years.

Although goats provide less than 3% of the worldwide milk supply, it is believed that more people consume goat milk than cow milk. This is because the highly populated countries of Asia and Africa account for 93%

Figure 17.10 Milking dairy goats in a rotary type milking system (from Dairymaster).

of the world's goats. Of course, per capita milk intake is low in those nations.

Worldwide dairy goats annually produce about 15.5 million metric tons (34.2 million kg) of milk, accounting for about 2.2% of the total milk produced by the world's livestock species. Developing countries produce approximately 83% of the total amount. India and Bangladesh far out-produce all other nations. Kilograms of goat milk produced worldwide increased from about 22 billion in 1990 to about 34.2 billion in 2008, partly because of a trend towards self-sufficiency by rural people, especially in developing countries where goat milk can help to improve their nutrition. However, marketing of goat milk and its products has not been done on a broad scale. Table 17.3 shows the amount of goat milk produced by the top 10 countries in the world, along with the total number of dairy does and the average milk produced per doe.

Table 17.3 The World's Top 10 Goat Milk Producing Countries.

Country	Goat Milk Production (million kg)	Dairy Goat Numbers (millions)	Milk Produced per Dairy Doe (kg)
India	8,820	30.2	132.5
Bangladesh	4,853	27.1	80.0
Sudan	3,309	43.1	34.8
Pakistan	1,544	4.9	141.9
Spain	1,324	1.4	422.3
France	1,324	0.8	703.8
Greece	1,103	4.1	123.9
Iran	882	13.7	29.9
Somalia	882	6.6	59.7
China	662	1.4	194.8

Source: Food and Agriculture Organization of the United Nations, Statistics Division (http://faostat.fao.org).

Three European countries, Spain, France, and Greece, produce considerable and similar amounts of goat milk. France has organized programs for selection, processing, and commercialization of goat milk, produced mainly from Saanen and Alpine breeds. France leads in annual milk production per dairy doe. China has the largest total number of goats in the world, but they are kept mainly for meat production.

17.6 Major Diseases of Goats

Goats can become infected with various viral and bacterial diseases such as caprine arthritis, encephalitis, and mastitis. They can transmit certain **zoonotic** diseases to people such as tuberculosis, brucellosis, and Q-fever (cf. chapter 13). Goats, especially kids and yearlings, are more susceptible than cows to parasites and must be well protected. Parasitism is particularly a problem among goats when they are allowed to pasture with sheep, which are also susceptible to parasites.

As with other farm mammals, monitoring body temperature of goats during illness is recommended. The normal body temperature of goats is from 101 to 104°F (38.3 to 40°C); that of a kid averages about 1°F (0.556°C) higher.

17.6.1 Mastitis in Goats

Goats are susceptible to intramammary infections by the same mastitis-causing bacteria as cows and respond by releasing white blood cells (leukocytes, phagocytes, somatic cells) into infected glands to fight the infecting microorganism (cf. chapter 14). Common infecting bacteria include *Staphylococcos aureus*, *Streptococcus agalactiae*, and *Escherichia coli*. Mastitis occurs more frequently in goats raised under intensive management systems than in family-type settings. Factors contributing to incidence and severity are high numbers of infected glands in the herd, poor hygienic practices, bruising of udder tissues, and stresses such as extreme temperatures, muddy and wet living conditions, or a sudden change in diet. Under stress a doe's immune system is compromised and has a difficult time fighting off invading microorganisms. The dipping of teats in 1% iodophor or 4% hypochlorite after milking kills most bacteria on the exterior of the teat.

Somatic cell counts (SCC) can be used to detect subclinically infected mammary glands and for monitoring udder health at the herd level (Koop et al., 2009), but the SCC in goats' milk is strongly affected by stage of lactation (figure 17.11).

Udder **edema** with congestion is common in high-producing does during the late dry period and early lactation. Control measures include limiting consumption of high-energy feedstuffs, such as corn meal, to less than 20% of the ration, sodium to 0.3%, and potassium to 0.7%.

17.7 Management

Goats display an intensely inquisitive and intelligent nature; they will explore anything new or unfamiliar in their surroundings. Goats will test fences, either intentionally or simply because they are handy to climb on. If any of the fencing can be spread or pushed down, the goats will escape. Due to their high intelligence, once they have discovered a weakness in a fence they will exploit it repeatedly. Goats are very coordinated and can climb and hold their balance in most precarious places.

Dairy goats are generally pastured in summer and may be stabled during winter. Because dairy does are milked daily, they are generally kept close to the milking shed. Stabled goats may be kept in stalls or in large group pens.

The usual practice in the United States is to rebreed does annually. In some European commercial dairy systems, does are bred only twice and are milked continuously for several years after the second kidding. In much of Asia goats are kept largely for milk production in both commercial and household settings. In Africa and the Middle East goats typically run with flocks of sheep, which maximizes production per acre since goats and sheep prefer different food plants. Many goats are allowed to wander the homestead or village. Others are kept penned and fed in what is called a cut-and-carry system. This type of husbandry is also used in parts of Latin America. Cut-and-carry refers to the practice of cutting grasses, corn, or cane for feed rather than allowing the animal access to the field where the crop could be destroyed easily by trampling.

17.7.1 Feeding Dairy Goats

Contrary to popular belief, goats do not eat tin cans! Aside from sampling many things, goats are quite particular in what they consume, preferring to browse on the tips of woody shrubs, vines, and trees as well as broad-leaved plants, which are sometimes toxic (figure 17.12). Goats are ruminants; therefore, like the cow, buffalo, and sheep, they respond favorably to feeding regimes that include large quantities of high-quality forages. To maintain a high level of milk production, dairy goats should be fed the same type and quality of feeds as dairy cows. Of course, because of differences in body size, goats do consume substantially less than cows, approximately one-sixth as much. Typically their feed from grazing is supplemented with hay and **concentrates**.

The adult size of a particular goat is a product of its breed (genetic potential) and its diet while growing (nutritional potential). A diet containing 10 to 14% protein and sufficient calories is needed during the prepuberty period for the animal to attain its potential size. Large-framed goats reach mature weight at a later age (36 to 42 months) than small-framed goats (18 to 24 months) if both are fed to their full potential. Large-framed goats need more calories than small-framed ones for maintenance of daily functions.

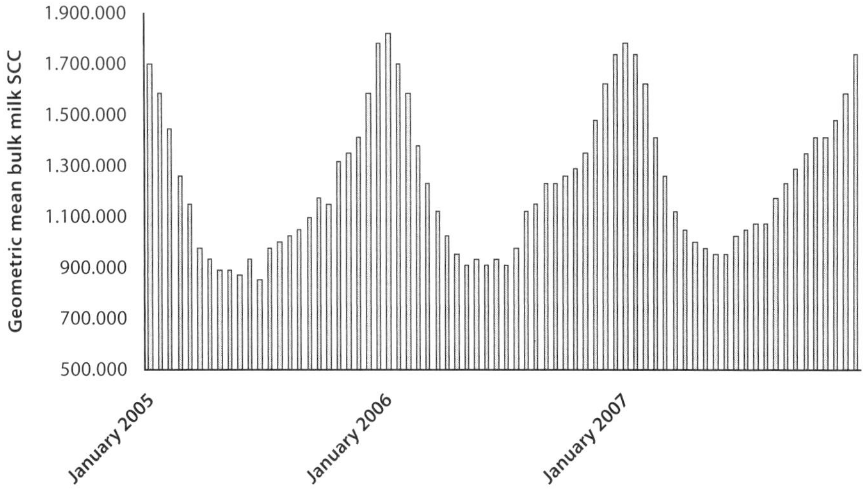

Figure 17.11 Bulk milk somatic cell count shows a constant pattern over the years in about 90% of all (~300) Dutch dairy goat herds (courtesy of Gerrit Koop, Faculty of Veterinary Medicine, Utrecht University, Utrecht, The Netherlands).

Figure 17.12 A domestic goat, feeding in capeweed. Swifts Creek, Victoria, Canada (Fir0002/Flagstaffotos.com.au) (http://www.gnu.org/licenses/old-licenses/fdl-1.2.en.html).

For its size, the goat commonly consumes more feed than cows or sheep—5 to 8% of its body weight in dry matter compared with 3 to 4% for cattle and sheep. However, since the goat ingests two to three times as much feed daily on a body weight basis, it has proportionately more nutrients available than the cow for making milk. The overall efficiency of milk production in dairy goats is about the same as that of dairy cows. With most grasses and silages an 18 to 20% protein-grain mixture is needed to supplement the forage since corn and/or sorghum silage and average-quality grass hays are low in digestible protein. With good-quality hay (15 to 20% protein) or the use of small grain pastures such as oat, rye, and ryegrass, a 14 to 16% protein-grain mix is generally adequate.

17.7.1.1 Feeding Offspring

It is important that baby goats receive colostrum (cf. chapter 7). Orphan goats are often raised using a bottle and nipple for the first two months. Many goat breeders feed milk replacers (the same fed young dairy calves) to young goats. This is a sound economic practice, since goat milk can be marketed at a much higher price than the cost of milk replacer. The digestive physiology of a very young kid (like the young of other ruminants) is essentially the same as that of a monogastric animal. Milk digestion begins in the abomasum, the milk having bypassed the rumen via closure of the reticuloesophageal groove (reticular groove) during suckling. At birth the rumen is undeveloped, but as the kid begins to consume solid feed the rumen soon increases in size and in its capacity to absorb nutrients.

17.7.1.2 Feeding Nonlactating Goats

The **dry** goat should not be neglected at feeding time. More than two-thirds of fetal growth occurs during the last two months of pregnancy. Thus, to help assure the birth of a strong, well-developed kid and to assure the proper body condition of the mother to yield maximal milk, the nonlactating goat must receive a balanced diet. Dry does can be fed a ration lower in protein than that fed lactating goats, and it is advisable always to keep trace-mineralized salt available.

17.7.2 Determining Ages of Goats

The age of goats may be determined by their front teeth on the lower jaw. There are no teeth on the upper jaw. Both the growing kid and the adult have eight teeth. These are small and sharp in the kid. At about 12 months of age the center pair drops out and is replaced by two large permanent teeth. When the goat is two years of age, two more small teeth, one on each side of the first pair, are replaced by two more large teeth; at about three years, the goat has six permanent teeth, and at about four years, the goat has its complete set of eight permanent teeth. Following this, the degree of wear provides an approximate indication of age. Teeth of old goats spread apart, finally become loose, and some may drop out. When this occurs, goats begin losing their usefulness as grazing animals.

17.7.3 Behavioral Aspects of Goats

Goats are affectionate, playful mammals and suffer greatly when left alone or when separated from herdmates. Thus, a doe that is producing well will often decrease milk production greatly when moved to another herd. It is unlikely she will achieve her production potential again until the next lactation. This is one reason for purchasing two or more animals from a herd at any given time.

Goats are well known for their agility and extraordinary physical activity, especially climbing and jumping.

They can climb steep cliffs and slanted tree trunks to reach tender leaf tips. Their mobile tongue and lips enable them to select tasty portions of available feeds and forages as well as discriminate against the less desirable ones. As shown in figure 17.13, they may demonstrate impressive jumping ability by two weeks or less of age.

Goats display social order, even if confined. Strong, aggressive goats commonly dominate weak and submissive ones for feed, water, and preferred locations. Dominance is displayed by biting, butting, body blows, and mounting. The mature male goat is often vicious and should be respected accordingly. This is true especially if he has horns.[3] Grouping goats by similar body size and age is recommended. Where frozen semen is available, the use of artificial insemination makes it unnecessary to keep a buck.

Figure 17.13 At two weeks of age, this kid had learned to jump onto its mother's back, jump down, and repeat the feat numerous times before being inclined to stop (courtesy of Miranda A. Wall, Chesterfield, Missouri).

17.8 Dairy Goat Industry and Organizations

Some of the challenges facing the dairy goat industry regarding growth as a commercial industry include: (1) collection of complete and reliable data on production through developing nationwide strategies, (2) identifying superior and proven bucks that accelerate genetic improvement, and (3) solving the seasonality of production problem to ensure a consistent supply of goat milk. Seasonal breeding and the resulting annual fluctuations in goat milk supply have made development of new markets difficult and have impeded the adoption of milk yield as a breeding goal.

The International Goat Association is a global network of people and organizations with the mission of promoting research and development of socially just, environmentally sound, and economically viable goat farming that will benefit humankind, help alleviate poverty, and improve the quality of life. The Animal Genomics and Improvement Laboratory at USDA conducts research to discover, test, and implement improved genetic evaluation techniques for economically important traits of dairy goats.

In the United States, each dairy goat breed has its own organization; however, registrations for all breeds are handled by the ADGA and the American Goat Society (AGS). Coordinating efforts of the respective goat breeds is done by the ADGA, which was organized in 1904 as a pedigree-registry association. Its activities today represent all phases and programs of the dairy goat industry (e.g., registry, milk production testing, and official dairy goat shows). An annual ADGA handbook is provided to its members. AGS maintains the Dairy Herd Improvement Program as a tool for members to use in improving production of purebred dairy goats. The program provides production information on herds and individual does. AGS Dairy Herd Improvement Registry (DHIR) testing involves monthly tests that follow the uniform operating procedures, the code of ethics of the National Dairy Herd Improvement Program, and the AGS Dairy Herd Improvement (DHI) rules and procedures. DHIR testing involves monthly visits to the farm by a DHI technician to observe the day's milkings, weigh the milk, record individual milk weights, and take samples from each doe. Samples are sent to a lab that tests for fat and protein. Monthly reports sent to the herd owner include each doe's production data to date.

17.8.1 *Dairy Goat Journal*

Serving the dairy goat industry since 1916, the *Dairy Goat Journal* is published monthly and circulated nationally in the United States. It includes articles on uses of goat milk, answers by competent individuals to timely questions, reports of goat sales and breed shows, advertisements indicating where dairy goat stock may be purchased, and suggestions for the improvement of goat keeping.

17.9 Summary

Although dairy goats do not provide a major portion of milk for humans in the United States, they are major suppliers for people of many other lands. Of all milk-yielding animals, the goat provides milk for more people worldwide than any other domestic mammal. This is because the goat population is large in mountainous areas in which goats thrive on grass and browse on brush, whereas the limited feed in these areas will not adequately support dairy cows. The goat, then, is especially useful to people of hilly, dry countries. Goats are especially important sources of milk for the peoples of India, Africa, southern Europe, and western Asia.

The only goat native to the United States is the Rocky Mountain goat, but it has never been domesticated. The snow-white Saanens from Switzerland; the colorful French Alpines; the sleek-coated Nubians with droopy ears, rich colors, and aristocratic noses; and the productive Toggenburgs are especially popular in the United States.

Goats are alert, agile, curious, and often playful mammalian friends of humans. They quickly become lonely when left alone and respond favorably in milk production to gentle care. Goats, like sheep and cattle, have no front teeth on the upper jaw. Each year they obtain new teeth until four to five years of age, when they have their permanent set of eight teeth.

Goat milk is no more susceptible to off-flavor than cow milk. When properly handled, cooled, and stored, goat milk is slightly sweet, nutritious, and tasty. Moreover, because of its minute fat globules and fragile curd, it can be easily digested.

STUDY QUESTIONS

1. In which regions and nations of the world are dairy goats most abundant?
2. Name three major challenges to commercialization faced by the dairy goat industry.
3. Describe the differences in the milk produced by the goat and the cow.
4. What color is the milk of goats? Why?
5. Why does goat milk form cream very slowly?
6. What major type of casein present in cow's milk is virtually absent in goat's milk?
7. Why is it important to separate bucks and lactating does?
8. How can one cool goat milk quickly on the farm? To what maximal temperature should it be stored? Is this important? Why?
9. Name the eight dairy goat breeds recognized in the United States, name their country of origin, and give their breed characteristics.
10. In what major way does the appearance differ between Saanen and Sable goats? How is this related to suitability for life in a hot and sunny climate?
11. Be able to identify and/or name the anatomical parts of a goat.
12. Describe the desired conformation of a dairy goat.
13. At what minimal age and percentage of adult body weight should goats be bred?
14. What is the normal length and frequency of estrous periods among goats? When is the preferred time to breed during estrus? What are the usual signs of estrus in goats?
15. What is the length of an average gestation? Are most births singles or multiples? What is the average weight of newborn goats?
16. Which of the eight goat breeds is known for high reproductive performance, as well as very high fat content of its milk?
17. Discuss the merits of artificial insemination of goats.
18. Do goats commonly consume more or less dry matter per pound of body weight than cows? Which species—cattle or goats—has higher body maintenance costs on a per pound basis? Which is the more efficient producer of milk?
19. Describe the types of forages preferred by goats. Name some plants goats may eat that cows will avoid.
20. Discuss briefly the feeding of lactating goats and dry goats. Why is it an economically sound practice to feed milk replacer to baby goats?
21. How many mammary glands do goats have? Is the principle of milking goats the same as that of cows?
22. Describe the infection and development of mastitis in dairy goats. What management steps should be taken to prevent this disease?
23. Are the productive lives and lengths of lactation of cows and goats similar?
24. How many teeth would one expect to find in a five-year-old doe? Discuss briefly how one may use teeth to estimate age in goats.
25. Why should one not leave a goat alone? Why should goats be purchased and/or moved in pairs?
26. Discuss briefly the unpleasant odor of bucks. Can this affect the flavor of milk? What precautions should be considered and taken to prevent an off-flavor in goat milk?

NOTES

[1] It is interesting that milk of thyroidectomized goats has a yellow color. This may be explained by the fact that the thyroid is necessary for conversion of carotene to vitamin A.

[2] According to the USDA, doe milk is not recommended for human infants (as the only source of nutrients) because it contains "inadequate quantities of iron, folate, vitamins C and D, thiamin, niacin, vitamin B_6, and pantothenic acid to meet an infant's nutritional needs." It may cause harm to an infant's kidneys and could cause metabolic damage. The Department of Health in the United Kingdom has also stated that goats' milk is unsuitable for babies, and infant formulas based on goats' milk protein have not been approved for use in Europe. Also, according to the Canadian Federal Health Department, Health Canada, most of the dangers or counterindications of feeding unmodified goat milk to infants are similar to those incurring in the same practice with cow's milk, namely in the allergic reactions.

[3] It is interesting that after a thousand years of effort to breed the horns off goats, it has not been accomplished completely.

REFERENCES

Dohner, J. V. 2001. *The Encyclopedia of Historic and Endangered Livestock and Poultry Breeds.* Topeka, KS: Yale Agrarian Studies Series.

Koop, G., M. Nielen, and Y. van Werven. 2009. Bulk milk somatic cell counts are related to bulk milk total bacterial counts and several herd-level risk factors in dairy goats. *Journal of Dairy Science* 92:4355–4364.

McNulty, R. W., A. D. Aulenbacher, E. C. Loomis, N. F. Baker, and R. R. Bushnell. 1997, April. *Your Dairy Goat*. 4-H Goat Project Development Committee, University of California Cooperative Extension.

US Food and Drug Administration. 2013. *Grade "A" Pasteurized Milk Ordinance.* Washington, DC: Author.

Websites

Alpines International Club (http://www.alpinesinternationalclub.com)
American Dairy Goat Association (http://www.adga.org)
American Goat Society (http://www.americangoatsociety.com)
American LaMancha Club (http://www.lamanchas.com)
American Nigerian Dwarf Dairy Association (http://www.andda.org)
Dairy Goat Journal (http://www.dairygoatjournal.com)
International Goat Association (http://www.iga-goatworld.com)
International Nubian Breeders Association (http://www.i-n-b-a.org)
International Sable Breeders Association (http://www.sabledairygoats.com)
National Saanen Breeders Association (http://nationalsaanenbreeders.com)
National Toggenburg Club (http://www.nationaltoggclub.org)
Oberhasli Goat Club (http://www.oberhasli.us)

18

Dairy Sheep

> As with new industries, the US dairy sheep industry owes its existence
> to the hard work of a few pioneer producers and processors.
> *National Research Council, 2007*

This chapter was contributed by David L. Thomas, PhD, Professor of Sheep Management and Genetics, University of Wisconsin–Madison, who has an appointment in research, teaching, and outreach. He leads the only dairy sheep research program in North America and was awarded the first Distinguished Service Award from the Dairy Sheep Association of North America. He was recognized by the American Society of Animal Science with the Rockefeller Prentice Memorial Award in Animal Breeding and Genetics and the Animal Management Award. In 2009 he received the Spitzer Excellence in Teaching Award and in 2010 he received the Animal Industry Service Award from the American Society of Animal Science. He has worked on livestock development projects in Kenya, Bulgaria, Indonesia, Pakistan, Ethiopia, Kazakhstan, Kyrgyzstan, Uzbekistan, and Armenia and was awarded the Bouffalt International Animal Agriculture Award from the American Society of Animal Science in 2005. He is a Fellow of the American Society of Animal Science.

18.1 Introduction

The relationship between humans and sheep has always been beneficial for both. Sheep are unsurpassed at converting grass and other herbage into the highest quality food products, meat and milk, and unmatched textiles and hides.

Professor C. V. Ross, University of Missouri

The intact male sheep is called a *ram,* the female a *ewe.* A sheep less than one year old is called a *lamb*, and a castrated male is called a *wether.* There is no special term commonly used in North America to refer to a female sheep prior to giving birth to her first lamb, such as the term heifer for young female cattle or gilt for young female swine. The young female sheep is simply referred to as a *ewe lamb.*

Sheep and goats, and perhaps reindeer, were the first **ruminants** domesticated by humans—only the dog was domesticated earlier (Zeuner, 1963). Domesticated sheep (*Ovis aries*) are thought to have descended from the wild mouflon of Europe and Asia, since both have the same number of chromosomes (54 or 27 pairs) (Ryder, 1983), but some domesticated breeds may owe some or all of their ancestry to the wild urial and argali found in Asia east of the range of the mouflon (Zeuner, 1963). Domestication most likely occurred in southwestern Asia in the area south of the Black and Caspian Seas in modern day eastern Turkey and the northern portions of Iran, Iraq, and Syria. Sheep were domesticated in this region approximately 11,000 years ago (Zeder, 2008).

Today, the primary products of sheep throughout the world are meat, wool, and skins; especially in North America, South America, Australia, and New Zealand. However, milk and milk products have been a part of human sustenance for centuries in regions of Asia where sheep were first domesticated and in areas of their early expansion into the Middle East, North Africa, and Europe.

Dairy sheep production was practiced as a subsistence agricultural enterprise on small farms and among nomadic herders through the first half of the twentieth century and continues in much this same way today in many parts of the world. However, from the 1950s to the present, dairy sheep production and sheep milk processing has been commercialized and is an important industry in many countries; especially those in the Mediterranean basin, including Spain, France, Italy, Greece, Turkey, and Israel. Commercial dairy sheep production did not appear in North America, South America, Australia, and New Zealand until the 1980s, and although it is still a minor agricultural industry in these regions, it is growing.

18.2 World Sheep Milk Production and Trade

Sheep milk represents a small percentage of total milk produced annually in the world. World production of sheep milk in 2010 was estimated at 22 billion lb and represented only 1.4% of milk produced from all species (cow, buffalo, goat, sheep, and camel). The 11 leading countries for sheep milk production in 2010 are listed in table 18.1.

China is the leading producer of sheep milk and produced over twice as much as Greece. The southern European countries bordering the Mediterranean Sea are among the major sheep milk producing countries of the world and have the best developed commercial dairy sheep industries. Total sheep milk production statistics for the United States are not collected because of the industry's small size; production in 2011 was estimated as no more than 5 million lb.

The primary sheep milk commodity of the world is cheese. Approximately 100 million lb move among countries each year. Presented in table 18.2 are the leading countries for sheep milk cheese exports and imports in 2009. Italy and France account for 60% of total world exports. Italian dairy sheep production is concentrated on the island of Sardinia, where milk from the Sarda breed of sheep is used to produce the hard cheese, pecorino romano. The dairy sheep industry of France is primarily in its south-central part where the famous blue-veined Roquefort cheese is made from the milk of Lacaune ewes.

The United States, Germany, and the United Kingdom are the leading importers of sheep milk cheese, accounting for 86.6% of world imports in 2009. The United States is the destination of over half of the world trade in sheep cheese (60 million lb). American consumers have developed a desire for the unique taste of sheep milk cheeses, and imports have increased by 50% from 1989 to 2009. This indicates a large potential market for domestic sheep milk cheeses since annual US production is estimated at less than 1.0 million lb.

Table 18.1 Leading Countries for Sheep Milk Production in 2010.

Country	Production (million lb)	Percent of Total Milk Production[a]
China	3,793	4.2
Greece	1,881	40.7
Turkey	1,797	6.0
Romania	1,433	12.9
Syria	1,415	28.7
Italy	1,320	5.3
Somalia	1,299	20.1
Spain	1,287	8.4
Iran	1,054	6.3
Sudan	995	5.8
France	570	1.1
World	22,102	1.4

[a] Sheep milk production as a percentage of total milk production from all species (cattle, buffalo, goats, sheep, and camels).

Source: Food and Agriculture Organization of the United Nations, Statistics Division (http://faostat.fao.org).

Table 18.2 Leading Countries for Exports and Imports of Sheep Milk Cheese in 2009.

Exports		Imports	
Country	Amount (million lb)	Country	Amount (million lb)
Italy	38.0	United States	60.0
France	22.7	Germany	16.4
Bulgaria	11.7	United Kingdom	10.2
Luxembourg	7.5	The Netherlands	4.8
Greece	5.1	Luxembourg	4.7
The Netherlands	2.1	Spain	4.6
Germany	1.5	Sweden	4.0
Romania	1.3	Greece	2.8
Spain	1.0	France	2.7
Denmark	0.5	Belgium	2.3
Belgium	0.2	Austria	2.0
United Kingdom	0.2	Italy	1.1
World	92.3	World	119.1

Source: Food and Agriculture Organization of the United Nations, Statistics Division (http://faostat.fao.org).

18.3 Dairy Sheep Breeds of North America

Access to foreign dairy sheep genetics is a priority of US dairy sheep producers.

Professor Dave Thomas, University of Wisconsin–Madison

There are no dairy sheep breed associations in the United States to maintain pedigree records and to promote breed improvement. Likewise there is no national entity to organize performance recording and to calculate across-flock estimates of genetic value for lactation traits. These are major deficiencies in the US dairy sheep infrastructure that need to be remedied for the long-term viability of the dairy sheep industry.

18.3.1 East Friesian

The East Friesian breed originated in East Frisia in northern Germany. The breed is generally white in color, with some black individuals. The head is **polled** in both rams and ewes. The head, long ears, and legs are free of wool. Their most distinctive feature is their "rat-tail," which is free of wool. The body is completely covered with wool; although it is not considered to have great meat conformation, East Friesians are more muscular than Lacaunes. Ewes have larger udders that tend to get "baggy" and less desirable in later lactations. East Friesians are very docile; at maturity a ewe weighs approximately 175 lb (79 kg), rams weigh 230 lb (104 kg). Sheep are seasonal breeders and begin mating in the fall; the average litter size of East Friesians is two lambs per ewe. The breed has a high conception rate (95+%), while out-of-season (spring) breeding success is average. Lambs can reach puberty in 7 to 10 months; ewes may start reproducing at 1 or 2 years old. East Friesians have one of the highest rates of milk production and longest lactation period of sheep breeds. A mature ewe can produce 800 lb (363 kg) of milk per lactation (210 days), yielding 5.4% milkfat and 4.8% protein (figure 18.1).

Figure 18.1 (A) East Friesian ram (United States); (B) East Friesian ewes (England); and (C) crossbred East Friesian ewe lambs (United States) (courtesy of Department of Animal Sciences, University of Wisconsin–Madison).

The first importations of East Friesian genetics into North America for dairy sheep production were by Hani Gasser, British Columbia, Canada (semen from Switzerland in 1992); Chris Buschbeck and Axel Meister, Ontario, Canada (embryos from England in 1994); and Peter Welkerling, Alberta, Canada (semen from England in 1994) (Berger, 2007). It appears there were no importations of East Friesian genetics into Canada after 1996. The East Friesian breed is recognized by the Canadian Sheep Breeders' Association, and pedigrees are maintained and registration papers provided by the Canadian Livestock Records Corporation.

The first East Friesian genetics in the United States were a group of 50% East Friesian and 50% Rideau Arcott ram lambs imported in 1993 from Hani Gasser of British Columbia, Canada. They were imported by the University of Minnesota, the University of Wisconsin–Madison, and Hal Koller, a dairy sheep producer from Amery, Wisconsin. Small importations of purebred East Friesian genetics into the United States from England, Belgium, the Netherlands, and New Zealand continued through 1996, and then continued from Canada until 2002, when the US–Canada border was closed to the importation of breeding sheep due to the presence of bovine spongiform encephalopathy (BSE) in Canada. There is no association of East Friesian breeders in the United States for the maintenance of pedigree records. Concentrations of East Friesian breeders are located in Wisconsin, New York, and parts of New England.

18.3.2 Lacaune

The Lacaune breed originated near the town of Lacaune in south-central France, where it is still the most numerous breed of sheep. The breed is almost always white, while a few individuals are black. Both rams and ewes are polled, while the head, long ears, and legs are free of wool. They have very little, if any, wool on the belly, bottom of the sides, and the underside of the neck. Several animals have just a cape of wool across the back. Their tails may or may not be covered in wool. Lacaune ewes have a tighter, more desirable udder conformation than the East Frieisan. Lacaunes are less docile, like a meat sheep. A mature ewe can weight 160 lb (72.5 kg), while a ram can weight 210 lb (95.3 kg). The average litter size of a mature ewe is 1.80 lambs. The breed has a high conception rate (95+%), while out-of-season (spring) breeding success is good. A mature ewe can produce 750 lb (340.2 kg) of milk per lactation (210 days), yielding 5.6% milkfat and 5% protein (figure 18.2).

Milk of the Lacaune is used to manufacture Roquefort, the famous French blue-mold cheese. The Lacaune breed has been selected in France for its high milk production, high milk solids, and desirable udder conformation. The sophisticated Lacaune selection program in France has incorporated artificial insemination, milk recording, and progeny testing longer than for any other

Figure 18.2 (A) Lacaune rams (France); (B) Lacaune ewes (France); and (C) crossbred Lacaune ewe lambs (United States) (courtesy of Department of Animal Sciences, University of Wisconsin–Madison).

dairy sheep breed in the world. By 2006 the genetic gain for milk yield in the Lacaune breed was an impressive 13 lb/yr (5.9 kg/yr). Fat and protein percentages increased as milk yield increased (Barillet, 2007).

Josef Regli of Ontario, Canada, was the first person to import Lacaune genetics into North America when he imported Lacaune embryos from Switzerland in 1996. He remains the only breeder of purebred Lacaune sheep in North America. The University of Wisconsin–Madison imported the first Lacaune genetics into the United States in 1998 via semen from three Lacaune rams of England and two Lacaune rams from Josef Regli in Canada. A few Lacaune rams were subsequently imported by other US producers from Josef Regli in Canada.

Other than the 1996 importation from Switzerland and the 1998 importation from England, there have been no new Lacaune importations into Canada or the United States due to strict import regulations in both countries. There is a significant need for new Lacaune genetics in the United States. The Lacaune is a recognized sheep breed by the Canadian Sheep Breeders' Association. Pedigrees are maintained by the Canadian Livestock Records Corporation. There is no association of Lacaune breeders in the United States for maintaining pedigree records.

18.3.3 East Friesian Compared to Lacaune

The University of Wisconsin–Madison operates the only dairy sheep research farm in North America at its Spooner Agricultural Research Station. Starting in 1998, a comparison of the performance of the two breeds was conducted. Lacaune ewes produced 8% less milk and gave birth to 8% fewer lambs than East Friesian ewes. However, fat and protein content of the milk was greater for Lacaune ewes than for East Friesian ewes, which resulted in approximately the same amount of fat and protein yield from the two breeds. The East Friesian breed also has a greater susceptibility to respiratory disease than does the Lacaune breed.

18.3.4 Other World Dairy Sheep Breeds

Several other improved dairy sheep breeds exist throughout the world in addition to East Friesian and Lacaune, but they have not been imported into the United States or Canada due to strict animal health import regulations. Some additional prominent dairy breeds of Europe and the Middle East are the Manchega and Latxa of Spain, Manech of France, Sarda of Italy, Chios of Greece, improved Awassi of Turkey and Israel, and Assaf of Israel.

18.4 Reproduction of Dairy Sheep

Sheep are seasonal breeders and become sexually active with decreasing hours of daylight. On the North American continent, dairy sheep will readily mate from September to March. The average estrous cycle duration of a ewe is 17 days. Sheep in estrus do not ride each other as do cows and goats, and there are no obvious physical signs of a ewe being in estrus, but ewes in estrus will seek out the ram.

Artificial insemination (AI) is not common in sheep. The cervix of the ewe has several folds and blind pockets that do not easily allow passage of an insemination pipette through the cervix to deposit semen in the uterus. The AI that is done infrequently requires a minor surgical procedure called laparoscopic artificial insemination (LAI). This procedure requires restraint of the ewe on her back and two small incisions in the abdomen—one for the laparoscope to view and position the uterus and the other for the insemination pipette with a small needle attached at the end. The semen is injected directly into the lumen of each uterine horn. LAI is the only alternative with frozen semen, but the procedure is too expensive for regular use.

The normal gestation period of sheep is 147 days—about five months. Ewe lambs should be bred first at seven to eight months of age to produce their first lambs when they are at or slightly more than one year of age. Healthy dairy ewe lambs with a minimum weight of 110 to 125 lb (49.9 to 56.7 kg) when mated will have conception rates of at least 85%.

The most common birth type is twins, but triplets and quadruplets are not uncommon. Number of lambs born per pregnancy increases with age of the ewe. Average numbers of lambs born per year per ewe at the University of Wisconsin–Madison were: age 1 (1.59), age 2 (1.77), age 3 (1.97), age 4 and older (2.04). Average birth weights of lambs born alive were: singles (12.1 lb [5.5 kg]), twins (11.4 lb [5.2 kg]), triplets (10.4 lb [4.7 kg]), and quadruplets (9.0 lb [4.1 kg]). Male lambs weighed slightly more than females (11.6 versus 10.8 lb [5.3 versus 4.9 kg]) with a male to female ratio of approximately 1.1:1.0.

A well-grown ram as young as seven to eight months of age may be mated naturally with up to 20 ewes during his first season. During his second and subsequent seasons (18 months and older), he may breed 50 to 60 ewes per breeding season. If mating occurs during the height of the breeding season (September to November), 85 and 95% of the ewes should become pregnant by the end of the first and second estrous cycles, respectively. Rams should undergo a breeding soundness exam prior to the start of the breeding season. This is done by a veterinarian or a qualified sheep reproduction specialist. The testes are palpated for the absence of scar tissue and the penis examined for signs of infection. A semen sample should be taken and examined for sperm number and quality. Questionable rams should be replaced.

During the breeding season, rams should be fitted with a marking harness. This harness fits over the shoulders and between the front legs of the ram and has a colored grease chalk attached to the harness between the front legs. When the ram mounts a ewe, a colored mark from the grease chalk is left on the rump of the ewe. The color of the grease chalk is changed every 14 to 17 days. If there are no chalk marks on the ewes, or if most of the ewes are marked a second time with a different color, the ram lacks libido or is either sterile or of low fertility and needs to be replaced with a more active and fertile ram.

18.5 Composition and Characteristics of Sheep Milk

Cheesemakers found that sheep milk responded differently in the cheesemaking procedure. It was more sensitive to rennet, coagulated faster, produced a firmer curd and yielded more cheese per unit of milk than cow milk.

Professor Bill Wendorff, University of Wisconsin–Madison

Most sheep milk produced in the world is transformed into cheese or yogurt, and fresh sheep milk is rarely consumed. For this reason, Bencini and Pulina (1997) refer to the quality of sheep milk as its capacity to be transformed into high-quality products and to produce a high yield of these products. The processing performance of the milk is mainly dependent on its clotting properties—a combination of renneting time (cf. chapter 30), rate of curd formation, and consistency of the curd. The clotting properties of the milk are affected by its composition.

Composition of sheep milk varies with several factors including nutrition, stage of lactation, milk yield, and breed. Average composition of sheep milk compared to goat and cow milk is presented in table 18.3. Striking differences are the higher total solids, fat, and protein of sheep milk compared to goat and cow milk. Goat milk is more similar to cow milk than to sheep milk in gross composition. As a result of the high total solids of sheep milk, only 4 to 5 lb (1.8 to 2.3 kg) of sheep milk are required to produce 1 lb (.45 kg) of cheese, which is almost double the cheese yield from goat or cow milk. In comparison, the higher casein content results in a shorter rennet coagulation time and a firmer curd for sheep milk (Jandal, 1996).

Table 18.3 Average Composition of Sheep, Goat, and Cow Milk.

Composition	Sheep	Goat	Cow
Solids, not fat, %	12.0	8.9	9.0
Fat, %	7.9	3.8	3.6
Total protein, %	6.2	3.4	3.2
Casein protein, %	4.2	2.4	2.6
Lactose, %	4.9	4.1	4.7
Ash, %	0.9	0.8	0.7
Calories/100 mL	105.0	70.0	69.0

Source: Park et al., 2007.

18.5.1 Milkfat Content

Fats (lipids) are vital components of milk because of the physical and sensory characteristics they impart to cheese and other products made from sheep milk.

The average diameter of fat globules of sheep milk (3.3 μm) is similar in size to the fat globule of goat milk (diameter of 3.5 μm), but smaller than the average fat globule of cow milk (4.5 μm) (Park et al., 2007). The comparably smaller fat globule size of sheep and goat milk results in their faster digestibility by humans than the fat of cow milk.

Approximately 20% of the fatty acids of sheep and goat milk is the metabolically valuable short- and medium-chain saturated fatty acids (C4:0 to C12:0) compared to approximately 12% in cow milk (table 18.4). The higher proportion of short-chain fatty acids, especially caproic, caprylic, and capric, in sheep milk gives sheep milk cheeses their pleasant but "sheepish" flavor. These three fatty acids are even higher in goat milk and are largely responsible for the distinctive flavor of goat milk cheeses. Lipases attack the ester linkages of the short- and medium-chain fatty acids more rapidly than those of long-chain fatty acids, so these differences may contribute to more rapid digestion by humans of goat and sheep milk compared to cow milk.

Conjugated linoleic acid (CLA) encompasses a family of isomers of linoleic acid found primarily in the meat and milk of grass-fed ruminants. CLA has been found anticarcinogenic and to increase lean body mass. There is evidence that compared with products of cow and goat milk, products made from sheep milk may be higher in CLA (Prandini et al., 2007). However, these studies were not controlled for diet among species, and differences observed may be due to more forages

Table 18.4 Major Fatty Acids and Their Content in Sheep, Goat, and Cow Milk.

	Percent of Total Fatty Acids by Weight		
Fatty Acid	Sheep	Goat	Cow
C4:0 Butyric	4.0	2.6	3.3
C6:0 Caproic	2.6	2.9	1.6
C8:0 Caprylic	2.5	2.7	1.3
C10:0 Capric	7.5	8.4	3.0
C12:0 Lauric	3.7	3.3	3.1
Total: C4–C12	20.3	19.9	12.3
C14:0 Myristic	11.9	10.3	9.5
C16:0 Palmitic	25.2	24.6	26.5
C16:1 Palmitoleic	2.2	2.2	2.3
C18:0 Stearic	12.6	12.5	14.6
C18:1 Oleic	20.0	28.5	29.8
C18:2 Linoleic	2.1	2.2	2.5
Total: C14–C18	74.0	80.3	85.2

Source: Jandal, 1996.

included in the diets fed sheep (resulting in higher CLA) compared to more grain in the diets fed goats and cows (lower CLA). For example, Jahreis et al. (1999) reported large differences in CLA content of sheep milk due to season—1.28% in summer when ewes were on pasture compared to 0.54% at the end of the indoor winter feeding period.

Both sheep and goats convert ingested β-carotene to vitamin A, which results in their milkfat being whiter than milkfat of cows, thus making their milk white.

18.5.2 Milk Proteins

Sheep milk contains approximately 80% caseins and 20% whey proteins. Casein content is positively related to cheese yield and curd firmness. Since sheep milk contains more casein than either goat or cow milk, cheese yield is higher and curd forms faster and more firmly.

The caseins of sheep milk are α_{s1}-CN, α_{s2}-CN, β-CN, and κ-CN. A fifth casein, γ-CN, is a product of the breakdown of β-CN by plasmin. Both genetic and nongenetic variants of these caseins have been identified in various sheep populations throughout the world. There are only a few studies that have evaluated these variants for their effects on cheese and its manufacture. For example, the most common α_{s1}-CN variants in the Sarda breed of dairy sheep in Sardinia, Italy, are the C and D variants. Pirisi et al. (1999) collected milk from Sarda ewes of CC, CD, and DD genotypes. The CC milk had better renneting properties and curd formation than the DD milk, with the CD milk intermediate in function.

The major whey proteins of sheep milk are α-lactalbumin and β-lactoglobulin. Three variants of α-lactalbumin have been identified, but there are no reports of the effects of the variants on cheesemaking. Three β-lactoglobulin variants are known (A, B, and C), with the C variant currently occurring only at a low frequency in nondairy sheep. Conflicting results have been reported regarding the superiority of the A or B variant for cheesemaking (Moioli et al., 2007).

18.6 Milking Dairy Ewes

The principles of milking cows discussed in chapter 11 also apply to ewes, except ewes have only two mammary glands. Gentleness at milking time cannot be overstressed.

Dairy ewes are machine-milked in an array of different types of parlors, from elevated platforms for very small flocks (figure 18.3) to pit-type parlors for medium-sized flocks (figure 18.4 on the following page) to rotary parlors for very large flocks (figure 18.5, also on the next page). Parlors must be constructed to meet state and federal regulations. Generally, all equipment and surfaces must be nonporous (no wood) so that the parlor can be cleaned easily and effectively. Parlors need to be designed for efficient movement of sheep in and out and for fast milking. For example, a well-designed double-12 pit parlor (12 ewes on each side of a central pit occupied by the milkers) with 12 milking machines can milk more than 140 ewes/hr. This type of throughput is necessary for even medium-sized flocks of 200 to 300 ewes when using family labor. If throughput is less than 100 ewes/hr, family members will spend most of their waking hours milking sheep.

Milking parlors can have either a low line or a high line (figure 18.6 on p. 255). Low lines have the advantage that less vacuum pressure is needed, resulting in less agitation of the milk and a lower probability of damage to the teat during milking. In a double-sided pit parlor, the high line has the advantage that the same set of milking units can be used on both sides of the parlor, cutting in half the number of milking units that must be purchased and maintained.

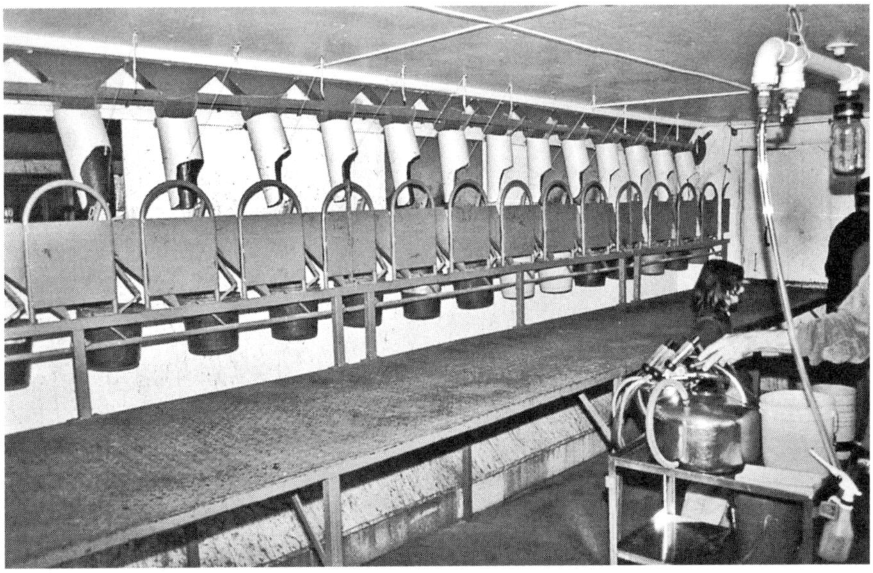

Figure 18.3 Metal platform milking parlor in Vermont. The milking device is a "cow bucket" modified for sheep (courtesy of Department of Animal Sciences, University of Wisconsin–Madison).

Proper settings on milking machines are important for efficient milking and maintenance of udder health (Billon, 2004). The pulsation rate on machines for dairy ewes is higher than for cows and should be 150 to 180 cycles/min; with a rate at the higher end of this range generally used. The pulsation ratio generally is 50:50, i.e., the amount of time for the suction stroke is equal to the amount of time for the compression stroke. The vacuum level is dependent upon the milking system: for a low line or milking into buckets, 34 to 38 kPa (10.2 to 11.4 in Hg) is sufficient, for a high line it should be 38 to 40 kPa (11.4 to 12.0 in Hg).

Ewes are milked from the rear. Udder conformation is an important factor in the milking ability of dairy ewes, and the scoring system in figure 18.7 can be used to describe udder conformation. A very desirable udder would have the following scores: udder height = 5, teat angle = 9, teat length = 5, and udder shape = 9. These scores describe an udder with good support that will not become pendulous and "baggy" in later lactations and with teats of moderate length located at the bottom of the udder.

Lactating ewes on pasture or in pens with clean bedding seldom require their udders to be washed or

Figure 18.4 The double-12 pit parlor at the Spooner Agricultural Research Station, University of Wisconsin–Madison (courtesy of Department of Animal Sciences, University of Wisconsin–Madison).

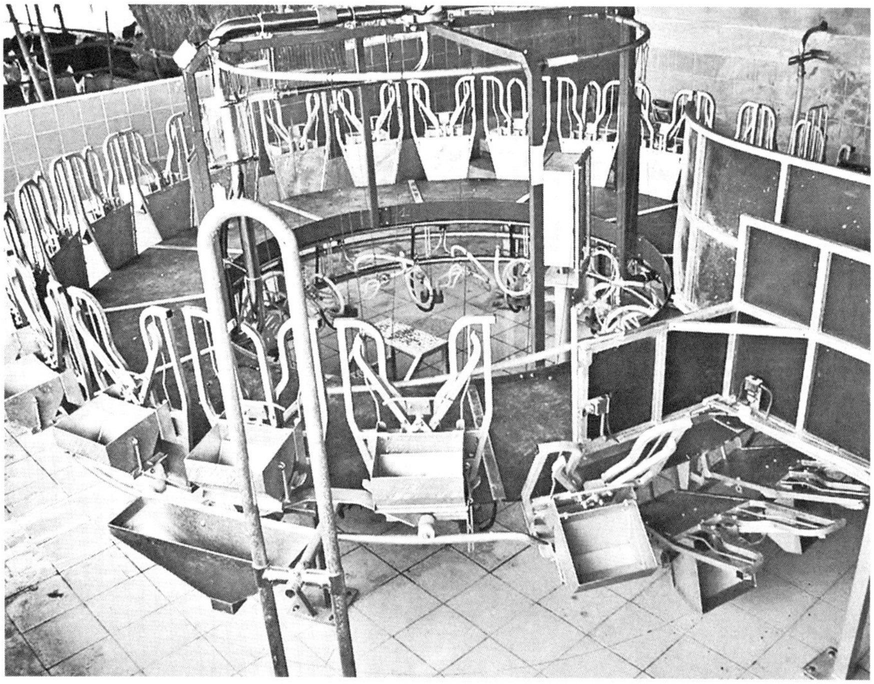

Figure 18.5 Rotary sheep or goat milking parlor (courtesy of Greenoak Dairy Equipment, Kitchener, Ontario).

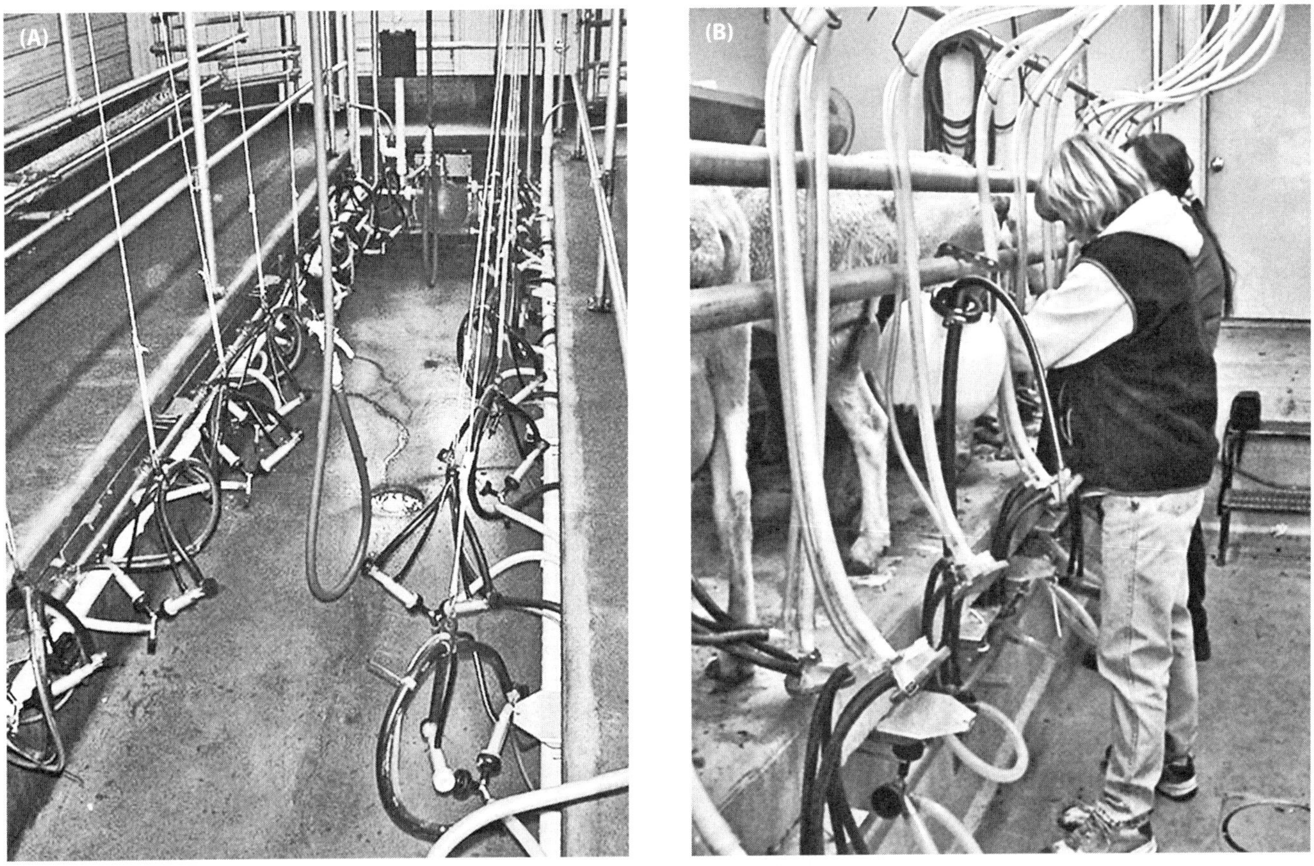

Figure 18.6 (A) Low line system—milk and vacuum lines attach to pipes below the level of the ewes. (B) High line system—milk and vacuum lines attach to pipes (not shown) above the ewes and milkers (courtesy of Department of Animal Sciences, University of Wisconsin–Madison).

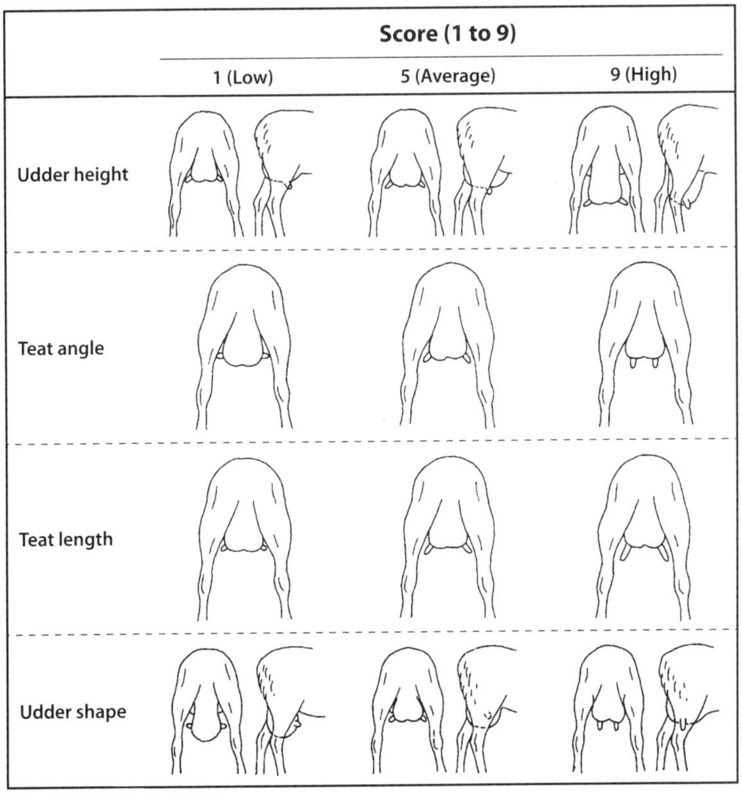

Figure 18.7 Linear scores for the evaluation of the main morphological traits in dairy sheep (Caja et al., 2000).

prepped prior to milking. Sheep feces are much drier than those of cows so there is very little, if any, fecal material on ewe udders. The general practice on most sheep dairies is to clean only udders that are obviously dirty prior to milking. All teats should be dipped in an approved antimicrobial teat dip after milking to assist in preventing mastitis.

Use of the pressure method is the preferred means of drying off a ewe. When milking is abruptly stopped, intramammary gland pressure increases, and milk secretion stops in accordance with the principles discussed in chapter 10.

18.7 Milk Production of Dairy Ewes

There is not a long tradition of dairy sheep production in North America, so the first dairy sheep milk producers in the United States in the 1980s milked meat-type sheep. Given the very low milk yield of domestic sheep breeds that had been developed for meat and wool production, early dairy sheep producers in North America realized from the start that they needed access to specialized dairy sheep breeds if they were to have profitable operations.

Professor William Boylan at the University of Minnesota established the first dairy sheep research program in North America with construction of a sheep milking parlor in 1984. He and his students compared domestic breeds of sheep for milk production in a commercial-type dairy system. As the research in sheep milk production progressed, industry interests became focused on the opportunities and economic potential of a dairy sheep industry in the United States.

Dairy ewe flocks have a large variation in average commercial milk yield; the variation is determined primarily by nutritional status and the date milking starts relative to lambing date. Flock averages can easily range from 450 to 800 lb (204.1 to 362.9 kg) of milk/ewe/lactation. Table 18.5 summarizes the 527 lactation records collected in 2010 and 2011 from dairy ewes at the University of Wisconsin–Madison that were put to milking 24 to 36 hr after lambing. Ewes had an average lactation length of about seven months and averaged 3.4 lb/d (1.5 kg/d) of milk. The best ewe in the flock averaged 5.7 lb/d (2.6 kg/d) of milk. Ewes in their second and later lactations produced 1 lb/d (.45 kg/d) more milk than first lactation ewes. The best single test day milk yield for a ewe during these two years was 13.2 lb (6.0 kg) (data not shown).

Peak milk production of ewes is commonly achieved at three to four years of age. The average productive life of a ewe is four to five years, with very few ewes older than eight or nine years found in dairy flocks.

18.7.1 Storage, Transport, and Value of Sheep Milk

In the early years of the industry in the United States, which was characterized by small flock sizes, low milk production per ewe, and few producers in a region, raw milk typically was frozen at the farm until sufficient quantities were accrued for transport to a processing plant. Freezers typical for a home that reached a minimum temperature of about 5°F (–15°C) were used initially, and cheese processors experienced problems with milk quality and stability of this frozen milk. Research conducted at the University of Wisconsin–Madison showed that after six months of storage at 5°F (–15°C), thawed milk samples exhibited destabilized protein, which settled to the base of containers. However, milk stored in commercial type freezers at –16.6°F (–27°C) exhibited good protein stability throughout 12 months of storage (Milani and Wendorff, 2011). It is now recommended that sheep milk be rapidly frozen in 40-lb plastic bags on racks (figure 18.8) and stored at temperatures of –4°F (–20°C) or lower in commercial-type freezers for no more than 6 to 12 months.

As flock sizes, milk production per ewe, and number of dairy sheep farms in a region have increased, there is sufficient sheep milk production in many areas for economical fluid milk pickup. However, production is generally not high enough for daily pickup. Milk is cooled, stored, and held in bulk tanks on the farm and generally picked up by milk tanker trucks every two to four days and transported to a processing plant. Even with pickup of fluid milk in many regions, most sheep dairies maintain a commercial-type freezer in order to freeze small quantities of milk produced early and late in the milking season when only a few ewes are milking.

Table 18.5 Summary of 527 Ewe Lactation Records (2010 and 2011).

		Lactation Yield (lb)			
Group	**Lactation Length (days)**	**Milk**	**Fat**	**Protein**	**Milk/Day (lb)**
All ewes	212	713	41	35	3.4
First lactations	186	490	27	23	2.6
Second+ lactations	221	789	46	39	3.6
Top 10% lactations	239	1,103	64	53	4.6
Best single lactation	235	1,335	75	64	5.7

Source: Spooner Agricultural Research Station, University of Wisconsin–Madison.

The price paid farmers for sheep milk is two to three times the price paid for goat milk and three to five times the price paid for cow milk. A price twice that of cow or goat milk is justified based on the increased cheese yield from sheep milk relative to cow and goat milk. The additional price advantage of sheep milk is due to the unique flavor and texture it imparts to specialty dairy products. Although some sheep milk is sold on a weight basis only, most is now sold based on quality and content of fat and protein.

Figure 18.8 (A) Fluid sheep milk in 40-lb plastic bags being frozen rapidly on a rack in a commercial-type freezer. (B) Previously frozen bags of sheep milk (courtesy of Department of Animal Sciences, University of Wisconsin–Madison).

18.8 Major Diseases and Parasites of Sheep

Disease and mortality problems increase as we confine, intensify, and concentrate our operations. We must seek ways of minimizing this.

Dr. C. C. Beck, Michigan State University

Sheep are more susceptible than cows, but less susceptible than goats, to internal parasites. Internal parasites are a potential problem if sheep are grazed. Sheep that are housed or penned for the entire year will have insufficient internal parasites to cause health or production problems. The most susceptible sheep are lambs that have not had previous exposure and an opportunity to develop natural immunity. Lambs on pastures heavily infested with internal parasite larva can die from internal parasitism. Although adult animals generally do not die from parasitism, heavy parasite loads will result in reduced milk production. Most chemical dewormers cannot be administered to lactating animals due to possible residues in the milk. Therefore, parasite control of pastured dairy ewes requires the deworming of the ewes after the end of milking season in autumn. A grazing program that uses a rotation of pastures with long intervals between grazing of individual paddocks may also help in the control of internal parasites.

Most sheep have a naturally long tail that hangs down. A wool-covered tail can collect feces and moisture and becomes an excellent environment for flies to lay their eggs. The fly larvae (maggots) that hatch from the eggs greatly irritate and can feed on the flesh of sheep. This condition is called "fly strike." To prevent fly strike, the tails of most lambs are removed within a few days to a few weeks of birth. The procedure is called docking, and the remnant of the tail that remains on the sheep is called the dock.

Enterotoxemia (overeating disease) is a clostridial disease of lambs and adult sheep on high grain diets. Lambs should receive two vaccinations with *Clostridium perfringens* type C & D toxoid prior to two months of age, and ewes should receive the vaccination annually one month prior to lambing.

Ecthyma (sore mouth) is a viral disease causing lesions on the lips of lambs that can be spread to the udders of ewes. Lambs should be vaccinated once before one month of age. Vaccination imparts lifetime immunity.

Johne's disease, caseous lymphadenitis, and ovine progressive pneumonia all result in the progressive loss of weight and eventual death in adult sheep. Ovine progressive pneumonia is especially detrimental to dairy sheep operations because the retrovirus causing the disease can occupy the soft tissues of the udder and significantly reduce milk production. Testing of purchased animals to ensure that the diseases are not being introduced and periodic testing of the flock for these diseases, combined with the culling of positive animals, are the best ways to control these diseases.

18.8.1 Mastitis in Dairy Sheep

Milk secretion in ewes is largely of the **apocrine** type, where cells of the udder alveoli are damaged in the normal secretion of milk, and cytoplasmic particles (similar in size to somatic cells) and epithelial cells are regular components of milk from healthy udders. This is different from the **merocrine** secretion in cows, where there is little damage to the milk secretory cells in the secretion of milk. However, milk from cows and ewes with healthy udders have a similar somatic cell count (SCC) and a similar presence of cell types, which is very different from goats that have a naturally higher SCC than either cows or ewes (Souza et al., 2012). Therefore, the methods used to measure SCC in cow milk in certified laboratories should produce accurate SCC measurements for ewe milk.

High SCC values in ewe milk are associated with udder infection. In a study with dairy ewes in Wisconsin, ewes with three or more monthly SCCs of 400,000 or higher during the previous lactation were 5.6 to 7.5 times more likely to test positive for mastitis pathogens compared to ewes with lower SCCs (Spanu et al., 2011). Infected udders result in reduced milk production, greater probability of culling, and poorer processing properties of the milk. In a study where cheese was made from ewe's milk with three different SCCs, milk containing > 10^6 cells/mL had less casein, a lower casein to true protein ratio, and slower coagulation time. The cheese made from high SCC milk had a higher proportion of fatty acids with undesirable flavor profiles, and rancid flavors were detected by cheese graders (Jaeggi et al., 2003).

Somatic cell count tends to increase with parity and days in lactation. Since sheep are seasonal breeders and ewes in a flock are in a similar stage of lactation on any given date, bulk tank SCC is generally greater in the fall of the year toward the end of lactation than in the winter and spring at the start of lactation. In the United States, the *Grade "A" Pasteurized Milk Ordinance* requires the SCC of sheep milk not to exceed 750,000 cells/mL, the same as for cow milk.

18.9 Feeding Dairy Sheep

A little bit of grain, fed over a long period, does a sheep a world of good.

Art Pope, University of Wisconsin–Madison

Many production systems put ewes to milking as soon after giving birth as possible and raise the lambs on milk replacer. This is done to be able to market as much milk as possible. It is important that baby lambs receive colostrum (cf. chapter 7). Lambs can be allowed to nurse colostrum from their mother for the first 12 to 24 hr after birth, suckle pooled colostrum through a nipple on a bottle, or be administered pooled colostrum via a stomach tube. Once the lamb has adequate colostrum, it is hand-fed lamb milk replacer from a bottle with a nipple every 4 to 6 hr for one or two days until it is trained to the nipple. Then the lamb is transferred to a pen with free-choice access to milk replacer from nipples attached to a reservoir of milk replacer or a machine that automatically mixes milk replacer on demand. The milk replacer must be specially formulated for lambs. Calf and kid milk replacer cannot be used because they contain less fat and protein than lamb milk replacer.

Although lambs will eat only small amounts of feed while they are on the milk replacer, they need to receive a palatable, high protein (20 to 22% crude protein) diet so they can easily transition to a complete dry diet on weaning. This diet should be pelleted or coarsely ground for maximum intake. Lambs should be weaned from milk replacer at 30 to 45 days of age but continue to have free-choice access to the high protein diet until they are 60 to 75 days of age. Research at the University of Wisconsin–Madison has shown that a lamb weaned from milk replacer at approximately 30 days of age on the system described here will consume 18 lb of milk replacer powder.

At 2.0 to 2.5 months of age, lambs can be transitioned to a lamb grower diet of 16% crude protein, and ultimately to a 14% crude protein finishing diet. These diets are commonly provided as a high protein pellet (e.g., 34 to 36% crude protein with added minerals, vitamins, and medications) mixed with whole shelled corn in the ratio necessary to provide the desired level of protein. These diets are fed free-choice to the lambs. A small amount of legume or legume-grass hay should be provided also.

If lambs are to be raised on pasture after being weaned from milk replacer, they should be supplemented with grain during the entire pasturing period or at least until they are consuming enough pasture to support reasonable body weight gains.

Replacement ewe lambs can be removed from lambs being fed for market before they reach 100 lb and fed a higher forage and lower energy diet to save on feed costs. Some research suggests that ewe lambs fed for maximum body weight gain produce less milk as adults than do ewe lambs that have reasonable, but not maximum, growth rates. However, a three-year trial at the University of Wisconsin–Madison with approximately 250 ewe lambs found no difference in adult milk production of ewe lambs fed free-choice for maximum body weight gains compared to ewe lambs fed at 75% of maximum feed intake. Therefore, a producer should feel comfortable selecting replacement ewe lambs out of the lambs in the finishing pen if they have not been selected earlier.

Sheep are **ruminants**; therefore, like the cow and goat, feeding regimes for adults based on large quantities of high-quality forages should be the rule. Since most dairy sheep operations are small with little capability for the mixing of complicated diets, total mixed rations (TMR) that are common in dairy cow operations are generally not fed to dairy sheep. However, if size of the operation can justify the investment in equipment, a dairy cattle TMR, with minor modifications, will work well in a dairy sheep operation. Sheep select short fiber components of rations, leaving the long fiber components in the feed bunk. Therefore, forages must be chopped to a length of about 0.79 in (20 mm) to decrease this selective behavior. High-yielding ewes fed TMR should receive additional concentrates in the milking parlor.

Energy requirement related to body weight is higher in sheep than in cattle. Furthermore, since retention time in the rumen is not higher in sheep than in cattle, digestion of roughage tends to be lower in sheep so they are not able to consume as much roughage as cattle per kg of body weight. They must compensate for small body size by ingesting a relatively higher amount of dietary energy. Sheep can adapt well to diets high in starch. They are less sensitive than cattle to acidic conditions in the rumen. Their small intestine, where bypass residual starch is digested, is relatively longer than that of the cow. Thus, they digest grain more completely than do cattle and adapt well to diets rich in carbohydrates.

Presented in table 18.6 are the daily requirements of lactating dairy ewes for total digestible nutrients (TDN, a measure of energy), crude protein, calcium, and phosphorous. The lactating ewe requires a diet that has approximately 16% crude protein. Ewes reach their peak in milk yield in the third to fifth week of lactation, which is the reason the greatest nutrient requirements are in early lactation. Average lactation length of dairy ewes in the United States is 200 to 210 days.

Diets of lactating dairy ewes need not be complicated, but they must be balanced to meet requirements for maximum milk production and health of the ewes. Table 18.7 presents simple rations containing the common feedstuffs of alfalfa hay, whole shelled corn, and soybean meal that meet the daily nutrient requirements of lactating ewes. Since forage is generally the least expensive feedstuff, these diets are based on 4.5 lb of alfalfa hay, close to the maximum amount of dry forage that a ewe can consume in a day. Corn, an energy feed, is next added in the amount needed to meet the TDN requirement not met by the hay, followed by addition of soybean meal, a protein feed, in the amount needed to meet the shortfall of protein from the hay and corn. Finally, dicalcium phosphate is added to meet a shortfall of phosphorous from the feedstuffs. As the ewe progresses through lactation, her milk production decreases, and the amount of corn (for energy) and soybean meal (for protein) required to meet her requirements also decreases.

The **dry** ewe should not be neglected at feeding time, especially during late pregnancy. More than two-thirds of fetal growth occurs during the last six weeks of pregnancy. Thus, to help assure (1) the birth of strong, well-developed lambs, (2) bountiful and high-quality colostrum, and (3) adequate body condition of the ewe to yield maximal milk production, the nonlactating ewe must receive a balanced diet. Prior to late pregnancy, ewes can meet their nutrient requirements on high-quality forage alone—legume or grass-legume hay or haylage. During the last six weeks of pregnancy, the addition of 1 lb of grain to the high-quality forage will be adequate. Dry and pregnant ewes should be fed at a level that results in a 3.5 to 4.0 body condition score at the time of lambing (1 = very thin, 5 = very fat).

Table 18.6 Daily Nutrient Requirements of a 176-lb Dairy Ewe.

		Daily Nutrient Requirements (lb)			
Stage of Lactation	**Daily Dry Matter Intake (lb)**	**TDN**	**Crude Protein**	**Calcium**	**Phosphorous**
Early: Day 1–90	≤6.69	4.42	1.10	0.03	0.03
Mid: Day 91–150	≤6.45	3.67	0.88	0.02	0.02
Late: Day 151–end	≤6.20	3.30	0.65	0.02	0.02

Source: Adapted from National Research Council, 2007, with minor modifications.

Table 18.7 Simple Rations to Meet the Nutrient Requirements of Lactating Dairy Ewes.

	Feedstuffs, as Fed (lb/ewe/d)[a]			
Stage of Lactation	**Alfalfa Hay, Midbloom**	**Whole Shelled Corn**	**Soybean Meal**	**Dicalcium Phosphate**
Early: Day 1–90	4.50	2.50	0.49	0.05
Mid: Day 91–150	4.50	1.74	0.13	0.02
Late: Day 151–end	4.50	1.27	—	0.04

[a] Amount of feed required to meet the requirements of a 176-lb ewe.

All rams, ewes, and lambs should have free-choice access to a mineral that is formulated especially for sheep. A sheep mineral will have no added copper since sheep are very susceptible to copper toxicity. In many regions, selenium is deficient in the soil so a good sheep mineral will include selenium.

18.10 Management of Dairy Sheep

18.10.1 Determining Ages of Sheep

The age of sheep may be determined by their front (incisor) teeth on the lower jaw. (There are no incisor teeth on the upper jaw.) Both the lamb and the adult have eight incisors. These are small and temporary in the lamb. At about 12 months of age the center pair drops out and is replaced by two large permanent teeth. When the sheep is about two years of age, two more small temporary teeth, one on each side of the first pair of permanent teeth, are replaced by two more large permanent teeth; at about three years, the sheep has six permanent teeth, and at about four years, the sheep has its complete set of eight permanent incisor teeth. Following this, the degree of wear provides an approximate indication of age. Teeth of older sheep spread apart, finally become loose, and some may drop out. When this occurs, sheep begin losing their usefulness as grazing animals.

18.10.2 Behavioral Characteristics of Sheep

Sheep have a well-developed flocking instinct and wish to stay with other sheep. A sheep that is separated from its flock mates will become agitated and take extreme measures to rejoin the flock. This great desire to be part of the flock is useful when moving sheep. Once one or two sheep are moving in the correct direction, the rest of the flock will follow.

Shepherds often use herding dogs to assist in moving of sheep. The most common breed of herding dog is the border collie. Developed in northern England and Scotland with a genetic instinct for herding sheep, the natural flocking instinct of sheep allows the border collie to form them into a compact group and to move them forward by being behind them. A trained border collie can save a shepherd many hours of labor by moving sheep quickly and efficiently.

Mature rams can often become mean toward people and attempt to butt them. People have been severely injured by a rogue ram. All persons should be watchful of rams when in the same pasture or pen. Rams that are played with as lambs have a greater probability of becoming mean as adults. Don't make pets out of ram lambs.

18.11 Dairy Sheep Association of North America

The charter meeting of the Dairy Sheep Association of North America (DSANA) was held on November 7, 2002, in Ithaca, New York. The goals of DSANA are: (1) promote effective dairy sheep management by educating, supporting, and encouraging new and established sheep milk dairies, farmsteads, and artisanal sheep milk cheesemakers; (2) promote cooperation and exchange of information among producers of sheep milk and cheesemakers; (3) promote products manufactured from sheep milk; (4) help producers organize for the genetic improvement of dairy sheep; (5) inform and educate the public as to the merits and availability of sheep dairy products; and (6) foster international understanding and free exchange of ideas among producers internationally.

A major activity of DSANA is sponsorship of an annual symposium, which brings together dairy sheep producers, sheep milk processors, animal and food scientists, and industry and government representatives to share in the latest information regarding dairy sheep production and sheep milk processing. The printed proceedings from DSANA symposia contain the best source of current information for the North American dairy sheep industry. DSANA also publishes a quarterly newsletter for communication among its members.

18.12 Summary

Production of milk and cheese from sheep is an ancient enterprise in many countries of the world, especially in the countries of southern Europe, North Africa, and the Middle East. Dairy sheep production is relatively new to North America, but the industry is growing due to the large growth of specialty cheeses in recent years. Domestic sheep milk cheeses, not to be found in food stores 20 years ago, are now commonplace in most cheese stores and many regular grocery stores.

The common breeds of sheep found in the United States were selected for meat and wool production and are unsuitable for commercial milk production. Recent imports of European dairy sheep genetics (East Friesian and Lacaune breeds) resulted in increased milk production of US dairy sheep flocks and increased the probability of profit for dairy sheep farms. There are other foreign dairy sheep breeds that would be valuable to US producers, but strict animal health import regulations have kept those breeds out of the United States.

Sheep milk has a much higher percentage of fat and protein than milk from cows or goats and, as a result, a higher yield of cheese per unit of milk. Sheep milk products have a distinctive flavor but a much milder flavor than products made from goat's milk.

Dairy sheep are seasonal breeders and generally lamb in the winter and spring. They lactate for approximately seven months and average about 3.5 lb/d of milk. The milk is quite valuable; the price paid to producers is two to three times that for goat milk and three to five times that for cow milk.

Dairy sheep are fed simple diets of high-quality forage supplemented with grain and protein during the high production periods of late pregnancy and early and mid-lactation.

A major need of the US dairy sheep industry is a national organization to calculate across-flock estimates of genetic values for lactation traits. Without such an organization, genetic improvement of dairy sheep will be very slow.

Study Questions

1. In what regions of the world were sheep first domesticated?
2. In which regions of the world are dairy sheep most important? Which three countries are the largest producers of sheep milk?
3. What is a ewe? A ram? A lamb? A wether?
4. What is docking and what is its purpose?
5. In what ways do milks of the ewe, goat, and cow differ?
6. What color is the milk of ewes? Why is it that color?
7. Name the two dairy sheep breeds of the United States, their country of origin, and describe their breed characteristics.
8. Describe the desired udder conformation of a dairy ewe.
9. At what age and body weight can ewes be bred for the first time? Are ewes seasonal breeders? Explain.
10. What is the normal length and frequency of estrous periods among ewes?
11. What is the length of an average gestation? Are most births singles or multiples? What is the average weight of newborn lambs?
12. Is artificial insemination common in dairy sheep? Explain why or why not.
13. Why should you not feed calf milk replacer to artificially reared lambs?
14. Discuss briefly the feeding of (a) lactating ewes and (b) dry ewes.
15. How many mammary glands do ewes have? Is the principle of milking ewes the same as that of cows?
16. What is the average daily milk yield of dairy ewes?
17. Are the productive lives and lengths of lactation of cows, goats, and ewes similar?
18. How many permanent incisor teeth would one expect to find in a three-year-old ewe? Discuss briefly how one may use teeth to estimate age in sheep.
19. How do you prevent enterotoxemia in sheep?
20. Name three progressive wasting diseases of sheep. What are the best methods to control these diseases?
21. Are ewes more or less susceptible to internal parasites than cows? Than goats? Why?

For Further Study

Berger, Y. 2002. "Dairy." In *SID Sheep Production Handbook* (Vol. 7, pp. 100–136). Centennial, CO: American Sheep Industry, Inc.

Berger, Y., P. Billon, F. Bocquier, G. Caja, A. Cannas, B. McKusick, P. G. Marnet, and D. Thomas. 2004. *Principles of Sheep Dairying in North America* (US Pub. No. A3767). Cooperative Extension, University of Wisconsin–Madison.

Fuquay, J. W., P. F. Fox, and P. L. H. McSweeney (eds.). 2011. *Encyclopedia of Dairy Sciences*, Vol. 2 (2nd ed.). San Diego, CA: Academic Press.

(1) G. Molle and S. Landau. "Sheep: Feeding and Management" (pp. 848–856).

(2) C. Macaldowie. "Sheep: Health Management" (pp. 857–864).

(3) O. Mills. "Sheep: Milking Management" (pp. 865–874).

(4) J. Hatzminaoglu and J. Boyazoglu. "Multipurpose Management" (pp. 875–881).

(5) D. L. Thomas. "Sheep: Replacement Management" (pp. 882–886).

(6) E. Gootwine. "Sheep: Reproductive Management" (pp. 887–892).

Pulina, G. (ed.). 2004. *Dairy Sheep Nutrition*. Wallingford, Oxfordshire, UK: CABI Publishing.

References

Barillet, F. 2007. Genetic improvement for dairy production in sheep and goats. *Small Ruminant Research* 70:60–75.

Bencini, R., and G. Pulina. 1997. The quality of sheep milk: A review. *Australian Journal of Experimental Agriculture* 43:485–504.

Berger, Y. M. 2007. "The East Friesian Breed of Sheep in North America." In Proceedings, Third Biennial Spooner Dairy Sheep Day (pp. 10–16). College of Agricultural and Life Sciences, University of Wisconsin–Madison.

Billon, P. 2004. "The Designing of Small and Medium Sized Milking Machines for Dairy Sheep." In Proceedings, Tenth Great Lakes Dairy Sheep Symposium (pp. 28–54). Hudson, Wisconsin, Dairy Sheep Association of North America.

Caja, G., X. Such, and M. Rovai. 2000. "Udder Morphology and Machine Milking Ability in Dairy Sheep." In Proceedings, Sixth Great Lakes Dairy Sheep Symposium, November 2–4. Guelph, Ontario, Canada, Dairy Sheep Association of North America.

Jaeggi, J. J., S. Govindasamy-Lucey, Y. M. Berger, M. E. Johnson, B. C. McKusick, D. L. Thomas, and W. L. Wendorff. 2003. Hard ewe's milk cheese manufactured from milk of three different groups of somatic cell counts. *Journal of Dairy Science* 86:3082–3089.

Jahreis, G., J. Fritsche, and J. Kraft. 1999. Species dependent, seasonal, and dietary variation of conjugated linoleic acid in milk. In M. P. Yurawecz, M. M. Mossoba, J. K. G.

Kramer, M. W. Pariza, and G. J. Nelson (eds.), *Advances in Conjugated Linoleic Acid Research*, Vol. 2. Champaign, IL: American Oil Chemists Society.

Jandal, J. M. 1996. Comparative aspects of goat and sheep milk. *Small Ruminant Research* 22:177–185.

Milani, F. X., and W. L. Wendorff. 2011. Goat and sheep milk products in the United States (USA). *Small Ruminant Research* 101:134–139.

Moioli, B., M. D'Andrea, and F. Pilla. 2007. Candidate genes affecting sheep and goat milk quality. *Small Ruminant Research* 68:179–192.

National Research Council. 2007. *Nutrient Requirements of Small Ruminants: Sheep, Goats, Cervids, and New World Camelids.* Washington, DC: National Academies Press.

Park, Y. M., M. Juarez, M. Ramos, and G. F. W. Haenlein. 2007. Physio-chemical characteristics of goat and sheep milk. *Small Ruminant Research* 68:88–113.

Pirisi, A., G. Piredda, C. Papoff, R. Di Salvo, S. Pintus, G. Garro, P. Ferranti, and L. Chianese. 1999. Effects of sheep α_{s1}-casein CC, CD and DD genotypes on milk composition and cheesemaking properties. *Journal of Dairy Research* 66:409–419.

Prandini, A., S. Samantha, G. Tansini, N. Brogna, and G. Piva. 2007. Different levels of conjugated linoleic acid (CLA) in dairy products from Italy. *Journal of Food Composition and Analysis* 20:472–479.

Ryder, M. L. 1983. *Sheep and Man*. London, UK: Duckworth.

Souza. F. A., M. G. Blagitz, C. F. A. M. Penna, A. M. M. P. Della Libera, M. B. Heinemann, and M. N. O. P. Cerqueira. 2012. Somatic cell count in small ruminants: Friend or foe? *Small Ruminant Research* 107:65–75.

Spanu, C., Y. M. Berger, D. L. Thomas, and P. L. Ruegg. 2011. Impact of intramammary antimicrobial dry treatment and teat sanitation on somatic cell count and intramammary infection in ewes. *Small Ruminant Research* 97:139–145.

Zeder, M. A. 2008. Domestication and early agriculture in the Mediterranean basin: Origins, diffusion, and impact. *Proceedings of the National Academy of Sciences* 105:11597–11604.

Zeuner, F. E. 1963. *A History of Domesticated Animals*. New York: Harper & Row.

WEBSITES

Canadian Livestock Records Corporation (http://www.clrc.ca)

Dairy Sheep Association of North America (http://www.dsana.org)

Food and Agriculture Organization of the United Nations, Statistics Division (http://faostat.fao.org)

Proceedings of Great Lakes Dairy Sheep Symposium (http://fyi.uwex.edu/wisheepandgoat/dairy-sheep)

University of Wisconsin Sheep and Goat Extension (http://fyi.uwex.edu/wisheepandgoat)

19

Managing the Dairy Farm Enterprise

> If cows could talk, they would be heard all over this country
> calling for an improved breed of dairymen.
>
> *William Dempster Hoard (1836–1918)*

19.1 Introduction

Milk is an essential food and *management* of the dairy farm enterprise is the most important factor in its production. Regardless of how well cows are bred and fed, unless management is correct and conforms to best practices, a cow's production potential will never be realized. Management, then, ties it all together. Most of the difference between profit and loss—indeed, between success and failure—in the dairy farm enterprise is attributable to sound management.

Dairy farming is both a business and a way of life. As a business, dairy farm enterprises represent an estimated 90% of the total investment of the dairy industry. As a way of life, dairying allows one to be in close contact with nature and to have the satisfaction of producing the most nearly perfect food—milk. It allows one to develop closeness with animals, to gain experiences with conception, birth, growth, and development of dairy animals, to experience life and death and all the challenges presented therein. Dairy farming provides a number of individual freedoms, especially authority and responsibility; it offers an opportunity for creative work and continuing challenge.

Management is both an art and a science; it is most successfully achieved by those who evaluate facts, weigh alternatives, and arrive at economically sound solutions. Managing the dairy farm is a continuous decision-making process; a manager must provide an environment that fully utilizes the inherent genetic capability of cows to secrete milk in order to obtain maximum milk production. Dairy farm management entails care, use, and handling of animals, equipment, and facilities. It also involves production records and other business aspects. As dairy farms increase in size and mechanization, management becomes increasingly important. To attain profitable milk production, managers must possess skills in combining cows, land, labor, machinery, equipment, housing, and milk quality into an overall efficient system.

Absolute rules that will assure successful operations in dairy farming cannot be written. Guidelines and important considerations, yes, but in the final analysis it is the experience, judgment, and expertise of the manager that makes the dairy enterprise successful! Working with dairy cows that have desirable conformation and high production is a profound aspiration and often the favorite diversion of progressive dairy farmers. How can they profitably care for and feed cows so they will express their inherent genetic potential? What is the best management system for achieving profitable milk production? Although we cannot answer these and related questions precisely, many aspects of managing the dairy farm enterprise have been discussed in preceding chapters. In this chapter, we cite important considerations of dairy cattle management.

19.2 The Manager

The most important part of any dairy operation is the person or persons involved in management and care of the herd.

Clinton E. Meadows (1913–2003)

Whether considered a cowperson, herdsperson, director, or proprietor, the manager is the most important single factor in the profitable operation of a dairy enterprise. Everyone involved in profitably producing milk is necessarily a manager, whether the involvement is with calves, cows, feeding, breeding, milking, or care of the product. In addition, expertise is needed in management of human resources, information, and the production and utilization of records.

What characterizes an effective manager? First and foremost, it is an enjoyment of dairying and a love and respect of good cows. This love of cattle enhances the fundamental understanding of animal needs and the proper interpretation of their individual requirements.

In reviewing management practices within high-producing dairy herds, there is one common observation that surfaces consistently. Namely, considerable individual attention is given each cow. This, of course, requires additional labor. Therefore, a decision each manager must make is, "Will the extra labor be reflected in personal satisfaction and sufficiently higher milk yield to offset its additional cost?"

Dairy farm managers should get to know their cows. Knowing health histories of the animals, for example, can benefit both manager and veterinarian in treating animals and often prevents problems. Knowing production records and pedigrees aids in sales to prospective buyers of dairy cattle. Collection of data into the farm computer programmed to detect significant changes in milk produced, its conductivity, and other measurable parameters can assist personnel in identifying animals that need individual attention.

Patience is a precious attribute to those adept at managing the cow. When handlers display temper, animals become frightened and, naturally, refuse to cooperate fully. Fast movements and/or loud noises often excite cows. The "Cow Casanova" is likely to harvest more milk and smile all the way to the bank.

Recognizing that there are some members of society who are suspicious that some dairy producers do not respect animal rights and treat cattle cruelly, National Milk Producers Federation sponsors the National Dairy FARM (Farmers Assuring Responsible Management) Program. The program provides a national set of guidelines designed to demonstrate commitment by dairy farmers to outstanding animal care and high-quality milk.

Finally, no set of rules or instructions will ever replace experience and good judgment. A careful eye searching the needs of dairy cattle will be an important factor in successful management.

19.3 Recruiting and Managing Labor for Dairy Farms

Dairy farm managers accomplish their goals of profitability and long-term viability through workers. Successful managers hire competent, caring workers then orient, train, and motivate them with effective communication and incentives.

Steps involved in hiring include: discern labor needs, develop a job description, build a pool of applicants, review applications, interview applicants, and hire the best—including those with a positive attitude (Erven, 2011).

Most new employees do not come to dairy farm-related jobs with all the knowledge, skills, and abilities needed. Therefore, training is important to their successful performance and satisfaction with the job. Two-way communication plays a major role in employee-employer relations. Messages should be clear and understood, and feedback from the receiver is important to assure clarity and maintain commitment.

Motivated employees are those for whom the following categories of needs have been fulfilled: physiological, security, social, esteem, and self-competence. Detrimental factors associated with impeding motivation include: poor working conditions, excessively long work days and weeks, unfair pay, disagreeable supervisors, unreasonable rules and policies, and conflicts with coworkers. Employees welcome recognition for achievements and good work performance. They need to feel they are important members of the farm team.

19.4 Management Strategies

It should come as no surprise that high production is accomplished by more hours of attention. In any endeavor, excellence requires more effort.

John A. Speicher (1928–2008)

A dairy farm manager must manage more than just the farm. In order to become an effective business manager, there are several practices that are critical to success: planning, risk management, managing overhead, and applying management tools.

19.4.1 Planning

An important role of the dairy farm manager is to develop and perfect a business plan and to implement and monitor that plan effectively. That plan should include performance standards. When those standards are not met, action must be taken to enhance performance and/or modify the plan.

Dairy farms that are family owned often involve family members who have personal goals. Some of these goals will be financial in nature while others may be related to lifestyle, community service, and involvement in organizations. Any conflict among these goals must be reconciled before an effective business plan can be devised and implemented. Consideration must be given within the plan to the financial realities of potential profit, cash flow, and level of family income needed.

The plan should specify what and how much the enterprise will produce, where it will be marketed, and how farm finances will be structured to achieve the planned results. The plan will be affected by the economic, social, and political environment in which the business exists, especially the available resources. A target level of performance must be established in order to judge whether a successful and responsible operation has been achieved.

Routine planning in an established operation may involve day-to-day decisions as well as making financial plans and projections for the coming year. In contrast, setting the course for extended operations and new investments requires much more thorough analysis and decision. Benson (2011) listed the following steps to effective strategic planning: (1) setting clear business goals; (2) making an inventory of farm, financial, and human resources; (3) analyzing past and projected performance; (4) identifying alternatives; (5) assessing the external business environment; (6) evaluating the production, marketing, and financial feasibility of alternative courses of action, including risk management; (7) making decisions; (8) specifying how the plan will be implemented; and (9) specifying how farm performance will be evaluated.

Since there is substantial diversity among dairy farming systems, data from similar operations are best used in making reliable decisions during planning sessions. Dairy farmers who process their own milk for sale face additional challenges, especially those related to marketing.

19.4.2 Risk Management

Risks should be prioritized on the basis of probability of occurrence with highest priority given to those having greatest financial impact. The goal should be to reduce the probability that selected risks will occur. Examples include providing a backup power supply and an excess inventory of items that are critical to the operation, carrying adequate insurance, and using futures markets to manage price risks. Laws and regulations, especially those related to liability and taxation, can have significant effects on risks.

Effective record-keeping systems yield information for monitoring and evaluating farm performance and for making appropriate decisions. Both financial and production information must be included.

In a healthy business the positive net cash flow from operations must be large enough to offset the negative cash flows associated with investing, financing, and nonfarm spending (Benson, 2011).

19.4.3 Overhead

Dairy farm managers must seek to maintain the lowest feasible capital investment. Before expanding or adding new facilities, consideration should be given to expected return on investment. It is possible to have too high an overhead (feed, land, labor, cow depreciation, equipment and machinery, building, utilities, taxes, and insurance) in relation to herd size and amount of milk produced. Table 19.1 shows how income and expenses were affected by numbers of cows in herds producing milk on a national basis in 2014. Overhead was a major factor determining costs. Costs of production decreased markedly with increases in cow numbers. Only in herds with 500 or more cows did this analysis show a profit margin over total costs, even when unpaid labor was discounted. Overall, feed costs averaged 53%, operating costs 11%, and overhead 36%. Of course, these are gross data, and many factors affect individual situations including market conditions and production per cow. In this analysis, output per cow varied from more than 15,000 lb in the smallest herds to about 22,500 lb in herds of 500 to 999 cows. Nevertheless, overhead must be considered a major factor by owners/managers.

19.4.4 Management Tools

Records are the most dependable management tools available. These include records of milk production, feeding, breeding and calving, labor, health, sales (milk and dairy cattle), forage testing data, and farm accounting (cf. chapter 4). Effective record-keeping systems yield information for monitoring and evaluating farm performance and for making appropriate decisions. In a healthy business the positive net cash flow from operations must be large enough to offset the negative cash flows associated with investing, financing, and nonfarm spending (Benson, 2011).

19.5 Factors Affecting Profitability

A profitable dairy operation is dependent largely on three factors: (1) price of milk, (2) production of milk per cow, and (3) cost of producing milk. Individual milk producers can do little about the price of milk—it largely depends upon supply and demand (cf. chapter 20). They can, however, greatly affect costs of production. Through improved breeding, feeding, herd health, and use of records, milk production per cow can be maximized and the cost of producing milk minimized. The latter, of course, includes maximal use of labor-saving devices and equipment commensurate with the size of the operation and the number of cows milked.

19.5.1 Costs of Feed in Milk Production

Since the dairy farm enterprise is based on the sale of milk, productivity of individual cows and of the dairy herd is probably the most important single factor affecting profitability of the operation. Feed represents approximately half the total cost of producing milk. However, costs assigned to feed in 2010 among major areas of milk production in the United States ranged from 35% in Minnesota to 74% in California and Texas. The average for 23 major milk-producing states was 56% compared with 45% in 2005. Most dairy enterprise managers concur that as milk production per cow increases, feed cost per cwt of milk produced decreases.

19.5.2 Labor Management and Efficiency

Annual cow labor costs in milking, feeding, and handling manure vary from 30 hours with highly automated operations to more than 100 hours on farms where most work is done manually. Efficient stanchion and freestall housing systems require, respectively, approximately 60 and 35 hours per cow annually. Missouri studies found that annual labor averaged 48 hours per cow among participating dairy farmers; moreover, total labor per cow decreased as herd size increased. Because milking is the largest labor expenditure on dairy farms (it constituted 56.6% of the labor per cow in Missouri studies), practices and procedures used should be fully and continuously examined.

Research at Michigan State University showed that labor required for milking can be reduced 10% either by eliminating grain feeding in the milking parlor or by using

Table 19.1 Milk Production Costs and Returns per Hundred-Weight (cwt) Milk Sold by Group Size, 2014.

	Number of Cows					
Milk Production Costs	**< 50**	**50–99**	**100–199**	**200–499**	**500–999**	**> 1,000**
Gross value of production ($)	29.81	29.02	28.02	27.60	26.89	25.26
Total feed cost ($)	20.19	18.44	17.38	17.43	15.82	13.89
Total operating costs ($)	24.98	23.03	21.30	21.54	19.39	16.44
Allocated overhead ($)	25.35	16.87	11.22	8.64	6.02	4.56
Total costs ($)	50.33	39.90	35.52	30.18	25.41	21.00
Cost minus unpaid labor ($)	35.88	32.44	28.79	28.60	24.86	20.83
Output per cow (lb)	15,365	17,137	18,927	19,842	22,579	23,016

Source: USDA Economic Research Service, 2015.

milker units on both rather than only one side of a herringbone parlor. Group feeding allows better use of mechanization and automation. Feeding forage accounts for 13% of chore time when both hay and silage are fed, but 8% when only silage is fed. Thus, feeding hay and silage requires about 2.5 to 3.0 hours more labor per cow annually than feeding silage alone. Importantly, it is usually easier to automate the feeding of silage than that of hay.

19.5.2.1 Incentives for Employees

Most employees show greater interest in making a successful dairy farm enterprise even more profitable if they share in the profits.

Incentives may be based on any or all of the following: (1) milk production (a bonus based on progressive increases in milk production above breed average); (2) breeding efficiency (a bonus as calving interval becomes less than 14 months, or for each cow detected in estrus and/or declared pregnant); (3) calf raising (a bonus for low calf mortality);[1] and (4) total net income (a bonus that increases as net income increases).

19.5.3 Economics of Size

Several studies have shown that increases in herd size are not reflected in increased production per cow. However, labor efficiency commonly increases as herds are enlarged, therefore decreasing costs of producing milk (creating a lower investment per cow, for example).

Economic and technological forces continue to encourage adjustments to larger, more highly mechanized and efficient dairy management units. For maximal efficiency, buildings and equipment should be utilized to capacity. This is more nearly possible with large herds. Moreover, purchases and sales are accomplished more economically when quantity is large. Of course, large size, per se, guarantees neither increased labor efficiency nor greater profit.

19.5.4 Money Management and Related Aspects

That part of the holding of a farmer or landowner which pays best for cultivation is the small estate within the ring-fence of his skull.

Charles Dickens (1812–1870)

Successful managers are familiar with principles of economics and practices of sound business management. High net income is their ultimate concern and objective.

Investments for milking, housing, manure and feed storage, and handling systems for the same number of cows depend on the climate and variations in temperature and humidity. Regardless of location, however, successful managers utilize buildings and equipment to their capacity and make the enterprise large enough that labor-saving provisions are economically sound. This minimizes costs per cow. The largest percentage allocation of capital investments on selected Missouri dairy farms was found to be land and buildings.

Used assiduously, credit is a valuable tool in dairy farm management. It benefits owners to borrow money if the returns from borrowed capital are greater than the costs of borrowing and using it. Financial records serve as a means of analyzing past actions and as a basis for future decisions. Software programs have been designed to assist in estimating financial consequences of adjustments in the dairy enterprise. Most university dairy and farm management extension specialists can assist managers in making use of computer-assisted analyses.

Managers of dairy farm businesses must study, understand, and apply information concerning insurance (liability, storm, fire), taxes, credit, wages, land values, and agricultural law.

19.5.5 Cow Costs and Culling

Two important and related aspects of dairy farm management are cow costs (depreciation) and culling of unprofitable cows.

19.5.5.1 Cow Costs

Cow depreciation is the initial value of a cow when she enters the milking herd minus her salvage value when sold divided by her years in production. For example, assume a cow initially worth $1,800 brings $800 when sold. If her productive life is 4 years, her annual depreciation cost is $250 ([$1,800 – $800]/4 = $1,000/4 = $250). Using this example, a herd of 100 cows would have an annual depreciation cost of $25,000 (100 × $250). Of course, managers can reduce depreciation cost by raising herd replacements or, calculated another way, they can realize a handsome profit through the appreciation of calves and heifers raised.

19.5.5.2 Culling

Low-producing cows are seldom profitable. Cows in first lactation yielding 30% or more below herd average and older ones producing considerably below **herdmates** of the same age should be considered for culling. Cows that tend to go **dry** in six to eight months, those chronically infected with mastitis, those with disagreeable dispositions, and those of low reproductive efficiency are additional prospects for culling. Occasionally a cow is injured or has another reason for low production and deserves a second chance. Mississippi Agricultural Experiment Station researchers reported that cows with poor temperaments should be culled from the herd since they can affect the entire herd and reduce producer profits. Additionally, anxious and aggressive cows were more prone to illness, had more difficulty gaining weight, and more often damaged farm equipment. Their rowdy behavior may disturb cows that would otherwise be calm. Genetic inheritance was the primary factor affecting cow behavior. Finally, the researchers concluded that when a wild female is mated to a high-strung bull, the result will likely be a highly temperamental calf.

19.6 Practices That Contribute to Efficient Management

Management is the successful accomplishment of doing a number of little things well.
Jack L. Albright (1930–2014)

It is difficult to identify precisely those practices and procedures that assure a profitable dairy operation. Certainly management aspects of breeding, feeding, milking, and reproduction command a high priority of attention.

Usually sound management includes breeding cows and heifers so calving and peak production coincide with periods of highest milk prices (usually in the fall and winter months). Average persistency of lactation among cows commonly reflects feeding and management practices, and knowledge of persistency curves (cf. chapter 10) aids managers in synchronizing peak production with highest milk prices.

The more successful managers have high-quality feed available for cows at all times. They should regularly compare average feed cost to the value of 100 lb (45.4 kg) of milk produced, i.e., income over feed cost.

Average days in milk should be 85 to 90% for each period between calvings. A high conception rate (average days **open** less than 90) and minimal dry period (45 to 60 days) are required to achieve this goal.

Milk production per cow and milk sales per hour of labor and per dollar invested are meaningful measures of labor and management efficiency of dairy enterprises. In striving to maximize these measures, managers should attempt to increase pounds of milk produced per person-hour, per employee, per acre, per farm, and per dollar invested. Of course, complete and meaningful records are essential components of these important goals.

Major differences in management exist among dairies on which production is pasture-based, pasture-optimized, or **drylot**. Constraints imposed by availability and quality of pasture, as well as abilities of cows to graze pastures, can result in significant variability in milk production. However, there are regions of the world in which climate and availability of land afford the opportunity for economical production under pasture-based systems.

19.6.1 Pasture-Based Dairying

This system, which includes seasonal calving, has been used to produce much of the milk in southern Australia and New Zealand for many years. In recent years dairy farmers from those countries have established pasture-based dairies in selected areas of the United States, especially in Missouri, where Holstein–Jersey cows are found in numbers of 500 or more per farm. A spring-calving system, often used in Europe, calls for calving in the spring to provide lactating cows for milking during seven to nine months of pasture consumption.

Variables important in this system are quality of grasses and clover, fertility of land, length of growing season, and the ability of management to optimize pasture utilization without compromising growth or persistence of the sown species. Strip or rotational grazing is practiced. Ideally, grazing management matches pasture consumption with growth rate keeping the plants in productive state. Major factors that affect intake of pasture include (1) cow size, breed, milk yield, and stage of lactation; (2) disease and environmental stress; (3) pasture mass and composition; and (4) amount of pasture and supplementary feeds provided.

19.6.2 Pasture-Optimized Dairying

Traditional dairying in the United States and other areas having temperate to subtropical climates was of this type during the twentieth century. Pasture was the major source of nutrients during the growing season, whereas silage and hay provided the forage for other parts of the year.

Pastured dairy cattle typically experience greater postpartum condition loss than cows fed in confinement. This may be due to higher maintenance requirement, lower dry matter intake, reduced energy density, or poor energy utilization. However, the incidence of mastitis has been found lower among pasture-fed cows than those fed in confinement.

The optimal time for lactations to start and end depends on feed availability and prevailing seasonal milk prices. The nonseasonal, pasture-based milk production system provides desirable flexibility in this regard.

19.6.3 Drylot Dairying

This form of dairying confines the cows and brings forage to them rather than sending the cows to pasture. In general, the drylot system requires more labor and a greater investment than do pasture grazing systems, and this is reflected in higher costs of producing milk. However, it has the advantage of permitting harvest of more tonnage per acre of cropland. Of course, the price of land and taxes per acre influence the decision on whether drylot is more profitable than other alternatives. In certain areas (in the Los Angeles **milkshed**, for example), land is at a premium and is priced and taxed accordingly. Therefore, drylot dairying is a necessity. In other areas pasture operations require less labor and capital investment per cow and provide equally profitable dairy farm operations.

19.6.4 Milk Composition and Management Decisions

High numbers of cows in today's dairy herds coupled with high production per cow make vital the discovery of impaired productive capacity of each animal. Fat, protein, lactose, and urea of individual cow milk samples can be analyzed monthly as part of milk-recording systems. Daily measurements of these components, with results fed into a digital matrix for comparative evaluation, can be made at a reasonable cost; they can provide early detection of changes in cow health, thereby providing vital input and enabling important management decisions. This is possible when milk composition is analyzed online during milking.

The concentration and total production of various milk components can be used diagnostically to monitor individual cow health and nutritional status. For example, the content of urea and protein indicate the balance between protein and energy in the rumen and can be used to optimize the efficiency of nitrogen utilization. Some milk composition parameters could also be involved in the detection of metabolic disorders induced by an unbalanced diet. A high fat-to-protein ratio can be used as an indicator for cows with a negative energy balance and ketosis, displaced abomasum, ovarian cyst, lameness, mastitis, and body condition loss (Friggens et al., 2007). Milkfat depression may imply subacute ruminal acidosis (Plaizier et al., 2008). Changes in the milk composition (e.g., somatic cells, lactose, minerals, and enzymes) can also be attributed to a cow's response to mastitis, decreased secretory activity, and alteration of the blood-milk barrier (Mulligan et al., 2006).

Aernouts et al. (2011) measured milk composition using visible and near-infrared (Vis/NIR) spectroscopy utilizing both transmittance and reflectance techniques. They showed that Vis/NIR reflectance measurements produced accurate fat and protein prediction with a simple configuration highly suitable for online measurements, but it gave poor results for lactose prediction. Accurate measurements of lactose content in milk can be achieved by Vis/NIR transmittance spectroscopy through only 1 mm of milk. This would allow monitoring concentrations of fat, protein, and lactose with sufficient accuracy. However, the narrow light path of Vis/NIR spectroscopy could cause excessive vacuum fluctuations. Future developments should provide milk producers this type of important tool.

19.6.5 Fly Control

Flies and other parasites reduce milk production. Most common among flies that inhabit dairy farms are houseflies. Moisture is the most important factor in fly reproduction. Eggs and small larvae do not develop in dry or in very wet conditions. Thus, basic fly control consists of the regular removal of decaying organic matter (manure, feed, hay, straw) on a weekly basis, as houseflies have a 10-day life cycle. Various insecticides and larvicides may be used in fly control. However, since recommendations and regulations change, we will not include specific generic or common trade names. Instead, readers are referred to their entomology specialists who may be contacted at land grant agricultural experiment stations and colleges of agriculture.

Chemicals that can be absorbed through the skin and later secreted into milk should not be applied directly to dairy animals. In fact, apply only those substances for which the label specifically states that they can be used on lactating dairy cows. Above all, *read and heed the label*!

19.6.6 Trimming Hoofs

The udder is the most sensitive and critical part of a cow—followed by her feet. Thus, it is important that hoofs of cows be kept trimmed and in good condition. Long hoofs that give the appearance of sled runners cause an unbalanced foot. Such a condition causes the cow to shift additional weight onto the heel of the foot. This takes weight from the front of the foot and allows even more growth in front. Both length and depth of hoof affect foot balance. The shift of weight onto the heel puts pressure on the tender part of the foot and stretches tendons and leg muscles, resulting in a weakened leg. This is why problems with feet and legs often go together.

By trimming the feet properly and restoring foot balance, many foot and leg problems can be corrected. Therefore, a hoof-trimming program is an important aspect of dairy herd management. Cows with sore feet lack aggressiveness in making their way over enough pasture or across the lot to feed bunks frequently enough to sustain high milk production.

We reserve techniques and details of hoof trimming, as well as those of clipping, dehorning, tattooing, fitting for sales and shows, and other related management practices, for advanced dairy production and management courses and textbooks.

19.6.7 Keeping Updated

Successful dairy enterprise managers keep abreast of changes in the industry, especially those of economic significance. Such managers are knowledgeable about current and projected prices of crops and dairy products. They constantly monitor supply-and-demand projections. Reliable sources of information include the US Department of Agriculture, colleges of agriculture at land grant universities, extension specialists, breed journals, and numerous farm publications. The latter includes *Hoard's Dairyman*, *Farm Journal*, and *Dairy Herd Management*. Short courses related to breeding, feeding, and managing the dairy farm enterprise are offered periodically in most states. Participation in these is a worthwhile means of keeping current. Breed and AI organizations, product manufacturers, supply firms, colleges and universities, government entities, and others provide websites from which managers can gain valuable information. We have cited many of them in appropriate chapters of this book.

Each manager must know and keep abreast of changes in regulations affecting production of safe milk (cf. chapter 24). For grade "A" milk the applicable rules are published in the *Grade "A" Pasteurized Milk Ordinance* in the section entitled Standards for Grade "A" Raw Milk for Pasteurization, Ultra-Pasteurization, Aseptic Processing and Packaging or Retort Processed after Packaging.

One aspect of dairy farm management that is destined to demand an increasing amount of time and attention is waste management. Complete recycling of wastes, both solid and liquid components, may become an integral part of dairy cattle housing systems. Solid storage is less costly but presents significant odor and fly problems. Liquid storage in underground tanks requires large investments in construction, equipment, and labor for distribution. Anaerobic lagoons together with storage ponds or sprinkler systems for lagoon effluent are gaining popularity (cf. chapter 12).

19.6.8 Using Available Services and Information

Progressive dairy managers take advantage of services that complement their functions, including computerized least-cost feed formulation (cf. chapter 8) and feed analyses. In addition, managers who strive to improve their dairy farm enterprise will invest time in visiting operations recognized for high milk production and efficiency. Several artificial insemination organizations provide counsel regarding breeding programs (cf. chapter 6). Producer cooperatives and various suppliers provide field representatives who assist with management.

The eXtension Foundation has created DAIReXNET (http://www.extension.org/dairy_cattle), which is a national extension resource that provides science-based, peer-reviewed information about the dairy industry for professors, students, researchers, and the general public. DAIReXNET provides an interactive learning environment that includes social media, listservs, and webinars.

The sanitarian/inspector that represents the health authority in a designated area has the responsibility of assisting milk producers in following practices that result in safe milk. This person should be considered a valuable asset even though debits on the inspection sheet may suggest that the inspector is not a friend. As the producer regularly engages the sanitarian in discussions about milk quality and safety, both become better able to perform professionally.

19.6.9 Partnerships

A dairy farm partnership is a joint operating agreement in which members share in business management, investments, labor input, and profits. Profits are normally shared in proportion to contributions of land, labor, capital, and management that each member brings to the dairy enterprise. Written agreements and records are essential components of long-term, successful partnerships. As a minimum, an agreement should include (1) purpose and duration of the agreement, (2) record of inputs, (3) how the business is to be managed, (4) work responsibilities, (5) handling of receipts and expenses, (6) method of financial settlement, (7) procedures and methods of terminating the agreement, and (8) a means of arbitration.

Corporations and other types of business organizations are also available to owners who wish to develop plans to assure that their dairy farm enterprise will be maintained beyond their active participation and/or lifetime. In all such business matters, legal counsel should be obtained from an experienced attorney.

19.6.10 Photographing Dairy Animals

Photographs are often useful in selling animals for dairy purposes. Certain details must be cared for in getting high-quality pictures. For example, in posing heifers and bulls, the rear leg nearest the camera is set slightly farther back than the rear leg on the other side. Cows are posed just the opposite. The leg nearest the camera is set slightly ahead so both the fore and rear udder can be seen. Be certain the front teat is not covered by the rear leg (cf. chapter 3).

Front feet of cows and heifers should be elevated about 6 in higher than rear ones. Calves need a little less elevation. The front leg on the side opposite the camera is set slightly behind the leg nearest the camera. Trees, shrubs, or an open field make a good background for animal photographs. The animal should be photographed from an angle slightly to the rear with the camera held at a height just lower than the animal's topline. Naturally, the animal should be clean, clipped, and well groomed. Additionally, it is important to have the animal appear alert. Usually having a person move toward the front of the animal and wave a handkerchief will make the animal move its ears forward and present an alert appearance.

19.7 Using Simulation in Management

Treat your cow as you would want to be treated if you were a cow.

William Dempster Hoard (1836–1918)

A computer simulation is a beneficial tool that can guide the dairy farm manager's decision-making process by projecting the results of various managerial decisions. In order to illustrate the benefits of such a tool in the classroom setting, students can be divided into groups of five or six members who comprise a cooperative. The goal of each cooperative is to study the effects of various management inputs on milk production and enterprise profit. The first exercise will simulate dairy cattle breeding. Assume each cooperative is allocated $700,000 and a list of pedigrees and related milk production data of available cows and bulls. In chapter 6, we discussed the use of total performance index and estimated average transmitting ability to evaluate cows and predicted transmitting ability to evaluate bulls. With this knowledge we now select 100 females (at an average cost of $6,000 to $7,000 each) and the bulls to which we want to mate our new dairy herd. To make this phase more meaningful, students may actually analyze and debate pedigrees, study advertisements, and even hold a mock auction.

In the second exercise, each cooperative has $1,000,000 to invest in land. They must consider productivity, average rainfall, taxes, interest, nearness to a good milk market, and other factors related to the purchase and management of land for purposes of dairying.

The third exercise is devoted to feeding. Since feed represents approximately half the cost of producing milk, students should use a simulation program to determine which feed ingredients would provide a balanced ration at the least cost.

In the fourth exercise, various housing and milking arrangements and their effects on milk production, labor efficiency, and cost of milk production will be studied. Using simulation, decisions, situations, data, and models provide students the opportunity to explore possibilities in dairy cattle management.

19.8 Summary

Milk production per cow is a measure of the cow's genetic ability to secrete milk and of the manager's ability to provide the feed, care, and management needed for the inherent milk-producing ability of the cow to be expressed. The best of breeding and feeding shines dimly if management fails to put it all together.

The profitable dairy farm enterprise is managed on a business basis. Successful managers enjoy dairying and good cows. Indeed, they are quick to identify profitable cows and to determine which ones are prospects for the meat packer and not the milking parlor. There is no hormone, vitamin, antibiotic, or secret compound that will transform an average-producing dairy herd into a high-producing and profitable one. Instead, rewarding results come from the manager's attention to details and the use of good management tools. The latter include (1) maximal use of production testing and DHI reports, (2) working with veterinarians to build sound herd-health programs, (3) mating cows and heifers to bulls having high predicted transmitting ability, and (4) testing the nutritional value of forages and giving appropriate attention to feeding practices.

Management is especially important in developing completely mechanized systems that result in efficient and profitable milk production. Managing a profitable dairy enterprise includes (1) a breeding program that increases the genetic value of the herd, assures a high first-service conception rate, and produces 12-month calving intervals; (2) a herd size that exploits the fixed resources of land, labor, capital, and the managerial ability of the operator; (3) a feeding program that meets the daily nutritional needs of animals with respect to energy, protein, minerals, and vitamins to challenge the inherent productive potential of the herd fully; (4) the use of facilities that assure both low capital investment and efficiency of production; and (5) the use of a complete record-keeping system.

Through simulation students can face a constructed situation resembling actual problems, and they may be asked to clarify or solve a given problem, assemble data that bear on it, map out alternative solution strategies, and estimate the risks and probable outcomes of each. This technique, together with the more structured game playing, is frequently employed with the aid of a computer (Campbell, 1972).

When attention is given to these and other aspects of managing the dairy farm enterprise as discussed in this and preceding chapters, milk producers will more likely enjoy success in providing milk for humans and financial rewards for themselves.

Study Questions

1. Why is management the most important factor in the production of milk?
2. Identify traits common to successful managers of dairy farm enterprises.
3. Does an increase in herd size usually result in higher milk production per cow?
4. Identify three important factors affecting the profitability of a dairy farm enterprise.
5. What is the approximate distribution of the major costs of milk production?
6. List 10 or more factors affecting costs and quantity of milk produced per cow.
7. Identify factors affecting the amount of individual labor devoted annually to the average dairy cow.
8. Of the total average annual labor per cow, approximately what proportion is devoted to milking itself?
9. Are incentives for dairy farm employees considered worthwhile? Cite examples.
10. Identify potential economies and efficiencies of size that are often realized with large dairy farm operations.
11. Why is credit an important management tool in the dairy enterprise?
12. If a dairy farmer has a 120-cow herd and replaces one-third of them annually, what is average annual per cow depreciation if replacement heifers cost $1,600 and the salvage value of cows is $800?
13. Identify major factors affecting culling rate of profitable dairy herds.
14. Why should dairy managers consider seasonal variations of milk prices when establishing optimal calving dates?
15. Cite several practices that contribute to labor and management efficiency on dairy farms.
16. What time of year is optimal for calves to be born on farms practicing pasture-based dairying? Is this decision based on expected variations in milk prices?

17. Identify factors influencing the decision to operate the dairy farm enterprise on a drylot basis.
18. Why should dairy managers be concerned with flies and other insects? Where should managers go for information on insecticides approved and recommended for use with dairy animals?
19. Why is hoof trimming an important aspect of dairy herd management?
20. Why is it important that milk producers keep updated with information pertinent to the entire dairy farm enterprise?
21. What is a partnership? Review items that should be included in a written agreement.
22. In photographing dairy animals, would one position cows and heifers alike? Why?

Note

[1] Missouri studies found calf losses to be nearly twice as great on dairy farms when outside labor cared for calves as when the owner or family members raised the calves.

References

Aernouts, B., E. Polshin, J. Lammertyn, and W. Saeus. 2011. Visible and near-infrared spectroscopic analysis of raw milk for cow health monitoring: Reflectance or transmittance? *Journal of Dairy Science* 94:5315–5329.

Benson, G. A. 2011. "Business Management." In J. W. Fuquay, P. F. Fox, and P. L. H. McSweeney (eds.), *Encyclopedia of Dairy Sciences*, Vol. 1 (2nd ed., pp. 481–491). San Diego, CA: Academic Press.

Campbell, J. R. 1972. *In Touch with Students . . . A Philosophy for Teachers*. Columbia, MO: Educational Affairs Publishers.

Erven, B. L. 2011. "Labor Management on Dairy Farms." In J. W. Fuquay, P. F. Fox, and P. L. H. McSweeney (eds.), *Encyclopedia of Dairy Sciences*, Vol. 3 (2nd ed., pp. 9–14). San Diego, CA: Academic Press.

Friggens, N. C., C. Ridder, and P. Løvendahl. 2007. On the use of milk composition measures to predict the energy balance of dairy cows. *Journal of Dairy Science* 90:5453–5467.

Mulligan, F. J., L. O'Grady, D. A. Rice, and M. L. Doherty. 2006. A herd health approach to dairy cow nutrition and production diseases of the transition cow. *Animal Reproduction Science* 96:331–353.

Plaizier, J. C., D. O. Krause, G. N. Gozho, and B. W. McBride. 2008. Subacute ruminal acidosis in dairy cows: The physiological causes, incidence and consequences. *The Veterinary Journal* 176:21–31.

USDA Economic Research Service. 2015, May 1. Milk Cost-of-Production Estimates—2010 Agricultural Resource Management Survey Base Annual (http://www.ers.usda.gov/data-products/milk-cost-of-production-estimates.aspx).

Websites

Agricultural Resource Management Survey (http://www.ers.usda.gov/data-products/milk-cost-of-production-estimates.aspx)

DAIReXNET (http://www.extension.org/dairy_cattle)

University of Wisconsin, Department of Dairy Science, Dairy Management Tools (http://dairymgt.uwex.edu/tools.php)

20

Marketing Milk and Milk Products

The production of milk constitutes the most economic method of converting feeds into animal products, and the maintenance of our milk supply is the most important object of dairy farmers.

N. C. Wright

20.1 Introduction

From the time the early settlers arrived in the New World until post-Civil War days (1860s) in the United States, few family farms had the labor or the technical ability to produce more milk than they could consume. There was little need to think about a market for milk and milk products. With industrialization, however, came the opportunity for vocational specialization, and farmers began to follow the trend. As people migrated from farms to cities, dairy farmers found markets for their milk among their city cousins. Due to a lack of refrigeration and the high perishability of milk, suppliers were restricted on how far their products could be transported safely. Therefore, each town and city had its own milk suppliers that made frequent, usually daily, deliveries.[1]

With the development of mechanical refrigeration, pasteurization, **aseptic packaging**, and an excellent transportation system throughout the United States, the face of the dairy industry began to change. Marketing areas became larger, as did processing plants and areas from which they were supplied with raw milk. Improved equipment, extensive rural electrification, and better sanitation increased the keeping quality of milk. Thus, processors were able to secure greater economies of production by increasing the size of their operations.

These trends continue today. However, important limitations are imposed on the growth and development of the industry by the market situation. Of all the factors related to marketing of milk and milk products, supply and demand are most important. In fact, decisions and practices in milk marketing are based on current and expected supply and current and expected demand.

Producers and processors want to supply the quantity of milk that will satisfy demand in a way that they can maximize profit. Too great a supply depresses prices. If, however, demand pushes prices too high, consumers may seek substitute products and total revenue to the industry will be less than its potential.

Supply management has been an important issue among milk producers and governmental agencies. Capitalization for milk production and processing is costly and time for newborn calves to reach milk-producing maturity is more than two years. Furthermore, demand for the many dairy products varies with season. Finally, milk producers market milk independently, even though the majority participate in cooperative operations. Therefore, it is difficult for the industry to keep supply in line with demand so as to maximize profit. In this chapter, then, let us consider supply and demand and other factors important to the marketing of dairy products.

20.2 Market Prices

The total supply of milk brought to market by producers depends primarily on prices paid milk producers, expected future prices, present and expected costs of milk production, availability of financing, and alternative farm and off-farm opportunities. Volatility of prices, primarily of feed and milk, are primary factors affecting milk supply. An example of this volatility is found in the prices of corn/soybean meal mix, which averaged $4.56 in 2005 but rose to $9.97 in 2008. During the same time period, replacement cow values rose from $1,800 to $1,953. Nationwide uniform milk prices were $15.20 and $18.45, respectively. This meant that on a nationwide basis the ratio of the value of milk was 3.33 times the cost of concentrate in 2005 but only 1.85 times the cost of concentrate in 2008. At the same time the cost of replacement cows rose about 8%. Obviously, milk producers had incentive to increase production in 2005, but by the time calves born in 2005 came into production in 2008 the prospect of earning a profit had disappeared. Under usual conditions of supply and demand a cycle of high to low prices occurs about every three years.

Further evidence of price volatility is seen in prices averaged over all federal milk markets from 2006 through 2014 (figure 20.1). During 2013 average prices per hun-

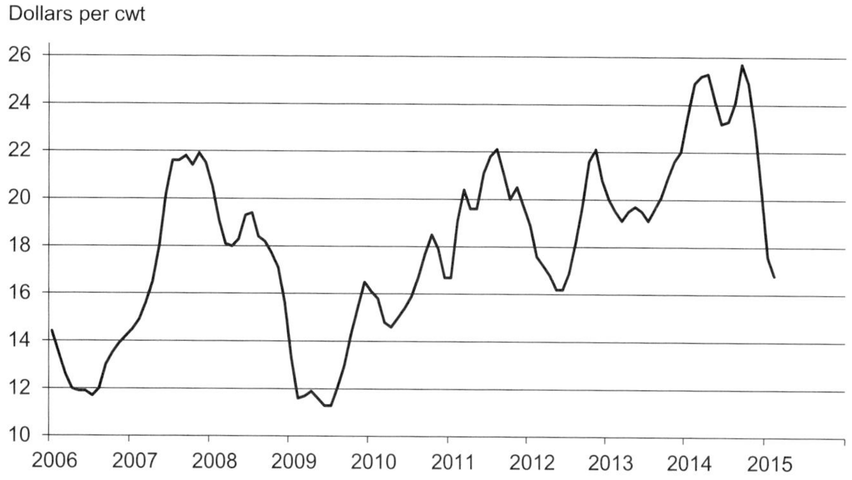

Figure 20.1 Price volatility from 2006 to 2014 (USDA National Agricultural Statistics Service, 2015).

dredweight (cwt) ranged from a high of $22.00 in December to a low of $19.10 in July. In addition, prices differ by location: for example, in January 2015 the price in Florida averaged $22.60 whereas the average in New Mexico was $16.40.

Finally, extreme volatility is reflected in an analysis of global prices as reported by the Food and Agriculture Organization of the United Nations (FAO). Average weighted world export prices of cheese, butter, casein, nonfat dry milk, and dry whole milk in 2002 to 2004 = 100. Since 2006 the index doubled, then halved, then doubled again. After falling to a cyclical low of 114 in February 2009, the index climbed continuously during the next 10 months to a peak of 216 then fell to 187 in the next three months.

20.2.1 US Dairy Exports

The world markets for dairy products are affected by a range of volatile drivers, including conditions for crop growth, income in the developing world, relative values of currencies, dietary shifts, and changes in regulations. A stronger US currency makes US exports more expensive on the international market. Changes in export prices and volume have a dramatic influence on domestic prices as well. Increased output from the world's top dairy suppliers (United States, European Union, Australia, Argentina, and New Zealand) can result in a volume of milk that is greater than importers can consume. World prices can plunge as a result, and competitors have to be willing to sell their milk at lower prices, or find other markets for their products. In the United States, prices have the ability to remain high because of strong domestic demand.

In recent years the US dairy industry has been increasingly competitive on world markets. The estimated total value of US dairy exports in 2014 was $7.1 billion while dairy imports were valued at $3.52 billion. During 2014, 3.963 billion pounds of milk solids were exported. On the basis of total milk solids produced, the United States exported 15.4% while importing 2.9 to 3.0%. The US Dairy Export Council (USDEC) leads overseas market development on behalf of the US dairy industry.

Whereas the United States is a major exporter of dry whey, whey protein concentrate, and lactose, it is a major importer of casein, caseinates, and milk protein concentrates. This reflects the high use of milk in the production of cheeses from which much whey is co-produced. A significant demand exists internationally for these whey products.

20.2.2 Reporting and Tracking Market Prices

The best way to track the supply and demand picture for the dairy industry is to go to the websites of the following three governmental agencies:

- Agricultural Marketing Service (AMS) (http://www.ams.usda.gov)
- Economic Research Service (ERS) (http://www.ers.usda.gov)
- National Agricultural Statistics Service (NASS) (http://www.nass.usda.gov)

As part of the US Department of Agriculture, all three agencies are active in various aspects of dairy marketing. AMS publishes *Dairy Market News Weekly Report*, which reliably indicates the supply and utilization of milk in the United States, milk cow numbers, governmental purchases and price support activities, imports and exports, and stocks of dairy products in storage.

ERS provides research and analysis on milk production, the marketing and consumption of dairy products, international dairy trade, and government programs. ERS has the lead role in USDA's short-run and long-run dairy market outlook activities. Monthly and quarterly dairy industry data and the dairy market outlook are provided in the monthly *Livestock, Dairy, and Poultry Outlook*.

NASS reports statistical information on agricultural products. The publication entitled *Milk Production, Disposition, and Income Annual Summary* contains the annual number of milk cows, production per cow, and production for the year by state and nation. Other reports include amount of milk used on farms and amount sold, cash receipts, dairy products produced, and selected prices.

Estimates of milk supply, use, and prices are also reported in the monthly *World Agricultural Supply and Demand Estimates* (WASDE) report, issued by USDA's World Agricultural Outlook Board. The WASDE estimates are developed by a committee of dairy analysts from the following USDA services and agencies: ERS, AMS, the Farm Service Agency, and Foreign Agricultural Service (FAS). The USDA's Agricultural Baseline Projections provide long-run dairy industry projections.

20.3 The Demand for Milk and Milk Products

Demand for milk and milk products depends largely on their prices, availability, and price of substitute products, consumer income, population numbers, and changes in consumer tastes and preferences.

About 38% of US milk production is used as Class I for **fluid milk** products. About 41% is used in making cheeses. Soft manufactured products, including ice cream, yogurt, and cream products, require about 12%. Finally, about 9% goes to making butter and dry milk products.

In 1909, Americans consumed a total of 34 gal of fluid milk per person—27 gal of whole milk and 7 gal of milks lower in fat than whole milk, mostly buttermilk. Back then buttermilk was the by-product of churning cream into butter, which was often done on farms. Today, the major by-products of the commercial butter-making process are nonfat dry milk and dry buttermilk, whereas cultured buttermilk is soured by addition of bacteria that produce lactic acid and flavorful substances in sweet milk. In 1909, more than half of the milk was consumed on the farms where it was produced compared with 10% in 1960 and 0.3% in recent years.

During World War II, fluid milk consumption shot up from 34 gal/person in 1941 to a peak of 45 gal/person in 1945. War production lifted Americans' incomes but curbed civilian production and the goods consumers could buy. Many food items were rationed, including meats, butter, and sugar. Milk was not rationed and consumption soared. After 1945, however, milk consumption fell steadily, reaching a low of just under 23 gal/person in the first decade of the twenty-first century, consisting of 36.5% whole milk, 47.5% lower fat milks, and 16% nonfat milk. These changes are consistent with increased public concern about cholesterol, saturated fat, and calories. However, the decline in per capita consumption of fluid milk also may be attributed to competition from other beverages, especially carbonated soft drinks and bottled water, a smaller percentage of children and adolescents in the United States, and a more ethnically diverse population whose diet does not normally include milk.

Americans are consuming about 33 lb of cheese per person, which is eight times more than they did in 1909 and more than twice as much as they did in 1975. Although some 300 varieties of cheese are available in the United States, American and Italian types make up about 82% of sales and are approximately equal in amounts consumed. Demand for time-saving convenience foods is a major force behind this growth in cheese consumption. More than half of the cheese now comes in commercially manufactured and prepared foods (including for food service), such as fast-food sandwiches and packaged snack foods. New products, such as resealable bags of sliced or shredded cheeses, have also raised consumption.

US per capita consumption of ice cream and other frozen desserts ranges between 26 and 27 qt, with ice cream representing 87% of the total volume. Regular ice cream, including economy, regular, premium, and super-premium types, represents 64% of that market, while lowfat and nonfat products make up some 25%. Over the past two decades Americans have been eating a little less ice cream overall but more of the higher priced, higher milkfat, premium and super-premium ice creams as well as frozen yogurt and other frozen dairy products.

20.4 Promoting the Benefits of Milk and Educating the Public

In the United States, milk producer and processor promotion programs have been established under the Dairy Production and Stabilization Act of 1983 (Dairy Act) and the Fluid Milk Promotion Act of 1990 (Fluid Milk Act). These two acts grant USDA the authority to oversee these programs, to ensure that program funds are properly accounted for, and that they are administered in accordance with the respective act. As required by law, the programs must reimburse the secretary of agriculture for all of USDA's costs of program oversight and for an independent analysis.

20.4.1 The Dairy Board

Established in 1984, the National Dairy Promotion and Research Board (Dairy Board) consists of 38 board members who are nominated by industry representatives or organizations and appointed by the secretary of agriculture. Its mission is to coordinate a promotion and research program that maintains and expands domestic and foreign markets for fluid milk and dairy products produced in the United States.

Qualified state or regional dairy product promotion, research, or nutrition education programs are certified annually by the secretary of agriculture. To receive certification the qualified program must (1) conduct activities that are intended to increase human consumption of milk and dairy products generally; (2) have been active and ongoing before passage of the Dairy Act; (3) be primarily financed by producers, either individually or through cooperative associations; (4) not use a private brand or trade name in its advertising and promotion of dairy products, and (5) not use program funds for the purpose of influencing governmental policy or action.

Expenditures by the Dairy Board and many of the qualified programs are integrated through a joint process of planning and program implementation so that the programs on the national, regional, state, and local level work together. The Dairy Board develops and implements programs to expand human consumption of dairy products by focusing on innovation, product positioning with consumers, and new places for dairy product consumption. In combination with the National Dairy Council, the Dairy Board instituted the 3-A-Day® Dairy program, a nutrition-based marketing and education program, as well as its New Look of School Milk campaign, which includes efforts to improve the school milk experience for the nation's children through improvements in packaging, flavors, and availability. In 2008, the Dairy Board paired up with the NFL in the Fuel Up to Play 60 program, the nation's largest in-school wellness program. With more than 73,000 schools enrolled in this program, 14 million children have access to better foods (including dairy) and get more physical activity in schools.

20.4.2 The Fluid Milk Board

The Fluid Milk Promotion Act of 1990 established the National Fluid Milk Processor Promotion Board (Fluid Milk Board). It administers a generic fluid milk promotion and consumer education program funded by America's fluid milk processors. The program is designed to educate Americans about the benefits of milk, increase fluid milk consumption, and maintain and expand markets and uses for fluid milk products. It is financed by a mandatory $0.20/cwt assessment on fluid milk processors who process and market commercially more than 3 million lb of fluid milk products per month in the United States. The secretary of agriculture appoints 20 members to the Fluid Milk Board. Fifteen members are fluid milk processors who each represent a separate geographical region and five are at-large members. Assessments generated $103.3 million in 2012.

The Fluid Milk Board also oversees the Milk Processor Education Program (MilkPEP), which supports federal nutrition goals, the US Dietary Guidelines for Americans, and USDA's Food Guide Pyramid. MilkPEP is a national marketing voice for milk in a marketing environment subject to a high degree of federal and state regulation, helping to maintain the strength and stability of the fluid milk industry. The Fluid Milk Board launched The Breakfast Project in 2012 in order to focus on highlighting breakfast at home as the ideal opportunity to drink milk. Through the use of a social media listening and engagement program, the Fluid Milk Board was able to engage with consumers and influencers and share why milk should be a part of breakfast each day.

20.4.3 Dairy Management, Inc.

Created to help increase sales and demand for dairy products, Dairy Management, Inc. (DMI) and its related organizations work to increase sales and the demand for dairy products through dairy research, education and innovation, and the maintenance of confidence in dairy foods, farms, and businesses. The National Dairy Council, Innovation for US Dairy, Dairy Research Institute, and the US Dairy Export Council are all registered trademarks of DMI. It works with state and regional promotion organizations and is a strategic consultant and resource to businesses and organizations in the food and beverage sector that seek to increase sales and raise their image through the innovative use of dairy and dairy ingredients.

DMI is partially funded by the Dairy Board and the United Dairy Industry Association, which serves as a federation of state and regional dairy producer-funded promotion organizations. DMI is also funded by nearly 47,000 dairy farmers in the United States, as well as dairy importers.

DMI's export enhancement program is implemented by the USDEC, which receives primary funding, about $17 million, from three sources: DMI, FAS, and membership dues from dairy cooperatives, processors, exporters, and suppliers.

20.4.4 The National Dairy Checkoff Program

DMI manages the Dairy Research and Promotion Program, also known as the national dairy checkoff program. The Dairy Act specifies that US producers of milk collect and remit an assessment of $0.15/cwt of milk used for commercial purposes. Producers who market solely organic milk are granted "organic exemptions," representing approximately 1 billion lb of milk. The Dairy Act provides that dairy farmers can receive a credit of up to $0.10/cwt of their assessment for contributions to qualified state, regional, or local dairy product promotion, research, or nutrition education programs. In 2012, mandatory assessments totaled $102.5 million. The Dairy Board portion of the assessment was $99.7 million and the importer assessment was $2.8 million. Qualified programs revenue was $197.2 million.

In 2011, USDA extended the national dairy **checkoff** to imported dairy products, with the assessment set at $0.075/cwt of milk or **milk equivalent**. Producer milk in 2006 to 2007 contained 12.45% milk solids (3.68% fat and 8.77% nonfat). Therefore, on a milk solids basis, the assessment rate was set at $0.00602/lb ($0.075 / 12.45). This action was based on provisions in the 2008 US Farm Bill to establish the assessment on imports, which was followed by an announcement by USDA of the proposed rule in May 2009. Following considerations of inputs from both domestic and foreign entities, the rule was implemented August 1, 2011.

20.4.5 Effectiveness of Promotion Programs

An independent econometric analysis of the effectiveness of the dairy and fluid milk promotion programs was conducted in 2007 by Cornell University. It was estimated that the generic fluid milk marketing efforts and activities sponsored by fluid milk processors and dairy farmers helped mitigate the decline of fluid milk consumption. The generic fluid milk marketing activities increased fluid milk consumption by 30 billion lb (13.6 billion kg) from 1998 to 2007, or 3 billion lb (1.4 billion kg) per year on average. Had there not been generic fluid milk marketing conducted by the two national boards, fluid milk consumption was estimated to have been 5.4% lower. The study concluded that these marketing efforts have had a positive and statistically significant impact on per capita fluid milk consumption.

20.5 Dairy Cooperatives

Like the milk processor, the dairy farmer must sell his/her product—milk. But unlike the milk processor, the producer has little chance as an individual to set a sale price. To gain bargaining power, milk producers have formed cooperatives. These fall into four types of categories:

1. *Bargaining only*—this type of cooperative assists members by negotiating prices, facilitating sales, and ensuring accurate weights and test results;
2. *Niche marketing*—typically these cooperatives use their members' milk to make specialty products such as organic dairy foods and artisan cheeses;

3. *Fluid processing*—typically these cooperatives process and package member milk into fluid milks and soft manufactured products such as cottage cheese, yogurt, and ice cream; and
4. *Diversified*—while selling a portion of their raw milk to other processors, cooperatives own and operate plants that produce a variety of products including cheese, butter, and dry dairy foods.

Most dairy cooperatives are organized on a centralized basis—farmers are direct members. A few are organized on a federated basis—members of the cooperative are other cooperatives or a combination of direct members and cooperative members. Many are organized to serve farmers in a local area or single state while others serve members in multiple states, regionally or nationally. Some dairy cooperatives have made additional business arrangements to increase outlets for members' milk through subsidiaries, partnerships, federations, marketing agencies in common with other cooperatives, and joint ventures with other cooperatives or investor-owned firms.

According to a survey conducted by USDA Rural Development every five years, in 2012 the United States had 133 dairy cooperatives. Data from 89 co-ops that handled 80% of the milk marketed through all co-ops revealed the following characteristics on a per hundredweight basis: assets of $10.90, total liabilities of $8.12, and member equity of $2.78. Assets per hundredweight of milk were lowest for bargaining co-ops ($2.31) while they were highest for diversified co-ops ($12.15).

Sales of milk through a cooperative are claimed to gain revenue for the milk producers by returning moneys that would normally go to a middleman. However, risk is involved in the cooperative business venture, as is the case with the private commercial firm. Investments in facilities and equipment are required and management and labor must be paid. Generally, the normal market price is paid the producer for milk and any profits from the operation are returned in the form of either ownership (equity) in the cooperative's assets or as yearly dividends, or both.

20.5.1 The Development of Cooperatives

Dairy farmers started bargaining with fluid milk processors through their cooperatives prior to World War I. **Classified pricing** systems were established in nearly all major markets in the 1920s. They provided that milk used for fluid purposes would be priced higher than that used in **manufactured milk products**. Moreover, these classified pricing systems reflected greater costs of handling and marketing perishable milk in fluid form. During the depression of the 1930s, however, these arrangements became unworkable in most markets. The Agricultural Adjustment Act of 1933 provided milk price stabilization through a licensing system, and through similar acts in 1935 and 1937 the system of federal milk marketing orders was established.

Until the late 1960s relatively small cooperatives functioned in individual **milksheds**. In some areas two or more cooperatives competed for milk from individual producers and for sales to the same **handlers**. Large regional organizations then began to be formed by the merger of individual cooperatives of both the manufacturing and marketing types.

Conversion from delivery of milk in cans to plants to the use on farms of bulk tanks with delivery by bulk milk trucks, which began after World War II, was completed by the 1970s. This required substantial capital investment and additional milk volume for efficient operations. This led to fewer, but larger, farms and processing plants. Dairy cooperatives adapted similarly with mergers and consolidations in the mid-1960s that markedly reduced cooperative numbers. These larger organizations put farmers in a better position to negotiate with large corporate milk processing firms. Managers of highly specialized investor-owned fluid processing plants found it more efficient to contract with cooperatives to provide the amount of milk they needed. Large-scale, multiplant cooperatives negotiated "full-supply" contracts with these fluid processors (and in some cases, manufacturers). Under a full-supply contract, a cooperative provides the milk volume the plants need and manufactures the excess into other products, especially cheese, butter, and nonfat dry milk.

The task of balancing the volume of milk supplied with the volume demanded is complicated because of the nature of milk production—seasonal variations with daily harvest. However, producers participating in a cooperative can be assured that the cooperative will market all of the milk they produce. The larger the volume under one cooperative's control the more the random variation of supply and demand is offset by one another.

Dairy cooperatives came to dominate the functions of supplying fluid milk markets, routing the movement of milk, and balancing supply with demand. In addition, their role in verifying weights and tests assures members of proper payment for their milk. In this way they increased efficiencies in milk marketing. Their guarantee to market all of their members' milk distinctly sets dairy cooperatives apart from proprietary milk handlers.

Another wave of consolidation in the dairy sector among investor-owned dairy firms and grocery retailers occurred as the twentieth century drew to a close. As the capital cost of keeping processing capabilities up-to-date continued to grow, the pace of mergers among dairy cooperatives picked up again. Some of this activity was to satisfy the needs of large, integrated food companies that increasingly looked for (1) milk suppliers with national reach, (2) the ability to provide entire lines of dairy products, or (3) the ability to meet certain product specifications. Other dairy cooperatives merged to address regional needs and to consolidate complementary or duplicate operations. In 1997 the top five dairy cooperatives handled 41% of the net milk handled by all cooperatives. By 2002 three of these had merged into a single entity, and the five largest cooperatives accounted for about 54% of the milk. Two of them were national in both membership and marketing. Legal action taken against a national cooperative resulted in a court decision that it had reduced competition for producer milk within an area to an unacceptable extent and caused it to revise practices.

20.5.2 Cooperatives Working Together

Cooperatives Working Together (CWT) was launched in 2003 by the National Milk Producers Federation (NMPF). It was developed and funded by dairy farmers to provide assistance to its members during economic crises. For example, until 2010 it offered the herd retirement program, where individual milk producer members could bid to sell their cows to slaughter, thus reducing the milk supply during a time of economic stress and instability. Successful bidders received partial payment for the milk they did not produce over a 12-month period, plus the value of animals sold to slaughter. They also agreed not to re-enter the market for at least one year. For example, in October 2009 the program accepted bids from 154 producers for 26,412 cows and 517 million lb (235 million kg) of milk.

Today, the focus of CWT is to assist its members in exporting their dairy products. The product to be exported is identified, along with its quantity, the end customer, the country of the customer, and the amount of assistance per metric ton of product. CWT and NMPF, with guidance from the USDEC, analyze each bid considering the market of the sale, time of delivery, and world and US product prices. If the request is reasonable, the bid is accepted. When requests are deemed unreasonable, the member is advised of the maximum assistance CWT will provide. Since the US federal government is not involved in CWT, responses to changes in market conditions can be made without lengthy legislative or bureaucratic processes.

20.6 Risk Management

Modern dairy farm operations are big business with large initial capital investments. Establishment and maintenance of such an enterprise is a long-term effort. Furthermore, milk cow numbers cannot be varied rapidly to meet changes in supply and demand because of the time lapse of more than two years required between birth (or conception) of an animal and its first lactation. Of course, some cows can be shifted from milk to beef production, and vice versa, but this is not a reasonable alternative for the highly specialized milk producer. Such factors make it important that raw milk prices remain relatively stable so that consumers will be assured of an adequate milk supply at reasonable prices.

The minimum price for raw milk in the United States is established indirectly by support prices set by the federal government. However, government programs may be cancelled, modified, or extended. Processors compete for supplies of **manufacturing milk** through price competition and many forms of nonprice competition. Normally, it is when raw milk supplies for manufacturing approach or exceed the commercial demand that government must provide price support through the purchase of butter, nonfat dry milk, and/or cheese. USDA is typically responsible for the price support program and is legally required to maintain the support price between 75 and 90% of **parity price**. Nevertheless, prices are not uniform within an area where several processors compete. Some transactions may occur at a price below the established support level and some above it.

20.6.1 Volatile Prices and Risk

As demand conditions change relative prices of manufactured products also change, affecting the ability of manufacturers of different products to compete for raw milk supplies. This situation has been particularly noticeable in competition between cheese and butter-powder plants. Although these products were equally profitable in 1960, cheese became increasingly more profitable in response to greater demand. In such a profitable market, cheesemakers frequently were willing to bid higher for milk to increase their output. Butter plants competing for the same milk supplies were under considerable pressure to raise their buying prices to compete with those of cheese plants, but they could not increase their product prices. Thus, butter-powder plants were often in a difficult cost/price squeeze.

Prices of milk and feed have varied widely, especially during the first decade of the twenty-first century (figure 20.2 on the following page); demand for corn to make ethanol had increased markedly, causing an increase in prices. Extremes in price volatility are evidenced by data on percentages of operating costs assigned to feed in 2010, which ranged from 35% in Minnesota to 74% in California and Texas, while the national average was 56%. The national average was 45% in 2005.

20.6.2 Insurance against Risk

In 2008, the USDA Risk Management Agency introduced a risk management tool called Livestock Gross Margin for Dairy Cattle (LGM-Dairy). It provides protection against loss of gross margin (market value of milk minus feed costs) on milk produced from dairy cows. It is available to producers in the 48 contiguous states and only for milk for commercial or private sale intended for final human consumption. There is no minimum number of hundredweights a producer can insure, however, the maximum amount of milk that can be insured is 24 million lb (10.9 million kg) per crop year.

Established by the Farm Bill of 2014, the Margin Protection Program (MPP) provides dairy producers

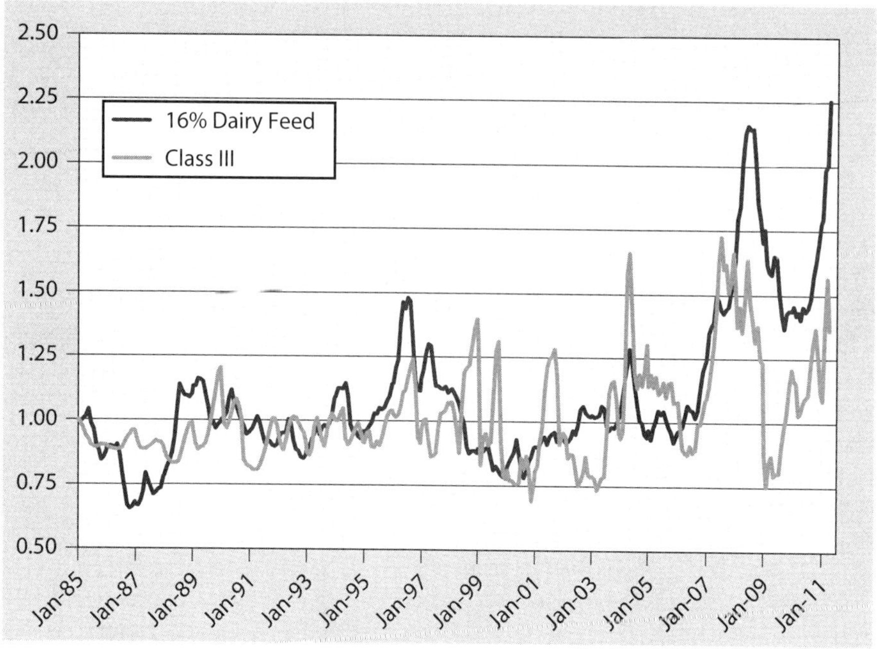

Figure 20.2 Price indexes for 1985–2011 demonstrate the high degree of volatility in prices of milk and dairy cow rations (University of Wisconsin, Dairy Marketing and Risk Management Program, Professor Brian W. Gould, Department of Agricultural and Applied Economics).

with payments when dairy margins are below the coverage levels a producer chooses each year. Its focus is to protect farm equity by guarding against destructively low margins, not to guarantee a profit to individual producers. The program supports producer margins, not milk prices, and is designed to address catastrophic conditions as well as prolonged periods of low margins. Simply defined, it is the all-milk price minus the average feed cost. Average feed cost is determined using a national feed ration that has been developed to reflect costs associated with feeding all dairy animals on a farm on a hundredweight basis.

All dairy operations producing milk commercially are eligible to participate in the MPP. Coverage is based on a producer's production history. Adjustments are to be made based on the national average growth in overall US milk production as estimated by USDA. Growth of an operation beyond the national average increase is not protected by the program. Producers may protect from 25 to 90% of their production history in 5% increments. They may select margin protection coverage from $4 to $8 per hundredweight in 50-cent increments.

The Federal Crop Insurance Act (1980) limits funds available to support livestock insurance plans to $20 million in a fiscal year. This includes both LGM-Dairy and MPP, both of which operate on a calendar year basis. The 2014 Farm Bill allows dairy farmers to participate in either MPP or LGM-Dairy, but they may not simultaneously use both programs.

20.7 Federal Milk Marketing Orders

Federal milk marketing orders (FMMO), administered by USDA AMS, were instituted in 1937 to establish rules under which dairy processors purchase fresh milk from dairy farmers supplying a marketing area. The orders are established and amended through the formal hearing process. Federal milk marketing orders regulate handlers that sell milk or milk products within an order region by requiring them to pay not less than an established minimum price for the **grade "A" milk** they purchase from dairy producers, depending on how the milk is used.

The 1996 Farm Bill (P.L. 104-127) required USDA to consolidate the number of federal milk marketing orders. USDA implemented these changes in 2000. Currently there are 10 milk marketing orders, down from 31 when the law was enacted (figure 20.3). In February 2015, three groups sent a proposal to USDA AMS to establish an order for the state of California. More than 50% of the cows not included in FMMO programs are located in California.

Federal milk marketing orders define the terms under which handlers, who engage in distribution of fluid milk in specified areas, purchase milk from producers. These orders are legal instruments by which a uniform system of classified pricing is established throughout the market. Producers are not controlled nor guaranteed a market. Although the orders apply almost completely to grade "A" milk, they do not impose sanitary restrictions on production or manufacturing practices. Other aspects not affected by federal orders are retail prices or maximum prices paid producers.

20.7.1 Establishing Federal Milk Marketing Orders

Development of an FMMO usually starts with a request by a cooperative association of milk producers that petitions the US secretary of agriculture. If the peti-

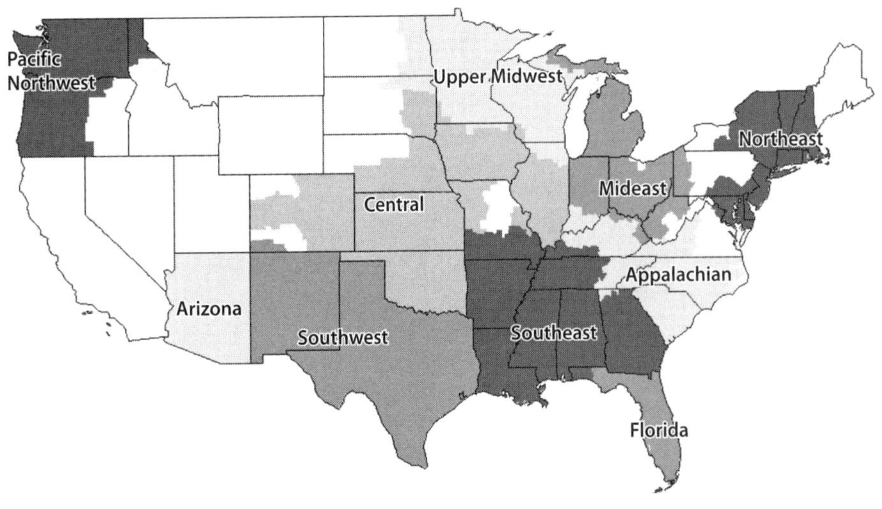

Figure 20.3 Areas covered by the 10 federal milk marketing orders.

tion has merit and the secretary decides to proceed, a public hearing is called at which all interested parties (producers, handlers, and the public) are given the opportunity to present related facts and opinions. If merited, a recommended order is developed and issued with time allowed for examination and filing exceptions. Exceptions are considered and they may be reflected in the final order published in the Federal Register. However, the order must be approved by at least two-thirds of the producers or producers supplying two-thirds of the milk for the area. A milk producer's cooperative may vote, as a block, for all of its members who deliver milk during a specified period. An order may be terminated at the request of more than 50% of the producers supplying more than 50% of the milk for the market, or by the US secretary of agriculture.

Federal orders are administered by a **federal milk market administrator** appointed by the secretary of agriculture. The administrator is responsible for (1) calculating and announcing minimum milk prices, (2) collecting reports from handlers showing quantities of milk received and used, and (3) receiving reports showing payments for raw milk by each handler. The administrator verifies these handler reports. Handlers are assessed for expenses incurred in the operation of the market administrator's office to a maximum limit established in the respective order.

20.7.2 Handlers

These are commonly individuals, partnerships, corporations, or cooperative associations that purchase milk for sale in fluid products. Because of the large interchange of milk among markets, handlers may be fully regulated, partially regulated, or exempt, depending upon the amount of their receipts disposed of as **Class I milk** in a given area. Exempt handlers are usually producer-distributors who use a low volume of milk.

20.7.3 Product Classification

Establishment of minimum prices on the basis of use of grade "A" milk is known as classified pricing. Milk, cream, and skim milk used in fluid milk products are Class I and receive the highest price. Milk used for soft manufactured products (e.g., cottage cheese, yogurt, and ice cream) is Class II. Class III milk is used for making cheese. Milk used for butter and concentrated and dry milk products falls into Class IV (table 20.1).

Uniform prices reported for 2013 ranged from $18.83 in the Pacific Northwest Order where Class I utili-

Table 20.1 Definitions of Classes of Milk Sold in Federal Milk Marketing Orders in the United States.

Classification	Uses
I	Milk, skim milk, lowfat milk, milk drinks, cultured buttermilk, filled milk, and milk shake and ice milk mixes containing less than 20% total solids. Products in this category contain not more than 9% milkfat and at least 6.5% nonfat milk solids or 2.25% true milk protein. For example, if a product contains 6% nonfat solids, 2.3% protein, and 9% milkfat, it is to be considered a fluid milk product.
II	Cottage cheese (all types), fluid cream products (containing 9% or more milkfat), yogurt, eggnog, frozen desserts, frozen dessert mixes containing 20% or more total solids, and bulk fluid milk or cream products disposed of to any commercial food-processing establishment.
III	Cheese (other than cottage).
IV	Butter, any milk product in dry form, evaporated or condensed milk in consumer-type package, and other products not classified in Classes I, II, and III.

zation was 26%, to $23.53 in the Florida Order where Class I utilization was 86%. The Upper Midwest Order had the lowest Class I utilization of 11% but other factors caused the uniform price to be $19.44. During 2013 percentages of milk marketed through federal orders were as follows: Class I = 32%; Class II = 12%; Class III = 47%; and Class IV = 9%.

California is a principal milk-producing area that does not fall under the jurisdiction of the FMMO. The California Department of Food and Agriculture (CDFA) has maintained California's own state milk marketing order with the passage of the Young Act in 1935. Pricing in California is based on a five-class system: Class 1 includes fluid milk products; Class 2 includes sour cream, yogurt, cottage cheese, etc.; Class 3 includes ice cream and frozen desserts; Class 4A includes butter and dry milk products; and Class 4B includes most cheeses.

20.8 Prices and Formulas for Class Products

The FMMO uses component values to calculate monthly class prices. The value of each component is based on its relationship to dairy product yield and its value. Component values are associated with each milk class. The value of a component is determined by the following equation:

milk component value =
(component relationship to product yield) ×
(product price – manufacturing cost)

The National Dairy Products Sales Report (Sales Report) provides current wholesale market prices for basic dairy commodities. It is based on survey data collected from reporting entities via the electronic Dairy Product Mandatory Reporting (DPMR) system. The DPMR program consists of 96 reporting entities selling one million lb or more of dairy products as defined by statute (7 USC 1637b). There are 16 plants reporting on 40-lb cheddar blocks, 11 plants reporting 500-lb cheddar barrels, 19 plants reporting butter, 31 plants reporting nonfat dry milk, and 19 plants reporting dry whey. Milk prices are given per 100 lb or cwt, rounded to the nearest cent; component prices are per lb, rounded to nearest one-hundredth cent; and cheese, dry whey, butter, and nonfat dry milk prices are weighted monthly averages of weekly Sales Report prices, rounded to the nearest one-hundredth cent.

The Mandatory Price Reporting Act (2010) requires USDA AMS to release dairy product prices and sales information on or before Wednesday at 3:00 PM (EST). This information is used to determine advance and announced prices that are used in class price formulas. Advance prices are released by the twenty-third of the month preceding their use. They are composed of two-week Sales Report average prices for butter, nonfat dry milk, cheese, and dry whey. Class I and Class II skim milk prices depend on advance prices before the pricing month begins. Therefore, the Class I milk price for June will depend on dairy product prices given in mid-May. Announced class prices are released by the fifth of the month using monthly average Sales Report prices. Component, Class II butterfat and 3.5% butterfat, Class III, and Class IV prices all use announced prices. Therefore, Class II, III, and IV prices will be based on the preceding month's prices.

20.8.1 Formulas for Calculating Class Prices

These formulas are subject to revision and may be affected by changes in the US Farm Bill, which commonly is rewritten on a five-year basis (table 20.2).

- **Class I Milk Price**: Class I milk is assumed to be composed of 3.5% fat and 96.5% skim milk. The Class I price (based on "higher of" advanced Class III or Class IV skim milk price) plus the Class I differential for each order determines Class I skim milk price. The mover equals (base skim milk price for Class I × 0.965) + (advanced butterfat pricing factor × 3.5). The Class I mover is announced on or before the advance prices. A detailed county-level Class I differential map is provided by the USDA AMS.
- **Class II Milk Price**: Most Class II products can be manufactured from Class IV products, so there is a relationship between the two prices. The Class II skim milk price equals the advanced Class IV skim milk price factor, plus $0.70. The Class II butterfat value is the announced Class IV/III butterfat price plus a differential of $0.007. Class II prices are tied to butter-powder markets (Class IV), not cheese markets (Class III). Class II skim milk uses advanced pricing formulas and is determined by the twenty-third of the prior month.
- **Class III Milk Price**: Class III milk is used primarily for cheese. The value of butterfat is based on the price of butter, protein on the price of cheese, and other solids on the price of dry whey. Weekly averages of Sales Report data are used as the basis for these calculations.
- **Class IV Milk Price**: The Class IV component values (butterfat and solids nonfat) are based on two dairy products (butter and nonfat dry milk). Class IV prices not only price butter and powder, but also affect the Class I mover. Weekly averages of Sales Report data are used as the basis for these calculations. The Class IV price is an announced price that is released by the fifth of the month.

As an example, figure 20.4 (on p. 284) illustrates the offsetting impact of high butter prices on milk protein prices. As the value of milkfat goes up with the butter price, the value of protein commonly goes down. Typically milkfat has a higher value when used in cheese

Table 20.2 Class Pricing Formulas.

Class I[a]	
Class I price	(Class I skim milk price × 0.965) + (Class I butterfat price × 3.5)
Class I skim milk price	Higher of advanced Class III or IV skim milk pricing factors + applicable Class I differential
Class I butterfat price	Advanced butterfat pricing factor + (applicable Class I differential ÷ 100)
Class II	
Class II price	(Class II skim milk price × 0.965) + (Class II butterfat price × 3.5)
Class II skim milk price	Advanced Class IV skim milk pricing factor + $0.70
Class II butterfat price	Butterfat price + $0.007
Class II nonfat solids price	Class II skim milk price ÷ 9
Class III	
Class III price	(Class III skim milk price × 0.965) + (Butterfat price × 3.5)
Class III skim milk price	(Protein price × 3.1) + (Other solids price × 5.9)
Protein price[b]	[(Cheese price – 0.2003) × 1.383] + {[((Cheese price – 0.2003) × 1.572) – Butterfat price × 0.9] × 1.17}
Other solids price	(Dry whey price – 0.1991) × 1.03
Butterfat price	(Butter price – 0.1715) × 1.211
Class IV	
Class IV price	(Class IV skim milk price × 0.965) + (Butterfat price × 3.5)
Class IV skim milk price	Nonfat solids price × 9
Nonfat solids price	(Nonfat dry milk price – 0.1678) × 0.99
Butterfat price	See Class III
Somatic cell adjustment rate	Cheese price × 0.0005, rounded to fifth decimal place. Rate is per 1,000 somatic cell count difference from 350,000.

[a] Advanced pricing factors are computed using applicable price formulas listed in the table, except that product price averages are for two weeks.
[b] [(cheese price – make allowance) × protein's yield factor] + {[((cheese price – make allowance) × fat's yield factor) – butterfat price × 0.9] × fat to protein ratio)}
Source: USDA Agricultural Marketing Service (http://www.ams.usda.gov/AMSv1.0/PriceFormulas2012).

than when used in butter. Prices of milk protein and fat have varied widely over a 15-yr period (2000 to 2015) (figure 20.4). Each component had low values of approximately $1.00/lb. The price of fat peaked three times above $2.00/lb while protein was valued at or above $4.00/lb from mid-2007 to early 2009. The inventory of butter is characteristically high in summer and low in winter months, primarily due to the **flush season** in milk production during spring and early summer. Protein value is determined by the following complex formula (values in 2013) that takes into account two factors: (1) the price for protein based on the wholesale price for cheese and (2) the value of milkfat when it is used in cheese rather than butter:

$$\text{Protein price} = [(\text{cheese price} - 0.2003) \times 1.383] + \{[(\text{cheese price} - 0.2003) \times 1.572] - \text{butterfat price} \times 0.9\} \times 1.17$$

Included in the formulas for calculating the prices of protein and milkfat are the so-called "make allowances." These values represent the calculated costs to manufacture 1 lb each of cheese (protein formula), butter (milkfat formula), and dry whey (other solids formula). In 2013 the make allowances per pound were $0.2003 for cheese, $0.1715 for butter, and $0.1991 for dry whey.

20.8.2 Differentials

Prices paid by handlers and received by producers may be adjusted to reflect differences in distance of the producer from the handler and differences in milk composition. Therefore, each producer will have a unique milk price based on the total value of their milk's components, plus a price differential.

20.8.2.1 Location Differentials

These represent the cost of transporting milk from the area of production to the processor. Zones are determined by distance of the producer, or of the **receiving station**, from the processing plant.

20.8.2.2 Milkfat Differential

In the past, fat was considered the most valuable component of milk, however, it is the most variable component. The traditional method of pricing milk provided a fixed value for milk containing 3.5% fat, regardless of its content of nonfat solids or of protein. Current pricing systems place a market-determined value on the skim milk portion of milk. In pricing Class IV milk, the prices of nonfat solids and milkfat depend on the prices of nonfat dry milk and butter, respectively. Class III prices reflect the value of protein in cheese, milkfat in butter, and "other solids" in dry whey. The higher of the Class III and

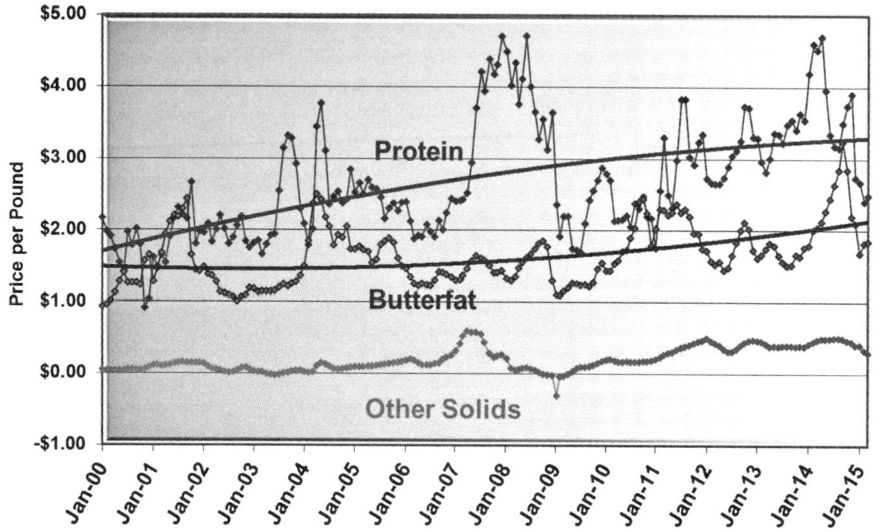

Figure 20.4 Component prices of butterfat (milkfat) and milk protein based on FMMO data (USDA Agricultural Marketing Service).

IV prices for skim milk becomes the basis of pricing skim milk for Class I; however, the value of skim milk used as Class II is derived from the Class IV skim milk price.

On an individual producer basis, a **milkfat differential** is paid or deducted for each 0.1% in fat content the producer's milk tests above or below 3.5%, respectively. In markets where component pricing is not used, the effect of such a system is to set a price based on fat content plus the value of skim milk. For example, if the price of 3.5% milk is $18.00/cwt and the milkfat differential is $0.20, milk testing 3.0% fat would sell for $17.00 [$18.00 – (5 × $0.20)]. Milk testing 4.0% fat would sell for $19.00 [$18.00 + (5 × $0.20)].

Such a pricing practice, however, places no real value on the concentration of its nonfat solids and encourages watering. (The reader is reminded that addition of water is illegal and is referred to chapter 25 for discussion of methods of detecting added water in milk.) In the above example the producer could actually profit by adding water to milk testing 4.0% fat. To make 4.0% milk test 3.0%, about 33 lb of water is added, producing 133 lb of watered milk. The value placed on the milk-water mixture under the above pricing plan would be $22.94 [(1.33 × $18.00) – (5 × 0.20)], which is an additional $4.94 ($22.94 – $18.00) in profit. This additional profit would more than pay the additional hauling cost. Payment for milk on the basis of both fat and nonfat solids or protein and other solids content negates this advantage.

A second reason for considering both fat and nonfat components in pricing relates to the variation in these constituents (cf. chapter 21). As fat content changes by 1%, average changes in nonfat solids are as follows: Holsteins (0.55%); Jerseys (0.25%); and Guernseys (0.36%). Component pricing has been made practical with the availability of automated instruments that accurately and efficiently determine milk's composition (cf. chapter 24).

20.8.3 Premiums

The sellers of raw milk, i.e., dairy farmers (producers) or their cooperative representatives, can bargain with regulated handlers to sell milk at prices higher than the FMMO minimums. These higher than minimum prices are called premiums, or over-order prices (the differences between the prices are called over-order charges or over-order payments). Sometimes the over-order price is referred to as the effective price. These premiums offset costs of balancing the supply with demand for Class I milk and are justified where full-supply agreements exist.

There are several implications from the use of premiums over federal order prices. They may entice increased production and reduce consumption, thus increasing government purchases under the price support program. Premiums may also alter intermarket price relationships so that handlers in markets where no premiums exist find it profitable to ship packaged milk into markets where substantial premiums are paid. Handlers who are able to purchase milk from nonmember sources gain a competitive advantage. This creates disorder, a situation that FMMOs are designed to remedy.

20.8.4 Pooling

An objective of the FMMO is to ensure that qualified producers within an order receive a uniform price for their milk and milk products within each class, no matter how those products are used by handlers. However, individual handlers vary in the percentage of producer receipts they use in various milk products, creating a discrepancy in prices paid to producers. Therefore, in order to finance a uniform price, a **producer settlement fund** (also called an equalization fund) is used. Handlers with higher than average utilization pay into the fund and handlers with lower than average utilization receive from this fund in proportion to the utilization value of the milk purchased. In general, dealers who have a high Class I utilization will pay into the pool, while dealers having high Class III and IV utilization will receive from the pool. For example, assume class prices are as follows:

Class I = $18.00
Class II = $17.00
Class III = $16.00
Class IV = $16.00

And utilization in the market is 25% for each class. The uniform price is then

$18.00 × 0.25 = $4.50
$17.00 × 0.25 = $4.25
$16.00 × 0.25 = $4.00
$16.00 × 0.25 = $4.00
Uniform price = $16.75

Each producer will receive $16.75/cwt. Any dealer whose utilization results in milk costing $16.74 or below receives from the pool while those having blend prices of $16.76 or above pay into the pool. For example, the milk purchased by a cheesemaker should cost only $16.00/cwt, but the producer must be paid $16.75/cwt. So, for every hundredweight of milk purchased for cheese, the cheesemaker must receive from the pool $0.75. Conversely, a dealer who purchases Class I milk will pay producers only $16.75/cwt for milk worth $18.00/cwt. Therefore, this dealer must pay into the pool.

As cooperatives have merged, resulting in their members living in several marketing areas, prices paid producers of respective cooperatives (cooperative pool) have been determined by a process called reblending. Reblending of returns under a number of different federal orders can result in blend prices that differ considerably from those established by individual orders.

20.8.5 Forward Contracting

Section 1502 of the Food, Conservation and Energy Act of 2008 (2008 Farm Bill) established the Dairy Forward Pricing Program, which allows producers and cooperative associations of producers to voluntarily enter into forward price contracts with handlers of milk used for Class II, III, or IV purposes under the FMMO. The program authorized milk handlers to pay producers or cooperative associations of producers a negotiated price rather than the federal order minimum blend price for producer milk, provided the volume of such milk does not exceed the handler's Class II, III, and IV utilization for the month. The program applies to producer milk regulated under federal milk marketing orders when it is not intended for fluid use. In accordance with the 2008 Farm Bill, forward contracts under the program were prohibited from being entered into after September 30, 2012, and could not be extended beyond September 30, 2015. However, with passage of the 2014 Farm Bill, the USDA issued a final rule extending the Dairy Forward Pricing Program to allow producers and cooperative associations to enter into forward price contracts through September 30, 2018. This final rule also extends all other requirements of the program until September 30, 2021.

20.8.6 Payment for Producers

The FMMO program has two systems for paying producers: the component pricing system and the skim milk/butterfat pricing system. The component pricing system is used in six orders (Northeast, Upper Midwest, Central, Mideast, Pacific Northwest, and Southwest). Under this system producer prices are based on (1) the values of butterfat, protein, and other solids in the milk each individual markets plus (2) the value of the producer price differential. The component values are those computed for the Class III price. The producer price differential is the weighted average value of the Class I and II differentials, the weights being the proportional volumes of milk used in Class I and II. The effective differentials are the actual differences in the applicable class prices for the month. Producers receive a price for each pound of milk component marketed plus the producer price differential per hundredweight. For informational purposes, a statistical uniform price is calculated that equals the Class III price plus the producer price differential.

The skim milk/butterfat pricing system is used in four orders (Appalachian, Florida, Southeast, and Arizona) in which much of the milk is used in fluid form. Under this system, producer prices are based on the uniform skim milk price and the uniform butterfat price. These prices are the weighted average values of each class's skim milk and butterfat prices. The weights are the volumes of skim milk and butterfat used in each class. Producers receive the uniform skim milk price for each hundredweight of skim milk marketed and the uniform butterfat price for each pound of butterfat marketed. The uniform milk price under this pricing system is the weighted average value of the uniform skim milk and butterfat prices in 3.5% butterfat content milk.

In four orders (Central, Mideast, Southwest, and Upper Midwest) producer payments are adjusted for the quality of the milk marketed based on the somatic cell count of the milk. The reason is that high somatic cell counts, due primarily to bovine mastitis, are associated with lowered protein content. As protein concentration decreases, so does cheese yield. In calculating the deduction in value of milk containing high numbers of somatic cells, the rate is determined by subtracting 350,000 from the observed count and the result is divided by 1,000. The cheese price is multiplied by 0.0005. The two derived numbers are then multiplied to provide the deduction. For example, if the cheese price = $2.00 and the somatic cell count = 550,000, then

($2.00 × 0.0005) × [(550,000 – 350,000) ÷ 1,000] = $0.20

In this example the penalty is $0.20/cwt of milk delivered during the applicable time period.

20.9 Production Controls

Forces that tend to increase the supply of grade "A" milk are relatively high Class I prices, negotiated premiums, sanitary restrictions on the production of manufacturing milk, and technical advances that make production of high-quality milk practical for larger numbers of milk producers. Increased production of all grades of milk is stimulated by price supports, economics of scale (such as spreading fixed costs over more units of production), and improvements in breeding, feeding, and management (more milk per cow).

A surplus of grade "A" milk is required to ensure its availability to consumers, but production of more than can be used as fluid milk reduces the uniform price.[2] Each producer would like the largest possible share of his/her milk used as Class I. Some producers are willing to reduce production when it means they will share a larger proportion of the Class I market.

In California, which does not participate in the federal order program, a system of allocating production bases and pool quotas has existed since 1967. Under this plan producers are allocated a base determined by a previous level of production and by the demand for Class I milk.

Problems involved in establishment and operation of a Class I base plan include (1) treatment of intermarket movement of milk, (2) establishment and ownership of bases, (3) deciding what action to take when producers deliver excess production outside the market, and (4) entry of new producers. The largest problem is achieving equity and integrity in all operations.

In 1984, the European Union (EU) imposed milk production quotas. Each member state had its own milk production quota, which was then distributed among dairy farmers. If a member state were to exceed its quota, it would have to pay a penalty to the EU. As of April 2015, the quota system was abolished. With the removal of the system, it is anticipated that the creation of a "milk belt" will be accelerated across northern Europe, along with a broad structural change across the dairy industry in the EU. In addition, high grain and oilseed prices have encouraged farmers in central Europe to focus on grain production instead of dairy production. In this case, geographic specialization and concentration may contribute to production controls in certain areas of the EU.

In Canada, the Market Sharing Quota (MSQ) is the national milk production target. The Canadian Dairy Commission monitors trends in Canadian requirements (demand) and production (supply) on a monthly basis. It then makes a recommendation to the Canadian Milk Supply Management Committee on the MSQ. The target is constantly monitored and is adjusted when necessary to reflect changes in demand, as measured in terms of butterfat, during the course of the year. Canadian requirements are defined as total domestic consumer demand plus planned exports for industrial dairy products.

The National Milk Marketing Plan establishes each province's share of the MSQ, and distributes any quota increase or decrease. Each province allocates its respective share of the MSQ to its producers according to its own policies, and in accordance with pooling agreements.

20.10 Problems in Intermarket Movement of Milk

With improved sanitation, refrigeration, and transportation, it became feasible to ship raw milk long distances. However, several factors have restricted intermarket movement; these are called **trade barriers**.

The earliest recognized trade barriers were sanitary restrictions. For many years such local restrictions were considered appropriate because they protected the health of consumers. However, upon the widespread adoption of uniform sanitary controls, reasons for restricting entry of milk from outside an area decreased. Some states and municipalities were reluctant to relinquish full control over milk sanitation, and it was left to the courts to rule that "a city may make reasonable regulations to insure purity of milk but cannot keep out milk unobjectionable for human consumption" (Krausz, 1960). In Arkansas the court held that an ordinance requiring a duplicate inspection of milk from out of state that was produced and processed under acceptable conditions and ordinances, plus fees, was unreasonable and placed an unreasonable burden on interstate commerce.

Bargaining activities of cooperatives that produce full-supply agreements (the cooperative provides 100% of handlers' demands daily but no more) and negotiated premiums often limit the handler to the use of local milk under control of the cooperative.

Shipments of excess milk to other markets, especially irregular shipments, may disrupt operations in local manufacturing plants. Such plants need a consistent milk supply to allow operation in the most efficient manner and to provide dairy products for customers.

Because costs of transportation are much less when shipments are large, handlers in the distant receiving market must be large enough to use tank-load quantities.

Finally, imperfect communication that prevents knowledge of prices and availability of milk may form a barrier to trade. Sufficient differences in price must exist for relatively long time periods to make it feasible for handlers in a milk-deficient area to do business with potential sellers in another area.

20.11 Hauling Milk

In most developed countries practically all milk is placed in mechanically cooled bulk tanks on farms and is transported to plants in insulated **bulk tank trucks** of capacities varying from 2,000 to 6,200 gal (7,600 to

23,500 liters). Over the past 60 years the dairy industry has become highly concentrated in both production and manufacturing aspects. This has resulted in increased distances of hauling both raw and finished products. However, efficiencies of operation in both farm and plant have greatly increased.

Ownership of the milk changes at the farm when milk becomes commingled with other milk in the truck tank. This factor places responsibilities on the truck operator to evaluate odor, flavor, and appearance of the milk and to measure and sample it accurately and properly (cf. chapter 26). Failure to evaluate the milk properly may result in contamination and rejection of the entire load. Improper sampling and measuring may result in economic losses to the producer or buyer.

Producers pay costs of hauling to most firms. Some pay on a flat-rate basis per hundredweight of milk. Others pay according to distance transported. A more equitable system of payment is reflected in a formula system for rate determination. Factors commonly employed are (1) distance from market by all-weather roads, (2) density of route—average miles between producers, (3) volume—discount per hundredweight if over the average volume per producer per pickup, and (4) a minimum charge per stop.

20.11.1 Localization

While increased localization of food supply chains provides some benefits, there has been uncertainty as to how it affects costs. A study by Nicholson et al. (2011) developed an economic model for the US dairy industry that included costs of assembly, interplant transportation, processing, and distribution for all dairy products including milk, yogurt, cheese, and butter. The average distance traveled for all US dairy products was about 320 miles from farm to market. Scenarios were developed to compare the impacts of increasing local sales focusing on reducing weighted average source distance, a unit of measurement comparable to food miles. Increased localization reduced assembly costs while increasing processing and distribution costs. The weighted average source distance for some products decreased at the expense of increases for other products. In one scenario, reducing the average distance traveled of beverage milk by 10% required a 30% increase in overall food miles for all other dairy products.

20.12 Sales and Distribution of Dairy Products

Processors of milk may choose from several alternative methods of getting their products to the consumer. They may sell direct, which is called retailing, or they may sell through an intermediate, which is called wholesaling. Within each category of sales several alternatives exist.

20.12.1 Retail Sales

Marketing dairy products directly to the consumer usually involves delivery of the product by company-owned vehicles to the consumer's doorstep. In the early years of development of the market milk industry, most milk was marketed in this manner. As late as 1930, 94% of all milk sold to Chicago consumers was distributed to homes. During the 1960s, about 30% of milk was supplied via delivery. By 2005, the last year for which USDA has data, only 0.4% was home delivered. However, early in the twenty-first century (among various concerns such as processed and contaminated foods) a resurgence began. Firms that have found home delivery a profitable business have minimized costs per unit and provided additional services, products, and flexible schedules for delivery, which are desired by the customers. To decrease costs chargeable to milk, other services such as sales and distribution of other foods or certain nonfood items is needed.

20.12.2 Wholesale Marketing

Neighborhood, convenience, and grocery stores are three types of wholesalers. Some are chain stores that handle only the brands of milk products processed in facilities owned by the chain. These are *captive* plants and all their output is sold to stores owned by the chain. This arrangement may result in certain efficiencies including (1) better inventory control, (2) handling in larger volume, (3) reduced costs of selling at the wholesale level, and (4) direct control (especially quality control) from processor to consumer.

Some chain stores and associations of independent stores operating on a district or regional basis have achieved several of these efficiencies by purchasing and receiving milk at the manufacturing plant. This is called *dock delivery*. Such products are often packaged in containers bearing the brand of the chain or association, a practice called *private labeling*. Distribution to stores is thus accomplished on a large-volume basis with efficiencies that may not be available to the processor. Many of these stores also sell dealer, independent, or national brands and this is especially true of specialty products such as whipping cream and yogurt. Private-label milk products are commonly sold at lower prices than are dealer brands, especially if more than the private-label brand is offered. The loss or gain of a contract to supply a food chain's stores in a region can have considerable effect on the dairy product sales volume and financial well-being of a processor. Private-label brands make it easier for food chains to change suppliers and exercise controls over pricing and merchandising.

20.12.3 Vending Machines

Convenience and reductions in labor costs are factors that promote sales through coin-operated vending machines and self-service refrigerators in areas of intense traffic such as service stations and office buildings.

The availability of raw milk in Europe has encouraged the spread of vending machines for unpasteurized milk. The United Kingdom, France, and Spain have permitted installation of self-service vending machines. Italy, Austria, Slovenia, Switzerland, and the Netherlands have begun to permit installations as well. Although the health benefits of raw milk are perceived more positively in Europe, regulatory agencies are warning consumers about potential health risks, as well as limiting its availability. In the United States the likelihood of raw milk vending machines becoming available is very low, as the FDA has warned against the consumption of raw milk because of its potential for mild to severe effects, such as vomiting, diarrhea, abdominal pain, and flu-like symptoms.

20.12.4 Marketing Margins and State Controls

A marketing margin is the difference between the price paid by consumers and that paid farmers for raw materials used in the final food product. Milk processors, distributors, and store operators must share this margin in attempting to cover costs and accumulate profits. Margins on fresh milk differ between milk sold in stores and on retail routes, among sizes and types of containers, and in some markets, with volumes purchased by stores or by consumers on retail routes.

Marketing margins of handlers must cover the costs of the following general categories: raw materials, transportation, salaries, wages and commissions, containers and other supplies, plant and office operations, taxes, licenses, insurance, advertising, depreciation and repairs, and miscellaneous expenses.

Some states have regulations that address marketing margins. The Maine Milk Commission establishes minimum producer prices, which are normally set at the prevailing southern New England price plus a premium. The Maine Milk Commission also establishes minimum processor and retailer margins based upon periodic cost studies (Doyon et al., 2008).

20.12.5 International and Online Trading

Trading commodity futures has become important to the dairy industry. Two special participants in futures trading are the Chicago Mercantile Exchange (CME) and New Zealand's Fonterra Cooperative. Offerings through CME include Class III and IV milk, nonfat dry milk, dry whey, butter, and cheese. Figure 20.5 summarizes prices of Class III milk in the CME during four months of 2015.

The Fonterra Cooperative of New Zealand sponsors GlobalDairyTrade (GDT) on a semimonthly basis in which they trade skim, sweet whey, and whole milk powders, anhydrous milkfat, butter and buttermilk powder, cheese, lactose, and casein products.

Since pricing programs may force prices of domestically produced milk and milk products above world prices, it is often necessary to restrict imports of milk products. In the United States the president administers import controls with the advice of appropriate agencies of the executive branch of government. Controls are generally imposed on semiperishable products such as cheese, nonfat dry milk, sodium caseinate, butter oil, and butter.

Canada uses supply-management systems to regulate its dairy industry. They include production quotas, producer marketing boards to regulate price and supply, and tariff-rate quotas (TRQs). These systems severely limit the ability of US producers to increase exports to Canada above TRQ levels and inflates the prices Canadians pay for dairy products. Under the current system, US imports above quota levels are subject to prohibitively high tariffs (e.g., 245% for cheese and 298% for butter).

Canada also applies import duties to certain commercial "food preparations" that contain cheese. For these particular food preparations, rather than classifying the product under a single tariff heading, the components are classified separately. As a result, cheese components of food preparations are now subject to prohibitively high tariff rates.

Figure 20.5 Prices of Class III milk offered and settled through the Chicago Mercantile Exchange during four months of 2015 (from CME Group).

In the United States the FAS administers the Dairy Import Licensing Program, which provides issuance of licenses to import certain dairy articles under TRQs. Licenses are issued on a calendar-year basis and provide low-tier tariff rates (in-quota rates) to the licensee for a specified quantity and type of a dairy article. Higher "over-quota" rates apply to imports in excess of that quantity. *Historical licenses* (inception in 1950s) are reissued annually only to importers who qualified during a base period, provided they remained qualified. *Nonhistorical (lottery) licenses* (first issued in the 1960s) are available to qualified applicants, but there is no guarantee of renewal year-to-year. *Designated licenses* are issued to importers nominated by a foreign government or entity to which the United States granted the right to designate an allocation. Of the 346 million lb (156.9 million kg) of dairy products imported under TRQs in 2012, 86% was cheese. Imports within the three categories amounted to: historical (43%), lottery (27%), and designated (30%). In 2013, FAS proposed review of the regulation and requested public comments on possible changes in its provisions.

20.13 Summary

Pricing of milk is no longer done by simple bargaining between buyer and seller. Prices are influenced by many agencies, as well as by the producer, processor, retailer, and consumer. Interjection of forces other than competition, per se, is done on the premise that the production of milk cannot be turned on and off at will in a short period of time. Consumers need a continuing, adequate, and reasonably priced supply of high-quality milk and milk products. Therefore, it is in the public interest to stabilize prices of raw milk at levels that will assure an adequate supply.

Marketing of milk and milk products is big business. Sales are made by the milk producer, producer cooperatives, processors, distributors, and retailers. Their sales depend upon consumer demands. Prices paid at each point of sale are influenced by factors other than demand. Raw milk prices are kept above certain minimums by actions of the federal government. Federal milk marketing orders provide minimum prices for raw grade "A" milk, with higher prices for milk used in fluid products but lower prices for **surplus milk**, which must compete with **manufacturing grade milk**. Producer bargaining with handlers through producer cooperatives often results in price changes. Production and import controls play lesser, but necessary, roles in price determination.

Handlers and retailers influence pricing by their marketing practices. Ownership of both processing plants and retail stores by the same company—the captive plant situation—provides certain economies of operation. Private labeling and dock delivery have been adopted because of economic advantages to segments of the industry. Cooperative ownership of manufacturing facilities provides flexibility in usage of raw materials. Merging of producer cooperatives has resulted in greater bargaining power for the producer, and economies have been realized in areas of milk collection, management of the routing and shipment of milk, and processing operations.

Study Questions

1. Describe the marketing of milk and milk products in the United States during the Civil War. How did it change during the process of industrialization? Describe the changes that occurred during the twentieth century.
2. Identify four or more factors affecting the supply of milk.
3. What agencies in the USDA are active in the marketing aspects of milk and milk products?
4. Is the total milk production in the United States increasing or decreasing? How do sales on a per capita basis influence milk production?
5. What factors influence the demand for dairy products in the United States?
6. What portion of the US milk supply is used for fluid milk products? For what is the balance used?
7. What is the per capita consumption trend in milkfat? Nonfat solids?
8. Why do cooperatives offer producers bargaining power in the marketplace? What services do they render producers?
9. What are federal milk marketing orders? Discuss their merits and limitations. How are they established?
10. What is a milkshed?
11. What is a full-supply agreement?
12. What are the merits and limitations of support prices? What is meant by parity?
13. What is meant by a milk handler?
14. How does classified pricing of milk function?
15. What is the purpose of pooling milk?
16. Explain how a utilization ratio is used to calculate the blend price of milk.
17. Cite an example of how the producer settlement fund affects prices paid to producers.
18. Which is more seasonal in nature—milk production or milk consumption? Why?
19. What is meant by premiums? How do they affect the price paid producers for milk?
20. How do location differentials affect the price paid producers for milk?
21. Discuss milkfat differentials. What variables are needed to calculate them?
22. Describe the advantages and disadvantages of the current methods of the pricing of milkfat and nonfat solids.
23. Identify factors that tend to increase the supply of grade "A" milk. What factors stimulate the production of all grades of milk?

24. What are milk production controls? Why and when are they useful?
25. What is meant by the term flush season?
26. Which dairy products are affected most by import controls?
27. Discuss the impacts, merits, and limitations of the intermarket movement of milk. What conditions tend to favor such movement of milk?
28. What responsibilities does the farm bulk tank truck method of transporting milk place on the truck operator?
29. What factors influence transportation costs to milk producers?
30. Producers can market their products via retail or wholesale sales. Which method is beneficial for small producers? Large producers?
31. What is meant by the term dock delivery? Why has the practice of private labeling increased in recent years?
32. What are the advantages and disadvantages of offering raw milk in vending machines?
33. What is meant by marketing margin? What factors influence it? Are there usually regional differences in it?
34. Under what conditions are retail prices of milk controlled?
35. Describe the national dairy checkoff program. How is the assessment paid by producers used?
36. Identify three organizations that aid in consumer education, research, and promotion related to milk and milk products.

Notes

[1] For a comprehensive review of the history of dairy marketing from its infancy in the Jamestown colonies to modern times see *Dairy Marketing in America* by Carl E. Zurborg (2005).

[2] In order to allow for imperfect distribution and for seasonal and other unavoidable decreases in production, a supply of milk for fluid use must be available beyond what consumers will purchase.

References

Doyon, M., G. Criner, and L. A. Bragg. 2008. Milk marketing policy options for the dairy industry in New England. *Journal of Dairy Science* 91:1229–1235.

Krausz, N. G. P. 1960, February. Proceedings, Second Agricultural Industries Forum, University of Illinois.

Nicholson, C. F., M. I. Gómez, and O. H. Gao. 2011. The costs of increased localization for a multiple-product food supply chain: Dairy in the United States. *Food Policy* 36(2):300–310.

USDA National Agricultural Statistics Service. 2015, March. Agricultural Prices (http://www.nass.usda.gov/Publications/Todays_Reports/reports/agpr0315.pdf).

Zurborg, C. E. 2005. *Dairy Marketing in America.* Columbus, OH: National Dairy Shrine.

Websites

Cooperatives Working Together Program (http://www.cwt.coop)
Food and Agriculture Organization of the United Nations (http://www.fao.org/home/en)
GlobalDairyTrade (http://www.globaldairytrade.info)
Milk Processor Education Program (http://www.milkpep.org/login)
MilkPrice (http://milkprice.blogspot.com)
US Dairy Export Council (http://www.usdec.org)
USDA Agricultural Marketing Service (http://www.ams.usda.gov/AMSv1.0)
- *Dairy Market News Weekly Report* (http://www.ams.usda.gov/AMSv1.0/WeeklyPrintedReportsMain)
- Marketing Orders, Current Price Formulas (http://www.ams.usda.gov/AMSv1.0/PriceFormulas2012)

USDA Economic Research Service (http://www.ers.usda.gov)
- *Livestock, Dairy, and Poultry Outlook* (http://www.ers.usda.gov/publications/ldpm-livestock,-dairy,-and-poultry-outlook.aspx)

USDA Foreign Agricultural Service (http://www.fas.usda.gov)
- Dairy Import Licensing Program (http://www.fas.usda.gov/programs/dairy-import-licensing-program)

USDA National Agricultural Statistics Service (http://www.nass.usda.gov)
- *Milk Production, Disposition, and Income Annual Summary* (http://usda.mannlib.cornell.edu/MannUsda/viewDocumentInfo.do?documentID=1105)

USDA Office of the Chief Economist (http://www.usda.gov/oce)
- World Agricultural Supply and Demand Estimates (http://www.usda.gov/oce/commodity/wasde/)

USDA Rural Development (http://www.rd.usda.gov)

21

Biological, Chemical, Physical, and Functional Properties of Milk and Milk Components

I would address one general admonition to all; that they consider what are the true ends of knowledge, and that they seek it not for pleasure of mind, or for contention, or for superiority to others, or for profit, or fame, or power, or any of these inferior things; but for the benefit and use of life, and that they perfect and govern it in charity. For it was from lust of power that the angels fell; from lust of knowledge that man fell; but of charity there can be no excess, neither did angel or man ever come in danger by it.

Francis Bacon, "The Great Instauration" (1618)

21.1 Introduction

We began in chapter 1 by showing the extraordinary complexity of milk. It consists of (1) an oil in water emulsion with fat globules dispersed in the continuous or serum phase; (2) a colloidal suspension of casein micelles, globular proteins, and lipoprotein particles; (3) a solution of milk sugar (lactose), soluble proteins, most of the minerals and the water-soluble vitamins, and (4) somatic cells. The major components are accompanied by various minerals (notably calcium and phosphorus), vitamins, enzymes, and various minor organic compounds (such as citric acid), some of them nitrogenous in nature. The characteristic opaque color of milk reflects primarily the dispersion of milk proteins and calcium salts. In this chapter we discuss salient aspects of the biological, chemical, physical, and functional properties of milk.

The following terms are used to describe milk's components:

- *Plasma* is milk minus its fat and is called **skim milk** or nonfat milk.
- *Serum* is plasma minus casein micelles and is called *whey.*
- *Nonfat milk solids* (*NMS*) or *solids-not-fat* (*SNF*) is milk minus fat and water and is composed of proteins, lactose, minerals, acids, enzymes, and vitamins.
- *Total milk solids* (*TMS*) is composed of fat plus NMS.
- *Milkfat* and *butterfat* are used interchangeably.

Only some 250 years have elapsed since it was proved that lactose is a true sugar. The constituents of milk listed in 1790 were butterfat, casein, lactose, a small amount of extractive matter, a little salt, and water. Only about 150 years have elapsed since lactalbumin, lactoglobulin, and iron were identified in milk. After 1900, however, the elucidation of milk's composition was rapid. Today, not only have the minor components of milk been identified, they are increasingly being harvested for specific purposes, both for function and nutrition. For example, milk's known bioactive components include certain amino acids and peptides.

21.2 Lipids

The lipids of milk are insoluble in the water phase (with the exception of traces of water-soluble fatty acids) and are present in a globular (minute droplet) form called an **emulsion**. Their primary component is the triglyceride fraction. Each triglyceride is composed of one **molecule** of glycerol (a trihydric alcohol) and three molecules of fatty acid.[1] Monoglycerides and diglycerides comprise only traces of fat. Glycerol and butyric acid can be **esterified** to form a monoglyceride (figure 21.1).

Fatty acids would be added to the two remaining hydroxyl (–OH) groups of glycerol in the same kind of reaction to form a triglyceride. Milkfat contains at least 400 different fatty acids (Jensen et al., 1991). Their kinds, numbers, and the manner in which they are associated with glycerol determine the properties of milkfat. About 98% of the total weight of the milk lipids occurs in the fatty acid **moiety** of the triglycerides (figure 21.2).

Compared with other food fats (table 21.1), milkfat contains a much higher percentage of the low molecular weight fatty acids. Being liquid at low temperatures, they favorably influence the melting point of milkfat (they lower it). The major fatty acids are those with straight chains and even numbers of carbon atoms (table 21.2). However, numerous fatty acids with branched chains and odd numbers of carbons have been isolated. The fatty acids with less than 10 carbons are all saturated, that is, they contain no double bonds between adjacent carbon atoms. The saturated acid in highest concentration in milkfat is palmitic acid (figure 21.3 on p. 294).

A single carbon-to-carbon bond may be unsaturated in acids with 10 through 24 carbons. These comprise about 35% of the fatty acids in milkfat, and oleic acid (18C) predominates, comprising about 30% of the total fatty acids. At least 60% of the triglycerides contain one or more double bonds. In molecules that have double bonds, the chain of carbon atoms has 90° angles on each side of each double bond (figure 21.4 on p. 294).

Molecules that have more than one double bond are **polyunsaturated**. Their content in bovine milk is lower than in human milk primarily because of hydrogenation in the bovine rumen. These acids are important in human nutrition, but only the cis, cis form is active as an **essential fatty acid**.[2] Milkfat contains about 3% trans fatty acids. The major portion (more than 60%) of the 18-carbon diunsaturated acids (dienes) in milk are in the cis, cis form, called linoleic acid. The dienes include carbon chains from 18 through 24 in length. The structural formula for linoleic acid of the cis, cis configuration (hydrogens are omitted for clarity) is found in figure 21.5 (on p. 294).

Three unsaturated bonds occur in linolenic acid, which also has 18 carbons and is of the cis, cis, cis config-

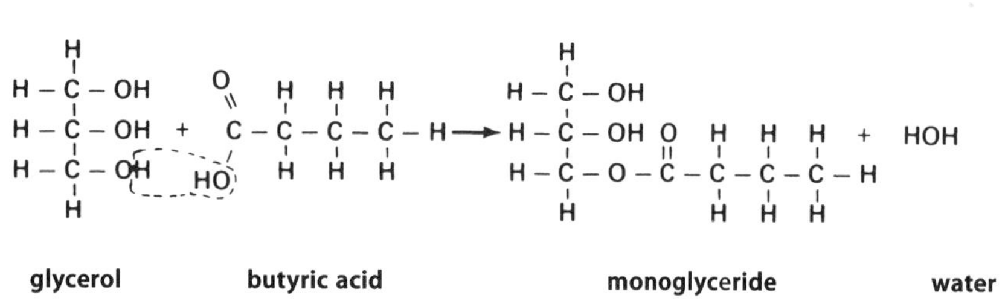

Figure 21.1 Esterification of glycerol and butyric acid to form a monoglyceride.

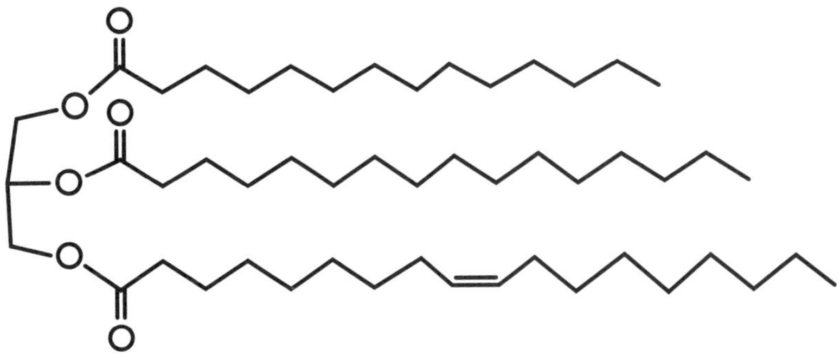

Figure 21.2 Milkfat is a triglyceride (fat) derived from fatty acids such as myristic, palmitic, and oleic acids, which contain 14, 16, and 18 carbons, respectively.

Table 21.1 Average Fatty Acid Composition (%) of Selected Commercial Fat and Oil Samples.

Fatty Acid	Milkfat	Lard	Tallow	Coconut	Soya Bean	Peanut	Safflower	Cottonseed	Corn
Butyric	3.5	—	—	—	—	—	—	—	—
Caproic	2.2	—	—	trace	—	—	—	—	—
Caprylic	1.0	—	—	8.0	—	—	—	—	—
Capric	1.8	—	—	7.0	—	—	—	—	—
Lauric	3.0	trace	0.5	53.0	—	trace	—	trace	—
Myristic	11.2	3.0	3.2	16.3	0.5	1.4	0.3	1.1	trace
Palmitic	25.5	28.4	28.9	8.5	12.2	8.3	4.0	21.2	11.8
Palmitoleic	2.2	4.0	2.8	—	trace	0.5	—	1.2	trace
Stearic	11.3	13.5	21.3	2.0	1.2	3.1	2.0	2.3	1.9
Oleic	30.4	43.0	38.8	4.2	22.3	52.0	17.0	22.8	30.4
Linoleic	2.8	7.9	2.0	1.0	60.7	28.3	74.0	48.0	54.7
Linolenic	0.5	0.2	trace	trace	3.1	2.8	2.7	3.4	1.2

Source: Kinsella, 1969.

Table 21.2 Major Fatty Acids in the Milkfat of Cows.

Class	Number of Carbons: Double Bonds	Mole Percent Range[a]	Percentage by Weight	Melting Point (°C)
Saturated				
Butyric	C4:0	8.1–10.5	2.8	-8.0
Caproic	C6:0	1.2–3.9	2.3	-4.0
Caprylic	C8:0	1.3–2.7	1.1	16.3
Capric	C10:0	1.5–3.7	3.0	31.3
Lauric	C12:0	3.0–8.4	2.9	44.0
Myristic	C14:0	8.6–24.2	8.9	54.0
Palmitic	C16:0	15.8–27.2	24.0	63.0
Stearic	C18:0	7.6–10.0	13.2	69.6
Monounsaturated (monoenoic)				
Myristoleic	C14:1	0.9–1.7	0.7	
Palmitoleic	C16:1	1.9–3.7	1.8	0.5
Oleic	C18:1	15.3–31.4	29.6	16.3
Diunsaturated (dienoic)				
Linoleic	C18:2	1.0–4.9	2.1	-5.0
Triunsaturated (trienoic)				
Linolenic	C18:3		0.5	-14.4

[a] Mole percent expresses percentage on a number of molecules basis. For example, the 18-carbon acid, stearic, weighs approximately 4 times as much as the 4-carbon acid, butyric, but is present in about the same mole percentage.

Source: Adapted from data of Jack and Smith, 1956; and data of Herb et al., 1962.

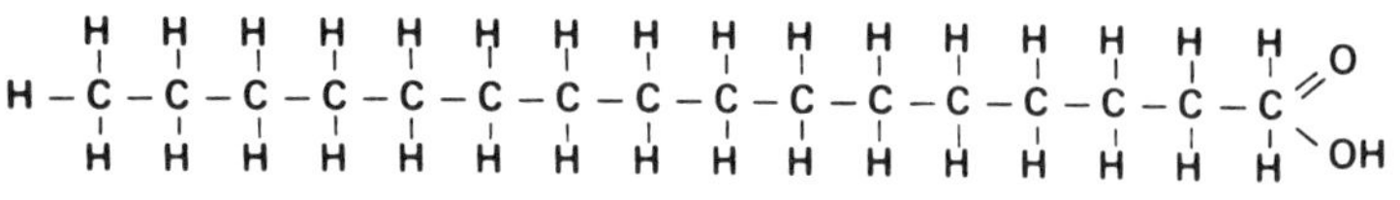

palmitic acid — — — saturated

Figure 21.3 The structural formula of palmitic acid.

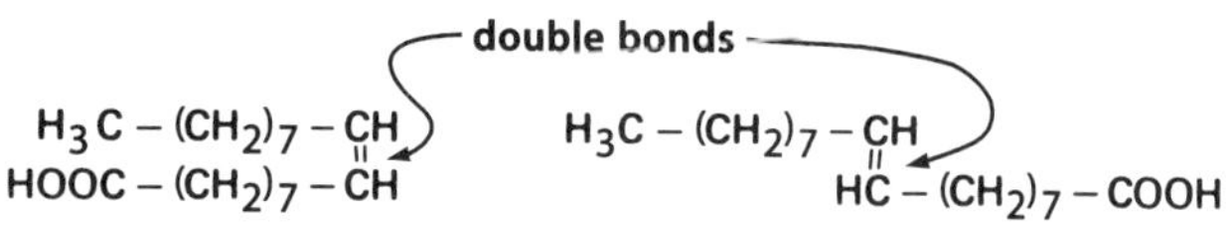

cis **oleic acid** trans **oleic acid**

Figure 21.4 Oleic acid in cis and trans configurations.

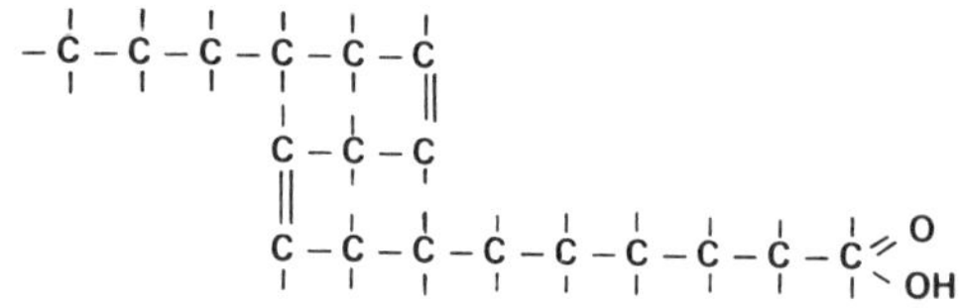

cis, cis **linoleic acid** — — — — **diunsaturated**

Figure 21.5 The structural formula for linoleic acid of the cis, cis configuration.

uration. Linolenic acid is the major triene (three double bonds) of milkfat; other trienes comprise about 0.14%. Four double bonds occur in arachidonic acid, which is 20 carbons in length, cis, cis, cis, cis in configuration, and comprises 0.14% of the total. There are traces of other tetraenes (four double bonds) and of pentatenes (five double bonds in 20- to 22-carbon acids). Most fatty acids have straight chains of carbons, but 2% or more of those in milk, by weight, are likely to have branched chains.

Three significant factors affect the melting properties of fat, namely (1) the fatty acid composition (molecular weight and degree of unsaturation), (2) the distribution of acids in glycerides, and (3) the **polymorphic** forms of fat crystals. Although the normal melting point of milkfat is 99°F (37°C), the temperature can vary from 86 to 106°F (30 to 41°C) depending on previous treatment of the fat, especially rate of cooling and physical form of the fat. The state of dispersion affects crystalline structure. Globular fat has a lower melting point than bulk fat, and globule size affects crystalline behavior. Globules formed by homogenization, typically less than 2 μm in diameter, are more uniform in composition (therefore in individual melting point) than are naturally produced globules that vary more widely in composition and in diameter from about 1 to 10 μm. The greatest mass is formed by globules of about 4 μm in diameter.

The fully saturated triglycerides in milkfat provide plasticity—the ability to form, at room temperature, self-standing structures, as in butter (Philhofer et al., 1994). This property results from the existence of large crystals of high-melting triglycerides. These form three-dimensional lattices that support a large volume of butter oil (Juriannse and Heertje, 1988). The relatively low concentration of solid crystals and low frequency of cross-linking in butter at room temperature permits it to be spread (Walstra, 1987). Proper working of butter during churning enhances spreadability (cf. chapter 32).

Milkfat has a synergistic effect on the cocoa butter of milk chocolate, in which it softens the otherwise brittle texture. Additionally, it tends to retard formation of bloom in the chocolate by interfering with polymorphic transitions in cocoa butter crystals (German et al., 1997).

A unique feature of bovine milkfat is the high content of butyric acid (4 carbons): over 3% by weight of the major fatty acids and, on a molar basis, approximately one-third of the amount of palmitic acid (16 carbons), the most abundant fatty acid by weight. Sheep milk contains about 4%. The average total content of fatty acids with chain lengths of 4, 6, and 8 carbons is reported as follows: sheep (9.1%), goats (8.2%), and cows (6.2%) (see table 18.4).

Short-chain and unsaturated fatty acids depress the melting point, whereas long-chain saturated acids elevate it. For example, melting points of individual triglycerides vary from −103°F (−75°C) for tributyrin to 162°F (72°C) for tristearin. Softer butter is produced from the milkfat of cows on spring and summer **pasture** because of increased dietary intake of unsaturated fatty acids. However, the microflora of the rumen is an active force in adding hydrogen to (thus reducing) the unsaturated acids. This strongly limits quantities of unsaturated acids that pass into the intestinal tract where they can be readily absorbed and transported into the bloodstream as precursors of milkfat. Researchers at the USDA Dairy Products Laboratory applied a coating of formaldehyde-treated casein to highly unsaturated droplets of safflower oil and fed this to cows. The coating protected much of the oil from hydrogenation in the rumen, allowing it to pass to the intestines. There the digestive process was completed, and the unsaturated fatty acids were transported to the mammary glands via blood. Thus they became part of milkfat, causing it to contain an unusually high amount of unsaturated fatty acids, 35% linoleic compared with 3% in control animals.

There is uncertainty as to the manner in which fatty acids of milkfat are distributed on the three reactive groups of the glycerol molecule. Probably there is a general random distribution with some biological modification. In strictly random distribution one-third of each fatty acid should appear in each of the three possible positions, but the distribution in milkfat does not achieve this status. Biological alteration may be necessary to prohibit production of fats with too high a melting point (solid at body temperature) as would occur upon the formation of triglycerides with three long-chain saturated fatty acids. It is interesting that there is a strong predisposition for the saturated fatty acids myristate (14:0), laurate (12:0), and stearate (18:0) to be located in the central position of the glycerol backbone (sn-2) (table 21.3).

Because carbon chains of fatty acids attached to glycerol in a glyceride are free to rotate on an axis perpendicular to the axis of the glycerol portion, various configurations can arise for the molecule. Thus, triglycerides will pack together differently when they crystallize, a phenomenon known as polymorphism. The crystals may, therefore, have different densities and melting points. Molecules that contain branched chain fatty acids impede formation of dense crystal structures. Thus, they decrease melting points. However, trans fatty acids increase melting points.

Collectively, monoglycerides and diglycerides comprise less than 0.5% of the total lipid fraction of milk, and diglycerides are approximately 10 times higher in concentration than monoglycerides. These compounds tend to concentrate at the fat globule surface because they have one (diglyceride) or two (monoglyceride) free hydroxyl groups, which tend to be associated with water. Certain mono- and diglycerides are added in some dairy products to help stabilize fat emulsions. Their basic structure is illustrated in figure 21.6.

Although minor in quantity (0.2 to 1.0%), phospholipids are important components of milkfat globules. They are comprised of fatty acids, phosphoric acid, and other groups, including glycerol and choline (in lecithin), ethanolamine and serine (in cephalin), and sphingosine (in sphingomyelin). Low molecular weight fatty acids are present only in trace amounts. Other complex lipids

Table 21.3 Positional Distribution of Fatty Acids in Milkfat (mol%).

Fatty Acid	sn-1	sn-2	sn-3
4:0	—	—	35.4
6:0	—	0.9	12.9
8:0	1.4	0.7	3.6
10:0	1.9	3.0	6.2
12:0	4.9	6.2	0.6
14:0	9.7	17.5	6.4
16:0	34.0	32.3	5.4
18:0	10.3	9.5	1.2
18:1	30.0	18.9	23.1
18:3	1.7	3.5	2.3

Source: Berner, 1993.

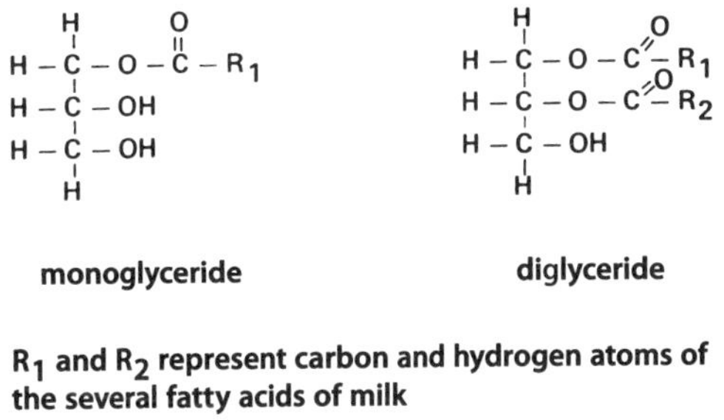

Figure 21.6 Structure of mono- and diglycerides added to stabilize fat emulsions in some dairy products.

include cerebrosides (sphingosine + galactose + long-chain fatty acid) and plasmalogens (glycerol + aldehydes + fatty acids + phosphoric acid + choline or ethanolamine). Neutral plasmalogens do not contain phosphoric acid.

The **hydrophillic** (water-loving) head of phospholipids contains the negatively charged phosphate group, and may contain other polar groups. The hydrophobic (water-hating) tail usually consists of long hydrocarbon chains of fatty acids.

Phospholipids exist in complex with proteins in milk. As triacylglycerols are synthesized within epithelial cells of the mammary gland, they assemble into small droplets. These acquire a coating of phospholipids and protein as they are secreted from the cell by a budding mechanism in which the globule is enveloped by the plasma membrane. The coat material, milkfat globule membrane (MFGM), is enriched with three proteins: the redox enzyme xanthine oxidoreductase, mammary gland specific butyrophilin, and adipophilin (Nielsen and Anderson, 1999). In addition, mucins, or proteins covalently linked to carbohydrates (about 30% by weight), are adsorbed to the MFGM (Patton, 1999). One of the two known mucins, MUC1, carries a strong negative charge because of its high content of sialic acid. This assists in keeping globules from clumping. However, cooling and agitation dislodge MUC1, thus favoring clumping and rising of fat globules. Proteins comprise about 1% of milkfat globules. Enzymes comprise a minor amount of adsorbed protein.

Cream, centrifugally separated from milk, contains about 65% of milk's lipid-bound phosphorus because of the high degree of adsorption of phospholipids to the fat globules they help stabilize (emulsify). In churning and in homogenization, phospholipids are dislodged from fat globules and are not completely readsorbed. They appear to be less susceptible to oxidation in the serum phase than at the fat globule surface, thus reducing the susceptibility of homogenized milk to copper-catalyzed oxidation.

The principal sterol of milk, cholesterol (0.3 to 0.4% of the fat), comprises part of the MFGM. Its structural configuration is shown in figure 21.7. It is a cellular constituent in animals, being especially abundant in nervous

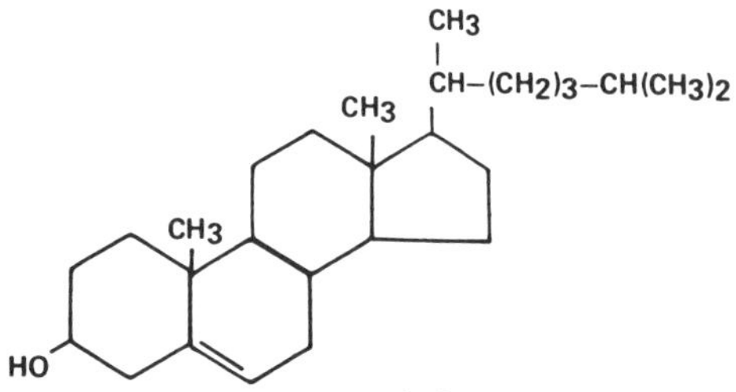

cholesterol

Figure 21.7 The structural configuration of cholesterol.

tissue. One form, 7-dehydrocholesterol, is the precursor of vitamin D_3 and can be activated (changed to the vitamin) by ultraviolet irradiation of milk (see section 21.6). Since cholesterol has no ester linkages, it cannot be saponified as can triglycerides and other esters. Other unsaponifiable lipid components include the fat-soluble vitamins A, D, E, and K.

For purposes of characterization, certain constants have been determined for the more common fats. The following describes those that apply to milkfat:

- *Refractive index*—the degree of bending of light waves passing through the liquid fat. For milkfat the number varies between 1.4538 and 1.4578 and is lower than for most other fats.
- *Saponification number*—the number of milligrams of KOH (potassium hydroxide) required to saponify one gram of fat; indicates the average weight of the fatty acids in the fat. For milkfat the value usually falls between 225 and 230, which is much higher than values for fats other than coconut and palm kernel oil.
- *Iodine number*—the number of grams of iodine absorbed by 100 g of fat; indicates the degree of unsaturation of the fat. For milkfat the values range from 26 to 35, which are generally lower than for other food fats.
- *Reichert–Meissel number*—the number of milliliters of 0.1 N alkali required to neutralize the volatile, soluble fatty acids distilled from 5 g of fat. It primarily measures butyric acid, and the value for milkfat, between 17 and 35, exceeds that for other fats and oils.

Milkfat is completely melted as it exits the mammary gland. However, on cooling the lower melting glycerides start to crystallize. Fat should be in this melted form for efficient homogenization of the globules. However, it is important to have fat at least partially crystallized in whipping, churning, and freezing ice cream mix. On churning, some fat globules coalesce (they irreversibly increase in size), while others may flocculate (reversibly cluster) without loss of identity as long as the crystalline structure is maintained. There may also be irreversible agglomeration in which the fat globules are held together by a combination of liquid and crystalline fat with some individual globules remaining intact. Coalescence will occur if the crystals melt. This state of partial coalescence exists in whipped cream, frozen dairy desserts, and aerated emulsions.

21.3 Proteins and Nitrogenous Materials

The major milk proteins, the caseins, β-lactoglobulin, and α-lactalbumin, are synthesized only by mammary epithelial cells. Immunoglobulin and serum albumin are absorbed from the blood, except that a minor amount of immunoglobulin is synthesized by lymphocytes within the plasma cells of mammary tissue. Lymphocytes provide local immunity to the mammary gland.

Milk proteins are composed primarily of amino acids and most of the nitrogen in milk is contained within them. These amino acids are connected to one another primarily by **peptide bonds**. In such bonding the **α-amino** ($-NH_2$) group of one amino acid is joined to the carboxyl (–COOH) group of the next (see figure 21.8). The pattern is repeated to form a long chain (polymer or polypeptide). Nitrogen contained in amino acids comprises about 16% of the protein. Hence, we determine the amount of nitrogen by the Kjeldahl method (cf. section 22.3) and multiply by 6.38 (for milk proteins) to calculate the amount of protein contained.

Figure 21.8 Peptide chain containing three α-amino acids (R = remainder of structure of each amino acid).

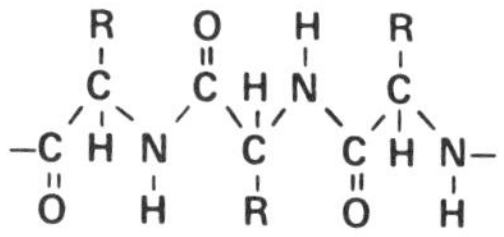

Proteins are constructed by a **synthesizing** mechanism of the cell, and each protein is characteristic of the cell (and animal or plant) that produces it. Material in the cell nucleus, called **deoxyribonucleic acid (DNA)**, is the ultimate determinant of the kind of protein that will be produced. However, an intermediate material, ribonucleic acid (RNA), serves as the carrier of essential information. This information is in the form of a template that establishes the sequence of amino acids in the protein chain. Nineteen amino acids are commonly found in milk proteins, including the **essential amino acids**. Tryptophan and lysine are especially plentiful. However, cystine and methionine, the amino acids that contain sulfur, are present in less than optimal concentrations.[3] The nutritional values of the proteins of corn, potato, or white bread are virtually doubled by consuming them with adequate milk to balance the essential amino acids (cf. chapter 1).

As proteolytic enzymes degrade milk's proteins, peptides of various lengths and composition are produced. Some have important biological functions. These **bioac-**

tive peptides are produced during digestion of milk and during fermentation. Beneficial effects on health include antimicrobial, antioxidative, antithrombotic, antihypertensive, and immunodulatory (Madureira et al., 2010). Activities of biofunctional peptides depend on composition and sequence of amino acids contained therein. Sizes of active sequences vary from 2 to 20 amino acids, and some have multiple functions.

In bovine milk, using a proteomic approach based on liquid chromatography connected with tandem mass spectrometry, Reinhardt and Lippolis (2006) identified more than 120 proteins with diverse functions such as trafficking, cell signaling, or immune response. Moreover, they characterized changes in the bovine MFGM proteome during transition from colostrum to mature milk (Reinhardt and Lippolis, 2008). However, except for mucins, MFGM proteins from goat milk are poorly characterized compared with their bovine or human counterparts. Mucins are high molecular weight proteins that contain more than 50% carbohydrates and have well-recognized anti-infectious effects. In the goat, MUC-1 and MUC-15 have been characterized either at the DNA or protein level.

21.3.1 The Micellar Proteins (Caseins)

Bovine casein, the major protein of milk, consists of individual casein fractions, namely α_{S1}-, α_{S2}-, β-, and κ-CN at an approximate proportion of 4:1:4:1.3 (Walstra, 1999). In casein, phosphate is esterified with serine and threonine. Both have free hydroxyl groups on the end opposite the carboxyl group, and therefore will react with phosphoric acid to form an ester. These phosphate groups are important to the structure of the casein micelle. Calcium binding by individual caseins is proportional to the phosphate content. Proteins that contain constituents other than amino acids are called **conjugated** proteins; the phosphoprotein casein is an example. The distinguishing property of all caseins is their low solubility at a pH of 4.6. Casein's biological function is to carry, in liquid form, large amounts of highly insoluble Ca/P to stomachs of mammalian young where it forms a clot for efficient digestion. More than 90% of the calcium content of skim milk is associated with the casein micelle. Casein (2.5% of milk depending mainly on breed) is the protein precipitated by acidifying skim milk to a pH near 4.6. Proteins remaining in whey after removal of casein are called whey proteins or milk serum proteins (0.6% of milk). Some whey proteins are precipitable by heat, and commercial lactalbumin is a mixture of these heat-labile, acid-precipitable whey proteins.

The most abundant components of casein are α- and β-caseins. However, κ-casein is especially important because it stabilizes α- and β-caseins up to 10 times its own mass in the presence of calcium. Milk protein fractions have been isolated in several forms called genetic variants (table 21.4). These variants arise when a deletion or substitution of a **nucleotide** occurs in a chromosomal gene. This leads to changes in the amino acid composition of the protein. Since there are two **alleles** in each chromosome, one or both may have experienced change in nucleotide sequence. For example, in the case of κ-casein, cows have been found having AA, AB, and BB genotypes. Several researchers have reported that milk from cows producing κ-casein BB is higher in protein and casein than milk from cows producing the AA or AB genotype.

Table 21.4 Summary of the Known Genetic Variants of the Major Milk Proteins.

Milk Protein	Genetic Variants[a]
α_{s1}-casein	A, B, D, (E), F
α_{s2}-casein	A, (B, C), D
β-casein	A^1, A^2, A^3, B, C, (D), E
κ-casein	A, B, C, E
β-lactoglobulin	A, B, C, D, (E, F, G), H, W, X
α-lactalbumin	A, B, (C)

[a] Parentheses indicate variant not found in European cattle.

Source: FitzGerald, 1997.

Comparisons of frequencies of various protein alleles among some dairy breeds revealed that the genotypic frequency for κ-casein BB is 2 to 4% for Holstein–Fresians, but 46 to 49% for Jerseys. Others report higher frequencies (cf. chapter 30). In cheesemaking, milks containing κ-casein BB have relatively short rennet coagulation time, faster **curd** formation, higher curd firmness, and fewer fines in the whey than milks containing κ-casein AA (see FitzGerald, 1997). For example Walsh et al. (1995) observed comparative rennet coagulation times of 20.3 and 16.7 minutes for milks containing AA or BB variants (also known as α and β variants), respectively, at pH 6.60. Higher yields of cheese were associated with the higher recovery of fat in cheeses made with κ-casein BB milk. The "carbon footprint" per unit of cheese produced was 20% less for Jersey than for Holstein cows (Capper and Cady, 2010).

Colloidal calcium phosphate binds together the hundreds or thousands of submicelles forming casein micelles (figure 21.9 on the next page). Binding may be covalent or electrostatic. Submicelles rich in κ-casein occupy surfaces, whereas those with less are buried in interiors. Walstra (1999) proposed a structure for casein micelles (CMP = caseinomacropeptide, a hydrophilic portion of κ-casein).

High numbers of proline residues in caseins cause bending of protein chains and inhibit formation of close-packed, ordered secondary structures. Caseins contain no disulfide bonds. The lack of tertiary structure accounts for the stability of caseins against denaturation by heat because there is little structure to unfold. Without a tertiary structure there is considerable exposure of hydrophobic residues, resulting in strong associations among casein molecules, rendering them insoluble in water.

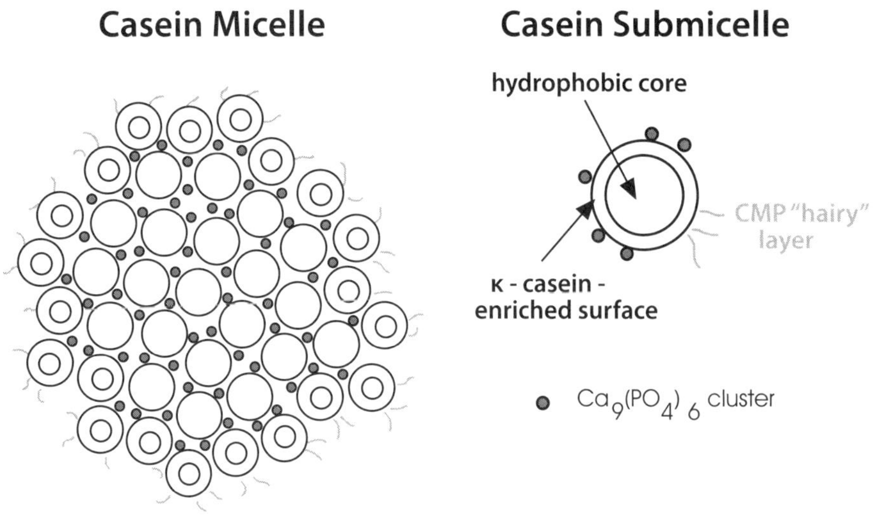

Figure 21.9 Proposed structure of casein micelle and submicelle (Department of Food Science, University of Guelph).

In-depth studies of the structure and nature of casein have been conducted in the laboratories of the USDA resulting in the pictures shown in figure 21.10 (Carroll et al., 1985; Farrell et al., 2006). Although studied for many years, the nature of the internal structure of casein micelles remains unresolved. Some models focus on the importance of protein structure, which is paramount during micelle biosynthesis; others focus on the importance of added calcium/phosphate.

To characterize a protein one must determine content of amino acids and other constituents, sequence of these components and their number in the molecule, and molecular configuration (arrangement of the molecule in space). Other valuable information about a protein includes its **isoelectric point**, heat stability, stability in the presence of salts such as sulfates of ammonium and magnesium, electrophoretic mobility (direction and rate of movement in an electrical field), and molecular weight. Tests of these properties indicated considerable diversity (polymorphism) among proteins in milk.

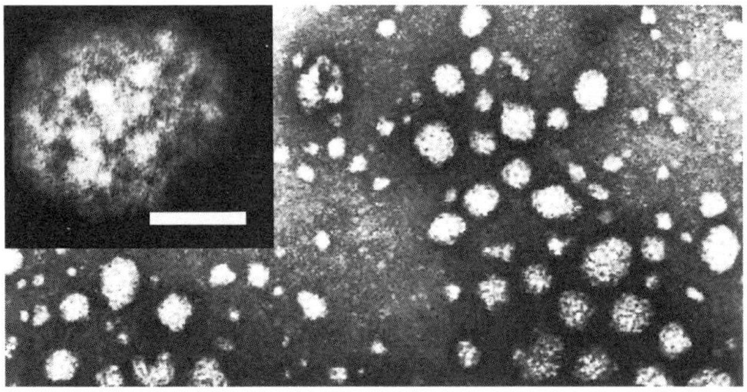

Figure 21.10 Negatively stained casein micelles from human milk; this stain emphasizes the internal protein structures of the micelles. The inset shows a potential submicelle with the bar equal to 30 nm (Carroll et al., 1985).

Casein is synthesized in secretory cells of mammary glands in a highly aggregated state making more or less spherical micelles ranging from 0.03 to 0.3 μm in diameter (1 μm = 10^{-6} m) in bovine milk. Inside the casein micelle are cavities or channels that admit relatively large molecules of at least 36,000 molecular weight. Molecules in the micelle are arranged in a definite structure, which determines many of the functional properties. However, a dynamic equilibrium exists among the micelles, submicelles, free casein molecules, and the dissolved calcium and phosphate. About 10,000 casein monomers comprise the average casein micelle. On cooling to 39°F (4°C), β-casein begins to dissociate from the micelle, and at 32°F (0°C), micelles do not aggregate.

Among casein monomers, β-CN has some unique characteristics that make it of particular interest to food producers. It is composed of 209 amino acids, has a flexible structure, lacks internal covalent cross-links, including disulfide bonds, and has a moderate charge with a distinct distribution of charged groups (Farrell et al., 2004). It has a hydrophilic (water-loving) phosphoseryl cluster at the N-terminal and hydrophobic amino acid residues, such as proline, at the C-terminal. This makes β-CN an excellent surfactant and stabilizer ingredient for functional food applications (Dickinson, 2006). Importantly, β-CN may serve as a precursor for production of bioactive peptides having antihypertensive, opioid, or mineral-binding properties (Phelan et al., 2009). Methods of isolating and concentrating β-CN for various uses are being developed.

21.3.2 The Soluble or Whey Proteins

Proteins remaining in the liquid portion of milk after precipitation of caseins at pH 4.6 are collectively called whey proteins. These globular proteins are water soluble and are subject to heat denaturation. Native whey proteins have good gelling and whipping properties. Denaturation increases their water binding capacity. The principle fractions are β-lactoglobulin, α-lactalbumin, bovine serum albumin (BSA), and immunoglobulins (Ig).

β-lactoglobulin contains only amino acids—162 of them. Unlike casein, it contains free sulfhydryl (–SH) groups in the form of cysteine (an amino acid) residues. These have been implicated in development of a cooked flavor when milk is heated. All ruminant milk contains β-lactoglobulin, whereas milk of nearly all nonruminants does not. The hydrophobic area (a barrel-shaped structure) within the molecule effectively binds retinol. Thus it is suspected of having a regulatory role related to vitamin A in the mammary gland. Below pH 3 and above pH 8, β-lactoglobulin exists as a monomer, between pH 3.1 and 5.1 and at low temperatures it forms an octamer, and at milk's pH (6.7) it exists as a dimer.

Certain hormones are required to synthesize α-lactalbumin: insulin, cortisone, estrogen, and prolactin; synthesis is inhibited by progesterone. Late in pregnancy progesterone concentrations in the mammary gland decrease and manufacture of α-lactalbumin is initiated. This protein is essential for the synthesis of lactose in milk. It is relatively heat stable because it interacts with calcium to form intramolecular ionic bonds that help it resist thermal unfolding, even to temperatures as high as boiling given favorable pH and calcium content.

The four classes of immunoglobulins, IgG1, IgG2, IgA, and IgM, are closely related in chemical composition and physical properties. Antibodies of milk are especially abundant in colostrum, providing passive immunity to the newborn calf. These antibodies have the highest molecular weight of milk proteins. Since they are not synthesized in the mammary gland, they must be transported via the blood into milk.

Blood serum albumin apparently passes unchanged through membranes of udder tissues into milk. Concentrations of blood serum albumin increase when such tissues become inflamed because of increased permeability. The authors have demonstrated that blood serum albumin binds and inactivates some antibiotics (notably penicillin G) used in mastitis therapy.

A large pool of minor proteins exists in milk in amounts varying with stage of lactation. They can make a substantial contribution to the total bioactivity related to the calf. Some of these are encrypted within the native amino acid sequences of the major proteins. Exposure of the neonate to the diverse and numerous proteins and peptides of milk regulates the animal's development and boosts its defenses against pathogens. Particularly significant is the presence of large amounts of hormones and growth factors in colostrum (Wynn et al., 2011).

Highly soluble, low molecular weight (short polymers or polypeptides) proteins are found in milk, as are enzymes that are composed primarily of proteins. Proteose-peptones are those proteins that remain in solution after milk has been heated to 203°F (95°C) for 20 minutes and then acidified to pH 4.7. They can be precipitated with 12% trichloroacetic acid. Enzymes are organic catalytic agents that participate in chemical reactions without being used themselves. They lower requirements for energy to make a reaction occur. They are vital in biological reactions.

Milk allergy is the body's response to an allergen-antibody reaction. An allergen is an ordinarily harmless substance present in the diet or environment, which in sensitized persons is able to produce disorders such as asthma and gastrointestinal upsets. Symptoms of milk allergy are asthma, rhinitis, vomiting, abdominal pain, diarrhea, urticaria, and **anaphylaxis**. Milk proteins are the **etiologic** agents in milk allergy. Almost all proteins are antigenic and must be considered as potential allergens. Although principal allergenic activity among milk proteins has been ascribed to β-lactoglobulin and α-lactalbumin, the question is controversial. Inconclusive but substantial evidence points to increased allergic reaction among sensitive persons to the condensation product of lactose and the ε-amino group of lysine (an amino acid in β-lactoglobulin). This is a product of the browning or **Maillard reaction** that results from extensive heating of milk.

Milk allergy occurs primarily in infants and children up to two years of age. Estimates of incidence vary widely from 30% in allergic children to less than 0.1 to 7.0% in nonallergic children. Antibodies to milk proteins are detectable within less than a month following the first ingestion of milk. From birth until the third month, the human gastrointestinal mucosa is relatively permeable to undigested proteins of foods. Hence, sites of antibody formation are more likely to be exposed to foreign protein, particularly in infants from families with a genetic background of allergy. Obviously, since the probability of developing milk allergy decreases markedly with increasing age, the practice of breast feeding would be recommended especially in those families with a history of unfavorable reactions.

Infants fed breast milk for different periods, before being switched to cow's milk, reacted with significantly lower levels of antibody than those fed pasteurized or dry milk immediately following birth, with the degree of reduction in response being directly related to the period of breast feeding (Gery, 1970).

21.4 Lactose

Lactose is the carbohydrate of milk and is usually the most abundant of the three major solid components of milk. Cows' milk contains about 5.0% lactose, whereas human milk averages 7.0%. For this reason lactose is often added to formulas for infants. In addition to providing caloric value (30% milk's food energy), lactose makes milk sweet, provides the substrate from which lactic acid is produced in fermented dairy products, adds body to milk and foods in which it is used, takes part with proteins in the caramelization (browning) reaction, and is related to texture and solubility in stored milk products. Milk also contains about 7 mg of glucose and about 2 mg of galactose/100 mL. Lactose is relatively insoluble so that it may crystallize in some dairy products, especially ice cream and sweetened condensed milk.

Lactose is synthesized only by cells of lactating mammary glands through the reaction of glucose with uridine diphosphate galactose. The enzyme galactosyl transferase is modified by α-lactalbumin, specifying it (modifying galactosyl transferase) so it catalyzes synthesis. In the nonlactating gland, glucose participates in a different reaction because α-lactalbumin is not present.[4]

Hydrolysis of lactose (12 carbons) yields glucose (6 carbons) and galactose (6 carbons). In the cells of the digestive human tract the enzyme lactase (β-D-galactosidase) is responsible for the hydrolysis of lactose. Lactase acts primarily in the jejunum, the second of three main sections of the small intestine, where villi (metabolically active, differentiated, nondividing cells) contain it in their brush-like border. Galactose and part of the glucose enter the bloodstream and are metabolized by the liver. Some glucose is fermented by intestinal microorganisms, thus lowering the pH. This, in turn, helps inhibit growth of undesirable putrefactive bacteria (Proudfit and Robinson, 1967).

While growing, desirable bacteria synthesize B vitamins for use by the host. Lactose in the intestine also favors absorption of calcium, magnesium, and phosphorus and the utilization of vitamin D. If the quantity of lactose present in intestinal cells exceeds the hydrolytic capacity of the available lactase, a portion of the lactose remains undigested. Some is absorbed, passes into the blood, and is metabolized by the liver or excreted in urine. In the jejunum the remainder exerts high osmotic pressure, causing water and sodium to be excreted into the intestinal lumen. In the colon the lactose is fermented into short-chain fatty acids, hydrogen, methane, and CO_2. Some short-chain fatty acids are absorbed into the bloodstream. The remainder is excreted in feces, resulting in acidic stools. The symptoms of lactose maldigestion are excess gas production and, in extreme cases, diarrhea. Of all the dietary sugars, lactose is hydrolyzed most slowly, about one-half the rate of sucrose hydrolysis.

Since some lactose is fermented in cultured dairy foods, the amount of lactase needed is less than for an equivalent amount of the nonfermented counterpart. Ripened cheeses have had all of their lactose hydrolyzed. Most individuals who have primary lactase deficiency do not experience intolerance to the amount of lactose contained in an 8-oz glass of milk (12 g).

Although infants generally produce sufficient lactase, a small number appear to be deficient, as indicated by breath hydrogen tests. Under select conditions, presence of hydrogen in the breath suggests fermentation of lactose in the colon. Primary lactase deficiency, as well as the decline of lactase activity with age, occurs at variable times after weaning, depending on racial/ethnic background (Miller et al., 1995). People of African and Asian descent have the earliest decline in lactase levels, beginning at about age three. As adults, about 80% of these people show some level of lactose maldigestion while persons of north European descent have the lowest occurrence at about 20%.

Consumption of lactose is reported to increase the ability to digest lactose, either because it induces the production of lactase or because it results in modification of the microbial flora of the intestine so the number of bacteria that metabolize lactose increases. The preferred test for lactose intolerance involves measurement of blood glucose after ingestion of lactose (figure 21.11). Intolerance means that, after a 12-hr fast, blood glucose concentrations fail to increase significantly after ingestion of a large quantity (50 to 100 g) of crystalline lactose. Conditions of this test differ significantly from ingestion of moderate quantities of milk[5] and the diagnosis of lactose intolerance by this method should not be equated with milk intolerance. Many persons who react to the clinical lactose tolerance test can consume a glass or two of milk daily with meals and have no apparent difficulty.

Lactase deficiency may be found in diseases of the small intestine wherein the enzyme is decreased due to

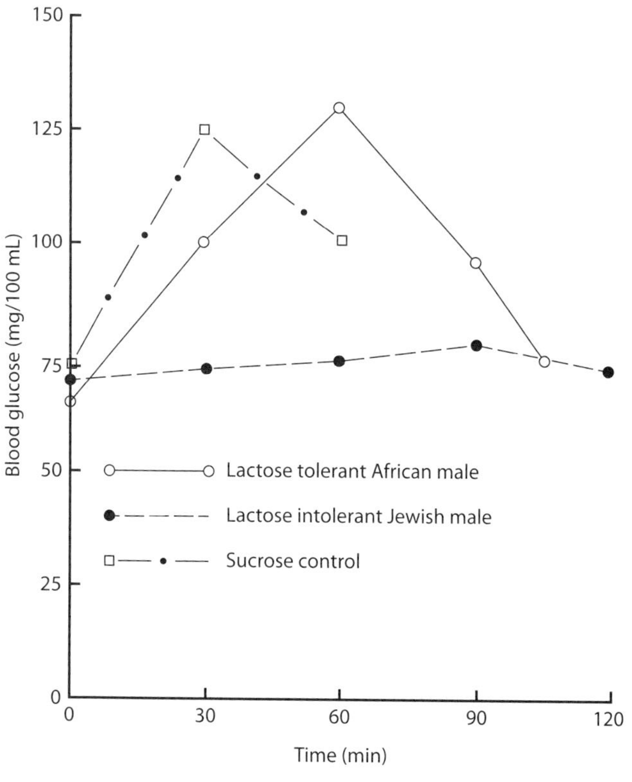

Figure 21.11 Blood glucose concentrations after ingestion of lactose and sucrose indicate lactase activity (adapted from Kretchmer, 1972).

structural abnormalities in the bowel wall. Such secondary lactase deficiency may occur transiently after an attack of gastroenteritis or, more chronically, in diseases of the intestinal mucosa such as celiac disease or tropical sprue. After successful therapy, lactase levels normalize and lactose may again be tolerated well. Environmental factors may play a role in some areas of the world since unsanitary conditions are often associated with recurrent gastrointestinal parasitosis. These conditions will injure the bowel wall, thereby inhibiting production or release of intestinal lactase. Some researchers have suggested that lactose intolerance in adults is the normal genetic state and that, among groups of people who began to consume milk of cows, sheep, and goats, a selective advantage was conferred on individuals who, because of chance mutation, had high enough lactase activity to digest lactose. This is, of course, speculation based on evidence of a genetic basis for lactose intolerance.

Partial fermentation of lactose in buttermilk, yogurt, and such products increases quantities of such food a lactose-intolerant person can ingest without experiencing adverse symptoms. More than 20% of the lactose of milk is fermented to raise the acidity to 1.0% lactic acid because the metabolic efficiency is less than 100%.

Two isomers of lactose exist, alpha (α) and beta (β). The alpha form normally crystallizes as a monohydrate (one molecule of water attaches to each molecule of lactose). However, on drying at a high temperature (above 212°F [100°C]) in vacuum, water is removed to form α-anhydride. Both alpha and beta forms exist in solutions at room temperature. Also, lactose glass (highly concentrated or supersaturated solution without crystals) can be formed by rapid removal of water from lactose solutions. No crystals form because high viscosity slows diffusion of lactose molecules to centers of crystal growth (nuclei).

Lactose is one-half to one-fifth as sweet as sucrose (table sugar) depending on concentration. It is relatively sweeter at higher than at lower concentrations. The sweetness of β-lactose is greater than that of α. However, the small difference is quickly eliminated in solutions due to mutarotation (conversion of β to α, and vice versa).

Compared with most sugars, lactose is relatively insoluble in water. At 77°F (25°C) about 18 g of lactose (racemic mixture of α and β forms) will dissolve in 100 g of water. Milk, concentrated 3:1 (as in condensed milk), approaches saturation with lactose at room temperature and the saturation point may be exceeded on cooling. If so, α-lactose hydrate crystals form. Solubility of α-lactose approximates 7 g/100 g of water at 59°F (15°C), whereas about 50 g of β-lactose will dissolve in the same amount of water. Tendencies of concentrated solutions to crystallize present both problems and opportunities to dairy technologists. Problems occur in the manufacture of frozen desserts, concentrated milks, dry milks, and dry wheys.

21.5 Minerals

Minerals of milk are not only important in nutrition; they also strongly affect the physical state and stability of milk proteins. They affect flavor and sometimes catalyze chemical reactions that occur in milk. The following minerals are considered dietary essentials for humans: potassium, calcium, chlorine, phosphorus, sulfur, magnesium, zinc, iron, copper, iodine, fluorine, manganese, molybdenum, and cobalt.

Minerals of milk are present in salt forms and they form the residue that remains after ashing. However, it is incorrect to equate minerals, salts, and ash in milk. Salts include the ionic constituents except for hydrogen ions and hydroxyl ions. Thus, they include metals, inorganic acid radicals (e.g., phosphates), and organic acid radicals (e.g., lactate). Even proteins can be considered salts since they carry a net negative charge and associate primarily with cations (e.g., calcium). Incineration destroys the organic portion of milk and modifies certain minerals. Carbonates, sodium, and chloride may be volatilized. However, it is possible to obtain an approximation of the mineral content of milk by an analysis of ash.

The major minerals in milk and their approximate percentages are potassium (0.14); calcium (0.12); chlorine (0.10); soluble and **colloidal** phosphorus (0.9); sodium (0.05); sulfur (0.02); and magnesium (0.01). In addition to these, milk contains minor amounts of zinc, iron, copper, iodine, bromine, fluorine, boron, nickel, manganese, molybdenum, and cobalt, plus at least 13 microelements. For some elements the concentration in milk depends on the amount consumed in the feed. Copper, iron, and nickel are dissolved in milk slowly, hence they can enter from equipment. As milk acidity increases, so does the tendency to dissolve zinc, tin, and aluminum. Amounts of minerals in milk are not constant. For example, significant differences by season and geographical location have been shown for calcium, with lowest concentrations occurring in late summer.

Potassium, sodium, chlorine, and sulfates of milk are dissolved (in true solution). However, the solubility of some salts is exceeded, and the excess is present in colloidal particles containing casein, calcium, magnesium, phosphate, and citrate. Two-thirds of the calcium is in the colloidal form as calcium caseinate, phosphate, and citrate. Phosphorus is present in five types of compounds of milk: inorganic salts (33%) and organic esters (7%) in solution; and inorganic salts (38.5%), proteins (20%), and lipids (1.5%) in colloidal dispersion.

Of course, colloidal salts are in equilibrium with dissolved salts. If water is added some of the colloidal salts will dissolve. Attempts to study milk salts are hampered because of the dynamic equilibrium (changing ionic and molecular state) of the ions in milk. Shifts in balance among the salt forms occur as milk is heated or cooled. The solubility of some components increases with temperature, but calcium phosphate becomes less soluble as temperature is increased. As milk is concentrated, calcium phosphate and calcium citrate accumulate in colloidal form and hydrogen ions are liberated from their salts. This lowers pH. Additions of acid cause pronounced shifts in salt equilibriums. Calcium and phosphorus are released from colloidal particles as pH is lowered until, at slightly above the isoelectric point (pH of 4.7), they are completely solubilized.

Considerable carbon dioxide (CO_2) is dissolved in fresh milk (about 20 mg/100 mL), but much of it is quickly lost to the air. The balance is present as carbonates and, as such, acts as an acid. However, exposure to heat and vacuum allows the removal of virtually all CO_2. Consequently, titratable acidity decreases (except as countered by release of H+ from colloidal calcium and phosphorus) and the pH and freezing point increase (less dissolved substance in milk).

Ionic constituents of milk, including salts, aid in *buffering*, which is a measure of the amount of acid or base (number of equivalents) needed to shift the pH of a liter by one unit. Major buffering components of milk are phosphates, CO_2, proteins, citrates, and lactate.

21.5.1 Calcium

Milk and milk products supply about 75% of the dietary calcium for humans in the United States. Of other foods, only leafy vegetables have sufficient calcium to meet dietary requirements, and calcium of vegetables is not as readily utilized by humans as is calcium of milk.

Calcium is an essential nutrient with multiple functions. The average adult human body contains 1.0 to 1.5 kg of calcium. About 99% is in the skeleton. The skeleton serves as the storehouse of calcium, a mineral that performs vital functions in coagulation of blood and regulation of muscle activity (including the heart) and nerve irritability. In the adult, new bone is continually being formed in certain areas and reabsorbed in others, but in normal healthy persons these processes are in equilibrium, so total mass of bone remains relatively constant. Unlike bone, teeth do not participate in the full cycle of cellular remodeling. It is important that proteins and minerals be present in optimal amounts during primary development in childhood.

The quantity of dietary calcium recommended varies with age and sex. For children 4 to 8 years of age, the recommend daily allowance (RDA) for calcium is 1,000 mg. For males and females 9 to 18 years of age it increases to 1,300 mg, but for males ages 19 to 70 it decreases back to 1,000 mg. For females ages 19 to 50 it decreases back to 1,000 mg, however, for females over age 50, requirements increase once again to 1,200 mg. Pregnant and lactating females also require between 1,000 to 1,300 mg of calcium, depending on their age (Institute of Medicine, 2011).

Continued low intake of calcium may result in failure of growing persons to attain their genetic potential in height. In adults osteoporosis often results. Osteoporosis is a disease of the bone in which the mass is seriously diminished and fragility results. It is estimated to affect about one-third of postmenopausal women. It appears to be a complex disorder with both endocrinal (hormonal) and nutritional factors involved in its development and treatment. Improvement in calcium storage can be obtained by elevating dietary calcium intake to well over the equivalent of a quart of milk daily.

Calcium and fat absorption are interrelated. Impairment of the absorption of either impairs absorption of the other. Lactose stimulates absorption of calcium, magnesium, phosphorus, barium, and other minerals in the intestinal tract of humans.

Calcium concentrations of milk remain stable despite changes in calcium content of feed. This is true because the cow can draw on skeletal reserves in times of dietary deficiency or can excrete calcium in times of dietary surplus. However, calcium concentrations can be affected sufficiently by feeding to influence the stability of milk products (Sommer, 1952). The authors have observed decreases in calcium and phosphorus content of milk associated with mastitis.

21.5.2 Phosphorus

About 85% of the phosphorus in the human body is combined with calcium in bones and teeth. Muscles, the brain, nervous tissues, and blood also contain considerable phosphorus. Phosphorus is of immense importance in enzymatic reactions that produce energy for the body because it is used in the formation of high-energy bonds in adenosine triphosphate (ATP). In blood it is needed especially in maintaining the acid-base balance.

Milk contains calcium and phosphorus in the highly favorable ratio of about 1.2:1.0. A quart of milk contains about 0.9 g of phosphorus, an amount sufficient for children and adults; teenagers and pregnant and lactating women need about 1.2 g.

Phosphorus is not likely to be deficient in diets that contain adequate protein and calcium. Human milk contains about 30% as much calcium and only 15% as much phosphorus as cows' milk. In some infants under two weeks of age the kidneys are unable to excrete the excess phosphorus absorbed from cows' milk, and phosphorus accumulates in the serum where it combines with calcium and can cause tetany.

21.5.3 Interrelationships among Minerals

Since milk contains all essential minerals (iron and copper in limited amounts), it protects against mineral imbalance. Many minerals are involved in the intricate processes of regulation of enzymatic reactions, balancing acid-base ratios, and facilitating transfer of essential compounds across membranes. Especially important are magnesium and potassium, which are counteracted by calcium and sodium.

21.6 Vitamins

Vitamins are substances found in low concentrations in plant and animal tissues (including milk), which are necessary for metabolic reactions within tissues. Each exerts control over specific reactions. Humans must ingest vitamins in foods, though some are produced in part by bacteria in the alimentary tract or by tissues.

Removal of fat from milk separates the fat-soluble from the water-soluble vitamins. This means, unless fortified, skim milk contains only traces of the fat-soluble vitamins. Table 21.5 shows vitamins of milk and several important characteristics.

Table 21.5 Selected Vitamins of Cow's Milk.

Vitamin	Quantity (per qt)	Percent of Adult RDA
A[a]	450 μg	50
D[b]	5–15 μg	100
Thiamine	0.4 mg	30
Riboflavin	1.5 mg	100
Nicotinic acid	0.8 mg	5
Ascorbic acid	15 mg	20

[a] Includes carotene.

[b] Vitamin D added milk (13 to 31 IU average for raw milks in winter and summer, respectively).

Quantities of vitamins in milk and milk products are determined by three assay methods: biological (feeding trials with animals), microbiological (growth studies with bacteria), and chemical (or physicochemical).

21.6.1 Fat-Soluble Vitamins

This group includes vitamins A, D, E, and K. The vitamin A content of milk varies with feeds and feeding practices. Provitamin A carotenoids occur abundantly in green, leafy plants. However, loss of potency occurs in haymaking and **ensiling**. Thus, milk produced from pasture usually contains higher concentrations of vitamin A. Generally, the more natural the color maintained, the higher the potential vitamin A in forage. Feeding of tocopherol (vitamin E) appears to improve assimilation and utilization of carotene, probably because of vitamin E's antioxidant properties. Summer milk has been estimated to contain 1.6 times as much vitamin A as winter milk.

21.6.1.1 Vitamin A

Cows differ in the amount of conversion of carotenoids to vitamin A. Guernseys and Jerseys, for example, convert a smaller proportion (about 60%) of β-carotene[6] to vitamin A, whereas Holsteins and Ayrshires accomplish a higher rate of conversion (about 80%). Conversion of carotene into vitamin A in the dairy cow occurs in the wall of the small intestine.[7] Humans readily convert β-carotene to vitamin A.

The quantity of vitamin A in milk is roughly proportional to the fat content, but there are indications that the carotenoids and vitamin A are present in both the fat globule membrane and in the protein complex.

Normal heat processing, such as pasteurization, evaporation, sterilization, and drying, causes little reduction in the vitamin A or carotene content of milk. However, prolonged heating in the presence of air results in reduced vitamin A activity.

Added vitamin A in fortified skim milk loses potency because the vitamin is no longer protected by fat, as it is in whole milk. Protection from light decreases the rate of deterioration, as does the addition of antioxidant. Palmitate is attached to the alcohol form of vitamin A (retinol) in order to make vitamin A stable in milk.

21.6.1.2 Vitamin D

Several structurally related compounds have vitamin D activity. These can be formed by ultraviolet irradiation of certain provitamin steroids. If irradiation is extended too far, vitamin activity is lost. The two principal forms of the vitamin, both of which appear in milk, are vitamin D_2 (calciferol), produced by irradiation of **ergosterol** (found in plants and yeast), and vitamin D_3, produced by irradiation of 7-dehydrocholesterol (found in higher animals and humans) (figure 21.12). Although in the purified state neither is stable to air and heat, solutions in oil or dispersions in milk keep excellently. Biological assays are required to determine activity.

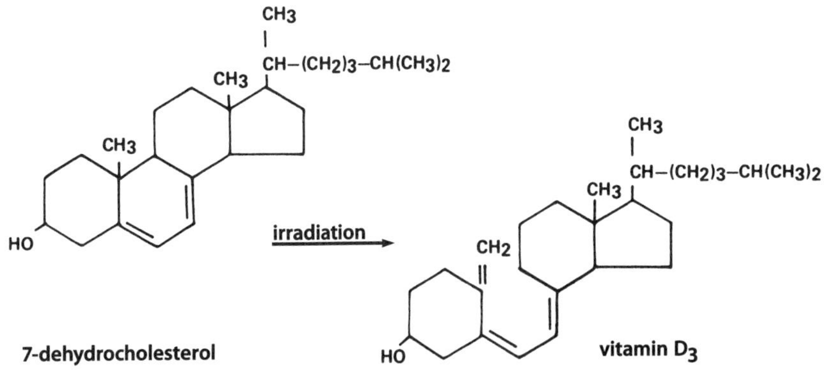

Figure 21.12 Vitamin D_3 is produced in humans and higher animals by irradiation of 7-dehydrocholesterol.

The main function of vitamin D is promotion of absorption and transport of calcium through the intestinal wall. It is converted in the kidneys to calcitriol, which has hormonal activity that promotes healthy mineralization of bone (Adams and Hewison, 2010). A deficiency results in rickets in children and osteomalacia in older people. Both diseases cause soft and misshapen bones. Calcitriol also acts in the immune system to defend the body against microbial invaders. Vitamin D also modulates neuromuscular function, reduces inflammation, and influences the action of many genes that regulate the proliferation, differentiation, and apoptosis of cells.

The content of vitamin D in milk, as in other natural foods, is low. Amounts present are related to the fat content of milk and depend both on diet and exposure of cows to sunlight. Milk has been selected as the logical food in which to add vitamin D since it furnishes the major supply of dietary calcium and phosphorus. It is usually added in the amount of 400 IU (10 μg) per quart (100 IU/cup). By law in Canada, milk is fortified with 35 to 40 IU/100 mL. Fortification is almost always by addition of vitamin D concentrates (irradiated ergosterol or concentrates from fish liver oil) to milk, though it can be accomplished by feeding substances rich in the vitamin (e.g., irradiated yeast). European milk was at one time fortified with vitamin D, but too much was used; the toxic effects on children led to a European government ban on vitamin D fortification of milk in the 1950s (Holick and Jenkins, 2005).

21.6.1.3 Vitamin E

Vitamin E activity is exhibited by a group of substances known as tocopherols. Of the four best known, α-tocopherol is most potent. It functions as an antioxidant that stops production of reactive oxygen species formed when fat undergoes oxidation. Thus it is important in efficient energy production from foods.

Milk contains relatively small amounts of vitamin E, almost solely in the alpha form, which mammals selectively absorb and deposit in their tissues. Quantities average less than 1 mg/qt; whereas the RDA for adults is 15 mg. Milk from cows on pasture or chopped green forage usually contains about twice as much tocopherol as when they are fed dry feeds. The efficiency of transfer of the vitamin from the ration was as low as 2% in one study. Feeding of α-tocopherol gives partial protection against development of copper-induced oxidized flavors in milk. Extensive losses in vitamin E result when milkfat becomes oxidized with the development of metallic to fishy flavors. Vitamin A and carotene are also lost. Alpha-tocopherol is stable to most treatments used with milk processing.

21.6.1.4 Vitamin K

The chief function of vitamin K is regulation and maintenance of normal concentrations of prothrombin in blood. Low levels of prothrombin lead to delayed clotting. Deficiencies occur most often in newborn infants prior to establishment of the intestinal microflora that synthesize vitamin K. Although animal products, including milk, contain little vitamin K, the intestinal flora synthesizes adequate amounts for most animals.

21.6.2 Water-Soluble Vitamins

This group includes the B complex and vitamin C.

21.6.2.1 B Complex Vitamins

Vitamins of the B complex are synthesized by the **rumen microflora** at a position in the alimentary tract from which they become available for absorption into the system. Quantities in milk, therefore, are independent of quantities consumed in cattle feeds. Concentrations of thiamine, riboflavin, vitamin B_6, biotin, pantothenic acid, nicotinic acid, and folic acid are all greater in ruminal contents than in feeds ingested by the cow. Nicotinic acid is also synthesized from the amino acid tryptophan in animal tissues. Inositol is probably synthesized in the rumen. The calf begins to synthesize thiamine and riboflavin upon commencing consumption of forage.

Thiamine (B_1) is essential for metabolism in all animals. In the phosphorylated form it is a coenzyme active in several biochemical reactions in the intermediary metabolism of carbohydrates. The body stores only enough thiamine for a few weeks' reserve. It is widely distributed in foods and is especially abundant in milk. One quart of milk supplies about 30 to 40% of the RDA for persons over 10 years of age (about 0.4 mg of the 0.9 to 1.2 mg recommended). Less thiamine is recommended for children (0.4 to 0.5 mg) and infants (0.2 to 0.3 mg). Thiamine is partially destroyed by heat, although losses in **high-temperature short-time (HTST)** or **ultrapasteurization** are less than 5%. Sterilization in cans, as with evaporated milk, results in losses of 20 to 60% of the activity. Modern drying practices result in retention of most thiamine activity. Most bacteria used in preparing cultured milks use thiamine, particularly during early hours of incubation. Some produce appreciable quantities later in the process.

In 1933 several investigators observed growth-promoting effects of the water-soluble yellow extracts of several foods. Depending on origin, they were known as lactoflavin, ovoflavin, hepatoflavin, verdoflavin, and uroflavin, but their biological activity proved to depend on a single substance, riboflavin. It is the component of whey that imparts a greenish-yellow color. The vitamin is relatively stable to heat in neutral and acid foods, but is highly sensitive to light, especially at high pH and temperatures. The pigment in chocolate milk protects against light degradation of riboflavin.

Riboflavin is a constituent of two coenzymes vital to metabolism of glucose, fatty acids, amino acids, and purines in humans and animals. Deficiencies are manifested as skin lesions, especially around the mouth and eyes. Milk, skim milk, and whey are rich sources of riboflavin. One quart of milk is a major source of the vitamin in diets of consumers. Quantities in milk are usually highest during spring and summer. It appears, however, that effects of alteration of the cows' diets must be indirect, being mediated through differences in the ruminal microflora. Jersey and Guernsey cows secrete milk with 25 to 50% more riboflavin than that of Ayrshires and Holsteins, and colostrum contains two to four times the concentration found in milk secreted later.

Riboflavin is associated with development of light-activated (oxidized) flavor in milk, which involves the photolysis of methionine (oxygen is required). In fermentations of milk some microbial **cultures** use riboflavin, especially in early incubation. Sometimes they secrete riboflavin on extended incubation.

Xanthine oxidase, which contains riboflavin, is part of the lipoprotein complex of milk. This complex is composed of phospholipids, nucleic acid, and enzymes. It is adsorbed to the surface of fat globules in fresh milk. Milk contains about 120 mg/L of xanthine oxidase.

Another B vitamin of great importance in tissue respiration is nicotinic acid (niacin). Nicotinamide is a closely related active form of the vitamin. Its deficiency results in pellagra, a disease that once caused thousands of deaths in the southern United States. Nicotinamide is a constituent of two important coenzymes, **nicotinamide adenine dinucleotide** (**NAD**) and **nicotinamide adenine dinucleotide phosphate** (**NADP**). They are involved in the glycolytic and citric acid cycles and in breakdown of fatty acids.

Although a quart of milk supplies only about 5% of the niacin requirement (13 to 20 mg), it is among the most valuable antipellagric foods. There are two reasons: (1) the amount present in milk is fully available, whereas much of it in certain other foods is bound and unavailable, and (2) milk is an excellent source of tryptophan, which can be used by humans to synthesize niacin (60 mg of tryptophan equals 1 mg of niacin).

Little variation in the concentration of niacin due to diet, breed, or stage of lactation occurs in milk. The vitamin is stable to heat and sunlight. In fermentations of milk and cheese most bacteria use niacin in early growth but synthesize and release it later.

Vitamin B_6 is important in metabolism of **essential fatty acids** and amino acids and is required in transamination, i.e., the transfer of amino nitrogen from one amino acid to the **keto analogue** of another. It has three forms: pyridoxine, pyridoxal, and pyrdoxamine. The RDA for humans is 1.0 to 1.5 mg/day, and milk supplies about 0.5 mg/qt.

It is probable that alterations in the ruminal microflora result in the higher concentration of vitamin B_6 often observed in milk produced when cows go to pasture in spring and early summer. Breed seems to have no effect. Pasteurization causes no significant decreases in vitamin B_6 activity but sterilization destroys about 50% of it. It is mildly susceptible to light inactivation.

Pantothenic acid forms part of coenzyme A, which plays a vital role in metabolism of carbohydrates, fatty acids, and nitrogenous compounds. The human requirement has been estimated as 5 mg/d, and milk supplies about 3 mg/qt. Neither breed nor feed is an important variable affecting the content found in milk. There is little destruction of the vitamin in pasteurization, sterilization, drying, or prolonged exposure to sunlight.

Choline is required by all animals in comparatively large amounts, chiefly to be used in lecithin and sphingomyelin. About half of milk's choline is water soluble. The remainder is bound, mainly in the phospholipids. One quart of milk supplies slightly more than 100 mg of choline; the RDA ranges from 375 to 550 mg for adults ingesting mixed diets. Heat has little effect on choline.

Milk is a dependable source of biotin and is a substance from which this vitamin was originally isolated. Average concentration in milk is about 29 mg/qt, about 60% of the RDA. Intestinal bacteria produce biotin except when sulfa drugs (and probably antibiotics) affect them adversely. Avidin in raw egg whites binds biotin, rendering it inactive. No more than 10 to 15% of the biotin is destroyed in pasteurizing or sterilizing milk.

Folic acid (pteroylglutamic acid) is necessary for development of normal red blood cells. Its **conjugates** are also active. Milk is a relatively poor source of this compound.

Vitamin B_{12} possesses the most complex structure of any vitamin and is the only one with a metallic element (cobalt) in its molecule. Milk contains about 4 μg/qt, which exceeds the RDA for adults. It is highly stable to pasteurization (HTST or **ultrahigh temperature** [**UHT**]) but not to in-can sterilization, such as in evaporated milk. **Deaeration** stabilizes the vitamin to heat.

21.6.2.2 Vitamin C

Scurvy is a disease that results from a deficiency of ascorbic acid (vitamin C). For hundreds of years many deaths occurred among sailors who were away from land for extended periods of time. In the sixteenth century citrus fruits were recognized as useful in preventing or curing scurvy. Vitamin C is needed for formation and maintenance of intracellular material, such as collagen of fibrous tissue and matrices of bone, cartilage, and dentin.

Immediately after milk is removed from the udder, vitamin C is in the reduced form, L-ascorbic acid. Slow oxidation occurs at a rate dependent on oxygen content, exposure to light, copper content, and temperature. The first oxidation product, dehydro-L-ascorbic acid, is biologically active. Copper catalyzes this reaction, which is reversible. Further oxidation to diketo-L-gulonic acid renders vitamin C inactive, and the reaction is irreversible. Fresh **raw milk** contains an average of about 20 mg/qt of vitamin C. However, dehydroascorbic acid is completely oxidized during preheating in HTST pasteurization, and by the time milk reaches the consumer only about 25 to 50% of the vitamin is active, sometimes less, depending on treatment. (Vacuum treatment removes oxygen and slows deterioration.) The RDA for adults ranges from 75 to 90 mg. Therefore, milk is not an important source of vitamin C. The cow requires no external source of vitamin C; she synthesizes it in her body tissues, especially in the liver.

The vitamin C content of milk appears to be practically independent of the cow's diet. When milk is held in storage, however, vitamin C is more stable if cows have been fed grass silage or pasture than if they have been fed dry winter feeds. Presumably there are more antioxidants in the milk produced from silage or pasture. Fermentations in milk increase the stability of vitamin C, probably because they lower the amount of dissolved oxygen and add acid. Vitamin C is stable under both of these conditions.

There is little or no ascorbic acid in ripened cheese. The dispersal of air into ice cream in freezing appears to contribute to oxidation of the remaining vitamin C. Either addition of large amounts of ascorbic acid to milk or its complete destruction by rapid oxidation retards or prevents an oxidized flavor. Riboflavin participates in the oxidation of ascorbic acid and removal of it prevents the reaction.

21.6.3 Hypervitaminosis

Americans are nutritionally privileged yet they spend millions of dollars annually for supplemental vitamins. Hypervitaminosis A has become a recognized clinical syndrome. Intake of 50,000 IU (15,000 μg) of vitamin A daily for several weeks usually results in symptoms.[8]

Vitamin D toxicity (hypercalcemia) in children may occur with intakes of 4,000 IU (100 μg) per day. Toxic effects develop in adults receiving 100,000 IU (2,500 μg) daily for several months. The National Institutes of Health has set upper daily intake levels at 25 μg for infants and 50 μg for persons one year of age and older. These levels are well above quantities that would be ingested through milk and milk products with vitamins A or D added.

There is little, if any, detrimental effect from ingesting large quantities of the water-soluble vitamins because the excess is readily excreted. Folic acid presents an exception in that large doses may mask the presence of **pernicious anemia**.

21.7 Enzymes

The organic catalytic components of milk that participate in reactions without getting themselves used are called enzymes. All consist wholly or in part of proteins, but they form a negligible amount of milk's protein. Since enzymes are proteins, they tend to be inactivated by heat and to have specific pH and other conditions under which they act most favorably. Enzymes are important for two reasons. First, they may modify the chemical makeup of the substrate so the flavor, aroma, appearance, or reaction of the product is different. Secondly, they may be used as evidence of quality or of treatment given the product. For example, the presence of alkaline phosphatase in milk after pasteurization is evidence of insufficient heating or recontamination with raw milk (cf. chapter 25).

Fresh raw milk contains enzymes that gain entrance from the cow. Other enzymes are produced in milk or milk products by microorganisms, some inside and some outside the mammary gland. Many enzymes are in low concentrations in milk but have specific functions, including protection of the calf digestive system and promotion of absorption and digestion of milk. These enzymes can have desirable or deleterious effects on milk and milk products. Enzymes in bovine and human milks are tailored for needs of the respective species. Amounts of enzymes in milk vary with cow age, environmental conditions, diet, parity, and other physiological factors.

Most enzymes can be classified broadly according to the substrates they act upon; for example, carbohydrases act on carbohydrates, proteases act on proteins, and lipases act on lipids. However, the International Union of Biochemistry and Molecular Biology has divided enzymes into six major classes:

1. Oxidoreductases, which catalyze oxidation or reduction reactions;
2. Transferases, which catalyze the transfer of specific chemical **moieties**;
3. Hydrolyases, which hydrolyze substrates with concomitant uptake of water molecules;
4. Lyases, which remove or add specific chemical moieties to their substrates;
5. Isomerases, which catalyze isomerization; and
6. Ligases, which catalyze the synthesis or bonding together of substrate units.

Most of the important enzymes of milk are oxidoreductases (peroxidase and catalase) and hydrolases (amylase, protease, phosphatase, and lipase) (Shahani et al., 1973). Let us now consider those having practical importance. Their major characteristics are summarized in table 21.6.

21.7.1 Phosphatases

These catalyze hydrolysis of esters of phosphoric acid. Alkaline phosphatase is always present in raw milk, although its concentration is variable. Most of it is associated with microsomal particulates, which are adsorbed to fat globules. Destruction of alkaline phosphatase is practically complete after heating at minimum temperature and time combinations for pasteurization. However, a slight decrease in either time or temperature leaves some of the

Table 21.6 Major Enzymes of Milk and Selected Characteristics.

Enzyme	Function	Sensitivity to Pasteurization	Optimal pH	Importance in Milk
Alkaline phosphatase	Hydrolyzes esters of phosphates	Inactivated	9.8	Used as determinant of adequacy of pasteurization
Phosphoprotein phosphatase (formerly acid phosphatase)	Hydrolyzes esters of phosphates	Stable	4.0	Apparently unimportant
Catalase	Decomposes hydrogen peroxide	Mostly inactivated	7.0	Quantity in milk roughly correlates with number of leukocytes
Peroxidase (lactoperoxidase)	Catalyzes transfer of oxygen from peroxides to other substances	Stable	6.8	Indicator of heating milk at high temperatures
Xanthine oxidase	Catalyzes addition of oxygen to a substance or removal of hydrogen from it	Stable	6.9	Milk is an excellent source
Esterases and lipases	Hydrolyzes ester linkages, especially fats to fatty acids and glycerol	Inactivated	8.5 9.0	Produce rancid flavor and odor
Proteases	Hydrolyzes peptide linkages of proteins	Moderately stable	8.5	Relatively unimportant compared with microbial proteases in stored milk and milk products

enzyme active. Reactivation may occur in high fat products pasteurized at ultrahigh temperatures and stored at relatively high temperatures (cf. chapter 25). This enzyme is activated by the magnesium ion. Its optimal pH is 9.6. Phosphoprotein phosphatase is much more heat stable, being inactivated at 190°F (88°C) after 30 min of heating. It has a pH optimum of 4.0. It acts on aromatic phosphates, casein phosphorus, and pyrophosphates.

21.7.2 Esterases

These catalyze the hydrolysis of esters to form alcohols and acids. Lipases are a specific type of esterase that break emulsified fats into fatty acids and glycerol. Freshly drawn milk contains two lipases (Tarassuk and Frankel, 1957). Upon cooling, membrane lipase is irreversibly adsorbed to fat globule membranes. If the concentration of membrane lipase is excessive, certain milks develop rancidity with no apparent treatment other than cooling. Milk from cows in late lactation and on dry feed is most susceptible to this spontaneous lipolysis (Deeth, 2011). Plasma lipase must be activated by agitation, homogenization, or temperature manipulations. These treatments apparently cause a transfer of lipase from casein to fat globules. However, it is also known that violent agitation and homogenization disrupt the protective fat-globule membrane. Such treatments allow greater access of lipases to the substrate (fat) and, for this reason, would be expected to cause increased lipolysis. Because of the increased susceptibility of homogenized fat to lipase activity, it is imperative that homogenization be preceded or followed immediately by pasteurization. No lipolytic activity remains after pasteurization (Chandan and Shahani, 1957). However, after heating at ultrahigh temperatures (about 194°F) [90°C]) for 3 sec, lipase was reactivated on storage for 48 hr.

Bacteria may produce lipases in milk, some of which are not inactivated by pasteurization. However, the frequency of occurrence of rancidity in market milk due to microbial lipases is low.

Aryl esterase (A-esterase) activity is high in both colostrum and mastitic milk. It apparently gains access to milk from blood serum when tissue permeability increases during inflammation.

21.7.3 Proteinases

These attack peptide linkages (an amino group linked to a carboxyl group):

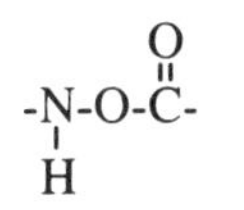

adding water (hydrolysis) and splitting the protein into large peptide fragments. Given the proper protease and peptidase, which produces small fragments and individual amino acids, a protein can be degraded completely to its constituent amino acids. These enzymes act at optimal pH and temperature levels. Hydrolysis occurs between certain amino acids, and some require presence of certain metal ions.

Bacteria produce proteases in milk, thus making it difficult to demonstrate the presence of natural milk protease. Nevertheless, proteolysis has been demonstrated in milk incubated at 97°F (37°C) when bacteria counts were negligible and growth was controlled with inhibitors. The active fraction of protease is precipitated with casein when milk is acidified. *Pseudomonas* bacteria sometimes produce proteases that are not inactivated by pasteurization (Mayerhofer et al., 1973; White and Marshall, 1973). One microbial protease studied in the authors' laboratories produced defects, especially bitterness, in ultrapasteurized milk, in cheddar cheese, and in cottage cheese after storage of these products at 41°F (4°C).

21.7.4 Catalase

This enzyme decomposes hydrogen peroxide (H_2O_2) to water (H_2O) and oxygen (O_2). In milk it is primarily associated with leukocytes that are elaborated in mastitic milk. Therefore, tests based on catalase content have been used to screen milk for high leukocyte content. However, correlation of the quantity of oxygen liberated to the leukocyte content is relatively low and other screening tests are preferred (cf. chapter 14). Some microorganisms produce catalase in milk and this may complicate the interpretation of results of screening tests based on this principle. Catalase in milk is destroyed by pasteurization.

21.7.5 Peroxidase (Lactoperoxidase)

This enzyme catalyzes the transfer of oxygen from peroxides to other substances. Milk contains relatively high concentrations of peroxidase (about 1% of total serum protein content). It is a heme protein with an iron content of about 0.07%. Pasteurization at minimal temperatures does not inactivate it.

21.7.6 Xanthine Oxidase

This enzyme is present in bovine milk in high concentrations but is absent from human milk.[9] It catalyzes oxidation-reduction reactions. Examples of substances oxidized by xanthine oxidase are xanthine, aldehydes, purines, pyrimidines, and reduced NAD. Examples of oxygen acceptors are molecular oxygen, methylene blue, cytochrome C, ferricyanide, nitrate, and quinone. Riboflavin is a component of xanthine oxidase. Pasteurization does not destroy it, but it is less stable to heat than is peroxidase.

21.7.7 Other Enzymes

Other enzymes of milk include, but are not limited to, alpha-amylase (attacks starch), aldolase (hydrolyzes fructose, 1,6-diphosphate), and carbonic anhydrase (catalyzes both the hydration of carbon dioxide [CO_2] and the reverse reaction, dehydration of carbonic acid [H_2CO_3]).

21.8 Cellular Constituents

Cellular constituents of milk include leukocytes (white blood cells), epithelial cells, and, in some instances, erythrocytes (red blood cells). Collectively these are classified as somatic (body) cells. Their numbers in milk are related to infection and trauma of mammary glands. Bacterial cells from both the interior and exterior of the mammary gland are also present. Numbers within the gland are determined by the ability of infecting bacteria to reproduce in the glands and produce toxins, by the amount of tissues infected, and by the animal's defensive capabilities. The latter include phagocytic activities of leukocytes and the immunity conferred by antibodies in milk. In uninfected animals, numbers of bacteria in **aseptically** drawn milk usually number less than 300/mL. In milk from infected glands numbers may reach several million/mL, but they are more commonly in the range of 10,000 to 100,000/mL. Streptococci tend to produce higher bacteria counts than do staphylococci.

21.9 Physical Properties of Milk

21.9.1 Acidity and pH

The normal pH of milk at room temperature is 6.5 to 6.7. Colostrum is more acid in reaction whereas mastitic milk usually has a higher pH, ranging to 7.3. Infrequently, streptococci produce sufficient acid in the mammary gland to drop the pH below normal. As temperature increases, pH decreases because calcium phosphate becomes insoluble. Phosphates are the main constituent of milk that buffer (tend to stabilize) its pH. Milk has a low buffering capacity in the region of the phenolphthalein end point (pH 8.3), which is used in the test for titratable acidity (cf. chapter 25). An electrometric pH meter with a glass electrode is the most satisfactory instrument for measuring the pH of milk.

In the United States the acidity of fresh milk (apparent acidity) is expressed in terms of the equivalent percentage of lactic acid. In European countries it is expressed as milliliters of alkali (usually 0.1 is normal) per given volume of milk. Most fresh milk samples fall within the range of 1.5 to 1.8 milliequivalents (meq) of acid per 100 mL, or a titratable acidity of 0.15 to 0.18%. Lactic acid is seldom present in fresh milk. The apparent acidity (**titer**) results from carbon dioxide, proteins, phosphates, citrate, and minor constituents. All these act as buffers. Therefore, as the percentage of nonfat solids in milk increases, titratable acidity increases. The authors have been consulted by processors who were unnecessarily concerned about high titratable acidities in incoming raw milk when, in fact, they were receiving milk of unusually high solids content. The titratable acidity test was reflecting the high content of buffering components.

It is the *developed* (real) acidity of milk with which milk processors and distributors are most concerned. This represents bacterial action in breaking down lactose to lactic acid. The average person will declare milk "sour" when its titratable acidity is about 0.25%. This represents about 0.07 to 0.10% developed acidity, which added to 0.15% (apparent acidity) equals about 0.25%.

21.9.2 Oxidation-Reduction Potential

Oxidation is a reaction that causes the removal of electrons. Fresh milk, drawn without exposure to air, has a negative oxidation-reduction potential (E_h). If methylene blue is added to such milk, it is reduced and becomes colorless. Exposure of the milk to oxygen causes oxidation of the methylene blue, and it turns blue. The reversible oxidation-reduction reaction involving the ascorbate-dehydroascorbate system (see section 21.6.2.2) is the principal one stabilizing the potential of milk containing oxygen in the zone of +0.2 to 0.3 volts (Webb et al., 1974). As milk undergoes fermentation, the potential usually decreases. Dissolved oxygen is used by bacteria, which transfer electrons to it. After it is virtually consumed, other substances become oxidants. Dyes such as methylene blue and resazurin function in this manner. Heating milk sufficiently to produce a cooked flavor liberates sulfhydryl groups by denaturing protein (primarily β-lactoglobulin), and the oxidation-reduction potential is lowered. Deaeration (vacuumization) also decreases the potential. Such treatments tend to stabilize dried milks against oxidative degradation. Contamination of milk with copper raises its E_h. Copper is a potent oxidizing agent that catalyzes production of metallic and oxidized flavor defects in milk and milk products.

21.9.3 Density and Specific Gravity

The weight of a given volume of milk or milk product can be important in quantitative analysis for composition and in packaging and sale. Density is weight per unit volume at a given temperature. For example, the density of water is essentially 1.000 at 39°F (4°C). Specific gravity, however, is the density of a substance divided by the density of water at the same temperature.

The density of milk decreases as temperature is increased, but the specific gravity remains relatively constant at about 1.032. The overall density is affected by densities of the constituents. Milkfat has a density that approximates 0.9 g/mL at 140°F (60°C). Densities of lactose, proteins, and other solids are collectively about 1.616.

Specific gravity is used to control the composition of concentrated milks during manufacture. It is used to estimate nonfat solids and the total solids content of milk and to screen samples for added water.

21.9.4 Viscosity of Milk

The viscosity of normal milk varies between 1.5 and 2.0 centipoises (units of viscosity) at 68°F (20°C). Casein is the major component affecting viscosity; however, fat contributes considerably depending on amount and sizes of globules and the extent of their clustering. Lactose, whey proteins, and salts make small contributions to viscosity.

Viscosity tends to increase as temperature increases to the point of coagulation of milk proteins because proteins aggregate. Acidity, salt balance, and the activity of certain enzymes are interacting effects. Homogenization increases milk's viscosity by increasing the surface area of the fat and the amount of protein adsorbed onto fat. Increases tend to parallel homogenization pressure.

21.9.5 Surface Tension and Interfacial Tension

Milk has a lower surface tension (about 50 dynes/cm at 68°F [20°C]) than water (72.75 dynes/cm at 68°F [20°C]) because proteins, phospholipids, milkfat, and free fatty acids tend to concentrate at interfaces. The important interfaces in milk are those between the liquid and air (the location of surface tension) and between the aqueous and lipid phases (the location of interfacial tension).

Tension is developed at these interfaces because some molecules are mutually attracted to each other, and numbers of mutually attractive molecules are much greater on one side of the surface than on the other. In air the concentration of molecules is relatively small. Fat and water are not **miscible**, and molecules in each have a low degree of attraction for the other. Therefore, the two phases remain separated. However, certain molecules move into surfaces in which they relieve stress (lower tension) in the process of adsorption. Other molecules tend to leave surface films (negative adsorption), thus increasing surface tension.

Factors that tend to lower surface tension of milk significantly are the development of rancidity (releases fatty acids) and churning (releases phospholipids). Surface tension increases upon heating sufficiently to denature and coagulate proteins.

21.9.6 Refractive Index

Light passing at an oblique angle from air into water, a denser medium, is bent or refracted. The magnitude of bending is the *refractive index*. Dissolved substances increase the refractive index above that of water, which is 1.33299 at 68°F (20°C) at a wavelength of 589.3 nm. Neither colloids (casein in milk) nor emulsoids (fat in milk) affect the refractive index, but in milk they interfere with its measurement. Therefore, at least fat must be removed to perform an accurate analysis. The refractive index of the aqueous phase generally ranges from 1.3440 to 1.3480 nm. An approximation of the total solids in condensed skim milk can be obtained by refractometry, which can also be used to estimate casein content after precipitation and resuspension. Detection of added water by refractometry is not as satisfactory as by cryoscopy (freezing point determination).

21.9.7 Freezing Point

The freezing point test (cf. chapter 25) has its basis in the fact that the **colligative properties** of milk, including osmotic pressure and freezing point, are controlled by the amount of dissolved substances in milk. The latter is, in turn, controlled by concentration of dissolved substances in blood, which is maintained within close limits. Consumption of large amounts of water by cows can cause a limited decrease in osmotic pressure of the blood with a consequent increase in the freezing point of milk. Other factors reported to influence freezing point are season, feed, ambient temperature, breed, and time of milking. Lactose and chlorides account for about 75% of the freezing point depression, and each tends to compensate for changes in concentration of the other.

Pure water freezes at 32°F (0°C). Most milk samples freeze within the range of –0.507 to –0.527°C. The addition of 1% water to milk will increase the freezing point by slightly more than 0.006°C. **Thermistor** cryoscopes are sensitive to changes in temperature of 0.001°C. (A word of caution: Some cryoscopes are calibrated in degrees Hortvet. To convert °H to °C, multiply by 0.96231 and subtract 0.00240.)

The commonly accepted upper limit of the freezing point in the United States is –0.507°C (–0.530°H). However, evidence of added water is best established when the freezing point of the suspect sample is compared with that of an authentic sample. An authentic sample is one taken by a representative of a control agency, such as the health department, in a way that precludes contamination with water.

Freezing point can be affected by vacuum treatment, by sterilization, and by freezing and holding samples before testing. Addition of milk solids, salts, or sugar will significantly lower the freezing point, as will development of acid by fermentations within milk.

21.9.8 Boiling Point

Composition and pressure determine the boiling point of milk. Normal milk boils at approximately 100.5°C at an atmospheric pressure of 14.696 psi (101.3 kPa). Doubling or tripling the concentration of milk solids increases the boiling point by 0.5°C and 0.75°C, respectively.

21.10 Summary

Chemical, physical, biological, and functional properties of milk are numerous and complicated because of the wide variety of components in milk, biological variation, cow diets, the environment, and types of treatment milk receives in production, handling, and processing. Biological variation causes each cow's milk to differ from the milk of every other cow and even from her own from milking to milking.

Lipids are dispersed as an emulsion in milk. Uniquely, they contain relatively large quantities of short-chain fatty acids. Proteins and nitrogenous components are divided into 11 fractions in contemporary nomenclature. Caseins comprise about 80% of these proteins, and they, combined with calcium and phosphorus, are colloidal in milk. Serum proteins of milk also combine with calcium, but they do not form colloids. Therefore, they are found in the aqueous phase. All essential fatty acids and amino acids are found in milk. Lactose is the sugar of milk. It is less soluble than glucose or sucrose, and this low solubility has importance in its digestion and in the processing of milk and milk products.

There are seven major minerals in milk (including calcium and phosphorus), 11 minor minerals, and at least 13 microelements. Some occur partially in colloidal complexes, whereas others are completely in true solution.

Milkfat contains vitamins A, D, E, and K. The skim milk portion is a particularly good source of riboflavin and contains each of the other vitamins of the B complex as well as vitamin C. In addition to the biological significance of each vitamin, a few enter into reactions that affect the quality of milk and milk products. Riboflavin, ascorbic acid, and vitamin E are involved in oxidative reactions.

Enzymes in milk are those involved in synthesis and secretion of milk as well as some enzymes produced by microorganisms in the mammary gland. Some are destroyed by heating, and testing for them can provide information about specific heat treatments. Others, notably lipases, degrade milk lipids under certain conditions.

Cellular constituents are generally undesirable in milk. They consist mainly of leukocytes and bacteria. However, were it not for the phagocytic activities of leukocytes in milk, infecting organisms would often overwhelm the cow's body defenses.

Milk has a buffering capacity that maintains its pH at about 6.5 to 6.7 and its titratable acidity at 0.15 to 0.18% (apparent acidity). Its oxidation-reduction potential is dependent on exposure to oxygen, growth of microorganisms, metabolism of leukocytes, heat treatment, and deaeration or vacuumization. The density of milk and milk products and the refractive index are dependent on composition. Fat affects the former markedly but does not affect the latter. Casein is the major milk component affecting viscosity. Surface tension and interfacial tension are affected by proteins, phospholipids, fat, free fatty acids, and treatments that change concentrations of these substances.

In processing, quality assurance, product development, and in nutrition, it is imperative that workers have knowledge and basic understanding of the chemical, physical, biological, and functional properties of milk and milk components.

Study Questions

1. Compared with the fats of other foods, how does milkfat rate in terms of length of fatty acids, molecular weight, and digestibility?
2. What are essential fatty acids? Are these found in milk?
3. Differentiate between saturated and unsaturated fatty acids.
4. What is the temperature range within which milkfat melts? Identify factors affecting the melting point of milkfat.

5. Why is butter made from milkfat of cows on pasture softer than that made of milkfat secreted when cows are fed dry feeds?
6. Can diet affect the amount of unsaturated fatty acids of milk? Cite an example.
7. Does the percentage of milkfat affect the cholesterol content of milk? Why? Would you suspect any seasonal variation in the cholesterol content of milk? Why?
8. How many amino acids are commonly found in milk? Does milk include all of the essential amino acids needed by humans?
9. What is the major protein of milk? Name milk's minor proteins.
10. What is the isoelectric point of casein? Of what practical significance is this in cheese manufacture?
11. What is a milk allergy? What are the etiologic agents in a milk allergy? Describe its effects on the human body.
12. How do cows' and women's milks differ in percentage of lactose? Is this of significance in formulating infant diets?
13. Which is sweeter, lactose or sucrose? Do α- and β-lactose have the same degree of sweetness?
14. What is lactose intolerance? Does there seem to be a genetic basis for lactose intolerance? Cite examples.
15. Does fermentation of lactose in products such as yogurt and cultured buttermilk affect the quantity of such products a lactose-intolerant person can ingest? Why?
16. Name the major and minor minerals of milk and note which ones are considered dietary essentials for humans.
17. Does a cow's diet affect the amounts and types of minerals found in her milk? Cite examples.
18. Name the major buffering components of milk.
19. What is the average Ca/P ratio of cows' milk? Does this differ from that of human milk?
20. Cite important roles of calcium in human health and well-being. Cite examples of interrelationships between calcium and other components of milk.
21. What major contributions does milk phosphorus make to humans?
22. Which vitamins are most abundant in skim milk? In whole milk?
23. Does a cow's diet affect the amounts and types of vitamins found in her milk? Cite examples.
24. Which vitamins are affected by oxidation of milk?
25. Does heating affect the vitamin content of milk?
26. Which breeds of dairy cattle secrete milk of the highest vitamin content? How can this be explained?
27. Of what significance is the phosphatase enzyme in milk processing?
28. Which enzyme is present in cows' milk but curiously is absent in women's milk?
29. May aseptically drawn milk contain bacteria? How can this be explained?
30. What is the normal pH of fresh milk? Does temperature affect the pH of milk?
31. Differentiate between apparent and developed acidity in milk.
32. What is the percent titratable acidity of fresh milk?
33. What is the specific gravity of milk?
34. Does the density of milk increase or decrease as temperature is increased?
35. Is milk more or less viscous than water? Why? Which components contribute most to the viscosity of milk?
36. What are the effects of temperature and homogenization on the viscosity of milk?
37. Compare the surface tensions of milk and water. Which components of milk contribute most to its surface tension?
38. What factors affect the refractive index of milk?
39. What are the freezing and boiling points of milk? Compare them with the freezing and boiling points of water.

NOTES

[1] Fatty acids are alike in one respect—the carboxyl group (–COOH) that gives them their acid property. The hydrogen atom in this group ionizes (becomes free of the remainder of the molecule). It can be replaced by metals (such as sodium) to form soaps or by glycerol to form fats. Another atom or group of atoms must be attached to carboxyl groups to complete the molecule. In fatty acids these vary from a single hydrogen atom (in formic acid) to long chains of carbons and hydrogens.

[2] Functions of essential fatty acids are incompletely defined. They may be of importance in structures of cell membranes, in normal transport of lipids **in vivo**, and in connection with enzyme systems. They accelerate both the synthesis and catabolism of cholesterol.

[3] Casein contains about 0.8% sulfur, whereas β-lactoglobulin and α-lactalbumin contain about 1.6% and 1.9%, respectively.

[4] The only mammals that have no lactose—or any other carbohydrate—in their milk are certain members of the Pinnipedia: seals, sea lions, and walruses of the Pacific basin. If these animals are given lactose in any form, they become sick. No α-lactalbumin is synthesized by the California sea lion. However, when α-lactalbumin from cows' milk is added to a preparation of sea lion mammary gland in a test tube, lactose is manufactured (Kretchmer, 1972).

[5] One would have to ingest about a liter of milk to obtain 50 g of lactose.

[6] The yellow color in butter results from the presence of carotenoids. Estimates of amounts of inactive or nonpotent carotenoids in milk range from 5 to 25%.

[7] Calves less than two weeks old apparently do not convert carotene to vitamin A and there is inefficient conversion in calves to age four months.

[8] Symptoms include hard lumps in the extremities and cortical thickening of underlying bone. Some patients exhibit fissures of the lips, loss of hair, dry skin, jaundice, and liver enlargement.

[9] Absence of xanthine oxidase in human milk provides a basis for a test to distinguish between human and bovine milks.

REFERENCES

Adams, J. S., and M. Hewison. 2010. Update on vitamin D. *Journal of Clinical Endocrinology and Metabolism* 95(2):471–478.

Berner, L. A. 1993. Defining the role of milkfat in balanced diets. *Advances in Food and Nutrition Research* 37:131–257.

Capper, J. L., and R. A. Cady. 2010. A point-in-time comparison of the environmental impact of Jersey vs. Holstein milk production. *Journal of Animal Science* 88, E-Suppl:569.

Carroll, R. J., J. J. Basch, J. G. Phillips, and H. M. Farrell Jr. 1985. Ultrastructural and biochemical investigations of mature human milk. *Food Structure* 4:323–331.

Chandan, R. C., and K. M. Shahani. 1957. Milk lipases: A review. *Journal of Dairy Science* 40:418–430.

Deeth, H. C. 2011. "Lipolysis and Hydrolytic Rancidity." In J. W. Fuquay, P. F. Fox, and P. L. H. McSweeney (eds.), *Encyclopedia of Dairy Sciences*, Vol. 3 (2nd ed., pp. 721–726). San Diego, CA: Academic Press.

Dickinson, E. 2006. Structure formation in casein-based gels, foams, and emulsions. *Colloids and Surfaces A: Physiochemical and Engineering Aspects* 288:3–11.

Farrell, Jr., H. M., R. Jimenez-Flores, G. T. Bleck, E. M. Brown, J. E. Butler, and L. K. Creamer. 2004. Nomenclature of the proteins of cows' milk—Sixth revision. *Journal of Dairy Science* 87:1641–1674.

Farrell, Jr., H. M., E. L. Malin, E. M. Brown, and P. X. Qi. 2006. Casein micelle structure: What can be learned from milk synthesis and structural biology? *Current Opinion in Colloid and Interface Science* 11:135–147.

FitzGerald, R. J. 1997. "Exploitation of Casein Variants." In R. A. S. Welch, D. J. W. Burns, S. R. Davis, A. I. Popay, and C. G. Prosser (eds.), *Milk Composition, Production and Biotechnology* (pp. 153–171). Wallingford, United Kingdom: CABI Publishing.

German, J. B., L. Morand, C. J. Dillard, and R. Xu. 1997. "Milk Fat Composition: Targets for Alteration of Function and Nutrition." In R. A. S. Welch, D. J. W. Burns, S. R. Davis, A. I. Popay, and C. G. Prosser (eds.), *Milk Composition, Production and Biotechnology* (pp. 35–72). Wallingford, United Kingdom: CABI Publishing.

Gery, I. 1970. *Final Report* (USDA Grant Number FG-IS-237; Project UR-AIO-(60)-61). The Hebrew University, Jerusalem, Israel.

Herb, S. F., P. Magidman, F. E. Liddy, and R. W. Riemenschneider. 1962. Fatty acids of cows' milk. A. Techniques employed in supplementing gas-liquid chromatography for identification of fatty acids. *Journal of the American Oil Chemists' Society* 39:137–142.

Holick, M. F., and M. Jenkins. 2005. *The UV Advantage.* New York: iBooks, Inc.

Institute of Medicine. 2011. *Dietary Reference Intakes for Calcium and Vitamin D.* Washington, DC: The National Academies Press.

Jack, E. L., and L. M. Smith. 1956. Chemistry of milk fat: A review. *Journal of Dairy Science* 39:1.

Jensen, R. G., A. M. Ferris, and C. J. Lammi-Keefe. 1991. The composition of milk fat. *Journal of Dairy Science* 74:3228–3243.

Juriannse, A. C., and I. Heertje. 1988. Microstructure of shortenings, margarine and butter: A review. *Food Microstructure* 7:181–188.

Kinsella, J. E. 1969. Butterfat complexity results from variety of fatty acids. *American Dairy Review* 31(3):82–84.

Kretchmer, N. 1972. Lactose and lactase. *Scientific American* 227:70.

Madureira, A. R., T. Tavares, A. M. P. Gomes, M. E. Pintado, and F. X. Malcata. 2010. Invited review: Physiological properties of bioactive peptides obtained from whey proteins. *Journal of Dairy Science* 93:437–455.

Mayerhofer, H. J., R. T. Marshall, C. H. White, and M. Lu. 1973. Characterization of a heat-stable protease of *Pseudomonas fluorescens* P26. *Applied Microbiology* 25:44–48.

Miller, G. D., J. K. Jarvis, and L. D. McBean. 1995. *Handbook of Dairy Foods and Nutrition*. Boca, FL: CRC Press.

Nielsen, R. L., and M. H. Anderson. 1999. Isolation of adipophilin and butyrophilin from bovine milk and characterization of a cDNA encoding adipophilin. *Journal of Dairy Science* 82:2543–2549.

Patton, S. 1999. Some practical implications of the milk mucins. *Journal of Dairy Science* 82:1115–1117.

Phelan, M. A., R. J. Aherne, R. J. FitzGerald, and N. M. O'Brien. 2009. Casein-derived bioactive peptides: Biological effects, industrial uses, safety aspects and regulatory status. *International Dairy Journal* 19:643–654.

Philhofer, G. M., H. C. Lee, M. J. McCarthy, P. S. Tong, and J. B. German. 1994. Functionality of milk fat in foam formation and stability. *Journal of Dairy Science* 77:55–63.

Proudfit, F. T., and C. H. Robinson. 1967. *Normal and Therapeutic Nutrition* (13th ed.). New York: Macmillan Company.

Reinhardt, T. A., and J. Lippolis. 2006. Bovine milk fat globule membrane proteome. *Journal of Dairy Research* 73:406–416.

Reinhardt, T. A., and J. Lippolis. 2008. Developmental changes in the milk fat globule membrane proteome during the transition from colostrum to milk. *Journal of Dairy Science* 91:2307–2318.

Shahani, K. M., W. J. Harper, R. G. Jensen, R. M. Parry, Jr., and C. A. Zittle. 1973. Enzymes in bovine milk: A review. *Journal of Dairy Science* 56:531–543.

Sommer, H. H. 1952. *Market Milk and Related Products* (3rd ed.). Milwaukee, WI: The Olsen Publishing Company.

Tarassuk, N. P., and E. N. Frankel. 1957. The specificity of milk lipase. IV. Partition of the lipase system in milk. *Journal of Dairy Science* 40:418–430.

Walsh, C. D., T. Guinee, D. Harrington, R. Mehra, J. Murphy, J. F. Connelly, and R. J. FitzGerald. 1995. Cheddar cheesemaking and rennet coagulation characteristics of bovine milks containing κ-casein AA or BB genetic variants. *Milchwissenschaft* 50:492–495.

Walstra, P. 1987. "Fat Crystallization." In J. M. V. Blanshard and P. Lillford (eds.), *Food Structure and Behavior* (pp. 67–84). Orlando, FL: Academic Press.

Walstra, P. 1999. Casein sub-micelles: Do they exist? *International Dairy Journal* 9:189–192.

Webb, B. H., A. H. Johnson, and J. A. Alford. 1974. *Fundamentals of Dairy Chemistry* (2nd ed.). Westport, CT: AVI Publishing.

White, C. H., and R. T. Marshall. 1973. The reduction of shelf-life of dairy products by a heat-stable protease from *Pseudomonas fluorescens* P26. *Journal of Dairy Science* 56:849–853.

Wynn, E. H., A. J. Morgan, and P. A. Sheehy. 2011. "Minor Proteins, Blood Serum Albumin, Vitamin-Binding Proteins." In J. W. Fuquay, P. F. Fox, and P. L. H. McSweeney (eds.), *Encyclopedia of Dairy Sciences*, Vol. 3 (2nd ed., pp. 795–800). San Diego, CA: Academic Press.

Websites

American Journal of Clinical Nutrition (http://ajcn.nutrition.org)

Dairy Science and Food Technology (http://www.dairyscience.info/index.php)

National Institutes of Health, Office of Dietary Supplements (http://ods.od.nih.gov)

University of Guelph, Department of Food Science (https://www.uoguelph.ca/foodscience)

22

Determining the Composition of Milk and Milk Products

Life is short and the art long; the occasion fleeting;
experiment dangerous; and judgment difficult.
Hippocrates

22.1 Introduction

Milk is complex—not only because it contains more than 250 individual components, but also because of variation in ratios of its components. In buying, selling, and processing milk it is important to know its composition. There are many reasons why the dairy industry may need to determine the composition of milk and milk products. For example, buyers and sellers establish prices based on content of valuable characterizing components (usually fat and protein or nonfat solids). Federal milk market administrators depend on accurate tests, as they monitor and account for milk marketed within federal orders. Dairy Herd Improvement laboratories require reliable methods as well, as they serve milk producers. The nutritional and sensory qualities that concern consumers depend on composition. Researchers monitor compositional values in efforts to improve efficiencies and effectiveness of production, processing, and marketing. In new product development, composition is the center of attention.

Methods and procedures discussed in this chapter are those most frequently used in purchasing, selling, manufacturing, and assuring high-quality milk and milk products.

22.2 Sampling

Proper collection and delivery of milk samples to the laboratory is critical if results of tests are to be accurate. The perishable nature of milk compounds the problem. Since the sample must be representative of the entire mass, adequate mixing of liquid samples and proper randomized selection of subsamples from solid products is imperative. Preventing contamination of samples is essential. Proper temperature must be maintained. Ideally, testing should be done within 24 hr when there is the possibility that chemical, physical, or biological changes may affect test outcome.

Collectors of samples must assure that all pertinent information is recorded and delivered to the laboratory. This includes identity, date, time, temperature, and who collected the sample. When collecting samples for regulatory purposes, a chain of custody should be established. Collection and testing of multiple samples increases the statistical probability that results will accurately reflect the condition of the mass.

Physical nature of the mass (fluid, viscous, semisolid, solid, dry) dictates how the sample will be taken, held, and transported. For full coverage of the topic see chapter 3 in *Standard Methods for the Examination of Dairy Products* (Wehr and Frank, 2004). Additionally, the *Grade "A" Pasteurized Milk Ordinance* (US Food and Drug Administration, 2013) contains several sections related to samples and sampling.

22.3 Protein

The protein content of milk averages about 3.0% and varies much less than does that of fat. The Kjeldahl method has been the standard method of testing for protein since 1883. The test is based on digestion of protein in concentrated sulfuric acid and potassium sulfate forming ammonium sulfate from the released nitrogen. By adding sodium hydroxide, ammonia is freed and is distilled into excess boric acid. Titration with hydrochloric acid follows, and protein is calculated based on amount of ammonia times the milk protein conversion factor of 6.28.

Because not all nitrogen in milk is combined with protein, a test for true protein can be done by precipitating the protein with trichloroacetic acid. The precipitate is then subjected to the Kjeldahl procedure. Casein content can be determined by precipitating it from milk by lowering the pH to 4.6 using acetic acid and sodium acetate. The filtrate is then tested by the Kjeldahl method, and casein protein is calculated as the difference between total nitrogen and noncasein nitrogen. Alternatively, the content of nitrogen, thus of protein, of the casein precipitate can be determined by the Kjeldahl test.

Modern laboratories, especially those performing large numbers of samples, depend on instrumental analyses for protein and other major components of milk (Barbano and Lynch, 2006). Those instruments are calibrated by chemical tests, including the Kjeldahl test (see section 22.6).

22.4 Milkfat

It is vital that milk processors know the fat content of the raw milk they receive so producers can be paid appropriately. Knowledge of fat content is needed for many other purposes. Traditionally, milkfat has been the most valued and variable component of milk. Scientists recognized early on that a simple and inexpensive but accurate test for milkfat would be of significant value. Only in standardizing fat content in manufacturing is there the possibility to meet label requirements and maintain operational efficiencies. For example, in cheddar cheese 50% of the dry matter must be fat. Fat tests permit standardization of milk for production of cheese and most other dairy products.

22.4.1 Volumetric Analyses

Fat content is determined volumetrically as a percentage of the milk sample. It is important that a proper amount of milk be introduced into the test bottle, that conditions of chemical and physical reactions are appropriate, and that the volume of the fat column is determined in a precisely calibrated neck of a test bottle at the prescribed temperature. The quantity of milk introduced is normally determined by volume using a pipet that delivers 17.5 mL of milk (specific gravity × milliliters = grams; thus, 1.032 × 17.5 = 18.06, or about 18 g).

The weight of milk per unit volume is called density. Specific gravity is the density of milk divided by the density of water at the same temperature. If, for example, the density of milk is 1.032 g/mL and that of water is 1.000 g/mL, the specific gravity of milk is 1.032. The density of milk varies inversely with temperature, primarily because water and fat expand as they are heated. However, specific gravity of milk changes little with temperature because water is its major ingredient; thus, water density largely determines **skim milk** density. Figure 22.1 indicates only a slight decrease in the specific gravity of skim milk when the temperature increases, probably because proteins become more hydrated (bind more water) at higher temperatures.

Dr. S. M. Babcock developed a volumetric test for milkfat in 1890, and it has become the official test in much of the United States and in many other countries.[1] The Babcock test, when applied to milk or cream, has been given a Class A1 rating in *Standard Methods for the Examination of Dairy Products*, which means it has been collaboratively studied and subjected to thorough evaluation. Furthermore, it has been given official final action by AOAC International. With slight variations in protocol it is applicable to a wide range of milk products.

Both chemical and physical forces are used in the Babcock test to break the fat-in-serum emulsion, free the fat, and force it into the calibrated cylindrical neck of the test bottle (figure 22.2). When added to milk at a ratio of about 0.8:1, concentrated sulfuric acid produces an exothermic chemical reaction that hydrolyzes colloidal proteins while melting and freeing fat from its emulsified state. Prior to adding acid, the aqueous portion of milk has a specific gravity of about 1.036, and the fat is approximately 0.93. The comparatively heavy sulfuric acid (specific gravity 1.825) combines with the aqueous phase (making specific gravity about 1.43) and exerts a pronounced physical force to separate the aqueous and fat phases. The fat, which has a much lower density than the remaining mixture, is separated by centrifugal force into a graduated neck of the test bottle, which is calibrated to express the fat content as a percentage by mass.

When applied to ice cream, cream, and cheese, the Babcock method requires samples to be weighed into the test bottle.

Because the Babcock method does not yield the total amount of milkfat in some dairy products, the Pennsylvania and Roccal modifications have been developed. The Babcock test for skim milk will usually read low by up to 0.10%, whereas the test for buttermilk will read 0.32% low. Results of tests on frozen desserts, chocolate milk, and whey also tend to be improved by this modification, which involves addition of ammonium hydroxide and/or alcohol to assist in freeing fat.

A similar volumetric test is the Gerber test, which is used primarily in Europe. It has been given a Class O rating in *Standard Methods for the Examination of Dairy Products*, which means it has been subjected to thorough evaluation and wide use but not formally validated. Principles of the Gerber test resemble those of the Babcock, but the sulfuric acid is weakened slightly by the added isoamyl alcohol. Moreover, the test bottle is considerably different (figure 22.3). Sulfuric acid digests protein and frees fat concurrently, producing heat. The alcohol aids

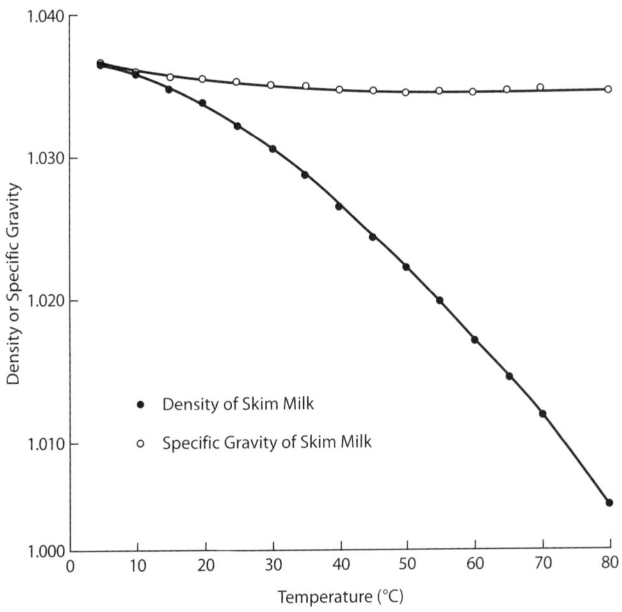

Figure 22.1 The relation of density and specific gravity of skim milk to temperature (Whitaker et al., 1927).

Figure 22.2 Babcock test bottles for (from left to right) skim milk, cream, and cheese.

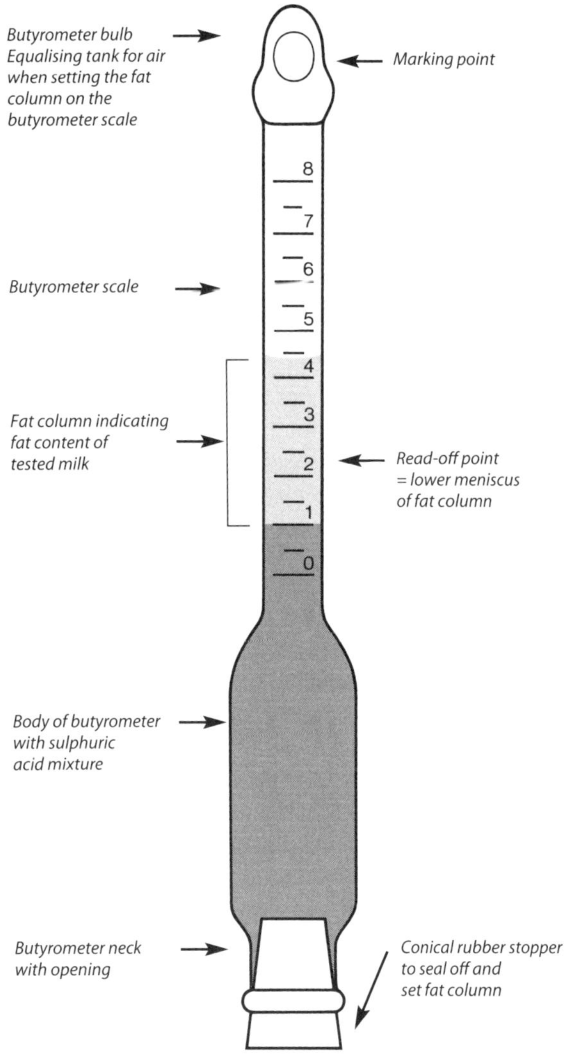

Figure 22.3 Butyrometer used in the Gerber test for quantity of fat in milk (courtesy of Funke-Gerber).

in fat separation. Centrifugal force is used to move fat into the graduated flat tube of the test bottle (butyrometer) where quantity is determined by visual observation.

22.4.2 Gravimetric Analysis

Gravimetric analyses of milkfat are more precise than volumetric ones. However, they require more time, expensive equipment, and technical skill. The Mojonnier method (Wehr and Frank, 2004) requires that samples be weighed to the nearest milligram into an extraction flask. Ammonium hydroxide is added to dissolve the casein and ethyl alcohol is added to prevent formation of a gelatinous mixture when ether is added. Both ethyl ether and petroleum ether are added to extract fat. Chemicals should leave no residue after extraction and evaporation. The aqueous and fat solvent layers are then formed by moderate centrifugation. The lighter solvent layer, containing fat, is decanted into a clean and preweighed dish, and a second extraction is made with alcohol and the two ethers. The solvent is evaporated from extracted fat on a hot plate. The dish containing the fat is then cooled and weighed precisely. Fat percentage in the sample is readily calculated by dividing the weight of fat (minus the weight of any residue found in the "blank" control) by the weight of the sample and multiplied by 100. Test results by the Mojonnier method average 0.03 to 0.06% lower than results by the Babcock method.

Repeatability of the Mojonnier test on raw whole milk, i.e., the absolute difference between two independent single test results obtained on the same material in the same laboratory by the same operator, should be no more than 0.044%. The interlaboratory reproducibility (difference) should not exceed 0.056%.

22.5 Lactose

Milk contains approximately 4.9% lactose (12 carbons), which can be hydrolyzed to two 6-carbon molecules, glucose and galactose, by microbial enzymes. Fresh milk seldom contains the hydrolyzed moieties. Two commonly used chemical methods to quantify lactose are presented below.

22.5.1 Enzymatic Method

This method is based on the ability of β-galactosidase to hydrolyze lactose to glucose and galactose plus water. The resultant β-galactose is then quantified when it is oxidized by NAD to galactonic acid, forming an equivalent number of NADH molecules. This reduced form of NAD is then measured by the amount of UV light (340 nm) absorbed and a conversion factor is used to calculate the amount of lactose.

22.5.2 HPLC Method

Lactose in raw and pasteurized milks can be determined by high-performance liquid chromatography (HPLC). The apparatus contains an NH_3 analytical column (carbohydrate column) and a refractive index detector. The mobile phase is acetonitrile/water, and the flow rate is 2 mL/min. Before injection into the apparatus, proteins of the samples are removed by precipitation with sulfuric acid followed by filtration. Standard deviations for groups of samples should not exceed 0.05%.

22.5.3 Older Chemical Methods

Because lactose is a polarizing sugar, the content of lactose can be determined when fat and protein are removed and the clear filtrate is tested in a polarimeter. Another property of lactose is that it is a reducing sugar. This facilitates testing by determining the amount of cupric salt reduced to cuprous oxide. After fat and protein are removed, the filtrate is heated with copper sulfate. The precipitate of cuprous oxide that forms is dried and weighed. Lactose content is calculated from Munson–Walker tables as modified by Hammond. Other reduction-based tests use picric acid, chloramine-T, or ferricyanide as reagents.

22.6 Minerals

Atomic absorption or flame emission methods are most often used for quantifying minerals in milk and milk products. Samples are dry-ashed or digested while wet to destroy organic components. The dissolved sample is then sprayed in a fine mist into a flame and the decrease in intensity of radiation from a hollow cathode lamp, due to absorption by specific atoms in the flame, is measured. Concentrations of the minerals are then calculated by comparison to the absorption of standard solutions of the different minerals of interest.

Atomic absorption instruments are capable of quantifying down to 0.001 μg/mL of some minerals, but the more common limit is 1 μg/mL (Pomeranz and Meloan, 1987).

Since costs and technical skills needed to make analyses for minerals are expensive, tests are usually conducted in dedicated service laboratories rather than in dairy laboratories. Furthermore, data used for nutrition labeling are determined for each product based on real data from product tests.

22.7 Total Solids and Nonfat Solids

Tests for total solids have been based on the principle of heating a precisely measured quantity of sample to evaporate any moisture, then the weight of sample residue is determined. The percentage of total solids is calculated using the formula:

$$\text{Total solids (\%)} = \frac{\text{weight of dry residue (g)} \times 100}{\text{weight of sample (g)}}$$

For total and nonfat solids, the forced draft oven method (Class A1) is recommended. It is applicable to milk, cream, evaporated and condensed milks, natural cheeses, and frozen desserts. Samples are weighed to the fourth decimal into a dried, preweighed dish. Drying occurs at 212°F (100°C) and cooled samples are again weighed. The within-laboratory standard deviation of the method used on milk is 0.019.

A microwave oven or an infrared heater can be used for evaporating moisture from samples when testing moisture (Class B). However, variability of test results is high compared with the forced draft oven method.

Estimates of total solids in milk can be made using the lactometric method (Class O). **Lactometers** are designed to determine the specific gravity of milk tempered to 102°F (39°C), which assures that all fat is melted and has a uniform density and constant volume. At a temperature below 93°F (34°C), milkfat begins to crystallize and occupy a smaller volume. Amount of crystallization depends on temperature and time of storage.

This simple test requires only a tall cylinder, a lactometer (Watson type), and a water bath. The lactometer reading is converted into total solids by the following formula:

$$\text{Milk, \% total solids} = 1.33\text{F} + \frac{273\text{L}}{\text{L} + 1{,}000} - 0.40$$

$$\text{Skim milk, total solids} = 1.33\text{F} + \frac{273\text{L}}{\text{L} + 1{,}000}$$

Where F = % fat, and L = lactometer reading in degrees.

22.8 Multicomponent Methods

22.8.1 Infrared Analyses

Development of instrumental methods for laboratory testing typically has proceeded based on the following objectives: (1) cost savings by reducing labor required, (2) improved accuracy by removing technician error, and (3) increased speed of determination. Development of infrared-based analytical equipment made possible concurrent testing for multiple major components. In addition, total solids may be determined by adding an experimentally determined factor to the percentage of fat, protein, and lactose. This factor accounts for the nonfat, nonprotein, and nonlactose solids present in the milk. Therefore, mid-infrared (MIR) spectroscopy (3 to 8 μm) has been a widely used method of determining compositional characteristics of fluid milk.

Infrared (IR) analysis of milk is based on absorption of IR energy at specific wavelengths that differ in maximal absorption by component:

1. *Fat*—carbonyl groups in ester linkages (these bonds between fatty acids and the glycerol backbone absorb at 5.72 to 5.76 μm) as well as the carbon-to-hydrogen bonds in the fatty acid carbon chains, which absorb at 3.48 to 3.51 μm.
2. *Protein*—peptide bonds between amino acids, which absorb at 6.40 to 6.60 μm.
3. *Lactose*—hydroxyl groups, which absorb at 9.47 to 9.61 μm.

For example, in the MilkoScan™ Minor by Foss, the sample is diluted and homogenized. The mixture then passes through a flow cuvette where fat, protein, and lactose are measured by their infrared absorption at appropriate wavelengths. The value for water is calculated on the basis of the sum of the values for fat, protein, and lactose plus a constant value for mineral content. The instrument requires precise calibration and must be thermostatically controlled.

A broad range of products can be tested using certain models of IR equipment, and multiple calibrations can be stored to provide tests on a full range of sample types. Calibration against chemical methods is required and, when conducted properly, test results are accurate and reproducible. Typical reference methods are the Mojonnier ether extraction method for fat, the Kjeldahl method for total protein, and the enzymatic or HPLC tests for lactose. Independent laboratories provide calibration samples to multiple users, which allow multilaboratory comparisons of data for validation of testing accuracy.

An important characteristic of IR testing is that the absorption of light at any specific wavelength is not solely by the targeted bonds. Water absorbs strongly at most wavelengths and each individual milk component can absorb some light at most wavelengths. Intercorrection factors must be used to correct for variances in ratios of components: fat, protein, lactose, and water.

As samples enter equipment, fat is homogenized to a consistent particle size since absorbance is highly affected by differences in diameters of fat globules. Therefore, homogenizer function must be monitored closely.

22.8.2 Fourier Transform Infrared Spectrometers

Newer instruments use Fourier transform infrared (FTIR) analysis. They scan the full infrared spectrum, enabling the measurement of multiple parameters even in complex dairy products. Suppliers include Denmark-based Foss with its MilkoScan™ FT2 and Minnesota-based Bentley Instruments with its FTS component analyzer. FTIR spectrometers capture the complete infrared absorption spectrum of the milk sample for component analysis, thus allowing calibration to be based on all spectral characteristics of each particular component (figure 22.4). They simultaneously measure milk composition including fat, protein, and lactose. The sampling, sequencing, and identification of the sample vials are functions performed by an autosampler. Instruments are available in models for analysis of up to 500 samples per hour.

The following are design and operational requirements to assure high reliability of FTIR milk analyzers (Chalmers and Griffiths, 2002).

- *Calibration*: At the beginning of each testing day, a set of calibration samples should be tested and a pilot sample included throughout the test run to check on instrument stability. To test for repeatability of an instrument, 18 samples of the same raw milk are tested. The result of the first test is discarded and the remaining results are checked to see that the range is no more than +/–0.04%.
- *Sample cells*: The analyzer must (1) have a light path length of 30 to 50 μm, (2) be resistant to buildup of chemical residue, (3) have a refractive index window that is the same as that of the sample, (4) have a high degree of mechanical stability, and (5) include easy flushing sample to sample.
- *Fat globules* must be of uniform size (~1 μm) so the diameter does not exceed one-third of the shortest IR wavelength.
- *Materials used* in the homogenizer must resist high sheer—balls are made of ruby (AlO_3) and seats of ceramic (ZnO_2).
- *Pressure* within the flow system is kept >1,400 kPa to prevent volatilization of dissolved CO_2 and to force any free air to become dissolved.
- *Temperature* must be controlled within 1°C to avoid thermal expansion of the flow cell.
- *Humidity* outside the flow cell and within the beam pathway must remain constant to avoid beam drift.
- *Vibrations* within the unit, especially of the mirror, must be dampened.

Another component of milk that is of interest to dairy producers is somatic cell count. Laser-based flow cytometry can be used to determine somatic cell counts in raw milk. Flow cytometry is an extremely powerful and versatile technique. It is the method of choice in the medical field for detecting, analyzing, and sorting cells. For milk, the sample is first treated with a buffer solution that stains somatic cells with a fluorescent dye. This suspension is then injected into the flow cytometer, where hydrodynamic focusing ensures that stained somatic cells intersect an intense laser beam, which causes the cells to emit fluorescent light. This fluorescent light is then collected and detected (figure 22.5). Analysis of the histograms (which show the heights and widths of the electronic pulses) provide the total somatic cell count.

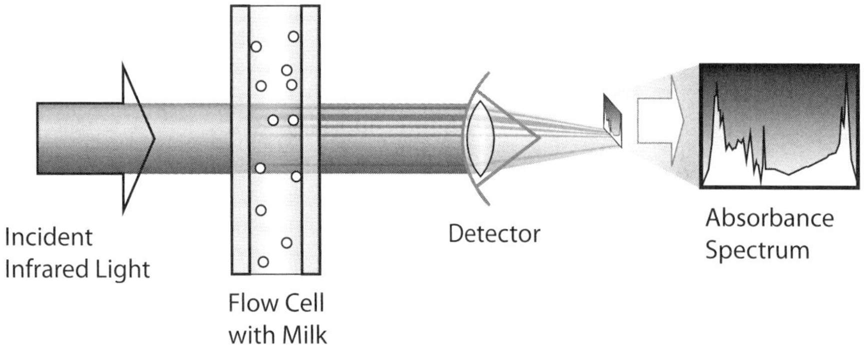

Figure 22.4 Principle of analysis of fat, protein, and lactose in milk by FTIR spectroscopy (courtesy of Bentley Instruments, Inc., Chaska, Minnesota).

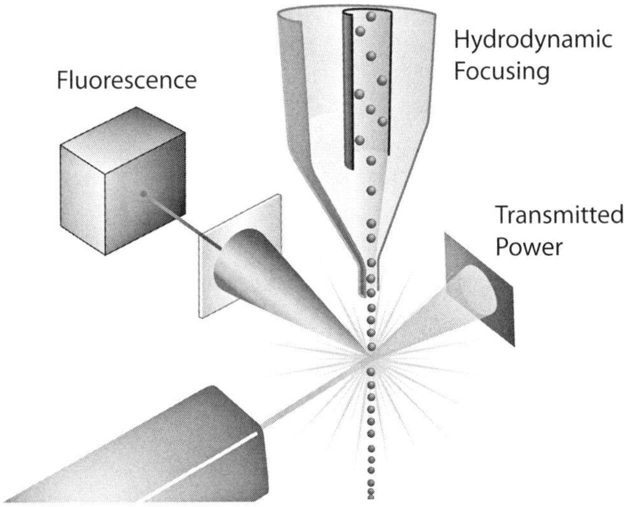

Figure 22.5 Illustration of the application of flow cytometry in enumeration of somatic cells in milk (courtesy of Bentley Instruments, Inc., Chaska, Minnesota).

The histograms can be recorded and archived so they may be recalled in the future.

The Somacount FCM by Bentley Instruments consists of two identical flow and optical channels that provide redundancy (figure 22.6). Should a problem arise in one channel, the operator can simply shut down that side and continue to test samples at a reduced speed of up to 300 samples per hour. The FCM exceeds the International Dairy Federation 148A standard and International Committee for Animal Recording requirements for somatic cell counting and uses AOAC approved methodology.

22.8.3 Near Infrared Analysis

ProFoss™ employs near infrared (NIR) based technology known as high-resolution diode array analysis in tests for fat and moisture content in butter, cheese, frozen desserts, and dry milk products. The NIR measurement system uses wavelengths in the "overtone region" (1400 to 2400 nm for reflectance and 650 to 1050 nm for absorbance) of the mid-IR spectrum. Analyses of the signals depend on application of sophisticated statistics, which is accomplished by built-in computers.

In-line compositional control of fluid milks is facilitated by NIR milk analyzers. Since NIR penetrates deeper than MIR, it is also useful for solid products such as cheese. This method is based on overtone and combination vibrations within the wavelengths of 0.7 to 1.4 μm. Bands observed are very broad, requiring careful development of calibration samples and application of multivariate calibration techniques.

22.8.4 The Kohman Method

This method is applicable to butter, margarine, and similar high-fat spreads that contain no water soluble emulsifiers, gums, or gelatin. It is based on the principle that a weighed sample can be heated to drive off moisture. The moisture lost is calculated after cooling and reweighing the sample. After extraction of the fat with petroleum ether, the remaining solids are weighed to determine fat content. Finally the remaining solids are dissolved in water and salt content is determined by titration with standardized silver nitrate, with potassium nitrate as the color indicator.

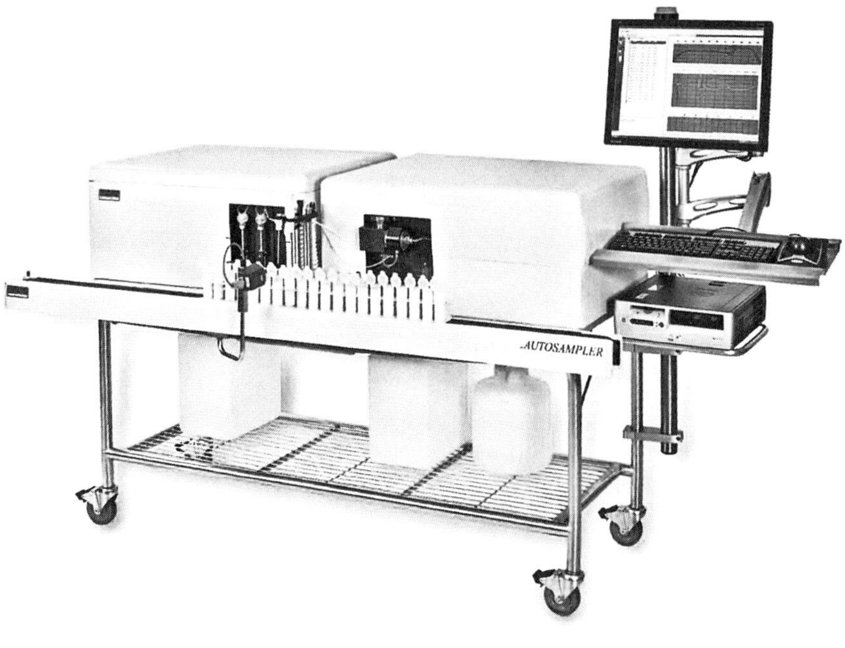

Figure 22.6 Bentley FCM instrument for simultaneous analysis of milk composition and somatic cell count (courtesy of Bentley Instruments, Inc., Chaska, Minnesota).

22.9 Summary

Methods presented in this chapter are those commonly used in determining the gross composition of milk and of some milk products. Much of the dairy industry has adopted instrumental methods, but instruments must be calibrated with reference samples that have been prepared by the chemical method. Some firms make a business from selling reference samples to industry laboratories. Equipment is expensive but high throughput and good transportation make possible the centralization of laboratories. Furthermore, concentration of the industry has reduced costs per test.

Persons who perform or use data from compositional analyses are obliged to know the capabilities and limitations of methods used. Technicians often are the primary sources of error in analytical testing. Standardization of technique, simplification of procedures, and adequate training are essential to obtain and maintain an efficient and reliable laboratory. Users should understand the cost factors of test performance and keep abreast of new developments in quality requirements and testing methods.

Study Questions

1. Describe the characteristics of a properly obtained and reliable sample of milk.
2. Cite three reasons why it is important to know the composition of milk.
3. What chemical component of protein is measured by the Kjeldahl test?
4. What parts of the Babcock method cause fat to rise to the neck of the test bottle?
5. Calculate the percentage of total solids in a 150 g sample of fluid milk that had a dried weight of 19 g.
6. Why is it important that chemicals used in extracting fat in the Mojonnier test leave no residue?
7. What enzyme is used to hydrolyze lactose in the enzymatic method for quantification?
8. Name two types of flame photometry used to quantify minerals in dairy foods.
9. Discuss differences between density and specific gravity of fluid milk products.
10. What factors cause user laboratories to analyze dairy products using instrumental rather than chemical methods?
11. What are the major functional groups of fat, protein, and lactose that absorb infrared energy?
12. Why is the homogenizer a critical component in an infrared milk analyzer?
13. Explain how it is possible for FTIR instruments to simultaneously measure milk's fat, protein, and lactose content.
14. How do fluorescent dye and a laser beam fit into the instrumental counting of somatic cells found in milk?
15. Name the three components of butter that can be determined using the Kohman method.
16. What techniques can be used to determine the calcium content of milk?

Note

[1] Babcock could have patented the test, which would have made him wealthy. He recognized this, but he also knew how much milk producers and processors needed the test; therefore, he forfeited a potential fortune and gave the test to the dairy industry.

References

Barbano, D. M., and J. M. Lynch. 2006. Major advances in testing of dairy products: Component and dairy product attribute testing. *Journal of Dairy Science* 89:1189–1194.

Chalmers, J. M., and P. R. Griffiths (eds.). 2002. *Handbook of Vibrational Spectroscopy.* Chichester, United Kingdom: John Wiley.

Pomeranz, Y., and C. E. Meloan. 1987. *Food Analysis: Theory and Practice* (2nd ed.). Roslyn, NY: AVI Publishing.

US Food and Drug Administration. 2013. *Grade "A" Pasteurized Milk Ordinance.* Washington, DC: Author.

Wehr, H. M., and J. F. Frank (eds.). 2004. *Standard Methods for the Examination of Dairy Products* (17th ed.). Washington, DC: American Public Health Association.

Whitaker, R., J. M. Sherman, and P. F. Sharp. 1927. Effect of temperature on the viscosity of skim milk. *Journal of Dairy Science* 10:361–371.

Websites

American Journal of Public Health (http://ajph.aphapublications.org)

AOAC International (http://www.aoac.org)

Bentley Instruments (http://bentleyinstruments.com)

Foss North America (http://www.foss.us/industry-solution/dairy)

ISO Standards (http://www.iso.org/iso/home/store/catalogue_tc.htm)

23

Microbes, Milk, and Humans

> Dairy microbiology is a specialized branch of dairy science that deals with microorganisms of importance to the economy of the dairy industry and to the health of humans.
>
> *H. L. Russell (1866–1954)*

23.1 Introduction

Microorganisms bring both benefit and harm to the dairy industry. In fact, some organisms can be either beneficial or harmful depending on the circumstances of their presence and activities. For example, the same bacteria that cause milk to sour and become undesirable for drinking may be used to manufacture desirable cultured buttermilk and similar cultured products. There are instances in which large numbers of microorganisms may be present without having a significant effect on the quality or wholesomeness of milk and milk products or on the well-being of dairy animals or the humans who are in contact with them. At other times only a few microorganisms can be highly significant when given the opportunity to reproduce, either in foods or in a dairy animal or a human. Fortunately, humans have learned to control most microorganisms; they can be destroyed, preserved, or caused to grow. They can have their enzymes extracted, their activities modified, and some can even be used as food. Therefore, an understanding of the science of microbiology is essential to those who would successfully pursue a career in the dairy or food industries. Even consumers need to understand how microbiological factors influence quality, shelf stability, and healthfulness of foods.

In this chapter we discuss the most important microorganisms in milk and milk products. Pathogenic types are discussed in chapter 24 and reference will be made to others used in the manufacture of cultured products, cheeses, and butter in chapters 29, 30, and 32, respectively.

23.2 Understanding Microbiology

Preserve the enthusiasm inspired by bacteriology—
which has acquired such might from its recent conquests
that it has captivated the minds of all.
Louis Pasteur (1822–1895)

The world of the microbe is so small it taxes the human imagination. Single cells of bacteria are measured in micrometers (1 μm = 0.001 mm or 0.000039 in). To see such a cell clearly, magnification of about 1,000X is necessary. Viruses are much smaller, rickettsia are intermediate in size between viruses and bacteria, whereas molds and yeasts are several times larger than bacteria.

Since bacteria are the most important microorganisms in milk and milk products, our discussion focuses on them.

23.2.1 Morphology

This refers to shape and usually applies to single cells, though it may refer to organization of a group of cells, as in a chain, cluster, or even a colony. Single cells are generally spherical (cocci) or rod-shaped (bacilli). A few are spiral-shaped or rodlike with distinctly bent ends. These are called spirochetes. Representative cells are shown in figure 23.1.

Some cells may have one or more fine, long, protruding, whip-like filaments. These *flagella* wave back and forth to propel the cell in liquids. Mucilaginous or gummy envelopes (capsules) that cling to the cell are secreted by some bacteria. They play a protective role against **phagocytosis** by white blood cells (phagocytes) in animals. Some encapsulated bacteria used in fermentations of milk products can increase the viscosity of those products. Capsules may provide protection for bacteria as they travel through the acidic environment of the stomach.

23.2.2 Gram Reaction

A convenient method for separating bacteria into large groups is the Gram stain.[1] A film of bacteria is fixed on a slide and stained with crystal violet. Iodine and alcohol are applied in succession, followed by a red stain, safranine. Alcohol removes the primary (violet) stain from certain

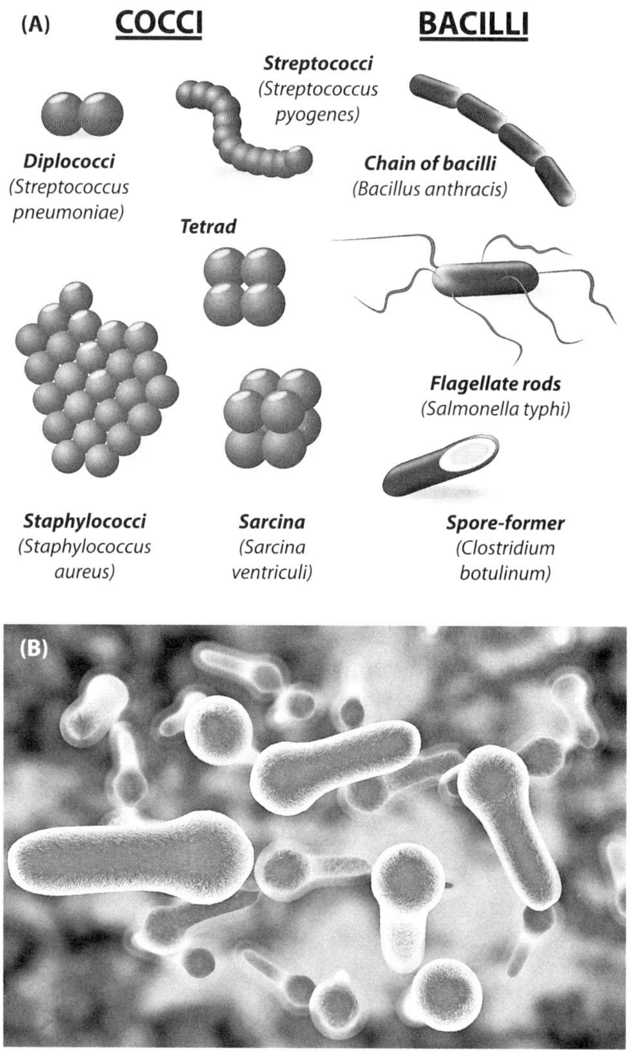

Figure 23.1 Typical shapes and arrangements of common bacteria: cocci, bacilli (rods), and spore-forming rods. (A) Designval/Shutterstock.com. (B) Kateryna Kon/Shutterstock.com.

bacteria, the Gram-negative group. Gram-positive bacteria have a thick, mesh-like cell wall made of peptidoglycan (50 to 90% of cell envelope) and are stained purple by crystal violet, whereas Gram-negative bacteria have a thinner layer (10% of cell envelope) and do not retain significant amounts of the violet stain. Gram stain reactions of bacteria important in milk and milk products are listed in table 23.1.

23.2.3 Spores

Two major genera of bacteria, *Bacillus* and *Clostridium*,[2] have the ability to form spores within **vegetative** cells (these are called *endospores*). Spores function to preserve the bacterium under adverse circumstances such as desiccation and heat. Therefore, conditions required to sterilize milk and milk products are determined by the number and types of spores in the product.

23.2.4 Oxygen Requirements

Most bacteria important in milk and milk products require oxygen for respiration. Some require large or moderate amounts, whereas others, called *facultative anaerobes*, require only small amounts. Clostridia, however, are poisoned by oxygen and their vegetative cells, but not their spores, soon die in its presence. They are, therefore, important in canned dairy products and in ripened cheese in which oxygen tension is low.

23.2.5 Growth Temperature

Rates of reproduction among microorganisms vary in response to changes in temperature at which the **culture** is held. Therefore, the optimal temperature for bacterial growth can be used in characterizing particular microorganisms. Those with high, medium, and low optimal temperatures are known, respectively, as *thermophilic*, *mesophilic*, and *psychrophilic*. Members of another important group are called *psychrotrophic*; these grow well at refrigerated temperatures, about 45°F (7°C), but their optimal temperature is not necessarily low. Pathogens and organisms of the **rumen** and other parts of the animal body tend to grow best at body temperature, 98.6 to 104°F (37 to

Table 23.1 Selected Characteristics of Major Genera of Bacteria of Milk and Milk Products.

Genus	Gram Reaction	Shape	Major Reasons for Importance in Milk or Milk Products	Need for Oxygen (O_2)
Streptococcus	+	cocci	Produce acid; some cause mastitis	low
Lactococcus	+	cocci	Produce acid in cheeses and cultured milk products	low
Leuconostoc	+	cocci	Produce diacetyl in cultured milk products	low
Staphylococcus	+	cocci	Some cause food poisoning; some cause bovine mastitis	low
Micrococcus	+	cocci	Large numbers occur in **milkstone** on dairy equipment	moderate
Enterococcus	+	cocci	Gastrointestinal habitat, thermoduric, facultative anaerobe	low
Lactobacillus	+	rods	Produce acid and ripen cheese	very low
Propionibacterium	+	rods	Ripen Swiss cheese	very low
Corynebacterium	+	rods	Inhabit streak canal of teats of cows	moderate
Microbacterium	+	rods	Heat resistant	moderate
Listeria	+	rods	Pathogenic psychrotroph	moderate
Bacillus	+	rods	Produce spores, many of which are heat resistant; some coagulate milk under alkaline conditions by means of rennin-like enzymes; proteolytic	usually high
Clostridium	+	rods	Anaerobic spore formers; many are heat resistant; includes food-poisoning bacteria; gas producers	none
Escherichia	–	rods	From intestinal tract of humans and animals; indicators of contamination after pasteurization; some cause food-borne infections; gas producers	moderate
Enterobacter	–	rods	Indicators of postpasteurization contamination; sometimes cause bovine mastitis; gas producers	moderate
Proteus	–	rods	Hydrolyze protein, producing unpleasant odor	moderate
Salmonella	–	rods	Some produce food-borne infection; from intestinal tract of humans and animals	moderate
Shigella	–	rods	Some produce food-borne infection; from intestinal tract of humans and animals	moderate
Pseudomonas	–	rods	Many are psychrotrophic; attack fats and proteins readily; one species causes bovine mastitis	high

40°C). Collectively, these are called *mesophiles.* Thermophilic (heat-loving) bacteria thrive at temperatures of 113 to 149°F (45 to 65°C). Few bacteria in milk grow at 149°F (65°C), but the lactobacilli of yogurt, acidophilus milk, and Swiss cheese and the *Streptococcus salivarius* ssp. *thermophilus* (*S. thermophilus*) of yogurt and Swiss cheese prefer temperatures near 113°F (45°C). Psychrophilic (cold-loving) microorganisms are sometimes found in milk, but **psychrotrophs** usually predominate. Thermoduric (heat-enduring) types survive the high temperatures employed in pasteurization. Vegetative cells of most species of spore-forming bacteria are sensitive to such heat treatment, but spores of the same species are often thermoduric.

23.2.6 Rate of Reproduction

The time required for a single cell to produce two cells is called *generation time* or *doubling time.* Depending on growth phase, nutritional factors, environmental factors, and genetic capabilities, generation time may vary from as little as 20 minutes to several hours, or even days. Bacteria multiply fastest when in the logarithmic or exponential phase of growth, the phase during which the viable population increases at a constant rate. This phase is preceded by the lag phase, a period during which there is minimal cellular division. Dairy microbiologists are concerned with keeping undesirable microorganisms in the lag phase, but they strive to ensure logarithmic growth of desirable microorganisms in many processing situations. Obviously, manufacturing and distribution schedules are highly dependent on rates of microbial reproduction.

23.2.7 Nutritional Requirements

To grow, all microorganisms require energy and the building blocks of cells, particularly carbon, hydrogen, oxygen, and nitrogen. There are also requirements for certain minerals. However, specific needs for the growth of different microorganisms vary widely. Many that exist in water, such as coliforms, grow on simple substances, including inorganic nitrogen (such as ammonium ion). Others, especially the lactobacilli, require preformed amino acids and/or vitamins. Their nutritional needs are met only in a complex medium such as milk, which is capable of supporting the growth of all but the most **fastidious** of bacteria.

23.2.8 Inhibition of Growth and Destruction of Microorganisms

We have discussed how oxygen, temperature, and nutritional factors relate to microbial growth. Other environmental factors of importance include pH or acidity, moisture availability (water activity, a_w), osmotic pressure, and microbial inhibitors.

Most bacteria grow best when the pH is between 6 and 8. Molds and yeasts tolerate high acidity. Proteolytic bacteria are sensitive to acids. Thus, increasing the acidity of cultured milk products improves their keeping quality. For example, cottage cheese with a pH of 5.0 is likely to keep significantly longer than cottage cheese with a pH of 5.2. It is well to remember that when pH changes by one unit the hydrogen ion concentration has changed tenfold.

Bacteria require water for metabolic processes. Thus, an important method of preventing growth is to drain dairy equipment quickly and completely after cleaning. Microbial growth in several milk products is inhibited by drying (as in, for example, dry milks). However, drying may also preserve bacteria for long periods. Freeze-drying, or **lyophilization**,[3] is used to preserve lactic **starter cultures** that become active upon inoculation into milk (cf. chapter 29).

In a manner similar to drying, the addition of large amounts of sugar or salt to solutions inhibits bacterial growth and kills microorganisms. However, the concentrations required vary widely. Water passes from the cell into the medium because of high osmotic pressure. This principle is responsible for preservation in sweetened condensed milk and in selective culturing of staphylococci, which tolerate as much as 10% sodium chloride (NaCl) in a **nutrient broth**.

Sodium chloride influences microbial growth by a toxic effect as well as an osmotic effect. Differences in salt tolerance provide a useful means of controlling the development of microorganisms in and on certain cheeses.

Free fatty acids may be inhibitory or stimulatory to microorganisms. The short-chain fatty acids, especially acetic acid (two carbons), inhibit and even destroy salmonellae. However some bacteria, particularly lactobacilli, lactic streptococci, and corynebacteria, are stimulated by oleic acid (18 carbons).

Antibiotics are substances produced by microorganisms that, in extremely small concentrations, inhibit other microorganisms. Microorganisms vary in sensitivity to antibiotics. For example, penicillin G is highly effective against streptococci but not very inhibitory to coliforms and pseudomonads. Sanitization of dairy equipment depends on destruction of microorganisms by chemicals (cf. chapter 27).

23.3 Groups of Bacteria Important to the Dairy Industry

Species of bacteria that have been isolated from milk number in the thousands. However, there are about 19 genera of noteworthy significance in milk and milk products (see table 23.1). We discuss them here according to type of activity in which they participate.

23.3.1 Lactic Acid Producers

Microorganisms in this group add tremendous value to milk when grown in it or fractions of it under controlled conditions.[4] They degrade lactose to lactic acid, thus lowering the pH (cf. section 29.2 and 30.4.3). They also produce small amounts of **volatile fatty acids** (mainly acetic and propionic), diacetyl, and acetaldehyde, which contribute the aroma that complements the tartness or sourness imparted by lactic acid. They belong to the families Lactobacillaceae and Streptococcaceae of the order Lactobacillales. The **strains** used depend on conditions of manufacture (the reverse is also true) and the end product desired. These lactic acid producers are Gram-positive, nonmotile, microaerophilic, or anaerobic. Some are cocci and others rods, but all form pairs or chains because they divide in only one plane, like rods. They have complex nutritional requirements, including amino acids and minerals.

Most often used are *Lactococcus lactis* ssp. *cremoris* and *Lactococcus lactis* ssp. *lactis*. They are similar in function and grow well at temperatures of 68 to 91°F (20 to 33°C), producing up to 1% lactic acid. *S. thermophilus* can grow at temperatures up to about 124°F (51°C); therefore, it is well adapted for manufacture of Swiss and brick cheeses and yogurt. Lactobacilli, especially *Lactobacillus acidophilus*, *Lactobacillus delbruéchii* ssp. *bulgaricus* (*Lb. bulgaricus*), and *Lactobacillus casei*, produce and tolerate more acid than do streptococci. The bacteria listed above produce primarily lactic acid from carbohydrates and are known as *homofermentative*. Several other organisms in the same family (Lactobacillaceae) produce acetic acid, ethanol, carbon dioxide, and other substances. They are, therefore, *heterofermentative*. *Leuconostoc cremoris* (family Leuconostocaceae) produces diacetyl (2,3-butanedione) from citrate in cultured milk products. Diacetyl provides the nut-like flavor note in cultured buttermilk, sour cream, and cottage cheese. *Lactococcus lactis* ssp. *lactis* var. *diacetylactis* is a potent producer of diacetyl, carbon dioxide, and other compounds (Hassan and Frank, 2001).

Lactobacillus helveticus is widely used as a starter culture to manufacture yogurt and certain Swiss and Italian cheeses and as a flavor-enhancing adjunct culture for other cheese types (cf. section 30.4.11). Abilities to reduce bitterness and to accelerate flavor development in cheese can vary widely among strains. Moreover, milk fermented with certain *Lb. helveticus* strains has been found to be enriched with antihypertensive and immunomodulatory bioactive peptides. These attributes are associated with enzymes involved in proteolysis and/or catabolism of amino acids. Broadbent et al. (2011) determined that genes for peptidases and amino acid metabolism were much the same among strains within the species, whereas genes for proteinases associated with the cell wall varied widely. Some genetic differences may help explain variability observed among *Lb. helveticus* strains regarding their functionality in cheese and fermented milks.

23.3.2 Probiotic Microorganisms

Live microorganisms that confer a beneficial health effect on the host when administered in sufficient amounts are called probiotics. Anukam and Reid (2007) provided an excellent review of probiotic microorganisms and their characteristics. O'Flaherty and Klaenhammer (2010) provided a detailed list of benefits enabled by probiotic bacteria. There are many factors to consider in developing and labeling dairy foods containing probiotic microorganisms. The following are guidelines for developing and marketing a probiotic-type food:

1. Microorganisms should be identified to the strain level; genomic sequencing recommended.
2. Each probiotic strain should be able to be identified and enumerated from the product. DNA-based approaches often are applicable.
3. Efficacy should be substantiated for probiotic strains in specific products.
4. Product formulation should be based on the type of food, dose, and choice of strain(s) used in clinical studies.
5. Viability of all probiotic strains in the product should be maintained above the minimal target level through the end of shelf life.
6. Product labels and any supplementary communications should provide clear, accurate information pertaining to types and numbers of probiotics, any documented health benefits, and the amount of product that must be consumed for a favorable effect.
7. Probiotic strain(s) must be safe for their intended use (Sanders et al., 2007; Tong, 2010).

The genera *Lactobacillus* and *Bifidobacterium* are major contributors to the reservoir of probiotic bacteria (table 23.2 on the following page).[5] To render a lasting effect, the organism must be capable of surviving and growing in the intestinal tract. These microbes are generally bile and acid resistant as are most species of the natural flora of the gastrointestinal tract.

Although the usual cultures for yogurt (*Lb. bulgaricus* and *S. thermophilus*) cannot grow and survive in the intestinal tract, they provide benefits by contributing enzymes important in digestion including lactase (β-d-galactosidase; cf. chapter 21). Furthermore, milk contains a high concentration of calcium phosphate, a natural buffering substance that moderates the low pH of the stomach, thereby assisting the survival of probiotic bacteria as they pass through.

The potential benefits of probiotic microorganisms, which vary from strain to strain, include (1) controlling the growth of undesirable organisms in the intestinal tract, (2) enhancing the immune response, (3) improving lactose digestion, (4) enabling anticarcinogenic activity, and (5) influencing control of serum cholesterol. Cholesterol is incorporated into the cellular membrane of lactobacilli and can be reduced to coprostanol, which is less easily absorbed from the intestine than is cholesterol (Liu et al., 2007). Microorganisms in the intestine communicate with their host and vice versa (Moller and de Vrese, 2004; O'Flaherty and Klaenhammer, 2010).

Some manufacturers add prebiotic ingredients to cultured foods, especially yogurts. Prebiotics beneficially affect the host by stimulating the growth and/or activity of bacteria inherent in the colon. To be classified as a prebiotic, the ingredient must be (1) neither hydrolyzed nor absorbed in the upper digestive tract, (2) a selective substrate for one or a few beneficial bacteria of the colon, and (3) able to alter the colonic flora to induce luminal or systemic effects that are beneficial to the host. Inulin-type fructo-oligosaccharides (FOS) and galacto-oligosaccharides (GOS) are known as prebiotics and are recognized as meeting these criteria as food ingredients (Chandan, 1999). Fermentation of these soluble fibers in the large intestine produces short-chain fatty acids that are important to the host in lipid metabolism. As growth of these organisms is enhanced, an antagonistic effect on undesirable bacteria is expected (see Gilliland, 2001).

Table 23.2 Effects in Humans of Some Probiotic Microorganisms.

Strain	Scientifically Established Effects
Lb. johnsonii LJ1	Mucosal adherence, improved oral vaccination
Lb. acidophilus NCFB 1748	Reduction in mutagenicity, improvement in bowel movements during constipation
Lb. acidophilus NCFM	Alleviation of lactose intolerance symptoms
Lb. rhamnosus GG	Shortened duration of rotavirus diarrhea, mucosal adherence, prevention and treatment of antibiotic-associated diarrhea, reduced risk of atopic diseases in infants
Lb. rhamnosus HN001	Increased killer cell and phagocytic activity, reduced risk of atopic diseases in infants
Lb. casei Shirota	Lowering fecal enzyme activities, normalizing intestinal microbiota, reduced recurrence of superficial bladder cancer
Lb. gasseri (ADH)	Altering intestinal metabolic activity
Lb. paracasei F19	Shortening duration of rotavirus diarrhea
Lb. reuteri	Shortening duration of rotavirus diarrhea, reduced crying of infants with colic
B. lactis Bb12	Shortening duration of rotavirus diarrhea, treatment of atopic eczema
Saccharomyces boulardii	Prevention of antibiotic-associated diarrhea, treatment of *C. difficile* colitis

Source: Adapted from Salminen et al., 2011.

Food manufacturers must design their claims regarding the benefits of probiotics with the regulatory environment in mind and must develop their research plans to provide evidence that satisfies substantiation requirements. The law allows that, in addition to nutrient content claims, manufacturers of dietary supplements may make structure/function or health claims for their products. A structure/function claim describes the process by which the dietary supplement or conventional food maintains normal functioning of the body. For dietary supplements, premarketing demonstration of safety and efficacy and approval by the US Food and Drug Administration (FDA) are not required; only premarket notification is required. The FDA requires that the evidence of effectiveness is accepted by experts in the disciplinary field and that the claim is truthful and not misleading. In general, the level of substantiation and the quality of evidence needed to make a structure/function claim are less than that needed to make a health claim. When a structure/function claim is made, the manufacturer must state in a disclaimer that the FDA has not evaluated the claim and that the product is not intended to "diagnose, treat, cure, or prevent any disease"; such a claim can legally be made only with regard to a drug (*Federal Register*, 2000; Heimbach, 2008).

There are no officially recommended dosages for health effects of probiotics. Nor are there established limits for safe consumption as measured in colony-forming units (CFU). However, experts on the subject suggest that at least one billion CFUs are needed for meaningful benefits. *Consumer Reports* (2005) reported 15 to 155 billion CFUs per serving in yogurt products tested. This was compared with about 20 to 70 billion CFUs per daily dose of probiotic supplements.

The kefir lactobacillus, *Lb. kefiranofaciens* M1, was recently demonstrated to have an anticolitis effect in mice. Based on these results it was suggested as having the potential for use in fermented dairy products as an alternative therapy for intestinal disorders.

Lactobacillus plantarum strain Tensia was found suitable for generally recognized as safe (GRAS) and qualified presumption of safety (QPS) criteria (Songisepp et al., 2012). It had no undesirable characteristics as a component in regular semihard Edam-type cheese at a daily dose (50 g) or in excess (100 g) and with a probiotic daily dose of 10^{10} CFU for three weeks.

Although acidified dairy products are the primary carriers of probiotic microorganisms, in 2012 an Australian company announced development of nonfermented milk and juices containing probiotic microorganisms and/or omega-3 fatty acids encapsulated with alginate. Encapsulation slows or prevents growth of the bacteria. Additionally, it may slow the rate of oxidation of the unsaturated fatty acids while masking any fishy flavor that may exist.

Freeze-drying is a common method for preservation of probiotics, including bifidobacteria, depending on the freeze-drying method and subsequent storage conditions. Effects of temperature, water activity, and atmosphere on microbial viability were evaluated during storage of the freeze-dried cells of a commercially used *Bifidobacterium* strain resilient to stress, *Bifidobacterium animalis* ssp. *lactis* Bb-12 (figure 23.2), and an intestinal strain more sensitive to stress, *Bifidobacterium longum* DJO10A (Celik and O'Sullivan, 2013). *B. animalis* ssp. *lactis* survived under all conditions tested, with optimum survival at temperatures up to 70°F (21°C), water activities up to 0.44, and all three atmospheres tested (air, vacuum, and nitrogen). The intestinal-adapted strain *B. longum* was much more sensitive to the different storage conditions, but could be maintained in an atmosphere of nitrogen with water activity between 0.11 and 0.22.

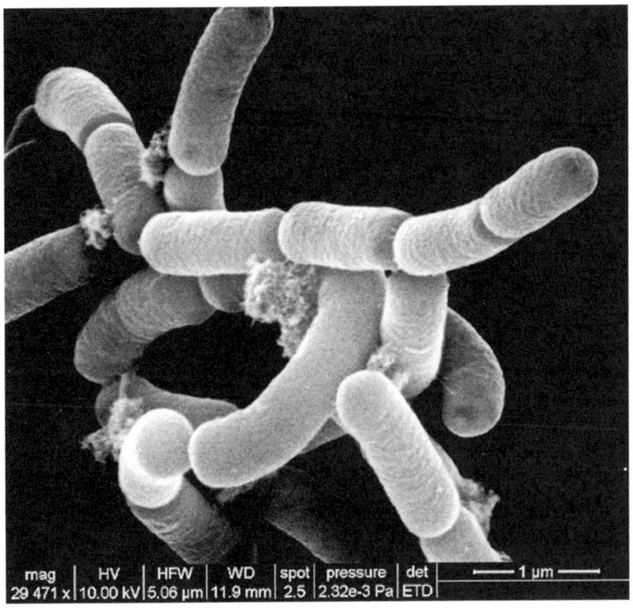

Figure 23.2 Scanning electron micrograph of *Bifidobacterium animalis* (courtesy of Ami Yoo, Food Science Department, University of Missouri, Columbia).

23.3.3 Proteolytic Microorganisms

Without protein hydrolysis, ripened cheese would lack its fine characteristic flavor and elastic body. However, ripening must be accomplished under controlled conditions or defects develop. Microorganisms and enzymes that are either naturally present or added must be only mildly proteolytic. Weakly proteolytic bacteria in dairy foods include streptococci, lactobacilli, propionibacteria, and leuconostoc.

Milk proteins may act as precursors of biologically active peptides with different physiological effects. Among these biological activities, the inhibition of the angiotensin-converting enzyme (ACE) is one important target of antihypertensive drugs to prevent hypertension. One way to liberate bioactive peptides from proteins in fermented dairy foods is by using proteolytic strains of lactic acid bacteria. During growth, lactic acid bacteria not only generate peptides by the action of cell wall proteinases, but also produce free amino acids by the action of intracellular peptidases. *Lactobacillus* and *Lactococcus* are the genera commonly used to generate ACE-inhibitory peptides. The question remains as to conditions necessary for production of sufficient ACE-inhibitory peptides to have a clinically positive effect in humans.

Many other microorganisms produce enzymes that rapidly degrade proteins to amino acids. This destroys the protein and causes off-flavors and other defects. Highly proteolytic bacteria occur predominantly in the genera *Bacillus*, *Clostridium*, and *Pseudomonas*. There are also several genera of highly proteolytic molds. Several bacteria and molds produce rennin-like enzymes that cause sweet coagulation, that is, coagulation of proteins without acid. Most such enzymes hydrolyze proteins excessively and induce bitter flavors. However, a few induce mild proteolyses and therefore can be used as **rennet** substitutes.

23.3.4 Lipolytic Microorganisms

Psychrotrophic bacteria are the most active lipase producers in milk. Species of the genus *Pseudomonas* predominate. Several molds and yeasts are lipolytic also and these sometimes spoil refrigerated milk products. Lipolysis is vital in ripening cheeses, especially soft-ripened types.

23.3.5 Enteric Bacteria

Intestinal tracts of humans and animals are habitats of numerous bacteria. Several are of importance as contaminants of milk and milk products including *Escherichia coli*, *Clostridium perfringens*, *Streptococcus fecalis*, and many species of salmonellae and shigellae. *E. coli* belongs to a group of Gram-negative bacteria that ferment lactose, producing gas. Members of this group, called coliforms, are destroyed by pasteurization. Their presence, then, in pasteurized milk and milk products indicates contamination after pasteurization and is evidence of unsanitary conditions. Therefore, the coliform test (cf. chapter 25) is widely used in the dairy industry to detect and enumerate these organisms. It is also used to check the quality of water intended for drinking and for cleaning dairy equipment.

23.3.6 Psychrotrophic and Psychrophilic Microorganisms

Cold-tolerant and cold-loving bacteria constitute the most troublesome group of microorganisms to the dairy industry because they reproduce relatively rapidly at temperatures of 45°F (7°C). The predominant genus of this group is *Pseudomonas*. Other genera include *Achromobacter*, *Alcaligenes*, and *Flavobacterium*. Several researchers have isolated psychrotrophic spore-forming organisms (*Bacillus* spp.) from raw milk.

23.3.7 Microorganisms Associated with Food-Borne Disease

Microorganisms of a few species produce **toxins**. Those formed from simple substances and then excreted in foods are called exotoxins. Endotoxins are held within microbial cells from which they must be released to affect humans adversely.

The exotoxin produced by some strains of *Staphylococcus aureus* is a potent enterotoxin (a toxin that acts within the intestinal tract). It is heat stable, even after boiling for 30 min, and staphylococcal cells need not be present for it to be active. Although toxin-producing strains of *S. aureus* are sometimes associated with mastitis, the possibility of their producing effective amounts of toxin in fresh milk are extremely remote for three important reasons: (1) they do not reproduce at temperatures below 39°F (4°C), (2) other bacteria compete with them and usually outgrow them at higher temperatures, and (3) the number of cells necessary to produce critical amounts of toxin exceeds two to three million/mL. In cheesemaking, however, possibilities exist for staphylococcal food poisoning. Conditions most often necessary are starter failure, resulting in insufficient acid production in milk that has been heated sublethally or recontaminated with an enterotoxigenic strain of *S. aureus*. On ripening of cheeses, many staphylococci die, but the toxin remains active.

The most potent toxin known to humans is the exotoxin produced by *Clostridium botulinum*.[6] These bacteria are anaerobic spore-formers. The toxin is destroyed by boiling for 10 min, but spores are heat resistant and are of concern in canned high-protein foods. Of 659 botulism outbreaks reported in the United States from 1899 to 1969, only four were traced to milk or milk products. *C. botulinum* is particularly susceptible to lactic acid and unsaturated fatty acids. It is generally agreed that growth is inhibited by pH of 4.5 or lower. Its outgrowth in natural cheese is highly unlikely. However, in processed cheese, a favorable environment for Clostridia may be created by additives, selective heat treatments, higher pH, and packaging conditions. Canned milk products must be heated to at least 239°F (115°C) for 15 min to assure destruction of *C. botulinum* spores.

Listeria monocytogenes has been associated with food-borne illnesses among consumers of unripened soft cheeses. In food plants the bacterium is found in cold and wet environments, especially in floor drains. It commonly resides in biofilms, which protect it from cleaners and sanitizers, and can grow at temperatures as low as 39°F (4°C). Since listeriae are killed by pasteurization, it is the use of raw milk or postpasteurization contamination that pose the risk of illness.

Dairy farms are major reservoirs of food-borne pathogens and unpasteurized milk consumption has been implicated as a major risk factor for *Campylobacter jejuni* human infection. The sale of raw milk for human consumption by self-service automatic vending machines has been allowed in Italy since 2004, but several outbreaks of disease caused by *C. jejuni* have since been reported. Serraino et al. (2013) isolated 52 pathogens from 49 of 196 milk filters taken from 14 farms. The isolates were identified as *C. jejuni* (6), *C. hyointestinalis* ssp. *hyointestinalis* (8), *C. concisus* (1), *C. fetus* ssp. *fetus* (1), *Arcobacter butzleri* (22), and *A. cryaerophilus* (14). The presence of *C. jejuni* in in-line milk filters confirms that raw milk consumption is a significant risk factor for human infection. The other isolates appear to be of low pathogenicity, except to infants.

In July 2010 the National Advisory Committee on Microbiological Criteria for Foods reported their assessment of the importance of food as a source of exposure to *Mycobacterium avium* ssp. *paratuberculosis* (MAP). MAP is the causative agent of Johne's disease, which affects primarily the small intestine of ruminants. Milk, particularly raw milk, may be a likely food source for human exposure to MAP. However, the significance of MAP as a human pathogen is unknown and is being investigated by several research groups.

Mycotoxins are exotoxins produced by molds and are **carcinogenic** to humans. Among this group are the aflatoxins produced by species of the *Aspergillus* genus. They are polycyclic, unsaturated lactones and are quite heat stable. Mycotoxins are of concern in milk and milk products because cows may consume them in moldy feeds[7] and transfer them to milk; they may be produced by molds growing on products such as cheese or butter, or may be contained in nuts added to frozen desserts. Prevention of their direct addition to feeds and foods involves the protection of grains and nuts to prevent mold growth and the exclusion of affected materials. Most molds that grow on milk products are harmless. Fortunately, toxin produced by toxigenic molds diffuses slowly into cheese. The depth of diffusion depends on the quantity produced, water content, physical properties, and storage time and temperature. In one research study toxin was not detected in cheddar cheese more than 1.5 cm from the toxigenic mold after seven weeks of growth (Lie and Marth, 1967).

Bacteria that produce endotoxins must be ingested to cause disease. Important ones in milk and milk products are the salmonellae, of which there are some 1,400 **serotypes**. Dry milk has been an important vehicle of salmonellae. Recontamination of a dried product after processing is usually responsible, as salmonellae are killed during milk pasteurization. Bacteria of the related genus *Shigella* may be transmitted from person to person via milk.

Salmonellae are commonly named based on serological reactions. Only three species exist and there are six subspecies: *enterica*, *salamae*, *arizonae*, *diarizonae*, *houtenae*, and *indica*. New methods of classifying these organisms are based on analyses of the genome and include pulsed gel electrophoresis and polymerase chain reaction, a technique used to amplify a single or few copies of a piece of DNA generating thousands to millions of copies of a particular DNA sequence.

23.3.8 Micrococcaceae

Members of this family are Gram-positive, spherical or elliptical, and divide in two or three planes forming tetrads, packets, and irregular masses of cells. Two genera important to the dairy industry are *Micrococcus* and *Staphylococcus*.

Among the micrococci are species that survive pasteurization. These may affect the ripening of cheeses and may cause high bacteria counts in pasteurized products. However, few of them grow at refrigerated temperatures.

The *Staphylococcus* genus includes at least 40 species. The taxonomy is based on sequences of nucleic acids in the ribosomal 16s RNA (rRNA). Most are harmless and reside normally on skin and mucous membranes of humans and other organisms. However, *S. aureus* is quite important to humans and animals, and we have already introduced this species in the section on food intoxication and in chapter 14. This organism is characterized by an ability to hemolyze red blood cells, to produce coagulase, to grow in the presence of 10% NaCl and 0.02% potassium tellurite, and to hydrolyze egg yolk. Tests for the organism often are based on these characteristics. *S. aureus* produces the enzyme catalase, which converts hydrogen peroxide (H_2O_2) to water and oxygen. Therefore, the catalase test is useful to distinguish staphylococci from catalase-negative enterococci and streptococci.

Micrococci and coagulase-negative staphylococci are frequently isolated from the exterior of the cow's udder, and sometimes from its interior where they induce much less inflammation than do strains of *S. aureus*. Both species grow best at body temperature (98.6°F [37°C]) and are killed by pasteurization.

23.4 Viruses and Bacteriophages

Viruses and bacteriophages are similar in size and composition; in fact, bacteriophages (phages) are viruses that attack bacteria. A living cell is required for viruses to reproduce. Bacteriophages are adsorbed to the cell into which they inject their nuclear components and alter the metabolic system of the cell to reproduce them. Finally they lyse the cell to free themselves. Thus, many phages, even hundreds, may be produced in one bacterial cell. Lactic cultures that produce capsules have a greater resistance to phage attachment and bind more water than nonencapsulated strains. Culture manufacturers typically test their cultures for resistance to phages found present in manufacturing plants of their clients, thus permitting selection for resistance to the isolated strains.

Phages are adsorbed only to organisms that produced them and to closely related strains. In chapter 29 we discuss the importance of phages in the manufacture of cultured milk products in which they may prevent acid production and typical flavor development. Bacteriophages that attack psychrotrophic bacteria have been isolated from pasteurized milk (Whitman and Marshall, 1971). However, probabilities are low that the shelf life of milk is extended by their activities since relatively high numbers must be present (Ellis et al., 1973).

Viruses of the enteric group (including poliovirus and infectious hepatitis virus) may be transmitted in foods, including milk and milk products. Such viruses gain entrance to the body via the oral route, establish themselves in the intestinal tract, and are excreted in the feces. Numbers of carriers and asymptomatic cases exceed clinical cases. Therefore, improper hygienic and sanitary practices contribute to contamination. Since pasteurization kills viruses and since contamination must ultimately come from an infected person, the prevention of virus transmission in pasteurized milk and milk products depends on sanitary equipment and care on the part of personnel to avoid exposure of product, equipment, and packages to themselves or anything they may have contacted. Unless their bacterial host is present, viruses cannot reproduce in foods and few enteric types of bacteria have hosts other than humans.

23.5 Molds

Molds are multicellular microorganisms. They usually resemble cotton because of the threadlike strands of protoplasm they produce. Most molds produce asexual spores on stalks that rise vertically. Such spores are usually green, black, or brown (figure 23.3 on the next page).

Molds are aerobic and grow over wide ranges of pH, osmotic pressure (many require little moisture), and temperature. Mold spores are reproductive; they resist drying but are not especially resistant to heat. Because of their light weight, mold spores are easily spread by air currents, insects, and animals. Molds are found in soil, dust, feeds, and feces. Even compressed air can carry mold spores (and bacteria) when it is not protected or filtered properly. Compressed air is used in a broad range of applications in the dairy processing industry, such as mixing of ingredients, cutting, **sparging**, drying of product, transporting/propelling product through processing systems, and packaging of final product. In many of these applications, compressed air is in direct contact or indirect contact with and can contaminate the product or package.

Few molds grow rapidly in milk. They can be especially destructive, however, in cheeses, butter, sweetened

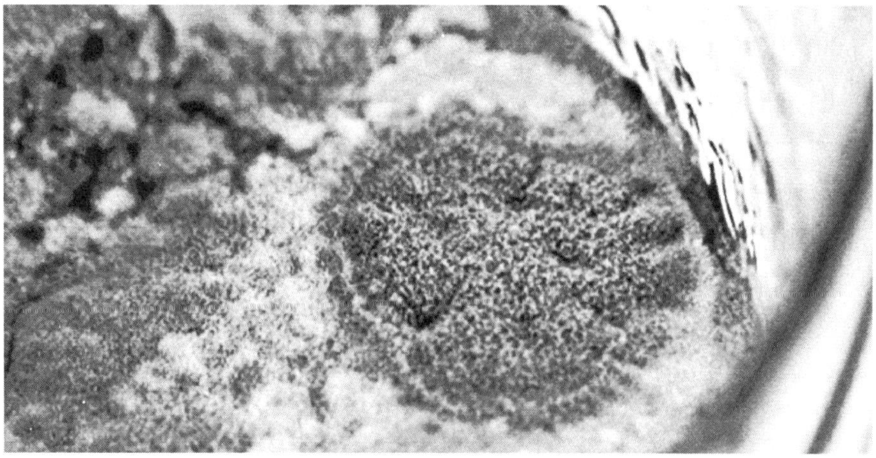

Figure 23.3 Surface growth of *Penicillium* mold on flavored cream cheese (courtesy of Food Science Department, University of Missouri).

condensed milk, and in cultured products such as cultured (sour) cream. The species *Penicillium roqueforti* (figure 23.4) is used in ripening blue mold cheeses. Molds, like yeasts, can be selectively cultured on malt agar or potato dextrose agar acidified to pH 3.5 to inhibit bacterial growth. Chloramphenicol and chlortetracycline hydrochloride are employed as bacterial inhibitors in some culture media.

23.6 Yeasts

Yeasts are likely among the earliest domesticated organisms. People have used yeast for fermentation and baking throughout history. Archaeologists digging in Egyptian ruins found early grinding stones and baking chambers for yeasted bread as well as drawings of 4,000-year-old bakeries and breweries. In 1680 the Dutch naturalist Anton van Leeuwenhoek first observed yeast with his microscope but considered them to be globular structures rather than living organisms.

The ovoid or elliptical, Gram-positive, nonmotile cells of yeasts usually measure 3 to 15 μm in diameter—several times larger than bacterial cells. However, yeasts are similar to bacteria in appearance of colonies on solid media and in biochemical activities. They are generally quite active in the metabolism of carbohydrates. Yeasts that ferment lactose usually produce carbon dioxide and ethyl alcohol as well as lactic acid. Most yeasts reproduce by budding (figure 23.5), the extension of protoplasm from one or more locations on a cell. Each bud grows into a large mature cell capable of forming new cells. The nucleus of the parent cell splits, forming a second nucleus that migrates into the daughter cell. A few yeasts reproduce by fission (mitosis), as do bacteria. Budding and fission are asexual processes. "True" yeasts (*Ascomycota*) reproduce sexually (as well as asexually) forming ascospores, whereas "false yeasts" (*fungi imperfecti*) produce no sexual spores. Yeasts are especially abundant in the soil of orchards and vineyards.

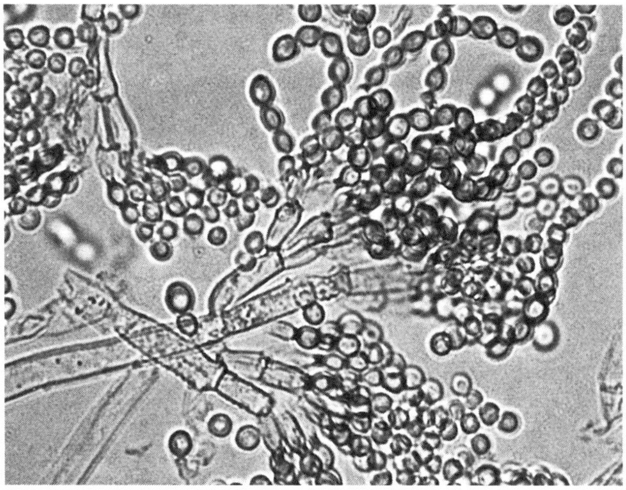

Figure 23.4 *Penicillum roqueforti*. Spores (conidia) are spherical bodies borne on branching structures at the top of conidiophores, the vertically oriented portion of the mycelial structure.

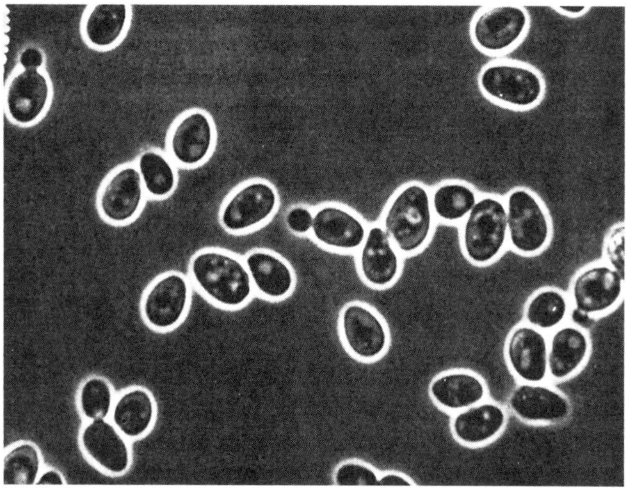

Figure 23.5 Cells of *Saccharomyces cerevisiae* that are reproducing by budding (courtesy of Anheuser-Busch, Inc., St. Louis, Missouri).

Important genera of yeasts in milk and milk products include *Saccharomyces*, *Schizosaccharomyces*, *Candida*, and *Torulopsis*. Yeasts are harmful in fresh milk products. They produce about 1 to 2% alcohol in kefir and koumiss (fermented milk drinks) and assist in ripening of some cheeses, but produce gassiness and off-flavors in others.

23.7 Effects of Microorganisms on the Shelf Life of Dairy Products

We define *shelf life* as the length of time a dairy product remains wholesome and edible. Microorganisms account for most deterioration of dairy products. Thus, there is a close relationship between microbe numbers and shelf life.

Shelf life of dairy products is both economically and nutritionally important. The economics involved include, for example, product loss, increased transportation and labor of handling, and dissatisfied consumers who reject an inedible product. Spoiled dairy products result in a loss of important dietary nutrients.

To maximize the shelf life of milk and milk products it is important to begin with high-quality milk having a low bacterial count. Processing, handling, and storage procedures must be such that contamination will not occur and conditions unfavorable to the growth of microorganisms will be maintained.

23.8 Summary

Without beneficial activities of microorganisms, consumers could not enjoy many fine-flavored, nutritious milk products. Microorganisms are chosen for manufacture of specific products because of their abilities to ferment sugars, hydrolyze proteins or fats, produce flavorful substances, or alter physical characteristics. Their requirements for growth must be considered in the design of processing techniques and in the provision of substrates in which they are to grow or not grow.

Indirectly, microorganisms also benefit humans by producing antibiotics useful in the control of animal diseases; by competitive growth (often resulting in suppression of microorganisms that may cause food-borne disease); and, of course, in the degradation of waste materials from farms, processing plants, and homes. Milk products, especially cultured ones, have become favorite vehicles for providing probiotic bacteria and prebiotic additives in diets of humans and certain animals.

Detrimental aspects of microorganisms command significant amounts of the time of milk producers, milk haulers, sanitarians, field representatives, quality assurance personnel, processors, wholesalers, and retailers. At each point in this chain care is taken to prevent the entry of microorganisms and the growth of those that are present. Several processes are designed to destroy microorganisms, thus preventing transmission of disease and increasing storage stability. Persons associated with these professions and activities benefit by knowledge of microbiology and its application to milk and milk products.

Study Questions

1. Cite an example of how the same microorganism can be either beneficial or harmful depending on the circumstances of its presence and activities.
2. Which collective grouping of microorganisms contributes most to the value of milk products?
3. Of what value is the Gram reaction? Do the spherical bacteria (cocci) of milk stain Gram-negative or Gram-positive?
4. What is the significance of spore-forming bacteria to the dairy industry?
5. Is it possible to control bacterial growth in certain milk products by controlling the oxygen concentration? Cite an example to support your answer.
6. Differentiate among thermophiles, mesophiles, and psychrophiles.
7. What is generation time as applied to microbes? Differentiate between the lag phase and the logarithmic phase of bacterial growth.
8. Why do most microorganisms grow well in milk?
9. Which of the following grows best in a high acid (low pH) environment—bacteria, yeasts, or molds? Apply your answer to cheese manufacture.
10. Cite several environmental factors affecting bacterial growth.
11. What principle is involved in the use of sugar and/or salt in the preservation of certain dairy products?
12. Cite examples of the beneficial contributions to the dairy industry of the bacteria that produce lactic acid.
13. What contributions do proteolytic microorganisms make in the ripening of cheese?
14. What is sweet coagulation? Which component of milk participates in the reaction?
15. Cite an example of a contribution that lipolytic microorganisms make to cheese.
16. What are coliforms? Why are they undesirable in milk and milk products? Are they killed by pasteurization?
17. Why are psychrotrophic microorganisms of considerable concern to the dairy industry?
18. Why might Clostridia be of more concern in processed cheese than in natural cheese?
19. Can mycotoxins be transferred from feed to milk?
20. Are most molds that grow on milk products harmful or harmless?
21. Are salmonellae killed by pasteurization?
22. What are bacteriophages? How do they reproduce? Does pasteurization affect them?

23. Do molds require oxygen to grow? Why are they easily spread? In which dairy products are molds most destructive? Cite an example of a mold used in ripening a particular cheese.
24. Cite examples of contributions made by yeasts to the dairy industry.
25. Why should persons associated with the production, handling, processing, distribution, and consumption of milk and milk products have knowledge of the application of microbiology?
26. What is meant by the shelf life of dairy products? How may it be extended?

Notes

[1] Named after the scientist who developed it, Hans Christian Gram.

[2] Bacteria in the genera *Bacillus* and *Clostridium* are **aerobic** and **anaerobic**, respectively.

[3] Freshly produced cells are concentrated and frozen, then exposed to a high vacuum while in the frozen state. Moisture passes directly from the solid (frozen) to the gaseous (vapor) state.

[4] Fractions of whole milk are skim milk, lowfat milk, cream, whey, and others.

[5] Henry Tissier, a French pediatrician, observed that children with diarrhea had in their stools a low number of bacteria characterized by a peculiar Y-shaped morphology. These "bifid" bacteria were, on the contrary, abundant in healthy children (Tissier, 1906). He suggested that these bacteria could be administered to patients with diarrhea to help restore a healthy gut flora.

[6] Seven types of toxin (A, B, C, D, E, F, and G) are produced by different strains of the species *C. botulinum*.

[7] Aflatoxins were found in the milk of cows fed toxic peanut meal.

References

Anukam, K. C., and G. Reid. 2007. "Probiotics: 100 Years (1907–2007) after Elie Metchnikoff's Observation." In A. Méndez-Vilas (ed.), *Communicating Current Research and Educational Topics and Trends in Applied Microbiology* (pp. 466–474). Badajoz, Spain: Formatex Research Center.

Broadbent, J. R., H. Cai, R. L. Larsen, J. E. Hughes, D. L. Welker, V. G. De Carvalho, T. A. Tompkins, Y. Ardö, F. Vogensen, A. De Lorentiis, M. Gatti, E. Neviani, and J. L. Steele. 2011. Genetic diversity in proteolytic enzymes and amino acid metabolism among *Lactobacillus helveticus* strains. *Journal of Dairy Science* 94:4313–4328.

Celik, O. F., and D. J. O'Sullivan. 2013. Factors influencing the stability of freeze-dried stress-resilient and stress-sensitive strains of bifidobacteria. *Journal of Dairy Science* 96:3506–3516.

Chandan, R. C. 1999. Enhancing market value of milk by adding cultures. *Journal of Dairy Science* 82:2245–2256.

Consumer Reports. 2005, July. Probiotics: Are enough in your diet? *Consumer Reports* 34–35.

Ellis, D. E., P. A. Whitman, and R. T. Marshall. 1973. Effects of homologous bacteriophage on growth of *Pseudomonas fragi* within milk. *Applied Microbiology* 25:24.

Federal Register. 2000. Regulations on Statements Made for Dietary Supplements Concerning the Effect of the Product on the Structure or Function of the Body (21 CRF Part 101). *Federal Register* 65(4):999–1050.

Gilliland, S. E. 2001. "Probiotics and Prebiotics." In E. H. Marth and J. L. Steele (eds.), *Applied Dairy Microbiology* (2nd ed.). New York: Marcel Dekker, Inc.

Hassan, A. N., and J. F. Frank. 2001. "Starter Cultures and Their Use." In E. H. Marth and J. L. Steele (eds.), *Applied Dairy Microbiology* (2nd ed.). New York: Marcel Dekker, Inc.

Heimbach, J. T. 2008. Health-benefit claims for probiotic products. *Clinical Infectious Diseases* 46 (Suppl 2):S122–S124.

Lie, J. L., and E. H. Marth. 1967. Development of aflatoxin in Cheddar cheese. *Journal of Dairy Science* 50:955.

Liu, Z., Z. Jiang, K. Zhou, P. Li, G. Liu, and B. Zhang. 2007. Screening of bifidobacteria with acquired tolerance to human gastrointestinal tract. *Anaerobe* 13:215–219.

Moller, C., and M. de Vrese. 2004. Review: Probiotic effects of selected acid bacteria. *Milchwissenschaft* 59(11/12):597–601.

National Advisory Committee on Microbiological Criteria for Foods. 2010. Assessment of food as a source of exposure to *Mycobacterium avium* subspecies *paratuberculosis* (MAP). *Journal of Food Protection* 73:1357–1397.

O'Flaherty, S., and T. R. Klaenhammer. 2010. The role and potential of probiotic bacteria in the gut, and the communication between gut microflora and gut/host. *International Dairy Journal* 20:262–268.

Salminen, S., W. Kenifel, and A. C. Ouwehand. 2011. "Probiotics, Applications in Dairy Products." In J. W. Fuquay, P. F. Fox, and P. L. H. McSweeney (eds.), *Encyclopedia of Dairy Sciences*, Vol. 1 (2nd ed., pp. 412–419). San Diego, CA: Academic Press.

Sanders, M. A., G. Gibson, H. S. Gill, and F. Guarner. 2007, October. *Probiotics: Their Potential to Impact Human Health* (Issue Paper Number 36). Ames, IA: Council for Agricultural Science and Technology.

Serraino, A., O. Florio, F. Giacometti, S. Piva, D. Mion, and R. G. Zanoni. 2013. Presence of *Campylobacter* and *Arcobacter* species in in-line milk filters of farms authorized to produce and sell raw milk and of a water buffalo dairy farm in Italy. *Journal of Dairy Science* 96:2801–2807.

Songisepp, E., P. Hütt, M. Rätsep, E. Shkut, S. Kõljalg, K. Truusalu, J. Stsepetova, I. Smidt, H. Kolk, M. Zagura, and M. Mikelsaar. 2012. Safety of a probiotic cheese containing *Lactobacillus plantarum* Tensia according to a variety of health indices in different age groups. *Journal of Dairy Science* 95:5495–5509.

Tissier, H. 1906. Traitement des infections intestinales par la methode de transformation de la flore bacterienne de l'intestin. *Comptes Rendus Social Biology* 60:359–361.

Tong, P. 2010. Opportunities beyond yogurt. *Dairy Foods Magazine* 111(5):62.

Whitman, P. A., and R. T. Marshall. 1971. Isolation of psychrophilic bacteriophage host systems from refrigerated food products. *Applied Microbiology* 22:220–223.

Websites

Council for Agricultural Science and Technology (http://www.cast-science.org)

Food Safety Magazine (http://www.foodsafetymagazine.com)

Frontiers in Microbiology (http://journal.frontiersin.org/journal/microbiology)

National Center for Biotechnology Information (http://www.ncbi.nlm.nih.gov)

24

Milk Quality and Safety

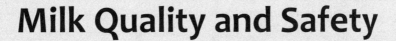

The rule is well established in American law that a vendor of food warrants that it is fit for human consumption and is liable for damages if the food causes physical injury to a consumer.

J. A. Tobey[1]

24.1 Introduction

Considering total quality and balance of components, no other single natural food meets the nutritional requirements of humans better than milk. However, hundreds of microorganisms have similar nutritional needs so milk provides an excellent medium for their growth. This means, given a favorable temperature, microorganisms can reproduce in milk and spoil it rapidly. Furthermore, if a disease-producing organism is present, it may infect the milk consumer. In this chapter we focus on effects of microorganisms and common adulterants of milk that may affect its safety and quality.

24.2 Historical Background

In the United States, it became evident early in development of the fluid milk industry that certain practices were necessary to preclude disease transmission through milk. Physicians in New Jersey recognized the importance of milk in transmission of disease. They studied methods of hygienic milk production, established rigid detailed standards, and selected dairy producers who agreed to produce milk under supervision of a medical milk commission. The first **certified milk** was produced in 1894 in Essex County, New Jersey. Medical milk commissions were subsequently established throughout the United States. Many practices established by the certified milk industry were subsequently adopted for production of grade "A" milk.

In 1893, commercial pasteurization was introduced in Baltimore, Maryland; in its infancy, it was often ineffective. Generally, bacteria counts were reduced no more than 99% (e.g., from 10^6 to 10^4). Postpasteurization contamination and inefficient cooling permitted counts to increase rapidly following heat treatment. Problems were related to an overburdened inspection system that consisted of a city chemist and two inspectors. The wagon trade required most of their attention. The 300 stores selling 10% of the milk, primarily to tenement dwellers, received infrequent attention. The city ordinance regulated milk production within the city (5,000 to 6,000 cows) but most milk was produced on unregulated farms outside the city, and the quality was controlled only with regard to composition, bacterial count, and temperature.

Many improper practices were noted by Moore et al. (1911) in a report entitled *The Milk Problem in St. Louis*. Samples were collected from milk for infants in the crowded districts of that city. Sources were dairies, milk wagons, and stores. At that time, half of the milk was being dipped from cans at the point of retail sales. This was termed *loose* milk as opposed to bottled milk. Loose milk was of comparatively low quality and sold for $.05 compared to about $.08/qt for bottled milk. According to the report, 35% of this loose milk had been "flagrantly watered" and another 33% was "suspicious" based on specific gravity tests. In those days bacteria counts often exceeded several million/mL, especially those of loose milk. In Chicago, early standards were 1,000,000/mL in summer months and one-half that in other months. Over 60% of the samples taken from vendors in St. Louis in the summer of 1910 had counts exceeding 3,000,000/mL, and over 40% were above 10,000,000/mL. It is reasonable to assume that similar conditions existed in most cities in the early 1900s. However, the movement to make milk safe was under way.

24.3 Diseases Transmissible to Humans through Milk

Some disease-producing microorganisms are capable of infecting both human and bovine species. Others are specific to humans but can be transmitted through milk that has been contaminated with them.

24.3.1 Tuberculosis and Brucellosis

Tuberculosis and brucellosis of cattle can be transmitted to humans through milk. During the early to mid-1900s tuberculosis (*Mycobacterium tuberculosis*) and brucellosis (*Brucella abortus*) were endemic in cattle in the United States. Governmental programs of testing and eliminating infected animals plus calf-hood vaccination for brucellosis were successful in virtually eliminating these serious diseases. Vaccination of females of dairy breeds against brucellosis should occur between the ages of three to six months.

The USDA Animal and Plant Health Inspection Service (APHIS) provides nationwide surveillance of both tuberculosis and brucellosis and produces monthly reports of their results. For tuberculosis, when infected animals are identified, officials investigate these cases to determine if additional animals or herds of animals are infected. Infected animals are removed to slaughter and annual testing is conducted in areas where positive tests have occurred.

Brucellosis is also known as contagious abortion or Bang's disease (cf. chapter 13). In humans it's known as undulant fever because of the severe intermittent fever accompanying human infection. In 1934 the Cooperative State–Federal Brucellosis Eradication Program was established to eliminate the disease from the country. The program's Uniform Methods and Rules set forth minimum standards for states to achieve eradication, as control of brucellosis was limited mainly to individual herds. States are designated brucellosis free when none of their cattle or bison have been found to be infected for 12 consecutive months under an active surveillance program. Brucellosis in cattle is detected by blood test or the milk ring test. Both are based on antigen-antibody reactions that depend on the presence of the specific antibody to the causative organism, *Brucella abortus*.

24.3.2 Diseases of Humans Transmissible through Milk

Before World War II diseases commonly associated with raw milk included diphtheria, scarlet fever, and typhoid fever. Although vaccines were developed to provide immunity to these diseases, adoption of the *Grade "A" Pasteurized Milk Ordinance* (*PMO*) and pasteurization of milk were also major reasons they became infrequent in the United States.

The organisms of major public health concern today are serotypes of *Salmonella* (especially Typhimurium and Dublin), *Listeria monocytogenes*, pathogenic (enterohaemorrhagic) *Escherichia coli*, and *Campylobacter jejuni*. These bacteria reside primarily in the intestinal tract of the bovine and enter milk as a result of fecal contamination. Therefore, keeping them out of milk depends on prevention of intestinal infections and contamination of milk with environmental substances. Some strains of *Staphylococcus aureus*, *E. coli*, and *L. monocytogenes* can cause bovine mastitis, making it important to exclude milk obtained from infected mammary glands.

Although they are potentially capable of causing serious illnesses, rickettsiae, viruses, and parasites are responsible for few milk-related illnesses. For a comprehensive review of milk-borne diseases see Ryser (2001) and chapter 13.

24.3.3 Consumption of Raw Milk

In recent years small, but increasing, numbers of individuals have sought to buy and consume raw milk (Yilmaz et al., 2009). Raw milk and raw milk products (such as cheeses and yogurts made with raw milk) can be contaminated with bacteria that can cause serious illness, hospitalization, or death. These harmful bacteria include *Brucella, Campylobacter, Listeria, Mycobacterium bovis, Salmonella,* Shiga toxin-producing *E. coli, Shigella, Streptococcus pyogenes,* and *Yersinia enterocolitica*.

Because of the potential for serious illness, federal law prohibits dairies from distributing raw milk across state lines in a package form so that it can be consumed. Raw milk can only be distributed across state lines if it is going to be pasteurized or used to make aged (over 60 days) cheese before being sold to consumers. Each state makes its own laws regarding selling raw milk within its own borders. In about half of states, sale of raw milk directly to consumers is illegal. In the remaining states, raw milk may be sold directly to consumers.

An outbreak of campylobacteriosis was reported in Michigan and Indiana in 2010 in which there were eight confirmed cases traceable to a cooperative in which buyers owned "cow shares." That cooperative was organized to allow raw milk to be available to persons who were not permitted to purchase it through retail channels. In the United States from January 2010 to March 2012, 23 disease outbreaks resulting in 300 illnesses, but no deaths, were assigned to consumption of fluid raw milk (20) and aged raw milk cheese (3).

24.4 Regulations and Responsible Organizations

The first law on milk quality in the United States was a Massachusetts act in 1856 that prohibited adulteration of milk. The first regulations for the production of milk were prepared by the Board of Health in England and were known as the Dairy, Cowshed and Milk Shop Order of 1879. In 1896, New York City established a permit system governing the sale of milk. The right to enforce the system was upheld by an opinion of the US Supreme Court in 1905. Between that year and 1920 city ordinances of many types were adopted.

Variations in methods and effectiveness prompted the US Public Health Service (USPHS) to develop a standard ordinance in 1924 and to publish it in public health reports. To provide for uniform interpretation of this ordinance, an accompanying code was published in 1927. It provided administrative and technical details pertaining to satisfactory compliance. Published by the USPHS and the US Food and Drug Administration, it is now known as the *Grade "A" Pasteurized Milk Ordinance.* The *PMO* was developed with the assistance of milk regulatory and rating agencies at every level of federal, state, and local government, including departments of health and agriculture; representatives of milk producers, processors, equipment manufacturers, and trade associations; educational and research institutions; and individual professional sanitarians. It incorporates provisions governing the processing, packaging, and sale of grade "A" milk and milk products, including buttermilk and buttermilk products, whey and whey products, and condensed and dry milk products. It is a "model ordinance" since adoption of milk control regulations is delegated to states.

Control of milk's quality and safety in the United States is the responsibility of the states and municipalities. The final seat of responsibility varies from state to state. In general, however, most milk is regulated under rules written in the *PMO.* More than 90% of milk in the United States is produced under these rules.

In 1946 the Conference of State and Territorial Health Officers requested the USPHS to develop a plan to certify interstate milk shippers. This request was made because shippers were encountering widely different regulations in various receiving states. Additionally, inspectors were traveling among states to inspect and enforce these varying regulations. The goal was to find agreement among states pertaining to requirements for production and processing of grade "A" milk products. As a result, the Surgeon General of the United States invited all states to send a representative to St. Louis, Missouri, in 1950 to participate in the first Interstate Milk Shippers Conference. Out of this meeting came the National Conference on Interstate Milk Shipments (NCIMS) and adoption by the states of the *PMO* for control of the safety of the nation's milk supply. By using "reciprocal agreements," agencies now accept inspections and testing by officials in the shipping states because the rules of production, processing, inspection, testing, and enforcement are the same among states. And no state may impose a requirement that is in excess of those required by the *PMO*. Furthermore, the NCIMS meets biennially to update the *PMO* with the concurrence of the USPHS/FDA.

As an example of state enforcement, in Missouri the agency responsible for milk safety is the State Milk Board. The office and employees operate under a budget approved by state government but technical and managerial decisions are under the Board of Directors, members of which represent producers, processors, health departments, and consumers.

24.4.1 Plant and Equipment Inspections

Typically, qualified sanitarians inspect producing farms at least twice annually and processing plants quarterly. An inspection form is used as a guide in checking for adherence to regulations. Deficiencies are marked on the form and time is given in which a deficiency must be corrected. Appendix P of the *PMO* describes a system of assigning farm inspection frequencies of one to four times per year based on performance. Dairy farms are categorized at least every three months based on the previous 12-month farm inspection and milk quality data. For example, in Wisconsin, among producers meeting other requirements, those having no monthly somatic cell counts (SCCs) over 500,000/mL qualify for an annual inspection; whereas those who have 4 of 13 monthly counts exceeding 750,000/mL must be inspected twice yearly. To qualify for a single yearly inspection no more than one bacteria count may exceed 25,000/mL and it shall be less than 100,000/mL. Other violations can increase inspection frequency to quarterly.

Periodically, each state or local health unit must be checked by an FDA Rating Officer to determine whether the sanitarians and laboratories within the unit are correctly enforcing the requirements of the *PMO* and that the producers and processors are complying with published regulations.

Inspectors collect samples for testing and deliver them to approved laboratories that follow procedures published in *Standard Methods for the Examination of Dairy Products* (*SMEDP*) by the American Public Health Association and in *Official Methods of Analysis* published by AOAC International. An excellent reference for compositional and labeling standards for milk and milk products is found in chapter 16 of *SMEDP* (Wehr and Frank, 2004).

FDA monitors proficiencies of laboratories involved in testing under the *PMO*. For example, FDA periodically sends "split samples" to official laboratories. These are tested for such properties as bacteria count, somatic cells, antibiotics, pesticides, and freezing point. Within a specified period results must be returned from each participating laboratory to the FDA for statistical analysis. Laboratories that fail to produce results within an acceptable tolerance must correct their procedures or face losing their certified status.

Equipment manufactured in conformity with guidelines outlined by 3-A Sanitary Standards, Inc. (3-A SSI) complies with the sanitary design and construction standards of the *PMO*. 3-A SSI is an independent, not-for-profit corporation dedicated to advancing hygienic equipment design. Additionally, there are 3-A Accepted Practices, including number 607 (which applies to spray drying systems) and number 604 (which applies to supplying air under pressure in contact with milk, milk products, and product-contact surfaces).

These sanitary standards are promulgated jointly by the Sanitary Standards Subcommittee, which is composed of: the Dairy Industry Committee (represents equipment manufacturers), the Committee on Sanitary Procedures of the International Association for Food Protection (represents milk sanitarians/inspectors), and the Milk Safety Team of the FDA (represents government regulatory agencies).

Food contact surfaces of equipment must meet specific design and fabrication requirements to ensure cleanability. Contact surfaces are those that have direct contact with a food or its residue, including those onto which food residue can drip, drain, diffuse, or be drawn. All multiuse containers, utensils, and equipment exposed to milk and milk products or from which liquids may escape into milk products must be made from smooth, impervious, nonabsorbent, nontoxic materials. Stainless steel is the most common material used in fabrication of equipment. 3-A sanitary standards specify AISI 300 series stainless steel, with 304 and 316 used commonly. These nonmagnetic materials are composed of alloys in which chromium and iron predominate. Chromium oxide forms a passive layer on the surface that protects the inner layer, containing iron, from corrosion. If the passive layer is compromised, the inner layer becomes vulnerable to corrosion by materials such as hypochlorites. The smoothness level desired for most dairy equipment is known as no. 4 finish (roughness average of 0.8 μm or less). Surfaces must be free of cracks, crevices, and sharp angles and be sloped to drain.

Since 1956 the 3-A symbol has been used to identify equipment meeting 3-A sanitary standards for design and fabrication. These standards provide processors with assurance that equipment meets sanitary standards and establishes guidelines for uniform evaluation and compliance by milk sanitarians. Value of the 3-A symbol has been enhanced by the third-party verification (TPV) inspection requirement by an independent, accredited professional. Instituted in 2003, successful TPV, applied in the United States and 22 other countries, authorizes equipment manufacturers to place the 3-A symbol on their product.

24.4.2 Hygiene and Equipment Materials

The European Hygienic Equipment Design Group (EHEDG) is composed of representatives from research institutes, equipment manufacturers, the food and pharmaceutical industry, and legislative bodies. Their objective is to provide guidance on hygienic and aseptic design.

Of particular interest to organizations such as the EHEDG is equipment that has product contact surfaces. In the dairy industry the needs for sanitary design and operation are particularly stringent. Both alkaline and acidic cleaners are used; therefore, materials used to make equipment must withstand their corrosive effects. The major material used is stainless steel, which is an alloy of chromium, nickel, and iron. The most commonly used combination is 18% chromium and 8% nickel with iron. Chromium produces a stable passive film, up to 5 nm thick, on exposed surfaces. An addition of 2 to 4% molybdenum improves corrosion resistance in the presence of acids. Extended exposure to hypochlorite sanitizers in high concentrations is the major risk factor in corrosion of stainless steels.

Plastic materials offer the advantages of low cost, easy fabrication, low weight, and corrosion resistance. However, they have low mechanical strength and poor resistance to heat and fatigue. They must be abrasion resistant and free from constituents that can migrate into dairy foods. Food grade polymers must be specified. These elastic materials must be nonabsorbent. Nylon and Teflon are often used as seals in storage vessels, piping, and pumps.

24.4.3 Hazard Analysis & Critical Control Points Program

The FDA is responsible for application of the Hazard Analysis & Critical Control Points (HACCP) program in the food industry. In 1999, the NCIMS initiated a voluntary pilot program to test whether the HACCP is an acceptable alternative to current plant performance ratings. The following points are made regarding the program.

1. HACCP is a science-based system used to ensure that food safety hazards are controlled to prevent unsafe food from reaching the consumer.
2. HACCP places a high level of ownership of responsibility on the food processor to identify and control hazards and to document effectiveness of the system. Additionally, it requires constant verification that the system is working properly.
3. The regulatory agency retains its authority and responsibility for oversight to verify that food is manufactured according to the firm's HACCP plan and is handled in such a way that its safety is assured.
4. Safety is enhanced by a proactive approach that continuously monitors food safety controls and documents results and corrective actions. This monitoring occurs in "real time" rather than as a reactive, after-the-fact approach.
5. HACCP requires monitoring to reveal when food safety limits have been violated. This results in taking corrective actions to reinstate control and through documented procedures to prevent recurrence. Operation of the system is verified constantly.

24.4.4 Reportable Food Registry

In 2007, the Food and Drug Administration Amendments Act required the FDA to establish the Reportable Food Registry, an electronic portal that allows responsible parties or public health officials to report possible food-related public health risks. Such submissions provide early warning to FDA of potential public health risks from reportable foods or food ingredients. Responsible persons may be at any level of the food production/distribution chain, both domestic and international. Reports during the first year of operation encompassed seven categories of food safety hazards. Of 18 reports related to dairy products, eight involved contamination with *L. monocytogenes* or with undeclared allergens/intolerances.

24.4.5 Grade "A" Milk

The scope of the *PMO* is described in detail in the following quotation from the introduction to the 2013 revision:

> An ordinance to regulate the production, transportation, processing, handling, sampling, examination, labeling, and sale of Grade "A" milk and milk products; the inspection of dairy farms, milk plants, receiving stations, transfer stations, milk tank truck cleaning facilities, milk tank trucks and bulk milk hauler/samplers; the issuing and revocation of permits to milk producers, bulk milk hauler/samplers, milk tank trucks, milk transportation companies, milk plants, receiving stations, transfer stations, milk tank truck cleaning facilities, haulers, and distributors; and the fixing of penalties.

Producers are issued permits to ship grade "A" milk. If the producer is found in repeated violation of requirements of the *PMO*, this permit is suspended or revoked. In brief, the requirements include (1) freedom from adulterants and of milk from diseased animals; (2) proper construction and cleanliness of the milking barn and milk room; (3) safe and adequate water supply; (4) sanitary waste disposal and adequate treatment of human wastes; (5) milk utensils and equipment properly constructed, cleaned, and sanitized; (6) milking accomplished with minimum risk of contamination from the animals, equipment, and personnel; (7) cooling within a prescribed time to a temperature that will satisfactorily restrict microbial growth; (8) control of insects and rodents; and (9) satisfactory condition of milk as it is presented to the processing operation.

Likewise, processors are issued permits and are required to meet standards that pertain to the facilities, equipment, personnel, operations, process and storage parameters, cleaning and sanitizing, packaging, waste disposal, and transportation. Of highest importance are proper design, installation, and operation of the milk pasteurizer. To ensure lethal heat treatment is given to pasteurized milk products the *PMO* provides detailed specifications for construction and operation as well as constant monitoring of pasteurization temperature and a series of test procedures to confirm proper operation.

Regardless of care taken that raw materials, processes, equipment, and packages meet sanitary standards, failure to manage flow of personnel within a facility can lead to product contamination from the environment. Access to areas within a plant must be limited to those who have responsibilities therein and those persons must be dressed suitably for the operation.

24.4.6 Manufacturing Grade Milk

The US Department of Agriculture published the General Specifications for Dairy Plants Approved for Inspection and Grading (7 CFR 58, Subpart B) involved in their grading and inspection service. These are plants that produce butter, cheeses, and dry milk products that bear a grade label, for example, grade "AA" butter. Milk produced for use in these products must either be from grade "A" supplies or from farms that are subject to the requirements of the USDA's *Milk for Manufacturing Purposes and Its Production and Processing: Recommended Requirements*. This document, revised in 2011, sets forth requirements similar to, but less stringent than, those for producing grade "A" milk. The quality classification of raw milk for manufacturing purposes from each producer is based on an organoleptic examination for appearance and odor, a drug residue test, and quality control tests for sediment content, bacterial estimate, and somatic cell count. The standard for maximal bacterial count is 500,000/mL.

24.5 Food Additive Regulatory Program

The US Congress enacted, and President Eisenhower signed into law, the Food Additives Amendment of 1958 (Public Law 85-929, 72 Stat. 1784). It is the basis for the US food additive regulatory program, which oversees most substances added to food (Neltner et al., 2011). Substances covered range from drugs and pesticides in animal feed to additives used to cultivate, preserve, and process food.

The term *food additive* means any substance the intended use of which results or may reasonably be expected to result, directly or indirectly, in its becoming a component or otherwise affecting the characteristics of any food. This includes any substance intended for use in producing, manufacturing, packing, processing, preparing, treating, packaging, transporting, or storing food and includes any source of radiation intended for any such use.

The term food additive includes all common ingredients, for example, emulsifying salts added to processed cheeses. These common ingredients, some of which have been used for centuries, are not normally thought of as food additives. The term also includes substances often called "indirect food additives" or "food contact substances," such as substances used during the food packaging process. Therefore, "food additive" may include substances in or on packaging, such as components of clear plastic film that cover cheese slices, as well as substances used in or on food processing equipment, including lubricating oil or cleaning chemicals, to the extent their components may migrate into food. Expanding the meaning of "food additive" further is the definition's inclusion of any source of radiation used to inspect, heat, or treat certain foods or packaging even though radiation per se is not a substance.

There are three broad exclusions from the food additive definition: (1) substances *generally recognized as safe* (*GRAS*) among experts who are qualified by scientific training and experience to evaluate safety under conditions of intended use; (2) substances that were sanctioned by the FDA or USDA before the Food Additives Amendment of 1958 was enacted; and (3) certain categories of substances for which distinct programs and standards for managing them were established (e.g., pesticides and their residues in raw agricultural commodities, color additives, drugs in animal feed, and dietary supplements).

FDA's Office of Food Additive Safety (OFAS), via its agency's Center for Food Safety and Applied Nutrition, regulates food additives, GRAS substances, color additives, and prior-sanctioned substances used in human food. FDA's Center for Veterinary Medicine (CVM) regulates drugs and additives used in animal feed and pet food. The US Environmental Protection Agency (EPA) sets standards for pesticide chemicals or residues in food, which FDA enforces.

24.6 Safety of Milk Products on a Global Scale

Assuring the safety of milk and milk products for human consumption has become extremely complex because of increased concentration of the industry and introductions of new products, processes, materials, and marketing patterns, each of which continue to be evaluated in terms of their public health significance.

As electronic (digital) communications and global transportation have been developed, exports and imports of dairy foods have increased. Local firms have become interstate and interstate firms have become international. Increasingly it has become important to harmonize regulations, composition, labels, and inspection practices. The ability to trace back to the source of raw materials and **handlers** is now demanded by many regulatory agencies, dealers, and consumers.

Coding of products with a "sell by" date has been used for several years. The need to trace back to its origin an outbreak of food-borne illness has prompted regulatory agencies and industry to standardize 14-digit codes, which each trader in the supply chain records as cases are received and when sent onward. If contamination is discovered, going back through the databases will identify the source and route of the product, thus enabling a direct approach to the problem. Traceability to the item level is a goal in some parts of the food industry. Use of technology such as the smart phone to scan codes is gaining favor.

Several international organizations are involved in harmonization efforts on a global scale. We will discuss some of them here.

24.6.1 Codex Alimentarius Commission

This commission was created in 1963 by the Food and Agriculture Organization of the United Nations (FAO) and the World Health Organization (WHO) to develop food standards, guidelines, and related texts, such as codes of practice under the Joint FAO/WHO Food Standards Programme. Important purposes of this program are protecting consumers' health, ensuring fair practices in the food trade, and promoting coordination of all food standards work undertaken by international governmental and nongovernmental organizations. The Codex Alimentarius is a collection (14 volumes) of international food standards adopted by the Codex Alimentarius Commission, which has over 165 member countries. It includes standards for all principle foods whether processed, semiprocessed, or raw that will be distributed to consumers. Codex standards are referenced by the World Trade Organization during trade disputes. Without scientific justification a country cannot refuse entry of products that meet Codex standards.

The Codex Code of Ethics stipulates that:

> No food should be in international trade which: (a) has in it or upon it any substance in any amount which renders it poisonous, harmful or otherwise injurious to health; or (b) consists in whole or part of any filthy, putrid, rotten, decomposed or diseased substance or foreign matter, or is otherwise unfit for human consumption; or (c) is adulterated; or (d) is labeled or presented in a manner that is false, misleading or deceptive; or (e) is sold, prepared, packaged, stored or transported for sale under insanitary conditions.

Codex Commission operates through committees. The most important ones to the dairy industry are committees on Food Hygiene, Milk and Milk Products, and Methods of Analysis and Sampling. Codex uses a deliberative eight-step procedure to elaborate on and adopt standards. The Codex Committee on Milk and Milk Products develops standards for various dairy products including cheese, butter, milk and cream powders, sweetened and condensed milk, and milkfat products. Each standard contains basic compositional requirements and references other Codex texts for requirements related to pesticide residues, food additives, food hygiene, labeling, and methods of analysis. Although Codex standards are voluntary, countries have an obligation to base their food safety standards on the work of Codex.

24.6.2 Food and Agriculture Organization of the United Nations (FAO)

The FAO has adopted the following principle: *Producing a food supply that is safe and of good quality is a prerequisite to the successful domestic and international trade in food and a key to sustainable development of national agricultural resources. All consumers have the right to expect and demand safe, good-quality food.*

24.6.3 World Trade Organization (WTO)

The WTO has adopted the Agreement on the Application of Sanitary and Phytosanitary Measures, which sets forth basic rules for food safety and animal and plant health standards. It allows countries to set their own standards. But it also says regulations must be based on science. They should be applied only to the extent necessary to protect human, animal, or plant life or health. And they should not arbitrarily or unjustifiably discriminate among countries where identical or similar conditions prevail. Member countries are encouraged to use international standards, guidelines, and recommendations where they exist.

24.6.4 International Dairy Federation (IDF)

The IDF, a science-based organization, represents the dairy industry worldwide. It provides a global source of scientific expertise and knowledge to support production of safe, nutritious, and high-quality milk and dairy products. IDF is a nonprofit private sector organization representing the interests of various international stakeholders in dairying. It was founded in 1903 and is represented in 53 countries. Its members account for approximately 85% of the world's total milk production. IDF members are organized in national committees composed of representatives of dairy farmers, dairy suppliers, academics, governments/food control authorities, and from the dairy processing industry.

IDF aims to identify, elaborate, and disseminate best practices at the international level to guide the dairy sector and to harmonize members' work on issues along the dairy production chain, including animal health and welfare, protection of the environment, nutrition, food safety and hygiene, and food standards. Among its services is validation of methods of analysis of dairy foods.

24.6.5 International Organization for Standardization (ISO)

The world's largest developer and publisher of international standards, the ISO is a network of national standards institutes from 164 countries with a Central Secretariat in Geneva, Switzerland. ISO is a nongovernmental organization that forms a bridge between the public and private sectors. Many of its member institutes (one per country) are government based while others belong to the private sector, having been set up by national partnerships of industry associations.

ISO enables a consensus to be reached on solutions that meet both the requirements of business and the broader needs of society. ISO standards (1) make development, manufacturing, and supply of products and services more efficient, safer, and cleaner; (2) facilitate fair trade; (3) provide governments with a technical base for legislation; (4) share technological advances and good management practices; (5) disseminate innovation; and (6) safeguard users of products and services.

Most ISO standards are specific to a particular product, material, or process. However, ISO 9001 and ISO 14001 are "generic management system standards," meaning that the same standard can be applied to any organization whatever its product or service, in any sector of activity, and whether it is a business enterprise, a public administration, or a government department. ISO 9001 contains requirements for implementing a *quality* management system and ISO 14001 for an *environmental* management system. ISO, IDF, and AOAC jointly sponsor numerous methods applicable to the dairy industry.

24.6.6 Additional Agencies and Programs

Along with the international aspects of control and standards for dairy foods, the US dairy industry is indirectly affected by other programs and issues related to providing a safe and reliable food supply. These include FDA-sponsored Current Good Manufacturing Practices (CGMP). The recent emphases on sustainability and traceability of the food supply will undoubtedly impact productivity, quality, and safety of dairy foods.

Many firms choose to have independent auditors evaluate their operations, and retailers often require reports of such audits. The Global Food Safety Initiative had its beginning in Europe in 1998 where retailers, through the British Retail Consortium, sought uniformity in audits of food-producing firms.

The program is based on the need for uniformity among agencies that audit food manufacturers when their auditors evaluate the status of operations and management thereof regarding food safety. The initiative focuses on standardization on an international basis of effective programs among certified auditing organizations, known as certifying bodies. A major one in the United States is the Safe Quality Food Institute.

The following Key Obligations of Food and Feed Business Operators were issued by the Health & Consumer Protection Directorate-General of the European Commission:

- **Safety**: Operators shall not place on the market unsafe food or feed.
- **Responsibility**: Operators are responsible for the safety of the food and feed which they produce, transport, store, or sell.
- **Traceability**: Operators shall be able to rapidly identify any supplier or consignee.
- **Transparency**: Operators shall immediately inform the competent authorities if they have reason to believe that their food or feed is unsafe.
- **Emergency**: Operators shall immediately withdraw food or feed from the market if they have a reason to believe that it is not safe.
- **Prevention**: Operators shall identify and regularly review the critical points in their processes and ensure that controls are applied at these points.
- **Cooperation**: Operators shall cooperate with the competent authorities in actions taken to reduce risks.

Effective in 2012 the European Union (EU) required firms shipping dairy foods to EU countries to comply with the European Union Health Certification Program and maintain an updated Certificate of Conformance with somatic cell count (400,000/mL) and standard plate count (100,000/mL) maximums based on three-month moving averages. Testing at the producer level is required. The USDA Agricultural Marketing Service (AMS) has the responsibility for reviewing records to assure compliance.

In the United States, a need for educational programs related to milk quality and safety prompted organization of the Northeast Dairy Practices Council (NDPC) in 11 northeastern states in 1970. In 1994, NDPC became a national organization named the Dairy Practices Council (DPC). An additional constitutional change in 2000 opened membership internationally. The intent of the organization is to evoke and enhance cooperation of representatives of industry, education, and regulatory bodies.

The objectives of DPC are twofold: (1) Develop and disseminate educational guidelines for the dairy industry related to proper sanitation and quality of dairy products. The DPC library consists of over 100 guidelines. (2) Provide mutual assistance internationally among countries, states, provinces, and territories for continued improvement of procedures involved in production, processing, and distribution of milk and dairy products.

The main strength of DPC is a large core of dedicated members and key sanitarians who work hard and volunteer many hours to author and review quality guidelines. DPC membership includes representatives of all phases of the dairy industry. Task Force Directors prepare task force sessions, coordinate the writing of many guidelines, and are responsible for the systematic review of guidelines to ensure each one is current.

24.7 Regulatory Standards for Grade "A" Milk

Standards described in this section are those from the FDA's *Grade "A" Pasteurized Milk Ordinance.*

24.7.1 Temperature

On farms, milk shall be cooled to 45°F (7°C) or less within 2 hr after milking, and the temperature of milk blended after the first and subsequent milkings shall not exceed 50°F (10°C). Milk processors shall maintain temperatures at 45°F (7°C) or less, except when processing.

24.7.2 Bacterial Limits

Standard plate counts (SPCs), which estimate numbers of aerobic bacteria, shall not exceed 100,000/mL for an individual producer's milk. The count may be as high as 300,000/mL in commingled milk (milk from multiple herds). The rationale is that pumping milk during transport operations breaks clumps and chains of bacterial cells, resulting in increased numbers of bacterial colonies when counts are made. The legal SPC limit for raw milk in European countries is 100,000/mL. The SPC limit for pasteurized milk is 20,000/mL. In addition to the SPC, tests for coliform bacteria are applied to pasteurized milk. A selective medium is used on which visible colonies of coliforms develop within 24 hr. Nearly all other bacteria are inhibited by the medium and coliforms, being producers of acid from lactose (milk sugar), form deep red colonies as they lower the pH and change the color of the violet red dye. Since coliform bacteria are easily killed by pasteurization, their presence in pasteurized milk indicates contamination after pasteurization. They are commonly present in the environment of dairy plants and on improperly cleaned and sanitized equipment. The count limit for coliforms is 10/mL.

24.7.3 Somatic (Body) Cell Count Limit

Counts of somatic cells in individual producer's milk shall not exceed 750,000/mL. European standards set a maximum of 400,000/mL, and any product exported to a European country is required to be made from milk that meets that standard. SCCs of milk from uninfected mammary glands consistently fall below 100,000/mL. A higher limit of 1,000,000/mL applies to somatic cells in goat milk. The reason is that the normal apocrine secretory system of goats (merocrine system in cows) releases cytoplasmic particles into milk, and these are enumerated along with phagocytes when cells are counted.

SCCs on milk taken from farm bulk tanks are usually conducted on electronic instruments using chemicals to disperse fat globules. Electronic counters take advantage that somatic cells stained with a fluorescent dye can be enumerated by flow cytometry or with an electrically monitored microscope focused on a rotating disc to which stained cells adhere.

24.7.4 The Three-Out-of-Five Provision

During any consecutive six months, at least four samples of raw milk for pasteurization and pasteurized milk product shall be collected from each producer or processor under direction of the regulatory agency. Because there can be significant variations in numbers of bacteria and somatic cells in milk due to sampling and laboratory errors, as well as normal biological variations, the standards are applied on counts made from the last four samples.

Whenever two of the last four consecutive bacterial counts (except those for aseptically processed milk and milk products), somatic cell counts, coliform determinations, or cooling temperatures, taken on separate days, exceed the standard for milk and/or milk products as defined in the *PMO*, the regulatory agency shall send a written notice thereof to the permit holder. This notice shall be in effect as long as two of the last four consecutive samples exceed the standard. An additional sample will be taken within 21 days of the sending of such notice, but not before the lapse of three days. Immediate suspension of permit and/or court action shall be instituted whenever the standard is violated by three of the last five bacterial counts (except those for aseptically processed milk and milk products), somatic cell counts, coliform determinations, or cooling temperatures.

All violations of bacteria, coliform, confirmed somatic cell counts, and cooling temperature standards should be followed promptly by inspection to determine and correct the cause. Following suspension of a permit, tests must show that effective remedy has been applied before the permit to sell can be reinstated.

24.7.5 Phosphatase

Alkaline phosphatase (EC 3.1.3.1), which is an enzyme that is abundant in raw milk, is inactivated almost completely by the minimal temperature and time of heating during the pasteurization process (cf. chapter 26). Because the heat stability of alkaline phosphatase is greater than that of any of the pathogens that may be present in milk, the presence or nonpresence of this enzyme serves as an indicator of product safety. Therefore, this test is applied as an added assurance that milk and milk products have received sufficient heat treatment. Both chemical and instrumental tests are available to detect the enzyme (see Wehr and Frank, 2004). When an accredited laboratory confirms a sample positive for phosphatase, the pasteurization process shall be investigated and corrected.

Products high in milkfat, especially those heated well above normal pasteurization temperature, may show a negative test immediately after heating but a positive test after a few hours of storage, especially if the storage temperature is unusually high and the holding time in the pasteurizer is minimal. This is called reactivated phosphatase and it can be differentiated from residual phosphatase remaining after inadequate pasteurization. This involves adding magnesium in the test protocol. This mineral will enhance the activity of reactivated but not residual phosphatase.

False positive phosphatase tests may occur if microbial cells have produced heat-stable alkaline phosphatase in the milk. The test has limited value in acidic dairy foods, such as cultured buttermilk and yogurt, because the enzyme is inactivated in acidic conditions.

The presence of phosphatase in milk results in exclusion of that milk from the supply. Application of the three-out-of-five rule is forbidden.

24.8 Adulterants of Public Health Significance

Adulterants most commonly found in raw milk are antibiotics, extraneous matter, pesticides, and chemicals used in cleaning and sanitizing equipment or udders and teats of cows. These get into the milk during operations not carefully controlled by the milk producer or processor. Usually they are present in traces only and are not harmful to consumers. However, important reasons exist to avoid contamination of milk with any one of them.

24.8.1 Antibiotics

Although milk cows are sometimes treated with antibiotics for other diseases, these drugs are most often used to treat cases of bovine mastitis (cf. chapter 14). Antibiotics are usually infused (injected through the teat canal) into the quarter from a syringe, but sometimes they are injected into muscle tissues (intramuscular) or, in severe cases, into the bloodstream (intravenous). Tests have determined the time milk must be withheld from the market after a cow receives antibiotic therapy. This withholding time, usually 72 hr after the last treatment, is given on the drug label. Since during this withholding period milk will be lost from sale, managers are reluctant to use antibiotics on lactating cows unless the infection is serious. Instead, they place much of their efforts on preventing infections and on treating infected cows during the dry period. Nevertheless, many cows receive antibiotic therapy during lactation. Therefore, all tank loads of milk are tested for beta-lactam type antibiotics before they are off-loaded at the milk processing plant. In the event of a positive test, samples must be taken from individual farms that make up that load and tested for antibiotics to identify the offending producer. Heavy penalties or fines are commonly levied on such producers.

24.8.2 Drug Residues

Data from the National Milk Drug Residue Data Base show that only a small number of positive tests for drug residue occur, and that number is very small in pasteurized milk and milk products. However, it is evident that dilution accounted for this small number since producer samples were positive in about one of every 750 tests, whereas tanker loads, representing mixed milk from multiple herds, were positive in about one of 2,300 tests. Removal of milk from tankers that were found positive at the plants accounts for the very low incidence of positive tests of pasteurized products, which amounted to only one for each 14,500 tests. Furthermore, the data show the program has reduced the incidence of drug residue positive samples in raw milk. This practice of careful testing has resulted in the incidence of positive drug tests of bulk tank milk decreasing markedly to only 0.014% of the samples tested in 2014 (table 24.1).

24.8.3 Extraneous Matter

The major sources of extraneous matter in milk are the udders and teats of cows that have not been cleaned adequately before milking. Occasionally a milking unit may fall off a cow. When this occurs, vacuum within the machine may draw extraneous matter from the environment into the milk line. An estimated 80% of the extraneous matter is dissolved in milk. It is a major source of bacteria.

Table 24.1 Drug Residue-Positive Milk Samples for Selected Years.

Source of Sample	Year	Total Samples	Number Positive	% Positive
Bulk milk tanker	1994	3,213,220	2,024	0.063
	2004	3,589,082	1,571	0.044
	2009	3,958,455	861	0.026
	2010	3,204,371	802	0.025
	2011	3,182,972	671	0.021
	2014	3,147,302	429	0.014
Producers	1994	824,132	1,634	0.198
	2004	677,507	895	0.132
	2009	510,760	508	0.099
	2010	545,148	431	0.079
	2011	478,533	395	0.083
	2014	445,233	266	0.060
Pasteurized products	1994	61,775	3	0.005
	2004	57,875	4	0.007
	2009	45,601	0	0.000
	2010	44,777	1	0.002
	2011	48,566	0	0.000
	2014	37,707	0	0.000

Source: GLH, Inc., 2015.

Many milk producers insert filters in the milk line to remove sediment from milk before it enters the **bulk milk tank** for cooling. This process does not remove all insoluble matter, but the materials that accumulate on filter pads include insoluble extraneous matter, clots of milk from mastitic mammary glands, somatic cells, and bacteria. Insoluble extraneous matter is called sediment because it tends to fall to the bottom of the storage vessel.

The insoluble matter can be detected by the sediment test. This test is normally applied to milk samples taken from the farm bulk milk tank after agitation to suspend the insoluble materials. The procedure calls for straining 470 mL (1 pint) of milk at a temperature of 95 to 100°F (35 to 38°C) through a cotton lintine filter disc with a filtering diameter of 1.02 cm. The amount of extraneous matter collected is estimated by comparison with photographs of standard discs prepared by the AMS. Presently the test is seldom used at the receiving plant. Instead, processors rely upon tests for microorganisms in the milk to control milk's quality.

24.8.4 Pesticides

Pesticides and herbicides are toxic to humans as well as pests. They can get into milk from direct application to animals or from contaminated feeds or water. Animals exposed to toxic vapors may inhale them. The term pesticide includes many kinds of ingredients in products such as insect repellants and weed killers, which are designed to prevent, destroy, repel, or reduce pests of any sort. Pesticides are found in nearly every home, business, farm, school, hospital, and park in the United States. Under the Food Quality Protection Act of 1996, the EPA must evaluate pesticides thoroughly before they can be marketed and used in the United States. Pesticides meeting requirements are granted a license or "registration" that permits their distribution, sale, and use according to specific use directions and requirements identified on the label.

Whereas some use of pesticides is necessary in milk production, emphasis is placed on prevention of fly breeding by removal of manure and applications of larvacides. Control of pesticide use on dairy farms and plants is monitored by public health officials who check that all pesticides stored on the premises are approved for use. They are detected in dairy products by chromatographic methods described in FDA's *Pesticide Analytical Manual*. Regulatory agencies can routinely detect residues as low as 0.01 ppm for many chlorinated organic pesticides.

When a pesticide residue test is positive in milk, an investigation shall be made to determine the cause and the cause shall be corrected. An additional sample shall be taken and tested for pesticide residues. No milk or milk products from that source shall be offered for sale until shown by a subsequent sample to be free of pesticide residues or below the actionable levels established for such residues.

24.8.5 Water

The most common adulterant of milk is water. When milk is priced according to weight with only fat content and class of use determining the price, it is usually profitable for the milk producer to add water (see chapter 20). Such an undeserved gain in value cannot be obtained by producers when milk is purchased on a "component price" basis. In such pricing, milk is tested for protein and fat with values being set for each component based on commercial prices of butter and cheese. A factor is added for "other solids" based on the value of whey solids in the marketplace. Thus, the amount of water in the milk is of no consequence in pricing. Of course, added water is of consequence to the manufacturer of dairy foods, especially cheeses and dry milk products, for which yield depends on concentration of solids in milk.

The unlawful adulteration of milk with water can be detected by testing for freezing point. Milk has a rather constant freezing point of about –0.517°C. Variability in this number is affected by certain environmental factors and bovine mastitis. Freezing point is one of the colligative properties of fluids. The others are boiling point, vapor pressure, and osmotic pressure. These properties are directly affected by the amount of dissolved substances in a solution. As osmotic pressure goes up, freezing point goes down. Since the animal must control the osmotic pressure of the bloodstream closely, and since blood must circulate through the mammary gland in the synthesis of milk, milk's freezing point is closely and biologically regulated by the concentration of blood's dissolved substances.

Freezing point tests conducted using a thermistor-type cryoscope can deliver results in 5 min or less. In enforcement of the freezing point standard, it is accepted that once a test shows a high value an inspector will visit the farm, make certain there is no water in the milking system, then take a milk sample from the bulk tank after all cows are milked. The result from testing this "authentic" sample will then be compared to the result of the suspicious test. The freezing point of the original sample is usually permitted to be 0.004°C above the freezing point of the authentic sample to allow for normal variance in repeatability of the test. Freezing point increases about 0.006°C with each 1% added water.[2]

24.8.6 Mycotoxins

Certain toxins produced by microorganisms of the environment have been found in milk and milk products. The molds *Aspergillus flavus* and *Aspergillus parsiticus* produce a group of highly toxic secondary metabolites called aflatoxins. These toxins belong to a broader group called mycotoxins (i.e., toxins of molds). Aflatoxins can be produced on seeds, such as corn, that are used in livestock feed. Of the common forms (B_1, B_2, G_1, and G_2), B_1 is most toxigenic. Cows that consume contaminated feeds convert B_1 toxin to M_1 by hydroxylation reaction. Milk is then contaminated with toxin M_1. The toxin disappears from the milk within three to four days following removal of a contaminated feed source.

The probability of the presence of aflatoxin is greatest in years of drought and delayed harvest of corn or other grains by wet weather. Feed manufacturers should be alert to the possible presence of aflatoxins and monitor incoming grains for fluorescence under ultraviolet light (black light). Fluorescence within the grain indicates possible contamination. Fluorescence is caused by kojic acid, which *Aspergillus flavus* produces along with the aflatoxin. Best results are obtained on cracked or ground kernels. Black light positive samples usually contain some aflatoxin, but this method cannot quantify aflatoxin levels and can give false positive responses. Further testing is then required to determine whether significant concentrations of aflatoxins are present. All corn exported from the United States must be tested for aflatoxin.

Tests for aflatoxins can be done by quantitative affinity chromatography with fluorometric detection, by enzyme-linked receptor binding assay, or by competitive binding assay with antibodies to aflatoxins B_1 and M_1. These tests are described in *SMEDP*. The USDA Grain Inspection, Packers and Stockyards Administration provides testing service for aflatoxins. The FDA will consider action if aflatoxin levels exceed 20 ppb for corn and other grains intended for immature animals or for dairy cattle.

24.8.7 Melamine

An estimated 300,000 Chinese children were sickened by melamine-tainted milk during an episode of adulteration of milk in 2008. Scientists from Peking University in Beijing found signs of renal damage during examination of ultrasound images of almost 8,000 children under the age of three living near the rural headquarters of Sanlu Group, the company at the center of the 2008 scandal. Six children died as a result of consuming dairy products laced with melamine. The industrial chemical was added to tons of milk to which water had been added to fool inspectors testing for protein content. Melamine is a nitrogen-containing compound that would be recognized, falsely, as protein by the Kjeldahl test for protein. By adding the melamine the perpetrators would be able to dilute the milk with water and still pass the test for protein content. The increase in volume of the adulterated milk would increase profits. Melamine used for this purpose contains cyanuric acid, an analogue that complexes with melamine to form crystals that accumulate in the kidneys, frequently causing kidney stones and renal failure. Renal abnormalities appeared in about 12% of children exposed to this milk.

Early in 2010 reports from China indicated some firms that kept supplies of the adulterated dry product were releasing it into markets. This prompted action by the Chinese government to establish a new public health regulatory system.

24.8.8 Allergens

Some foods cause serious allergies in many people. An allergen is an antigen capable of stimulating type I hypersensitivity (immediate hypersensitivity), a reaction provoked by re-exposure to a specific type of **antigen**. Sensitive individuals respond by producing large amounts of immunoglobulin E (IgE). Most humans produce significant IgE only as a defense against parasitic infections. However, some individuals may respond to many common environmental antigens. The FDA recognizes eight foods as being common for allergic reactions in a large segment of the sensitive population: peanuts, tree nuts, eggs, milk, shellfish, fish, wheat (and derivatives), and soy (and derivatives) as well as sulfites (often found in flavors and colors in foods) at 10 ppm (mg/L). In other countries the allergen list differs because of differences in genetic profiles and levels of exposure to specific foods. Canada adds sesame seeds and mustard to the list. The European Union additionally recognizes celery.

Milk and milk products containing eggs or nuts are most common on the avoidance list of hypersensitive individuals. The following practices are recommended in facilities in which potential allergens are used: store allergens segregated from other ingredients, schedule production separated from that of normal products, use dedicated equipment or assure cleaning before use, and verify package labels as meeting requirements.

24.8.9 Hormones

Consumers have expressed concern about possible hormone residues in milk. Vicini et al. (2008) compared 334 samples of whole milk from organic systems not using bovine somatotropin (bST) to treat cows, milk labeled "rbST-free," and conventional milk purchased at the retail level within the 48 contiguous United States. Minimal differences were found. Conventional milk had lower concentrations of estradiol and progesterone than organic milk. There were no differences in the level of bST in the three milks. Approximately 72% of the somatotropin values were less than the limit of detection (0.010 ng/mL) for the assay. Levels of insulin-like growth factor-1 (IGF-1) were similar in conventional milk and rbST-free milk and slightly lower in organic milk.

Another aspect of the hormone issue and one not appreciated by the general consumer is that, like other proteins, bST is digested in the human gut so any of it in the diet would be broken down and would no longer possess hormonal activity.

24.9 Clean and Sanitary Cows and Equipment

Major sources of spoilage bacteria in milk are dirty cows and unclean equipment. Bacteria that come from these sources include spore-forming bacilli, which have the potential to survive pasteurization, and **psychrotrophs**, which can grow well at temperatures used in milk storage, 34 to 41°F (1 to 5°C). When these two sources of microorganisms are well controlled, bacteria that get into milk will come primarily from interiors of mammary glands and are of types that grow poorly at refrigeration temperatures.

Therefore, it is highly important that cows be kept clean and prepared well before milking. Cows on pasture tend to be clean as they enter the milking barn; whereas those confined in lots or barns may have soil and manure attached to their skin and hair. Therefore, it is important to keep hair well clipped and the cow's environment clean. Cows entering the milking facility with dirty udders must be cleaned and sanitized before milking. Various methods are used, but certain principles apply to all of them. Firstly, all soil must be removed that will have a chance of entering the milk. Secondly, the udder must be practically dry because water carries bacteria into the milk. Finally, the teats should be disinfected with an effective sanitizer such as iodophor. If the udder appears clean, it is most effective to simply disinfect and then to dry the teats and the area immediately above them to remove excess sanitizer. Bacteria clinging to dry skin and hair above the area covered by the inflations are immobile unless released by water or sucked into the milk by unintentional exposure to vacuum through the teat cup.

Milking equipment can be a major source of psychrotrophic bacteria. The major species of these bacteria in milk is *Pseudomonas fluorescens*. Therefore, it is important that milking equipment is designed and constructed to be easily cleaned.

Equipment must be cleaned after each use. Most equipment is **cleaned-in-place** and minimal take down of the milking machines and milk lines is needed. As soon as milk has been drained from the lines into the bulk cooling tank, the pipe leading into the bulk tank is relocated to become the return line for cleaning solutions, and rinsing with water follows. Following the rinse, a solution of hot chlorinated alkaline cleaner is circulated through milking machines, receiver jars, and milk lines. This treatment lasts several minutes and is terminated with a water rinse. Rate of flow through the equipment is highly important because turbulence provides scrubbing action. The minimal rate of flow is 5 ft/sec. Periodically, if not daily, an acidified rinse is used to remove residual minerals that may precipitate from the alkaline cleaning solution. Equipment is then allowed to drain dry. This is important because some bacteria still remain in the equipment and they will grow given small amounts of nutrients and water.

Fat and proteins of milk are removed from equipment with alkaline-type cleaners whereas minerals are best dissolved and removed with acid-type cleaners. Milkfat, which is at about 99°F (37°C) when removed from an udder, starts to solidify on cooling at about 93°F (34°C), and solid fat is more difficult to remove from equipment surfaces than is liquid fat. Furthermore, detergents are more effective in hot than cold water. Therefore, cleaning is normally done with hot water. Rinse waters do not need to be warm. Alkaline detergents suspend and **saponify** fats so they are easily removed from surfaces. Mineral deposits on dairy equipment are composed of insoluble salts of calcium, magnesium, and other minerals plus proteins. This residue, when visible, is called **milkstone** (also water spots). Hot acid cleaner should be used to remove it. Milkstone tends to be deposited on equipment when hard, rather than soft, water is used for cleaning and rinsing, when alkaline cleaners are used without an acid rinse or intermittent application of acid cleaners occurs, and when insufficient detergent is used.

All milking equipment must be sanitized just prior to milking. The sanitizing solution is most often composed of sodium hypochlorite at 150 mg/L (ppm) in tap water. Minimal concentration of chlorine in the solution at the end of use is 50 mg/L. Sanitizer must be drained well from equipment before use, but rinsing after sanitizing is not permitted because most water is not sterile and may recontaminate surfaces. Hot water at 180°F (82°C) or higher is often used to sanitize milk lines and filling equipment in milk processing plants to destroy spoilage bacteria that may be hidden in junctions between two components inaccessible to chemical sanitizers. This is not practical on dairy farms. Hypochlorite sanitizers tend to be inactivated by residues of soil on equipment. Furthermore, bacteria, being so minute, are protected from a sanitizer by soil. This means equipment must be clean for chemical sanitizers to be effective.

To maintain proper fluid motion in cleaning and processing, all connections to equipment must be "close coupled" such that no dead ends or dead spaces exist. To avoid dead ends at connections of ancillary equipment (e.g., gauges and thermometers), the length of the connecting pipe cannot exceed its diameter.

24.10 Summary

Milk may become contaminated with disease-producing microorganisms that can come from the cow, the environment, and humans involved in its production, handling, and processing. Furthermore, certain adulterants may enter the milk supply. Nevertheless, consumers can be confident that safeguards are in place to ensure a safe and wholesome milk supply. These include the all-important nationwide adoption, application, and enforcement of the *Grade "A" Pasteurized Milk Ordinance* across the industry, from producing farms through processing facilities/equipment and distribution. This assures an extremely low probability that consumers will contract illness from the consumption of grade "A" milk products. An additional benefit consumers enjoy in these products is consistent superior flavor and keeping qualities.

The Interstate Milk Shippers Agreement and oversight by the FDA has enabled producers to cooperatively ensure that milk being shipped among states has met equivalent sanitary requirements. Importantly, too, the dairy industry requires skilled, knowledgeable, and conscientious workers at all levels to provide dairy food products of unquestionable quality and safety.

Study Questions

1. What is "loose milk"? How did elimination of it improve quality and safety of milk?
2. What were the main diseases of animals transmitted through milk before their control programs and pasteurization became customary?
3. Name four bacteria that are threats to humans who consume unpasteurized milk and/or cheese.
4. In what year did the first edition of the *Grade "A" Pasteurized Milk Ordinance* appear?
5. Explain the significance of reciprocal agreements among states as applied in inspection and testing of grade "A" milk moving in interstate commerce.
6. What functions do FDA rating officers perform related to the *PMO*?
7. What is the meaning of the acronym HACCP?
8. Note the nine basic requirements placed on producers of grade "A" milk in order to maintain their permits to sell milk for drinking purposes.
9. What federal agency is charged with the responsibility of regulating manufacturing grade milk used to make dairy foods sold with a US grade on the label?
10. What is the Codex Alimentarius Commission and how is it related to WHO and FAO?
11. What international organization is responsible for the Agreement on the Application of Sanitary and Phytosanitary Measures? How are the guidelines formulated?
12. What are the requirements as applied to grade "A" milk for temperature, bacterial count, somatic cell count, freezing point, antibiotics, and phosphatase.
13. What must happen in the event of a positive screening test for antibiotics in milk from a bulk milk truck containing milk from multiple farms?
14. What must happen when a test for pesticide in milk is positive?
15. Describe component price basis as applied to the price paid for milk. How does the addition of water to milk affect the pricing and purchasing of milk?
16. Name four colligative properties of solutions and describe the application of one of these to control adulteration of milk.
17. What organisms produce mycotoxins and through what pathway do they sometimes get into milk?
18. What is an allergen? How would one get into a dairy food?
19. What are the conditions under which milkstone is likely to form on dairy equipment?

Notes

[1] Tobey, J. A. 1947. *Public Health Law* (3rd ed.). New York: Commonwealth Fund.

[2] Some cryoscopes used to determine freezing point yield values in degrees Hortvet (°H), the reason being that the thermometer was miscalibrated when Hortvet developed the cryoscope. To convert one value to the other use the following formulas: °C = 0.96231°H – 0.00240; °H = 1.03616°C – 0.00250.

References

GLH, Inc. 2015, February. *National Milk Drug Residue Data Base, Fiscal Year 2014 Annual Report*. Lighthouse Point, FL: Author.

Moore, E., M. D. Weiss, and G. B. Mangold. 1911. *The Milk Problem in St. Louis*. St. Louis, MO: St. Louis School of Social Economy, Department of Research.

Neltner, T., N. Kulkarni, H. Alger, M. Maffini, E. Bongard, N. Fortin, and E. Olson. 2011. Navigating the U.S. Food Additive Regulatory Program. *Comprehensive Reviews in Food Science and Food Safety* 10(6):342–368.

Ryser, E. T. 2001. "Public Health Concerns." In E. H. Marth and J. L. Steele (eds.), *Applied Dairy Microbiology* (2nd ed.). New York: Marcel Dekker, Inc.

US Food and Drug Administration. 2013. *Grade "A" Pasteurized Milk Ordinance*. Washington, DC: Author.

Vicini, J., T. D. Etherton, P. Kris-Etherton, J. Ballam, S. Denham, R. Staub, D. Goldstein, R. Cady, M. McGrath, and M. Lucy. 2008. Survey of retail milk composition as affected by label claims regarding farm-management practices. *Journal of the American Dietetic Association* 108:1198–1203.

Wehr, H. M., and J. F. Frank. 2004. *Standard Methods for the Examination of Dairy Products* (17th ed.). Washington, DC: American Public Health Association.

Yilmaz, T., B. Moyer, R. E. Macdonell, M. Cordero-Coma, and M. J. Gallagher. 2009. Outbreaks associated with unpasteurized milk and soft cheese: An overview of consumer safety. *Food Protection Trends* 29:211–222.

Websites

3-A Sanitary Standards, Inc. (http://www.3-a.org)

Center for Food Safety and Applied Nutrition (http://www.fda.gov/AboutFDA/CentersOffices/OfficeofFoods/CFSAN)

Codex Alimentarius Commission (http://www.codexalimentarius.org/standards/en)

Dairy Practices Council (http://www.dairypc.org)

Food and Agriculture Organization of the United Nations (http://www.fao.org/home/en)

Food Quality Protection Act (http://www.epa.gov/pesticides/regulating/laws/fqpa)

Global Food Safety Initiative (http://www.mygfsi.com)

International Organization for Standardization (http://www.iso.org)

National Conference on Interstate Milk Shipments (http://www.ncims.org)

National Milk Drug Residue Database (http://www.kandc-sbcc.com/nmdrd)

Safe Quality Food Institute (http://www.sqfi.com)

USDA Animal and Plant Health Inspection Service (http://www.aphis.usda.gov)

25

Assuring High-Quality Milk for Humans

The quality of the finished product
can be no better than the quality of its ingredients.
William Henry Eddie Reid (1920–1964)

25.1 Introduction

A primary objective of producers, processors, distributors, and retailers of milk and milk products is to provide consumers with products of unquestionable quality. The establishment and continuation of a successful business enterprise by members of these groups depends upon their collective abilities to meet this objective.

Food manufacturers have an obligation to food quality and safety. Consumers depend on the diligence and discipline that goes into maintaining high standards. Lack of a good quality system not only results in failed audits but leaves a firm vulnerable to recalls and loss of brand equity. Programs for basic pest control, sanitation, and good manufacturing practices (GMPs) to document control, training, and product recovery are essential. The international marketplace and global auditing programs are expanding these requirements.

Milk of supreme quality has the following characteristics: (1) a bland, mildly sweet, and slightly salty flavor; (2) a low bacteria count (a few thousand/mL in raw milk and a few hundred/mL in pasteurized milk); (3) freedom from pathogens; (4) a low content of somatic cells; (5) no evidence of significant enzymatic activity; (6) freedom from poisonous, harmful, or extraneous substances; and (7) possession of essentially all its natural nutritive and functional properties.

Milk and its products are tested and evaluated as are few other foods because of their perishability and their importance in the diet of humans. The tests for quality that are discussed in this chapter apply primarily to fluid milk, both raw and pasteurized. Certain quality tests used with specific milk products are discussed in chapters that deal with those products.

The important quality tests of fluid milk may be categorized as (1) microbiological, (2) chemical, (3) tests for adulteration, (4) tests for abnormal milk, and (5) sensory evaluation. Most procedures for these tests can be found in *Standard Methods for the Examination of Dairy Products* (*SMEDP*), published by the American Public Health Association (Wehr and Frank, 2004). Methods are classified to reflect level of evaluation and degree of reliability.

- Class O methods have been subjected to thorough evaluation, are widely used, and have demonstrated to have value by extensive application.
- Class A1 methods have been subjected to thorough evaluation, have demonstrated application to a specific purpose on the basis of extensive use, and have been verified reliable by collaborative study.
- Class A2 methods meet the requirements of Class A1 but there is no record of extensive use.
- Class A3 methods have been evaluated as prescribed by the National Conference on Interstate Milk Shipments (NCIMS) and have been approved for use on milk that has been produced and shipped under provisions of the *Grade "A" Pasteurized Milk Ordinance* (*PMO*).
- Class A4 methods have been subjected to evaluation as prescribed by AOAC International Research Institute and have been given Performance Tested^{SM} status.
- Class B methods have one or more of the following characteristics: used successfully in research or other situations; developed or modified explicitly for the routine examination of dairy product samples; published in a peer reviewed scientific journal; and/or used widely in the dairy industry. However, there has been limited evaluation and no collaborative study has been applied to them.

Accurate and precise results of test methods are needed for (1) internal production control, (2) agencies that use test information to assure public safety, and (3) buyers and sellers of the product. Many of the methods published in *SMEDP*, in *Official Methods of Analysis of AOAC International* (*OMA*) (Latimer, 2012), and in FDA's *Bacteriological Analytical Manual* (*BAM*) have been subjected to measurements for repeatability and reproducibility. *Repeatability* is a measure of variance among multiple tests with the same method within a laboratory while *reproducibility* reflects variation of test results within and among laboratories. Acceptance of test methods depends on low variance among test results. Furthermore, each laboratory should be managed so its results fall within precision parameters published for the test method. To this end, *SMEDP* features a quality management system based on work by Early (1995).

Analysts must follow approved protocols as they perform each test. Analytical equipment must be properly calibrated. Samples for calibration of automated testing devices are available commercially as are validation and proficiency testing services for results of both microbiological and chemical tests. Precision parameters are provided in chapter 1 of *SMEDP*, while chapter 2 is an excellent reference on quality assurance and safety in the dairy laboratory.

25.2 Related International Organizations

We discussed in chapter 24 how the International Dairy Federation (IDF), the Codex Alimentarius Commission (CAC), and the International Organization for Standardization (ISO) operate on a global scale to develop and harmonize methods and practices used in the dairy industry to assure production of safe, nutritious, and high-quality milk and dairy products. The following shows how AOAC International (AOAC) functions globally in the realm of laboratory methods.

AOAC facilitates development, use, and harmonization of validated analytical methods and laboratory quality assurance programs and services. AOAC also serves as the primary international resource for knowledge exchange, networking, and laboratory information.

To meet these goals, AOAC has two methods validation programs, the Official Methods[SM] Program and the Performance Tested Methods[SM] Program. International laboratory accreditation is a requirement for participation in the global marketplace. Members of the AOAC Laboratory Accreditation Criteria Committee have provided laboratory managers with guidance needed to meet ISO 17025 requirements. The AOAC's Laboratory Proficiency Testing[SM] Program, among one of the first accredited proficiency testing programs, provides laboratories with a means for proving the accuracy and reliability of their test results. The program is the only one of its kind in the United States.

AOAC also provides key publications, hosts technical meetings and conferences, and offers training courses in laboratory management, quality assurance, accreditation, statistics, and measurement uncertainty. Publications include the *Journal of AOAC INTERNATIONAL*, a scientific, bimonthly online and print publication containing peer-reviewed articles and the *OMA*. This compendium, containing over 3,000 analytical methods, is distributed throughout the world and is considered the most authoritative volume in its field.

Now let us consider important tests performed in dairy laboratories.

25.3 Microbiological Tests

Microorganisms are responsible for most spoilage of milk and milk products. Usually the microorganisms are bacteria but occasionally yeasts or molds cause damage. Additionally, some pathogenic microorganisms can be transmitted via milk to consumers. It is important, then, in the prevention of spoilage and of disease transmission that the microbiological content of milk and milk products be assessed. This is a broad subject that can be treated only in principle here. Its importance is attested by the fact that development and application of microbiological tests for dairy products preceded their development and application in practically all other areas of the food industry.

25.3.1 Standard Plate Count (Class O)

The standard plate count (SPC) has long been the primary test for quality of fresh raw or pasteurized grade "A" milk. The process was described as early as the first edition (1910) of *SMEDP*, although the method has been modified several times in succeeding editions.

The SPC estimates total numbers of **aerobic** microorganisms. Therefore, an important first step is to optimize conditions for growing the types of such microorganisms commonly found in milk. Of greatest importance are nutrients contained in the plating medium, temperature and time of incubation, and proper dilution of the sample to avoid overcrowding of colonies on plates. Samples must be representative of the product, collected without contamination by other microorganisms, and kept under conditions that will not allow bacterial growth or destruction.

Samples, usually of two tenfold dilutions, are placed in transparent petri dishes (plates). A melted nutrient medium at 113°F (45°C) is poured into the dish and the sample and medium are mixed thoroughly. As the medium, standard methods agar (SMA), cools, the **agar** (1.5%) within it gels like gelatin. This prevents movement of microbial cells but allows them to reproduce in a single location. The reproduction rate of most bacterial species is adequate to produce visible colonies of several billion cells each within the prescribed incubation time of 48 hr at 90°F (32°C). Colonies that develop are counted and the number per mL is calculated by multiplying the number of colonies by the reciprocal of the dilution (figure 25.1).

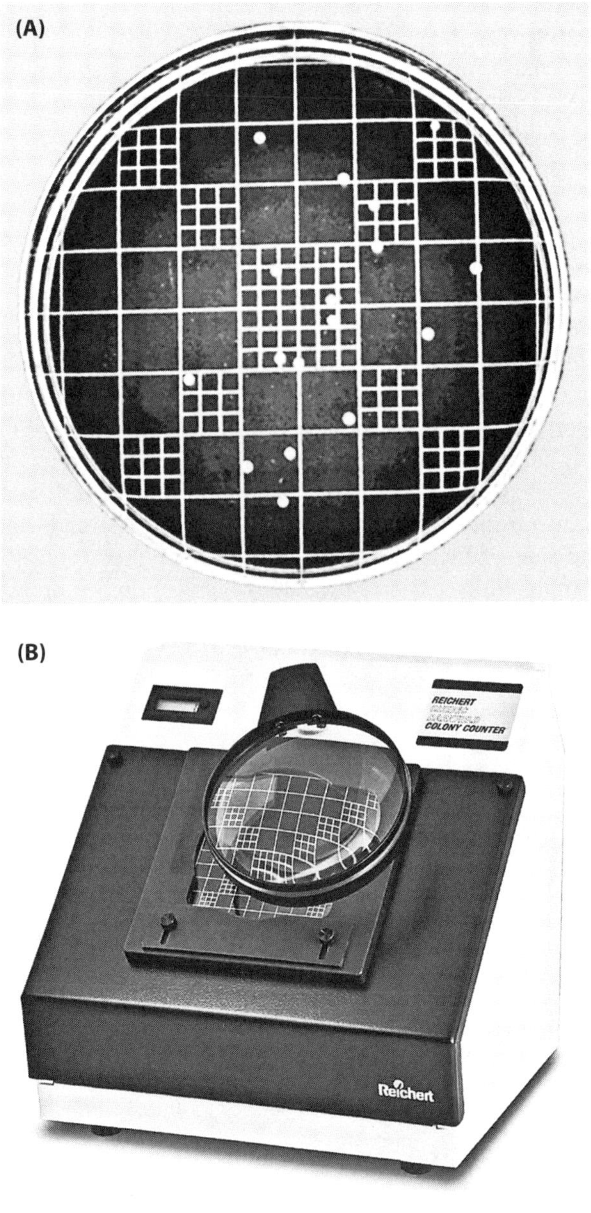

Figure 25.1 (A) Bacterial colonies on a plate (University of Missouri) and (B) a typical colony counter (courtesy of Reichert Technologies). The number of colonies on the plate is multiplied by the reciprocal of the dilution in determining the standard plate count.

If there is concern that the microflora of a sample could be inactivated at the 113°F (45°C) temperature of the fluid plate count agar, the Redigel™ method can be used. This method employs pectin, a cold water soluble gelling agent, instead of agar. In other aspects this method mirrors the SPC.

The area of the plate accommodates up to 250 discreet colonies, but since many milk products contain more microorganisms than 300/mL or g, it is necessary to dispense a diluted sample in the petri dish. For raw grade "A" milk the usual dilutions are 10^{-2} and 10^{-3}. The latter dilution allows a reliable analysis of milk with a count as high as 250,000/mL (the number of colonies in the dish multiplied by the reciprocal of the dilution, or, 250 × 1,000 / 1). Standards for numbers of bacteria detected by the SPC have been set for milks, dry milks, ice cream, and other dairy products. The usual maximum counts for raw and pasteurized grade "A" milk are 100,000/mL and 20,000/mL, respectively.

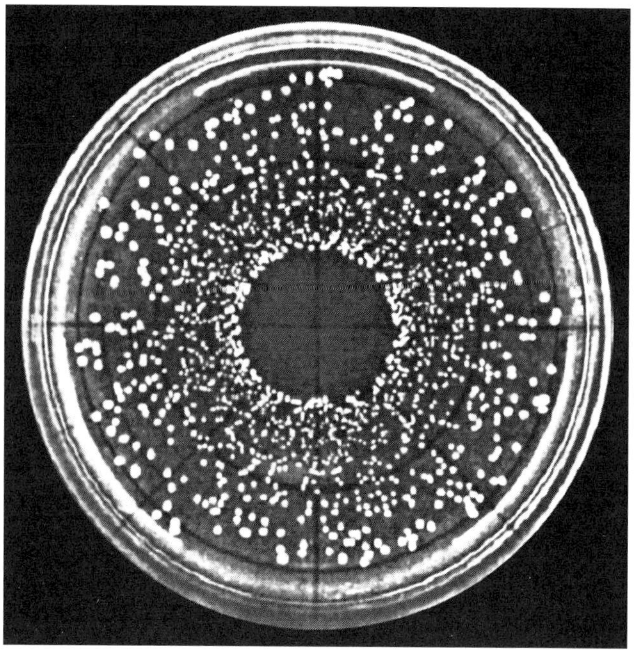

Figure 25.3 Example of plating done using FDA's spiral plating device (Maturin and Peeler, 2001).

25.3.2 Spiral Plate Count (Class A2)

Scientists of the FDA developed a mechanized system for plating liquid samples, especially milk (figure 25.2). An undiluted or diluted sample is spread from a Teflon tube in an **Archimedes spiral** on the surface of a prepoured petri dish containing SMA medium. The plunger of the syringe delivering the sample is moved at a decreasing speed as the dispenser tube is displaced from the center to the outer edge of the agar. Bacteria are restricted to the spiral path. Colonies develop in highest concentrations near the center and become sparse toward the side of the plate (figure 25.3). Since the amount of sample deposited and the rate of deposition are known, the volume contained on any selected portion of the plate is constant. By using a conventional colony counter with a counting grid, an estimate of the number of bacteria in the original sample can be made from any countable portion of the plate. A counter using a laser has also been developed to count and record colonies automatically. The spiral plating method is AOAC approved and studies have shown the method proficient in analyses of both raw and processed milk products. Where the method is used for official tests in lieu of the SPC, it must be validated using the protocol provided in *SMEDP.* The method is illustrated in the FDA's *BAM.*

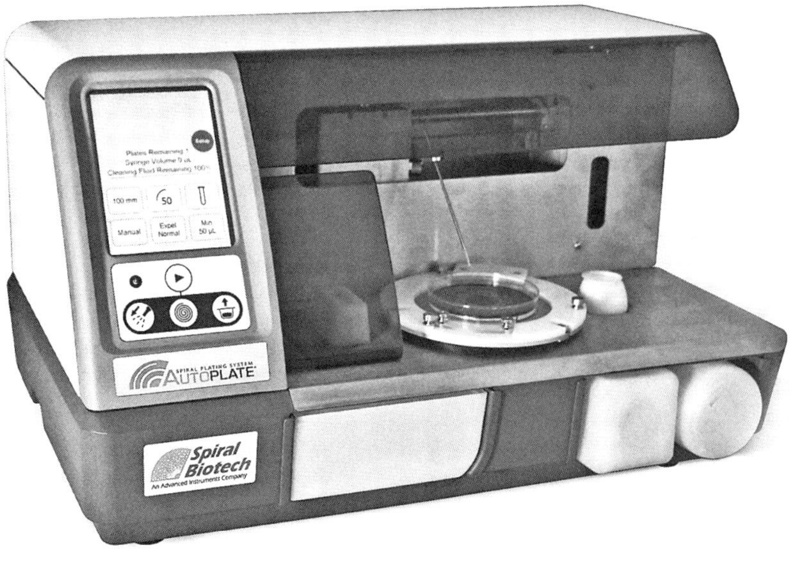

Figure 25.2 Device for delivering liquid sample to the surface of an agar plate in an Archimedes spiral and in a decreasing concentration from center to outer edge of the plate (courtesy of Advanced Instruments, Inc., Norwood, Massachusetts).

25.3.3 Petrifilm™ Aerobic Count (Class A1)

The Petrifilm™ aerobic plate count (PAC) method is applicable to all dairy products, including raw and pasteurized milks, rehydrated dry products, cheese, butter, and frozen desserts. With specific media it is adaptable to enumeration of selected groups of microorganisms. This method employs a ready-made culture medium system that consists of two plastic films coated with adhesive, dry standard methods nutrients (or selective media for other tests), and a dehydrated cold water soluble gelling agent. Indicator dye, triphenyltetrazolium chloride (TTC), is included to help visualize colonies for counting. After peeling back the top layer of film the medium is inoculated by pipetting 1 mL of sample directly onto the bottom film. The top film is dropped into place and the inoculum is distributed over the circular surface of the plate using a plastic spreader. The nutrients and gelling agent dissolve to initiate growth and hold developed colonies in place. Plates are incubated aerobically at 90°F (32°C) for 48 hr. The circular area of the plate covers about 20 cm^2 and up to 250 colonies can be counted on a single plate (figure 25.4).

Plates can be counted automatically with the Petrifilm™ Information Management System (PIMS). This device has been evaluated through the NCIMS Laboratory Committee to be an equivalent enumeration device for use with the Petrifilm™ aerobic count plates.

25.3.4 Plate Loop Count (Class O)

The plate loop count (PLC) method of estimating numbers of viable bacteria in milk is an alternative to reduction tests, direct microscopic counts (DMC), and the SPC. It provides a more reliable estimate of bacterial populations than do reduction tests or the DMC and is less expensive and time-consuming than the SPC.

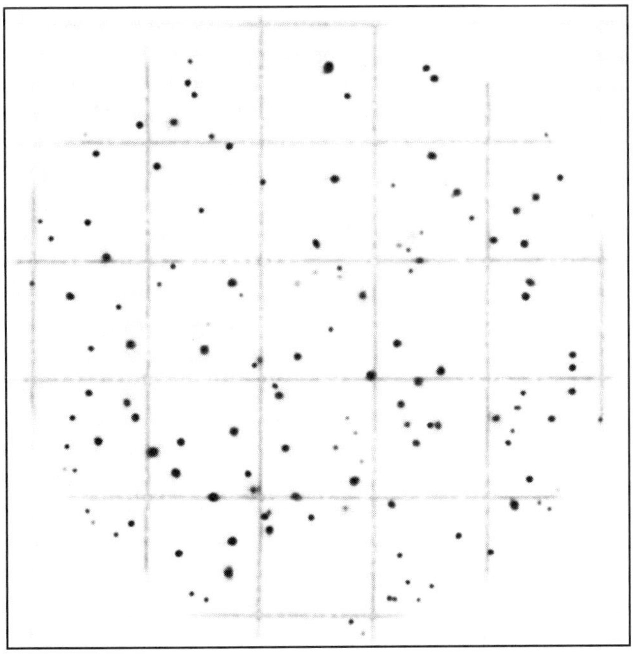

Figure 25.4 Typical example of bacterial colonies produced on a Petrifilm™ plate used for making a total aerobic plate count (© 3M 2015. All rights reserved.).

The principles of the SPC are followed, but samples are transferred using a platinum-rhodium loop instead of a pipette. The loop is standardized to transfer 0.001 mL and is sterilized by **flaming**. The loop is attached to a Cornwall continuous pipet syringe through a 13-gauge Luer-Loc hypodermic needle. The sample is washed off the loop into a petri dish with sterile buffered water from the pipet. The remainder of the procedure is the same as in the SPC. Colonies that arise during incubation at 90°F (32°C) for 48 hr are counted. The amount of milk delivered, 0.001 mL, limits counting accuracy to the range between 25,000 and 250,000/mL, so the method is applicable to raw milk. *SMEDP* provides a procedure for in-lab validation of the plate loop method in a comparison with the SPC.

25.3.5 Psychrotrophic Bacteria Count (Class O)

Rapid bulk cooling of milk and use of cleaned-in-place equipment on farms have made **psychrophilic** and **psychrotrophic** (also called psychrotolerant) bacteria the primary organisms in raw milk. Since these organisms may not grow or may grow poorly at the warm temperature of 90°F (32°C), they may be poorly enumerated by the SPC procedure. Thus, it may also be desirable to have a count of only the bacteria that will grow at refrigerator temperature (45°F [7°C]).

This is termed the *psychrotrophic count* and is made in the same manner as the SPC except plated samples are incubated at 45°F (7°C) for 10 days. Australian researchers observed counts comparable to those observed in this method when they incubated plates at 59°F (15°C) for 24 hr, then at 45°F (7°C) for three days.

Psychrotrophic bacteria commonly associated with dairy products belong to the following genera: *Pseudomonas, Flavobacterium, Alcaligenes, Acinetobacter, Klebsiella, Paenibacillus, Bacillus,* and *Lactobacillus.* Those most damaging to dairy foods are oxidative, Gram-negative rods belonging to the genus *Pseudomonas.* Various strains produce abundant amounts of enzymes that hydrolyze proteins, fats, phospholipids, and other large molecules of milk. Although pseudomonads are easily killed by pasteurization, some of their enzymes, especially proteases, are not inactivated by pasteurization and can spoil milk even when the producing organism is no longer viable.

The psychrotrophic bacterium most frequently observed in ultrapasteurized (long shelf life) fluid milk products is the spore-former *Paenibacillus.* It produces β-galactosidase, thus enabling it to make lactic acid. Most species of *Bacillus* do not produce this lactose-hydrolyzing enzyme and, therefore, do not produce lactic acid in milk. Ranieri et al. (2011) developed a real-time PCR assay to detect *Paenibacillus* spp. in fluid milk and to discriminate between these and other closely related spore-forming bacteria. They reported a detection limit of 1,000 cfu/mL. The assay was useful in screening colonies from plate count agar, thus providing a tool to rapidly identify *Paenibacillus* spp.

25.3.6 Preliminary Incubation Test (Class B)

A short-cut method that gives an indirect estimate of numbers of potential spoilage organisms in processed milk products is called the *preliminary incubation test*. Milk samples are incubated at 70°F (21°C) for 18 hr prior to plating, incubating, and counting by the SPC procedure.

25.3.7 Laboratory Pasteurization Count (Class O)

Bacteria that survive specific heat treatments are usually said to be *thermoduric* (heat-tolerant). High counts of these bacteria are consistently associated with unhygienic production practices. The laboratory pasteurization count is used to test for them. A raw milk sample is heated, usually at 145°F (62.8°C), for 30 min, the minimum required for **vat pasteurization** of milk. Following cooling the sample is subjected to the standard plate count. Some milk processors have standards for maximum numbers of thermoduric bacteria in raw milk, usually varying from 100 to 1,000/mL. Spores of psychrotrophic bacteria survive most pasteurization treatments, thus providing "seed" that can germinate at cold temperatures and shorten the shelf life of pasteurized milk.

25.3.8 Coliform Bacteria Count (Class O)

Coliform bacteria in milk and milk products have significance because (1) they are easily killed during pasteurization and (2) many originate in intestinal tracts of humans and animals. Therefore, the presence of them in pasteurized milk products suggests unsanitary conditions or practices during production, processing, or storage. The *PMO* sets a maximum of 10/mL (g) in pasteurized milk and milk products. However, the presence of any of these bacteria in pasteurized milk is strong evidence of postpasteurization contamination and a prediction of short product shelf life.

The coliform count differs from the SPC only in the medium used and incubation time. The medium, violet-red bile agar, contains inhibitors to most noncoliform bacteria. Coliforms grow rapidly, so only a 24-hr incubation is necessary. These factors prevent outgrowth of bacteria that might cause a false positive reaction. Furthermore, coliform bacteria ferment the lactose contained in the medium (producing lactic acid), so a red zone of precipitate forms in and near each coliform colony, making identification easy. Results are reported as coliform count per/mL or g. The Petrifilm™ coliform count method is applicable to milk (figure 25.5).

When necessary to **assay** milk products for small numbers of coliforms, tests of relatively large samples can be made using liquid media and the multiple tube, most probable number technique (MPN, Class O) as described in *SMEDP* (Wehr and Frank, 2004).

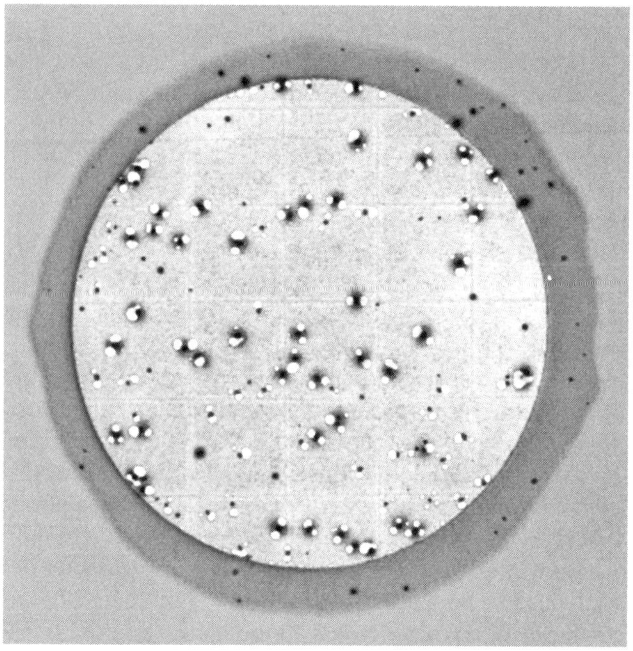

Figure 25.5 Petrifilm™ coliform plate showing typical red colonies surrounded by entrapped gas bubbles (© 3M 2015. All rights reserved.).

25.3.9 Direct Microscopic Count (DMC)

When bacteria counts are likely to be high in milk samples, two methods are satisfactory for estimating their concentration, for establishing a grade for the milk, or for determining the milk's acceptability. These are the DMC and the dye reduction tests.

The DMC is performed in practically the same way as the direct microscopic somatic cell count (see section 25.6). There are minor differences in the type of slide and counting procedure. Of course, bacteria (and yeasts and molds, if present) are counted instead of somatic cells. Bacteria tend to form clumps and chains and because these would produce only one colony in the SPC, such clumps or chains are counted as one colony-forming unit in the DMC procedure. The DMC method offers a rapid technique for determining the extent of bacterial contamination in raw milk or cream supplies taken directly from the producer or from pooled supplies in tankers or storage tanks. Individual samples may be examined within 10 to 15 min. However, high variability is associated with the method, especially among analysts. The morphology (size, shape, and arrangement) of cells in the stained film may suggest a cause of a high count. The method should not be used for the grading of pasteurized fluid milk or cream because dead cells may continue to stain for some time after they are killed by heat. This staining ability of dead cells is advantageous in the examination of dry milks, however, because it allows use of the DMC in attempts to determine the history of the milk used in the product, which normally contains few viable bacteria.

Results of the DMC are only estimates of the bacterial content of milk because there are numerous sources of error. The training and experience of the analyst are prime factors in determining accuracy, reproducibility, and significance in the results. However, the presence of large numbers of bacteria uniformly distributed is generally indicative of low-quality milk, whereas films from high-quality milk contain few or no bacteria in the field-wide strip (area viewed through the microscope). This strip crosses the entire central portion of the stained sample requiring that a single strip factor (SSF) be calculated to multiply by the clump count to provide the DMC/mL.

25.3.10 Dye Reduction Tests

Most bacteria of milk use oxygen in their growth. They transfer electrons to molecular oxygen in the process of respiration, thereby forming water (H_2O). When the concentration of molecular oxygen becomes sufficiently low, bacteria will utilize other compounds as electron acceptors. Methylene blue and resazurin are two such electron acceptors. These substances are reduced upon acceptance of electrons and, as a result, change in color. Oxidized methylene blue is blue in color, but when reduced it is colorless. Resazurin is purple-blue in color, but changes to pink and then becomes colorless as it accepts electrons. Hence, these substances are called *reduction indicators*. When placed in milk they give it color, but as bacteria grow the color changes occur. Rates at which molecular oxygen is used from milk and at which indicators are reduced depend upon the number of bacteria, their rates of growth, and their reducing capacities. Conditions for estimating numbers of bacteria by a reduction test are standardized, but there is high variability in the method. In particular; both thermoduric and psychrotrophic bacteria are less active in reducing than many other common milk contaminants. However, reducing activity is not diminished by clumping, whereas the SPC and DMC are lowered by clumping. Therefore, it is not justified to report results of reduction tests in terms of bacterial numbers. Instead, results are reported as reduction times, and milk is usually placed in a class or grade depending upon this time.

The method, briefly stated, involves mixing 9 mL of raw milk with 1 mL of a solution of methylene blue or resazurin (prepared by mixing a tablet of specially produced dye with boiled water). Tubes are incubated at 36±1°C and are examined periodically to determine when color changes have occurred. The methylene blue test is at its end point, methylene blue reduction time, when at least four-fifths of the sample is decolorized. Resazurin reduction time is determined by comparing color in the tubes to a set of color standards.

Both tests are applied almost solely to raw milk of manufacturing grade. Reduction rates are so slow for high-quality milk that results have limited correlation with bacterial content. Primary reasons for use of reduction tests over the DMC are (1) only a small amount of equipment and supplies is needed, (2) the method is simple, and (3) a large number of samples can be tested simultaneously. The methods, however, reveal little about the cause or causes of high bacterial counts. Because of the high microbiological quality of raw milk in the United States, dye reduction tests are no longer described in *SMEDP*.

25.3.11 Keeping Quality or Shelf Life Tests

Processors, distributors, retailers, and consumers have reasons for concern about the length of time fresh fluid milk will retain its good flavor and aroma characteristics. Usually the US dairy industry does not sterilize and aseptically package products other than whipping cream, coffee cream, and half-and-half. Therefore, microbial spoilage is expected to occur eventually in most fluid milk products (cf. chapter 23), and it is vital that limitations on time and temperature of storage be known. These limitations dictate methods and procedures of distribution, sales, and marketing.

The favorite test of most quality assurance personnel involves incubation of representative samples of finished product for five days at 45°F (7°C). Incubated samples are then evaluated for sensory qualities (taste, odor, and appearance) or for bacterial count (SPC) or both. Counts from this test should be compared with counts made on the fresh sample.

Usually, bacterial numbers remain static during the first several hours of storage (lag phase of growth). Then the count increases at an exponential rate (logarithmic phase of growth). Time required for counts to reach a fixed value depends upon (1) initial concentration of bacteria, (2) length of the lag phase, and (3) generation time or doubling time of the bacteria (likely there will be more than one species of bacteria in the sample). Figure 25.6 on the following page shows how much more rapidly the count will increase in a sample in which generation time is 6 hr compared with times of 12 and 18 hr. The count in each instance is 1/mL on the fifth day of storage. In this example, if the number of bacteria at 10 days is less than 1.7×10^7/mL with a generation time of 6 hr, the lag time is three days and the initial count is 1 in 256 mL (about 1 cup). Other similar conclusions are easily drawn from this illustration.

The 5-d/45°F (7°C) shelf life test is used because (1) this temperature is at or near the maximum that milk is expected to be stored; (2) psychrotrophic bacteria reproduce at this temperature; and (3) meaningful data can be

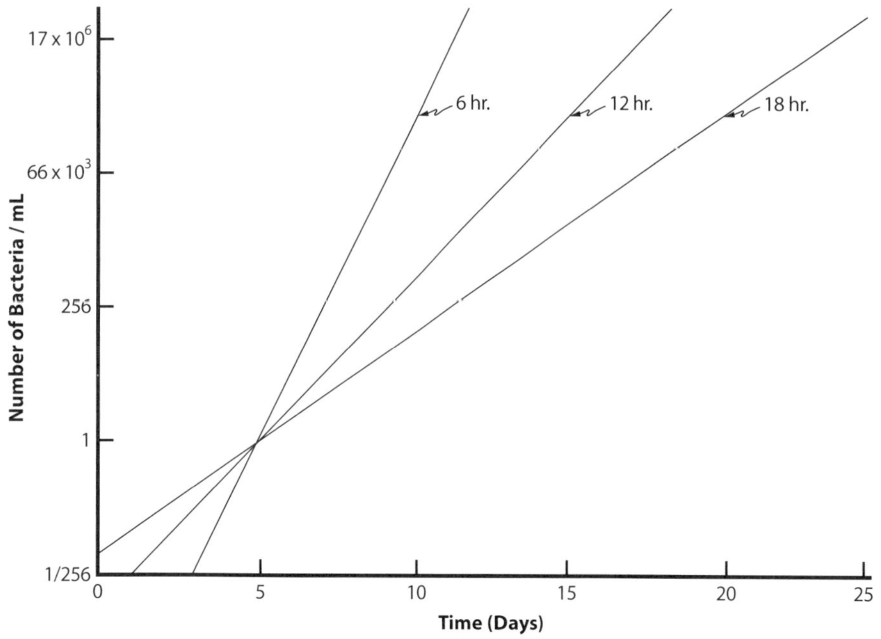

Figure 25.6 Changes in bacterial numbers during storage when generation times are 6, 12, and 18 hr (adapted from Marshall and Appel, 1973).

available for control applications in a reasonable time. However, variability among cartons of milk processed can be high and interpretations must be based on repeated sampling and testing. Greatest variability (counted in days) is associated with samples that are slightly to moderately contaminated, whereas those that are highly contaminated spoil so quickly that only hours separate the time of spoilage of the first sample from that of the last in a lot.

The 5-d/45°F (7°C) shelf life test can provide information on important sources of contamination in the milk processing system. Samples are taken at points in the processing line, incubated, and subjected to plate count analysis. Data in table 25.1, for example, show that for whipping cream the slide spigot was an important source of contamination.

25.3.12 Tests for Groups of Microorganisms

Proteolytic bacteria produce enzymes that degrade proteins, forming bitter peptides and other soluble substances that impart off flavors. A simple test involves adding sterile skim milk to SMA and using this medium in performing the SPC. To increase sensitivity for protease producers, plates are incubated at 70°F (21°C) for 72 hr. Plates are then flooded with acid to precipitate the casein, making the clear zones of hydrolysis more discreet.

Lipolytic bacteria produce extracellular lipases that hydrolyze fatty acids from the glycerol backbone of milkfat. Short-chain fatty acids of milkfat, especially butyric acid, are highly volatile, causing a rancid flavor. Lipolytic bacteria can be detected by plating them in a medium that contains emulsified tributyrin and spirit blue dye. Lipase positive colonies produce a clear zone and/or a deep blue color under and around them.

Lactic acid bacteria ferment lactose to form lactic acid, making milk sour. Although they are desirable in certain dairy foods, they cause defects in others. The organisms usually belong to the genera *Streptococcus*, *Lactococcus*, or *Lactobacillus*. In general these bacteria are differentiated from nonacid producers by enhancing the nutrients in SMA and adding bromcresol purple indicator. Acid producers will form yellow colonies on this medium.

Enterococci, spherical-shaped bacteria that are primarily of fecal origin, are sometimes enumerated as a check on the sanitary quality of churned butter since they survive well in the presence of salt in salted butter and in

Table 25.1 SPC/mL of Refrigerated Samples Showing Increasing Contamination during the Manufacturing Process.

Product	Above Bowl	At Filler Bowl	At Manifold	At Slide Spigot	In Carton
Homogenized milk	3,000	3,000	3,000	3,000	3,000
Skim milk	3,000	3,000	3,000	78,000	320,000
Half-and-half	3,000	18,000	24,000	180,000	240,000
Whipping cream	3,000	3,000	3,000	2,300,000	2,600,000
Multivitamin milk	3,000	90,000	90,000	160,000	280,000

Source: Data from Elliker, 1968.

the high acidity of yogurt. A selective medium, citrate azide agar, is used for plating samples and positive colonies on incubated plates are blue in color.

Aerobic bacterial spores, mostly of the *Paenibacillus* and *Bacillus* genera, can be enumerated from milk that is heated at 176°F (80°C) for 12 min to destroy vegetative cells. After cooling the tubes of heated milk, plates are prepared using SMA and incubation at 70 or 45°F (32 or 7°C) in order to visualize mesophilic or psychrotrophic colonies arising from the spores.

Anaerobic bacterial spores, primarily of the genus *Clostridium*, are selected from dairy foods by heating samples at 176°F (80°C) for 10 min then planting samples in reinforced clostridial medium in a series of tubes that have been heated to drive off oxygen. Tops of this selective medium are sealed with a layer of agar containing the reducing agent sodium thioglycolate. Development of gas in tubes incubated at 98.6°F (37°C) for seven days is a positive test. By selecting appropriate dilutions and numbers of tubes for each, MPN tables can be consulted to estimate populations of target bacteria.

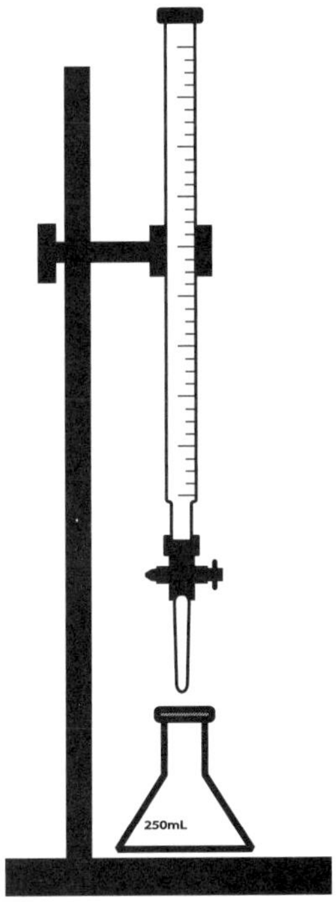

Figure 25.7 A ready-assembled acid titration set (Lalephat Aksomkouit/Shutterstock.com).

25.4 Chemical Tests

Many chemical tests are applied to milk and milk products; let us discuss some of the more common ones.

25.4.1 Titratable Acidity

An end product of the growth of fermentative microorganisms is lactic acid, which causes the sour taste of milk. The titratable acidity test is a simple acid-base reaction in which alkali of standardized **normality** (0.1 N) is added from a buret (figure 25.7) to 17.5 mL (18 g) of milk in the presence of a pH indicator (2 mL phenolphthalein) until the indicator changes from colorless to its first permanent shade of pink. This color change occurs at pH 8.3 when the concentration of phenolphthalein is 10 mg in 18 g of milk. This method allows a calculation of percent titratable acidity by dividing milliliters of sodium hydroxide (NaOH) used by 20.

The formula for calculation of titratable acidity follows:

$$\%\ \text{acidity} = \frac{\left(\text{mL NaOH} \times 0.1\ \text{N} \times (0.09) \times 100\right)}{\text{milk weight in grams}}$$

where 0.1 N is the normality of the NaOH, 0.09 is the milliequivalent weight of 0.1 molar lactic acid, and 100 converts the result to percentage.

Fresh milk rarely contains lactic acid, and then only because of bacterial growth within the udder. However, fresh milk produces a **titer** (i.e., requires alkali to increase the pH to the phenolphthalein end point) in the acidity test. This acidity of fresh milk (**apparent acidity**) is due to casein, whey proteins, carbon dioxide, and citrates and phosphates. Each of these is a component of the serum or aqueous phase of milk. The apparent acidity of normal fresh milk is 0.15 to 0.16%. However, milk containing a high quantity of nonfat solids may have an apparent acidity as high as 0.23%. Therefore, to determine whether a value higher than 0.16% represents **developed acidity**, more information may be necessary. Previous recent tests of milk from the same herd will be of value as will an estimate of the bacterial count. Several million bacteria per milliliter are necessary to produce detectable developed acidity. For these reasons the test has limited value in analysis of grade "A" milk. However, it is a useful tool for screening **tanker loads** of milk and in testing production of acid during manufacture of cheeses and fermented milk products (cf. chapters 28 and 30). If titratable acidity of raw milk is unusually high, the milk is suspect and its condition should be determined by taste and odor and by a DMC prior to receiving the milk.

25.4.2 Acid Degree Value (Class O)

All but traces of the fatty acids of milkfat are combined with glycerol in freshly drawn milk. However, lipolytic enzymes, or esterases (cf. chapter 21), may hydrolyze ester linkages between the fatty acids and glycerol, thus freeing fatty acids. Most of these free fatty acids are soluble in the fat phase. They contribute insignificantly to the titratable acidity of milk. However, the short-chained free fatty acids (especially butyric, caproic, and caprylic) produce a detectable rancid off flavor when present in low concentrations. Therefore, it may be desirable to test for free fatty acids in milk and cream. The acid degree value is defined as the milliliters of 1N base required to neutralize the acids of 100 g of milkfat. The result is expressed as milliequivalents of free fatty acids per 100 g of milkfat.

In the acid degree value test, fat is extracted from milk in a Babcock test bottle with a solution of 3% Triton X-100 and 7% sodium tetraphosphate at boiling in a water bath. Addition of methyl alcohol and centrifugation complete the separation. One milliliter of fat is removed from the neck of the test bottle and dissolved in a fat solvent. The solution is then titrated to the phenolphthalein end point with a dilute solution (0.02N) of potassium hydroxide (KOH) in absolute methanol. Fatty acids of high molecular weight hydrolyze considerably in water primarily because of their insolubility in it. An alcoholic solution is used to prevent hydrolysis, thus allowing the ionization and subsequent titration of each acid molecule.

Results are expressed as the acid degree value (ADV) and are interpreted as follows: normal = 0.4; borderline = 0.7 to 1.1; rancid to many people = 1.2; unsatisfactory = 1.5 and above.

25.4.3 Confirming Pasteurization

To prevent transmission of disease by milk the cardinal rule is that milk be pasteurized (cf. chapter 24). Not even a small amount of raw milk should be mixed with pasteurized product, nor should inadequate heating be allowed.

In 1933, Kay and Graham demonstrated that alkaline phosphatase (ALP), an enzyme of bovine raw milk, is almost completely inactivated at minimal times and temperatures approved for pasteurization. Furthermore, a simple and sensitive test is available for ALP. This test is used to verify records that are required to be kept on time and temperature of exposure of milk and milk products in the pasteurizer (cf. chapter 26). For a detailed history of development and application of phosphatase tests, see chapter 14 of *SMEDP*, where five tests are described. These tests involve detection by visual, spectrophotometric, fluorometric, and chemiluminescent methods. Properly applied, current test procedures permit detection of process temperature as little as 1.8°F (1°C) below the legal minimum and are sensitive to 0.01% added raw milk.

Here we describe the principles of the Scharer rapid visual ALP method (Class O), developed in 1938. The milk or milk product in question is mixed with disodium phenyl phosphate substrate and a buffer (pH 9.6). The mixture is incubated at 104°F (40°C) for 15 min. If alkaline phosphatase is present in a significant concentration, phenol is released. Free phenol combines with 2,6-dichloroquinone-4-chloroimide, which is added after incubation. The sample is then extracted with butanol and any blue color found is proportional to the concentration of phosphatase. The amount of blue compound can be estimated visually by comparing the unknown to standards or by colorimetric measurement.

Interfering substances, microbial phosphatase, and reactivated phosphatase pose problems for the analyst. Therefore, controls are necessary. A positive control is included to ensure that reagents are satisfactory. The positive sample is made from boiled and cooled milk plus 0.1% raw milk. Boiled and cooled milk serves as the negative control giving evidence that reagents and equipment are free of phenol. Interfering substances are detected by substituting buffer for the buffered substrate. Microbial phosphatase is more heat-resistant than milk phosphatase; therefore, pasteurization of the sample in question at 145°F (62.8°C) for 30 min, followed by retesting, allows discrimination of the two. If there is no significant reduction in the phenol equivalent following repasteurization, the original positive result must have been caused by microbial phosphatase.

In 2009, the NCIMS approved an alkaline phosphatase test developed by Charm Sciences, Inc. The test provides results for fluid milk products in less than 1 min (chocolate types in 90 sec). In 2007, the European Union approved the fluorimetric method (Advanced Instruments, Inc.) as an "official reference" (see ISO 11816-1:2013 [IDF 155-1]). This method has a high level of specificity and color or fat content in cheeses or flavored milk products will not affect test results.

25.5 Tests for Adulteration

Let us discuss briefly tests for adulterants of milk: antibiotics, water, whey, sediment, pesticides, and radionuclides.

25.5.1 Detection of Antibiotics

Several tests have been developed to detect antimicrobial inhibitors in milk and milk products. The oldest of these is based on the principle that growth of bacteria in a nutrient medium is inhibited by antibiotics. In the widely applied disc assay procedure, a suspension of spores of *Geobacillus stearothermophilus*, about 10^6/mL, is added to a thin layer of melted agar medium. (Lower concentrations of inhibitors are detectable with a thin rather than a thick agar layer.) Milk samples, 90 μL, are added to 0.5-in filter paper discs resting on the agar surface. After incubation at 147°F (64°C) for about 3 hr, a positive test is indicated by a clear zone around the periphery of the disc (called the zone of inhibition). This test has the advantage of a short time for the spores to germinate and grow rapidly. The resulting acid-producing vegetative cells change the bromcresol purple indicator from purple to yellow. Furthermore, the incubation temperature is so high that only the thermophilic cells of the test organism can grow, eliminating error from potential contaminating bacteria.

Penicillin or the semisynthetic penicillin antibiotics can be identified by mixing beta-lactamase (penicillinase) with a portion of the heated sample or by adding the milk sample to discs impregnated with the enzyme and repeating the test. (The beta-lactamase enzyme degrades antibiotics that have the beta-lactam ring structure of penicillin.) A positive test is indicated if a zone of inhibition forms around the disc with milk alone, but does not form around the disc with milk and beta-lactamase. Some milk products must be solubilized before the sample is added to the filter paper disc.

A modification of this test is called the Delvotest®. (Class A4 for raw milk and goat milk.) Suspect milk samples are added to small vials of nutrient agar containing bromcresol purple pH indicator and suspended spores of *Geobacillis steraothermophilus*. The vials are incubated at 147°F (64°C). Yellow color develops in negative samples within about 2.5 hr, whereas samples containing inhibitor remain purple. Addition of beta-lactamase to control samples assists in identification of the inhibitor.

Screening tests have been developed for use at the farm or the dairy plant receiving platform. They vary in applicability and sensitivity to various antibiotics. In general, they are designed to provide tentative results within 15 min. An example is the Penzyme® milk test (Class A4 for raw milk). It is based on the principle that beta-lactam antibiotics inhibit, on an equimolar basis, release of d-alanine from a substrate by DD-carboxypeptidase. If released, d-alanine is oxidized to pyruvic acid by d-amino acid oxidase, simultaneously forming hydrogen peroxide (H_2O_2). H_2O_2 oxidizes an organic redox indicator, the amount of which is related to the concentration of beta-lactam in the milk. Thus, in the absence of beta-lactam antibiotics, d-alanine is released in abundance, providing the oxidative force to activate the redox indicator. A positive test, then, is indicated by no change in the color of the indicator.

Other special tests have been developed for sulfonamides, tetracyclines, and chloramphenicol. When confirmatory identification of a drug residue is needed, high-performance liquid chromatography (HPLC) is the choice method.

25.5.2 Detection of Added Water

Milk containing added water is adulterated and illegal. Fortunately, there is a simple test that can detect small quantities of added water under carefully controlled conditions. Called the freezing point test, it is conducted with a cryoscope.

Hortvet introduced the cryoscopic method. His instrument uses a vacuum to evaporate ether from a container surrounding a tube containing the milk sample. Evaporation rapidly cools the sample to well below the freezing point without the formation of ice (supercooled). Ice crystallization is then induced, and the latent heat of crystallization is released, raising the temperature to the freezing point of the sample, which is read on a large glass thermometer.

Thermistor cryoscopes, introduced about 1956, employ a small thermistor, instead of a large thermometer, and a scale is provided from which to read the temperature. Tests take only 2 to 4 min per sample. Samples are supercooled in a refrigerated glycol bath to a precise point, usually 26.6°F (–3°C). Cooling is stopped and the sample is allowed to isothermalize (come to the same temperature throughout) as it is stirred gently. Crystallization is initiated by violent agitation within the sample tube. Latent heat of crystallization is given off causing the temperature to rise precisely to the freezing point, where it stabilizes. The instrument (figure 25.8) is calibrated against solutions of 7 and 10% sucrose, which freeze at –0.408°C and –0.600°C, respectively.

The freezing point test is based in the fact that **colligative properties** of milk, including osmotic pressure and freezing point, are controlled by the concentration of its dissolved substances. The latter is, in turn, controlled by the osmotic pressure of blood, which is maintained within close limits. Consumption of large amounts of water by cows can cause a limited decrease in osmotic pressure of blood with a consequent increase in the freezing point of milk. Other factors reported to influence the freezing point are season, feed, ambient temperature, the particular breed, mastitis, addition of colostrum, and the time of milking. Lactose and chlorides account for about 75% of the freezing point depression and each tends to compensate for changes in concentration of the other.

Pure water freezes at 32°F (0°C). Most milk samples freeze within the range –0.507 to –0.527°C. (Some cryoscopes read in degrees Hortvet, so results need to be converted to °C with the following factor: °C = 0.96231°H –

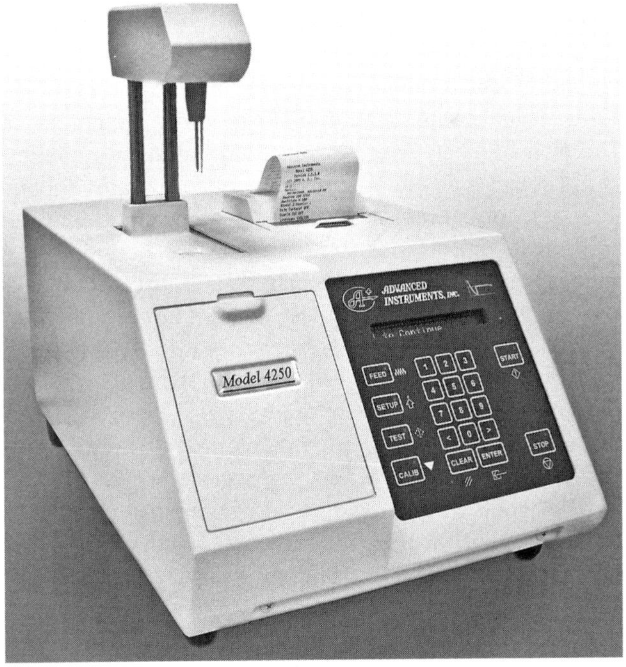

Figure 25.8 Thermistor cryoscope, single sample unit. A tube containing the milk sample is placed on the combined thermistor/stirrer then the unit is dropped into the freezing chamber (courtesy of Advanced Instruments, Inc., Norwood, Massachusetts).

0.00240.) Addition of 1% water to milk raises the freezing point slightly more than 0.006°C. Thermistor cryoscopes are sensitive to changes in temperature of 0.001°C.

The commonly accepted upper limit of freezing point in the United States is –0.530°H or –0.507°C. However, evidence of the addition of water is best established when the freezing point of the suspect sample is compared with that of an authentic sample. An authentic sample is one taken by a representative of the control agency (e.g., health department) while avoiding contamination with water.

In the dairy plant, freezing point can be affected by vacuum treatment, sterilization, and freezing and holding samples before testing. Addition of milk solids, salts, or sugar will lower the freezing point significantly, as will development of acid by fermentations within milk.

25.5.3 Added Whey

Fraudulent addition of cheese whey to milk is sometimes practiced. Low cost of cheese whey and high cost for fraud detection are among causative factors. The method for detection of cheese whey added to milk is determination of the caseinomacropeptide (CMP) index by HPLC. However, the cost is high for routine screening. CMP is a 7 kDa phosphoglycopolypeptide released from κ-casein during milk digestion and during chymosin-induced renneting of caseins.

Screening may be done using infrared-type milk testers with comparisons of ratios of fat to protein and protein to lactose.

25.5.4 Sediment Tests

Sediment is the insoluble portion of extraneous matter that may contaminate milk from cows, equipment, or the environment. Most extraneous material that enters milk, however, is soluble and does not appear as sediment.

A sediment test is performed by filtering 470 mL of milk through a white sediment disc of 1.02 cm diameter and observing the amount of residue. Discs of smaller diameter can be used with lesser quantities of milk. Test discs are compared in appearance with standard discs containing known quantities of sediment. Thus, milk may be placed in classes or grades. Acceptable milk contains no more than 1.50 mg of sediment in the sample (USDA Agricultural Marketing Service, 1977). Rate of flow of milk through the disc varies with temperature, solids content, abnormal milk content, and amount of sediment. All manufacturing grade milk received in cans must be subjected to sediment testing.

Milk producers can remove most extraneous material by filtering at the farm. Therefore, absence of sediment is not proof that milk was produced under clean conditions. However, presence of a large amount of sediment is strong evidence of carelessness in the production or handling of milk.

25.5.5 Detection of Pesticides

Methods for qualitative and quantitative analysis of pesticides in milk and milk products are based on chromatography and several residues can be detected simultaneously. Pesticides are commonly removed from sample materials in the following steps: (1) extracting fat in which the pesticide is solubilized, (2) partitioning from the fat (dissolved in petroleum ether) using acetonitrile (CH_3CN), and (3) column chromatography. Identity of residues may be confirmed by thin layer chromatography, element-specific gas chromatography, derivatization, and other means. The methods of analysis are described in the FDA (1999) *Pesticide Analytical Manual*.

25.5.6 Radionuclides

Milk and milk products are potential sources of radionuclides in the event of accidental spills, fallout from nuclear power plants, or weapons discharge. Most have a relatively short half-life or are not metabolized by the cow. The following have been found in measurable amounts in milk: strontium 89, strontium 90, barium 140, cesium 137, and iodine 131. Strontium may be absorbed on ion exchange columns, desorbed, purified, and its radioactive emissions counted to determine its concentration. A second method involves curdling milk with trichloracetic acid, precipitating alkaline earth elements (including strontium) from the filtrate, separating and purifying strontium by chemical means, storing for two weeks to allow conversion of some strontium 90 to yttrium 90, and finally counting. The gamma ray emitters, radiobarium, radiocesium, and radioiodine, are quantified by gamma ray spectrometry and no preparation of milk is required.

25.6 Abnormal Milk

The *PMO* defines milk as the "lacteal secretion of one or more healthy cows practically free of colostrum." Most states and municipalities include the same wording in their definition of milk. It is on this basis that milk from animals having mastitis is considered abnormal. The definition of abnormal milk has been somewhat arbitrarily established, though it is based on the scientific principle that infection produces inflammation within the mammary gland (cf. chapter 14). Inflammation is characterized by infiltration of the mammary gland by white blood cells (leukocytes) in numbers proportionate to the severity of infection. It is the authors' experience that milk from uninflamed glands of animals at first lactation will contain no more than 100,000 somatic (body) cells/mL. Numbers may be higher in older animals. Concentrations in excess of 300,000/mL are generally considered indicative of mastitis or other abnormalities. In the United States, most control programs for abnormal milk are based on control limits of 750,000/mL. European limits are 400,000/mL.

25.6.1 Screening Tests

To limit costs of control programs for abnormal milk and to provide information to producers, screening tests are regularly applied to fresh raw milk. The California Mastitis Test, described in chapter 14, is useful on milk from individual mammary glands and bulk tank samples. Screening can be done also by measuring conductivity of milk, which increases in response to inflammation of the mammary gland. Lactose synthesis is decreased in mastitic glands, and, since the osmotic equilibrium of milk with blood needs to be maintained, salts are transferred from the blood into the milk. Lactose is a nonconductor but serum salts are conductive. This results in increased electrical conductivity measurable with small conductivity meters.

25.6.2 Laboratory Tests

Enforcement of control standards of abnormal milk in the US milk supply is achieved by testing bulk tank milk. The *PMO* requires that no more than two of the last four tests indicate that the number of somatic cells is above control limits.

The direct microscopic somatic cell count (DMSCC) is adapted from the microscopic procedure for examining milk films first applied by Prescott and Breed (1910). A precisely measured sample of 0.01 mL volume is spread in a uniform thin film over a circular area of 1 cm^2 outlined on a glass slide. The film is dried then stained with acidified methylene blue dissolved in ethanol and tetrachlorethane. Basic dyes such as methylene blue, when at a low pH (acidic), react with nucleic acids contained in somatic cells staining them blue. Low pH and organic solvents prevent ionization of carboxyl groups of proteins so proteins combine poorly with the stain, hence the milk film forms a light blue background when excess dye is rinsed away. The film is examined under the oil immersion objective of a light microscope. A specific area of film is examined so a statistically valid estimate is obtained of the number of somatic cells in 1 mL of milk. Procedures for calibration and use of the microscope are given in *SMEDP.*

Somatic cell counts of producer milk generally are made using electronic instruments that provide high speed and accuracy. Sample size is approximately 100 times that examined in the DMSCC method and analysis requires about 15 sec after preparation of sample. However, cost of the instrument is significant and it must be carefully calibrated and standardized.

In performance of electronic somatic cell counts (ESCC), the preheated milk sample is mixed with buffer and dye. Part of the mixture is transferred to a rotating disc or into the channel of a flow cytometer (figure 25.9). Cells are illuminated by light that excites fluorescent dye that stains the nuclear material of the cell. Counts made of the electrical pulse of individual cells are transmitted to a computer.

Methods applicable to goat's milk differ from those used on cow's milk because goats shed cytoplasmic particles as well as nucleated cells into their milk. Therefore, Pyronin Y-methyl green stain, which differentiates between nucleated and cytoplasmic particles, is used for goat's milk examined by the DMSCC. Since counts by electronic methods are usually high, over 10^6/mL, the normal actionable level by control agencies should be confirmed by DMSCC.

25.7 Sensory Evaluation

Sensory science has been developed as a tool for documenting and understanding human responses to external

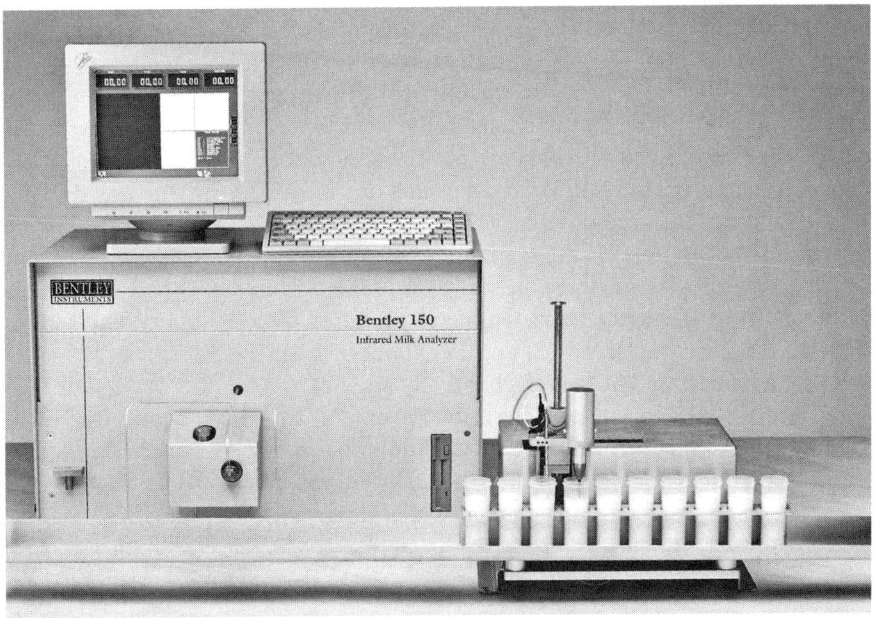

Figure 25.9 Instrument for counting somatic cells by flow cytometry (courtesy of Bentley Instruments, Inc., Chaska, Minnesota).

stimuli. This field of science now broadly encompasses psychophysics (understanding relationships between stimuli and human responses) and chemometrics (relating sensory perception to chemical measurements).

The term *sensory evaluation* denotes types of examinations that depend upon our senses of taste, odor, sight, and touch. Taste and odor are used primarily in evaluating milk and milk products but some attributes are evident to sight and/or touch.

Acceptance and preference are greatly dependent on how foods appeal to all senses. Therefore, the dairy industry depends on analysts who have developed expertise in sensory evaluation for examination of representative samples during receipt of raw materials, product development, processing, storage, and distribution. Expert sensory analysts are capable of recognizing both desirable and undesirable characteristics and can relate sensory perception to causative factors. For example, analysts detecting a rancid flavor in raw milk should relate it to factors favoring lipase activity. Furthermore, the analyst should be able to advise field service personnel, producers, and processors on procedures to follow in preventing defects. Personnel in product development need to predict as well as evaluate changes resulting from formulation, processing, and distribution practices. Therefore, these persons need specialized training in the functional properties of food ingredients and in experimental methods of sensory analysis. The latter includes design and analysis of experiments. Food texture is a key property consumers evaluate in determining food quality and acceptability (Foegeding et al., 2011). Changing composition, such as decreasing fat content, is often associated with changes in texture and flavor.

As a product is being consumed, volatile flavors are released within the mouth and pass back up through the nasopharynx and on to the olfactory epithelium. Aroma must be released into this area to be detected. Some components of dairy products may bind to aromatic ingredients, reducing their volatility. Perception of a volatile flavor note therefore depends on the physical structure of the compound and the matrix into which it is added.

Objectives of sensory evaluation may be to describe a flavor and/or aroma **profile** or to determine one or more of the following: (1) presence of typically desirable characteristics, (2) presence and magnitude of undesirable characteristics, (3) whether one sample differs from another, and (4) whether one sample is preferred over another. Objectives (1) and (2) are combined in USDA grading of cheese and butter (cf. chapter 30 and 32).

Here we discuss major defects of flavor and odor that occur in milk and milk products. Additional reference is made to sensory qualities peculiar to specific products in the respective chapters in which those products are discussed.

25.7.1 Preparation for Evaluation

The following are prerequisites to making evaluations: adequate facilities and supplies, personal training and experience, and representative samples.

A suitable room has adequate space and is comfortable and free of odors and dust. One should have high-quality water for rinsing the mouth and a sink or waste can for **expectoration** of samples. Occasionally it is necessary to examine samples in undesirable surroundings. It may be impossible, for example, to prevent odors in the milk room at times during the year; yet the milk hauler must perform sensory evaluation by both sight and smell before pumping milk from the farm bulk tank. Usually, however, it is possible to remove samples and to take them into a suitable environment, especially when a critical judgment is to be made.

Personal preparation is needed in both physical and mental respects. The analyst should not smoke and should refrain from eating spicy and highly flavored foods for a few hours prior to examining samples. Young people are usually more sensitive than older ones.

Mental preparation involves learning. One must first learn the sensory characteristics of typical desirable samples. This is best learned from study of descriptive information plus examination of samples chosen by an expert. Then one must learn what defects may occur and characteristics of each. Each individual should practice with samples containing known defects, being careful to note flavor, odor, and tactual sensations and their order of occurrence. Repetition helps to reinforce memory and is essential for those who would become highly proficient. Group practice with known samples allows discovery of taste blindness and insensitivity (table 25.2). A good technique is preparation of dilutions of a particular flavor in coded containers. Those samples are then presented in random order, along with one or more controls, and participants are asked to identify the sample or samples that contain the particular flavor. With adequate repetition, degrees of sensitivity to various flavors become apparent. It may be possible for a person who understands personal limitations to discover secondary characteristics useful in identifying specific flavors and odors. For example, one who has difficulty sensing the characteristic odor of rancidity may be able to identify the defect by the feeling it leaves in the mouth.

There are four primary taste sensations: sweet, sour, salty, and bitter. They are produced by sugar, lactic acid, table salt, and caffeine or quinine, respectively.

Confidence is important and is developed by adequate practice with representative samples and a thorough knowledge of sensations that should be experienced. The analyst must refrain from prejudging and allow the senses to identify the important characteristics. Imagination must not dictate decisions.

Table 25.2 Methods of Preparation of Typical Off Flavors in Milk and Cottage Cheese Samples.

Flavor	Method of Preparation[a]
Acid	Add about 2% cultured buttermilk.[b]
Bitter	Add 0.02% caffeine or a few drops of a 1% solution (or 0.02 g/qt) of quinine sulfate.
Coarse	To simulate the coarse (excessive diacetyl) flavor of cottage cheese, add about 0.01% diacetyl.
Cooked	Heat milk to about 167°F (75°C) and cool immediately.
Feed	Bubble-steam volatile extract of silage or chopped forages through milk.
Flat/watery	Add 5 to 10% water.
Foreign/ disinfectant	Add hypochlorite sanitizer or laundry bleach (flavor dissipates with time).
Fruity/ fermented	Add pineapple juice and vinegar (6:1) or select from samples held at 41 to 44.6°F (5 to 7°C) for several days.
Garlic/onion	Add garlic powder, onion juice, or macerated sections of onion.
Malty	Add malt extract or soak 2 tsp of Grape-Nuts™ cereal in 1 pt milk for 2 hr, then filter.
Metallic/ oxidized	Add 2 drops of 1% cupric sulfate ($CuSO_4 \cdot 5H_2O$) to 1 pt milk and expose in direct sunlight for 15 to 30 min, or add 10 drops copper sulfate (1%) to 1 pt milk and store at 41°F (5°C) for 48 hr. Different flavors will be developed by the two methods.
Musty	Store cottage cheese (or butter) in a small container along with an agar slant culture of *Streptomyces odorifer.*
Rancid	Add 1 part raw milk to 9 parts warm homogenized milk and refrigerate overnight or add 2 drops butyric acid per pt.
Salty	Add 0.5 g table salt/pt.
Unclean	Select from several samples of milk stored at 41 to 44.6°F (5 to 7°C) for several days. Typical samples will have a somewhat putrid odor and a slightly astringent to bitter flavor. Dilute as necessary.
Yeasty	Add baker's yeast to cream and hold at room temperature overnight then add to dry cottage cheese or mix with milk.

[a] Each substance should be added to high-quality milk unless otherwise noted.
[b] The diacetyl present in most cultured buttermilk may confuse the inexperienced analyst.

25.7.2 Defects in Flavor and Odor and Their Causes and Prevention

Off flavors and odors of milk and milk products can be placed in categories based on their causative factors: feed, microbial, chemical, and production or processing. Those *absorbed* from the cow, e.g., feed, garlic/onion, or cowy (when cows have ketosis), are present when milk is drawn. *Chemical* changes taking place after milk is drawn cause rancidity and metallic/oxidized flavors. Several off flavors result when *microbial* counts reach millions/mL.

25.7.2.1 Acid

This sour flavor develops when microorganisms ferment lactose and other carbohydrates. Lactic acid is the major acid in milk and milk products.

Fresh milk contains no lactic acid unless infecting bacteria ferment lactose within the udder, which is an unusual occurrence. Lactic acid is odorless but other compounds produced simultaneously create the typical odor of sour milk. The primary bacteria that make milk sour are streptococci and lactococci. They grow well at room temperature but slowly at a refrigerator temperature of 45°F (7°C). Therefore, souring of raw milk usually results from inadequate cooling. The acid flavor in butter results from churning sour cream. Acid is necessary to the production of most cheese but excessive acidity is detrimental to flavor, body, and texture.

25.7.2.2 Astringent

Astringency has been described as a "puckery" mouthfeel. It may be associated with feed flavor and slight rancidity or bitterness.

25.7.2.3 Bitter

Bitterness may result from protein breakdown, from consumption of certain weeds by cows, and development of rancid flavor. In grade "A" fluid milk the most important cause is proteolytic bacteria that grow well at 45°F (7°C). Proteases produced by these bacteria may withstand heating in cheese manufacture and produce bitterness in cheese during ripening. Prevention of bitter flavors often depends on prevention of bacterial contamination and growth. In addition, some bacteria in **starter cultures** produce proteases that release bitter peptides.

Bitterness is detectable by taste but not smell; however, bitterness is usually accompanied by other off flavors such as unclean flavor. Taste receptors adapt slowly to bitterness, so the flavor is usually stronger on repeated sampling, or it may appear when the mouth is rinsed with water. Bitterness is perceived on the sides and back of the tongue.

25.7.2.4 Cheesy

This flavor occurs in butter and results from breakdown of proteins and lipids in milk, cream, or butter itself. Microorganisms produce enzymes that cause proteolysis and lipolysis, giving butter a flavor like that of American cheese. Prevention, therefore, involves prevention of bacterial contamination and growth.

25.7.2.5 Coarse

This word is used to describe flavors in butter and cottage cheese. However, the meaning is different with each. Coarseness in butter denotes lack of the sweet, pleasant, delicate flavor. Instead there may be a slight hint of acidity and lack of freshness. In cottage cheese, coarse flavor has been used to describe the presence of excessive diacetyl, a desirable component in moderate concentration, therefore a balance of flavor is lacking.

25.7.2.6 Cooked

This flavor develops when sulfur-containing groups are released by heat from amino acids of certain milk proteins. The higher the temperature and longer the time of heating, the greater the cooked flavor. There is less damage, by equivalent heating, to the flavor of cream products than to skim milk. This flavor dissipates during the first day of storage. It is not particularly objectionable unless it has developed to the extent of being scorched. In fact, a slight to moderate cooked flavor is considered desirable in butter and vanilla ice cream. A pronounced cooked flavor is found in evaporated and dried milk products. The taste reaction time is short, and the aftertaste is pleasant. The aroma is particularly noticeable. Seldom is oxidized flavor found when cooked flavor is present.

25.7.2.7 Feed

This is the most prevalent off flavor found in fluid milk freshly drawn from the cow. The skunkweed flavor is an exception. It does not appear in milk but develops in butter or cheese from affected milk. Flavorful compounds found in feeds that cows consume are transferred via the bloodstream to milk. The rate of transfer is especially high when volatile odors of feeds the cow **regurgitates** are swept into the lungs as the cow inhales. Blood passing through the lungs for oxygenization picks up volatile compounds, and they are soon conveyed to tissues of mammary glands that synthesize milk. Inhaled odors may appear in milk in as few as 20 min. An experiment by Bradfield (1962) showed that milk absorbed a feed flavor when cows breathed the odor of silage, but the effect was greater when the silage was eaten. Some substances that produce flavor are absorbed by blood from the digestive tract.

Small-grain pasture, especially rye, is particularly offensive. Pasture grasses, silage, and some hays may also produce the flavor. Weeds such as ragweed, bittersweet, and peppergrass can be troublesome. The ensiling process destroys substances causing ragweed or peppergrass flavors but does not affect substances causing onion flavor. Removal of cows from access to offending feeds for 2 to 4 hr prior to milking usually reduces the flavor intensity to acceptable levels. This is because blood can reabsorb flavors from milk and partially transfer them to the air via the lungs. However, flavors from some weeds linger as long as 12 hr.

Feed flavors are characteristically aromatic, so they are detectable by the sense of smell. One can become accustomed to them by preparing typical samples using the following steps. Suspected feed is chopped into small pieces or crushed to allow escape of juices then placed in a side arm flask with water. The flask is closed with a stopper. A hose attached at the outlet of the flask is submerged in milk. Heat applied to the suspension drives aromatic components of feed into milk where they condense, imparting the feed flavor. (**Caution is advised:** The hose going into the milk must be removed before heat is removed from the flask to avoid a vacuum that will suck the milk back into the flask, which may cause it to burst.) This aromatic property is the basis for vacuum treatment of milk. Heat and vacuum are combined to remove volatile flavors and odors (cf. chapter 26).

25.7.2.8 Flat/Watery

The name describes the characteristic of this flavor well. It implies a low concentration of solids in milk, which most often occurs when water is added or when fat is removed to produce skim milk. Milk adulterated with water lacks characteristic sweetness. In cheese and other milk products the typical flavor is decidedly lacking. Most individuals are unable to detect flatness when less than 5 to 10% water is added to milk, and few can differentiate between samples containing 2 and 3.5% fat when the same milk is used to prepare samples.

25.7.2.9 Foreign/Disinfectant

Milk and milk products may become tainted by foreign materials either by absorption or by direct contamination. The former is rare in today's operations for, although milk readily absorbs odors, most health authorities limit the use and storage of detrimental substances around milking and processing facilities. Careless use of disinfectants, however, may result in off flavors. Failure to drain sanitizer from equipment is the most frequent cause, especially when its strength is excessive. Use of phenolic substances in salves and ointments applied to cow teats may provide sufficient phenol to interact with chlorine in sanitizers. The complex formed is detectable by some individuals in concentrations of a few parts per billion (ppb/mcg/L). Reaction to foreign flavor depends on the causative agent. Hypochlorite sanitizer is commonly used to prepare samples for practice in training analysts because it is frequently used in dairies. However, the intensity of this flavor dissipates rapidly, making it difficult to prepare a suitable sample.

25.7.2.10 Fruity/Fermented

This class of flavors results from bacterial growth and is usually associated with cultured products, especially cheese, and with pasteurized milk that has been stored refrigerated for 10 or more days. Fruitiness results from the combination of fatty acids with alcohols (both produced by bacteria) and the odor resembles that of ripe bananas, pineapple, or similar fruits. In such milk, the most abundant esters with fruity odors are ethyl butyrate and ethyl hexanoate (Reddy et al., 1968). Dr. M. E. Morgan of Oregon State University prepared a mixture of ethyl butyrate (0.35 ppm) and ethyl hexanoate (0.50 ppm) in 1,2-propanediol (essentially flavorless) for use in producing fruity flavor. Fermentation may be characterized by a vinegar-like taste and odor. These defects are especially undesirable and result in rejection of any sample exhibiting a distinct intensity of either flavor.

25.7.2.11 Garlic/Onion

The highly odoriferous garlic and onion plants are most offensive in early spring and late fall when they grow quickly and often become accessible to cows in large quantities. The flavor they produce is particularly undesirable and is not easily removed, even by steam-vacuum treatment of milk (cf. chapter 28). The flavor may be retained in milk products, especially butter and cheese. Prevention of the defect may require removal of cows from offending pastures at least 4 hr before milking. The garlic/onion flavor has a pronounced odor and persistent aftertaste.

25.7.2.12 Lacks Freshness

When milk or cottage cheese has been stored several days, bacteria counts increase and subtle chemical reactions occur that reduce concentrations of desirable flavorful compounds and initiate buildup of undesirable ones. For a period of several hours the product has no particularly identifiable off flavor, but neither is its flavor desirable. Products in this condition are described as lacking freshness. The condition is difficult to duplicate but mixing a small portion (perhaps 5 to 10%) of unclean milk with high-quality bland milk or cottage cheese usually is satisfactory.

25.7.2.13 Malty

Some streptococci that produce lactic acid also produce certain aldehydes that impart a flavor resembling malt extract or Grape-Nuts™. There is sometimes a hint of walnut or maple. Bacteria used in the manufacture of cottage cheese and cultured buttermilk produce off flavors in this category infrequently. The defect is seldom found in fluid milk because bacteria must number several million/mL before the concentration of offensive aldehydes reaches a detectable level. The flavor is detected by either odor or taste. It is often accompanied by high acidity. Adding about 0.34 ppm of 3-methylbutanal to milk will produce a malty flavor (Dr. M. E. Morgan, Oregon State University, personal communication).

25.7.2.14 Metallic/Oxidized

Milk, dry milks, frozen dairy desserts, butter, and certain other milk products are susceptible to oxidation. Unsaturated fatty acids, phospholipids, and some proteins are involved in the reaction. The flavor varies with both product and cause. Sunlight, fluorescent light, copper and iron, small amounts of ascorbic acid (vitamin C), riboflavin, and excessive agitation favor or catalyze the oxidative reaction, whereas antioxidants, such as tocopherol (vitamin E), and large amounts of vitamin C retard or prevent it. The flavor develops after milk is drawn from the cow.

The oxidation of unsaturated lipids leads to the formation of volatile compounds, off flavors, and off odors. Riboflavin (vitamin B_2) has been regarded as the major photosensitizer responsible for photooxidation in milk (Borle et al., 2001). Dairy products contain several photosensitizers: riboflavin, protoporphyrin, hematoporphyrin, tetrapyrroles, and a chlorophyll-like compound (Wold et al., 2005). All these sensitizers absorb light in the UV and violet region, whereas some of the tetrapyrroles absorb throughout the visible region, meaning that any wavelength in the visible region can initiate photooxidation in dairy products.

Flavor descriptors vary but include burnt, cabbage, oily, cardboard, tallowy, **stale**, medicinal, and metallic. The latter is characterized by distinct flatness and the sensation produced by a nail in the mouth. Variations of the class of flavors can be produced by (1) adding 2 drops of 1% cupric sulfate ($CuSO_4 \cdot 5H_2O$) per pint (473 mL) of milk and exposing to direct sunlight for 15 to 30 min, (2) adding 10 drops of $CuSO_4$ solution (or a copper penny) to a pint of milk and storing at 41°F (5°C) for 48 hr, or (3) adding a special mixture of vinylamyl ketone and octanal (both complexed with urea). Vinylamyl ketone used alone produces a metallic flavor.

This group of off flavors is of considerable economic importance. Surveys in retail milk markets determined that oxidation occurred frequently in homogenized milk that was packaged in translucent plastic (polyethylene) containers and stored near high-intensity fluorescent lights. Flavor breakdown may be detected 2 to 4 hr after exposure begins, and the milk is usually strongly oxidized in 12 hr. Riboflavin and ascorbic acid concentrations decrease in direct proportion to the amount of light exposure. Gold lights or gold and green filters offer protection from light waves that cause oxidation.

Milk from different sources is not equally susceptible to oxidation. Because stored **roughages** contain lower quantities of antioxidants than do green feeds, oxidized flavor is more likely to develop during winter months, especially if the growing season is short. Individual cows, apparently for genetic reasons, produce milk of varying susceptibility to oxidation, even when kept in the same environment and under the same management. Supplementing cows' rations with dl-α-tocopherol (1 g/cow/d) and with ethoxyquin was reported to increase the oxidative stability of milk (Dunkley et al., 1967; Oskarssen and King, 1968). Antioxidants apparently exert their effect by reacting with **free radicals**, thus terminating the chain reaction between unsaturated fatty acids and oxygen. Copper content of forages is of probable significance.

Pasteurization, especially at temperatures above 170°F (76.6°C) in the high-temperature, short-time method, releases sulfhydryl groups (of β-lactoglobulin), which have antioxidant properties. Free oxygen is removed by steam-vacuum treatment, thus reducing the probability of oxidation. Growth of bacteria in milk, though undesirable, uses oxygen dissolved in milk, thereby reducing the likelihood of oxidation. Homogenization retards development of light-induced oxidized flavor. The iron sometimes added to multivitamin mineral milk may promote oxidation.

25.7.2.15 Musty

This defect occurs infrequently in butter, cottage cheese, and other milk products that can absorb odors or can support growth of causative microorganisms. Mustiness is caused by certain molds and filamentous bacteria, especially *Streptomyces odorifer*, some of which may grow on stored milk products. Feeding musty forages or grain may allow absorption of the flavor by milk.

Mustiness is quickly detected by smell, the aroma resembling a poorly ventilated room with high humidity. There is a delayed reaction time when the sample is placed in the mouth, but the flavor lingers after expectoration.

25.7.2.16 Neutralizer

Because little cream is now separated on the farm, and as methods of handling milk and cream have reduced the incidence of souring, there has been little need to neutralize the acid in cream for butter making. Hence, the incidence of neutralizer flavor in butter is extremely low. It does not occur in other milk products unless neutralized cream or butter are used in them. The defect results from addition of too much neutralizer or improper methods of neutralization. Sensations of odor and taste are present, the taste being most apparent after expectoration of the sample. Baking soda produces some of the flavor characteristics of soda-type neutralizers. Lime [$Ca(OH)_2$] neutralizers produce more of a bitter taste.

25.7.2.17 Old Cream / Unclean

These defects are classified together because of similarities in their causes and characteristics. Modern production techniques have virtually eliminated the type of unclean flavor caused by absorption of unpleasant odors either in the milking barn or during storage.

Cowy flavor may be included in this group. It is absorbed from the bloodstream of cows experiencing **ketosis (acetonemia)**. The cow's breath carries the odor. One cow with ketosis may cause an off flavor strong enough to be detected when diluted with the milk of 50 to 60 healthy cows (cf. chapter 13).

The need to store raw and pasteurized milk refrigerated for extended periods allows growth of bacteria, which produce a variety of enzymes that will degrade milk and milk products. Certain combinations of these enzymes, especially the proteases, are responsible for unclean and old cream flavors. Thus, prevention of this group of flavors usually depends upon prevention of bacterial growth and enzyme activity. Sensory reaction to these flavors is usually delayed and they persist long after the sample is eliminated from the mouth (the mouth fails to "clean up").

25.7.2.18 Rancid

One of the most objectionable, yet one of the most prevalent, off flavors is hydrolytic rancidity. The pungent odor and somewhat bitter and soapy flavor resemble the odor and flavor of spoiled nutmeats and of blue and Roquefort cheeses.

Rancidity results from hydrolysis of the ester linkage (addition of water to the bond between a fatty acid and glycerol) in milkfat. Some of the released fatty acids are unobjectionable. However, the short-chain acids (butyric, caproic, and caprylic) are especially odorous. Of the common fats, only milkfat contains appreciable quantities of these fatty acids, which are not always undesirable. They add flavor to cheeses and are especially desirable in the blue-veined varieties (mold-ripened cheeses) and some Italian cheeses.

Lipases and esterases are enzymes responsible for hydrolysis of fats. Raw milk contains lipase synthesized by the cow. Since about 90% of it is associated with casein micelles, it is called lipoprotein lipase (LPL). Furthermore, fat is in globular form with a layer of phospholipid and protein surrounding each globule. This protective layer inhibits fat hydrolysis. Factors that weaken or destroy the protective layer promote development of rancidity. These include (1) excessive agitation at temperatures between 50 and 86°F (10 to 30°C), (2) homogenization, (3) warming raw milk to 86°F (30°C) and re-cooling, and (4) freezing and thawing. Milk secreted during advanced lactation has increased susceptibility to lipase activity.

Pipeline milkers and bulk coolers must be properly designed and operated to prevent rancidity. Opportunities for agitation are afforded when milk rises vertically in pipelines or milk hoses because air moves milk by displacing it in most pipeline systems.

Churning and foaming are indicative of treatment that may lead to rancidity. Both may be obvious when milk is observed in the bulk tank. Centrifugal pumps are used with most milk releasers on milk pipelines. These are operated intermittently to move milk to the bulk tank. Continuous operation results in foaming and churning and often leads to rancidity. Leaks in pipelines have a similar effect by promoting turbulence at the point where air enters. Uncontrolled agitation with air in silo storage tanks frequently causes rancidity. Controllers that regulate volume of air should be installed so excessive agitation does not occur as the tank is about to be emptied.

Bulk milk coolers should be designed and selected to prevent a rise in temperature of blended cold and warm milk to above 50°F (10°C). In this way activation of lipase by warming and re-cooling is prevented.

Pasteurization inactivates natural lipase of milk and prevents further fat hydrolysis. However, some bacteria produce lipases and not all of these are completely inactivated by pasteurization. Since homogenized milkfat is readily attacked by lipase, such milk must be pasteurized before or immediately after homogenization and must not be subsequently exposed to raw milk, even in trace amounts. Hydrolysis of 1 to 2% of the triglycerides of milk results in detectable lipolyzed flavor. The di- and monoglycerides remaining after hydrolysis of triacylglycerides are surface active, displacing foam-stabilizing proteins in foam bubbles. Consequently, foam stability is decreased when milk or cream is whipped, as in making butter, whipped cream, or cappuccino coffee.

25.7.2.19 Salty

Saltiness is a defect of chemical composition; the milk is salty as it comes from the cow. The condition is indicative of mastitis or the last stages of lactation. In mastitis, synthesis of lactose is impaired and chlorides increase to balance the osmotic pressure of milk with that of blood. Saltiness is seldom detected in mixed milk from large herds because of dilution.

The inexperienced analyst may overlook the salty flavor because it is sensed quickly but tends to fade with time. This flavor is perceived only by taste. Saltiness in milk products usually results from addition of too much salt rather than being carried over from the milk itself.

25.7.2.20 Storage

A group of flavors and odors, especially staleness, falls in this category. The cause may be either absorption or change in chemical composition or both, and several dairy products may be affected. The storage defect, however, is most detectable in bland or slightly flavored products such as butter, dry milk, and vanilla ice cream. Butter in quarter-pound prints is particularly susceptible. With this product, the high surface area exposed to the atmosphere enhances oxidation, absorption, and growth of microorganisms that degrade proteins, causing surface taint. A combination of these causative factors may be involved in production of certain types of storage flavor.

25.7.2.21 Yeasty

Yeasts are less prominent than bacteria in milk and milk products and they grow more slowly. They most frequently produce flavor defects in acid products, especially cheeses. Their growth in ripened cheeses such as cheddar may be accompanied by gassiness and holes may appear in the cheese. Yeasts often produce a sour, fermented-type flavor or a flavor like bread dough leavened with yeast. Yeastiness is the primary defect in orange juice pasteurized and packaged in fluid milk dairies. The alcoholic and sour note commonly develops 14 to 21 days after packaging when yeasts are introduced from equipment after pasteurization.

25.7.3 Scoring Dairy Products for Sensory Qualities

To communicate the quality of dairy products, it is sometimes desirable to assign numerical scores. Typically, the highest score of a range is assigned when a product has no defect, and lower scores are given for samples that fail to meet the ideal.

Two factors considered in assigning scores are (1) the relative importance of the defect in determining acceptability of the product and (2) the relative intensity of the defect. Scoring can be accomplished using a 5- or 10-point scale with the highest score being assigned when no criticism is made (table 25.3). Flavor and odor are marked on a scale of 1 to 10, whereas body and texture and color and appearance are marked on a 5-point scale. Some score cards are designed to give a total score of 10 to a product having no defects (table 25.4). Prac-

Table 25.3 Scoring Information Used in the National FFA Milk Quality and Products Career Development Event.

Flavor and Odor of Milk

Score (10-point scale)	Quality	Defects
10	Excellent	No defects
8–9	Good	Slight feed, flat
5–7	Fair	Slight bitter, feed, slight malty, slight metallic/oxidized, salty, slight unclean
2–4	Poor	Bitter, slight foreign, malty, metallic/oxidized, rancid, unclean, garlic/onion
1	Unacceptable	Foreign, strong garlic/onion, high acid, strong metallic/oxidized, strong rancid

Source: National FFA Organization, 2014.

Table 25.4 Scoring Guide for Flavor and Odor of Milk as Recommended by the American Dairy Science Association Committee on Evaluation of Dairy Products.

	Intensity of Defect		
Flavor Criticisms[a]	Slight	Definite	Pronounced[b]
Flat	9	8	7
Cooked	9	8	6
Feed	9	8	5
Lacks freshness	8	7	6
Astringent	8	7	5
Salty	8	6	4
Oxidized	6	4	1
Bitter	5	3	1
Foreign	5	3	1
Garlic/onion	5	3	1
Malty	5	3	1
Metallic	5	3	1
Rancid	4	1	—
High acid	3	1	—
Unclean	3	1	—

[a] A sample with no discernable defect is scored 10.
[b] A dash (—) indicates product of unsalable quality. Official rules prohibit use of such products in contests.

tices used in grading specific dairy products on a sensory analysis basis are given in chapters in which those products are discussed.

25.7.4 A Positive Approach to Flavor Control

Quality of processed milk and milk products cannot be better than the quality of raw materials from which they are made. A combination of good management factors is essential. Cows should be clean, healthy, and properly fed. Milking facilities should be clean, dry, and well-ventilated. Utensils and equipment must be clean, sanitized, smooth, free of copper, and constructed and installed properly. Milk must be quickly cooled and constantly protected throughout each stage of production, manufacture, and distribution.

25.8 Summary

The word *quality* is a neutral noun needing an adjective (modifier) to describe it. Milk producers and processors should never be content unless their milk products can be described as being of supreme quality. Costs of achieving excellence of quality are often less than they may seem because so many of the practices that improve quality also improve efficiencies of production.

Tests for quality are necessary primarily because practices of production and processing are not performed correctly. The better production and processing practices are performed, the less the necessity for quality assurance tests.

It is essential that the quality of milk and milk products be determined by tests for (1) microorganisms, (2) chemical components, (3) adulterants, (4) abnormalities, and (5) sensory characteristics. Generally the tests are expensive and require expertise in both performance and interpretation. Trained technicians and managers are essential for assuring high-quality milk for consumers.

Study Questions

1. What are the characteristics of high-quality milk?
2. Why are so many quality tests applied to milk and milk products?
3. How do international organizations benefit the dairy industry? Name four of them.
4. Why are microbiological tests of importance in assessing milk quality?
5. What is the standard plate count and how is it used in assessing the quality of milk and milk products?
6. What are the bacterial standards (number per milliliter) for grade "A" raw and grade "A" pasteurized milk?
7. Of what significance are psychrotrophic bacteria in milk?
8. Of what use is the preliminary incubation test in determining milk quality?
9. What are thermoduric bacteria?
10. Why are coliform bacteria especially undesirable in milk?
11. Of what use is the plate loop count method of estimating the number of viable bacteria in milk?
12. Which tests are useful in determining bacterial counts of milk suspected of having a large number of bacteria? How does their accuracy compare with the SPC?
13. The reduction tests for bacterial activity are based on what principle?
14. What bacteriological tests are made to estimate the shelf life of milk and milk products? Is sample variability important? Why?
15. Differentiate between apparent and developed acidity.
16. Identify factors affecting the titratable acidity of milk.
17. Name dairy products in addition to milk in which it is important to know the titratable acidity.
18. What laboratory test is used to detect free fatty acids in milk and cream?
19. Why is it important that milk be pasteurized? What chemical is the target of tests used to confirm pasteurization?
20. Of what significance are antibiotics in milk?
21. What is a zone of inhibition in an antibiotics test?
22. Which test is effective in determining if water has been added to milk? What physical property of milk serves as a basis for this test?
23. What is the freezing point of milk? Identify factors that affect it.
24. Why is the sediment test of limited use in determining milk quality?
25. Of what significance are pesticides to the dairy industry? Are they allowed in milk?
26. Why are somatic cells undesirable in milk? What does their presence indicate? Which tests are used to estimate somatic cell count numbers in milk?
27. What is meant by sensory evaluation of milk and milk products? What are the primary objectives of sensory evaluation?
28. Why are preparations of self, samples, and environment important to proper sensory evaluation?
29. Briefly outline a technique that can aid an assessor when learning sensory evaluation aspects of assessing milk quality.
30. What are the primary taste sensations? How may they be produced?
31. Name the four main categories of off flavors and odors of milk. What are the flavors within each category and their causes?
32. Which undesirable flavors of milk can be detected by odor?
33. Is a cooked flavor always undesirable in all dairy products? Why or why not?
34. Describe conditions in which milk may acquire a feed flavor. What is the most effective way to avoid a feed flavor in milk?

35. Which vitamins and minerals can affect oxidized flavor in milk?
36. What type of milk container can help to create an oxidized flavor? How is the light source in the storage environment important in this process?
37. How may dairy equipment be involved with creating a rancid flavor in milk?
38. What is the effect of pasteurization on lipase of milk? How is this related to milk flavor?
39. What is the relationship between mastitis and salty flavor of milk?
40. Outline a positive approach to milk flavor control.

References

Borle, F., R. Sieber, and J. O. Bosset. 2001. Photooxidation and photoprotection of foods, with particular reference to dairy products—An update of a review article (1993–2000). *Sciences des Aliments* 21:571–590.

Bradfield, A. 1962. Vermont Agricultural Experiment Station Bulletin 624.

Dunkley, W. L., M. Ronning, A. A. Franke, and J. Robb. 1967. Supplementing rations with tocopherol and ethoxyquin to increase oxidative stability of milk. *Journal of Dairy Science* 50:492–499.

Early, R. 1995. *Guide to Quality Management Systems for the Food Industry.* New York: Blackie Academic & Professional.

Elliker, P. R. 1968, September. *American Dairy Review.*

Foegeding, E. A., C. R. Dauberti, M. A. Drake, G. Essick, M. Trulsson, C. J. Vinyard, and F. Van de Velde. 2011. A comprehensive approach to understanding textural properties of semi- and soft-solid foods. *Journal of Texture Studies* 42:103–129.

Kay, H. D., and W. R. Graham. 1933. Phosphorus compounds in milk. *Journal of Dairy Research* 5:63–74.

Latimer, G., Jr. (ed.). 2012. *Official Methods of Analysis of AOAC International* (19th ed.). Rockville, MD: AOAC International.

Marshall, R. T., and R. A. Appel. 1973. Sanitary conditions in twelve fluid milk processing plants as determined by use of the rinse filter method. *Journal of Milk and Food Technology* 36:237–241.

Maturin, L., and J. T. Peeler. 2001, January. *Bacteriological Analytical Manual* (Chapter 3, Aerobic Plate Count). Washington, DC: US Food and Drug Administration.

National FFA Organization. 2014. *National FFA Career Development Events.* Indianapolis, IN: Author.

Oskarssen, M. M., and R. L. King. 1968. Effect of ethoxyquin on the development of oxidized flavor in milk. *Journal of Dairy Science* 51:928.

Prescott, S. C., and R. S. Breed. 1910. The determination of the number of body cells in milk by a direct method. *Journal of Infectious Diseases* 7:362.

Ranieri, M. L., W. R. Mitchell, R. A. Ivy, N. Martin, M. Wiedmann, and K. J. Boor. 2011. Development of a real-time PCR assay for rapid detection of spoilage *Paenibacillus* spp. in fluid milk. *Journal of Dairy Science* 94:292 (Abs. T78, E-Supplement 1).

Reddy, M. E., D. D. Bills, R. C. Lindsay, L. M. Libbey, A. Miller III, and M. E. Morgan. 1968. Ester production by *Pseudomonas fragi.* I. Identification and quantification of some esters produced in milk cultures. *Journal of Dairy Science* 51:656–659.

Scharer, H., 1938. A rapid phosphomonoesterase test for the control of dairy pasteurization. *Journal of Dairy Science* 2121–2134.

US Food and Drug Administration. 1999. *Pesticide Analytical Manual*, Volumes I and II. Washington, DC: Author.

USDA Agricultural Marketing Service. 1977. *United States Sediment Standards for Milk and Milk Products.* Washington, DC: USDA AMS Dairy Division.

Wehr, H. M., and J. F. Frank (eds.). 2004. *Standard Methods for the Examination of Dairy Products* (17th ed.). Washington, DC: American Public Health Association.

Wold, J. P., A. Veberg, A. Nilsen, V. Iani, P. Juzenas, and J. Moan. 2005. The role of naturally occurring chlorophyll and porphyrins in light-induced oxidation of dairy products. A study based on fluorescence spectroscopy and sensory analysis. *International Dairy Journal* 15:343–353.

Websites

Advanced Instruments, Inc. (http://www.aicompanies.com)

AOAC International (http://www.aoac.org)

Bentley Instruments, Inc. (http://bentleyinstruments.com)

Codex Alimentarius Commission (http://www.codexalimentarius.org)

International Dairy Federation (http://www.fil-idf.org)

International Organization for Standardization (http://www.iso.org)

National FFA Organization (http://www.ffa.org)

North Carolina State University, Food Safety Education and Training (http://foodsafety.ncsu.edu/about-us/publications)

Reichert Technologies, Inc. (http://www.reichert.com)

3M Food Safety, Dairy Industry (http://solutions.3m.com/wps/portal/3M/en_US/Microbiology/FoodSafety/industries/one)

US Food and Drug Administration, Food Contaminants and Adulteration (http://www.fda.gov/Food/FoodborneIllnessContaminants/default.htm)

US Food and Drug Administration, Science & Research (Food) (http://www.fda.gov/Food/FoodScienceResearch/default.htm)

USDA Agricultural Marketing Service, Grading & Standardization Division (http://www.ams.usda.gov/AMSv1.0/DairyStandardizationPublications)

26

Processing Milk and Milk Products

There is always a lag between the conception of an idea and its execution on one hand and its public appearance on the other.

Samuel Brody (1890–1956)

26.1 Introduction

Were it not for the risk of contracting disease, raw milk would likely be sold widely. However, there are other reasons for processing milk and milk products. **Pasteurization**, for example, destroys bacteria that cause deterioration of milk, as well as those that can cause disease. It also inactivates enzymes, including natural milk lipase, which can be detrimental to quality. **Homogenization** stabilizes the emulsion of fat in **milk serum**, so the last portion of milk removed from a container contains essentially the same concentration of fat as the first. **Clarification** and filtration remove insoluble extraneous materials and body cells. **Centrifugal separation** provides a means of adjusting milk composition so the components of fat can be readily standardized. **Vacuumization** removes volatile substances that may be offensive to smell and taste. Refrigeration is essential to control bacterial growth and enzyme activity.

Processes discussed in this chapter are those used in most milk processing operations. Bulk milk hauling and milk plant receiving operations are included. Processes unique to particular dairy products are discussed in subsequent chapters.

Before discussing these processes it will be interesting to observe how various mechanical, physical, chemical, and biological treatments are applied to milk and milk components in producing and marketing milk products. Let us review an abbreviated matrix for this purpose. The matrix will be based on typical raw cow milk: 3.7% fat, 3.1% protein, 4.9% lactose, 0.7% minerals.

Matrix: What major *processes* are applied in making selected milk products?
Centrifuge milk >> cream + nonfat milk (skim milk)
Standardize cream + nonfat milk >> half-and-half; milk; reduced fat milk; lowfat milk
Homogenize milk or cream >> stable emulsion of fat in milk product
Pasteurize liquid forms >> destroy microorganisms
Inoculate liquid milk product and *incubate* >> sour cream; cultured buttermilk; yogurt
Churn cream >> butter + buttermilk
Melt butter and *separate* lipid and aqueous phases >> butter oil + water and nonfat solids
Combine cream + concentrated milk + sugar + flavoring + stabilizer >> ice cream mix
Freeze ice cream mix while agitating under pressure >> ice cream
Drip ice cream mix into liquid nitrogen >> Dippin' Dots™ frozen dessert
Evaporate water from liquid milk products >> evaporated milk or concentrated milk
Dry concentrated milk product >> dry milk; nonfat dry milk; dry whey
Precipitate protein with acid and/or rennet (chymosin) >> cheese curd + whey
Ripen cheese curd >> cheese
Heat and filter cheese whey >> whey cheeses (e.g., ricotta)
Combine cream and nonfat cheese curd >> cottage cheese
Ultrafilter cheese whey >> whey proteins and permeate
Microfilter milk >> remove spores for longer shelf life
Condense and *cool* whey permeate >> lactose + mineral water

26.2 Bulk Milk Hauling

The bulk milk hauler occupies a key position in milk handling and to this end licensing is required. The hauler is responsible for examining milk to determine if its temperature, appearance, flavor, and odor are satisfactory (cf. chapter 25). She/he collects a sample to be used for a variety of compositional and quality tests. This sample, representing all milk in the tank, must be: (1) taken in a manner preventing contamination with chemical or biological substances, (2) protected from sunlight, (3) held at 34 to 39°F (1 to 4°C), and (4) delivered to the testing laboratory the day it is collected.

The hauler measures quantity of milk in the bulk tank. The common method involves insertion of a graduated stainless steel rod in the bulk tank (figure 26.1) to measure the milk depth. Each farm bulk tank is calibrated and a chart is placed in the milk room. Using this chart the hauler translates depth of milk into pounds and records this on a duplicate receipt, one copy each for the producer and the buyer. To assure accuracy in measurement several conditions must be controlled. The bulk tank must be (1) structurally stable so its walls will not move, (2) accurately calibrated to deliver the amount of milk measured, and (3) installed and maintained on the calibration plane (level) with a reliable device to verify maintenance of this plane. Movement of tanks, uneven or settling floors, and shims on legs not sealed in place are factors that can cause calibration problems. Milk must be motionless when measured, and this usually requires at least 10 min to elapse after agitation ceases. The gauge must be clean, dry, and fat-free. A measuring gauge taken from cold milk and wiped dry with a sanitary towel easily becomes a "wet stick" from moisture condensing from the air. The result is

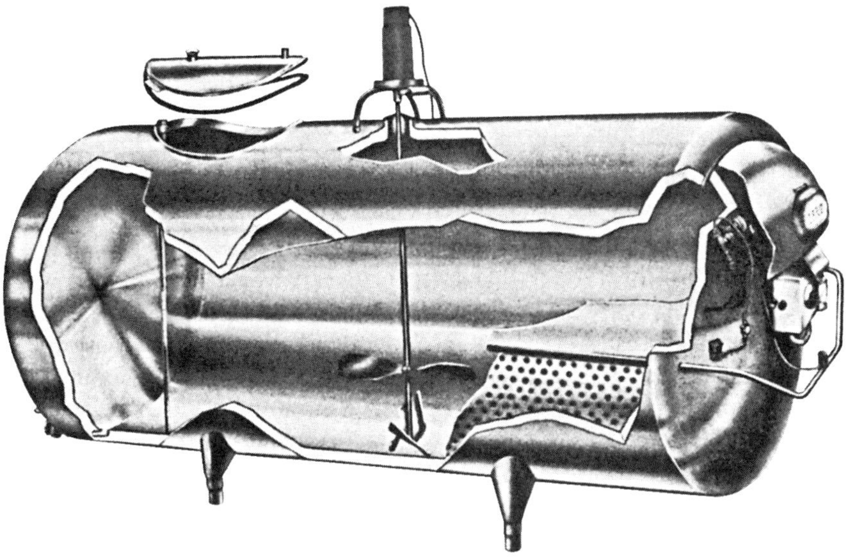

Figure 26.1 Cut-away drawing of a vacuum-type refrigerated bulk milk tank with graduated measuring rod (left), agitator (center), and refrigeration coils (right).

a high reading. Warming the gauge with warm water then drying it will alleviate this problem. High readings also result from foam, churned, or frozen milk and the presence on the gauge of wetting agents from detergents or towels impregnated with chemicals used in cleaning udders. A meter must be used with an air eliminator because air in milk produces error in volume measurement. The National Institute of Standards and Technology (2015) *Handbook 44: Specifications, Tolerances, and Other Technical Requirements for Weighing and Measuring Devices* includes details regarding the proper codes for meters used for measuring quantities of milk pumped through a milk line.

Duties of the milk hauler at the farm must be approached systematically. Objectives of the bulk hauler should be to

1. evaluate the condition of milk correctly so that
   a. acceptable milk will not be left on the farm, costing the producer unfairly, and
   b. unacceptable milk will not be loaded onto the truck, likely degrading thousands of pounds of milk from other producers;
2. sample milk correctly;
3. efficiently transfer milk to the transport tank; and
4. deliver milk and samples directly and expeditiously to the point of off-loading.

Milk becomes the property of the buyer once it is in the transport tank since it loses its identity when mixed with milk of other producers.

The following is an abbreviated version of the hauler's procedure at the dairy farm according to appendix B of the *Grade "A" Pasteurized Milk Ordinance* (US Food and Drug Administration, 2013).

1. Examine the milk by sight and smell for off odor or abnormalities that would class it unacceptable. Wash hands thoroughly and dry with a clean, single-service towel immediately before measuring and/or sampling the milk.
2. Measure the milk with the cold stick before agitation. If the agitator is running upon arrival, take the measurement only after the milk surface is quiescent.
3. The milk shall be agitated a sufficient time to obtain a homogeneous blend.
4. Identify sample container and collect sample, placing it in a sample case held at 32 to 40°F (0 to 4.4°C).
5. Record milk's temperature. Sanitize the hauler's thermometer when used to check accuracy of tank thermometer or before use when tank thermometer is inoperable.
6. After pumping milk into tank truck, disconnect, cap hose, and replace hose in truck. Be sure porthole cover is closed. Rinse tank and porthole thoroughly with tempered water.
7. When a route collection has started, complete it without unnecessary delay.

When a bulk tank truck has been cleaned and **sanitized** it must have a tag attached showing the date, time, place, and signature of the operator doing the work. This may not be required if the truck delivers to only one receiving unit where responsibility for cleaning and sanitizing can be definitely established. The tag is to be removed at the first stop on route and kept on file for the regulatory agency. Milk should never be collected at a producer's farm during the milking operation, and partial collections of milk from a bulk cooling tank, storage tank, or both are prohibited without special permission from the regulatory agency. Hauling toxic materials in bulk milk tank trucks is prohibited.

Milk is not to be collected unless it has been cooled to 45°F (7°C) within 2 hr after milking or if the blend temperature after the first and subsequent milkings has exceeded 50°F (10°C). Abnormal milk shall not be collected.

An accurate and complete record of each producer's milk pickup shall be kept. This shall include producer number and name; milk measurement, weight, or both; date and time of collection; temperature of the milk; remarks related to condition or quality; and the hauler's signature or initials.

Scientists at the University of Kentucky introduced the Bulk Milk Transportation Security and Traceability System in 2010. The system enables users to track milk from the farm to the processor. A data server, interfaced with a handheld computer, is used to store information on who, when, where, and why a tanker's door, dome, or valve was accessed; milk's quantity and temperature; tanker wash cycles; and condition of samples taken at farms. Operational limits set in the computer processor cause alerts to be communicated when these limits are violated.

26.3 Receiving Milk at the Plant

Upon arrival at the receiving dairy, milk is usually examined for odor and then agitated with sterile air or a mechanical agitator. The time necessary to provide complete mixing depends on the design of the agitator, size of tank, and quantity of milk. The milk is immediately tested for antibiotic residue. If there are questions concerning quality, an additional sample may be tested for titratable acidity and direct microscopic count. (Merits and limitations of these tests are discussed in chapter 25.) Other applicable tests are the freezing point, standard plate count, and the fat test. However, results of these are not available prior to unloading.

Dairy employees usually pump milk from the tanker (figure 26.2) then clean and sanitize it. The quantity received is determined by a metering device with an air eliminator. An alternative method is to weigh the truck before and after removal of milk.

After off-loading is completed, a spray device is lowered into an open manhole in top of the tank (figure 26.3). Tap water is used to rinse away most milk solids. Cleaning solution pumped through the sprayer flows from the bottom outlet of the tank and is returned to the solution reservoir for recirculation. After cleaning with a dairy cleaner, the tank is rinsed, sanitized, and drained. A seal is then placed on the tank indicating it is ready for receiving milk. Many health authorities require that tanks be cleaned after each load is delivered, even if the truck is to return immediately to farms for another load. This is necessary because the previous load may have contained excessive numbers of bacteria and because the temperature inside the tank may rise quickly when the tank is empty, permitting bacteria to begin rapid growth.

26.4 Transfer and Receiving Station Operations

In some milksheds it may be advantageous to collect milk from farms in small tank trucks (about 2,900 gal/11,000 L capacity) then transfer this milk to large trailer trucks (about 5,000 gal/19,000 L capacity) for hauling to a distant market. Two types of facilities are used. In the transfer station, overhead protection is provided for trucks during transfer of milk and for cleaning and sanitizing, whereas at a **receiving station** milk is pumped into stationary tanks. It may be clarified, cooled, and even standardized or separated before it is loaded onto another truck or trucks for transfer to the processing plant. Receiving stations must be constructed to meet the same sanitation requirements as milk plants.

Figure 26.2 Pumping milk from a transport truck in a milk receiving room (courtesy of Hiland Dairy, Springfield, Missouri).

Figure 26.3 A heavy-duty stainless steel sanitary dual head power washer is being electronically lowered into the raw milk tanker's top opening. Note the safety rails and personnel harness in place (courtesy of Prairie Farms Dairy, Springfield, Missouri).

26.5 Storing Raw Milk

Raw milk is normally processed within 24 hr of the time received at a plant, except for weekends and holidays. Many processors schedule operations daily, except on Wednesday and Sunday. Thus, they reduce storage time and meet the increased demand for weekends. In most large plants raw milk is stored in silo-type tanks (figure 26.4). These are equipped to agitate milk with filtered compressed air that is admitted at the bottom and stirs the milk as it rises. They are cleaned when emptied and must be cleaned at least every 72 hr.

Grade "A" raw milk for pasteurization should be stored at no higher than 45°F (7°C). The maximum bacterial count of raw milk in the plant is 300,000/mL compared with 100,000/mL at the farm. The reasons are that (1) during pumping, clumps of bacteria are broken; (2) during passage over additional equipment, limited contamination may occur; and (3) during the additional time of storage, a limited amount of growth of **psychrotrophic** bacteria may occur. The biological factors limiting the time raw milk may be stored are bacterial growth, metabolism, and enzymatic activity (microbial and natural milk enzymes). Chemical reactions (nonbiological) are of considerably less importance. These include oxidative reactions catalyzed by metal ions, especially copper.

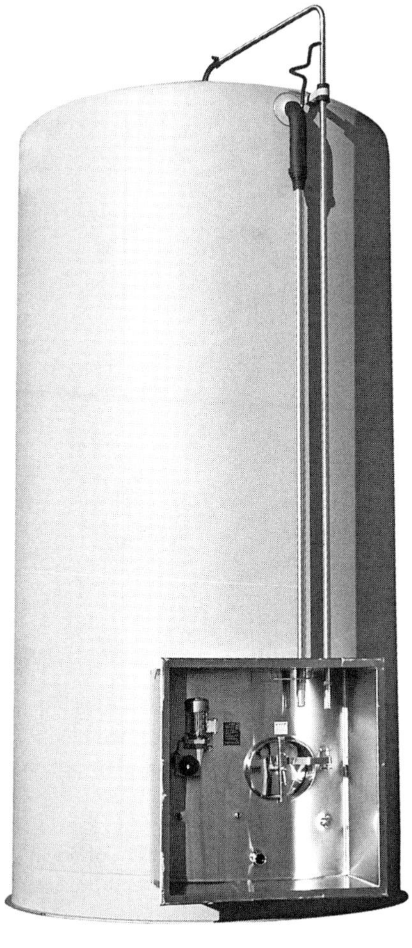

Figure 26.4 Silo-type storage tanks for raw milk. Capacities range from 5,000 to 70,000 gal (19,000 to 265,000 L) (courtesy of Paul Mueller Company, Springfield, Missouri).

26.6 Clarification and Filtration

Clarification and filtration are performed to remove insoluble suspended materials from milk. Clarification is considerably more efficient than filtration because small particles pass through filters that permit an acceptable rate of flow.

Clarifiers operate on the principle that the rate of separation of suspended materials from a suspending medium can be increased by applying centrifugal force and by reducing the distance through which the **suspensoid** must travel. The greater the difference in density between suspended particles and the suspending medium, the faster the separation.

A clarifier is actually a continuous centrifuge (figure 26.5A). Milk is pumped through it and centrifugal force is applied for a time that is sufficient to remove suspended particles, including somatic cells. These move to the bowl's periphery, which is designed with a large open space to collect and hold solid materials. There is some separation of fat globules from the serum but these are remixed as they leave the area of centrifugal activity via the central (tubular) portion of the bowl. Discs are used in the bowl to reduce the distance heavy particles must travel to become separated from milk. As they separate from the liquid these particles impinge upon the bottom of the disc immediately above them and centrifugal force carries them outward in a thin surface film until they reach the collection space in the outer bowl. Few microorganisms are removed.

Once the collection space in a clarifier bowl is filled, sedimented particles reenter milk and clarification is rendered ineffective until the clarifier is cleaned. This is a manual operation unless the machine is equipped with a hydraulic device that opens the bowl. Sediment can be flushed out as the bowl continues to spin (a self-cleaning clarifier).

Clarifiers may be operated with cold (39 to 45°F [4 to 10°C]) or hot (about 140°F [60°C]) milk. Because of its lower viscosity, hot milk is more efficiently clarified; hence the capacity of a given machine is increased. Unless milk is to be processed immediately it should be re-cooled. This results in exorbitant costs for heating and cooling. Therefore, no heat should be applied if clarification is to be accomplished before storing raw milk. Also, clarification of cold milk results in fewer deposits of casein in the bowl thus allowing longer periods of operation.

Sediment also can be removed by filtering milk through a layer of specially designed cotton fiber. This process is often called straining when performed on the farm. Leukocytes and other somatic cells pass through the filter. The operation is simple and comparatively inexpensive but much less efficient than clarification.

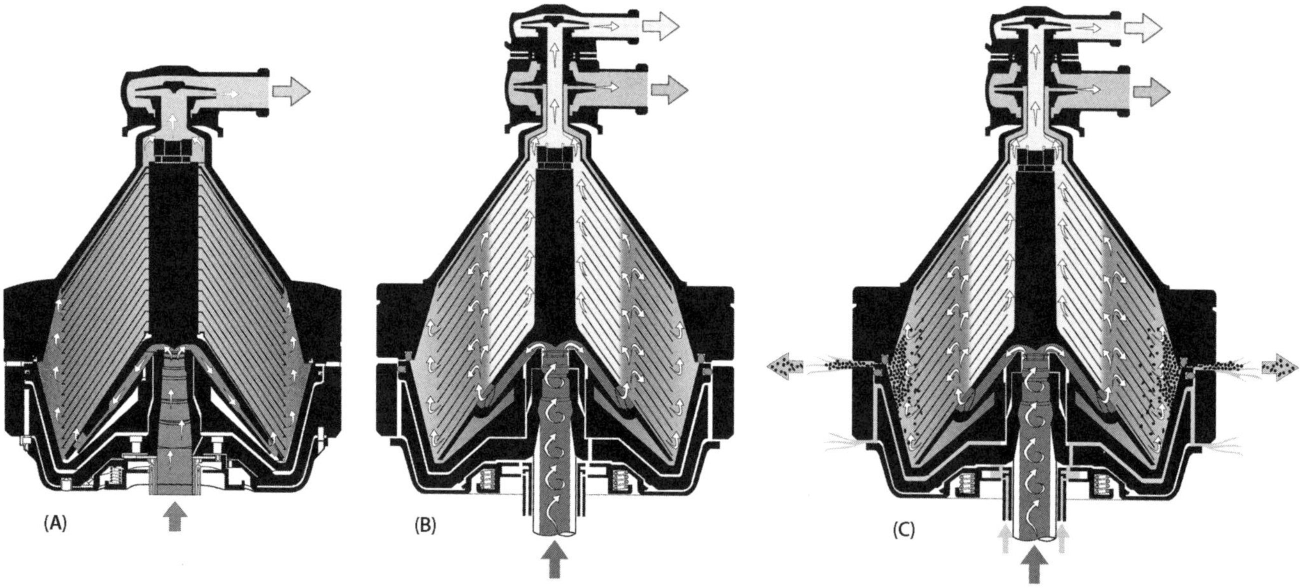

Figure 26.5 Bowls of a self-unloading centrifugal clarifier (A) and separator (B and C). Sediment and cellular materials are centrifuged from milk in the clarifier and incoming milk displaces clarified milk toward the center and top of the bowl, where it exits. In the separator, milk moves upward through distribution holes then outward. Skim milk is forced outward against the bottom of the disc above it, while cream moves inward along the top of the disc below it. Particulates collect in the slime storage space at bowl perimeter. Skim milk is displaced upward along the side of the bowl and exits at the top. Cream is displaced to the center and upward to its exit. Hydraulic pressure holds top and bottom portions of bowls together. On release of pressure, the bottom of the bowl drops and solids are discharged (C), thus cleaning the bowl so operations may continue (courtesy of Tetra Pak, *Dairy Processing Handbook*).

26.7 Separation and Standardization

Today's consumers enjoy milk products that contain several concentrations of fat. Processors produce these products because they can quickly and economically remove fat (in the form of cream) from milk. The process is called separation. Standardization is the adjustment (up or down) of fat concentration to a specific value. It is desirable that fat content of all dairy products be closely controlled. Fat is an expensive component; efficient operation requires that managers plan and accomplish standardization within practicable limits. Consumers are entitled to receive all the fat they pay for, but many also want the assurance that excess fat is not included.

Fat, emulsified in the serum phase of milk, is comparatively easy to remove and minimal physical damage results. Fat will rise slowly when milk is stored quiescently. The top portion containing concentrated fat is cream; the bottom portion with little fat is skim milk. Individual suspended fat globules of milk behave according to Stoke's law, where the viscosity of the fluid, diameter of the fat globules, and densities of fat and serum determine the rate of separation. Density of fat is lower than that of milk's other constituents, which leads to natural separation of large globules.

26.7.1 Centrifugal Separation

The first continuous cream separator was invented by Antonin Prandtl of Germany in 1875. Early separators were hollow bowl machines. Gustaf de Laval added discs about 1891. Improvements have since been made, the major ones being airtight construction and self-cleaning capability. Separators are constructed much like clarifiers. Discs are mounted on a shaft inside a sealed bowl (see figure 26.5). Milk is pumped in near the center where it enters a series of vertical holes in discs. As the discs spin, heavier skim milk is forced against the bottom of the disc above it and lighter fat globules are displaced inward against the top of the disc beneath them. Each forms a stream and the two streams move in opposite directions (the skim milk outward, the cream inward).

The purpose of discs in a bowl is to divide milk into thin layers so centrifugal force can separate fat globules from milk easily. This is true because the globules need to travel only a short distance before being separated from milk. Once skim milk reaches the outer disc edge, it is forced upward toward its outlet. Similarly, cream is displaced by other cream toward its outlet. Efficiently operating separators should produce skim milk with no more than 0.01% fat by the Babcock test or 0.06 to 0.10% by the Mojonnier test. Self-cleaning separators (figure 26.6), like self-cleaning clarifiers, are opened hydraulically during operation. When closed down they can be completely **cleaned-in-place**. Practically all separators and clarifiers are now airtight so foaming is no problem, microbial contaminants are not exhausted from the bowl in the form of a mist, and skim milk and cream may be piped to the desired location without additional pumping (the lack of such a feature created disadvantages for earlier models).

In pumping milk to a separator it is necessary to prevent entry of air at or before the pump (on its suction side). Air causes foaming and churning in the separator, the capacity is decreased, and the fat content of the resulting cream varies.

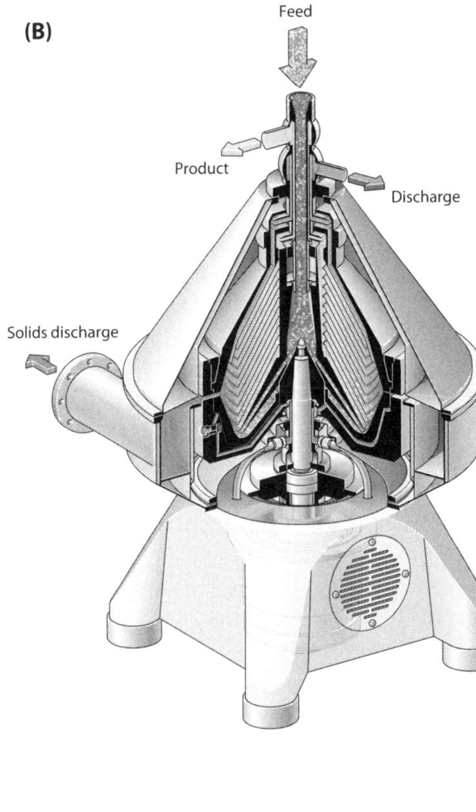

Figure 26.6 (A) Self-cleaning cleaned-in-place (CIP) separator installed in a fluid-milk processing line capable of processing 60,000 lb/hr (27,300 kg/hr). (B) Cut-away view of a milk separator (from GEA Westfalia Separator, Burlington, Ontario, Canada).

26.7.2 Membrane Processing

Membrane processing involves separation of milk components from a fluid stream, based primarily on component size, using a porous membrane under pressure to serve as a selective barrier. The pressure gradient across the membrane forces solute and smaller molecules through pores in the membrane, while larger molecules/particles are retained. Thus, one feed stream is split into two product streams. The retained stream (the "retentate" or "concentrate") will be enriched in retained larger macromolecules. The fraction going through the membrane (the "permeate") is depleted of those macromolecules. The retentate will also contain some permeate solutes. Membrane processing can produce milk products with reduced microorganisms, minerals, and lactose, along with increased protein. Conditions differ for producing each product. Membrane processes applicable to the dairy industry are, in order of smallest to largest membrane pore sizes, reverse osmosis (RO), microfiltration (MF) and ultrafiltration (UF).

Pores in RO membranes are permeated only by water. Therefore, they are useful in concentrating milk or whey prior to evaporation or transportation. RO technology can be used to recover milk solids from water used to flush milk lines, caustic from the cleaning of evaporators or dryers, and brine used in the processing of certain cheeses. In each case the water permeate can be used in cleaning operations while the caustic and brine can be recycled into their respective supply tanks.

Microfilters with ceramic membranes can retain 99.99% of milk's microorganisms while permitting skim milk or whey to pass through the pores. When applied concurrently with pasteurization (e.g., to produce extended shelf life milk), keeping quality is improved primarily because spores and thermotolerant bacteria are removed, some of which remain viable when only pasteurization is applied. Other applications include pretreatment of cheese milk to remove spores of gas-producing bacteria and production of dry milk products with very low bacteria and spore counts. To avoid fouling of the membrane with fat, only the skim milk or whey portion is microfiltered. Before pasteurization, cream is recombined with microfiltered nonfat milk in ratios needed in standardized milk products.

Researchers at the Wisconsin Center for Dairy Research recently developed a process using polymeric membranes for microfiltration, thus producing permeate streams of beta-casein and whey protein plus a retentate stream containing the other milk components. The retentate is suitable for making cheese or casein concentrates. Another recent development involves placing a positive charge on UF membranes to separate beta-lactoglobulin and glycomacropeptide from cheese whey.

In common dairy applications using UF, protein is concentrated in the retentate. Skim milk can be concentrated to a maximum of about 42% solids. The protein in cheese whey can be concentrated up to 80% (Cheryan, 1998) or higher if fat is removed by MF. A major application of UF is in increasing the protein content of milk for cheesemaking and in concentrating whey proteins (cf. sections 30.4.2 and 33.3).

The FDA has tentatively determined that the appropriate name for UF milk is "ultrafiltered milk." It is advantageous to use ultrafiltered milk to increase the percentage of protein in fluid milk. This allows the flavor and mouth feel enhancing properties of milk proteins to be achieved naturally as opposed to adding nonfat dry milk (NFDM), which often leaves a cooked flavor in the fluid milk as well as increased sweetness from excess lactose in NFDM. The resulting nonfat or lowfat varieties have the flavor and mouth feel of a whole milk product without the higher fat.

26.7.3 Labeling/Usage

FDA regulation 21 CFR 101.4 requires ingredients of a food to be declared by their specific common or usual name in the ingredient statement of the finished food. This regulation permits use of a collective term for a few specific ingredients. For example, skim milk, concentrated skim milk, reconstituted skim milk, and nonfat dry milk may be declared as "skim milk" or as "nonfat milk" (see section 101.4(b)(3)). This specific provision, however, does not extend to include ultrafiltered milk. Therefore, when used in foods, ultrafiltered milk must be declared by its specific common or usual name, i.e., "ultrafiltered milk."

The standard for the identification of milk permits the addition of certain specific milk-derived ingredients for the purpose of adjusting milk solids not fat and milkfat content of the milk. However, the standard for milk does not encompass a food that is prepared by separating the nonfat portion of milk into its various components and by subsequently remixing these separated components to create a specific profile. Similarly, the separation and subsequent remixing are not modifications to prevent nutritional inferiority, nor are they covered under provisions that permit minimal deviations in the ingredient. Therefore, a product prepared by passing milk through various filtration processes, including ultrafiltration, and then remixing the separated components, does not comply with either the standard for milk in 21 CFR 131.110 or the minimal modifications provided under the general standard in 21 CFR 130.10.

26.7.4 Standardization

The average fat percentage of milk received by most fluid milk processors is 3.6 to 4.0; whereas, milk is defined as having at least 3.25% fat. Therefore, it is necessary to remove some fat to provide a uniform and economically competitive product.

Standardization is generally accomplished by (1) removal of fat on a continuous basis with a standardizing clarifier, (2) removal of all fat from a portion of milk with the return of the correct amount of skim milk to the original storage tank, or (3) the addition of cream to milk or skim milk.

Manufacturers have modified the design of clarifiers and separators to enable them to standardize milk continuously. All three operations can be performed by some machines. For example, cream and skim milk are separated but only a fraction of the cream is removed. The balance is remixed with skim milk in the machine and milk of a lower fat content is discharged. Careful control of operating conditions is necessary. Unless there is online monitoring with feedback for control of fat content, the fat concentration in milk entering the standardizer must be kept constant for any particular setting of the standardizing valve. Without such control, milk must be agitated in the storage tank to avoid creaming and, when milk from another tank is admitted to the line, a new setting of the standardizing valve will be necessary if the fat concentration differs. Increased back pressure on the cream discharge port of the standardizer will cause a higher fat content in standardized milk, whereas increased back pressure on the milk discharge port will reduce the fat test. The reason is that the amount of back pressure determines residence time for each component in the separator. The longer milk, or skim milk, stays in the separator (higher back pressure on milk discharge), the more fat is separated, whereas the longer cream is retained, the more of it will escape in milk. Wide variations in temperature (5°C or more) will affect the fat content of standardized milk.

Computer-controlled online blending of cream and skim milk from the separator has proved successful. Input data entered by the operator are batch size, composition of raw materials, and desired composition of the finished product. Ancillary equipment required includes (1) turbine-type meters of CIP design, (2) air-operated throttling valves, (3) continuous in-line detection of fat content, and (4) a digital-based control system. In some applications an in-line analyzer (based on near-infrared technology) can be placed in the flow stream to monitor composition and signal the controller to adjust the system.

When standardizing by mixing skim milk with milk of a high fat content, it is necessary to calculate the quantities of each product that will be needed. A convenient method is the Pearson square, and is best explained by an example (figure 26.7). Assume unlimited quantities of skim milk (containing essentially no fat) and of 3.7% milk are available to make 10,000 lb of milk testing 3.3% fat. In order to calculate how many pounds of skim milk and 3.7% milk are needed to make 10,000 lb testing 3.3% fat, use the following procedure:

1. Draw a square, placing the desired fat content in the center and the percentages (or test) of fat content of the two available ingredients at the top left and bottom left of the square.

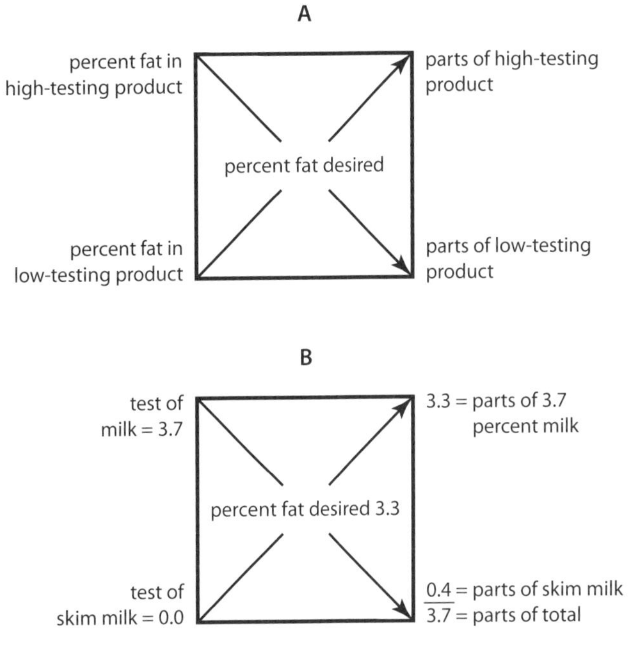

Figure 26.7 Example of application of the Pearson square in the standardization of milk.

2. Subtract the test of standardized milk (desired % fat) from test of milk, placing the result at the corner diagonally opposite. Then subtract the test of skim milk from that of standardized milk and place it at its diagonally opposite corner.
3. The numbers at the upper and lower right of the square represent, respectively, parts of 3.7% milk and of skim milk needed to obtain the total quantity to be produced. Therefore:

   10,000 / 3.7 × 3.3 = 8,919 lb of 3.7% milk
   10,000 / 3.7 × 0.4 = 1,081 lb of skim milk

The same procedure is adaptable to situations in which a fixed amount of one ingredient is to be mixed with an unlimited amount of another (cf. chapter 8). For example: 1,000 lb of 40% cream are to be mixed with 3.7% milk to produce half-and-half testing 10.5% fat. Below is a completed Pearson square.

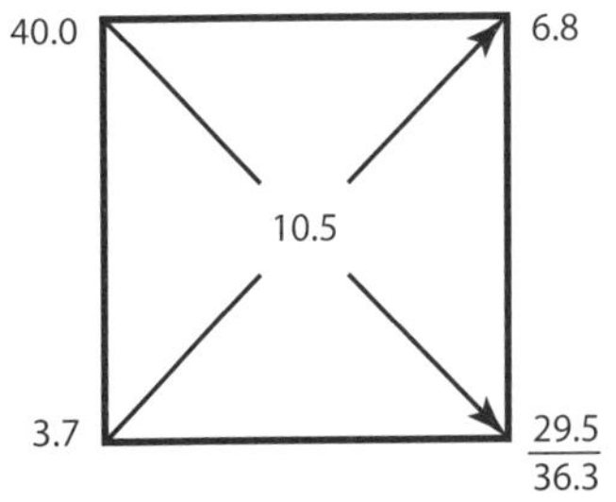

The ratio of cream to total product is 6.8:36.3. In order to determine the resulting lb of total product, solve for X:

$$\frac{6.8}{36.3} = \frac{1{,}000}{X}$$

$$6.8X = 36{,}300 = 5{,}338$$

$$5{,}338 - 1{,}000 = 4{,}338 \text{ lb of } 3.7\% \text{ milk}$$

To prove the solution:

1. Multiply lb of cream times its fat test in percent
   $1{,}000 \times 0.40 = 400$ lb
2. Multiply lb of milk times its fat test in percent
   $4{,}338 \times 0.037 = 160.5$ lb
3. Add the results of (1) and (2) = 560.5 lb
4. Compare to total lb of half-and-half times its fat test in percent
   $5{,}338 \times 0.105 = 560.5$ lb

Since the quantities of fat in proofs (3) and (4) are equal, the solution is correct.

In producing specialty milk products changes in composition caused by addition of ingredients must be considered in standardization. For example the formulas of some milks are adjusted to enhance nutritional properties: (1) vitamin-mineral milk containing 25% of the Daily Value (DV) of eight vitamins, the minerals calcium and phosphorus, plus 10% of the DV of vitamins A and E plus magnesium and zinc; (2) protein enhanced milk containing 12 g/serving; (3) calcium milk containing 33% more calcium than regular milk; and (4) fiber milk that is enhanced by 2.5 g/serving. Even greater effects on formula can occur when flavorings are added.

26.8 Pasteurization

The primary purpose of pasteurization is to destroy microbial pathogens. A secondary purpose is the destruction of microorganisms and enzymes that might impair product quality. With certain products (e.g., ice cream) heating adds a flavor that is preferred by many consumers.

Although Louis Pasteur pioneered in studies of heat treatment in preservation, pasteurization of milk was first attributed to Dr. Soxhlet of Germany in 1886. The first milk heater to be used in the United States was imported from Denmark in 1893. Commercial pasteurization began in the United States about 1900, sometimes being used secretly to aid in preserving milk. Early pasteurization equipment was designed to heat milk to 160°F (71.1°C) for 30 to 60 sec, but design was crude and temperature control was variable. In 1900 Russell and Hastings, and soon thereafter Rosenau (1909), established thermal death points of a variety of microorganisms that might be present in milk. They established that holding milk for 20 min at 140°F (60°C) was adequate to kill *Mycobacterium tuberculosis* and other pathogens, and in 1907 the vat pasteurization method was introduced. In 1908 the first compulsory pasteurization law was enacted in Chicago for milk from cows that had not been tuberculin tested. The minimum temperature for vat pasteurization was increased to 143°F (61.7°C). This temperature was increased to 145°F (62.8°C) in 1956. The process, accomplished by heating milk in a vat, is called *low-temperature long-time pasteurization* (holding method). The process is slow and several vats are required to process large quantities of milk.

Pasteurization in the home was first recommended in 1824 by William Dewees, professor of obstetrics at the University of Pennsylvania. According to Hall and Trout (1968), Dewees recommended heating "quickly to the boiling point; so soon as this is perceived, it should be removed and cooled as speedily as possible." However, boiling causes denaturation of part of the serum proteins, which coagulate and float on the surface as a **pellicle**. For home pasteurization, controlled heating at 145°F (62.8°C) for not less than 30 min is recommended. Heating in a double boiler avoids the "burn-on" flavor that occurs when too much direct heat is applied. Stirring is necessary. Because of the time and attention required, many persons who produce or can purchase raw milk are reluctant to pasteurize it. However, the risk of disease, though decreased markedly in the United States in recent years, is sufficient to require pasteurization, especially for milk fed to infants and geriatrics.

Early American families learned that pouring boiling water into milk enhanced its keeping qualities. Although unaware of it, they were actually performing a crude form of pasteurization. In many countries, boiling of milk is still commonplace. Lysine and cystine are sensitive to heat so that prolonged boiling of milk tends to reduce its **biological value**.

It is possible to destroy undesirable microorganisms in a comparatively short time by using a higher temperature. For example, at 161°F (72°C) only 15 sec are required for pasteurization. These conditions are employed in **high-temperature short-time (HTST)** pasteurization.[1] Even higher temperatures are often used to provide greater killing of spore-forming bacteria. Table 26.1 lists temperature and time combinations approved as adequate to kill pathogenic microorganisms. Because of a protective effect on microorganisms produced by fat and sugar, the temperature must be increased in the pasteurization of chocolate milk, cream, ice cream mix, eggnog, and similar products (except for **ultrahigh temperature** pasteurization, in which times and temperatures are the same for all dairy products).

Various types of heat exchangers are used to pasteurize or sterilize. These include:

1. Indirect heat exchangers
   a. Vat type—water jacketed and spray equipped
   b. Plate type
   c. Tube type
   d. Scraped surface or swept surface

2. Direct steam heaters
   a. Steam injectors
   b. Steam infusers

The major type presently used in fluid milk plants is the plate heat exchanger. Unit capacities range from a few hundred to 60,000 lb/hr. Many operators heat milk to near 176°F (80°C) and hold it about 25 sec in an attempt to destroy practically all viable **vegetative** forms of bacteria.

In achieving efficient heat exchange, four factors are important: (1) a high temperature gradient (temperature difference between heating and heated mediums), (2) high thermal conductivity of metal walls separating the two mediums, (3) a large amount of heat transfer area, and (4) rapid renewal of surface films. To provide these conditions in a plate-type pasteurizer cold milk is caused to flow on one side of a plate in the direction opposite the flow of hot milk or water on the other side (the *heating cycle*), and hot milk is pumped in the direction opposite the flow of cold milk or water (the *cooling cycle*). Stainless steel plates are relatively good heat conductors. Surface area is high because plates are spaced so only a thin film of milk passes between them. Also, plates are designed with indentations and with waffle-like surfaces (see figure 26.8) to increase surface area. This design also promotes turbulence, providing rapid renewal of surfaces in contact with the plate.

Table 26.1 Minimum Time and Temperature Combinations Used in Pasteurization.

1. Every particle of milk or milk product is heated in properly designed and operated equipment, to one (1) of the temperatures given in the following chart and held continuously at or above that temperature for at least the corresponding specified time:

Batch (Vat) Pasteurization	
Temperature	**Time**
63°C (145°F)*	30 min
Continuous Flow (HTST and HHST) Pasteurization	
Temperature	**Time**
72°C (161°F)*	15 sec
89°C (191°F)	1.0 sec
90°C (194°F)	0.5 sec
94°C (201°F)	0.1 sec
96°C (204°F)	0.05 sec
100°C (212°F)	0.01 sec

* If the fat content of the milk product is 10% or greater, or a total solids of 18% or greater, or if it contains added sweeteners, the specified temperature shall be increased by 3°C (5°F).

2. Eggnog shall be heated to at least the following temperature and time specifications:

Batch (Vat) Pasteurization	
Temperature	**Time**
69°C (155°F)	30 min
Continuous Flow (HOST) Pasteurization	
Temperature	**Time**
80°C (175°F)	25 sec
83°C (180°F)	15 sec

Source: US Food and Drug Administration, 2013.

Regenerative heating and cooling are important features of plate heat exchangers. In regeneration, hot product in one set of passages is cooled by cold product in the opposite passages (product-to-product regeneration). This warms the cold product and cools the hot. Regeneration recovers heat; consequently, less energy is needed for final heating. It also saves refrigeration. Importantly, efficiencies of regenerators commonly range from 80 to 90%.

The ability of HTST pasteurizers to assure a safe, finished product depends on maintenance of appropriate time-temperature-pressure relationships during operation of the system.

Milk usually enters the HTST plate pasteurizer at 39°F (4°C). The sequence of flow and approximate tem-

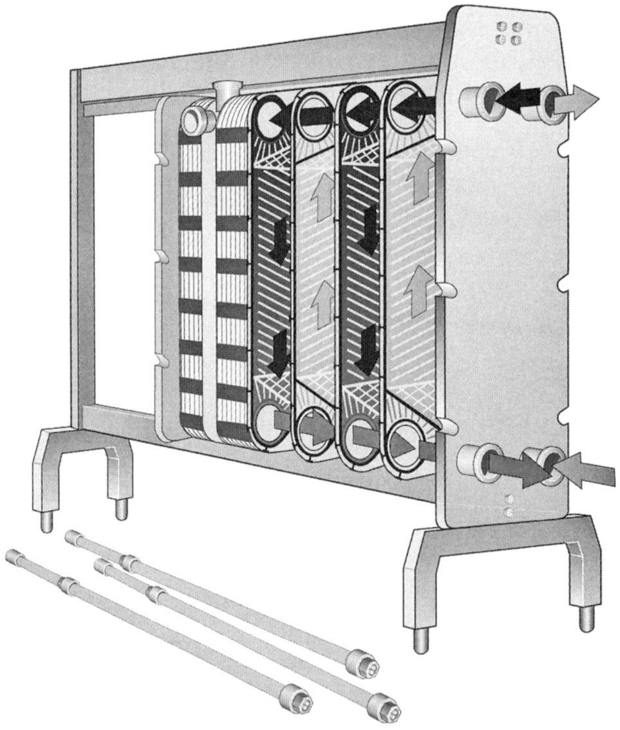

Figure 26.8 Cut-away view of a plate pasteurizer showing waffle-like surfaces of heat exchange plate (courtesy of Tetra Pak, *Dairy Processing Handbook*).

peratures attained are: raw milk regenerator, 136°F (58°C) → heater, 165°F (74°C) → pasteurized milk regenerator, 68°F (20°C) → cooler, 39°F (4°C). In the process, product is conveyed through the raw milk regenerator by suction force of a **positive-type pump**. This pump pushes milk at a constant speed through the remaining parts of the system, keeping a positive pressure on the pasteurized side of the regenerator. This causes milk to be held at pasteurization temperature for at least 15 sec in the holding tube that originates after the heater section and terminates at the flow diversion valve (figure 26.9). This tube must slope upward in the downstream direction at a rate of not less than 2.1 cm/m (0.25 in/ft). This prevents entrapment of air (foam) that could decrease the effective diameter and increase the rate of flow of the hot milk. The maximum velocity of milk product passing through the holding tube is affected by speed of the timing pump, diameter and length of the tube, and surface friction. The smaller the tube diameter, the higher the friction per unit volume. Holding tubes must be installed so sections of pipe cannot be left out, resulting in a shortened holding time. A temperature sensing and recording device (recorder-controller) is placed at the end of the holding tube and immediately upstream of the flow diversion valve (FDV). The FDV, developed in 1938, and its temperature sensor and control system (together they are the flow diversion device, or FDD) function to detect possible sublegal temperature at the end of the holding process. Should that occur, the valve is instantly moved by air pressure to a position that diverts milk back to the raw milk supply tank (balance or constant level tank) to be recirculated through the pasteurizer. Specifications for construction and operation of FDVs minimize the probability that unpasteurized product will enter packaging equipment. Furthermore, specifications for the pasteurizer are of the fail-safe type. That is, should any malfunction occur the equipment is designed to preclude mixing raw milk with pasteurized milk and the passage of raw milk out of the pasteurizer system.

Separators in HTST pasteurization systems must be installed and operated so they will not adversely affect regenerator pressures, create a negative pressure on the FDD during operation, or cause milk or milk product flow through the holding tube during times when such flow would compromise a required public health safeguard.

26.8.1 Higher Heat Shorter Time Pasteurization and Extended Shelf Life Milk

Extension of the shelf life of fresh milk products has long been an objective of milk processors. The industry recognizes the term "extended shelf life" (ESL) as indicative of products with a refrigerated shelf life of 21 to 28 days. These products are treated at temperatures well above HTST minimums and extreme care is exercised to prevent contamination after heating. Filling machines are available with enclosed filling valves surrounded by filtered air, treated by spraying with sanitizers, and/or exposed to UV light. ESL milk can be produced also by microfiltration, a process by which concentrations of bacteria can be reduced by 4 to 5 $\log_{10}$. This involves passing skim milk through a membrane with pore diameters small enough (up to 5 μm diameter) to retain practically all microbial cells while allowing skim milk solids and water to pass through. The retentate, a relatively small amount containing microbial cells, is then heated at elevated temperature and used for alternate products. The skim milk permeate is then standardized to the

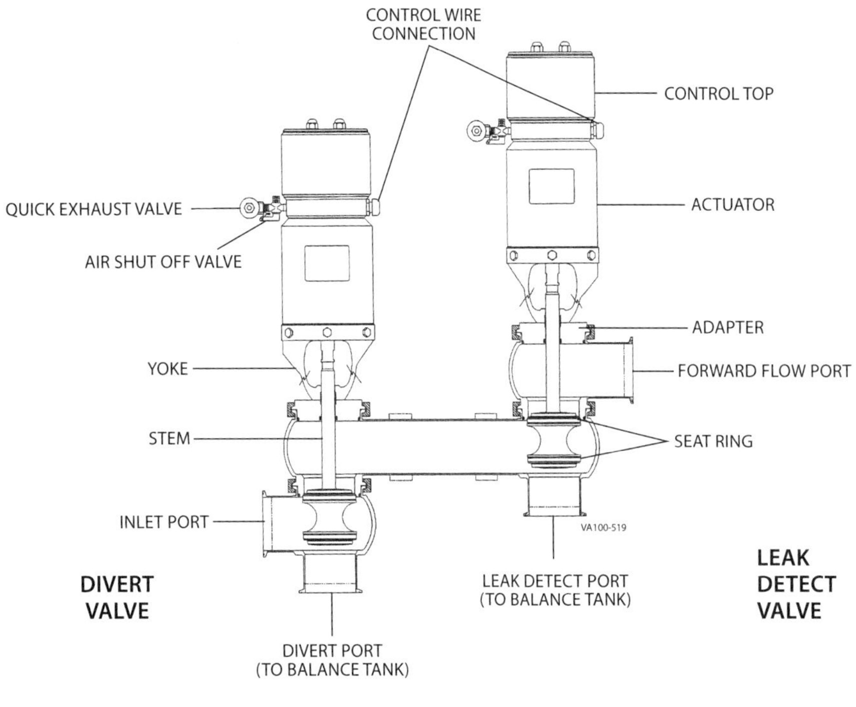

Figure 26.9 Cleaned-in-place (CIP) flow diversion valve that assumes three positions, divert, leak detect, and forward flow. A temperature sensing device and related electrical components control valve positions. Air opens the valves, which are spring-loaded to the divert position (as shown in this drawing). On reaching pasteurization temperature the leak detect position is assumed for about 10 sec. The divert valve (left) closes downward while the forward flow valve (right) remains seated upward allowing the valve body and leak detect port to be flushed with pasteurized product. The forward flow valve is positioned downward only when the product is at or above the minimal pasteurization temperature. During cleaning, the pistons are positioned to allow detergent solution to pass through all lines leading from the unit (courtesy of Waukesha Cherry Burrell).

desired fat content and pasteurized at minimal temperature and time to produce milk with slight to moderate cooked flavor. Another approach to producing ESL milk products is to reduce microbial numbers by high speed centrifugation, a process called bactofugation (see section 26.13). Pasteurization is required because the process does not remove all microbial cells.

Higher heat shorter time (HHST) pasteurization, developed in the 1950s, involves temperatures from 190 to 270°F (88 to 132°C) and holding times of 2 sec or less. Because holding tubes are quite short, timing must be done by measuring the rate of flow. This requirement applies as well to **ultrapasteurization**. To prevent vaporization in the holding tube, which may substantially reduce residence times, HHST systems operating above 212°F (100°C) must include a pressure limit switch set at 69 kPa (10 psi) above the boiling pressure of the product at its maximum temperature in the holding tube. The switch is inter-wired with the FDD to move it to the divert position if pressure on the milk falls below a prescribed value.

26.8.2 Ultrapasteurization

Federal standards of identity stipulate that to be labeled *ultrapasteurized* (UP) a product must have been heated to 280°F (138°C) or above and held at that temperature for at least 2 sec. The process significantly extends the shelf life of milk and requires that UP products be refrigerated (see 21 CFR 131.3) because they may contain small numbers of viable microorganisms. Temperatures and times may go as high as 302°F (150°C) and 9 sec if sterility is required, as in producing milk that is shelf stable at common room temperatures. Very high temperatures do comparatively little damage to milk flavor during the short exposure time, provided heating and cooling are accomplished quickly. Fortunately, they are highly detrimental to bacteria.

A problem that delayed adoption of ultrapasteurization was lack of ability to obtain quick response of controllers to temperature change. When holding times are 1 sec or less, the time between a temperature drop to below minimum and closure of the FDV is comparatively long. For the HTST pasteurizer the maximum control response time is 1 sec. As a remedy, flow diversion valves for UP pasteurizers have been placed at the end of the cooling section, but the temperature-sensing element is placed at the end of the holding tube. This increased distance provides ample time for the flow diversion valve to close if temperature in the holding tube becomes substandard. However, regenerating and cooling sections become contaminated with raw milk, and forward flow, through the diversion valve, must not occur until all product surfaces downstream from the flow diversion valve have been sanitized. This is accomplished by turning off the coolant to the cooling section and recirculating the dairy product until all surfaces have been heated to pasteurizing temperature. Special controls are required to accomplish this.

Another problem associated with short holding tubes is lack of a means to measure residence time in the tube. In HTST units, residence time is measured by injecting salt water through an electrical conductivity cell at the upstream end and detecting, with a second conductivity cell, the time it takes for it to pass to the other end. In UP equipment, tubes are sized to hold the product the minimum time under the most unfavorable condition, that of true **laminar** flow. In fully developed laminar flow, the maximum fluid velocity (at the tube centerline) is twice the average velocity (Knudsen and Katz, 1958). The more viscous the product, the more closely it approaches laminar flow, but even the most viscous dairy product will not achieve it. The maximum velocity of condensed skim milk is 1.7 times greater than its average velocity (Dickerson et al., 1968).

Design and operation of the equipment must prevent pressure in the pasteurized milk side of the regenerator from falling below 6.9 kPa (1 psi) of the pressure in the raw side of the regenerator at all times, including during shutdown. To accomplish this, a pressure relief valve is located in the line on the pasteurized side of the FDV. After the relief valve and before entrance to the pasteurized side of a regenerator, all milk or milk product must rise at least 12 in (30.5 cm) higher than the highest raw milk or milk product in the system. This provides the pressure differential needed in case of shutdown of the system.

Whereas holding time at pasteurization temperature in the conventional pasteurizer is set by limiting the rate of flow based on size of the holding tube, it is permissible to use magnetic flow meter-based timing systems. Flow through these systems is developed by a combination of flow-promoting devices, including booster and stuffer pumps, separators and clarifiers, homogenizers, and positive displacement pumps. The sanitary magnetic flow meter must be accurate and reliable, which can be demonstrated by its production of six consecutive measurements of holding time within 0.5 sec of each other. Flow recorders are used to continuously produce records of flow rate within the holding tube and an alarm system is used to stop flow and warn operators of shutdown.

26.8.3 Ultrahigh Temperature Processing with Aseptic Packaging (UHT/AP)

Aseptically processed milk and milk products are being packaged in hermetically sealed containers and stored for long periods of time under nonrefrigerated conditions. These conditions favor growth of many types of bacteria including pathogenic, toxin producing, and spoilage organisms. All ultrahigh temperature (UHT) processes are designed to achieve commercial sterility. This calls for destruction of microorganisms in the fluid product with heat plus treatments to sterilize the equipment and packages so any remaining microbial cells are unable to reproduce in the food under normal conditions of storage and distribution.

Subsequent handling, packaging, and storage processes must exclude any opportunity for recontamination of the milk or milk product. The selected process must conform to 21 CFR 113 requirements for low acid canned foods.

When bacteria and/or their spores are exposed to heat treatment, not all are killed at once. Instead they die in a logarithmic fashion. If in a given period of time at a certain temperature 90% are killed, continuing to treat in the same manner for the same time will destroy another 90%, making a total of 99% destroyed. On this basis, the lethal effect of heat can be expressed mathematically as a logarithmic function:

$$K \cdot t = \log N / Nt$$

Where

N = number of microorganisms/spores originally present

Nt = number of microorganisms/spores present after a given time of treatment (t)

K = constant

t = time of treatment

A logarithmic function can never reach zero, which means that sterility, defined as the absence of living bacterial spores in an unlimited volume of product, is impossible to achieve. Therefore, the concept of "sterilizing effect," expressed as the number of decimal reductions achieved in a process, is commonly used. A sterilizing effect of 9 indicates that out of 10^9 bacterial spores in the heated material, only 1 (10^0) survived.

Bacterial spores can survive heat treatments used to make ESL milk products. Several can grow at refrigerator temperatures. In one study, after heating at 261°F (127°C) for 5 sec, 85% of the isolates were of the *Bacillus* genus (Mayr et al., 2004).

Spores of *Bacillus subtilis* or *Geobacillus stearothermophilus* are normally used as test organisms to determine the efficiency of UHT systems because their spores are heat resistant.

The destructive effect of heat on biological materials increases rapidly with increases in temperature. The Q_{10} value is an expression of the increase in speed of reactions and specifies how many times the speed of a reaction increases when the temperature is raised by 10°C. The Q_{10} for changes in flavor is in the order of 2 to 3, meaning that a temperature increase of 10°C doubles or triples the speed of chemical reactions.

The Q_{10} value expected for killing bacterial spores ranges from 8 to 30 depending on sensitivity of the strain exposed to the heat treatment. *D-value*, the decimal reduction time, is the time required to reduce the number of microorganisms to one-tenth the original value, i.e., a reduction of 90%. *Z-value* is the temperature change that gives a tenfold change in the D-value.

High temperature processing of milk and other foods results in complex chemical reactions, causing changes in color (browning), development of off flavors, and formation of sediments. These unwanted reactions are largely avoided through heat treatment at a very high temperature for a very short time. The optimum time/temperature combination must completely inactivate the expected load of spores while limiting damage to the product and, thereby, enhancing consumer acceptance.

Direct heating with steam may be done by injection, i.e., forcing of steam into the milk flow through a nozzle (figure 26.10), or by infusion in which product flows into an atmosphere of steam in a chamber.

A small area is required for a steam injector and the equipment is simple. However, maintenance of stability is difficult. Unless steam is completely condensed (transformed from vapor to water) inside the injector, temperatures will vary in the holding tube. Three design criteria are required: (1) product and steam flows must be isolated from pressure fluctuations inside the injection chamber, (2) pressure on the product in the holding tube must be sufficient to condense steam and keep the heated product in liquid phase,[2] and (3) steam should be practically free of noncondensable gases, which will displace product in the holding tube and impair milk flavor. The latter results from the pronounced effect of heat on colloidal substances that tend to migrate to the **interface** formed by gas bubbles.

Steam injection tends to produce considerable noise, whereas steam infusion is relatively noiseless. Infusers can be operated with relatively lower steam pressures than injectors so there is less temperature difference between steam and product. There is also greater flexibility of temperature and rate of product flow in an infusion-type system. Product generally enters the top of a cylinder (figure 26.11) and is conveyed through a tube to the distributor, which causes it to fall in a fine film. Steam of culinary (food) quality enters near mid-cylinder and heats the falling product. Temperatures up to 302°F (150°C) may be obtained.

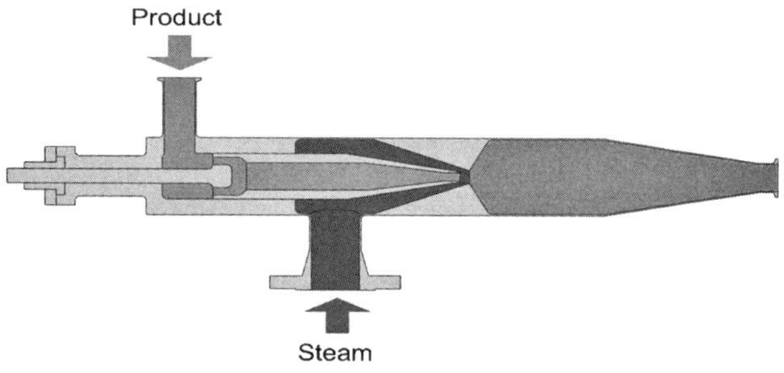

Figure 26.10 Steam injector nozzle by APV, a Division of SPX Corporation, Charlotte, North Carolina.

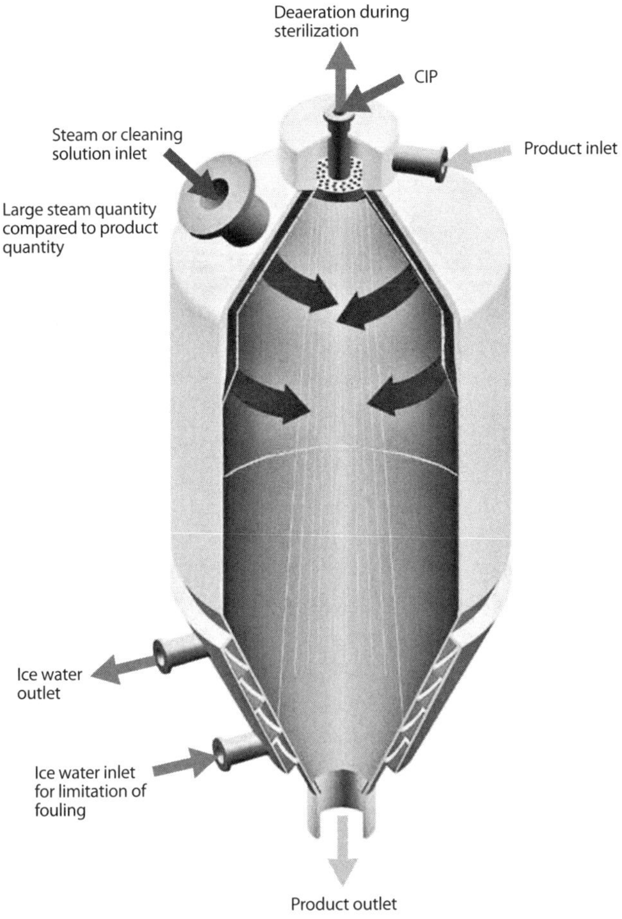

Figure 26.11 Steam infusion heater (from APV, Division of SPX Corporation, Charlotte, North Carolina).

26.8.4 Evaluation of Computerized Systems for Grade "A" Public Health Controls

The following is taken in abstracted form from part VI of appendix H of the *Grade "A" Pasteurized Milk Ordinance* (*PMO*) (US Food and Drug Administration, 2013).

Computer systems are commonly used to manage functions of public health control devices (valves, pumps, others) that operate milk pasteurization systems. These computer systems may be programmed for monitoring and controlling instruments of HTST and HHST pasteurizers. They may also control operational states of devices such as FDDs and booster pumps. Although this technology can furnish numerous advantages throughout the manufacturing process, the public health computer system should essentially replace its hard-wired counterpart. Computer systems differ from hard-wired controls in three major categories, and public health controls must address these differences.

First, unlike conventional hard-wired systems that provide full-time monitoring of public health controls, the computer performs its tasks sequentially, and the computer may be in real-time contact with the FDD for only 1 millisecond (msec). During the next 100 msec, or the time required for the computer to cycle once through its tasks, the FDD remains in forward flow, independent of temperature in the holding tube. Normally, this is not a problem, because most computers can cycle through the steps in their program many times during 1 sec. The problem occurs when (1) the public health computer is directed away from its tasks by another computer; (2) the computer program is changed; or (3) a seldom used JUMP, BRANCH, or GOTO instruction diverts the computer away from its tasks.

Second, in a computerized system, the control logic is easily changed because the computer program is easily changed. Sealing access to the public health computer's programming function can solve this problem.

Finally, for public health controls, the computer program must be made error-free. This is accomplished by keeping that program simple and of limited control scope.

26.9 Vacuumization

Vacuumization is exposure of hot milk or milk product to reduced pressure (vacuum) to remove volatile substances, especially those entering milk from feed. Varying intensities of treatment are necessary to remove volatile components. Some are removed under conditions in which no vaporization of product occurs, others (especially onion and garlic) are removed only by steam-vacuum distillation, and some are not removed. Fat soluble components are especially difficult to remove.

Units that employ vacuum and indirect heat are classified as deaerators and vaporizing units. Deaerators usually operate at temperatures below 140°F (60°C) and with 23 in of vacuum (159 kPa); therefore, there is no product vaporization (figure 26.12 on the next page). There is loss in volume of product (shrinkage) when temperature or vacuum is increased above these levels because vaporization occurs. Vaporizing units remove off flavors more effectively but if vapors are condensed and returned to milk, water soluble substances will be returned as well. Product shrinkage is approximately 1% for each 10°F (5.5°C) temperature drop in the vacuum chamber (expansion chamber).

Because vaporization causes a temperature decrease, vacuum chambers can be used as coolers in steam-vacuum units. These are employed with direct steam heaters, infusors, or injectors. The steam-vacuum unit is commonly placed downstream from (after) the FDV where the temperature exceeds 161°F (72°C). The temperature is instantaneously increased by direct steam heating to as high as 285°F (140°C), and the product is directed into a large chamber in which vacuum is produced. Thin films or small droplets are produced in the chamber so a maximum amount of surface is exposed to vacuum. If the

Figure 26.12 Milk deaerator (from APV, Division of SPX Corporation, Charlotte, North Carolina).

product leaves the steam-vacuum unit at the same temperature it enters, its composition will be essentially the same as when it entered the unit. Only the small amount of heat lost in radiation must be considered in engineering the mechanism (ratio controller) that controls temperature and vacuum, thus preventing dilution or concentration. In situations requiring comparatively high nonfat milk solids (NMS) in milk it may be desirable to concentrate milk slightly in this process. Whereas the federal minimum NMS in fluid milks is 8.25%, the state of California requires the following NMS minimums: milk, 8.7%; reduced fat milk, 10%; lowfat milk, 11%; and nonfat milk, 9%.

Steam used in direct heating must be of culinary quality. Strict limitations are placed on types of compounds permitted for use in controlling scaling in boilers producing steam. When a limited quantity of steam is required, it can be produced by indirectly heating potable water in a heat exchanger with steam from the plant supply. The heat transfer is made through a coil.

With direct steam heating processes, holding time is reduced because the milk or milk product volume increases as steam condenses to water during heating in the injector. This surplus water is evaporated as the pasteurized product is cooled in the vacuum chamber. For example, with a 120°F (66°C) increase by steam injection, which is probably the maximum temperature rise that will be used, a volume increase of 12% will occur in the holding tube. The measurement of the average flow rate, at the discharge of the pasteurizer, does not reflect this volume increase in the holding tube. Therefore, this volume increase (i.e., holding time decrease) must be considered in the calculations.

26.10 Homogenization

Fat globules of milk range in diameter from about 0.1 to 16 μm, and approximately 80 to 90% are from 2 to 6 μm. The number in milk has been estimated as 1.5 to 5 × 10^9/mL. The light weight of fat globules in comparison with milk serum causes them to rise; the larger the globule the faster the rate of separation. Individual globules rise at velocity rates proportional to the square of their radii (following Stokes' law). Sommer (1952) calculated the velocity of rise for globules of 1, 5, and 10 μm diameter to be 0.08, 2.09, and 8.34 mm/hr, respectively (based on a temperature of 50°F [10°C], milk serum density of 1.036, viscosity of 0.0247 poise, and fat density of 0.930). Homogenization disrupts fat globules so 98% or more are 2 μm or less in diameter. Globules of this size rise little because of forces of fluid friction and electrical charges at their surfaces.

After 48 hr of quiescent storage at 45°F (7°C), **homogenized milk** must have no visible cream separation, and the fat percentage of the top 100 mL in a quart of milk (or of proportionate volumes in containers of other sizes) shall not differ more than 10% from the fat percentage of the remaining milk as determined after thorough mixing.

Milk homogenizers are usually piston-type pumps that force product through one or two specially designed valves (1-stage or 2-stage homogenizers). The area through which the product flows is restricted sufficiently to produce pressures of 12.4 to 17.2 mPa (1,800 to 2,500 psi). Pressures decrease precipitously downstream from the valve (figure 26.13). These pressures, the high velocities, the turbulence they create, and their impact in the valve disrupt fat globules.

Physical state and concentration of the fat phase at the time of homogenization are major determinants of size and dispersion of fat globules formed. Cold milk, in which fat is mostly solidified, cannot be effectively homogenized. Fat must be in liquid form at the time of homogenization (140°F [60°C] or more preferred), and homogenized milk must be pasteurized to prevent accelerated activity of lipase. Because ultrapasteurized milk may stay on shelves for months, homogenization pressures of 24.13 mPa (3,500 psi) may be required.

When the small globules formed during homogenization coalesce, viscosity of the product increases, especially in high-fat products such as cream and ice cream mix. Dispersion of the lipid phase increases with increas-

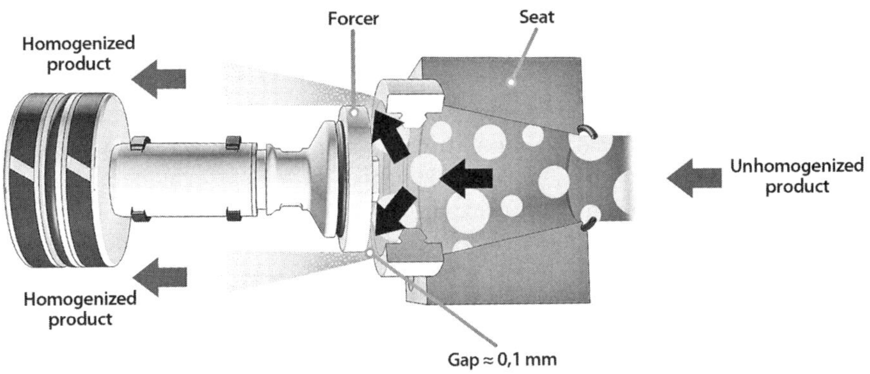

Figure 26.13 Drawing of a typical homogenizer valve showing fat globules being reduced in size (courtesy of Tetra Pak, *Dairy Processing Handbook*).

ing temperatures of homogenization because viscosity decreases with increasing temperatures.

Clumping is promoted by homogenization at a high pressure and relatively low temperature, especially in products having high fat content in which clusters are formed as a result of insufficient membrane material (casein). Satisfactory homogenization requires a casein to fat ratio of approximately 0.2:1. Passage of product through a second homogenizing valve at comparatively low pressure (3.45 to 6.9 mPa [500 to 1,000 psi]) disrupts the clumps (figure 26.14). This is called *2-stage* homogenization.

Homogenized milk tends to have a richer flavor than untreated milk because of increased number and surface area of fat globules. For example, a globule 5 µm in diameter will yield 125 globules 1 µm in diameter. As a result, the surface area of these globules will be five times that of the original globule. Also, homogenization increases the digestibility of milk by decreasing its **curd** tension so a softer, more fragile curd is formed in the stomach. The physical condition of protein is modified by homogenization so coagulation by heat and acid occur more readily. The lowered curd tension and protein stability shorten the time required for cooking homemade custards and increase the probability that milk may curdle in such foods as soups and gravies. Homogenization makes milk more susceptible to the off flavor induced by sunlight but less susceptible to the copper-induced oxidized flavor.

It is possible to increase the capacity of a processing operation if homogenizer capacity is the limiting factor. This is accomplished by separating cream from milk and homogenizing it before it is remixed with the skim milk portion.

It is vital that milk be pasteurized before or immediately following homogenization to inactivate native milk lipase. Homogenization disrupts the protective milkfat globule membrane, permitting access of lipase to its lipid substrate.

26.11 Cooling and Refrigeration

Heat always passes from a warm body to a cooler one,
the exchange being capable of continuing
until the two bodies are at the same temperature.
A. W. Farrall (1899–1986)

Development of the fluid milk industry depended directly on availability and use of refrigeration by producers, processors, distributors, retailers, and consumers. Home refrigeration is especially important because it allows purchase of pasteurized milk on a weekly basis. The excellent rate of cooling and low temperature of storage afforded by bulk milk tanks allows economical production because high-quality raw milk can be preserved adequately to be picked up at farms every other day.

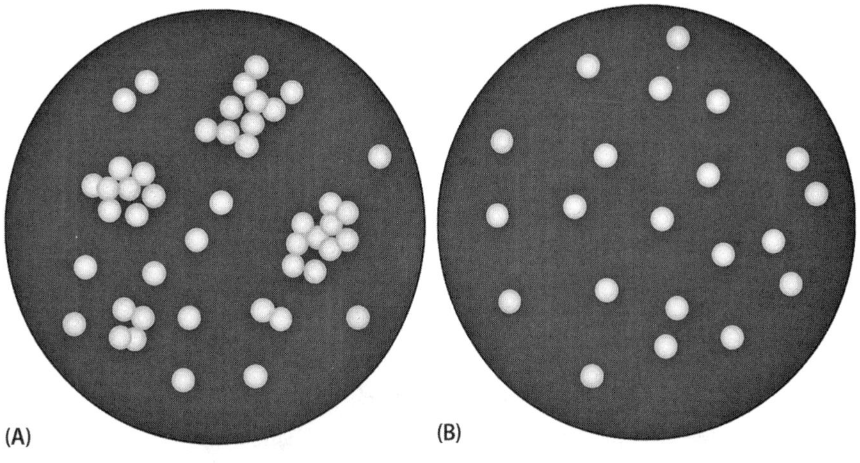

Figure 26.14 (A) Clumps of fat globules produced in the first stage of homogenization are disrupted by (B) second stage treatment (courtesy of Tetra Pak, *Dairy Processing Handbook*).

Every day pickup is usually necessary when milk is cooled in cans because of less efficiency in cooling and a generally lower level of sanitation/hygiene in production.

Low temperatures reduce the rate of deterioration of milk quality because rates of chemical reactions are directly dependent on temperature. This principle applies to both purely chemical and biochemical reactions, the latter being associated with biological or living matter. Enzyme-catalyzed reactions, such as hydrolysis of protein and fat, are particularly sensitive to temperature. Bacteria obtain energy for growth and metabolism from temperature-dependent biochemical reactions. Enzymes of some microorganisms require temperatures near 98.6°F (37°C) to provide optimum rates of reactions—these microorganisms grow slowly at 77°F (25°C)—but enzymes of psychrotrophic types function well at 45 to 59°F (7 to 15°C). Minimal growth of bacteria occurs at 32 to 39°F (0 to 4°C). However, hydrolytic enzymes secreted from bacteria (extracellular) often continue to function at these low temperatures.

Regardless of size (home refrigerator or industrial ice builder), refrigeration systems operate on the same principle (figure 26.15). A condensable gas is (1) compressed mechanically (compressor), (2) directed through a condenser to remove heat and change it to a liquid, (3) passed through a liquid control (expansion) valve into a receiving tank, and (4) piped into an evaporator where it accumulates heat (latent heat of evaporization). As it boils and becomes a gas, refrigerant is conveyed back into the compressor and the cycle is repeated. Heat required to change the liquid to a gas is taken from the evaporator and its surroundings. The evaporator, then, is the refrigerating element of the system. Heat is removed from one material at a low temperature level and transferred to another at a higher temperature. For example, from raw milk at 95°F (35°C) heat is transferred first to refrigerant then to air at an **ambient** temperature in the bulk-cooling operation.

Ammonia is the major refrigerant (gas) in large systems but non-halogenated hydrocarbons are usually contained in bulk milk tanks, store display cases, and home refrigerators. The capacity of a system depends on the quantity of refrigerant it can evaporate, compress, and condense. Components of a system must be balanced in size to achieve efficient operation. Ammonia boils at –28°F (–33.3°C) at atmospheric pressure.

The ultimate purpose of any refrigeration system is to cool or freeze dairy products. When the cooling system is of the *direct* type, the cooling medium is the refrigerant and the evaporator is the heat exchanger. Most refrigerated (cold wall) milk tanks are so designed. Home freezers and refrigerators are also of this direct expansion type. If the refrigerant cools water, glycol, or salt brine and this in turn cools milk or product, the system is an *indirect* type. Indirect cooling is applied in plate heat exchangers and in some farm bulk tanks.

Water that is reused in a closed water cooling system is called *sweet water.* If cooling to be done by sweet water

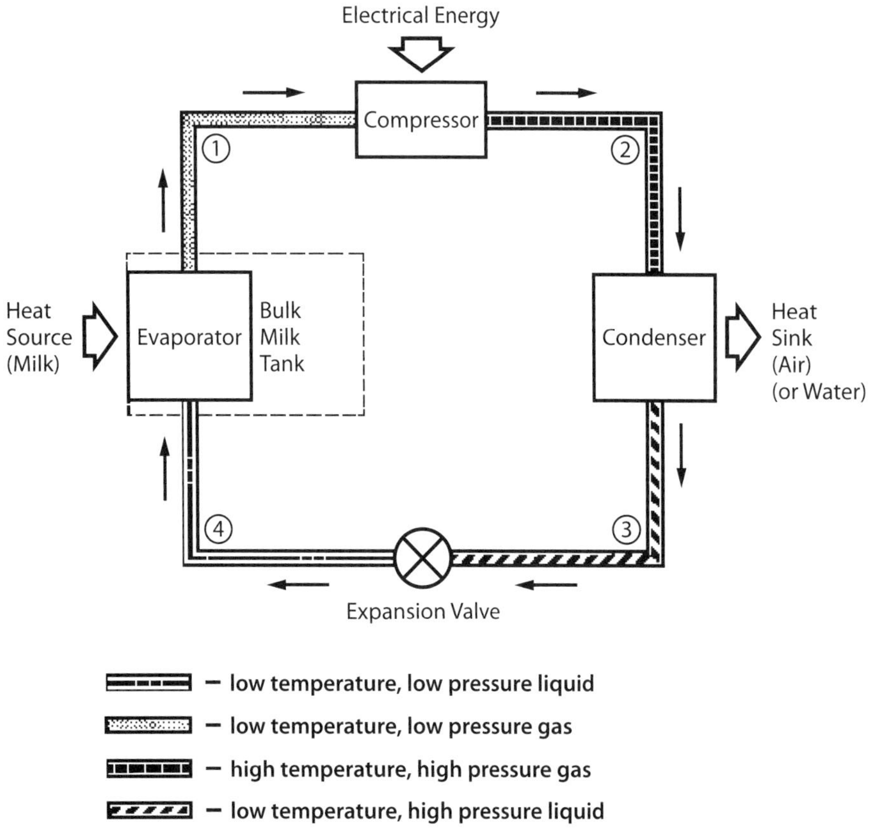

Figure 26.15 Schematic diagram of a simple refrigeration system (courtesy of Ontario Ministry of Agriculture, Food and Rural Affairs).

is considerable and of short duration, an ice storage tank is useful. Associated refrigeration equipment can be relatively small because it builds (accumulates) ice over a long period of time to be used in a short period.

If desired temperatures are lower than can be obtained with water as the cooling medium, brine may be used. *Brine* is any liquid cooled by a refrigerant and used for transmission of heat without a change of state. In common terminology, brine is considered to be a mixture of water and soluble material, usually sodium chloride or calcium chloride, that will lower the freezing point. Both of these salts are corrosive, especially at a low pH. The most acceptable brine in dairy plants is propylene glycol. It is widely used in cooling fluid milk to 34 to 36°F (1 to 2°C) immediately before packaging because cooling to this temperature range with sweet water is impractical.

Heat is transferred from product through container walls to air then to refrigerant much more slowly than from product to refrigerant through metal walls of heat exchangers. Therefore, thorough chilling of milk before packaging reduces probabilities of microbial growth in packages in cold storage rooms because cooling is accomplished earlier and faster.

Freezers for ice cream are swept surface heat exchangers. The evaporator is a tube through which product flows. A thin film of product freezes on the wall and is swept off by scraper blades. In some ice cream hardening chambers, air is chilled when it is blown over pipes in the evaporator. It passes over filled containers, removing heat, and is recirculated to the evaporator. In other systems cartons are placed in direct contact with refrigerated surfaces and heat passes from product to the colder surface. Desserts are often kept frozen with dry ice, which is solid carbon dioxide (CO_2), and truck bodies are often quick-chilled by dispensing compressed CO_2 gas in them. Carbon dioxide boils at –108.4°F (–78°C).

Materials that are poor conductors of heat (insulators) are used to maintain low temperatures in refrigerated equipment and cold rooms. Thus, rapid absorption of heat from the environment is prevented and costs are minimized. Corkboard, Styrofoam, and foam glass are examples of insulation used commonly. These materials contain a myriad of individual air cells connected to each other by only a thin membrane of solid material. Each cell forms a minute dead air space and air that is not moving is one of the best heat insulators.

26.12 Clarifixation

A clarifixator is a continuous centrifuge that clarifies *and* homogenizes milk. The bowl operates as an ordinary separator, but separated cream flows to a cream paring chamber from which it is pumped by the stationary cream paring disc (homogenization paring disc). This disc partially disintegrates fat globules. Cream is remixed with entering milk but only large globules are separated; small ones remain in milk and leave the centrifuge. Only when fat globules have become sufficiently small do they leave with the milk.

The small space needed, its dual performance as both clarifier and homogenizer, and low investment costs are advantages of the clarifixator. It is probable that cavitation (the rapid formation and subsequent collapse of bubbles in a liquid) causes breakup of fat globules in the machine. Surface energy released upon collapse of the bubbles is high and is apparently responsible for globule breakup.

The clarifixator can be used also to produce anhydrous milkfat. In this application the disc breaks down the emulsion of fat so that liquid leaving the machine is in a continuous oil phase in which buttermilk is dispersed.

26.13 Bactofugation

The process of centrifugal removal of bacteria and other microorganisms from milk is called *bactofugation* (Simonart et al., 1962). The objective is removal of bacteria to avoid the effects of heat on functional properties of milk, including destruction of certain enzymes that assist in ripening cheese. However, at its best bactofugation removes only about 99% of the bacteria in milk. Thus, it cannot be relied upon to remove all pathogenic bacteria. It probably has no significant effect on viruses or *Rickettsia* (which is intermediate in size between bacteria and viruses).

In a typical installation, two continuous centrifuges are set in a series and are fed with milk preheated to 129°F (54°C). The force exerted on milk is 9,000 times the force of gravity. Therefore, the relatively heavy microorganisms are thrown to the outside of the centrifuge bowl along with some casein. About 2.5% of the milk volume is removed in the form of sludge. This results in significant yield loss in cheesemaking. It is possible to pasteurize this sludge, however, and reincorporate it with milk.

26.14 Piping and Pumping Systems

26.14.1 Piping and Fittings

Milk and milk products are transferred via sanitary milk lines (pipes), usually constructed of stainless steel or glass. Energy in the form of motion is added to force milk through lines. Resistance to flow, called *head*, is provided by gravity and friction. The head caused by gravity—*static head*—equals the vertical distance (ft/m) the fluid is above a reference level. *Friction head* is the resistance to the flow of water through pipes and fittings expressed in ft/m. Friction head increases with increased flow rate but decreases with increased pipe size. In the table on the next page we have provided an example of the flow rates and friction caused by 100 ft of straight pipe at various diameters.

Pipe Diameter (in)	Flow Rate (gal/min)	Friction Head (ft)
1	20	28.3
1	25	43.7
2	20	0.97
2	25	1.43
3	20	0.138
3	25	0.231

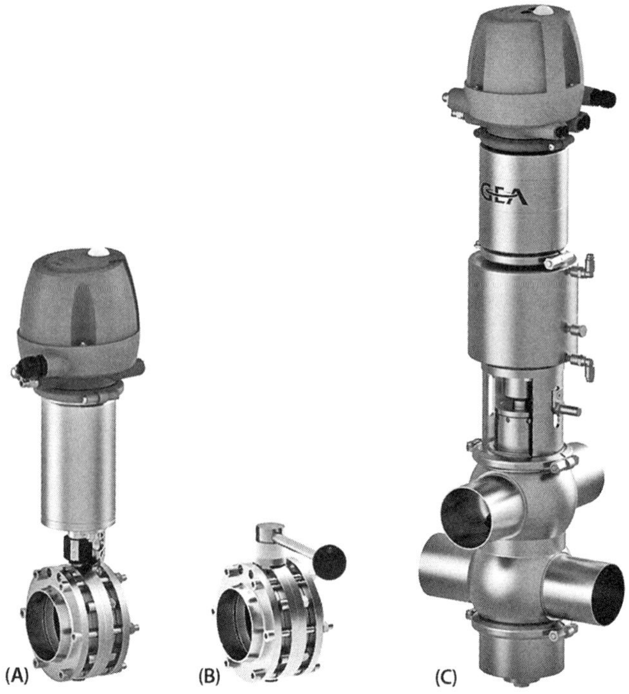

Figure 26.16 Examples of valves used in milk pipelines. (A) Remotely operated butterfly valve. (B) Hand-operated butterfly valve. (C) Remotely operated mix-proof valve through which two different fluids can flow simultaneously with no mixing (courtesy of GEA Tuchenhagen North America).

Elbows and other fittings produce more friction head than an equal length of straight pipe. For example, a 2-in elbow produces friction head equivalent to 7 ft of 2-in pipe.

The size of piping to be installed depends on the quantity of product to be pumped through it in a given time. Therefore, fittings on processing equipment are sized to accommodate pipes that will convey adequate product at acceptable pressure. Pipelines are easier to clean and keep in good repair when joints are welded. Welds are ground and polished to provide a smooth surface, which facilitates effective cleaning.

To effectively clean piping systems, detergent solutions should flow turbulently. The degree of turbulence that is needed to be effective is characterized by the Reynolds number (Re), which is calculated by the following formula: Re = (pipe diameter × velocity) (fluid density/fluid viscosity). Thus, Reynolds number is an expression of the ratio of the inertia force to the viscous force on an element of fluid. When Re is less than 2,000, laminar flow exists; values of 4,000 or more indicate turbulent flow.

The dairy industry was the first to introduce sanitary fittings and pioneered the practice of standardizing guidelines and practices to be used within an industry in the 1920s. The following list is representative of the current 3-A Sanitary Standards for piping and fittings:

1. Easily accessible and readily cleanable. Removable parts are readily demountable.
2. Product contact surfaces are self-draining.
3. No pipe or fitting threads are exposed to the product.
4. Permanent joints are welded on metallic surfaces that contact product.
5. Threads are wide, open, and easy to clean.
6. Interior surfaces are smooth and even, especially at joints, thus eliminating crevices and projections where milk residues might lodge and support bacterial growth.
7. Gaskets are eliminated where possible.
8. Gaskets are smooth, nontoxic, and product-resistant.

26.14.2 Valves

Valves are important components of piping systems. In some small systems it may be possible to use only hand-operated butterfly-type valves (figure 26.16). However, in most systems valves are opened electronically from remote locations, often by programmed or situation-initiated devices. Development of the double-seated mix-proof valve has made it possible to simultaneously direct both product and cleaning fluids through two channels in the same valve while protecting against cross-contamination.

Development of the air-actuated valve (figure 26.17) has made automation of milk processing possible. Such valves, as well as motors, can be operated from remote control centers and can be opened or closed at predetermined intervals by timers or other controllers.

26.14.3 Pumps

Pumps commonly used in the dairy industry are of the centrifugal (nonpositive) and positive displacement types.

In centrifugal pumps (figure 26.18), a rotating impeller (lobe) operates at relatively high speeds producing nonpulsating pressure. The impeller is usually connected directly to the shaft of an electric motor, thus providing simple construction. Excessive pressures will not build if flow is restricted or stopped downstream from the pump; hence variations in flow rate are obtained by means of a throttling valve in the discharge line. However, some pumps are not self-priming and air leaks on the suction side cause problems, especially if centrifugal pumps are used to move milk to packaging equipment or through high-temperature heat exchangers. In the former, entrained air may produce erratic fill of containers and in the latter, milk solids are likely to adhere and burn onto heated surfaces.

Positive displacement pumps operate at high pressures without significant loss of capacity and do not need priming. There are three types: piston, gear, and lobe (or impeller). Homogenizers are piston-type pumps that produce very high pressures. However, they produce pulsating flow that is undesirable in some applications. Gear pumps have intermeshing teeth and exert pressures on product as gears are rotated. Gear pumps are commonly used on freezers for frozen desserts. The most commonly used positive pump is the lobe type in which impellers are usually dynamically balanced and do not mesh (figure 26.19 on the following page). One impeller rotates clockwise, the other counterclockwise. For moving cheese curd only one large lobe is attached to each impeller shaft, not dynamically balanced. This allows a large space into which cheese can move, minimizing curd breakup. Both steel and rubber rotors are available for various pumping applications.

The capacity of positive pumps is determined mainly by speed. Variable speed drives are attached when the application necessitates a change in the flow rate. A major application is in timing pumps, which control flow rate through pasteurizers. Once the flow rate is set, the health officer places a seal on the rate control mechanism to prevent adjustment to a faster rate of pumping. The pump can be slowed, however.

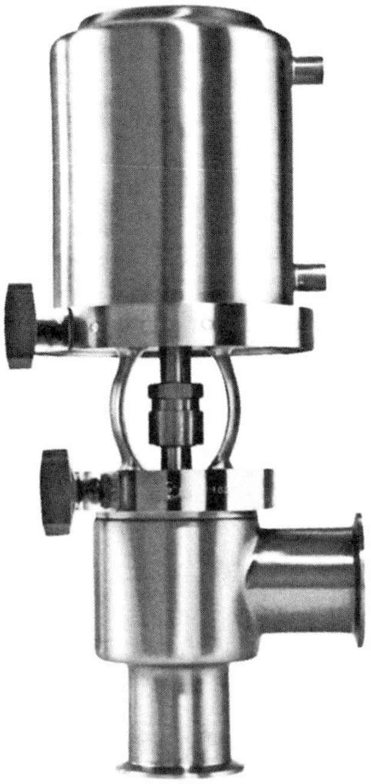

Figure 26.17 Basic parts of an air-actuated sanitary valve: valve body (bottom), valve stem (middle), and air actuator (top) (courtesy of Waukesha Cherry Burrell).

26.15 Compressed Air

Compressed air is used in a broad range of applications in milk processing: mixing ingredients, cutting, sparging, drying product, transporting/propelling product through processing systems, and packaging. In many applications compressed air contacts the product, potentially contaminating it.

Compressed air contains water vapor, particulate matter, oil vapor and droplets, and microorganisms. After compression raises its temperature, air is cooled, thus condensing water vapor into aerosols and droplets. Water in the compressed air can produce rust and corrosion in piping, which may flake off and be carried downstream,

(A)

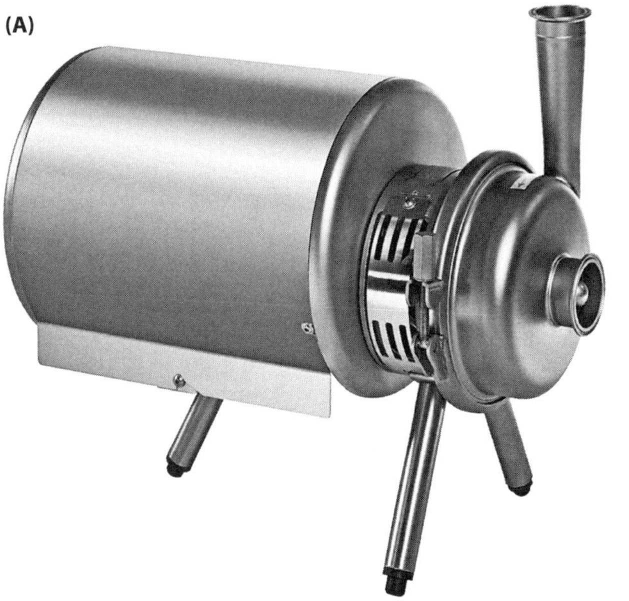

(B)

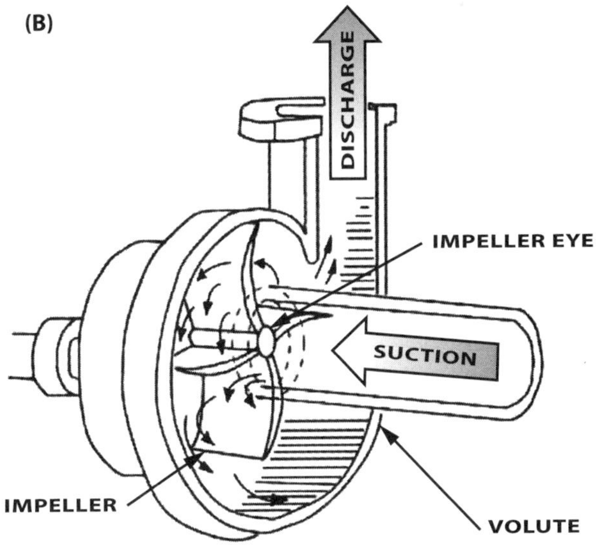

Figure 26.18 (A) Self-priming centrifugal-type milk pump (from APV, Division of SPX Corporation, Charlotte, North Carolina). (B) Product enters through the center pipe and exits through the vertical pipe.

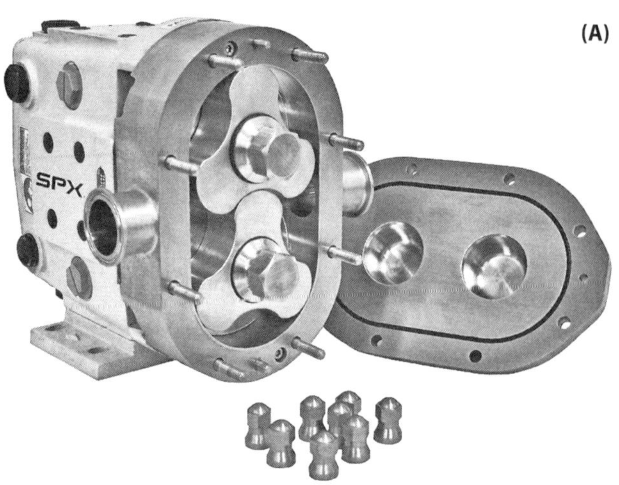

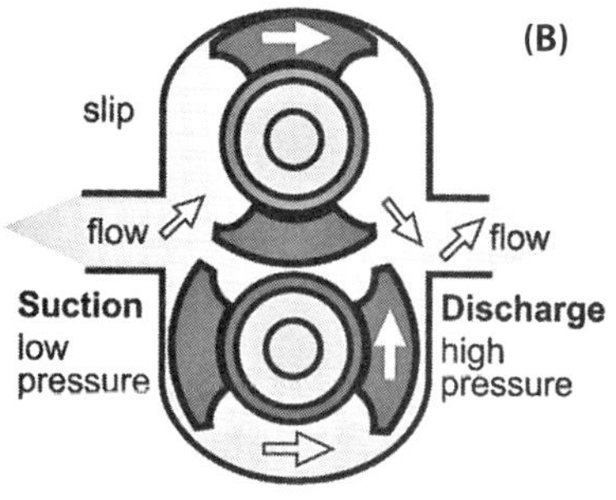

Figure 26.19
(A) Positive displacement pump showing three lobes per shaft for pumping liquids (from Waukesha Cherry Burrell) and (B) two lobes per shaft for pumping particles.

potentially contaminating the product. In addition, water condensate and warm compressed air provide a favorable environment for microbial growth.

The *PMO* requires that air compressing equipment be designed to preclude contamination of the air with lubricant vapors and fumes and describes the type of equipment permitted. Air under pressure in excess of 1 bar (i.e., 103.5 kPa [15 psi]) must have the moisture removed. Removal may be achieved by condensation and coalescing filtration or absorption or equivalent. When necessary to cool the compressed air, an after-cooler shall be installed between the compressor and the air storage tank.

As Hazard Analysis and Critical Control Points analysis is pursued, every location in which compressed air is used becomes a critical control point. Standard ISO 8573-1:2010 is commonly used in conjunction with The Code of Practices jointly developed by the British Compressed Air Society and the British Retail Consortium Trading, Ltd. Section 6 of this code defines three categories of compressed air used in the food industry based on their contact with food: (1) direct, (2) indirect, and (3) noncontact—no risk. Compressed air dryers should achieve a –40°F (–40°C) dew point specification and be monitored for performance. A two-stage filter system should be installed near the point of compressed air use. A typical system consists of a first stage, high efficiency coalescing filter with a rating of at least 99.99% at 0.01 μm, and a second stage sterile air filter with an efficiency rating of at least 99.9999% at 0.01 μm.

26.16 Automation

Hall and Trout (1968) define automation as automatic control of mechanical operations whereby output variables are sensed to control input variables. For example, a sensor of electrical conductivity or resistivity can be used to control the quantity of detergent added to a cleaning solution, and a sensor of temperature can be used to control the amount of heat applied. A process must be mechanized to be automated. The usual benefits of automation are increased productivity per worker-hour and less variation in processing performance. However, investment costs are commonly higher and increased skill by labor is required in operation and maintenance of such equipment.

26.16.1 Processing Systems

Units in an automated processing system are classified as *operating* or *controlling* types. A controlling unit energizes an operating unit through one or more electrical components. Typical operating units in a milk-processing system are clarifier, heat exchanger, homogenizer, centrifugal pumps, timing pumps, and air-operated valves. Some processing units are controlled mechanically using microswitches that open and close electric circuits.

Today's dairy industry makes wide use of computers to receive data from multiple sources to control production processes. For example, a digital in-line blending system uses an algorithm-based control scheme that integrates the demand and measured values of components and subcomponents of a fluid milk product within an ingredient stream. Using instruments such as flow meters to measure process parameters, it manipulates the component flows that have a significant effect on each other. The system then couples these measurements with the floating-point math capabilities of programmable logic controllers to execute algorithms to produce the desired composition of milk products.

26.16.2 Switches

There are many components that comprise a dairy processing system; in order for the system to function effectively, each component must be monitored and controlled to ensure the quality and safety of the product that is produced. For example, pressure on a processing line is monitored by a pressure switch that can detect a decrease or increase in pressure or an unacceptable difference in pressure such as might occur in certain types of pasteurizers. Temperature switches are important in the control of heat exchangers. Temperature changes of a medium are monitored by a sensor and are transmitted to a controller (figure 26.20). The controller, in turn, may energize or de-energize a unit and/or move a needle on a recorder indicating a change in temperature.

A conductivity switch is used in a controlling unit to detect the presence of a liquid (such as to detect the presence of water in a tank) or variations in the conductivity of a liquid (such as to determine the amount of detergent in a cleaning solution). In the example shown (figure 26.21), two sensing probes are immersed in a solution of detergent. A controller sends a low-voltage signal through one probe and detects the signal through another, providing there is sufficient detergent in the solution to produce adequate conductivity. If electrical current does not pass, the controller will correct the situation, for example, by opening a valve that permits detergent to flow into the tank.

(A) (B)

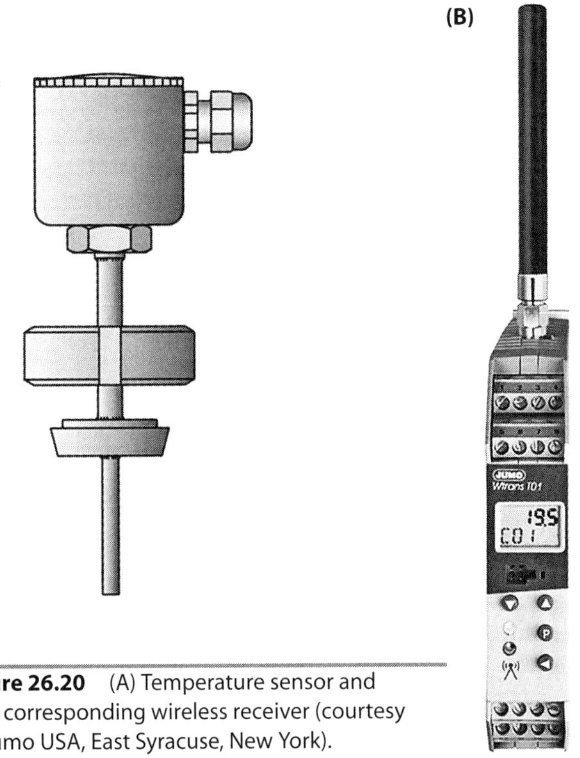

Figure 26.20 (A) Temperature sensor and (B) a corresponding wireless receiver (courtesy of Jumo USA, East Syracuse, New York).

A load cell is used in a controlling unit to monitor the weight of an object, such as a milk storage tank. A controller sends a low-voltage signal through a thin metal strip attached to the load cell. Voltage passing through the strip varies in proportion to weight applied to the cell. The controller responds to change in voltage by sending a signal, which may be fed into a readout device and/or a controlling unit, for example, the two solenoid-operated valves (A and B) that control the air-operated valves (C and D) in figure 26.22 on the next page.

26.16.3 Automated Management Systems

Some milk processors make products with little or no process automation, however, they do need automated systems to manage inventories, supply chains, and customers because of complex product mix and distribution variables. In addition, management and labor efficiency decisions can benefit from such systems. Enterprise resource planning (ERP) involves the integration and automation of key portions of a firm's business. For a manufacturer, this means the application of computer software to electronically store "farm-to-fork" information so that a given lot of product can be traced to all customers who received material produced from that particular lot; interactively calculate online customer pric-

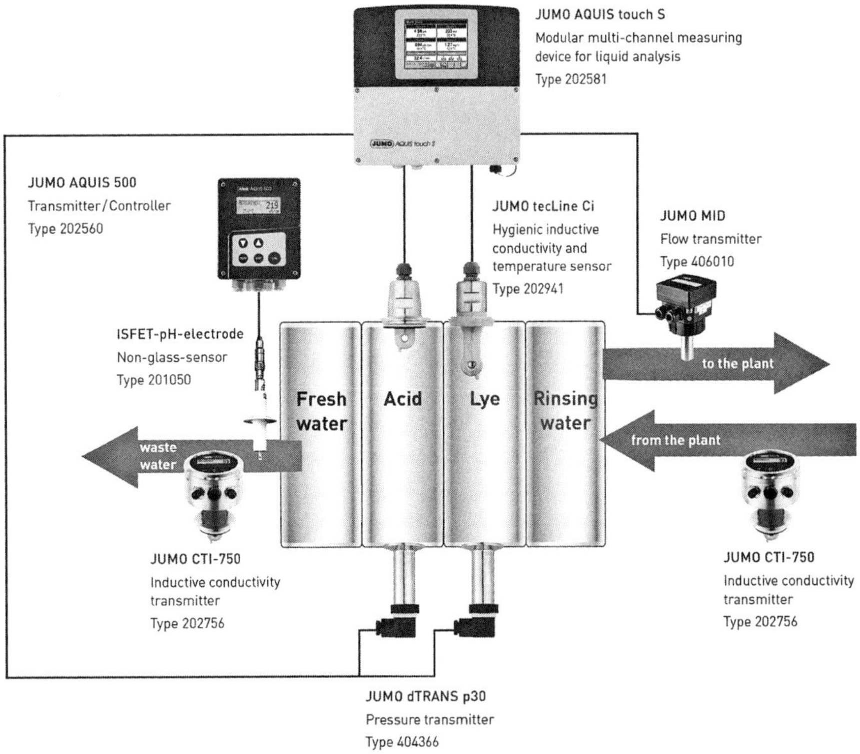

Figure 26.21 Conductivity probe and transmitter (in the two central tanks and the lines entering and leaving the station) used to control solute concentrations in a CIP cleaning system (courtesy of Jumo USA, East Syracuse, New York).

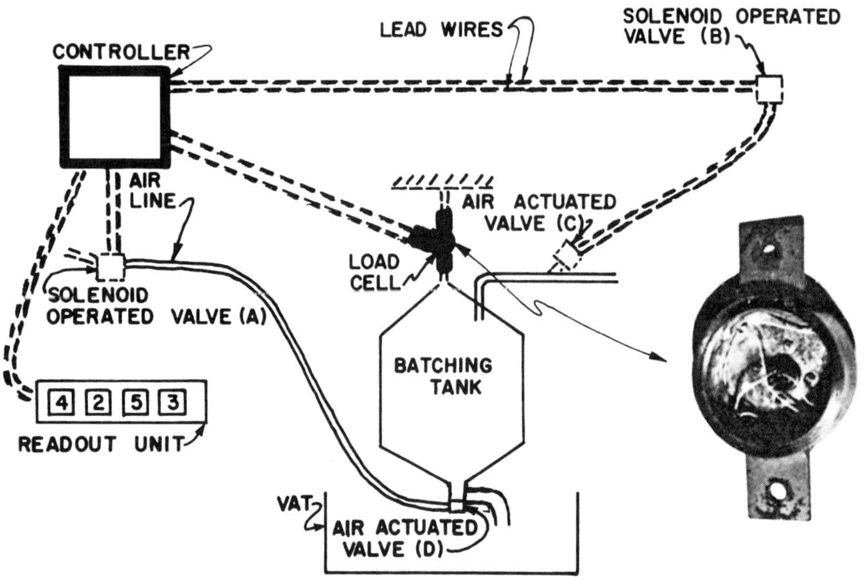

Figure 26.22 Load cell attached to a batching tank. The controller fills or empties the tank based on output of voltage from the load cell (courtesy of Missouri Agricultural Experiment Station).

ing based upon market price and customer pricing factors; and automatically generate customer-specific labels, which include not only customer logos and barcodes but also the retail customer's price. From inventory control to purchasing, an ERP system can perform key functions, including order management, manufacturing, general ledger, accounts payable, and accounts receivable.

Functions involved in software-based manufacturing execution systems (MES) include data acquisition/collection, process management, quality management, maintenance management, performance analysis, document control, product tracking and genealogy, and scheduling. Manufacturing, inventory, and plant scheduling need to work with the supply chain in real-time to yield the visibility required for a reliable supply chain.

Computer-centered wired and wireless technologies provide real-time data that facilitate meeting production goals. Data can be transferred between machines or among them in networks. Characteristically machines may be made to communicate with each other to maintain or enhance efficiencies related to the following: logic controller, visualization, input/output sensors, motion systems, and the safety system. Machinery communication technologies are moving toward Ethernet-based protocols.

26.17 Summary

The bulk milk hauler performs several vital functions in sampling, collecting, and transporting raw milk to the processing plant. Once there, the milk is sampled again, examined, and the quantity determined.

Processing operations include storage of raw milk, clarification, filtration, standardization, separation, pasteurization, vacuumization, homogenization, membrane processing, packaging, and refrigeration. Automation of these processes has developed rapidly in recent years. Air-operated valves allow for remote control of product flow. Various sensor elements and their electrical controls allow for processes to be monitored and adjusted remotely. Thus, centralized control is possible and efficiencies of operation are derived.

Each process to which milk is subjected increases cost. However, consumers desire and deserve milk and milk products of high hygienic quality, consistent and homogenous in composition, free of harmful microorganisms and chemical substances, free of off flavors and odors, and characteristic of the ideal product they seek. Therefore, processors must understand the functions, effects, and limitations of each process and must make proper choices in their management of milk-processing operations.

Study Questions

1. Why is the bulk milk hauler an important link in assuring high-quality milk for humans? What quality tests does he/she perform?
2. Identify possible causes of a high reading on the bulk milk tank quantity gauge.
3. When does raw milk become the property of the buyer?
4. What device is commonly used to clean bulk milk tank trucks?
5. What is a silo-type tank? How is it cleaned?
6. What is clarification? How is it achieved? Of what value is it in assuring high-quality milk?
7. What is filtration of milk? How does it differ from clarification?
8. What is standardization of milk? Of what significance is it? How is it accomplished?
9. Upon what principle are milk separators designed?

10. Using the Pearson square technique, determine the amount (lb) of skim milk (assume 0.0% milkfat) needed to standardize 10,000 lb of 4.0% fat milk to yield a lowfat milk of 1.5% milkfat.
11. Give two reasons for the need to pasteurize dairy products.
12. Differentiate between vat pasteurization and HTST pasteurization.
13. Discuss briefly how boiling of milk may affect its flavor and nutritional properties.
14. What is HHST? UHT?
15. Of what value is regeneration in milk processing?
16. Identify four important factors in achieving efficient heat exchange in milk processing.
17. Of what importance is a flow diversion value in the protection of human health?
18. Differentiate between steam injection and steam infusion.
19. What is vacuumization of milk? How is it accomplished? Does it affect product volume?
20. What is meant by culinary-quality steam?
21. Why and how is milk homogenized?
22. How does Stokes' law relate to the rise of fat globules in milk?
23. How does homogenization affect the flavor, curd tension, and digestibility of milk?
24. Discuss the significance of refrigeration to the dairy industry.
25. Differentiate between direct and indirect types of cooling media.
26. What is sweet water?
27. What is clarifixation? What are its advantages?
28. Upon what principle does bactofugation depend? Why might its use be desirable in processing milk intended for use in the production of cheese?
29. Differentiate between static head and friction head.
30. Describe the 3-A Standards for the design of piping and fittings used in milk equipment.
31. Differentiate between positive and nonpositive types of milk pumps. Identify factors that influence their capacity.
32. Of what significance are air-actuated valves in an automated milk-processing plant?
33. Cite one or more practical uses of a load cell in a milk plant.

NOTES

[1] A regenerative plate-type HTST pasteurizer was introduced in England in 1923 and in the United States in 1929. Standards for HTST pasteurization were first published in the US Public Health Service Milk Ordinance and Code in 1933.

[2] Vaporization of product in the holding tube reduces residence time of the product and may limit microbial destruction. Pressure on the holding tube must be at least 69 kPa (10 psi [0.69 bar]) above that at which the product boils. A pint of water flashed to steam can occupy as much as 100 cu ft.

REFERENCES

Cheryan, M. 1998. *Ultrafiltration and Microfiltration Handbook.* Lancaster, PA: Technomic Publishing.

Dickerson, R. W., Jr., A. M. Scalzo, R. B. Read Jr., and R. W. Parker. 1968. Residence time of milk products in holding tubes of high-temperature, short-time pasteurizers. *Journal of Dairy Science* 51:1731–1736.

Hall, C. W., and G. M. Trout. 1968. *Milk Pasteurization.* Westport, CT: AVI Publishing.

Knudsen, J. G., and D. L. Katz. 1958. *Fluid Dynamics and Heat Transfer.* New York: McGraw-Hill.

Mayr, R., K. Gutser, M. Busse, and H. Seiler. 2004. Indigenous aerobic sporeformers in high heat-treated (127°C, 5s) German ESL (extended shelf life) milk. *Milchwissenschaft* 59:143–146.

National Institute of Standards and Technology. 2015. *Handbook 44: Specifications, Tolerances, and Other Technical Requirements for Weighing and Measuring Devices.* Gaithersburg, MD: Author.

Rosenau, M. J. 1909. *The Thermal Death Points of Pathogenic Microorganisms in Milk* (Hygienic Laboratory Bulletin 56). Washington, DC: US Public Health and Marine Hospital Service.

Simonart, P., R. Poffé, and M. Weckx, 1962. Bacterial supercentrifugation and bacterial flora of milk. *Netherlands Journal of Dairy Science* 16:81–88.

Sommer, H. H. 1952. *Market Milk and Related Products* (3rd ed.). Milwaukee, WI: Olsen Publishing.

US Food and Drug Administration. 2013. *Grade "A" Pasteurized Milk Ordinance.* Washington, DC: Author.

WEBSITES

Au2mate (http://www.au2mate.com)

California Department of Food and Agriculture, California Milk Standards (http://www.cdfa.ca.gov/ahfss/Milk_and_Dairy_Food_Safety/Milk_Standards.html)

GEA Farm Technologies (http://www.gea-farmtechnologies.com)

National Institute of Standards and Technology (http://www.nist.gov)

3-A Sanitary Standards, Inc. (http://www.3-a.org)

University of Florida, IFAS Extension, Sanitary Design and Construction of Food Equipment (http://edis.ifas.ufl.edu/fs119)

US Food and Drug Administration, CFR Title 21 (http://www.accessdata.fda.gov/scripts/cdrh/cfdocs/cfCFR/CFRSearch.cfm?CFRPart=113)

27

Cleaning and Sanitizing Dairy Equipment and Containers

There is no single factor as important in the production of fine-quality dairy products as absolute and complete cleanliness.

R. B. Barrett, Research Chemist

27.1 Introduction

It is not enough to provide milk for consumers; it must be high-quality milk. It would be impossible to provide milk and milk products of superior quality if we were unable to remove **milk soil**[1] and to kill residual microorganisms. These two functions, respectively, are accomplished through the processes of cleaning and sanitizing. The *Grade "A" Pasteurized Milk Ordinance* (*PMO*) requires thorough cleaning and sanitization of all multiuse containers and utensils after each use and all equipment at least once each day if in continuous use. The exception is for milk storage tanks, which must be emptied within 72 hr of first use. Farm bulk milk storage tanks are commonly emptied after two days of milk storage. Evaporators in continuous use must be cleaned at least every 44 hr. In dairy plant operations a recording device is required to provide a continuous record of the results of monitoring the cleaning cycle time and temperature, cleaning solution velocity or cleaning pump operation, and presence or strength of cleaning chemicals for each cleaning cycle.

Those engaged in providing milk must understand the nature of soil and water, chemistry of cleaning, and fundamentals of sanitization. In this chapter we discuss these topics as well as choice of detergents and sanitizers and conditions of their use.

27.2 The Nature of Water

Water is the foundation of the entire cleaning and sanitization procedure.
Paul R. Elliker (1911–2005)

When filtering through the earth, groundwater flows through various mediums, dissolving and acquiring certain properties from them, such as carbon dioxide from decaying vegetation, minerals from deposits, and soluble substances from soil. Substances found in water that are objectionable in cleaning and sanitizing include suspended matter (clay, silt, and sand) and hardness (temporary and permanent). Suspended matter can be readily removed, but hardness is more challenging. The salts of magnesium and calcium, which are collected as water passes through the earth, constitute what is commonly called hardness. *Temporary hardness* (of bicarbonates), the usual cause of hardness difficulties, is precipitated by most alkaline materials and by heat. Components of water involved in *permanent hardness* are chiefly the sulfates and chlorides of calcium and magnesium. They are precipitated by many alkaline materials but not by heat (table 27.1). Additionally, soluble iron and manganese oxides in some well water are troublesome because they produce reddish-brown stains. In applications of water in washing foods such as cottage cheese these materials are especially objectionable.

Hardness in water is expressed as the amount of calcium and magnesium salts present in milligrams per liter (mg/L)[2] or grains per gallon (grains/gal).[3] The following tabulation, presented by the US Geological Survey, expresses hardness in relative terms:

	Concentration of Hardness Components	
Water Conditions	**mg/L**	**grains/gal**
Soft	0–60	0–3.5
Moderately hard	61–120	3.5–7.0
Hard	121–180	7.0–10.5
Very Hard	> 180	> 10.5

Constituents of water that precipitate under alkaline conditions and/or on heating cause problems associated with cleaning. Boilers and water heaters develop scale (liming), reducing efficiency of heat transfer with ultimate system failure. Lime deposits in heat exchangers (pasteurizers and coolers) decrease efficiency of heat

Table 27.1 Reactions Involved in Precipitation of Hardness Components of Water.

Reactions of Temporary Hardness

$Ca(HCO_3)_2$ Calcium bicarbonate	+	Heat	→	$CaCO_3$ (insoluble) Calcium carbonate	+	CO_2 Carbon dioxide	+	H_2O Water
$Mg\ (HCO_3)_2$ Magnesium bicarbonate	+	Heat	→	$MgCO_3$ (sl. soluble) Magnesium carbonate	+	CO_2 Carbon dioxide	+	H_2O Water

If pH of water is high, the slightly soluble $MgCO_3$ will precipitate as insoluble magnesium hydroxide, $Mg\ (OH)_2$.

Reactions of Permanent Hardness

$CaSO_4$ Calcium sulfate	+	Heat	→	No Change		
$CaCl_2$ Calcium chloride	+	Na_2CO_3 Sodium carbonate	→	$CaCO_3$ (insoluble) Calcium carbonate	+	2NaCl Sodium chloride
$Mg(NO_3)_2$ Magnesium nitrate	+	2NaOH Sodium hydroxide	→	$Mg(OH)_2$ (insoluble) Magnesium hydroxide	+	$2NaNO_3$ (soluble) Sodium nitrate

transfer. Finally, salts that precipitate on equipment leave films (waterstone) in which fat, protein, and other milk components combine to form **milkstone**. Therefore, it is imperative that water for use in cleaning be softened. Table 27.2 details six reactions in which water can be softened. In methods 1 to 3, calcium and magnesium are removed, whereas in methods 4 to 6 they are kept soluble by chemical treatment.

Choice of method of softening depends on amount of water used and its hardness. It is more economical to soften by removal of hardness when water is hard and quantities used are large. Other limitations may be imposed by soil type and conditions under which cleaning must be accomplished.

27.3 The Nature of Soil

Soil is any residue that must be removed from a surface in cleaning. In the dairy industry, soil consists mainly of constituents of milk and water. Other components include microorganisms, dust, lubricants, and certain ingredients of cleaners and sanitizers.

Milkstone and waterstone are white films that form on dairy equipment. On unheated surfaces they accumulate slowly because of ineffective cleaning or use of poorly conditioned water, or both. For example, calcium and magnesium salts precipitate when sodium carbonates are added to hard water. Some of this precipitate may adhere to equipment during cleaning, leaving a film of waterstone. Milkstone may form quickly on heated surfaces when proteins, denatured by heat, adhere to surfaces and other components adsorb to the proteins. Calcium phosphates are present in milk in large quantities and decrease in solubility at high temperatures. The nature of soil on heated and unheated surfaces usually differs in composition. Each type of soil requires a different cleaning procedure. Freshly deposited soil on an unheated surface is more readily dissolved than the same soil that has dried or has formed on a heated surface. Also, heavy deposits require more cleaner than light ones.

Practices that minimize soil deposition and maximize ease of its removal include (1) use of minimum time and temperature of heating, (2) cooling of product heating surfaces before and during emptying of heated processing vats, (3) rinsing of foam and product from surfaces before they dry, (4) keeping film of soil moist until cleaning is begun, and (5) rinsing with warm—not hot—water. Other factors affecting soil deposition in ultrahigh temperature (UHT) heaters include (1) air in milk, (2) high acidity, (3) age (fresh milk forms more deposits), (4) season (greater deposition in winter), and (5) preheating and holding at high temperature (reduces film deposition) (Burton, 1968).

Ease or difficulty of soil removal is determined by the nature of surfaces on which it exists. Pores of rubber parts, crevices in insufficiently polished surfaces, and

Table 27.2 Methods of Removing and of Maintaining Solubility of Calcium and Magnesium in Hard Water.

1. Removal with Zeolite Softener						
Calcium bicarbonate (hardness)	+	Sodium zeolite	→	Calcium zeolite (calcium removed by regeneration)	+	Sodium bicarbonate (remains in water)
2. Removal with Lime-Soda Softener						
A. Calcium bicarbonate (temporary hardness)	+	Calcium hydroxide (lime)	→	Calcium carbonate (insoluble)[a]	+	Water
B. Calcium sulfate (permanent hardness)	+	Sodium carbonate (soda ash)	→	Calcium carbonate (insoluble)[a]	+	Sodium sulfate
3. Removal by Boiling						
Calcium bicarbonate (hardness)	+	Heat	→	Calcium carbonate + Water + Carbon dioxide (insoluble)[a]		
4. Maintaining Solubility by Sequestration						
Calcium bicarbonate (hardness)	+	Sodium metaphosphate	→	Sequestered calcium complex (soluble)	+	Sodium bicarbonate
5. Maintaining Solubility by Chelation						
Calcium bicarbonate (hardness)	+	Chelating agent	→	Chelated calcium complex (soluble)	+	Sodium bicarbonate
6. Maintaining Solubility by Acidification						
Calcium bicarbonate (hardness)	+	Acid	→	Calcium salt + Water + Carbon dioxide (soluble)		

[a] Insoluble salts precipitate and are removed.

holes in corroded surfaces protect soil and microorganisms from effects of detergents and sanitizers.

27.4 Cleaning Compounds and Their Applications

There are two general types of dairy cleaning compounds, alkaline and acid, but within these types there are many combinations. Ingredients may be subdivided into five classes: alkalies, phosphates, wetting agents (surfactants, i.e., surface active agents), chelating agents, and acids.

27.4.1 Functions Required of Cleaner Ingredients

Because each ingredient has characteristic functional properties, it is possible to combine ingredients to form cleaners for removing specific soils. Major functional properties are:

1. *Emulsifying*—breaking fats and oils into small globules. Effective cleaners promote emulsification and stabilize emulsions, holding globules apart.
2. *Dispersing*—breaking aggregates or flocs into individual particles.
3. *Dissolving*—solubilizing inorganic and organic components of soil.
4. *Peptizing*—attacking and dispersing proteins by hydrolysis making them more soluble.
5. *Rinsing*—separating a solution or suspension from a surface by flushing with water or a solution.
6. *Saponifying*—combining an alkali with a fat to form soap. Soluble soap is easily removed.
7. *Sequestering*—removing or inactivating components of water hardness by forming a soluble complex. When the sequestrant is of an inorganic nature, such as polyphosphate, the term *sequestrant* is used; whereas when the sequestrant is of an organic nature, such as ethylenediaminetetraacetic acid (EDTA), the reaction is known as *chelation*.
8. *Wetting*—lowering the surface tension of a cleaning solution to increase its ability to penetrate soils.

The combination of these functions with a general purpose detergent will: (1) bring the detergent solution into intimate contact with the soil to be removed by means of good wetting and penetrating properties; (2) displace solid and liquid soils from the surface to be cleaned by saponifying fat, peptizing proteins, and dissolving minerals; (3) disperse soil in solvent by dispersion, deflocculation, or emulsification; and (4) prevent redeposition of dispersed soil on the clean surface by providing effective rinsing properties.

27.4.2 Properties of Major Components of Cleaners

27.4.2.1 Alkalies

Sodium hydroxide (NaOH) is the strongest alkali. It is the best for saponification but fulfills few other cleaner functions. It is irritating to skin and corrosive to tin, zinc, and aluminum. Frequently, strong solutions are used for circulation cleaning of heavy protein films from heaters and vacuum pans in milk-processing plants. Sodium metasilicate ($Na_2SiO_3 \cdot 5\ H_2O$) is a moderately strong alkali that inhibits corrosion. It has good emulsifying and deflocculating properties. Sodium carbonate (Na_2CO_3) (soda ash) is useful for hand-washing operations in conjunction with chemicals that soften water. Baking soda ($NaHCO_3$) and soda ash in combination form sesquicarbonate. Hypochlorites are often added to balanced alkaline cleaners to promote peptization of proteins. The activity of alkaline detergents is enhanced by hypochlorites, with efficiencies of some exceeding those of nonchlorinated counterparts by 40%. They also reduce tendencies for water spotting of equipment.

27.4.2.2 Phosphates

Trisodium phosphate ($Na_3PO_4 \cdot 12\ H_2O$) is a general utility cleaner and water softener; it softens by precipitation. Polyphosphates used in dairy cleaners are tetrasodium pyrophosphate ($Na_4P_2O_7$), sodium tripolyphosphate ($Na_5P_3O_{10}$), sodium hexametaphosphate ($(NaPO_3)_6$), and sodium tetraphosphate ($Na_6P_4O_3$). The latter two are called glassy phosphates because their crystals have the appearance of glass. Polyphosphates are excellent water softeners that prevent precipitation of calcium and magnesium. They rinse freely from surfaces, peptize, deflocculate, disperse, emulsify, and suspend. Sodium tetraphosphate is the most effective polyphosphate.

27.4.2.3 Surfactants

Anionic wetting agents commonly used in dairy cleaners are sulfated alcohols or alkyl aryl sulfonates (the active ingredients of California Mastitis Test reagent—cf. chapter 14). Nonionic wetting agents do not ionize, therefore they may be used with either **anionic** or **cationic** materials. Some are among the best emulsifying agents. Surfactants are organic materials of which one part is hydrophilic and one part is hydrophobic. Their dual nature makes them surface active. They lower surface tension, promoting wetting of surfaces, giving other detergent components access to soil.

27.4.2.4 Chelators

Chelation of heavy metals is accomplished with two types of organic chemicals: (1) sodium EDTA and (2) salts of organic acids. Such agents improve wetting of particles and penetration of soils by solutions. They tend to adhere to particles of soil, forming stable suspensions and emulsions.

27.4.2.5 Acids

Acids are usually combined with wetting agents. The most commonly used inorganic acid is phosphoric. It is relatively inexpensive but can be corrosive to metals. An inhibited type of hydrochloric acid is sometimes used for removing heavy scale, but it is not without the risk of serious corrosion. Nitric acid is less corrosive because it produces protective oxide films. In England, hot nitric acid has been highly recommended in circulation cleaning of pasteurizers.

Organic acids are much less corrosive than mineral acids and, though more expensive, are more widely used. The major one is gluconic acid (made from glucose).

For removal of milkstone or lime (scale), high concentrations of acids are needed, but prevention of the buildup of such soil on many surfaces may be accomplished by rinsing with acidified water following cleaning with an alkaline detergent and rinsing with tap water. Also, acid cleaners can be used alternately (every 3 to 7 days) with alkaline cleaners. Such a practice aids in removal of mineral deposits.

27.4.3 Principles of Cleaning

No single procedure or set of conditions and cleaning compounds will suffice for all cleaning situations in all dairy operations. We have already stressed how water quality, soil type, and type of cleaner influence cleanability. We now emphasize that temperature, physical action, and time determine how well cleaning is accomplished. If a cleaner is suited to the existing soil and water, and if it is of sufficient concentration, then these three physical factors become relevant variables. Decreasing the magnitude of any one requires an increase in the others. Higher temperatures produce faster chemical reactions and decrease viscosities, allowing greater turbulence. Both factors speed cleaning. Within the temperature range of 90 to 185°F (32 to 85°C), an increase of 18°F (10°C) will approximately double the cleaning efficiency. Because milkfat is liquid at temperatures above 92°F (33°C), cleaning solutions must be at least this warm.

Brushing by hand results in physical removal of soil. In circulation cleaning, turbulence of circulating solutions substitutes for brushing. Turbulence is determined by rate of flow and nature of surfaces and is expressed in terms of a **Reynolds number**. Velocity of flow of at least 5 ft/sec in **cleaned-in-place (CIP)** operations generally provides adequate turbulence. However, velocity and turbulence are not proportional under all conditions of flow. The higher the Reynolds number the greater the turbulence. In pipeline systems a desirable value is 30,000, though turbulent flow begins at about 3,000. Falling films of cleaning solutions on walls of storage vats are turbulent when the Reynolds number approximates 200.

Most soil is removed within the first or second minute of circulation cleaning, provided the cleaner and cleaning conditions are well chosen and the soil is not extremely heavy. However, it is a common practice to allow 15 to 30 min circulation time to provide a wide margin for error.

Prerinsing is the first step in cleaning. More than 99% of the soil is removed when unheated milk contact surfaces are properly prerinsed (figure 27.1). This reduces the soil loading of the detergent markedly; therefore, detergent solutions can be used to clean much more surface area when cleaning is preceded by adequate rinsing.

Cleaning may be done by hand or by in-place methods. The latter, known as CIP, involves recirculation of solutions through pipes, vessels, and tanks. Solutions are

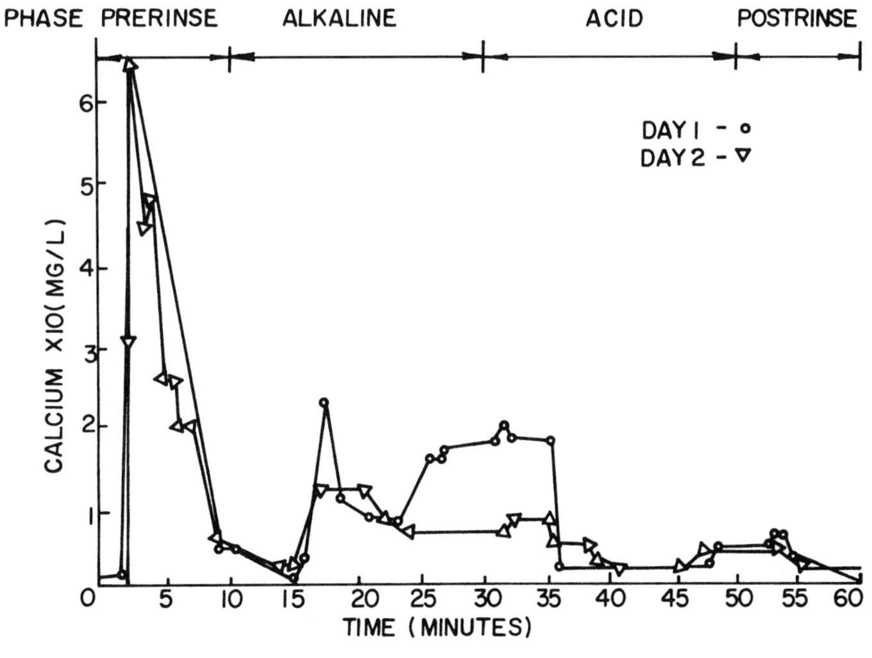

Figure 27.1 Rate of removal of milk soil from surfaces of stainless steel equipment as determined by quantity of calcium recovered from surfaces during cleaning-in-place (courtesy of Missouri Agricultural Experiment Station).

delivered from a cleaning station (figure 27.2) to large surfaces, such as walls, ceilings, and floors of tanks, by spray devices (figures 27.3), including orbital types, which rotate in both vertical and horizontal axes. Large silo tanks are sprayed with a special apparatus (figure 27.4). Even parts and sections of pipe may be cleaned by circulation in a vat called a parts washer (cleaned-out-of-place, or COP).

Cleaning by recirculation is used both on farms and in plants. It permits use of stronger cleaners and higher temperatures than can be used in hand washing. It virtually eliminates personnel as an important variable in cleaning, especially if systems are employed that automatically mix water and detergent in optimal concentrations, heat the solution to the proper temperature, and circulate it for the prescribed rate and time. It also releases workers for other important activities. Druce and Thomas (1972) found that for farm milk pipelines cleaned-in-place the incidence of bacteria counts greater than 50,000/ft^2 was 24% when the initial temperature of the detergent solution was >180°F (82°C), but was 68% when the initial temperature was < 140°F (60°C). These data stress the importance of temperature in effective cleaning.

Postrinsing should remove traces of cleaner and suspended soil from surfaces. However, to minimize mineral deposits, it is recommended that the last water used in postrinsing be acidified to near pH 5. Because chlorine corrodes stainless steel, acidified water should be used for rinsing after cleaning with chlorinated detergents.

Surfaces should be drained and allowed to dry after cleaning. Residual water provides a solution in which microorganisms can grow if sufficient organic matter remains. The hotter the final rinse water, the faster the drying.

The four steps in cleaning and sanitizing that should be applied to all types of equipment and multiuse containers are: (1) rinse with warm water; (2) wash with carefully selected cleaner dissolved in appropriately conditioned water at a temperature commensurate with the equipment, soil, and method of washing (hand, CIP, or COP); (3) rinse with hot water to promote drying (hot water may not be used in some refrigerated equipment); and (4) sanitize immediately prior to use.

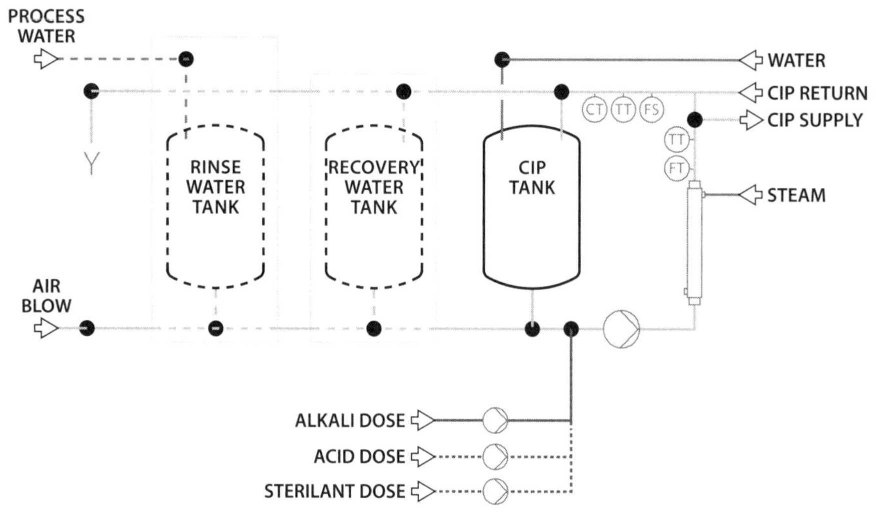

Figure 27.2 Line drawing of a typical system used to provide water, detergent, and sanitizing solutions for CIP spray cleaning of milk storage and transport tanks (from SPX Corporation).

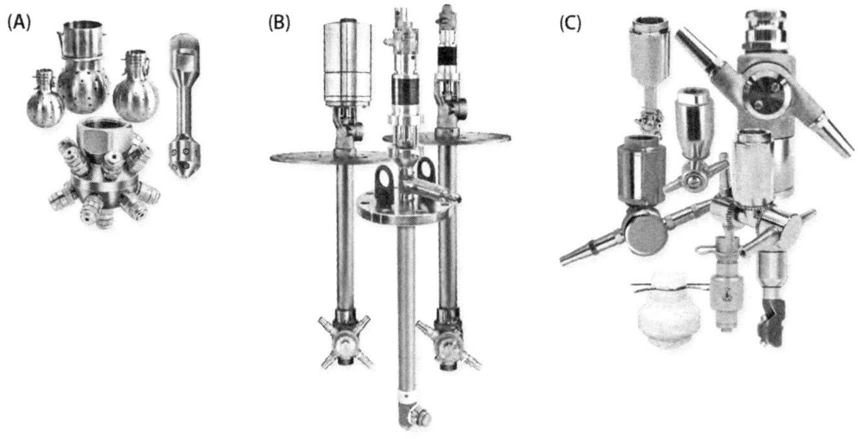

Figure 27.3 Three types of automated tank cleaning nozzles for delivery of sprays of cleaning solutions to equipment surfaces: (A) stationary, (B) motor driven, and (C) fluid driven (courtesy of Spraying Systems Company).

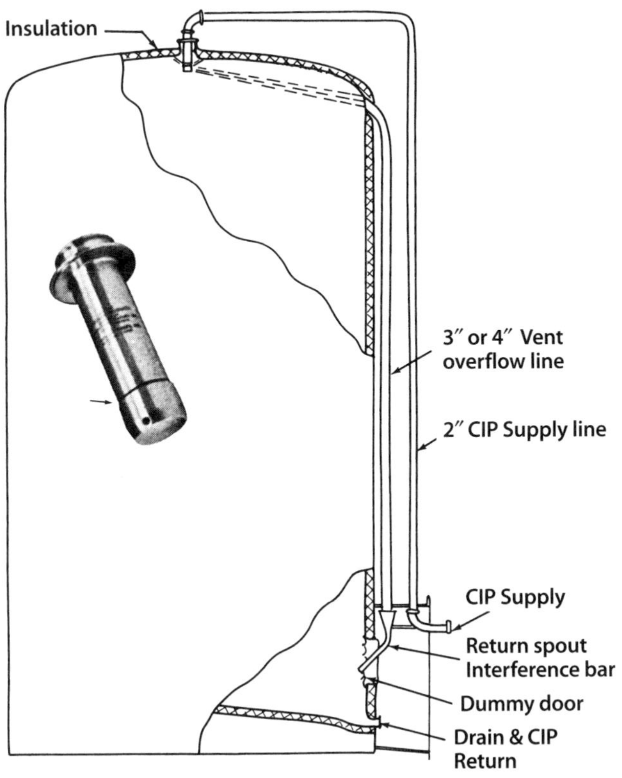

Figure 27.4 Cutaway diagram of a silo storage tank with inset showing a typical permanently installed spray device. Cleaning solutions are distributed to the top surface via the annular orifice (arrow). The tank vent line is cleaned by solution sprayed from the nozzle, a hole drilled in this instance at a 7° angle near the bottom. The spray device and the spud, through which it is inserted, are cleaned by solutions delivered via small holes at the top (courtesy of Equipment-Engineering Division, ECOLAB, St. Paul, Minnesota).

27.5 Sanitizers and Their Applications

Sanitization is the process of treating surfaces with physical or chemical agents that kill most, but not necessarily all, microorganisms present. *Sterilization*, the process of killing all microorganisms, is not necessarily accomplished through sanitization. Microorganisms may survive cleaning, even when surfaces appear scrupulously clean. Since the same solutions may be used to wash equipment exposed to raw and pasteurized milks, it must be assumed that pathogens may be transmitted via cleaning solutions. Insects and rodents may contact cleaned equipment, thus transferring microorganisms. Additionally, some water supplies contain bacteria that are either pathogenic or spoil food, therefore the postrinse can contaminate equipment. Many microorganisms can grow if moisture and food materials are present, so sanitization is usually performed just prior to the use of multiuse containers and equipment. This practice destroys residual microorganisms and it minimizes chances of recontamination and time of exposure to corrosive sanitizers.

To **sanitize** means to reduce numbers of microorganisms of public health significance to safe levels. In the dairy industry the additional aim is to destroy all microorganisms capable of spoiling the product. Chemical sanitizers may have either a **bactericidal** (killing) effect or a **bacteriostatic** (growth-inhibiting) effect. Effectiveness of chemical sanitizers depends on concentration, temperature, pH, and time of exposure. Heat, delivered as steam, hot water, or hot air, may be used but is more costly than chemical sanitization. Moreover, heat is often difficult to apply. The sanitization standard for destruction of contaminants on food-contact surfaces is generally accepted as 99.999% (a 5 $\log_{10}$ reduction) achieved in 30 sec.

When sanitizers are used below their lethal concentration, some bacteria may become increasingly resistant to their effects. Researchers at the National University of Ireland found that cultures of *Pseudomonas aeruginosa* exposed to sublethal concentrations of disinfectants adapted to more efficiently pump out of their cells antimicrobial agents, including antibiotics.

27.5.1 Chemical Sanitization

Chemical sanitizers of many types are available. These include compounds bearing chlorine, chlorine-bromine mixtures, iodine, quaternary ammonium compounds, and acid wetting agent sanitizers. Each has advantages and disadvantages, but compounds bearing chlorine are most widely used. Title 21, Part 178.1010 of the US Code of Federal Regulations lists the sanitizers permitted for use on food-contact surfaces.

27.5.1.1 Compounds Bearing Chlorine

Sodium hypochlorite, NaOCl, and calcium hypochlorite, $Ca(OCl)_2$, are inexpensive inorganic sanitizers. Due to their low stability, hypochlorites are very strong oxidizing agents. They react with many organic and inorganic compounds. Organic types include chloramine T (used infrequently because of slow action), dichlorodimethyl hydantoin, and di- or trichloroisocyanuric acid. Each of these compounds releases chlorine, which reacts with water to form hypochlorous acid, the bactericidal component believed to deactivate respiratory enzymes of microorganisms.

Sodium hypochlorite, which comes only in liquid form, is a widely used sanitizer in the dairy industry. It is economical, readily soluble, unaffected by minerals in hard water, and highly efficacious, especially in killing bacterial spores and **bacteriophages**. When calcium hypochlorite, a solid, is used, it is dispersed in water and sodium carbonate is added. Calcium precipitates as calcium carbonate, and hypochlorite combines with sodium in the clear supernatant. Inconvenience in use and in removal of precipitate from calcium hypochlorite is the major reason most users prefer sodium hypochlorite.

Items sanitized with sodium or calcium hypochlorite should be exposed for at least 2 min to at least 50 ppm (mg/L) available chlorine maintained at a temperature of at least 75°F (24°C) but not more than 110°F (44°C). Temperature, concentration, and time of exposure are dependent variables. As one is decreased, one or both of the others must be increased to maintain the same activity. For each change in temperature of 18°F (10°C) within the usual range of application, one can expect a twofold change in bactericidal effectiveness in the same direction. The concentration of chlorine must be at least 50 ppm after use.

Chlorine becomes inactive as it combines with microorganisms and residual soil. It has affinity for most types of organic matter, especially proteins. Therefore, it is advisable to start with a concentration of 150 to 200 ppm. When application is by spraying or fogging, initial solution strength should be 250 or 400 ppm, respectively.

Organic compounds bearing chlorine are used where their lesser corrosive nature is vital. Hypochlorites are corrosive and irritating to skin.[4] Alkaline conditions reduce the activity of all chlorine sanitizers, but acidity produces instability and rapid loss of strength. Therefore, buffering to maintain pH within the range of 6 to 8 provides a favorable balance between acidity and stability, except that the maximum pH for chloramine T and chlorinated hydantoin should be 6.4 and 7.0, respectively. Chlorine sanitizers tend to deteriorate when left in open containers or when exposed to light or heat. Also, they oxidize rubber parts when applied in high concentrations.

27.5.1.2 Bromine-Chlorine Mixtures

The combination of potassium bromide with sodium hypochlorite significantly increases bactericidal effectiveness of the latter. To be effective a solution need contain only 25 ppm of available halogen (combination of both hypobromide and hypochlorite, expressed as available chlorine). The pH of such a mixture varies from about 9.0 to 10.5, thus making the solution less corrosive to stainless steel than is hypochlorite.

27.5.1.3 Iodophors

Iodine has excellent antimicrobial properties but is sparingly soluble in water, tends to stain, and is unstable. Development of iodophors[5] overcame most of these undesirable properties. Iodophor contains iodine, a nonionic wetting agent, and buffered phosphoric acid. The wetting agent stabilizes iodine and increases its permeability toward bacterial cells. Buffered phosphoric acid maintains the acid pH necessary for iodine to be germicidal. Color of dilute solutions is proportional to concentration of iodine.

In many respects iodophors are similar to hypochlorites. Their activity increases proportionately with higher temperatures and more acidity but instability and corrosiveness also increase. Iodophor solutions should be maintained at pH 6.0 and below. Above pH 6.5, iodine is not released. Temperature should be kept below 120°F (49°C) to prevent iodine **sublimation**, which reduces the strength of the sanitizing solution. Organic matter reduces the activity of iodophors but not as greatly as that of hypochlorites. Water hardness constituents have little effect upon iodophors. A solution containing 25 ppm iodophor is effective against most microorganisms. Minimal exposure is 1 min for a solution containing at least 12.5 ppm available iodine at temperatures ranging from 75 to 110°F (24 to 44°C) and at pH 5 or below.

Free iodine, released from iodophors, inhibits respiratory enzymes of microorganisms. In comparison with hypochlorites, iodophors are more effective against yeasts but less effective against bacterial spores and bacteriophages. Otherwise, types of microorganisms affected are about the same.

27.5.1.4 Acid Wetting Agents

Mixtures of organic acids (e.g., dodecylbenzene sulfonic acid), anionic or nonionic wetting agents, and phosphoric acid provide detergency, solubilization of minerals, and germicidal action. Sanitizing effect is provided by low pH (2.0 to 2.2). Bactericidal action is slower than with chlorine- and iodine-type sanitizers; however, activity against bacteriophages is excellent. Since they foam readily, application by circulation is limited.

Because these compounds both clean and sanitize, there are limited situations in which they can be used as detergent-sanitizers to accomplish both tasks in one operation. The following precautions are suggested when detergent-sanitizers are used: (1) prerinse equipment thoroughly, (2) follow manufacturer's directions explicitly, and (3) resanitize prior to use.

27.5.1.5 Peroxyacetic acid

This acid has sanitizing activity against a wide range of microorganisms in most conditions, including water temperatures down to 40°F (4°C). It is active from acidic to neutral pH. It is also noncorrosive to 304 and 316 stainless steel and to aluminum surfaces when used at recommended concentrations. During use the active ingredients break down into water, oxygen, and carboxylic acids. Peroxyacetic acid sanitizers are often paired with stabilized hydrogen peroxide.

27.5.1.6 Quaternary Ammonium Compounds

These cationic wetting agents constitute a large class of sanitizers. They have the following basic structure:

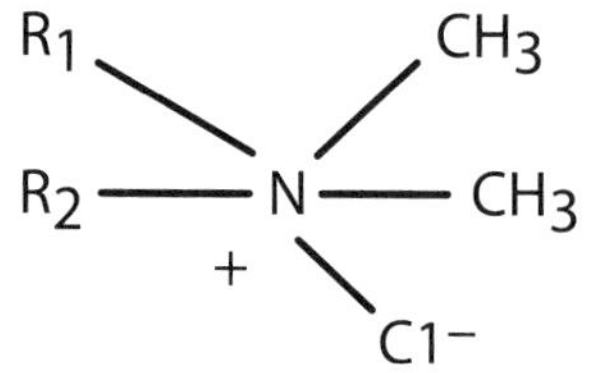

where R_1 and R_2 contain large numbers of carbons in straight chains and ring structures. The active portion of the molecule is the large cation, not the chlorine ion (Cl^-). However, most bacteria are only slowly destroyed and bacteriophages usually survive treatment. Additionally, quats are inactivated by water hardness components (salts of calcium and magnesium) and by anionic detergents. Use concentrations vary from 100 to 400 ppm (mg/L).

Advantages of these compounds are that they are noncorrosive, nonirritating, stable to heat, relatively stable in the presence of organic matter, and active over a wide range of pH.

27.5.2 Sanitization with Heat

Steam, hot water, or hot air may be used to sanitize equipment. The *PMO* specifies that each group of assembled piping shall be treated by maintaining steam flow from inlet to outlet for at least 5 min after the temperature at the outlet reaches 200°F (94°C). Exposure to hot water shall be for at least 5 min at 170°F (77°C). However, current dairy industry practice is to circulate water at 199°F (93°C) for 5 to 30 min.

Where pasteurized milk storage tanks are sanitized with hot water, they are specially constructed. Insulation and refrigeration coils are left off the tank, which is essentially made of a single layer of stainless steel. It is placed in a separate room, which may be heated during sanitization and refrigerated prior to and during storage of milk. This system, of course, requires considerably more energy in the form of heat and refrigeration than do conventional operations. However, reduction in postpasteurization contamination is significant. Tanks that are to be heat sanitized must be specially constructed so expansion and contraction do not result in the development of cracks in their walls or seams.

Small pieces of equipment may be sanitized in cabinets with steam or hot air. Exposure to steam at 171°F (77°C) for 15 min or at 199°F (93°C) for 5 min is minimal. Hot air is a less efficient destructive agent than steam or hot water, therefore a temperature of 180°F (82°C) for 20 min is required. If sterility is required, as in processes that sterilize and package aseptically, steam or hot water is employed under pressure. Equipment is exposed to temperatures of 248 to 302°F (120 to 150°C) for 30 min.

27.5.3 Sanitizing with Activated or Electrolyzed Water

Application of weak electrical current to water can cause ionization of water molecules, producing positively and negatively charged ions (activated water) that can attack organic substances and dislodge them from surfaces. The treatment creates charged bubbles that attach to soil particles leaving them charged so they tend to repel from surfaces to be washed away. The low-level electrical field produced during activation of water affects bacteria and viruses by creating pores in their cell membranes through which they can lose vital components. The treatment is, therefore, effective in both soil removal by flushing and killing microorganisms by **electroporation**. The electrically activated water returns to its native state in less than 1 min.

Electrolyzed water is produced by passing an electric current through water containing table salt (sodium chloride). Ionized components in the nontoxic (for humans) solution oxidize organic matter and can kill microorganisms within minutes. Oxidation damages cell membranes allowing water to infiltrate the cell, upsetting the osmotic equilibrium and causing the cell to burst.

Whereas activated water is effective when sprayed on surfaces, electrolyzed water requires a few minutes of contact time to produce lethality.

27.6 Recommended Practices in Cleaning, Sanitizing, and Plant Operations

Sanitary pipelines and holding tubes must be firmly supported to maintain desired slope. Pipe joints must be sealed with self-positioning sanitary gaskets that are monitored daily for leaks, checked monthly for condition, and replaced when worn. CIP circulation pumps must maintain turbulent flow, be free of leaks, and kept clean. Records of time, temperature, and detergency should be monitored to validate proper operation of each cycle. Any plug-type valves in the system should be dismantled and hand cleaned between the prerinse and CIP cycles. The valve body and adjacent lines should be cleaned inside and outside within a few inches of the valve. Outside surfaces of all equipment should be cleaned regularly.

Floors, walls, and ceilings should be included in a daily cleaning routine. Foaming chlorinated cleaners are recommended. Cracks should be carefully repaired to produce smooth surfaces. Dehumidifiers and air conditioning units must be sloped to drain outside the production area. Air handling units should produce a positive pressure while supplying filtered air that is not directed toward filler units. Some package fillers are fitted with high-efficiency particulate air (HEPA) filters. Particular attention should be given to cleaning and sanitizing floor drains to eliminate conditions for growth of *Listeria* bacteria. Personnel in the processing area should be only those needed and each should be dressed in clean clothing and footwear that are never worn outside the plant. Foot sanitizing baths should be in place at each entry of the processing area.

Workers should recognize that material safety data sheets (MSDS) are valuable references designed to protect them as well as consumers of the products. Included in the MSDS is information on the physical and chemical characteristics of the product, physical and health hazards, exposure limits, precautions for safe handling and cleanup, control measures, and emergency and first aid procedures.

Tests of cleanliness should be performed periodically. Screening surfaces for residues can be accomplished quickly using the adenosine triphosphate (ATP) bioluminescence procedure. ATP is an energy-containing molecule found in all microorganisms. The test involves swabbing a surface with a pen-like device or sponge disk, which is then inserted into a handheld reader (figure 27.5) that displays results within seconds in relative light units (RLU). The test indicates microbial presence but not identity. The method involves interaction of ATP with luciferin to form luciferyl adenylate, which is oxidized by luciferase (the firefly enzyme) and emits light. The amount of light emitted, read with a photodiode, is proportional to the amount of ATP present in the sample. This procedure is much faster than those involving swabbing surfaces and culturing the microorganisms.

Figure 27.5 Hygiena EnSURE Monitoring System with various Hygiena bioluminescent test devices (courtesy of Hygiena, LLC).

27.7 Summary

Only after thorough cleaning can one hope to sanitize equipment and containers with chemical sanitizers. Soil can be rendered sterile with heat but high cost and other considerations, such as reduced efficiency of heat exchange through soil films, give many reasons to support the practice of cleaning thoroughly.

Properties of efficient and effective dairy equipment cleaners are the ability to (1) soften water, suspend hardness, and prevent precipitates; (2) emulsify fats and certain proteins; (3) saponify fats; (4) penetrate (wet) soil; (5) disperse and suspend precipitates; (6) peptize proteins; (7) dissolve in water and rinse freely; and (8) be noncorrosive.

No single cleaning material has all of these desirable properties. Therefore, mixtures of components are formulated to remove specific kinds of soil when dissolved in water of known characteristics and used under specified conditions of temperature, concentration, time, and physical action. We see also, then, that no single formulation or set of conditions of use will suffice for all applications. In each application, however, the steps in cleaning are prerinse, clean, postrinse, and drain. Sanitization should immediately precede use of equipment and containers.

Chemical sanitizers combine with components of microbial cells and deactivate them, thus killing the cells. However, they may also combine with residual soil on equipment and become inactive. Hypochlorite sanitizers are especially sensitive to organic matter. Quaternary ammonium compounds are adversely affected by hard water.

Strength of sanitizer, time of exposure, and temperature in sanitization are principal determinants of effectiveness. Directions for use must be followed closely in all applications. Corrosiveness, irritation to skin and eyes, effect of pH, and spectrum of activity also play a role in choice of sanitizer.

Heat may be applied as steam, hot water, or hot air. The foremost advantage of these is that soil films cannot protect bacteria from the heat if exposure adequately heats equipment to a lethal temperature. Costs of sanitization with heat are markedly higher than those of chemical sanitization. Some processors, however, have concluded that today's fluid milk distribution systems require such long shelf life that the extra protection against postpasteurization contamination afforded by sanitization with heat justifies its cost.

Study Questions

1. In general, what do the processes of cleaning and sanitizing dairy equipment entail?
2. Identify substances that may enter water supplies used for cleaning dairy equipment.
3. Define hardness in water. Differentiate between temporary and permanent hardness. Why is hardness of water so important to proper cleaning of equipment? Cite an example of a dairy product in which direct contact is made with water.
4. What is milkstone? How may it be prevented? What is waterstone? What conditions enhance the accumulation of milkstone and/or waterstone?
5. What is soil? Does its nature commonly differ on surfaces of heated and unheated equipment? Why?
6. Identify factors that may minimize soil deposition and maximize ease of its removal.

7. Which would tend to accumulate more soil—rubber milking machine inflations or stainless steel pipelines used to convey milk? Why? Which of these two would be more easily cleaned?
8. What are the two general types of dairy cleaning compounds? What are their major functional properties? How are these functions combined into a general purpose detergent?
9. Describe the types, properties, and uses of the following for cleaning dairy equipment: alkalies, phosphates, surfactants, chelators, and acids.
10. How are temperature, physical action, and time related to cleaning dairy equipment? Does increasing or decreasing one of these physical factors affect the others?
11. What is CIP? Does turbulence affect this method of cleaning dairy equipment? Cite merits and limitations of CIP.
12. Is prerinsing of much significance in cleaning dairy equipment? What is the role of postrinsing?
13. Why should dairy equipment be stored dry when not in use?
14. Define the terms *sanitization*, *bactericidal*, *bacteriostatic*, and *germicidal*.
15. Identify four types of chemical sanitizers. Cite merits and limitations of each.
16. What times and temperatures are commonly employed in heat sanitization of dairy equipment? In sterilization?
17. What are the properties of efficient dairy equipment cleaners?

NOTES

[1] We define *milk soil* as those solids of milk that remain on surfaces after milk is removed from equipment and containers.

[2] Milligrams per liter (mg/L) means the same as parts per million (ppm).

[3] 1 grain/gal = 17 mg/L.

[4] Irritation seldom occurs when full-strength Clorox™ is used to disinfect teats after milking because the alkalinity in Clorox™ is low.

[5] Iodophor is a Greek word meaning "iodine carrier."

REFERENCES

Burton, H. 1968. Deposits from whole milk in heat treatment plant—a review and discussion. *Journal of Dairy Research* 35:317.

Druce, R. G., and S. B. Thomas. 1972. Bacteriological studies on bulk milk collection: Pipeline milking plants and bulk milk tanks as sources of bacterial contamination of milk—a review. *Journal of Applied Bacteriology* 35:253.

WEBSITES

Ecolab, Inc. (http://www.ecolab.com/market/dairy-farms)
Hygiena (http://www.hygiena.com/dairy-processors.html)
National Sanitation Foundation (http://www.nsf.org)
Paul Mueller Company (http://www.muel.com)
Spraying Systems Company (http://www.spray.com)
Title 21, Part 178.1010 US CFR (http://www.accessdata.fda.gov/scripts/cdrh/cfdocs/cfcfr/CFRSearch.cfm?fr=178.1010)
US Geological Survey, Water Quality Information (http://water.usgs.gov/owq/hardness-alkalinity.html)

28

Fluid Milk and Related Products

The German called it milch. The Frenchman called it le fait; the Spaniard called it la leche.
Foolish people! How could it be anything except milk?
B. L. Herrington (1904–1998)

28.1 Introduction

Fluid milk[1] and **fluid milk products** utilize approximately 35% of the total milk produced in the United States. These products are practically all made from **grade "A" milk** produced in accordance with the *Grade "A" Pasteurized Milk Ordinance* (cf. chapter 24).

In 2014, 206 billion pounds of milk were produced in the United States. In 2012, 332 fluid milk plants processed an average of 192 million pounds into the following products: whole milk (22%); **lowfat milk** (42%); nonfat milk (13%); flavored milks (7%); cream products (6%); yogurt (7%); cultured buttermilk (1%); and other products (2%).

As defined in the Code of Federal Regulations, Title 21, Part 131.110 (21 CFR 131.110), "Milk is the lacteal secretion, practically free from colostrum, obtained by the complete milking of one or more healthy cows. Milk that is in final package form for beverage use shall have been pasteurized or ultrapasteurized." Milk's composition may be adjusted by separating part of the milkfat therefrom, or by adding thereto cream, concentrated milk, dry whole milk, skim milk, concentrated skim milk, or nonfat dry milk. Milk may be homogenized. Standards for fat and nonfat solids for fluid milk and fluid milk products are published in the 21 CFR 131.

According to federal standards (21 CFR 131), fluid milk must contain a minimum of 3.25% fat and 8.25% nonfat milk solids (NMS). Food labeling regulations permit use of the "reduced fat" descriptor when fat content is reduced by 25%; therefore, milk containing 2% fat is so labeled. Milk containing 1% fat meets the standard (21 CFR 101.62) for the "lowfat" label. In each case the nonfat solids content is the same as for milk (8.25%). Higher fat milk products (21 CFR 131.3) contain the following percentages of fat: half-and-half = 10.5 to 18; light cream = 10 to 30; light whipping cream = 30 to 36; heavy whipping cream = 36 or more. Although there are no federal standards for nonfat solids content in the higher fat products, each one should contain an amount proportional to its content of skim milk.

Standards in California call for more nonfat solids than do federal standards: milk = 8.7%, skim milk = 9%; reduced fat = 10%; and lowfat = 11%.

Processes common to milk plants are described in chapter 26. Let us now make application of these to fluid milk operations. Figure 28.1 is an illustration of the flow of milk through a milk-processing facility.

28.2 Receiving and Processing

To achieve maximum efficiencies and economies in receiving operations, payment must be limited to quantities of milk and milk solids actually received. Therefore, accurate weighing or metering, plus sampling and testing, are imperative. Fat losses may range from 0.5 to 5.0%. The dairy manufacturing industry operates on a low profit margin. Assuming an annual net sales profit of 2% and raw milk cost of 50% (of sales), a loss of 1% of materials in the plant would equal 25% of yearly profit.

A close working relationship and a clear understanding of the division of responsibilities must exist among personnel who have receiving, processing, and quality assurance functions. This means that definite responsibility must be assigned for monitoring the following:

1. Scales, load cells, measuring rods, and metering pumps to assure accuracy;
2. Milk sampling techniques;
3. Analytical tests;

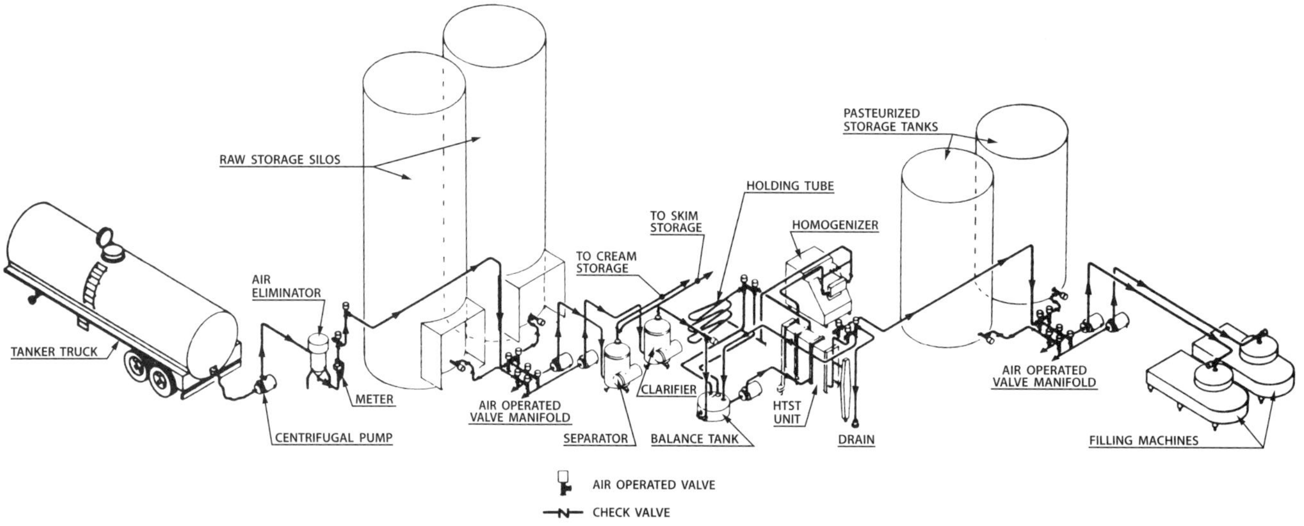

Figure 28.1 Product flow from tank trucks to filling machines of a fluid milk-processing facility.

4. Losses in standardizing and separating (e.g., fat in skim milk);
5. Leaks in valves, pipelines, and containers;
6. Entrainment losses and spillage;
7. Accuracy of fill;
8. Faulty stacker, unstacker, and conveyor operations;
9. Returned products; and
10. Accuracy and usefulness of records.

Processing and packaging schedules in most plants require that raw milk be available for standardization and pasteurization in early morning, a few hours before packaging begins. Therefore, some milk must be received the previous day.

For each product, the production manager must determine

1. Volume and composition needed;
2. Location for storage;
3. Processes (treatments) to which products will be subjected (rates of processing are generally determined by capacities of equipment);
4. Tanks for storage after processing;
5. Numbers of packages and sizes to be used;
6. Schedules for operation of package fillers;
7. Locations and procedures for placement in coolers and/or on trucks;
8. Assignments of personnel to operations; and
9. Schedule and personnel assignments to cleaning and sanitizing operations.

Standardization is accomplished before processing by the batch or, along with clarification, by the continuous method. Some plants separate into skim milk and cream then continuously recombine the two in correct proportions to produce the desired product. This process lends itself to automation. Following batch standardization, processes may occur in one of several orders. The more popular ones are:

1. Clarify, preheat, homogenize, pasteurize;
2. Preheat, clarify, homogenize, pasteurize;
3. Clarify, preheat, pasteurize, homogenize; and
4. Preheat, clarify, pasteurize, homogenize.

When standardization is done in-line, clarification becomes part of separation (since the separator removes insoluble extraneous matter) and separation is likely to come after preheating to maximize flow rate and efficiency in the process.

Vacuumization may be accomplished after preheating or pasteurizing or both. If steam injection or infusion is used, vacuumization must follow this heating process so water will be removed in proportion to the amount of steam that condenses in the product. Pasteurization is mandatory for all milk products but declaration of the term *pasteurized* is optional. However, if the product is ultrapasteurized, it must be so labeled (21 CFR 131).

Of course, cooling is necessary to a temperature which assures that the packaged product will reach storage at or below 41°F (5°C).

28.3 Making Milk Products Healthful and Desirable

Milk's bland flavor and near neutral pH make it an excellent carrier of fruits, flavorings, omega-3 fatty acids, probiotics, prebiotics, fiber, and other nutrients or functional ingredients. To enhance sales of fluid and cultured milk products, processors have developed specialty products that appeal to perceived needs of customers including nutrition, energy, low calories, and relaxation. Some nutritional benefits are defined through statutes or regulations passed by the US Food and Drug Administration (FDA). These health claims are outlined in the following categories: health, nutrient content, and structure/function. Under 21 CFR 104, fortified foods or foods that are labeled or advertised with a claim to their nutrition must be labeled to show content, in terms of percent of Reference Daily Intakes (RDI), of protein, vitamin A, vitamin C, thiamine, riboflavin, niacin, calcium, and iron. Serving size and quantities per serving must be stated for calories, fat, protein, and carbohydrates.

28.3.1 Added Vitamins

Milk and lowfat milk usually have 400 international units (IU) of vitamin D added per quart and are practically always homogenized, although this is not required. Skim milk (nonfat milk) is not homogenized but must contain vitamin A (2,000 IU/qt) and vitamin D may be added. In addition to vitamins A and D, multivitamin mineral milk typically has added, on a per quart basis, thiamine (1 mg), riboflavin (2 mg), niacin (10 mg), iron (10 mg), and iodine (0.1 mg). The FDA has issued the following two model health claims for calcium and vitamin D to describe the relationship between adequate calcium and vitamin D in reducing the risk of osteoporosis: (1) adequate calcium and vitamin D throughout life, as part of a well-balanced diet, may reduce the risk of osteoporosis and (2) adequate calcium and vitamin D as part of a healthful diet, coupled with physical activity, may reduce the risk of osteoporosis in later life (73FR56477).

Vitamins can be added at several points in the processing system, preferably after separation. These locations include the pasteurizing vat, the HTST constant-level tank, or on a continuous basis into the pipeline after standardization and prior to pasteurization. Addition by the batch or continuously with metering pumps can be used. The problem of underfortification is often related to the point in the system where vitamins are added. Vitamins A and D are fat soluble and will gradually become more concentrated in the milkfat portion of the milk. If vitamins are added in the proper amount before the product is separated and standardized, then the low-fat product will tend to be under fortified and the high-fat product over fortified. Water-soluble vitamin concentrates can minimize this problem if vitamins are added before separation. If this procedure is used, confirmatory assays should be done to ensure proper fortification levels of each product.

HTST systems are being used with in-line fat standardization, which also makes possible switching, without stopping, from milk and milk products being fortified with vitamin D to those being fortified with both A and D. These systems require metered injection of the proper vitamins at a point after standardization and before pasteurization. Sanitary positive-displacement pumps are available for this purpose.

Pumps should be calibrated based on the pasteurization system flow rate. If flow rates change for different milk products additional vitamin pumps may be needed. Routine calibration of metering pumps is recommended followed by verification of their accuracy.

28.3.2 Fortification with Other Ingredients

Fortification of milk with nonfat milk solids should be encouraged for two reasons: the flavor of most milk is improved and the nutritive value is increased. For NMS, trained taste panelists can detect changes as low as 0.6% to differentiate products, but a change of about 1.5% is required for fat. Differences in detectable lactose concentrations are even less, about 0.33%, whereas protein concentrations must differ by about 2.5% to be detected by sensory tests.

Processors cannot always market products that have the desired composition. Analyses of products of **handlers** in federal milk orders in eight Midwestern states illustrate these variations (table 28.1). The average fat percentage, 3.28, of homogenized milk was only slightly above the regulatory minimum of 3.25. However, percentages ranged from 2.84 to 5.98, meaning some standardizing operations were out of control. However, the overall variance was acceptable. Protein, which is not standardized in the nonfortified products, averaged 3.14% and had a much smaller variance than did milkfat. The average nonfat solids content of homogenized milk was 8.89%, well above regulatory minimum. Mean percentages of milkfat in the lowfat and skim milks were quite close to labeled values and variances were lower than was the variance of milkfat in homogenized milk.

Fiber-enriched milk products have recently entered the market. Soluble fibers, inulin, fructo-oligoaccharides, and galacto-oligosaccharides, which act as prebiotics, are favored in fluid products because they have little influence on product viscosity. Some soluble fibers, such as gums, absorb water and form viscous gels. They have the ability to trap cholesterol and bile acids while passing through the digestive tract, helping the body lower circulating cholesterol. Insoluble fibers, including lignin and cellulose, do not form viscous solutions and are associated with reduced risk of digestive disorders. The amount of fiber added to a product depends on the effects on the product and consumer and whether a health claim is merited.

Table 28.1 Percentages of Milkfat, Protein, and Nonfat Milk Solids in Fluid Milk Products Processed in Eight Midwestern States.

Product	Number of Observations	Mean	Low	High
Homogenized milk	541			
Milkfat		3.28	2.84	5.98
Protein		3.14	2.95	3.45
Nonfat solids		8.89	8.39	9.10
Lowfat 2%	538			
Milkfat		1.96	1.54	3.21
Protein		3.18	2.95	3.44
Nonfat solids		9.04	7.88	9.24
Lowfat (1%)	481			
Milkfat		0.98	0.82	1.38
Protein		3.21	3.07	3.50
Nonfat solids		9.17	8.79	9.34
Skim milk	173			
Milkfat		0.12	0.04	0.95
Protein		3.24	3.08	3.48
Nonfat solids		9.13	8.86	9.42

Source: Federal Order 032. Data from February 2010.

28.3.3 Reduced Sodium

Low sodium milk is beneficial for persons who suffer from high blood pressure, certain cardiac conditions, toxemias of pregnancy, certain kidney disorders, and other conditions in which **edema** is a problem. The sodium concentration in milk can be reduced from about 550 to 50 mg/L upon passage over an ion exchange resin bed. Sodium and some calcium are exchanged for potassium. In one process, part of the potassium is replaced by calcium in the passage of milk through a second resin bed. Pasteurization and homogenization follow ion exchange. There is some loss of B vitamins (and possibly calcium) but otherwise the product has the nutritive properties of regular milk.

28.3.4 Fatty Acids as Additives to Milk Products

The dairy industry has been moving toward production of milk, yogurt, and infant foods containing omega-3 fatty acids. These are known as alpha-linolenic acid (ALA), docosahexaenoic acid (DHA), and eicosapentaenoic acid (EPA). Omega-3 fatty acids are derived primarily from oils of fatty fish, certain algae, and flaxseed. Microparticles (< 1 nm) of the oils can be encapsulated to provide easy dispersability. These unsaturated fatty acids are easily oxidized but are known to provide nutritional benefits and are easily added to dairy foods containing milkfat. Although milkfat is rich in short-chain fatty acids, most of them are of the omega-6 type and are fully saturated (cf. chapter 1). Omega-3 fatty acids must be carefully processed to avoid oxidation of the unsaturated bonds that produces off flavors similar to those produced in milk by sunlight or fluorescent light. Protection from light, heat, and oxygen is imperative; therefore the fatty acids should be added near the end of processing.

A health claim has been authorized by the FDA for DHA and EPA: "Supportive but not conclusive research shows that consumption of EPA and DHA omega-3 fatty acids may reduce the risk of coronary heart disease." No minimum dose was provided, but a maximum of 3 g/d was indicated. In addition, structure/function claims applicable to EPA and DHA include:

> maintains a healthy cardiovascular system and supports normal development of the brain, eyes, and nerves. Foods containing at least 32 mg of EPA and DHA per serving may be labeled as being an "excellent source," "high in," or "rich in" omega-3 fatty acids.

If a product were to make a nutrient claim as an excellent source of omega-3 DHA and EPA, it must state: "Contains (xx) milligrams EPA and DHA combined per serving, which is (xx%) of the 160 milligrams Daily Value for a combination of EPA and DHA."

Conjugated linoleic acid (CLA) is the collective term used to describe the 18-carbon isomers of the dienoic essential fatty acid, linoleic acid (C18, n-6). This group of isomers is found in relatively high concentrations in milkfat. When consumed by experimental animals it is incorporated into cell membranes of phospholipids where it is an antioxidant. As little as 0.5% CLA, an amount found in most dairy foods, has an inhibiting effect on some cancers in test animals, especially mammary tumors (Council on Scientific Affairs, 1993).

Dairy products account for about 75% of the estimated 100 to 300 mg of CLA Americans consume per day. Numerous cis, trans and trans, cis isomers of CLA are contained in milkfat with double bonds positioned in the 18-carbon chain at Δ7,9; Δ8,10; Δ9,11; Δ10,12; and Δ11,13. They comprise about 0.7% of milkfat and most of it is the cis-9, trans-11 isomer (Huth et al., 2006). Much of this isomer is synthesized in the mammary gland by desaturation of vaccenic acid (18:1 trans-11) at the C-9 position.

Although the amount of CLA necessary to protect humans is unknown, extrapolation from amounts needed to reduce chemically induced tumors in rats indicates that 700 to 800 mg/d may be required for humans. Other estimates, which consider the ability of humans to convert vaccenic acid in milkfat to CLA, set the effective daily intake at 420 mg or less. Since the estimated daily human consumption of CLA in the United States ranges from 100 to 300 mg, the effective anticancer dose may be reached easily by enhancing the levels of CLA in dairy foods. Milk from grass-fed animals contains three to five times more CLA than that from cows fed the usual diet of 50% hay and silage plus 50% grain (Dhiman, 2001). Feeding oils from soybean, sunflower, peanut, linseed, and fish to dairy cows can increase the CLA content of milk. Feeding full-fat extruded soybeans and cottonseed to dairy cows doubled the CLA content of milk and cheese compared with feeding a normal diet. Other factors, such as feeding high grain diets and reducing the forage particle size, will decrease CLA in milk.

Although CLA is best known for its anticancer properties, researchers have also found that the cis-9, trans-11 form of CLA can reduce the risk for cardiovascular disease and help fight inflammation (Tricon et al., 2004). CLA is also known for its body weight management properties, which include reducing body fat and increasing lean muscle mass. A meta-analysis of over 30 clinical studies showed that CLA has a small impact on fat mass (Whigham et al., 2007).

Unlike other trans fatty acids, it is not harmful, but beneficial. In the United States, trans linkages in a conjugated system are not counted as trans fats for the purposes of nutritional regulations and labeling. In July 2008, CLA received a "no objection" letter from the FDA on its **generally recognized as safe (GRAS)** status for certain food categories including fluid milk, yogurt, meal replacement shakes, nutritional bars, fruit juices, and soy milk. With GRAS status, food companies are now able to add CLA to products in these food categories.

28.3.5 Probiotic Bacteria

The first dairy foods recognized to deliver probiotic bacteria (cf. section 23.3.2) to consumers were cultured products such as acidophilus milk and modern yogurts (cf. chapter 29). Recently, Ganeden Biotech developed the GanedenBC30 strain of probiotic spore-forming bacteria said to survive normal milk pasteurization. This strain of *Bacillus coagulans* is, therefore, suitable for use in fresh pasteurized milk as well as in cultured dairy products. Bacterial spores have advantages over vegetative cells in their abilities to survive the stomach's gastric acidity and bile of the intestinal tract. They must germinate and reproduce in the intestine to be effective. However, to possess these advantages when used in pasteurized fresh milk, they must not germinate to a significant degree in milk held refrigerated before consumption.

28.3.6 Flavored Fluid Milk Products

The most popular flavoring of milk is chocolate. The product may be called chocolate milk or chocolate-flavored milk only when the minimum fat content is 3.25%. Otherwise, it is given a fanciful name such as chocolate-flavored drink. Chocolate contributes fat: cocoa must contain at least 22% fat and chocolate liquor from 50 to 58% fat. In addition to milk, chocolate drinks contain about 6% sugar, 1.0 to 1.5% cocoa (or 1.25 to 2.25% chocolate liquor), 0.2% stabilizer (vegetable gum or sodium alginate), vanilla, and salt.

Most consumers prefer a flavor like that of milk chocolate. There should be little bitterness and mild sweetness. Sedimentation and unnatural color are the most frequent defects. Differences exist among markets and among individuals within a market as to preference for color and flavor intensity. Sedimentation is associated with solubility of chocolate, its fineness of grind, and viscosity of the finished product. Stabilizers are added to minimize sedimentation and commercial formulations appear to accomplish this objective. Over-stabilization may decrease consumption because products with a heavy body are more filling. Dutch-processed cocoa tends to have a higher solubility than its American-processed counterpart but it is darker in color and has a different flavor.

Cocoa is recognized as a good source of antioxidants due to its high content of polyphenols, which alter the blood concentrations of lipoproteins in favor of the high density versus low density types. However, cocoa treated with alkali has relatively low antioxidant activity.

Optimal processing conditions for chocolate-flavored drinks depend upon ingredients, especially stabilizer. For example, sodium alginate should not be added until the milk has been heated to at least 149°F (65°C). Creaming (rise of cream to surface) may occur if milk is not homogenized. However, homogenization of the milk-cocoa mixture results in wear on the homogenizer because of the abrasive effects of chocolate fibers. Heating the entire nonhomogenized but stabilized mixture at 156°F (69°C) for 15 to 30 min minimizes creaming.

Flavorings for other milk drinks include honey, strawberry, cherry, raspberry, pineapple, apple, orange, and banana. Because most of these fruits contain acid, stabilizers must be compatible with acids. An interesting sidelight is that the addition of 1 g/L (1,000 ppm) of terpineol, an essential oil from orange peel, slows bacterial growth and extends the shelf life of milk. Also, vitamin C from oranges and other citrus fruits prevents the development of oxidized flavor in milk.

Eggnog is a seasonal drink with maximum sales in November to January. US federal standards require at least 6% milkfat, 8.25% nonfat milk solids, and 1% egg yolk solids. Also, eggnog-flavored milk and eggnog-flavored lowfat milk must contain at least 0.5% egg yolk solids plus 3.25% and 0.5% milkfat, respectively. Most states permit 0.5% stabilizer in these products. Flavorings include nutmeg, cinnamon, vanilla, and rum concentrate. Artificial color and about 7% sugar are added also. Most suppliers offer prepared eggnog mix, which can be added to a mixture of plain homogenized milk, plain ice cream mix, and sugar. Spices tend to settle out if viscosity is low. For this reason, and to promote the appearance of richness, most processors make a somewhat viscous product (accomplished with stabilizers).

Fruit drinks can be profitable for fluid milk processors because the equipment and technology necessary for their production and marketing are available. These products often provide a greater profit margin than milk products. Most dairy plants sell orange juice and some offer orange drinks. Fruit drinks (also beverages or cocktails), products containing less than 100% of the expressed juice from the specific fruit, are described in 21 CFR 102.33. If prepared from concentrated juices, the label shall so indicate. Unless sterilized and aseptically packaged, the product must be kept refrigerated because yeasts, molds, and enzymes can spoil it. Chemical oxidation of oils, esters, aldehydes, and ketones, which are flavorful components, is also slowed by cold temperatures.

28.3.7 Specialty Products

Stabilizers are usually added to whipping cream to prevent coagulation during heating and to improve whipability, **overrun**, and stability of the whipped product. Because sales of high-fat products are relatively low compared with those of various milks, cream processing tends to be done in specially equipped plants in which large volumes are processed for wide distribution. Most producers use ultrapasteurization, 293°F (145°C) for 2 to 3 sec, and **aseptic packaging** to produce products with long shelf life. Contents of most packages are sterile. Nevertheless, refrigeration is employed throughout distribution and is recommended during use because sterility is not assured (1 to 2% of the containers may contain viable bacteria).

Few processors in the United States produce sterile milk or other lowfat products because costs of production exceed benefits of longer shelf life in distribution. In addition, heat-induced defects are likely to occur. However, aseptically packaged ultrahigh temperature (UHT) milk is produced widely in many countries of the world. Specialty products, such as lactose-free milk and specialty milk-based drinks, are usually UHT-processed for extended shelf life (ESL).

Milk protein concentrate (MPC) is produced by ultrafiltration (cf. section 26.7) of nonfat milk, the retentate being concentrated to 70% to 85% solids. Permeate produced in the process is high in lactose and minerals. It is dried forming an additive for baked goods, puddings, instant drink mixes, and more.

28.3.7.1 Lactose-Free Milk

Because they are malabsorbers of lactose, many consumers are unable to enjoy the benefits of milk. Lactose-free milk is widely available but at up to twice the cost of regular milk, primarily because of the cost of the lactose hydrolysis process and lower demand. Low lactose or lactose-free milks can be produced by addition of the yeast-derived enzyme lactase (β-galactosidase) to milk (cf. chapter 21). The enzyme hydrolyzes this 12-carbon (disaccharide) to its two substituent 6-carbon components, glucose and galactose. This process enables consumers who produce insufficient lactase in their small intestine to absorb the simple sugars.

Other solutions for consumers include in-home hydrolysis of lactose at cold temperatures, which can reduce the cost of purchasing lactose-free milk. Two brands of commercially available β-galactosidase hydrolyzed over 95% of milk's lactose in 24 hr at 4°F (2°C). However, it took twice the recommended dosage for two other brands to reach this objective. Alternatively, consumers can take a pill containing lactase as they consume the unhydrolyzed lactose in fluid milk products and unripened cheeses.

A common industrial process calls for in-line addition of microbe-free lactase to the milk immediately after UHT treatment. Lactose hydrolysis occurs in the final package. Since the active enzyme remains in the milk during storage, the enzyme preparation must be of the highest quality. Very high-quality raw milk is required when hydrolysis is done prior to pasteurization because treatment conditions, 43 to 46°F (6 to 8°C) for 24 to 30 hr, permit growth of psychrotrophic bacteria that can produce off flavors (cf. chapter 23).

Another way to produce lactose-free milk is to remove 50% of the lactose by ultrafiltration then hydrolyze the remaining portion with lactase. By splitting the lactose into its two monosaccharide components, the final product gains the same sweetness as standard milk. Ultrafiltration also permits concentration of proteins so products can be standardized for protein content. For example, one sports drink is reported to contain 18 g protein per 8 oz serving. This compares to 7 g in an equal quantity of milk.

28.3.8 Organic Milk

The Organic Foods Production Act of 1990 established the National Organic Program (NOP), which is implemented by the USDA Agricultural Marketing Service (AMS). The program facilitates domestic and international marketing of fresh and processed food that is organically produced and assures consumers that such products meet consistent, uniform standards. It established national standards for the production and handling of organically produced products, including a list of substances that are approved or prohibited from use in organic production and handling. Operations that wish to produce products with a "USDA organic" label are certified and regularly inspected by AMS.

The demand for organically produced foods has grown significantly during the past decade. Under the NOP, the label "organic milk" does not refer to the quality of milk produced, but to the elements that are used and fed to cows. The NOP provides specifics for livestock feeding and living conditions. The main components state that (1) animals must graze pasture during the grazing season, which must be at least 120 d/yr; (2) at least 30% of dry matter must be obtained by grazing pasture during the pasture season; and (3) producers must manage pastures to meet the feed requirements of the grazing animals and to protect the soil and water quality.

28.4 Packaging

There are still those who think that once packaging gets beyond the essentials of portability and labeling, it becomes an expensive triviality—something that should not be taken seriously unless society has more money than good fiscal judgment.

Faith Prior

The first glass milk bottle was invented by Dr. Harvey D. Thatcher in 1884. The paper carton was invented by John Van Wormer, who patented the "pure-pak" container in 1915. His inspiration came from dropping and breaking a bottle of milk and getting liberally splashed with the milk. It took ten years for this toy maker to perfect a machine for making the cartons, but by 1950 his company was producing cartons at the rate of 20 million per day.

Today's milk consumers have several choices of types and sizes of packages, including plastic-coated paper, blow-molded plastic, plastic bag in a box, and glass. Lightweight all-plastic bottles are made for single-service use; returnable types are considerably heavier. Plastic-coated paper containers, once the mainstay of the dairy industry, are found in sizes of one-half gallon and smaller. Those having a thin layer of aluminum foil between the paper and the inner layer of plastic are used with sterile and ESL products because of the necessity to limit oxygen penetration (figure 28.2 on the next page).

Blow-molded, all-plastic dominates the gallon-size market. Small plastic containers have become popular because of the versatility of production molds and resins, as well as marketing possibilities due to the availability of many printing styles and over-wrapping with heat-shrinkable plastic. In 1965, the percentages of fluid milk products packaged in paper, glass, and plastic were 65%, 29%, and 4%, respectively. By 2005, plastic dominated with 85% of the total; paper use had declined to 15%; and less than 0.5% was in glass. Table 28.2 on the following page shows the distribution of container sizes during 10-year intervals (1965–2005) for the month of November.

When selecting the appropriate type of package for a fluid product, the following factors should be considered: economy, convenience, appearance, disposability, safety, consumer preference, ease of filling and handling, durability, and protection afforded. A good consumer package (1) adequately protects quality during necessary shelf life in the given climatic region, (2) can be handled efficiently in production and distribution, (3) has high merchandising and consumer convenience values, and (4) minimizes packaging costs consistent with other essential requirements.

28.4.1 Paper Cartons

The first paper cartons were coated with wax through a dipping process, formed into shape by a mandrel, then filled with product and sealed shut with heat. Today's carton packaging material is a laminate of paper and polyethylene, and, for aseptic packages, aluminum foil (figure 28.3). Paper makes packages stiff, plastic renders them liquid-tight, and aluminum foil blocks light and oxygen. This combination of material is varied to suit product category, but in each case the only material to touch package contents is food-grade polyethylene.

Container materials may be sterilized with ethylene oxide in advance of shipment to users, however, for in-line processing it is impractical due to the time required to achieve sterilization. In order to create aseptic paper

Figure 28.2 An array of packages for aseptic filling (Tetra Pak; https://creativecommons.org/licenses/by-sa/2.0/deed.en).

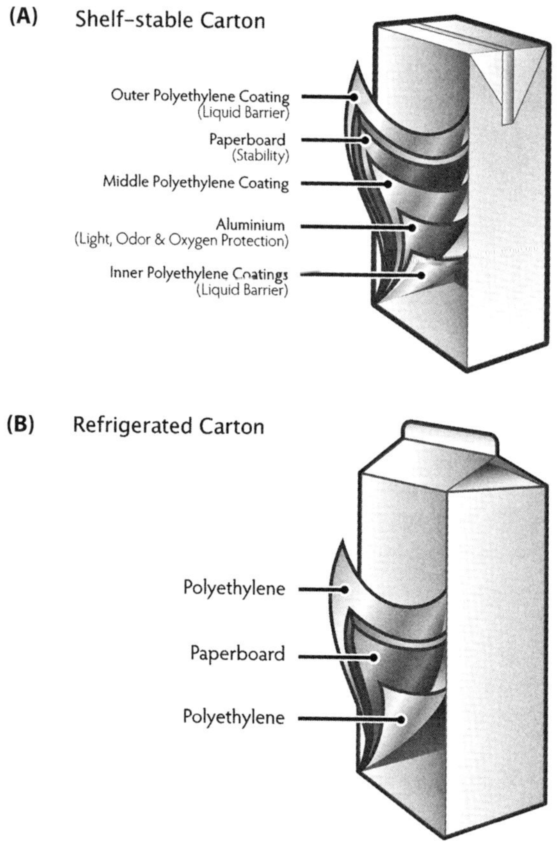

Figure 28.3 Construction of (A) shelf-stable carton and (B) refrigerated carton (courtesy of Carton Council).

Table 28.2 Percentage of Total Fluid Milk Products Sold by Size of Container.[a]

Year	Size of Container							Total
	Gallon	½ Gallon	Quart	Pint	Half-Pint	> 10 qt	Other	
	Percent							
1965	17	54	12	1	10	4	2	100
1975	43	34	7	1	11	3	1	100
1985	60	22	5	2	9	2	< 0.5	100
1995	64	18	4	2	10	2	< 0.5	100
2005	65	18	3	2	10	1	1	100

[a] Figures in the table are based on total sales volume in all federal milk orders combined.

Source: USDA Agricultural Marketing Service, 2005.

packaging, the paper used to make the container passes from a reel through a trough of hydrogen peroxide (H_2O_2). The paper is pressed clean and formed into a tube, then it passes electric heaters that elevate interior paper surfaces to 400°F (204°C) to volatilize the remaining peroxide and complete sterilization.

28.4.2 Plastic Containers

The plastic bag in a box has been available in multigallon sizes. The bag-in-box system improves space efficiency and reduces weight compared to rigid packaging. A modification of this type is the returnable plastic container into which a filled plastic bag is placed.

European companies pioneered the "pitcher-pak" idea in which plastic bags are formed from a roll of plastic film, sealed at the sides and bottom, filled, and heat-sealed through the column of milk at the top to eliminate headspace. The low-cost film is 3-mil polyethylene. Pouches for sterile packaging are made of laminated polyethylene-foil-polypropylene. Two or three individually filled pouches can be packed within a second polyethylene bag for multipackage sales. Plastic-coated paper and plastic film can be sealed with heat.

28.4.3 Packaging Equipment and Filling Processes

Machines are available that form, fill, and seal cartons aseptically (figure 28.4 on the following page). Some machines are capable of running three sizes of cartons simultaneously and can be preprogrammed to run selected amounts of product. Furthermore, programs are available for fault diagnosis concerning weight or volume of fill. Whereas fill is usually measured by weight, sales are based on volume. Most states that have legislation on volume measurement require testing after tempering to 68°F (20°C). A quart of milk at 68°F (20°C) weighs approximately 0.25% less than the same volume at 50°F (10°C). Therefore, in terms of total sales, the temperature at which the volume is determined can be significant.

Precise control in distributing quantity of product per package is vital. A sound, statistically based method of monitoring container fill should be pursued in every milk-processing plant. It is possible to have the average weight near that desired and yet observe deviation of individual weights outside established limits. In high volume operations, automatic devices are essential for checking package volume or weight then providing feedback to control mechanisms that, in turn, adjust quantity of product dispensed.

Overfill cannot be afforded, and underfill must be within regulatory limits. Several factors can influence fill: foam, head pressures in filler bowls, temperature variations in milk, and number of filling machines on a feeder line. The material of the container can also play a role; for example, blow-molded plastic containers may shrink during storage as much as 5%. Therefore, they should be used within 24 hr of production unless measures are taken to assure they are of proper size before filling.

Clean room or clean area packaging is essential in the manufacture of products with long shelf life. The area may consist only of an enclosed portion of the packaging machine into which sterile air is forced. Air is rendered essentially sterile by filtration through high-efficiency particulate air (HEPA) filters capable of removing 99.97% of particles measuring at least 0.3 µm in diameter. Filtered air is blown past the packaging site in **laminar** flow to prevent entrainment of microorganisms, which occurs readily in turbulent air. Operators wear special clean apparel that they change into just before entering the room.

28.4.4 Environmental Concerns

Increasingly important to many consumers and the dairy and food industry is the effect of packaging and transportation on the "carbon footprint" as well as biodegradability of food packaging materials. Since plastics are virtually nondegradable and require oil as the base material, researchers have sought to modify them by adding biodegradable components. "Bioplastics" describes plastics that are biodegradable, such as water soluble plastics like polyvinyl alcohol, or have been made from biodegradable materials (e.g., corn and sugar cane).

The ecological aspect of milk packaging cannot be disregarded. What are the interrelationships among humans, the materials used to package their food, and the environment? These must be considered from the standpoint of raw materials (availability, renewability, and alternate uses), processing (valuable resources required and pollution created), distribution (comparative energy requirements), and disposal (recyclability and pollution aspects). However, consumer needs and preferences must be part of these considerations.

Glass containers are reusable, except when they are broken. Raw materials for their manufacture are not renewable, but supplies are abundant. Several aspects of their use, however, are ecologically undesirable. Heat is required to raise the mixture of raw materials to 2,732°F (1,500°C). Glass is comparably heavy, requiring high inputs of energy in handling and transporting. This is of special significance in the return of bottles for reuse. Detergents used in washing must be disposed of, creating the potential for water pollution and additional demands for water.

Raw materials for paper containers are renewable. One acre of high-yield managed forest will produce about 80 tons of paper in 40 to 60 years. Additionally, the forest absorbs about 7.4 tons of carbon dioxide and discharges 5.4 tons of oxygen per year. However, cull material and trimmings left on land, when burned, pollute the air. Most chemicals used in processing wood into paper are recovered. Unusable parts of logs are burned, furnishing some of the power required for the process. The greatest problem in processing is air pollution by sulfhydryl compounds in the vicinity of the plant. Paper containers may be disposed of as solid waste and they are about 85% biodegradable in six months. However, plastic and metal found within laminated containers slow degradation of paper and are themselves not biodegradable.

Ethylene, the basic raw material for polyethylene, is stripped from natural gas, a limited, nonrenewable natural resource. Processing requires high energy input but there is no significant waste accumulation. Thermal pollution of water may be a problem and low molecular weight hydrocarbons may be discharged to the air in the vicinity of the plant. Some plastic (polyethylene) containers are returnable. These are heavier than single-service containers and transportation costs are higher. Moreover,

(A)

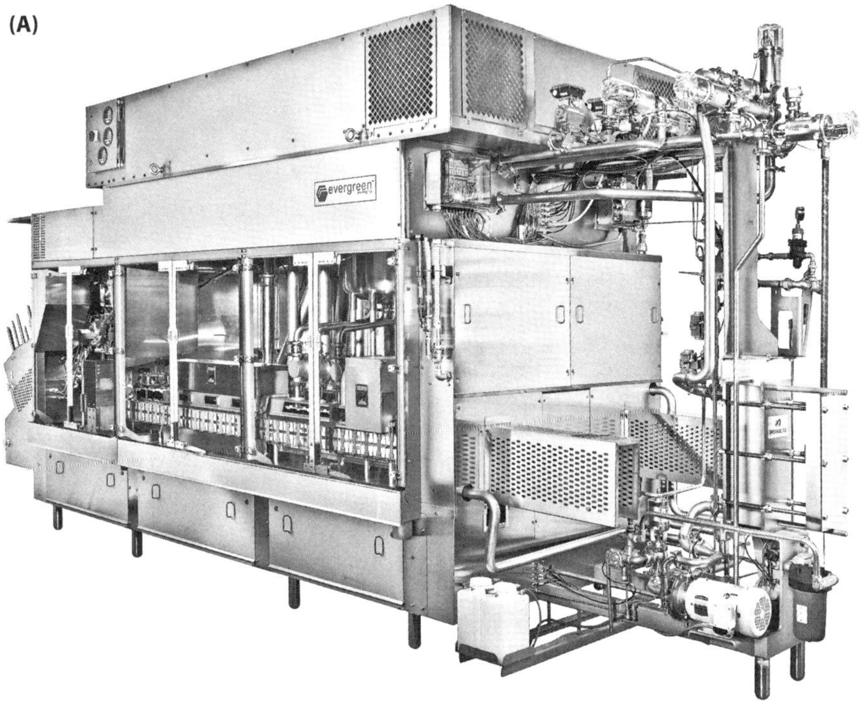

Figure 28.4 (A) EQ-70 gable top packaging machine including controlled environment features. (B) Mandrels for bottom forming. (C) Top heat sealing device (shown from the discharge end of the machine) (courtesy of Evergreen Packaging).

(B)

(C)

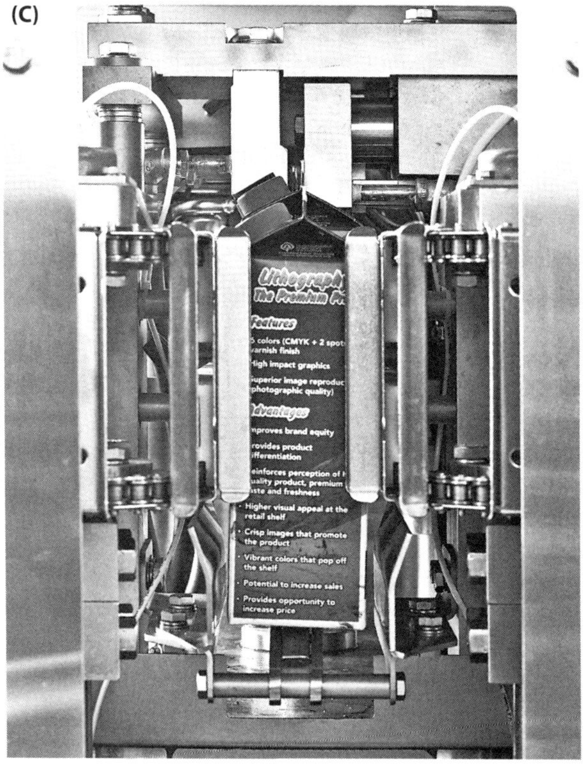

special incineration methods are required to burn plastic and it does not decompose in landfills.

Baribo (1972) summarized the situation interestingly: "How do you balance the different factors with different materials as they affect the ecosystem? There obviously is no simple answer. Trade-offs relate to demographics, consumer habits and preferences, the unique interrelationship of environmental problems in specific communities and regions, and the ways legislation and regulations are developed and interpreted."

28.5 Labeling

Section 4 of the *Grade "A" Pasteurized Milk Ordinance* (US Food and Drug Administration, 2013) states:

> All bottles, containers, and packages containing milk or milk products, except milk tank trucks, storage tanks, and cans of raw milk from individual dairy farms, shall be conspicuously marked with:
>
> 1. The identity of the milk plant where pasteurized, ultra-pasteurized, aseptically processed and packaged, retort processed after packaging, condensed, and/or dried.
> 2. The words "keep refrigerated after opening" in the case of aseptically processed and packaged low-acid milk and/or milk products and retort processed after packaging low-acid milk and/or milk products.
> 3. The common name of the hooved mammal producing the milk shall precede the name of the milk or milk product when the product is or is made from other than cattle's milk. As an example, "Goat," "Sheep," "Water Buffalo," or "Other Hooved Mammal" milk or milk products, respectively.
> 4. The words "Grade 'A'" on the exterior surface. Acceptable locations shall include the principal display panel, the secondary or informational panel, or the cap/cover.
> 5. The word "reconstituted" or "recombined" if the product is made by reconstitution or recombination.

Additionally, labels must meet applicable requirements of the Federal Food, Drug, and Cosmetic Act and the Nutrition Labeling and Education Act of 1990.

For milk to be labeled "antibiotic free," the dairy firm must provide evidence satisfactory to the regulatory agency that milk providers do not use antibiotics on cattle in their dairy herds. The fact exists that negative test results for antibiotics in milk does not mean that the milk tested is "free" of antibiotics. To assure by testing that milk is antibiotic free would require testing for every antibiotic that is or can be used to treat lactating dairy animals (US Food and Drug Administration, 2008).

28.6 Storage

Major considerations in storage of pasteurized milk products are maintenance of temperature between 34 and 39°F (1 and 4°C), protection of packages, rotation of inventory, efficiencies in product movement, space utilization, and inventory control. Systematic handling is a necessity. Requirements that buyers receive certain minimum quantities, preferably full pallets, result in operational efficiencies.

Materials handling systems are available that place all products in storage, pick orders from storage, and place these orders in case stacks at the truck-loading position. Each function can be accomplished mechanically. The system establishes product rotation and managerial control of inventory and of shipments. A code applicator and a code reader are essential in this operation. Manual transfer may be necessary for low-volume items.

28.6.1 Care of Milk in the Home, Restaurant, and Institution

No factor affects milk's keeping quality in the home as much as its storage temperature. Time during which its flavor remains acceptable decreases by about half for each increase in temperature of 5°F (3°C) above 40°F (4.4°C). Many refrigerators do not maintain temperatures consistently below 40°F (4.4°C). Other factors such as overloading with warm foods and leaving milk out of the refrigerator for prolonged periods are detrimental to keeping quality. When milk is purchased in large containers, small quantities are often withdrawn for table use. Portions withdrawn likely will be contaminated with small numbers of bacteria in the vessels. If such vessels are returned to the refrigerator for long-term storage, or if the same vessel is repeatedly filled without cleaning, spoilage is likely to ensue.

28.7 Sales and Distribution

The most valuable asset of a firm that sells milk and milk products is the good will and confidence of the consuming public.

W. H. E. Reid (Professor, 1919–1964)

About 84% of total fluid milk is sold through supermarkets (70%) and mass marketing supercenters (14%). Other outlets include convenience stores, dollar stores, club stores, and drug stores. Surprisingly over 5% is sold through federal school programs.

Costs of delivery per unit of product and per stop are determined by dividing total costs by the number of units (or points—some units are assigned more than one point depending on size or value) and number of stops, respectively. Because a large proportion of costs per stop is fixed, significant economies are realized when large volumes are delivered at each stop. Thus, reducing deliveries from three to two or one per week is favored. This has stimulated packaging in large containers for home delivery. Consumption is reportedly greater when larger containers are delivered.

Locations of sales of fluid-milk products, at homes or in stores, are determined primarily by relative costs, convenience, services rendered, keeping quality, and frequency of shopping. The economies associated with large-volume deliveries favor sales through superstores. However, this introduces a middle person and requires an additional marketing margin; thus, by efficient management, in many areas home delivery can be competitive.

Of course, many other factors are associated with economy in the delivery system. For example, route planning to optimize mileage and time on routes demands careful consideration; truck performance and fleet maintenance require management's attention; positive attitudes and salesmanship of drivers are keys to success. In relation to salesmanship, it is significant that in sales to retail stores, especially large superstore chains, upper-level sales personalities deal with store management. Therefore, store sales assume a different dimension than those at the consumer's residence. The distribution of the milk processor's sales proceeds to costs and income has been approximated as the following: raw materials = 55%; salaries and wages = 21%; facilities, depreciation, and insurance = 7.5%; containers = 7%; services and supplies = 5.5%; advertising = 1.5%; taxes and licenses = 1.5%; and net profit = 1%.

Table 28.3 Tests Used Frequently in Compositional and Quality Control of Fluid Milk.

Component	Test Methods
Fat	Babcock, Mojonnier, infrared and near-infrared
Nonfat solids	Mojonnier, lactometric (estimate)
Protein	Dye-binding, infrared, Kjeldahl
Lactose	Polarimetry, infrared
Lactic acid	Titratable acidity
Alkaline phosphatase	Fluorometric, chemiluminescent, Rutgers and Scharer methods
Sediment	Filtration
Added water	Thermister cryoscope (freezing point)
Inhibitors	Growth inhibition, enzyme substrate, receptor assay, ELISA, HPLC, and specific tests for sulfonamides, tetracyclines, and chlorampohenicol
Microorganisms	Standard plate count, direct microscopic count, psychrotrophic count, coliform count

28.8 Quality Assurance

Fluid milk processors market some of the most readily perishable foods. Both chemically and microbiologically induced deterioration threaten their shelf life. Tests must be performed on raw materials and finished products to assure quality standards. Additionally, compositional and sensory attributes of each product must be monitored regularly. We discussed tests of quality in chapter 25. Let us now make specific application to the fluid milk industry. Table 28.3 is a compilation of major tests employed by the fluid milk industry in compositional and quality control.

Antibiotic testing is always done before milk is off-loaded from farm bulk tank trucks. In receiving operations, composition is primarily controlled by periodic testing for fat content and added water. Quality is determined, usually monthly, by standard plate counts made on milk from individual producers and by sensory evaluation for flavor and odor. Other tests are applied less frequently to determine content of nonfat solids, protein, sediment, pesticides, and titratable acidity.

Each processor should maintain careful quality assurance of raw materials. When cooperatives that supply milk to processors monitor quality of milk from individual producers, processors usually need only maintain quality surveillance of tanker-loads of milk as they are received. However, when quality is unsatisfactory, coordination between the supplier and the field staff is imperative. Therefore, communication becomes highly important since there is at least one additional layer of personnel between producer and processor. Of course, the processor must communicate also with the federal milk market administrator (except in areas having no federal order; cf. chapter 20) and with health authorities (cf. chapter 24).

During processing, compositional control is of major concern. Fat is the component that must be most closely controlled because it varies more among different lots of milk. Tests should be used to control composition of products with added nonfat solids. Of course, fill control of containers deserves especially close attention.

After processing, the manufacturer is primarily concerned that products meet compositional specifications, have the desired flavor, aroma, and appearance, and have a long shelf life. Therefore, representative samples of each lot should be subjected to analyses for fat and nonfat solids, bacterial content, shelf life at 45°F (7°C), and, less frequently, to sensory evaluation in comparison with competitors' products. If steam-vacuum treatment has been used, freezing point determinations will provide data on probabilities of concentration or dilution (provided removal of carbon dioxide by vacuum treatment is considered a factor that will increase freezing point by as much as 0.008°C).

Periodic tests of sanitizer strength after use and of numbers of residual bacteria on equipment and in containers can provide vital data for use by perceptive quality-assurance personnel. It is important that tools/test kits be used to enable technicians to detect in real time potentially hazardous or undesirable conditions. Devices are available that enable rapid detection of microorganisms or invisible soil (fat, protein, minerals) on equipment. Also available is equipment to detect or eliminate metal fragments in certain foods.

No test should be performed if there is no definite purpose or a plan for use of the data. Too often management asks for tests to be performed with insufficient plans for their use.

The Public Health Security and Bioterrorism Prepardedness and Response Act of 2002 requires manufacturers to keep records of goods received and of shipped finished products. Manual record keeping can work well for bulk ingredients, but when numerous minor ingredients are used, automated data collection and analysis can reduce labor and increase accuracy. Such a system places necessary controls on the production floor to capture and control inventory lot numbers and ingredient usage. Lot numbers are captured at the receiving, weighing, mixing, portioning, and shipping stages, eliminating paperwork and administration. Usages at bulk, minor, and micro ingredient levels are recorded automatically with each lot number added to the mix. This system assures addition of correct ingredient weights and that the ingredient inventory will be used in correct rotation.

Finally, when samples are taken by regulatory agencies or for use of a second party in buying or selling, processors should obtain duplicate samples and have the same tests performed as are intended by that second party. These tests must be performed by the same standardized techniques by both parties. This means that laboratory personnel from the industry must be trained to perform as capably as personnel in any approved regulatory laboratory. To settle for less than precise, standardized techniques performed by trained and motivated technicians is to settle for mediocrity and questionable test results. From such results come undependable management decisions.

The Innovation Center for US Dairy initiated an industry-wide traceability project in 2013 that led to release of practices for processors to use in establishing programs to ensure traceability of products of their firms. This led to publication of *Guidance for Dairy Product Enhanced Traceability*. The document contains information needed to implement traceability standards. Included is a 21-point checklist of important components of a protocol critical to a traceability system. Three key elements are:

1. Identify and record lot IDs (key data elements): places where bulk products, ingredients, or packaging materials are added to make the final product.
2. Identify and record flows (critical tracking events): the main flow paths in the facility that products pass through.
3. Place a standard, human readable Lot ID on each product.

A good traceability program should allow a manufacturer to perform three key tasks: (1) find the source of an issue, (2) find the common point of convergence of products in an issue, and (3) find all the products that contain that common point of convergence across the corporation.

Guidance for Dairy Product Enhanced Traceability provides samples of typical entry points for manufacture of fluid milk products (figure 28.5), dry products, cheeses, cultured products, butter, and frozen desserts.

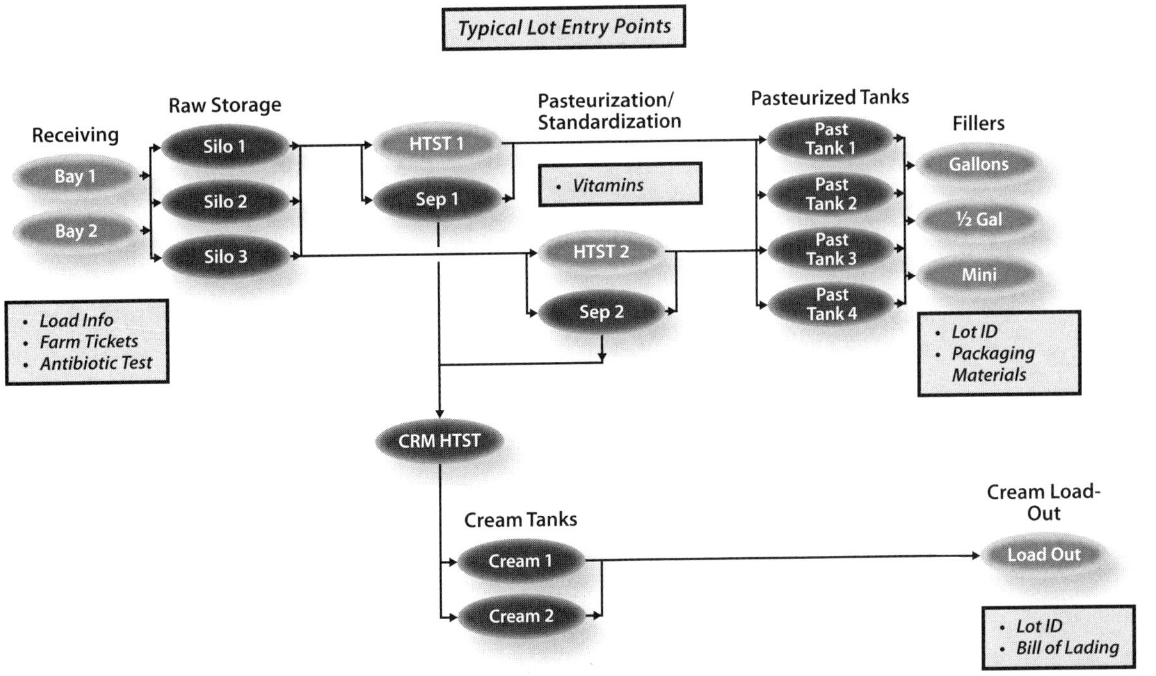

Figure 28.5 Typical lot entry points (text boxes) for ingredients, materials, and equipment in a fluid milk plant (Innovation Center for US Dairy, 2013).

28.9 Dating of Milk Containers

The term *milk dating* is used to designate the practice of conspicuously marking containers with a day of the week or date of the month that can be readily understood by consumers. Presumably, placing an identifiable date on the milk package enables buyers to distinguish milk by date of process and thus facilitates rapid movement of milk through distribution channels.

Historically, the premise for dating requirements in milk regulations of some state and municipal governments was the assumption that quality of pasteurized milk was directly related to its age. Research does not support this premise. A dating regulation that includes an expiration date may discriminate against the product by implying a deterioration of quality that actually is not occurring.

28.10 Management by Objectives

Management by objectives involves setting daily goals and then listing procedures to accomplish them. Each supervisor lists objectives and sets a timetable for their completion. From daily goals a formal program of weekly goals is obtained. Objectives are agreed upon by each foreman and the plant superintendent, who incorporates them with other objectives and shares them with the production manager. All objectives are written on an objective form and anticipated benefits are listed, together with estimated completion date. A good objective is (1) specific, (2) brief, (3) measurable (preferably in economic terms), (4) realistic, (5) consistent with total operational objectives, and (6) time-limited. Objectives can be selected from all areas of operations including (1) cost reduction, (2) management development and training, (3) sales, (4) quality assurance, (5) sanitation, (6) safety, and (7) labor relations.

Management by objectives has the further advantage of making it easier to select outstanding personnel for advancement and provides for improved communication among members of the management team.

28.11 Summary

Milk lends itself to fractionation and to the addition of nutrients and flavors, thus providing opportunities for manufacture of numerous fluid products. Various processes and packages are available. These allow dealers to differentiate their products from those of competitors. Careful managerial control must be exercised from receiving room through the distribution system to assure quality, conformity to compositional and quantitative specifications, efficient operation, and traceability.

Grade "A" raw milk is processed into a variety of fluid milk products that have been pasteurized, homogenized, standardized for content of major components, and packaged conveniently. Each step in both production and processing is regulated by the *Grade "A" Pasteurized Milk Ordinance.*

Major objectives of the processing firm are to produce products having desirable composition and sensory attributes, maximal keeping quality, unquestionable safety, and informative labeling. Concurrently, the firm must meet or surpass the financial expectations of the owners. Utilization of an objective system of management by expert personnel helps assure that this goal is met.

No processor can afford to neglect activities that assure consumers of safe, nutritious, and appetizing milk and milk products. One experience with an unsatisfactory product does more to alienate consumers than multiple satisfying experiences and much expensive advertising. It is obvious, then, that the milk processor must take seriously the key role played in providing safe, nutritious, and delicious milk for consumers.

Study Questions

1. Approximately what proportion of the US milk supply is utilized as fluid milk and fluid milk products?
2. Identify important aspects of accountability of milk solids in manufacturing fluid milk products.
3. What are some major responsibilities of the production manager in processing and packaging fluid milks?
4. Which vitamins are commonly added to milk? Why is it more important that vitamin A be added to lowfat and skim milks than to whole milk?
5. Can trained taste panelists more readily detect differences of fat or of nonfat solids in fluid milk? What are their relative sensitivities to concentrations of lactose and proteins?
6. What is the most popular flavoring of milk? Differentiate between chocolate milk and chocolate drink.
7. By law, what quantity of egg yolk solids must eggnog contain?
8. What is the approximate sodium reduction ratio in low-sodium milk (versus normal milk)?
9. Should most fruit drinks marketed by dairy foods processors be refrigerated? Why?
10. Why are stabilizers sometimes added to whipping cream?
11. What are the four basic types of packages currently used by the dairy industry?
12. Why are high-efficiency particulate air filters used in the packaging of products with potential long shelf life?
13. Identify important factors to be considered in selecting packages for fluid milk products.
14. What are important qualities and characteristics of good packages?
15. Identify factors that influence fill of milk packages.
16. Are milk containers commonly filled by weight or volume? Are sales by weight or volume?

17. What are major considerations in the storage of fluid milk products?
18. What is meant by quality assurance in the dairy industry? Outline steps of assuring high-quality dairy products for consumers.
19. What are the most commonly employed quality tests for milk?
20. What is meant by milk dating? What is a limitation of it?
21. Discuss briefly the proper care of milk in the home.

NOTE

[1] Also known as *market milk* because it is marketed in the liquid form.

REFERENCES

Baribo, L. E. 1972. The ecology of milk packaging. *Journal of Milk and Food Technology* 35:121–125.

Council on Scientific Affairs, American Medical Association. 1993. Report of the Council on Scientific Affairs: Diet and cancer: Where do matters stand? *Archives of Internal Medicine* 153:50.

Dhiman, T. R. 2001. Role of diet on conjugated linoleic acid content of milk and meat. *Journal of Animal Science* (Abstr.) 79:241.

Huth, P. J., D. B. DiRienzo, and G. D. Miller. 2006. Major scientific advances with dairy foods in nutrition and health. Centennial Issue, *Journal of Dairy Science* 89:1207–1221.

Innovation Center for US Dairy. 2013, September 10. *Guidance for Dairy Product Enhanced Traceability: Voluntary Practices and Protocols for Strengthening the US Dairy Supply Chain.* Innovation Center for US Dairy.

Tricon, S., G. C. Burdge, S. Kew, T. Banerjee, J. J. Russell, E. L. Jones, R. F. Grimble, C. M. Williams, P. Yaqoob, and P. C. Calder. 2004. Opposing effect of cis-9, trans-11 and trans-10, cis-12 conjugated linoleic acid on blood lipids in healthy humans. *American Journal of Clinical Nutrition* 80(30):614–620.

US Food and Drug Administration. 2013. *Grade "A" Pasteurized Milk Ordinance.* Washington, DC: Author.

US Food and Drug Administration, Center for Food Safety and Applied Nutrition. 2008, May 7. Questions and Answers from the Southeast Region Milk Seminar, Nashville, Tennessee (November 6–9, 2007) and FDA Training Courses Held in FY 2007. College Park, MD: Author.

USDA Agricultural Marketing Service. 2005, November. *Packaged Fluid Milk Sales in Federal Milk Order Markets: By Size and Type of Container and Distribution Method.* Washington, DC: Author.

Whigham, L. A., A. C. Watras, and D. A. Schoeller. 2007. Efficacy of conjugated linoleic acid for reducing fat mass; a meta-analysis. *American Journal of Clinical Nutrition* 85(5):1203–1211.

WEBSITES

Elopak (http://www.elopak.com)
Ganeden Biotech (http://ganedenlabs.com)
Tetra Pak (http://www.tetrapak.com)
US Food and Drug Administration, 21 CFR 131 (http://www.accessdata.fda.gov/scripts/cdrh/cfdocs/cfcfr/CFRSearch.cfm?CFRPart=131)
USDA Economic Research Service (http://www.ers.usda.gov/data-products/dairy-data.aspx)
USDA National Agricultural Statistics Service, Dairy Products Annual Summary (http://usda.mannlib.cornell.edu/MannUsda/viewDocumentInfo.do?documentID=1054)

29

Cultured and Acidified Milk Products

Bulgarian buttermilk gives a harsh, acid flavor which challenges those hardy souls
who would drink it for better health and, supposedly, longer life;
but the aromatic acid flavors of a good yogurt or
cultured buttermilk are as subtle as French perfume.

Frank. V. Kosikowski (1916–1995)

29.1 Introduction

It does not take long to discover that if milk from animals is not consumed soon after it is collected, it becomes sour and coagulated ("clabbered"). This results from contamination of milk by microorganisms followed by their growth and metabolic activities. Fortunately, the dominant bacteria produce lactic, acetic, and formic acids, thus lowering pH and inhibiting growth of bacteria that could produce disease and cause spoilage. Acid in fermented foods is important for this reason. In fact, certain lactic acid bacteria produce various bacteriocins: lacticin, nisin, diplococcin, thermophilin, plantaricin, propionicin, and pediocin. All these are protein or polypeptide antibiotics produced by bacteria used in milk fermentations. Salmonellae, listeriae, and staphylococci, important bacteria among food-borne disease producers, are inhibited by certain starter bacteria. However, in selecting strains of lactic acid bacteria for use in multistrain **starter cultures** it is important to avoid use of those that produce inhibitors to companion strains.

Throughout the world, people have discovered that fermented milks are edible, and in fact, delicious. In the United States and western Europe, cultured buttermilk, acidophilus milk, and **yogurt** are important (table 29.1). In eastern Europe and Russia, Bulgarian buttermilk, kefir, and koumiss are popular. The latter two contain from 1.0 to 2.5% alcohol produced by yeasts that ferment lactose. Koumiss is typically made from mare's milk, kefir is from the milk of goats, sheep, or cows. Dahi, a yogurt of the Indian subcontinent, is known for its characteristic taste and consistency.

Cultured or sour cream is prepared in the same manner as **buttermilk** but is semisolid because of its higher fat content (and the addition of stabilizers in some brands). Cultured milk products result from the souring of milk or its products by lactic acid producing microorganisms, whereas acidified[1] milks are obtained by the addition of food-grade acids to produce an acidity of not less than 0.20% expressed as lactic acid. The word *cultured* was chosen because pure bacterial cultures are used in commercial manufacture. It is also appropriate to use the words *fermented* or *sour* because lactic acid, which causes sourness, is produced by fermentation of milk sugar (lactose).

Of the genus *Lactococcus*, one species, *Lc. lactis*, contains the major mesophilic bacteria used in cultured dairy products. Its subspecies (ssp.) are *lactis* and *cremoris*. *Lc. lactis* ssp. *lactis* var. *diacetylactis* is able to convert citrate to diacetyl. Another diacetyl producer is *Leuconostoc mesenteroides* ssp. *cremoris*, a heterofermenative organism. The two subspecies of homofermentative *Lactobacillus delbrueckii* are *bulgaricus* and *lactis*. These organisms are commonly identified by their genus and subspecies names (see table 29.1).

Regardless of the type of cultured product, the same basic steps are necessary in processing through incubation. These steps are (1) starter culture preparation; (2) treatment of milk, skim milk, cream, or other product (e.g., pasteurization and homogenization); (3) inoculation; (4) incubation; (5) cooling; and (6) packaging. If the product is to be fluid for drinking, the **curd** must be broken and agitated to produce a smooth homogenous body. After incubation, and agitation when necessary, cooling to 44.6°F (7°C) or below is essential to stop acid development and to assure safety. The *Grade "A" Pasteurized Milk Ordinance* (US Food and Drug Administration, 2013) sets maximum (highest) pH limits of finished cultured products at: yogurt, 4.8 at filling and 4.6 within 24 hr of filling; cultured sour cream, 4.7; acidified sour cream, 4.6; and cultured buttermilk, 4.6. Products are then ready for packaging. Some processors package yogurt and cultured cream before incubation.

Most cultured products require addition of stabilizers to thicken and to prevent separation of whey from the coagulated mass. The food's pH is a factor in stabilizer selection. Acidic fruits and milk proteins often perform poorly together without the aid of stabilization. Pectins are choice stabilizers in products of low pH, including drinkable yogurts and smoothies. Extracts of plants, agar, locust bean gum, and guar gum may be certified organic. For more about stabilizers, see chapter 31.

Supplementing milk with nonfat dry milk or milk protein concentrate adds casein, which, after acidification, increases the viscosity of the product. Incubation time needed to reach a specific pH is increased because of buffering by the added protein. Adding whey protein concentrate increases water-holding capacity, conse-

Table 29.1 Selected Characteristics of Cultured Milk and Milk Products.

Product	Range of Fat Content (%)	Average Fat Content (%)	Range of Nonfat Solids (%)	Common Culture Bacteria
Buttermilk	0.1–4.0	1.0	8.5–11	*Lc. cremoris*[a] and *Le. cremoris*
Sour cream	15–20	18.0	7–10	*Lc. cremoris*[a] and *Le. cremoris*
Sour half-&-half	10–12	10.5	8–11	*Lc. cremoris*[a] and *Le. cremoris*
Yogurt	0.25–5.4	1.8	8.5–15	*Lb. bulgaricus* and *S. thermophilis*
Ymer	3–11	3.3	10–12	*Lc. diacetilactis, Lc. cremoris,* and *Le. mesenteroides*
Acidophilus milk	0.1–4.0	2.0	8.5–12	*L. acidophilus*

[a] *Lc. lactis* may be used instead of *Lc. cremoris*.

quently product viscosity. However, too much whey protein can result in a grainy texture. The high-temperature treatment given most milk used in cultured products causes interactions of whey protein and casein, thereby adding to product viscosity. By using high homogenizer pressures with fat-containing products, fat globules are made small. This improves the gel network by increasing globule surface area (cf. section 26.10) onto which casein can be adsorbed. However, high shear from stirring or pumping usually decreases product viscosity.

29.2 Cultures

We define a **culture** (starter) as a controlled bacterial population that is added to milk or milk products[2] to produce acid and/or flavorful substances that characterize cultured milk products. Table 29.2 includes information concerning organisms that are commonly used and their primary functions.

Together with their primary functions, most cultures make secondary contributions to the product. Each acid-producer contributes other flavorful substances. *Lactococcus diacetilactis* contributes more acid and more diacetyl than does *Leuconostoc cremoris.* Diacetyl is a major component of flavor and aroma in cultured products. This 4-carbon diketone (2,3-butanedione) imparts a nutlike flavor within the product to which it is added. Diacetyl is a degradation product of citric acid, hence *Le. cremoris* and *Lc. diacetilactis* are referred to as citrate fermenters. Citric acid is sometimes added to milk or cream to increase the **substrate** from which diacetyl can be formed.

For cultured buttermilk and sour cream, a mixture of strains of lactococci that produce acids (usually two strains) and leuconostocs that produce flavor usually is employed. Numbers of each strain must be balanced and strains must be compatible since some cultures antagonize others by producing inhibitory substances. Furthermore, strains selected should not be susceptible to the same **bacteriophages** (cf. section 23.4).

Dairies usually purchase cultures from suppliers either as frozen concentrate or in the lyophilized (freeze-dried) form. Frozen concentrates have become popular because they retain their original activity during several weeks of storage in liquid nitrogen (–320°F [–196°C]). When thawed they are immediately highly active and can be added directly to the milk from which the product will be made (called direct-to-vat set, or DVS), or they may be propagated as described below.

Figure 29.1 depicts commonly used alternatives in culture propagation. For preparing a few hundred gallons/liters of a product (such as buttermilk), frozen concentrated culture can be added directly to the processing vat. Some processors produce bulk culture from which several tanks of product can be made, particularly in cheese plants.

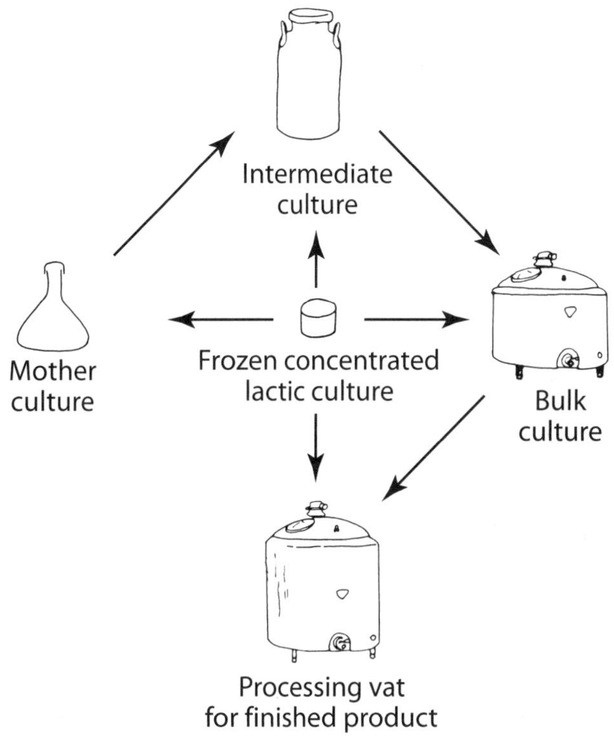

Figure 29.1 Alternative procedures for culture propagation: depending on the quantity of culture needed, frozen concentrated culture may be added to the milk (1) in the processing vat, (2) for bulk culture, (3) for intermediate culture, or (4) for mother culture (courtesy of Missouri Agricultural Experiment Station).

Table 29.2 Selected Characteristics of Bacteria Commonly Used to Produce Cultured Dairy Products.

Bacteria	Temperatures of Incubation (°C)	Maximum Titratable Acidity Normally Produced (%)	Primarily Produces
Leuconostoc cremoris	20–27	0.1–0.3	flavor
Lactococcus cremoris	21–30	0.9–1.0	acid
Lactococcus lactis	21–30	0.9–1.0	acid
Lactococcus diacetylactis	21–30	0.7–0.9	acid and flavor
Streptococcus thermophilus	38–45	0.9–1.1	acid
Lactobacillus acidophilus	38–45	1.2–2.0	acid and flavor
Lactobacillus bulgaricus	44–47	2.0–4.0	acid and flavor

Some processors prefer to propagate mother cultures whereby they keep a limited number of frozen cultures and transfer 0.5 to 1.0% of the mother culture to flasks of fresh media daily. Intermediate and/or bulk cultures (figure 29.2) are set as needed from the mother culture. Mother cultures should be evaluated for quality regularly. Extreme precautions are required to preclude their contamination, including the use of sterile containers and transfer devices as well as dedicated transfer hoods or clean rooms. Mother cultures are propagated in sterile media whereas milk for bulk cultures usually is not sterilized, but rather is heated at 167°F (75°C) for 30 min to 203°F (95°C) for 1 hr, depending on its intended use.

Media used for starter propagation must support rapid microbial reproduction. Milk is a good medium if free of antibiotics and if heated adequately to destroy its natural bactericidal components and its original bacteria. Overheating, however, may slow microbial growth because growth factors are destroyed. Heating at 250°F (121°C) for 15 min is the maximum temperature recommended.

Figure 29.2 Tank for cultivating large quantities of starter cultures (courtesy of Advanced Process Technologies, Inc.).

Commercial suppliers of cultures have recognized the need for stimulatory substances and bacteriophage inhibitors in starter media. Using a base of nonfat dry milk, pancreas extract is added as a stimulant and phosphate or citrate salts as bacteriophage inhibitors. The latter **sequester** calcium, which is needed by bacteriophages in the process of infecting bacterial cells. Not all phages are inhibited by the absence of calcium. Other practices are necessary, however, to prevent phage infection of bulk culture and the milk being used to make products. These practices include (1) effective cleaning and sanitizing, (2) rotation of cultures that have different sensitivities to bacteriophage, and (3) prevention of bacterial contamination.

Lactic acid bacteria vary widely in susceptibility to infection by bacteriophages (Hassan and Frank, 2001). Culture suppliers often combine strains of lactic acid bacteria that are resistant to different phages so if one strain of bacteria is attacked, a resistant strain will continue to grow and produce acid. These are called "mixed strain starter cultures."

The possibility of phage proliferation during culture propagation can be avoided by adding cell concentrates directly to the milk being used to make the product (DVS).

Control of pH and incubation temperature is important during production of bulk starter culture. Depending on the organism being propagated, specific amino acids and vitamins are needed. For example, lactococci require niacin, pantothentic acid, pyridoxine, and biotin, whereas *Streptococcus thermophilus* requires these plus nitroflavin. Neither lactococci nor *S. thermophilus* can synthesize branched chain amino acids (isoleucine, leucine, or valine) or histidine. Although milk contains these growth factors, concentrations are lower than needed for maximal growth.

29.3 Cultured Buttermilk

Buttermilk per se is the fluid from which fat is removed by churning cream into butter. It was formerly used as a beverage, but today is mostly dried for use by the baking industry. It has been replaced as a beverage by *cultured* buttermilk, which is prepared from skim milk or low-fat milk by fermentation with *Lc. cremoris* and *Le. cremoris*. Flavor components produced by these bacteria include lactic acid, diacetyl, volatile organic acids (formic, acetic, propionic, butyric, and valeric acids), acetaldehyde, and ethanol. Several factors, including microorganisms, milk, and processing, interact to give the end product its body and flavor characteristics. Therefore, flavor, aroma, body, texture, and appearance vary from plant to plant and from day to day. For this reason, processors should minimize variables so they can supply cultured buttermilk that is consistently of high quality.

When skim milk is used, nonfat solids should be added to increase total solids to 12.0 to 13.5%. Heat of pasteurization should be sufficient to denature whey proteins, thus increasing viscosity and freeing compounds containing sulfur, which reduce harshness of flavor. Adequate heating is accomplished by exposure to 178°F (80°C) for 30 min or 194°F (90°C) for 2 min. Milk is then cooled to 70°F (21°C) and pumped to a sterile stainless steel vat, inoculated with 1% starter culture and incubated for 12 to 14 hr, or until titratable acidity reaches approximately 0.85%.[3] The product is then cooled to 39°F (4°C) with mild agitation to minimize the incorporation of air. Butter granules may be added to enhance flavor and attractiveness. There are approximately 43 cal/glass (125 g) of cultured buttermilk made from skim milk (with no added butter granules). Because of the high heat treatment of processing and the effects of lactic acid, the protein of cultured buttermilk is especially easily digested.

29.4 Cultured Sour Cream, Crème Fraîche, and Dips

Consumers are fond of the smooth, custard-like body and delicate aromatic acid flavor of cultured sour cream, cultured half-and-half, and cultured dips. Cultured cream must contain not less than 18% milkfat, whereas sour half-and-half contains about 10% fat. Therefore, these products compare favorably calorie-wise with mayonnaise, salad oils, and other oil-base dressings, supplying only 25 to 50% as many calories in equivalent quantities. Dips made from sour cream or sour half-and-half are lower in fat because they are diluted by flavorings. Flavorings include herbs, garlic, onion, dill, blue cheese, honey, and fruits.

Procedures for the manufacture of cultured cream resemble those for cultured buttermilk. However, homogenization is an important extra step. Cream is pasteurized at 165°F (74°C) for 30 min or 185°F (85°C) for 25 sec, then immediately homogenized at 2,500 psi [17.3 mPa]. To increase its viscosity, it may be homogenized again at either 176°F or 104°F (80°C or 40°C). A mixed culture of bacteria that produce lactic acid and those that produce flavor, especially diacetyl, is added with thorough mixing. Incubation proceeds at 70 to 77°F (21 to 25°C) for 8 to 14 hr. If a vat is used for incubation, the finished product is pumped by a positive-type pump through a special valve that smoothes (homogenizes) the coagulum. Some processors prefer to incubate the inoculated cream in its final container.

Crème fraîche, which contains 28 to 42% milkfat, is manufactured in much the same way as sour cream. It is mostly used in sauces. The high fat content makes it resistant to curdling, whereas homogenization of the fat prevents churning. The product originated in Normandy and is the only cream to have an appellation d'origine contrôlée (controlled designation of origin). Similar products are used widely in Europe. European labeling regulations disallow any ingredients other than cream and bacterial culture.

Hot-pack sour cream is produced by introducing stabilizer into freshly prepared sour cream then heating, homogenizing, and packaging while hot in sealed glass or metal containers. This treatment extends **shelf life**.

29.5 Yogurt

Yogurt (yoghurt) originated in eastern Europe some 5,400 years ago. The oldest writings that mention yogurt are attributed to Pliny the Elder (23–79 AD). Yogurt was first introduced to the United States by Armenian immigrants Sarkis and Rose Colombosian, who started Colombo and Sons Creamery in Andover, Massachusetts, in 1929.

In recent years sales of yogurt in the United States have increased more than those of any other item in the dairy case. Based on federal order data for uses of Class II milk, yogurt production in the United States increased from slightly more than 1.8 billion lb in 2000 to 4.2 billion lb in 2010. This is attributed largely to the industry's production of a wide variety of fruit-flavored yogurts (introduced in the United States by Dannon in 1947) and to an increased appreciation among consumers of the low-calorie, highly nutritional aspects of yogurt. Added to this is the recent proliferation of yogurts containing probiotic microorganisms and to the increasing popularity of Greek-style (strained) yogurt. The most popular uses are as a between-meal snack, a dessert, and as part of a noon meal.

Yogurt (or yoghurt, yahourt, yaourt, leben, madzoon, naja, dahi, dudh, dadiah, and zabadi) is fermented by *Lactobacillus bulgaricus* and *S. thermophilus* in all countries except Turkey, where yeast that ferments lactose is included. Yogurt contains large quantities of protein, lactose, calcium, and B-vitamins and is tasty and highly digestible.

Dehydrated yogurt (yogurt powder or nonfat yogurt powder) has found many applications as a flavoring for soups, dips, chocolate brownies, snack bars, and other foods. Yogurt that has been cultured to a pH near 5.0 is dried yielding a protein content of 33 to 36% and a lactic acid content of about 6%. A related product, called cultured dairy solids, may contain, in addition to cultured nonfat milk, one or more of the following ingredients: cultured whey, cultured whey protein concentrate, nonfat dry milk, dry whey, lactic acid, natural and artificial flavorings, and silicon dioxide as an anti-caking agent (Gerdes, 2009).

There are two popular styles of fruit-flavored yogurt, Swiss and sundae. Fruit, flavoring, and color are distributed throughout Swiss-style yogurt. Fruit is placed in the bottom of the cup with yogurt on top to make sundae-style yogurt. Bacteria may attack portions of the fruit filling of Swiss-style yogurt.

29.5.1 Yogurt Production

Traditionally, the product was made from milk concentrated by boiling. Today yogurt is made from homogenized milk, lowfat milk, or skim milk with added nonfat dry milk. By increasing the milk solids, the body firmness of yogurt is increased. If milk with added nonfat dry milk is used, total solids will approximate 15%; however, if skim milk without added solids is selected, total milk solids will be only 8.5 to 9%. Most plain yogurt currently marketed in the United States contains from 11 to 15% milk solids and up to 2% fat. Rahmjoghurt, a creamy yogurt with a high fat content (10%), is available in Germany. "Drinking yogurts" containing probiotic bacteria, such as *Bifidobacterium* BB-12, are favored by many consumers.

Major variables affecting yogurt composition include (1) whether plain or flavored, (2) labeled fat content, and (3) whether sweetened with natural or artificial sweetener. Plain yogurts contain higher amounts of protein (4.5 to 7%) than flavored types (2.7 to 3.5%) and commonly contain no stabilizers. Naturally sweetened and fruit-flavored yogurts usually contain 2.7 to 3.5% protein, 15 to 18% sugars, and 19 to 22% total carbohydrates; total solids average about 23% and calories per serving about 170. In contrast, the artificially sweetened, nonfat, "light" fruit-flavored yogurts contain only 11 to 12% total solids whereas their calorie content per 170 g (6 oz) serving is about 80. To add viscosity to these low-solids products, stabilizers are added, including gelatin, pectin, carrageenan, xanthan gum, agar, and modified food starch. The standard of identity for nonfat yogurt (21 CFR 131.206) provides for use of stabilizers as optional ingredients. Sodium caseinate can function as a stabilizer. Therefore, it may be used to stabilize nonfat yogurt within good manufacturing practice at the amount necessary to accomplish the intended technical effect. The use of sodium caseinate for a purpose other than as a stabilizer (e.g., to increase the protein levels) is not permitted by the standard.

To improve the body and texture of nonfat yogurt, Nouri et al. (2011) successfully used renneted and regular skim milks mixed in a ratio of 3:7, then cultured. They reasoned that the treatment increased the protein's water-binding capacity.

High-heat treatment of milk, 185°F (85°C) for 10 to 30 min, used in manufacturing yogurt is essential to denature whey proteins, thus increasing the capacity of proteins to bind water and liberate amino acids that promote growth of *Lb. bulgaricus* (Tamine and Deeth, 1980). *Lb. bulgaricus* enzymatically hydrolyzes proteins providing amino acids, especially valine, stimulating growth of *S. thermophilus. S. thermophilus* uses oxygen, thus making the oxygen tension more favorable for *Lb. bulgaricus.* It also produces formic acid, which stimulates growth of *Lb. bulgaricus* in some milks. This is called **symbiosis**.

High incubation temperatures (106 to 113°F [41 to 45°C]) are required by these bacteria and their balance in the culture is determined by the temperature selected. Also, the ratio of one to the other in the inoculum must be controlled to produce favorable body and flavor characteristics. Usual ratios are 1:1 to 1:3 *Lb. bulgaricus* to *S. thermophilus.* A harsh acid flavor may occur when *Lb. bulgaricus* predominates. Lack of flavor is usually due to inactivity of *Lb. bulgaricus.* With 2% inoculum of active culture and incubation at 108°F (42°C), a firm gel is formed within 3 hr and the pH drops to 4.6 to 4.8. Manufacturing yogurt by direct addition of frozen concentrated cultures requires 5 to 8 hr of incubation. The 8-hr incubation time may provide the smoothest gel. Gradual cooling is then begun to arrest acid development and prevent **shrinkage** of curd. Quiescent storage is also required during early cooling to prevent whey separation in the fragile gel. Final pH varies from 3.8 to 4.4 depending on conditions of manufacture. Lactic acid, acetic acid, and acetaldehyde are important contributors to yogurt flavor. The symbiotic relationships between the two bacteria are vital to the production of sufficient acetaldehyde (at least 8 mg/L) for optimal flavor.

Some manufacturers **set** yogurt in a vat. When titratable acidity reaches 0.85 to 0.90%, the product is agitated and cooled to 68°F (20°C) to limit further acid development. A titratable acidity of at least 0.9% is required (21 CFR 131.200). Sugar and fruit flavoring (for flavored yogurt) are added and the mixture is pumped to the filler. Good reset occurs in the carton during refrigerated storage, especially if gelatin or another stabilizer is added.

Some suppliers of yogurt cultures claim to offer cultures of *Lb. bulgaricus* and *S. thermophilus* that do not produce undesirably high amounts of lactic acid during storage of finished yogurts. This is important in some parts of the world, including Asia, North Africa, the Middle East, and South America, where there is insufficient refrigeration to prevent significant continued growth of the culture.

29.5.2 Microflora and Probiotics in Yogurt

In 1905, Stamen Grigorov (1878–1945), a Bulgarian student of medicine in Geneva, first examined the microflora of Bulgarian yogurt. He described the bacteria as spherical and rod-like lactic acid producers. In 1907, the rod-like bacterium was named *Lactobacillus bulgaricus.* Nobel laureate biologist Élie Metchnikoff (born Ilya Ilyich Mechnikov in Russia in 1845) was influenced by Grigorov's work and hypothesized that regular consumption of yogurt was responsible for the unusually long lifespans of Bulgarian peasants. He believed *Lactobacillus* to be essential for good health and promoted yogurt as a food.

Many manufacturers add probiotic bacteria to yogurt (cf. section 23.3.2). These may include specific isolates of the following species:

- *Lactobacillus*: *acidophilus, brevis, casei, fermentum, johnsonii, paracasei, plantarum, reuteri, rhamnosus*, and *salivarius.*
- *Bifidobacterium*: *animalis, bifidus, breve, infantis, lactis*, and *longum.*

Bifidobacterium longum BB536 is a beneficial bacterium isolated from humans. It defends against digestive upsets (including constipation, diarrhea, abdominal discomfort, gas, and bloating) while improving the bodily ratio of beneficial to harmful bacteria and enhancing immunity. Although human isolates of beneficial bacteria are considered more compatible with the human system than animal species, they are also extremely sensitive, for example, to acids. However, BB536 was reported as highly stable in yogurt. Growth of bifidobacteria in the human intestine can be stimulated by addition of galacto- or fructo-oligosaccharides (FOS) or inulin to food. These additives, called *prebiotics*, are soluble dietary fiber and are not digested until they reach the colon. They bind water in the product, thus lowering the amount of fat needed to provide desirable texture. Prebiotics important to growth of beneficial intestinal bacteria include the oligosaccharides of human milk. These complex sugars exist in higher concentrations in human than in cow milk. Whey streams have the potential to be commercially viable sources of complex oligosaccharides that have the structural resemblance and diversity of the bioactive oligosaccharides in human milk (Zivkovic and Barile, 2011).

In addition to stimulation of growth of desirable intestinal bacteria, some studies indicate that inulin-type fructans improve calcium absorption and reduce amounts of circulating triacylglycerols.

Microencapsulation is an effective method for maintaining high viability of probiotic bacteria, as it protects probiotics both during food processing and storage as well as in gastric conditions. Several materials have been shown to help bacteria survive while passing through the acidic stomach and bile in the digestive tract.

Some thermophilic strains of yogurt bacteria can produce exopolysaccharides (EPS) that form filaments encapsulating the cells and linking them to casein micelles, thus increasing product viscosity. Some of these "ropy strains" are available commercially. The following functions have been claimed for EPS: (1) promote biofilm formation and adhesion to oral surfaces, (2) improve viability of cells after adhesion, and (3) promote resistance to nonspecific host immunity. Terms like slime, capsule, and mucoidal phenotype are often used interchangeably regardless of whether the condition is capsular (compact layer) or slime-like (diffuse layer). When the physical nature of the material has not been determined, the correct descriptive term is glycocalyx.

29.5.3 Greek Yogurt

Greek yogurt (strained yogurt) is relatively new to customers in the United States. It is called jocoque in Mexico, labneh in the Near East, and chakka in India and Pakistan, where it is concentrated from dahi. It is uniquely high in milk solids because, after coagulation of casein, whey is removed from the product by filtration. This, however, reduces the concentration of the essential amino acid cysteine, which is high in whey protein. The original process for whey removal was to place the curd in cheesecloth and hang it to drain. Other methods involve use of (1) fluid milk fortified with milk protein concentrate (MPC), (2) a mixture of MPC and whey protein concentrate, and (3) ultrafiltered milk. Original Greek yogurt was made from a mixture of milk and cream yielding a product containing about 225 calories and 15 g of fat per 6 oz serving. Numbers of culture bacteria are higher than in traditional yogurt because the solids are more concentrated.

Greek yogurt is sometimes made from milk that has been boiled to remove some of the water or is enriched by addition of cream and/or powdered milk. "Greek-style" yogurts are similar to Greek strained yogurt, merely made thicker with thickening agents.

Labels on popular brands of nonfat Greek yogurt in today's markets claim protein content of 11 to 15 g per 6 oz serving. On a 1 g of protein basis, Greek yogurts contain significantly less lactose than traditional yogurts since much of the lactose, which is dissolved in the whey, is removed during filtration. Yogurt fermentation usually converts about one-fourth of milk's lactose to lactic acid. Persons who absorb lactose poorly can consume more of milk's nutrients in yogurts than in fresh milk products without intestinal discomfort. Greek yogurt is a favored substitute for sour cream in many applications.

29.5.3.1 Growing Distribution of Greek Yogurt

In 1926, Athanassios Filippou opened a small dairy shop in Athens, Greece. Filippou's store became known for its delicious, creamy, one-of-a-kind strained yogurt. He brought in his sons and a family business named FAGE (pronounced "fa-yeh") was born. In 1954 the family created the first wholesale distribution network for yogurt in Greece. Until the mid-1970s FAGE was involved primarily in the small-scale production and distribution of traditional Greek yogurt. During that time retail outlets typically sold yogurt as a commodity product in bulk quantities and the consumer often was unaware of the manufacturer. However, in 1975 FAGE became the first company to introduce branded yogurt in small sealed tubs to the Greek market and soon expanded throughout Greece, and then internationally. Exports followed to the United Kingdom (1980); Italy (1983); the United States (1998); and others. With sales of over 2,000 tons/yr, a 220,000 sq ft processing plant for FAGE Total™ yogurt was built in Johnstown, New York, at a cost of $200 million. It started operations in 2008.

In 2005, another leading brand of Greek-style yogurt, Chobani, was founded by Hamdi Ulukaya (who was born and raised in Turkey). After two years of development, Hamdi and Kyle O'Brien launched Chobani. The pair decided on the name Chobani after the Greek word "chopani," meaning shepherd. It soon became the largest yogurt manufacturer in New York. Other well-known brands of Greek yogurt in the United States include Oikos and Yoplait.

New York now leads the United States in production of Greek yogurt. During 2012 to 2013, PepsiCo began a joint venture with Theo Muller Gmbh of Germany to build a $206 million plant to produce Greek yogurt in Batavia, New York. At that time there were 29 plants producing yogurt in New York.

29.6 Kefir

Traditional kefir, which is especially popular in Russia and eastern Europe, is an acidic and alcoholic fermented milk characterized by effervescent foaminess and a sour flavor. It commonly contains up to 1.1% lactic acid, 0.3 to 1.0% ethyl alcohol, and carbon dioxide (CO_2). The name kefir is thought to have been derived from the Turkish word "keif," meaning good feeling. Known as yogurt de pajaritos (bird's yogurt) in Chile, it has been consumed there for more than a century, where it may have been introduced by migrants from the former Ottoman Empire and eastern Europe. In recent years flavored kefir has been widely distributed in the United States.

The traditional method of making kefir involves direct addition of kefir grains (2 to 10%) to raw milk in a loosely covered acid-proof container and occasional agitation during incubation. Room is allowed in the container for expansion as CO_2 is produced. After fermentation for about 24 hr, ideally at 68 to 77°F (20 to 25°C), grains are removed by sieving and are reserved as starter for another batch. Kefir grains will ferment the milk from most mammals and will continue to grow in such milk. Typical milks used include cow, goat, and sheep, each with varying organoleptic and nutritional qualities.

In making kefir commercially, milk is heated to a high temperature (commonly 203°F [95°C] for 5 min), which denatures protein and increases viscosity of the finished product. A mixed culture of bacteria and yeast forms "grains" comprised of proteins, lipids, lactose, and microorganisms within the milk (figure 29.3). Kefir grains contain a water-soluble polysaccharide known as kefiran, which imparts a rope-like texture and feeling in the mouth; it ranges in color from white to yellow, and may grow to the size of walnuts. Lactic acid is derived from approximately 25% of the lactose in the starter milk, thus making kefir more readily consumable by lactose maldigestors than is the base milk product.

The fermented liquid containing living microorganisms may be consumed as a beverage, used in recipes, or kept aside for several days to undergo a slower secondary fermentation that further thickens and sours the liquid. However, without refrigeration the shelf life may be only three days. Grains enlarge during kefir production and eventually split. Grains can be dried at room temperature, lyophilized (freeze-dried), or frozen. Kefir fermented by small-scale dairies early in the twentieth century achieved alcohol levels up to 2%, but kefir made commercially with modern methods of production has less than 1% alcohol, probably because of reduced fermentation time since alcohol-producing yeasts multiply more slowly than do acid-producing bacteria.

Figure 29.3 Kefir grains (courtesy of F. P. Rattray, Chr. Hansen, Hoersholm, Denmark).

Numerous species of bacteria and yeast have been isolated and identified from kefir grains. Among the homofermentative bacteria are *Lactobacillus kefiranofaciens* ssp. *kefirgranum* and *Lb. delbrueckii*. Heterofermentative ones include *Lb. kefir* (an alcohol producer), *Lb. parakefir*, *Lc. lactis* (citrate fermenter), and *Le. mesenteroides* ssp. *cremoris* (diacetyl producer). Kefir yeasts include the lactose fermenters *Kluyveromyces marxianus* and *Debaryomyces hansenii* while nonlactose fermenters include *Saccharomyces cerevisiae* and *Pichia fermentans*. Complex interactions among these microorganisms occur within the kefir grains as each type may provide specific conditions including nutrients that support growth of the other. Even locations most favorable for growth within kefir grains vary with the microorganism.

Kefir can be produced using lyophilized cultures. A portion of the resulting kefir can be saved and used to propagate additional fermentations. Ultimately grains do not form and a fresh culture must be obtained. In preparing the so-called "kefir mild," kefir grains are no longer used. Instead the culture is composed of a mixture of selected bacteria and yeast, allowing the flavor to be kept constant. Kefir produced this way can have a shelf life of up to 28 days, whereas that produced with kefir grains may remain satisfactory for only 3 to 12 days. However, since this process does not provide the microbial diversity of kefir made with grains, the product may not possess the therapeutic and probiotic characteristics often associated with kefir.

The health benefits of kefir have recently been popularized in North America, Australia, and the United

Kingdom, where kefir can be found in pasteurized form in many stores and supermarkets. A brand of kefir popular in North America is marketed as containing the probiotic bacteria *Lactobacillus reuteri* and *Bifidobacterium lactis* as well as 10 other bacterial species. Made from lowfat milk with added flavoring and evaporated cane juice, an 8 oz (240 mL) serving is claimed to contain 2 g fat, 10 mg cholesterol, 20 g carbohydrate as sugars, and 11 g protein.

29.7 Ymer

Ymer is a creamed, high-moisture skim milk curd that is homogenized to a smooth and silky texture. Resembling a soft pudding, it is eaten as a breakfast, luncheon, or between-meal snack. This dairy food is similar to yogurt but has a higher content of diacetyl, less acetaldehyde, and more milk solids. Ymer contains an average of 85% moisture, 3.3% fat, and 75 cal/100 g.

Four steps in the manufacturing process make ymer unlike yogurt: (1) a different culture mixture is used—*Lc. lactis* ssp. *cremoris, Lc. lactis* ssp. *lactis* biovar. *diacetylactis*, and *Le. mesenteroides* ssp. *cremoris*—to assure strong gas production; (2) whey is removed, leaving 50% of the original volume; (3) cream is mixed with the broken curd; and (4) the mixture is homogenized.

29.8 Acidophilus Milk

Acidophilus milk is regular milk enriched with *Lactobacillus acidophilus.* It may also contain *Bifidobacterium bifidum*, the probiotic that's also found in yogurt. Sweet, or unfermented, acidophilus milk closely resembles regular milk in flavor and texture. Fermented acidophilus milk is more tangy and thicker than regular milk. While kefir comes in a variety of flavors, like yogurt, acidophilus milk is usually sold without additional flavorings. Yakult, which is similar to acidophilus milk, is produced in franchised plants in Japan, Taiwan, Hong Kong, and Brazil. The product is a pleasant-tasting, thin, slightly sour liquid containing an extract of *Chlorella* algae, vitamin C, and greater than 10^8/mL of viable cells of *Lactobacillus* (Shirota strain).

Early in the twentieth century, scientists debated the health-promoting effects of *Lactobacillus.* In 1907, Dr. Élie Metchnikoff published *Prolongation of Life*, a treatise based on his observation that the relatively large number of centenarians in southeastern Europe was correlated with the continuous use of milk fermented with lactobacilli (cf. section 23.3.2). He hypothesized that replacing or diminishing the number of "putrefactive" bacteria in the gut with lactic acid bacteria could normalize bowel health and prolong life.[4]

In 1900, Moro isolated *Lb. acidophilus* (figure 29.4) from the stools of infants; it has since been determined that this organism can colonize and grow in the intestinal tract. In cases of infantile diarrhea, pathogenic *Escherichia coli* were destroyed one to five days after initiation of therapy with acidophilus milk; the infants made a rapid recovery (Tomic-Karovic and Fanjek, 1962). In one study, feeding *Lb. acidophilus* to swine increased absorption of calcium and phosphorus (because of more lactic acid in the intestines), reduced numbers of *E. coli* in the intestinal tract, reduced enteritis, and increased weight gains (Sandine et al., 1972). *Lb. acidophilus* is used in therapy against yeast infections of the intestinal tract during or following systemic therapy with antibiotics.

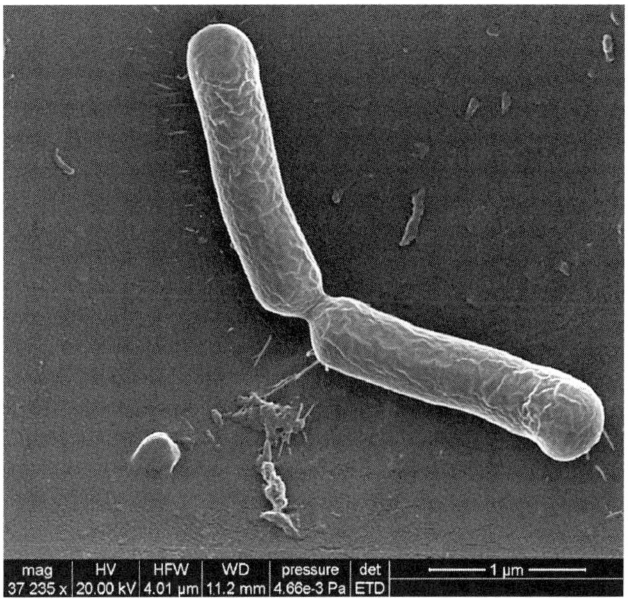

Figure 29.4 Scanning electron micrograph of *Lactobacillus acidophilus* (courtesy of Ami Yoo, Food Science Department, University of Missouri, Columbia).

Lb. acidophilus requires properly treated milk and elevated temperature for growth. Lowfat milk with 2% fat and 11 to 12% nonfat solids provides desirable characteristics for acidophilus milk. Unusually high heat treatment is recommended—250°F (121°C) for 15 to 20 min. Ultrahigh temperature treatment at about 293°F (145°C) for 2 to 3 sec appears ideal.

Incubation temperatures vary from 104 to 113°F (40 to 45°C) and the time of incubation from 18 to 48 hr depending on amount of inoculum, temperature of incubation, growth-promoting properties of the milk, and strain of *Lb. acidophilus* used.

The number of viable organisms is an important prerequisite for premium acidophilus milk. Numbers will approximate 5×10^8/mL at the end of incubation but decrease rapidly during storage at 41°F (5°C). Therefore, persons who drink acidophilus milk to obtain the bacteria must consume a freshly prepared product.

29.9 Other Fermented Milks

The origins of a wide variety of fermented milk products found throughout the world were dictated by local environments and resources, types of microorganisms present, prevailing temperature and humidity, availability of containers and utensils, and more. Presentations of the more popular of these fermented milks and the sciences and practices applied to them can be found in the *Encyclopedia of Dairy Sciences* (Fuquay et al., 2011). The origins and names of the most prominent of these products are:

- Scandanivia—Filmjölk (cow milk)
- Japan—Yakult (yacult) from skimmed milk
- Middle East—Zabady (buffalo milk) and gariss (camel milk)
- Central Asia—Koumiss (kumys) from mare's milk
- India—Dahi (cow or buffalo milk)
- Indonesia—Dadih (buffalo milk)
- Mongolia—Airag (horse milk) and tarag (goat milk)

29.10 Acidified Products

Propagation of cultures is expensive, time-consuming, and requires considerable training and expertise. Also, cultures are biological substances subject to effects of variables that may cause the end product to differ in characteristics from batch to batch. Knowing this, scientists and processors have endeavored to manufacture dairy foods that simulate cultured products by adding acidulants, flavors, and stabilizers to milk, to components of milk, and to other food materials. Products in the latter class are called dairy analogs.[5] Many of them contain vegetable fats as substitutes for milkfat. In some the only derivative of milk is sodium caseinate.

The common characteristic of cultured and acidified products is their acidic reaction. All cultured products contain lactic acid as the principal acid. For acidified products, lactic acid is the most desirable acidulant; however, it is also the most expensive. Citric acid is less expensive but the flavor is commonly characterized as tart. Phosphoric is the least expensive of the three acids but with it the product lacks desirable flavor. Gluconic acid is formed by hydrolysis of glucono-delta-lactone upon its addition to milk. This acid reduces the pH, producing coagulum. All four acids must be used with a stabilizer. In direct acidification, chemical and physical changes occur in a few minutes (longer with glucono-delta-lactone) rather than in several hours as in the culturing process.

In addition to acid, manufacturers add starter distillate or other flavoring compounds to acidified products. A major component of the flavoring is diacetyl. Stabilizers used include starch, vegetable gums, gelatin, carrageenan, mono- and diglycerides, and sodium phosphate. Basic steps in manufacture are (1) standardization of components within the mixture, (2) stabilization, (3) pasteurization, (4) homogenization, (5) cooling, (6) acidification, and (7) packaging. Acidified products include buttermilk, sour cream, sour half-and-half, dip bases, and imitations of these products.

The fine flavor and aroma that can be produced by culturing methods are often lacking in acidified products. However, these products have characteristics of their own that are preferred by some consumers. In particular, it often is easier to produce acidified products with long shelf life because cultured products can seldom be sterilized after incubation. Stabilized cream, for example, can be sterilized, mixed with sterile acid and flavoring, and aseptically packaged as acidified sour cream.[6]

29.11 Summary

Development and consumption of cultured or fermented milks probably resulted first from necessity because fresh milk became sour too quickly. A secondary benefit was the destruction of many pathogens by the products of the fermenting bacteria. Not insignificant, however, was the consumer's desire for more of the better-tasting products. Ultimately, scientists have implemented techniques and principles used first by artisans. Today, the manufacture of cultured and acidulated products is highly dependent upon scientific inputs. These relate to genetics of microorganisms (characteristics of cultures), food chemistry (chemical reactions that influence product characteristics), physiology (which organisms will grow at what rate and what they will produce), physical chemistry (how components of the formulation will react and interact), food engineering (kinds of fermenters and ancillary equipment needed), and sensory evaluation (sensory attributes of the product and whether people like them).

Cultured dairy foods in wide variety are available to consumers. Processors can vary product composition (including nonfat milk, milk, or cream), processing treatments, fermenting microorganisms, stabilizer and flavor additives, and packages. Thus, we see on grocers' shelves products such as cultured buttermilk, flavored and plain yogurt with and without probiotics, sour cream, and various cultured and acidified dips. Consumers of many nations do not need to depend on fermentations to preserve their milk or to make it safe to drink, but they do share in the benefits of these historical necessities that led to the development of cultured milks and milk products for their nutrition and pleasureful consumption.

Study Questions

1. Why does fresh milk deteriorate (become sour) quickly? What health-related aspect accompanies this phenomenon?

2. Name important cultured dairy products of the western world and of Europe and Russia. What types of milk are used to produce these products?
3. Differentiate between cultured milks and acidified milks.
4. What are the basic steps used in manufacturing cultured dairy products?
5. Define a culture. Of what significance are cultures to the dairy industry? Outline steps used in the preparation of cultures. What major advantage does a frozen culture hold over a lyophilized one?
6. What is a bacteriophage? Of what significance are antibiotics and bacteriophages in the preparation and use of cultures? What control measures can you recommend?
7. Differentiate between buttermilk and cultured buttermilk.
8. What effect does homogenization have upon the viscosity of sour cream?
9. What advantage does hot-pack sour cream hold over the more traditionally prepared sour cream?
10. Which cultured dairy product is currently enjoying the greatest consumption increase in the United States? What explanations can you offer for this?
11. What is ymer? To which dairy product of the United States is it most similar?
12. Of what nutritional and health-related significance is acidophilus milk? Yogurt containing probiotic bacteria?
13. Name and cite a limitation of the three principal acids used in the manufacture of acidified products.
14. What are the basic steps in manufacturing acidified products?
15. Why do acidified products often have a longer shelf life than cultured dairy products?

Notes

[1] The words *acidified* and *acidulated* may be used interchangeably.

[2] Cultures are also used in manufacturing other fermented foods, including sausages, pickles, and sauerkraut.

[3] The proper time to stop fermentation of cultured products is best determined by observing flavor, aroma, and acidity. Of course, titratable acidity is influenced by the quantity of milk solids contained. For skim milk containing 9% nonfat solids, acidity should be 0.85%, and a factor of 0.07% should be added or subtracted for each 1% increase or decrease in nonfat solids, respectively.

[4] In 1904 Metchnikoff became deputy director at the Pasteur Institute Laboratory in Paris, where he discovered the process of phagocytosis, thus demonstrating how specific white blood cells can kill harmful bacteria in the body. For this work he was awarded the Nobel Prize for Medicine in 1908.

[5] Dairy analogs are sometimes cultured also.

[6] Sour cream was the first acidified product produced on a commercial scale.

References

Fuquay, J. W., P. F. Fox, and P. L. H. McSweeney (eds.). 2011. *Encyclopedia of Dairy Sciences* (2nd ed.). San Diego, CA: Academic Press.

Abd El-Salam, M. H. "Middle Eastern Fermented Milks" (pp. 503–506).

Akuzawa, R., T. Miura, and I. S. Surono. "Asian Fermented Milks" (pp. 507–511).

Libudzisz, Z., and L. Stepaniak. "Buttermilk" (pp. 489–495).

Rattray, F. P., and M. J. O'Connell. "Kefir" (pp. 518–524).

Robinson, R. K. "Yogurt: Role of Starter Culture" (pp. 529–532).

Robinson, R. K. "Yogurt: Types and Manufacture" (pp. 525–528).

Roginski, H. "Nordic Fermented Milks" (pp. 496–502).

Surono, I. S., and A. Hosono. "Starter Cultures" (pp. 477–482).

Surono, I. S., and A. Hosono. "Types and Standards of Identity" (pp. 470–476).

Takano, T., and N. Yamamoto. "Health Effects of Fermented Milks" (pp. 483–488).

Uniacke-Lowe, T. "Koumiss" (pp. 512–517).

Gerdes, S. 2009. Exploring yogurt powder. *Dairy Foods* 110(10):88–90.

Hassan, A. S., and J. F. Frank. 2001. "Starter Cultures and Their Use." In E. H. Marth and J. L. Steele (eds.), *Applied Dairy Microbiology* (2nd ed.). New York: Marcel Dekker, Inc.

Nouri, N., H. Ezzatpanah, and S. Abbasi. 2011. Application of renneted skim milk as a fat mimetic in nonfat yoghurt. *Food and Nutrition Sciences* 2:541–548.

Sandine, W. E., K. S. Muralidhara, P. R. Elliker, and D. C. England. 1972. Lactic acid bacteria in food and health: A review with special reference to enteropathogenic *Escherichia coli* as well as certain enteric diseases and their treatment with antibiotics and lactobacilli. *Journal of Milk and Food Technology* 35:691–702.

Tamine, A. Y., and H. C. Deeth. 1980. Yogurt: Technology and biochemistry. *Journal of Food Protection* 43:939–977.

Tomic-Karovic, K., and J. J. Fanjek. 1962. Acidophilus milk in therapy of infantile diarrhea caused by pathogenic *Escherichia coli. Annales Paediatrici* 199:625.

Zivkovic, A. M., and D. Barile. 2011. Milk as a source of functional oligosaccharides for improving human health. *Advances in Nutrition* 2:284–289.

Websites

Dairy Foods Magazine
(http://www.dairyfoods.com/publications)

Dairy Science and Food Technology
http://www.dairyscience.info/index.php)

Journal of Dairy Science (http://www.journalofdairyscience.org)

Journal of Microbiology, Biotechnology, and Food Sciences
(http://www.jmbfs.org)

30

Cheeses

The poets have been mysteriously silent on the subject of cheese.
Gilbert K. Chesterton (1874–1936)

30.1 Introduction

The actual beginning of cheesemaking is unrecorded in history. Records of cheesemaking have been traced back to ancient times in Arabia, Egypt, India, Israel, and Greece. Archeologists working in cattle-rearing sites in present-day Poland discovered ceramic vessels replete with small holes. They were similar to sieves sometimes used to drain **whey** from cheese curds. Researchers of the Organic Geochemistry Unit of the University of Bristol, as well as colleagues at Lodz, Gdansk, and Poznan in Poland and Princeton University of the United States, extracted fatty residues from holes in this unglazed pottery. The fatty acids recovered were typical of those found in milkfat. The discoveries taken together provide strong evidence that cheese was being made by neolithic people about 7,000 years ago in this area. Cheese was vital to early nomadic tribes and became one of the prized mediums of exchange because it provided milk in a more portable and less perishable form. Rome was a rich market for cheese during the reign of the Caesars; crusaders brought treasured secrets of cheesemaking back to Europe. During the Dark Ages these secrets were guarded in monasteries. By the Middle Ages there were many famous European cheeses, including Swiss, sapsago, and Roquefort. Cheddar cheese originated in Somerset, England, around the twelfth century, taking its name from the Cheddar Gorge (caves) in the town where cheese was stored.

Immigrants from England, the Netherlands, and other European countries brought a taste for cheese and knowledge of its manufacture to America, especially to New England. The factory system of cheesemaking began at Rome, New York, in 1851 when Jesse Williams and his son pooled milk from their herds and those of their neighbors (Kindstedt, 2012). From New York, the factory system spread rapidly, especially to Wisconsin. Most plants were small. Typical factories in the 1920s seldom had more than 40 patrons (producers), and few lived more than two miles away. Most plants closed during the winter (Price, 1971). At that time, more than 4,000 factories made cheese in the United States. By 1971 there were only 920 plants producing all cheeses except full-skim American and cottage, and 548 produced cottage cheese. These plants produced a yearly average of nearly 2.6 and 1.4 million lbs, respectively. By 2014 annual production of cheeses (except cottage cheese) in the United States reached 11.45 billion lb. National production is divided into three regions: Central—5.062 billion lb in 229 plants (WI, MN, SD, IA, OH, IL); western—4.973 billion lb in 103 plants (CA, ID, NM); and Atlantic—1.415 billion lb in 204 plants (NY, PA, OH). Over the past 40 years the western states have become major cheese producers.

In 2014, 6.604 billion lb of American-type cheeses were produced, 68% of which was cheddar. Similarly, of the 6.011 billion lb of Italian type cheeses, 79% was mozzarella. Ohio, with nearly one-half of the production (46%) at 13 plants, made 138 million lb of Swiss cheese. Production of Hispanic cheese has grown markedly from about 85 million lb in 1999 to 250.3 million lb in 2014. Wisconsin (127 plants) led the nation followed closely by California (60 plants) in total cheese produced in 2014 with 25 and 21%, respectively. More milk was used for making cheese than for any other class of dairy foods. Major changes in relationships among percentages of the major types of cheeses produced in the United States over a 30-yr period are shown in table 30.1.

Table 30.1 Percentages of Cheeses Produced in the United States by Type in 1981 and 2011.

Cheese Type	1981	2011
Mozzarella	16.0	33.6
Cheddar	45.2	29.6
Other American	16.6	10.6
Other Italian	7.2	9.4
Cream/neufchatel	5.6	6.8
Swiss	5.0	3.1
Other	4.4	6.9

Per capita cheese consumption in the United States reached 33.7 lb in 2013; the two most popular varieties being total Italian cheeses, with mozzarella leading the way at 10.8 lb, followed by cheddar at 9.7 lb. Since 1970 per capita consumption of American cheeses has doubled while consumption of all other cheeses has increased fivefold.

A highly significant element in milk pricing is the value of cheese. In 2014, 7.1% of cheese produced in the United States was exported. Typical inventories of natural cheese equal a 4.5-week supply, and at normal production rates a 2% drop in exports for six months can add another one-half week to inventory levels, thus influencing farm milk prices (cf. chapter 20).

The financial health of the dairy industry in the United States rests heavily on cheese exports. Milk producers can contribute to competitiveness in world markets by producing higher protein concentrations needed for cheese manufacture and by keeping somatic cell counts low (high counts reflect lower percentages of protein). Whereas significant factors affecting exports are not controlled by milk producers, attending to those manageable is the best strategy.

Changes in cheese manufacture continue. Today's industry, having vastly improved facilities, equipment, quality, and volume, has experienced a revolution of manufacturing methods. Mechanized and automated processes have been developed, tested, and applied in large plants. Production of curd by direct acidification, a reality in the manufacture of Italian-type pizza cheese, promises to become part of the manufacturing process of other varieties. Immobilized enzymes, already used by some industries, may become useful for treatment of milk for cheesemaking—both for coagulation and for accelerated ripening.

Ultrafiltration (UF) of raw milk is practiced on-farm among some very large dairy herds. Ultrafiltered milk (30% solids) and UF skim milk (18% solids) are shipped to cheese makers for use in standardizing or supplementing existing milk supplies. Hauling costs are thereby reduced. Concentrations of protein and lipids increase during UF, but the concentration of lactose, which passes through the membrane, remains the same in both sides of the membrane. Therefore, **standardization** of normal milk with UF-concentrated milk enables production of increased amounts of cheese within the vat while preventing excessive acid development in the resultant cheese.

Milk protein concentrate (MPC) is often used to fortify cheese milk and increase cheese yield. However, when in the spray-dried form, long-term storage at high temperatures decreases its solubility. Casein micelles in MPC aggregate during high-temperature storage with average particle diameter increasing from 200 to 800 nm. Furthermore, whey proteins may attach to casein micelles. Such changes cause increased **rennet** clotting time and interfere with cheese matrix formation.

Desires of consumers for lower content of dietary fat and salt are leading researchers to develop methods of producing cheeses with desirable flavor, body, and texture that can be labeled reduced fat and reduced sodium. Included among the methods are additions of concentrated micellar casein and of selected strains of nonstarter lactobacilli.

30.2 Principles of Cheese Manufacture

It is serious business—
this turning of milk into a companion
of firm character and witty flavor.
Evan Jones (1846–1899)

The principle upon which nearly all cheesemaking depends is the ability of milk's casein to coagulate (form curd). Casein can be precipitated by lowering with acid the pH to its **isoelectric point** (the pH at which casein micelles have equal numbers of positive and negative charges; for casein it is pH 4.7 to 4.8). A few cheeses are made using proteins of whey.

A three-phase model for acid-induced gelation of milk was proposed by McMahon et al. (2009). The first phase involves temperature-dependent dissociation of proteins from the casein micelles with more dissociation occurring as temperature is decreased from 104 to 50°F (40 to 10°C). Dissociation continues as milk pH is lowered with the released proteins forming loosely entangled aggregates. The second phase of acid-induced gelation that occurs between pH 5.3 and 4.9 involves reassociation of caseins into compact colloidal particles. The third and final phase involves rapid aggregation of the colloidal casein micelles into a gel network at about pH 4.8. Different gel structures are formed based on temperature of acidification with a coarse-stranded gel network formed at 104°F (40°C) and a fine-stranded gel network at 50°F (10°C).

Milk's casein will coagulate when κ-casein, the stabilizing fraction of the casein micelle,[1] is sufficiently hydrolyzed from the surface of the casein micelle. Hydrolysis is accomplished with **rennin** or similar proteolytic enzymes. As the stabilizing power of κ-casein is destroyed by rennin, the caseinate micelle becomes increasingly sensitive to calcium ions. Calcium sensitivity may, in turn, be affected by pH, heat treatment, colloidal phosphate, and other ions. At pH 5.4 or lower, a small amount of Ca^{++} is needed to cause precipitation following treatment with rennin.

The simplest example of cheesemaking is the natural souring of raw milk[2] followed by stirring to break the coagulum, then heating (cooking) to shrink the curd, freeing whey. Curd is dipped from whey and drained to provide *fresh curd*, the basic component of cheese. Removal of whey after precipitation of protein is the fundamental act common in the manufacture of nearly all cheeses.

Despite the seeming simplicity of this method, complexity in cheesemaking reigns supreme. A great number of variables may affect the process. Stated in general terms, these variables include composition of milk, types and relative numbers of microorganisms present, types and quantities of enzymes present, temperature of incubation, degree of souring, fineness and uniformity of broken curd particles, temperature and time of cooking, and completeness of whey drainage.

Here we pause to note that our hypothetical cheese curd is not ripened. In fact, ripening of this curd (see section 30.4.11) is fraught with risk for we have only partial knowledge of the curd's microbial and enzymatic contents. Most natural enzymes of milk remain active in cheese from raw milk so if the pH is favorable, lipases and proteases will break down fat and protein. We are confident the curd contains a high amount of moisture because it has not been pressed or otherwise treated adequately. Salt, which would inhibit certain microorganisms, has not been added. The curd has numerous openings in which air is trapped and, with its high acid content, yeasts and molds would grow profusely if we allowed them a favorable temperature such as that often used in ripening (50 to 59°F [10 to 15°C]).

30.3 Classifications of Cheeses

There are about 18 distinct types or kinds of natural cheese. No two are made by the same method; therefore, there must be terms and classifications that are readily identifiable to producers and consumers. The method of classification of cheeses is based on the consistency or resistance of body and the method of ripening. Examples are:

1. *Very hard* (usually grated before use): 30 to 35% moisture
   a. ripened by bacteria—romano, parmesan, and asiago (aged)
2. *Hard*: 35 to 40% moisture
   a. ripened by bacteria, without eyes—cheddar, Colby, stirred curd, and provolone
   b. ripened by bacteria, with eyes—Emmentaler, Swiss, gruyere, and asiago (medium age)
3. *Semisoft*: 40 to 45% moisture
   a. ripened principally by bacteria—brick, Muenster, and asiago (fresh)
   b. ripened by bacteria and microorganisms on surface—Limburger and Port du Salut
   c. ripened principally by blue mold in the interior—blue, Roquefort, gorgonzola, and stilton
4. *Soft*:
   a. ripened usually by surface microorganisms, 45 to 52% moisture—Brie, Camembert, and bel paese
   b. unripened, 52 to 80% moisture—cottage, cream, neufchatel, mozzarella, and pizza

Fallacies exist in this method of classification. Hardness in cheese is controlled primarily by moisture content, but fat content and the degree of protein hydrolysis are also influential. No clear-cut objective measures place cheeses in various classes and moisture ranges. Nevertheless, this approach provides considerable information regarding the characteristics of cheese.

30.4 Fundamentals of Cheesemaking

Curious and inquisitive persons who focus on the scientific method prefer to look beyond the aspects of use and consumption of cheese into the hows and whys of turning liquid, bland-tasting milk into solid, aromatic, flavorful products. They want to know functions of the various milk components and effects of cheesemaking treatments on them. They want to understand how the following basic steps of cheesemaking are performed: acidification, coagulation, dehydration, gel synthesis, shaping, and salting. This section is devoted to fundamentals of the cheesemaking processes.

30.4.1 Milk and Its Treatment

Qualities of milk and its components are key determinants in converting milk to cheese. Milkfat, lactose, casein, whey proteins, enzymes, and minerals each have functional roles. Of course, the enzymes may be native to milk, come from microorganisms, or be added purposely. Understanding factors that determine cheese moisture content after pressing is essential for achieving optimal moisture, salt, and ripening. Control of cheese moisture is paramount to maximizing yield and profitability (Everard et al., 2011). Final cheese moisture is affected by many cheese-processing conditions, especially coagulation time, curd particle size, stirring, cooking time and temperature, time of dry stirring, cheddaring, salting, duration between salting and pressing, and ripening conditions.

Milk composition is the most important factor affecting cheese yields, of which approximately 93% of the dry matter is protein and fat. Major variables in cheese milk composition include (1) breed; (2) stage of lactation; (3) nutritional and health status of cows, especially udder inflammation/infection; (4) duration of cold storage as it limits proteolysis; and (5) efficiency of standardization.

Wendorff and Paulus (2011) compared milks from six breeds in terms of suitability for cheesemaking. As presented in chapter 21, milk containing the β-form of κ-casein has a shorter coagulation time, forms firmer curd, and has higher yields than that with the α-variant. Frequencies of the κ-casein β allele in milk from those six breeds were reported as follows: Jersey, 86%; Brown Swiss, 67%; Guernsey, 27%; Holstein, 18%; Milking Shorthorn, 11%; and Ayrshire, 7%. Accordingly, milk from Jersey, Brown Swiss, and Guernsey cows should be preferred, in this order, for making cheese. Nevertheless, no differences in cheese quality have been reported and economic factors, especially the raw milk pricing system, must be considered. Data from Wisconsin revealed the following average annual outputs: Jersey—16,800 lb milk, 4.84% fat and 3.59% protein; Brown Swiss—18,800 lb milk, 4.1% fat, and 3.37% protein; Guernsey—16,100 lb milk, 4.56% fat, and 3.33% protein; Holstein—24,000 lb milk, 3.7% fat, and 3.02% protein. When milk is purchased on the basis of content of fat and protein or on cheese yield, its value to the producer is highest among breeds containing the highest percentage of protein. Of course, yields of cheese per unit of time and equipment required to make it are important economic considerations.

The quality of raw milk is important because there can be at least five defects that result in problems. These include the following:

1. High bacterial count—off flavors, defective body and texture
2. High somatic cell count—inferior coagulum and, in Swiss cheeses, abnormal gas holes
3. Antibiotics and sulfa drugs—starter inhibition
4. Rancidity—starter inhibition
5. Off flavor—off flavor in cheese

The ideal eating quality of each variety of cheese can be obtained only if its components are in proper ratios, which differ among varieties. Additionally, standards are in effect for many varieties providing reasons for standardizing milk for cheesemaking: (1) ratios of components vary with the cheese variety, (2) regulatory standards must be observed, and (3) maximal recovery of solids from milk is an economic necessity. With milk low in casein, yield is correspondingly low because casein forms the structure inside which all other components are trapped or to which they are adsorbed. The optimum fat to casein ratio to provide maximum yield of high-quality cheddar cheese is 1:0.7. Despite these compelling reasons for standardization of this ratio, in some small cheese plants the benefits gained may not equal the cost of obtaining them. Protein to fat ratios of cheese milk ranging from 0.7 to 1.15 made major differences in cheese composition, cheese yield, and the percentage of milkfat recovered (Guinee et al., 2007).

UF raw milk, concentrated to 10 to 12% protein and containing 3.5% fat, often is used to fortify raw milk for cheese production. The process involves separating milk into cream and skim milk, concentrating the skim milk by UF approximately threefold, and adding back the cream. Lactose concentration remains at about 5% so its concentration in the fortified cheese milk is not significantly changed and the normal make procedure can be used. UF concentration reduces costs of shipment of treated milk to cheese plants and increases the yield of cheese per unit capability of cheesemaking equipment. Permeate, containing about two-thirds of the lactose, can be concentrated by reverse osmosis and then dried (cf. chapter 33).

Following or during standardization, milk intended for cheese manufacture is usually clarified (unnecessary if it has been centrifugally separated) then pasteurized (cf. chapter 26). Pasteurization (1) kills pathogens and microorganisms that produce defects, (2) inactivates certain enzymes, (3) enables greater uniformity in the product, and (4) increases yield of some cheeses. Minimal heat treatment is desirable for production of most ripened cheeses because if natural milk enzymes can be kept active, ripening progresses faster. This has prompted treatment of milk with hydrogen peroxide for some varieties of cheese, especially Swiss. The method calls for addition of about 0.02% hydrogen peroxide to milk as it enters the HTST pasteurizer for a 16-sec exposure to heat of about 133°F (56°C). Milk is immediately cooled to **setting temperature** and pumped to the cheese vat. Here catalase is added to deactivate residual peroxide prior to adding **starter culture**. This treatment selectively destroys coliforms, vegetative cells of clostridia, many staphylococci, and other bacteria that may cause defects in flavor, body, and texture. Most lactobacilli and streptococci survive. In addition, milk proteins are made more elastic and cheese ripens faster than when made from pasteurized milk. Cheeses sold in the United States not made from pasteurized milk must be ripened a minimum of 60 days. Importantly most pathogenic bacteria die during this time.

In 2012, Food Standards Australia New Zealand (FSANZ) approved the sale in Australia of hard to very hard cooked curd cheeses made from raw milk. The processing alternatives to pasteurization include thermization in combination with ripening or specific cooking of the curds in combination with controlled ripening and moisture content. Thermization involves heating milk at temperatures of 145 to 149°F (63 to 65°C) for 15 sec, whereas pasteurization involves heating milk at 161°F (71°C) for 15 sec. The lower heat of thermization, compared with heat of pasteurization, permits some of milk's natural lipase to remain active, thus assisting in development of typical flavor of blue-type cheeses.

Milk for some cheeses, especially blue, cream, and other soft cheeses, is homogenized to add smoothness and with some cheeses to decrease losses of fat in whey. In blue cheese varieties homogenization provides greater access of lipase enzymes to fat, thus allowing faster hydrolysis of the fat, as is necessary in releasing milk's short-chain fatty acids needed to produce the blue cheese flavor.

30.4.2 Preconcentration of Milk for Cheesemaking

Ultrafiltration (cf. chapters 26 and 33) can be used to concentrate milk by as much as 50% of the volume of the original milk before the setting step of the usual cheesemaking process. This is important economically because it saves equipment while potentially increasing yield. However, in producing cheddar cheese some modifications in the process are needed to overcome the increased buffering capacity of the **retentate**. These modifications include lower temperatures of setting, cooking, and cheddaring; addition of increased amounts of starter; and addition of rennet based on the original amount of milk (Sharma et al., 1989). Lowering the buffering capacity can be accomplished by acidifying the milk to pH 6.0 with acetic acid prior to ultrafiltration. In producing cheese for pizza, the retentate can be diafiltered to remove a portion of the lactose, which otherwise causes the pizza to brown excessively.

For some cheeses milk can be concentrated up to sevenfold producing "pre-cheese," to which starter and rennet are added. Only small amounts of whey are released so this pre-cheese can be put into molds to form cheeses such as Camembert. Other cheeses (e.g., mozzarella) can be produced after the coagulum is cut and cooked to remove small amounts of whey.

Advantages realized by concentration to near the chemical composition of the cheese include reduced processing costs and a significant increase in yield since more whey proteins and minerals are retained. However, addition of starter and coagulator are impaired somewhat by the high viscosity of the retentate. Additionally, functionality of the curd differs in that acidification is slowed because of the high buffering capacity of the retentate. Furthermore, the cheese may not conform to the standard of identity or the common definition.

30.4.3 Setting Milk

A series of steps leads to the formation of coagulum. The process usually involves adding lactic starter culture (cf. section 23.3.1 and 29.2) and coagulator (rennet or **chymosin**) to milk adjusted to a temperature favorable for the growth of starter. Often calcium ions (0.01 to 0.02% $CaCl_2$) are added to increase the rate of coagulation of pasteurized milk because pasteurization significantly slows the rate of coagulation. Quantity of starter added, its activity, quantities of rennet or other enzymes and $CaCl_2$ added, time of adding each, and characteristics of the milk are the primary determinants of the rate of coagulation and the properties of the coagulum. Therefore, cheese makers must be able to predict accurately how each factor will influence each vat of potential cheese.

Setting milk (coagulating the casein) is a relatively short process for ripened cheeses, requiring 15 to 60 min. The lactic culture needs to produce only limited amounts of lactic acid, usually less than 0.05%, to stimulate curd formation by rennet.[3] However, with unripened cheeses, such as cottage and cream, setting time varies from 4 to 16 hr depending on quantities of starter and rennet added and incubation temperature. For ripened cheeses, rennet is added after **titratable acidity** has increased about 0.02%.

Caseins are the proteins that form the curd during cheesemaking. Casein proteins have an open globular structure (cf. section 21.3.1), making them susceptible to proteolysis. Being phosphorylated, they bind the calcium of milk and exist as micelles. Chymosin (the protease of rennet) rapidly hydrolyzes the κ-casein component of the casein micelles. This protein, which is negatively charged, protrudes as a hairlike projection from the micelles, thereby stabilizing them. (Remember that like electrical charges repel each other.) After this hydrolysis, the bulk of the casein proteins precipitate under the influence of calcium ions. Chymosin specifically recognizes the sequence in the chain of amino acids from histidine to lysine, which are located in the 98th and 111th positions (His 98 to Lys 111), respectively (figure 30.1). Once attached to the chain it cleaves the peptide bond between the phenylalanine and methionine at positions 105 and 106 (Phe 105 and Met 106) in the κ-casein chain. This produces para-κ-casein and glycomacropeptide, the latter being released from the micelle.

Rennin (chymosin) is the proteolytic enzyme found in rennet, which is an extract found in the fourth stomach (**abomasum**) of calves. Since about 1960, the demand for rennet has exceeded supply for three important reasons: (1) US milk cow population has steadily decreased from 28 million in 1947 to 9.3 million in 2014, thus decreasing calf numbers, (2) more dairy calves are being raised as dairy beef and not being sold as vealers, and (3) cheese production has increased dramatically. To extend the supply of coagulator, manufacturers combine rennet and pepsin[4] in equal parts in a mixture. Additionally, stomachs from older animals may be extracted and used. Rennin content decreases as age of the animal increases but content of a similar enzyme, which some erroneously call bovine pepsin, increases. Normally about 90 mL (3 fl oz) of single-strength rennet are used per 1,000 lb milk for making cheddar cheese. For cottage cheese, from 0 to 1 mL of rennet is added per 1,000 lb of skim milk. To augment the supply of coagulator bovine chymosin is now produced recombinantly in *Escherichia coli* K-12, *Aspergillus niger* var. *awamori,* and *Kluyveromyces lactis.*

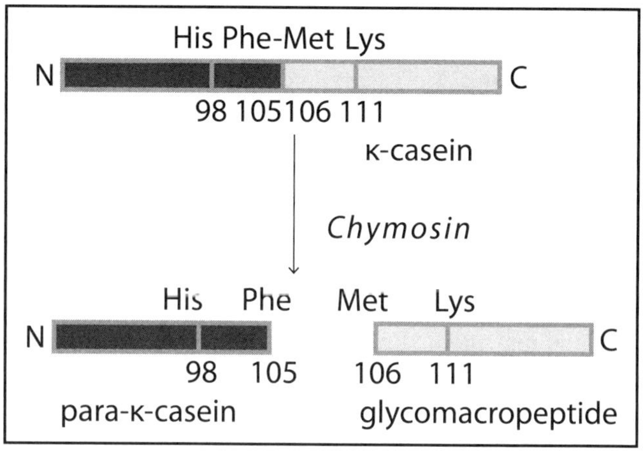

Figure 30.1 Site at which rennin (chymosin) hydrolyzes the κ-casein moeity from the casein micelle (David W. Brooks, University of Nebraska, Lincoln).

Some vegetarians find use of bovine rennet unacceptable. Microbially derived alternatives to bovine chymosin can be substituted for calf rennet. Suitable candidates are the proteases (microbial coagulants) from *Mucor miehei*, *Mucor pusillis*, and *Endothia parasitica*. These are acid proteases of similar structure to the bovine enzyme and are specific for κ-casein. However, they are unsuitable for long ripening cheeses because they have a broader range of nonspecific activities than chymosin and may produce off flavors, especially bitterness, on prolonged ripening. Additionally, yields may be lower when they are used.

Advances in bioengineering have resulted in efficient production of chymosin by selected fungi (fermentation-produced chymosin). For example, the bovine chymosin gene is inserted into the *Kluyveromyces lactis* chromosome. The yeast is grown by batch fermentation after which the yeast is killed by addition of benzoic acid and the chymosin is isolated by filtration.

Although recombinant bovine chymosin is the most frequently used coagulator in cheese production (about 85% in the United States and nearly 50% around the world), recombinant chymosins produced using genes of the goat, camel, or buffalo are available. In a comparative study, Vallejo et al. (2012) found that recombinant goat chymosin exhibited the best catalytic efficiency compared with the buffalo, bovine, or camel recombinant enzymes. Moreover, recombinant goat chymosin acted in a wider range of pH than other enzymes.

30.4.4 Coloring Cheeses

Extract of the annatto seed has been used to color cheese since the 1800s and is permitted under Good Manufacturing Practices according to 21 CFR 73.30. Annatto extract, a yellow/orange colorant consisting of the carotenoids bixin and norbixin, is commonly added to milk to produce orange-colored cheese such as cheddar. Differences in the carotenoid content of milk from various breeds, as well as seasonal variations in milk's color, make this desirable. There are no limits on usage of annatto in the United States, but in the United Kingdom the maximal amount to color orange, yellow, and broken white cheeses is 15 mg/kg. Approximately 20% of the added annatto escapes into cheese whey.

The color of whey may be detrimental to certain uses requiring that the colorant be bleached. This can be accomplished by oxidizing the double bond of the chromophore (color-emitting portion of the molecule). The dairy industry in the United States is permitted to use either hydrogen peroxide or benzoyl peroxide to bleach whey. Catalase is added after bleaching to destroy residual hydrogen peroxide. Both bleaching agents can affect cheese flavor. Since benzoyl peroxide is not permitted as a bleach of whey in Japan and China, they commonly test for the breakdown product, benzoic acid. However, test results vary because there are several sources of benzoic acid in cheese and whey.

Although not permitted to be used in the United States, two enzyme systems are effective in bleaching whey: a fungal peroxidase (trade name MaxiBright) and lactoperoxidase, a natural enzyme of milk. Benzoyl peroxide may be added directly to milk for making white cheeses such as asiago, blue, gorgonzola, parmesan, provolone, romano, Swiss, and Emmentaler.

30.4.5 Cutting or Breaking the Curd

This process is necessary to expedite whey removal. It is usually performed using "knives" made by stringing stainless steel wire through a stainless steel frame (figure 30.2). However, cheeses made in kettles, including Swiss, Emmentaler, and gruyere, are cut with a harp-shaped wire knife to produce particles the size of wheat grains. Gentle, uniform cutting minimizes breakup of coagulum and excessive losses of yield via whey. The smaller the particle of curd, the greater the surface area and the higher the surface to volume ratio. For example, a cube measuring 10^2 cm has a volume of 1,000 cm^3 and a surface area of 600 cm^2 for a surface to volume ratio of 3:5. Cutting the same mass of curd into 1 cm^3 yields the same 1,000 cm^3 volume. However, surface area increases to 6,000 cm^2 yielding a surface to volume ratio of 6:1. Thus, by reducing cube size by tenfold, surface area is increased tenfold. Considering

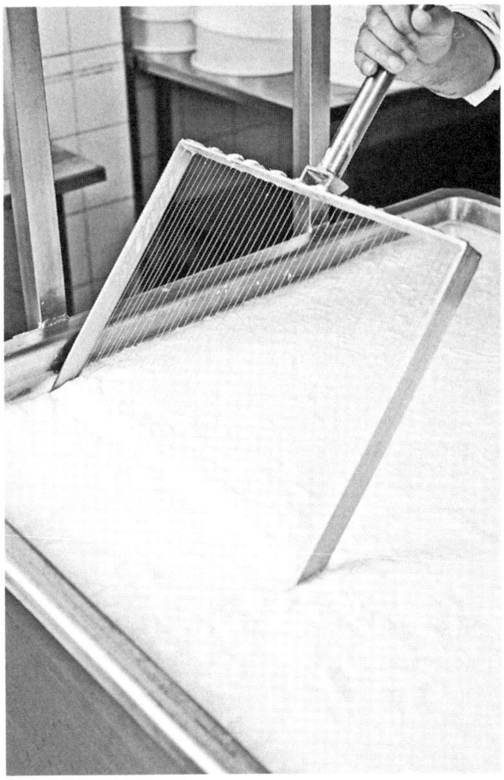

Figure 30.2 Curd knives are used in making vertical and horizontal cuts through cheese to form small cubes from which whey is easily expelled (racorn/Shutterstock.com).

that whey is more easily expelled from small curds (because moisture has a shorter distance to travel and more surface area from which to exit), it is readily apparent that curd particles are made smaller in the manufacture of hard than in the manufacture of soft cheeses. Of course, fine curd particles may escape in whey causing losses in yield.

Curd must have proper consistency and strength before cutting. For a rennet-set curd, this may be determined by the type of break that develops when a dairy thermometer is inserted at about a 30° angle from the surface into the coagulum and raised. If the curd breaks sharply and cleanly, cutting should commence. To increase moisture in cheese, coagulum should be firm at cutting; conversely, to decrease moisture, the curd should be cut when it is soft. The cheese maker should use the largest size cut that will result in the desired moisture. Fat losses occur at freshly cut surfaces and the probability of breaking off minute fragments of casein increases as surface area increases. Therefore, losses in yield commonly will be higher when curd sizes are smaller.

High-volume automated cheese plants employ closed vats in which each operation is computer controlled (figure 30.3). Consistency in the rate of fermentation is highly important since flow rates from raw milk storage to finished cheese packaging must accommodate product from multiple vats.

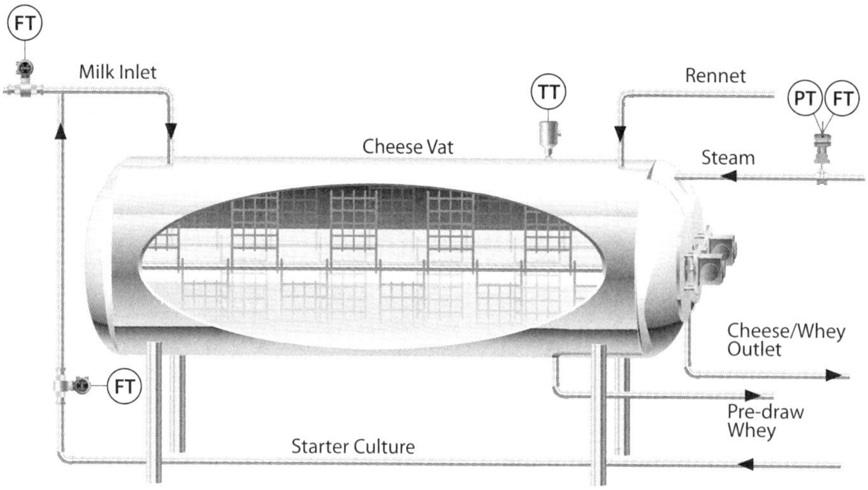

Figure 30.3 Cutaway view of an automatically and remotely controlled cheese vat containing blades that both cut and stir the curd. For cooking the curd, steam or water flows through channels behind the heating surface. The larger the vat diameter, the greater the speed of the agitator at the vat surface and the more small particles of curd produced during cooking (courtesy of Emerson Process Management).

Three methods may be used to determine when to cut cottage cheese coagulum: (1) pH measurement, (2) titratable acidity of whey, and (3) the A-C (acid-coagulation) test (Emmons and Tuckey, 1967). The correct pH must be determined by individual cheese makers for their milk supply and conditions of operation. It may vary from pH 4.6 to 4.9, it is often 4.8. The A-C test is based on the observation that normal skim milk, without rennet, will coagulate at pH 4.7. Therefore, after starter is mixed with skim milk and before rennet is added, a sample is removed into a beaker suspended in the vat to maintain equal temperature. After milk in the vat coagulates, milk in the beaker is observed closely by cutting with a knife or spatula. When thin lines of whey begin to appear at the cuts, the curd should be cut. Titratable acidity is a more variable test, primarily because of complex interrelationships of acids, salts, and proteins of nonfat solids that buffer (i.e., accept or contribute) protons (H^+) (cf. chapter 21). Insoluble calcium becomes soluble in milk during acidification thus reducing cross-linkages within the casein micelles leading to weaker, more flexible rennet gels (Choi et al., 2004). Higher concentrations of calcium in cheese tend to increase firmness while lowering meltability (Lucey and Fox, 1993).

These interrelationships are further complicated by growth of citrate fermenters, which use citrate, a buffering component, and produce carbon dioxide (CO_2). The latter increases titratable acidity as pH nears 8.3 in the titration. Therefore, unless solids content and extent of citrate fermentation are known, titratable acidity can be an unreliable estimate of time to cut the curd. In addition, variations in phosphate and citrate concentration, two important constituents, depend on the breed of cow whose milk is used, feed, stage of lactation, and evidence of mastitis. At pH 4.6, titratable acidity increases about 0.06% for each 1% increase in total solids above 9%.

30.4.6 Cooking

Cooking procedures vary widely in commercial practice depending on variety of cheese, methods of heating, and characteristics of each individual vat (no two vats of cheese are exactly alike). The primary objective of heating is to expel whey rapidly from curds, but acid development may be arrested and microorganisms destroyed if high temperatures are employed over an extended period of time. Cooking temperatures vary from 90 to 176°F (32 to 80°C). Lower temperatures leave more moisture in the curd, which results in a softer body. Average cooking temperatures for some cheeses are: blue, 91°F (33°C); brick, 95°F (35°C); cheddar, 102°F (39°C); pizza, 118°F (48°C); cottage, 120°F (49°C); Swiss, 136°F (58°C); and ricotta 176°F (80°C).

Heating may be accomplished directly by adding steam or hot water to curds, or indirectly by heating water with steam in a jacketed vat or by allowing steam to flow into the vat jacket. Temperature control is easiest when only steam is admitted to the jacket. However, local hot spots are avoided when water is also in the jacket.

Cooking should proceed slowly at first, then more rapidly. For example, for cheddar cheese the temperature is increased only 2 to 4°F (1 to 2°C) in the first 15 min, then raised 2°F (1°C) every 4 min with a total cooking time of 30 to 40 min.

When initial cooking proceeds too rapidly, major problems arise in moisture removal and defects occur in body, texture, and flavor. Curd particles become overfirm on the outside (case-hardened) and whey passes slowly through this membrane-like layer. Hence, curd interiors are mushy and high in moisture. Because lactose is retained in whey, considerable substrate is available for acid production and cheeses made from such curd are especially susceptible to defects related to high acidity.

30.4.7 Draining Whey or Dipping Curds

Whey can be removed by drainage from vats or kettles. Otherwise the mixture of curds and whey may be pumped or dipped from the vat to a drain table (figure 30.4), perforated molds (forms or hoops), or cheesecloth. The result is the same: whey and curds are separated (cf. section 33.6 for processing of whey).

Figure 30.4 (A) Curd being pumped from the vat to a drain table (Long Clawson Dairy, Leicestershire, United Kingdom). (B) Whey being drained from the curd through a stainless steel perforated cylinder (Agricultural Foundation, California State University, Fresno).

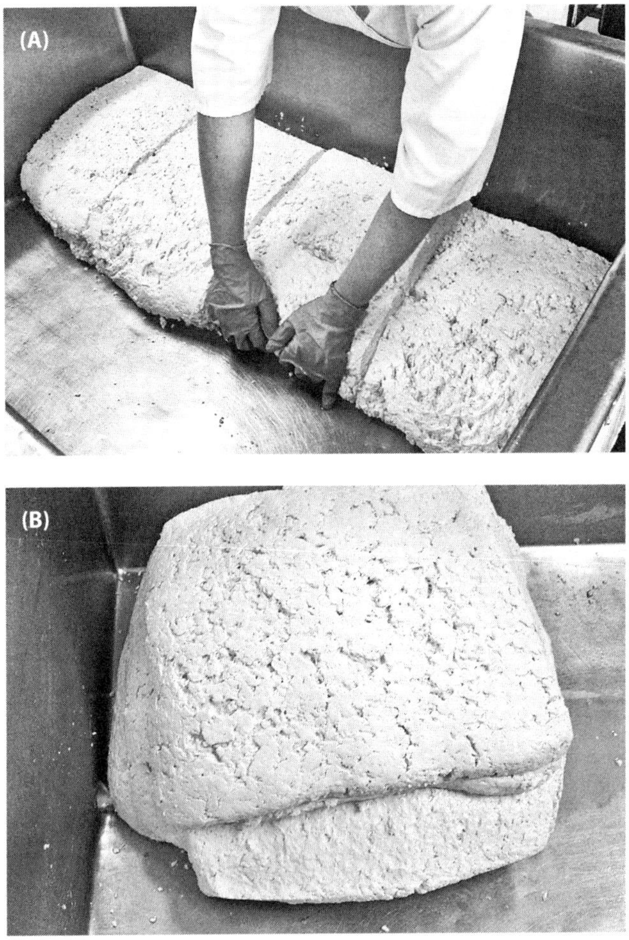

Figure 30.5 (A) In making cheddar cheese, curd is cut into slabs to allow piling. (B) Turning the slabs and repiling promotes whey drainage and knitting of curd (Agricultural Foundation, California State University, Fresno).

In making cottage cheese, curd must be prevented from sticking (matting) together. Therefore, only about half the whey is removed before water is added to wash and cool the curd. This water-whey mixture is drained and more water is added to wash the curd one or two additional times. Curd for most other cheeses is allowed to knit together. Acidity in the whey determines when drainage or dipping should begin for these cheeses.

30.4.8 Packing or Knitting

Differences in treatment at this point are important determinants of cheese characteristics. Curd for cheddar, washed curd, mozzarella, and pizza cheeses is allowed to knit together immediately after cooking; that for Colby cheese is washed before being packed in hoops; and that for Swiss, brick, and Muenster is placed in hoops immediately for pressing. Especially important factors that can be controlled in this step are moisture, acidity, body, and texture.

Since cheddar cheese is exceptionally popular in the United States, the cheddaring process deserves particular attention. After whey is drained and curds have knitted together sufficiently, slabs are cut from the curd (figure 30.5). These are turned and piled then repiled until sufficient whey has exuded and the body becomes elastic.

To obtain desirable elasticity of cheese, acidity must be controlled within narrow limits. Elasticity is related to the amount of calcium bound to casein. As more acid is produced in cheese, less calcium is bound. Early in cheesemaking the curd is tough, but as acid is produced it can be stretched considerably. However, if too much acid is produced, curd becomes brittle and crumbles when sliced. In cheddar, brick, Swiss, and similar cheeses in which the pH is 5.0 to 5.2, elasticity is desirable, whereas the more acid cheeses, such as Roquefort and Camembert (pH 4.6 to 4.8), are characterized by a *short* or crumbly body.[5] In such cheeses practically all calcium is free from casein. A curd-washing step is used in manufacturing Colby cheese to decrease the residual lactose content and thereby decrease the potential formation of excessive levels of lactic acid.

Cheddaring is followed by milling (figure 30.6), in which slabs of curd are reduced in size to facilitate salting. Acidity in the free whey at this time approximates 0.6%, an increase from about 0.2% at the time of draining the major portion of whey from curd (Wilson and Reinbold, 1965). This increase results from the favorable temperature for growth of bacteria-producing lactic acid, mostly lactococci. Desirable curd is free from mechanical openings and is silky smooth, fibrous, plastic, and yet firm.

Figure 30.6 (A) Milled curd, (B) salting, and (C) packing into a cheese mold (Agricultural Foundation, California State University, Fresno).

30.4.9 Salting

Salt is added to improve the flavor of cheeses, but it also provides other benefits. Some protein is dissolved, producing conditions for better matting in varieties such as cheddar; growth of several microorganisms capable of spoiling cheese is suppressed; lactic acid production is slowed; and whey is released at a faster rate. Efforts to reduce the concentration of salt needed in cheese may require increases in cooking, stirring, and drain times as well as increased temperatures compared with usual make procedures. Reduced-fat cheeses pose additional challenges in reduction of sodium. When fat is removed from cheese, it is replaced primarily with water, thus requiring an increase in salt to maintain the ratio of salt to moisture needed to inhibit microbial growth and favorably affect texture.

Salt is applied by one or a combination of three procedures: (1) adding it to curd particles or milled curds (as with Colby and cheddar), (2) rubbing dry salt on cheese surfaces during the first few days of ripening (as with blue and Roquefort), or (3) immersion of formed cheeses in brine for a few days (as with brick, Swiss, Edam, Muenster, and mozzarella). Quantities of salt added vary from 1 to 10% among varieties of cheese. If salt is not uniformly distributed, the cheese may acquire a mottled appearance.

30.4.10 Pressing

Most popular varieties of cheese, except for cottage and cream, are pressed in forms (molds) of wood, plastic, or metal (figure 30.7) or in cloth bags. Curd pressing, typically at 10 to 100 kPa, leads to a coherent mass of curd. Openings are closed, excess whey is removed, and the cheese is formed so it can be packaged easily. (Cheeses ripened by internal growth of molds must be open to allow the entry of needed air.) In some presses, vacuum is applied to cheese to draw out trapped air, thus promoting closure of mechanical openings. Temperatures during pressing should be high enough to promote fusion but low enough to inhibit growth of undesirable microorganisms. Pressure should be applied slowly at first to avoid early closure of exterior holes that allow whey to drain from cheese interiors. Pressing disrupts some milkfat globules within cheese so they coalesce, whereas free fat and fat globule membrane material can be released from the exterior of the cheese (Lopez et al., 2007). Fat globules become embedded in cheese cavities and the protein matrix, thus limiting casein aggregation by blocking the flow of whey through the cavities.

In some cheeses, curd grains of the outer layer fuse together forming a rind that reduces further expression of whey. The formation of a rind is accelerated by rapid

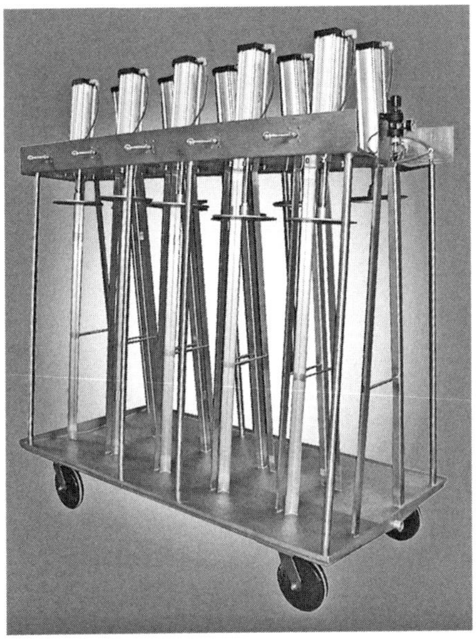
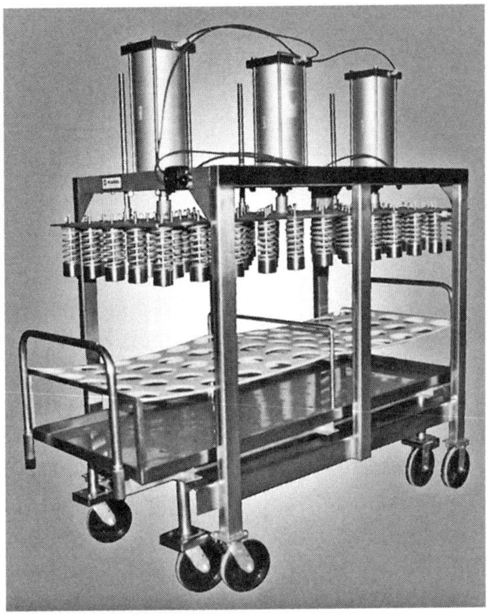
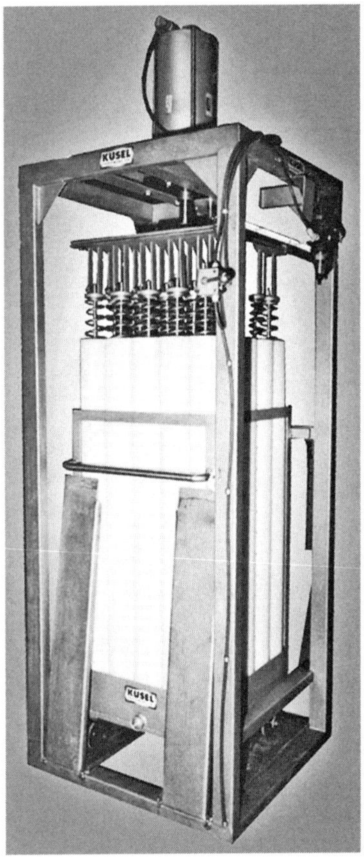

Figure 30.7 Cheeses are pressed in hoops varying in capacity from about 1 to 2,000 lb and in styles including blocks, cylinders, and barrels (courtesy of Kusel Equipment Company, Watertown, Wisconsin).

removal of moisture from this outer layer, especially when cheesecloth is applied. The higher the moisture content before pressing, the faster the formation and greater the thickness of the rind (Dejmek and Walstra, 2004).

As a replacement for draining cooked curd in a vat or on a drain table then forming it in hoops or barrels, Tetra Pak produces the Tetra Tebel Blockformer 6 in which cooked cheddar cheese curd is compacted in a series of vacuum and pressure relief cycles. Air and whey are gradually removed as the curd column moves down the tower (figure 30.8) by gravitational force. The fused curd is cut into blocks of uniform size and weight, and then gently ejected into plastic bags and onto a conveyor. Meanwhile, new curd is added to the top of the tower by vacuum-induced airflow.

30.4.11 Ripening

A complex transformation occurs during ripening of cheeses stored under controlled conditions after pressing (figure 30.9 on the next page).[6] Most fresh (**green**) cheese has little flavor except mild acid, and its body is firm, perhaps even tough. During ripening (also calling *curing* or *maturing*), enzymes from microorganisms and milk and those deliberately added (e.g., rennin, lipase) degrade components of curd, especially lactose, protein, and fat. Flavorful compounds produced include peptides, amino acids, amines, free fatty acids, aldehydes, ketones, and alcohols. These products of fermentation are much more water-soluble than their precursors; therefore, as ripening progresses, cheese body becomes softer and more pliable.

Inside the cheese, physical and chemical conditions change continually, influencing growth of microorganisms and activities of enzymes. This accounts for the succession

Figure 30.8 Tetra Tebel Blockformer 6 in which curd is formed into blocks of uniform weight, shape, and size then placed in plastic bags and sealed tightly (from Tetra Pak Cheese and Powder Systems B. V. Heervenveen, Netherlands).

Figure 30.9 Curing room containing ripening cheeses (PhotostockAR/Shutterstock.com).

of types of bacteria that dominate in many cheeses. In most cheeses, the oxygen supply is rapidly consumed and the interior soon becomes anaerobic. Thus, growth in the cheese interior is limited to microorganisms that require little or no oxygen. Because lactose is metabolized to other compounds in about two weeks, the supply of carbohydrates becomes limited. Subsequently the only organisms that can flourish are those that can gain energy from the metabolism of proteins, fats, and a few other compounds. For example, the lactococci that initially ferment lactose and produce acid that helps form the coagulum of cheddar cheese give way to the more nearly anaerobic lactobacilli as the lactose supply is exhausted. Major factors that influence the rate of ripening are the kinds and concentrations of enzymes, temperature, humidity, moisture in the curd, and time.

Temperatures for ripening are usually within the range of 4 to 60°F (2 to 15°C). Ripening at higher temperatures is associated with greater risk of developing defects, but time necessary to develop desirable properties is usually less at higher temperatures.

Producers of raw milk cheeses depend on the ripening process to destroy pathogenic microorganisms. It has been the general rule, and in some states a regulation, that hard raw milk cheeses, especially American types, be ripened at least 60 days before consumption. High salt to moisture ratio (low water activity), low temperature of ripening, and low pH are major factors preventing growth and causing death of most pathogens. Bacteria die more slowly in high moisture (soft) cheeses, and *Listeria monocytogenes* can grow at normal ripening temperatures. With these factors in mind regulators in Australia and New Zealand require that cheeses be made from pasteurized milk or milk treated with a process providing the equivalent reduction in bacterial numbers.

30.4.11.1 Lipases and Ripening

Catabolism of the triacylglycerides of milkfat by lipases and esterases produces free fatty acids. Lipases hydrolyze longer-chain fatty acids while esterases release shorter-chain ones. Oxidation of unsaturated fatty acids leads to production of alcohols, esters, thioesters (sulfur-containing), lactones, and ketones. Lactones are cyclic in structure with five- and six-sided rings.

Sources of lipases in cheese include lipoprotein lipase (LPL), the lipase of milk. It preferentially releases fatty acids from the sn_1 and sn_3 positions of the triacylglyceride. Since activity of LPL is greatly reduced by normal pasteurization, raw milk cheeses are likely to contain higher concentrations of free fatty acids than are cheeses made from pasteurized milk (El-Hofi et al., 2011). About 90% of LPL is associated with casein micelles. This association and the barrier presented by intact fat globule membranes limit lipolysis by LPL. Starter bacteria provide lipases that are weakly lipolytic and intercellular so that cells must lyse, releasing the enzyme, before substantial lipolysis occurs. Nonstarter bacteria contribute varying amounts of lipolytic activity. Lipases, especially rennet paste or lipase-producing microorganisms, are purposely added to some cheeses, especially Italian types. Rennet paste is harvested from abomasums of calves, kids, or lambs immediately after suckling. The pregastric esterase contained in rennet paste is released at the base of the tongue during suckling to enhance digestion of milkfat. Upon harvest the abomasum is partially dried then ground into a paste.

The following are sources of lipases for manufacture of selected cheeses: pregastric esterase for parmesan, romano, and feta; *Penicillium camberti* for Camembert; *Penicillium roqueforti* for Roquefort; and *Mucor miehei* for fontina.

Characteristic flavors produced by short-chain fatty acids include *rancid* and *cheesy* by butanoic acid (4 carbons), *pungent* and *blue cheese-like* by hexanoic acid (6 carbons), and *soapy, goaty, musty,* or *fruity* by octanoic acid (8 carbons). The shorter the carbon chain of the fatty acid, the higher the flavor produced per molecule of the acid.

30.4.11.2 Proteins and Ripening

Two fractions of casein, α_{s1}-casein and β-casein, are the major structural components of the protein fraction of cheese. During extended ripening of cheeses, α_{s1}-casein is hydrolyzed extensively (up to about 80%), whereas only about 25% of β-casein becomes hydrolyzed. In experiments in which a portion of β-casein was removed, increasing the ratio of α_{s1}-casein to β-casein, cheese melted more quickly than cheese with the normal ratio. Removal of some of the β-casein from cheese milk did not impair coagulation with rennet and improved stiffness of the gel, suggesting that α_{s1}-casein contributes more to gel stiffness than does β-casein (O'Mahony et al., 2009). Separation of bioactive β-casein from cheese milk may prove profitable provided the total concentration of casein is standardized to an optimal level.

If microbial growth is minimized on cheese surfaces, protein breakdown proceeds slowly. Seldom is more than 25 to 35% of the protein made soluble in hard cheeses such as cheddar and Swiss. The higher rate of protein and fat breakdown in surface-ripened cheeses is encouraged by (1) preparing cheese in small pieces to provide a high surface to volume ratio and a short distance for enzyme diffusion, (2) keeping moisture content high, and (3) deliberately planting desirable microorganisms on cheese surfaces. These aerobic microorganisms usually produce more highly proteolytic and lipolytic extracellular enzymes than do microorganisms that grow well inside cheese in an acid environment. In some surface-ripened cheeses, such as Camembert and Limburger, practically all protein is converted to water-soluble compounds.

By fortifying milk with the retentate of ultrafiltered milk, cheese yield is increased while times for curd formation and cutting are decreased because of a higher concentration of casein. Nonfat dry milk or condensed skim milk can be used but they add to the amount of lactose in the cheese thus increasing the amount of acid in the ripened cheese.

30.4.11.3 Lactic Acid Bacteria and Ripening

Lactobacillus helveticus was described by Orla-Jensen in 1919 as an isolate from Emmental cheese. This specie has the ability to grow at a relatively high temperature (130°F [55°C]) and is proteolytic. It is traditionally used in manufacturing aged cheeses including Swiss, gruyere, and provolone. Certain strains of the organism decrease bitterness and accelerate flavor development in cheese but the ability of different strains to affect these characteristics can vary widely. These attributes are associated with enzymes involved in proteolysis and/or catabolism of amino acids. Broadbent et al. (2011) determined that genes for peptidases and amino acid metabolism were much the same among strains within the species whereas genes for proteinases associated with the cell wall varied widely. Some of the genetic differences detected may help explain the variability observed among *Lb. helveticus* strains in regard to their functionality in cheese and fermented milks.

Lb. helveticus strain DPC4571 produced cheese of consistently high quality when used as an adjunct starter culture. This was attributed to its characteristics of rapid lysis and high proteolytic activity (Hannon et al., 2003). The early lysis of starter bacteria releases enzymes into the cheese matrix where they produce small flavorful molecules from the less flavorful proteins and fat. This results also in softening of the cheese and reduced incidence of bitterness. Although starter autolysis is vital for good flavor development, it serves primarily to enhance the effect of metabolic pathways already present in the starters and will not, in itself, result in enhanced flavor development (Kenny et al., 2006).

Because *Lb. helveticus* has a specific requirement for several amino acids to promote its rapid growth, it relies on a complex proteolytic enzyme system consisting of (1) an extracellular, cell envelope-associated proteinase that hydrolyzes casein into peptides; (2) specialized transport systems to take peptides and amino acids into the cell; and (3) intracellular endopeptidases and exopeptidases including many that are specific for proline-containing peptides, which degrade internalized peptides releasing amino acids. Although the biological role of the proteolytic enzyme system is directed toward provision of amino acids for cellular growth, the actions of these enzymes in cheese and fermented foods have important practical consequences on the sensory, functional (stretch, melt), and bioactive attributes of these products. Conversion of some free amino acids into volatile and nonvolatile compounds by lactic acid bacteria in cheese is believed to be rate-limiting in development of mature flavor and aroma. Broad differences in specificity and relative activity of individual enzymes exist, with lactobacilli generally showing higher proteolytic activity than other genera of lactic acid bacteria.

In experiments with soft cheese manufacture, addition of *Lactobacillus plantarum* I91 or *Lactobacillus casei* I90 as adjunct cultures with the starter containing *Streptococcus thermophilus* resulted in increased free amino acids, especially Asp, Ser, Arg, Leu, and Phe. Cheeses also contained significantly more diacetyl, a key flavor compound, than control cheeses. A triangle aroma test showed that addition of the lactobacilli strains significantly changed sensory attributes of cheeses, especially increasing the buttery flavor note, which is desirable in most soft cheeses. Both *Lb. plantarum* I91 and *Lb. casei* I90 performed well as adjunct cultures by influencing cheese aroma development and proteolysis (Milesi et al., 2010).

30.4.11.4 Probiotics in Cheeses

Refrigerated dairy products are considered ideal carriers for probiotic bacteria because they provide a suitable matrix for living microbes. A study showed that commercially produced cheese has the potential to provide immune health benefits when enhanced with *Lactobacillus rhamnosus* HN001 and *Lactobacillus acidophilus* NCFM. The intervention resulted in enhanced activity of phagocytic and natural killer cells, which was believed to be linked to immunity (Ibrahim et al., 2010).

30.4.11.5 Other Ripening Factors

In ripening Swiss cheese, early growth of *S. thermophilus* is followed in order by growth of *Lactobacillus bulgaricus* and *Propionibacterium shermanii*. The latter produces propionic acid and carbon dioxide (gas). Small bubbles of gas are joined by evolving gas, forming large holes (eyes). (Less energy is required to enlarge a gas hole already present than to form a new one.) Eye formation is virtually complete when the cheese is 5 to 7 weeks old; however, typical flavor is developed only upon continued storage at 39°F (4°C) for several more weeks (Reinbold, 1972).

In ripened, semisoft cheeses such as brick, Muenster, and Limburger, three microbial types predominate in succession. They are yeasts, micrococci, and *Brevibacterium linens*. Yeasts tolerate the high acidity of fresh cheese (pH as low as 3.5) and high salt concentrations at the surface (as high as 18 to 20%). As yeasts oxidize lactic acid, increasing the pH, micrococci begin growing. They, too, are salt tolerant. Yeasts also synthesize B vitamins needed for growth of *B. linens*. Together, these microorganisms produce several proteases that are important in developing typical flavors, especially in brick and Limburger cheeses (Olson, 1969).

30.4.11.6 The Rind

Rind develops on cheese that is allowed to dry considerably on the outside. In traditional manufacture cheesecloth bandages are placed inside the hoops when the cheese is pressed. The bandage is left in place when the cheese is dipped in hot paraffin to reduce **shrinkage** and to prevent surface growth of microorganisms. Moisture loss occurs slowly but is sufficient to form a rind during ripening. Because of considerable waste in rinds, manufacturers have developed rindless varieties. The cheeses are wrapped and sealed inside films that have low permeability to moisture and air, especially plastic and cellophane. Humidity in the ripening room is of limited significance for rindless cheeses.

Sometimes small white crystals of calcium lactate appear on surfaces of hard cheeses after two to six months of aging. Although harmless to consumers, they can cause a slight gritty mouthfeel. A patent was awarded in 2009 that claims addition of sodium gluconate at the time of salting the curd will inhibit growth of these crystals.

30.4.11.7 Cheese Mites

Artisan cheese makers frequently produce open-cured, cave-aged specialty cheeses using traditional bandage wrappings and aging in cool caves or cellars. Conditions therein favor breeding and transportation of cheese mites. These small arachnids are barely visible to the eye but their affinity for cheese and ability to bore holes in surfaces of rinds can add markedly to cost and decrease quality. Mites can be found on wood shelves used in cheese aging. Workers can distribute them since they attach to clothing and can be carried by air currents. Breeding is favored by unclean and moist conditions, especially where residue from cheese or whey exists. Mites can multiply profusely. They can be scraped from smooth cheese surfaces but only partially from open surface cheeses like gorgonzola or brick.

The first indication of cheese mites is a light gray or brown dust on cheese surfaces. Detection is assisted by sweeping of curing room floors onto white paper. After about an hour at room temperature mites will begin to move and become visible. Cheeses offered for sale through the USDA's Commodity Credit Corporation are ineligible for sale when the plant of origin is infested with cheese mites.

Ridding a cheese plant of mites is costly. Application of methyl bromide, an odorless, colorless gas, is effective but has been banned from use in cheese aging rooms. Aging of cheese at low temperatures of 40 to 45°F (4 to 6°C) minimizes reproduction and growth of cheese mites. Periodic cleaning of the room and equipment with detergent followed by sanitization is vital. Cheese mites can cause allergies, severe skin irritation, and gastrointestinal disorders.

30.4.12 Packaging and Preservation

Packaging materials for cheeses (figure 30.10) are designed to lock out oxygen, retain moisture, and protect cheese from contaminants. Some applications require films to shrink. Most barrier films are transparent to allow visibility of surfaces. However, they may be tinted to enhance appearance. Added features include grip and tear properties with reclosable openings for retail packages. In packages such as those used for shredded cheese it is possible to incorporate oxygen scavengers and preservatives encapsulated in cyclodextrins to reduce oxidation and growth of molds. Films are made by coextrusion or lamination processes. Large amounts of American-type cheeses, mostly used in manufacturing, are packaged in 500-lb barrels lined with barrier films.

Nisin, a bacteriocin, is useful as a food preservative that may be added to cheese or packaging materials (Molloy et al., 2011). This antibacterial polypeptide, discovered in 1928, contains 34 amino acid residues. Bacteriocins are ribosomally synthesized, extracellularly secreted, bioactive polypeptides that have bactericidal (killing) or bacte-

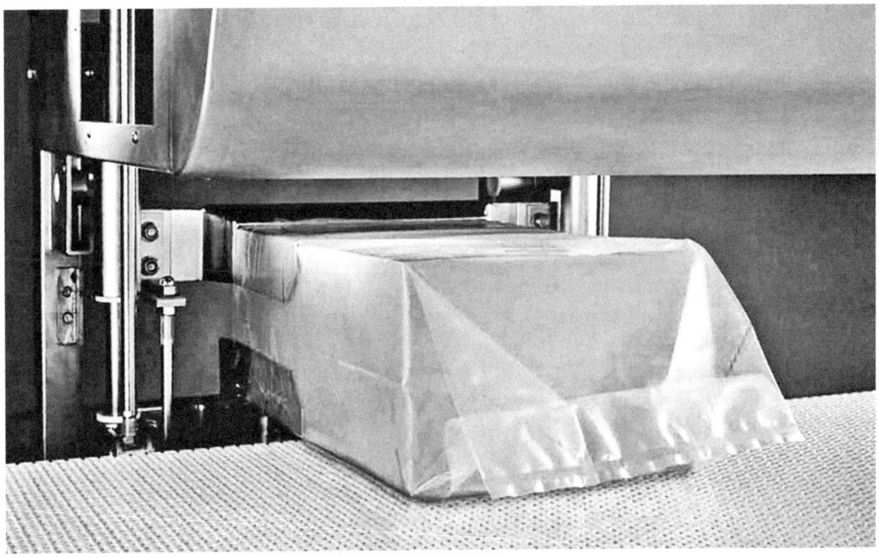

Figure 30.10 Packaging materials for 40-lb blocks of cheese (from Sealed Air Corporation, Duncan, South Carolina).

riostatic (growth inhibiting) effects on other bacteria. They are considered to act on the cytoplasmic membrane of the target cell, affecting cell permeability and dissipating ion gradients thereof. The producer cell is immune to its own bacteriocin.

Nisin is produced by fermentation using specific strains of *Lactococcus lactis*. It is a rare example of a "broad-spectrum" bacteriocin effective against many gram-positive bacteria, including *Clostridium tyrobutyricum* (which may produce gas in low-acid cheeses) and *L. monocytogenes* (a pathogen). When coupled with EDTA, a chelator, nisin also may inhibit gram-negative bacteria that possess an outer layer that prevents binding of nisin to the target lipid. Nisin is soluble in water and is permitted in cheese up to 12 mg/kg (European standard). It enhances the safety and shelf life of fresh mozzarella cheese when added to whey in which cheese is packed. It is used in processed cheeses to extend shelf life by suppressing gram-positive spoilage and pathogenic bacteria.

Researchers identified a culture of *Bifidobacterium longum* that produces a lantibiotic effective against gram-negative bacteria. Since this substance is produced in a mixed culture as a defensive metabolite, little of it is produced in pure culture. The task for researchers is to control the gene that turns on production in pure culture.

Lacticin is a second industrially important bacteriocin useful in controlling nonstarter bacteria during cheddar cheese ripening and preventing safety problems in cottage and mold-ripened cheeses and in meats. Lacticin-producing *Lc. lactis* 3147 strain can be incorporated directly in these foods where it produces lacticin during manufacture. Methicillin-resistant *Staphylococcus aureus*, vancomycin-resistant enterococci, and *Clostridium difficile* are each sensitive to lacticin.

30.5 Representative Varieties of Cheese

Many popular varieties of cheese originated in Europe and have a long history of production in a particular country or region. In order to provide those producers with a marketing shield from imitations, there are three types of protection in the European Union: (1) protected designation of origin (PDO), (2) protected geographical indication (PGI), and (3) traditional specialty guaranteed (TSG). The Database of Origin and Registration includes product names for foodstuffs registered under all three designations. In addition, a system of coding and control of packaging and shipment was designed to permit tracing of the cheese back to its origin. For example, in order for a product label to read Edam Holland or Gouda Holland, those products must be made in a region within the Netherlands.

There are about 1,000 product names protected by geographical indications. The estimated sales in 2010 of cheeses having PGIs in the European Union was €6.3 billion ($8 billion), and the sales volume was 866,000 metric tons (1.9 billion lb). The sources and types of cheeses that made up 88% of this sales volume include: Italy (grana padano and Parmigiano-Reggiano), France (Comté, Roquefort, and reblochon), and Greece (feta).

Premium products, such as cheese labeled with a PDO, are vulnerable to adulteration by substitution with less costly milk of another species (e.g., cow milk for sheep milk). This leads to substandard quality, loss of product identity, and unfair competition. DNA-based methods are appropriate for analysis of commercial products because DNA derived from somatic cells is present in dairy products. DNA is stable and retrievable for **polymerase chain reaction** analyses even after thermal treatment and acid coagulation.

The use of a PGI may act as a certification that the product possesses certain qualities, or enjoys a certain reputation, due to its geographical origin. Geographical indication laws restrict use of the PGI to the identification of a particular type of product unless the product or its constituent materials originate from a particular area and/or meet certain standards, including tests of quality. Geographical indications serve the same functions as trademarks because, like trademarks, they are source identifiers, guarantees of quality, and serve valuable business interests.

For definitions of cheeses defined in the United States, see US Code of Federal Regulations Title 21, Part 133. Table 30.2 shows how major nutrients are partitioned into six representative varieties of cheese. Now let us briefly discuss variations in the manufacture of 17 selected cheeses.

30.5.1 Cottage Cheese

This soft, unripened cheese is made from skim milk (or reconstituted dry skim milk in some countries) using lactic acid bacteria and a small amount, if any, of rennet. By law it must contain no more than 80% moisture (82.5% in lowfat cottage cheese). Creamed cottage cheese must contain at least 4% milkfat, whereas the lowfat variety may contain from 0.5 to 1.5% fat. Dry cottage cheese contains only a trace of fat. The protein of most cottage cheese is nearly all casein but one company has a patented process whereby skim milk is heated to denature whey proteins, which are then partially precipitated with casein. This process is called *coprecipitation*.[7]

Flavor, aroma, body, texture, and appearance vary widely among market samples of cottage cheese. The most common flavor is acid but its intensity may be mild to sharp. Diacetyl should be detectable by taste and smell but not so concentrated as to cause a coarse or bitter flavor. Body may be firm or soft; particle size may be small (less than 0.25 in [6.35 mm]) or large (0.5 in [12.7 mm]); cream may be well absorbed and stabilized or free from curd and of low viscosity. Each manufacturer should decide which characteristics are most desired by the most customers and then strive to produce that same type of cheese consistently.

Cottage cheese is the most perishable of all common cheeses. Psychotrophic bacteria and yeasts are the usual spoilage agents. However, both low pH (about 5.0) and compounds produced by culture bacteria (especially *Leuconostoc*) tend to inhibit their growth. Practices most effective in improving keeping quality are (1) prevention of contamination from water used in washing the curd and from equipment and (2) early reduction of temperature to about 37°F (2°C) and maintaining this low temperature. It is common to add potassium sorbate to cream dressing to improve shelf life of cottage cheese. However, overuse imparts a somewhat metallic flavor to the cheese. A patented treatment calls for addition of fine grains of food-grade calcium carbonate ($CaCO_3$) to cream dressing (about 0.06% of the finished cheese weight). At the pH of cottage cheese (4.8 to 5.2), $CaCO_3$ is slowly converted to carbonic acid that, in turn, converts to CO_2 and water. This produces an anaerobic condition in the cheese, slowing growth of aerobic spoilage bacteria and extending the shelf life from the usual 20 to 30 days to 30 to 40 days at normal refrigerator temperatures.

30.5.2 Cream and Neufchatel Cheeses

Cream cheese is smooth, white, and mild tasting. It contains at least 33% milkfat and not more than 55% moisture. Neufchatel cheese is similar but contains not less than 20% nor more than 33% milkfat and a maximum of 65% moisture. It has a slightly more acid flavor than cream cheese. Either may be made by hot-pack or cold-pack methods. In the former, cream is ripened by a lactic culture, heated to 149°F (65°C), stabilized, salted, separated (part of the whey is removed by centrifugal separation), homogenized, packed, and sealed. Shelf life usually exceeds 60 days.

30.5.3 American Cheeses

Cheddar, washed curd, Colby, and stirred curd cheeses constitute a group known as American cheeses. All must contain at least 50% fat in the solids (FS) (21 CFR 133.113). Modifications in the process for making cheddar led to development of three other varieties. *Washed curd* cheese is prepared in the same fashion as cheddar until the milling stage, when the curd is covered with cold water for 5 to 30 min. Washing increases moisture to a maximum of 40%. *Stirred curd* cheese has practically the same composition as cheddar[8] but has a more open texture and shorter (less elastic) body. It is manufactured like cheddar except that agitation of cooked curd particles is used to promote whey drainage and the ched-

Table 30.2 Percent of Recommended Daily Value of Six Cheeses.

Cheese Type	Total Fat	Sodium	Protein	Calcium	Vitamin A	Vitamin B_{12}	Vitamin D
Cheddar	13	7	14	20	5	3	1
Cottage	15	34	50	18	6	16	1
Mozzarella	7	0	16	20	2	5	1
Parmigiano-Reggiano	2	3	3	5	0	1	0
Queso blanco	44	34	48	81	13	35	8
Swiss	12	2	16	22	4	14	1

daring and milling steps are eliminated. *Colby* cheese is manufactured the same way as stirred curd except water is added to wash and cool the curd when most of the whey has been drained away, thus increasing moisture content to a maximum of 40%.

Both washed curd and Colby cheeses have shorter shelf lives (consequently they are not aged as long) than their corresponding unwashed counterparts, cheddar and stirred curd, because they contain more moisture and less acid. They also contain fewer soluble salts, have a more open texture, and are softer.

Much research has been conducted to develop lower fat American-type cheeses. The major problems are slow development of typical flavor and body. In 2009 lower-fat cheeses constituted less than 15% of total retail cheese sales.

30.5.4 Swiss Cheese Varieties

Swiss or Emmental cheese is one of the most difficult types to make. Only by having the correct combination of (1) milk, (2) three species of bacteria in an active culture, and (3) well-controlled conditions of manufacture can there be hope of developing typical Swiss cheese. The outstanding physical characteristic of Swiss cheese is its eyes, which should measure 3/8 to 13/16 in (0.4 to 2 cm) diameter. It is important that milk used to make this cheese be clarified to remove somatic cells and other particulates that can serve as nuclei for formation of excess eyes. Gas that forms eyes is produced by *Propionibacterium freudenreichii*, which also is the main agent of lipolysis, releasing free fatty acids during ripening.

Maximum moisture content is 41% and minimum fat content is 43% FS. Swiss cheese has a sweet, hazelnut-like flavor. The body is firm and meaty and the texture is **close** (that is, free of holes other than the normal Swiss-type gas holes). Once a hole begins to form, gas from the surrounding saturated mass moves into and enlarges it since that requires less energy than the formation of a new hole.

Gruyere, a related cheese with eyes, contains less moisture (39% maximum) and more fat (45% FS). It has smaller eyes and a more pungent flavor than Swiss because surface growth of microorganisms is permitted.

Besides the addition of three starter bacteria in optimal proportions, the method of Swiss cheese manufacture differs markedly from that for American cheese: (1) the curd is cut into finer particles; (2) cooking temperatures are approximately 50 to 60°F (10 to 15°C) higher; (3) there is no cheddaring, washing, or milling; and (4) salting is accomplished (by diffusion into pressed cheese) by immersion in a brine tank or by rubbing salt on surfaces of the formed cheese.

30.5.5 Italian Cheeses

The best known Italian cheeses are parmesan, romano, provolone, mozzarella, and pizza. In the United States this group is produced in slightly larger amounts than American cheeses.

Parmesan (grana padano is quite similar) is made from lowfat milk, thus producing a comparatively low fat content (32% FS). Moisture content is also low (32% maximum). Therefore, the cheese is very hard and usually must be grated before consumption. It must be ripened at least 14 months and commonly is ripened for two years. The low moisture content is obtained by cutting the coagulum finely and cooking at high temperatures (~135°F [57°C]). Benzoyl peroxide may be used to bleach milk for producing lighter-colored parmesan.

The coagulum for provolone, mozzarella, and pizza cheeses is formed primarily by the action of rennet (kids' rennet produces a piquant or sharp flavor). A mixture of heat-resistant lactobacilli and *S. thermophilus* is commonly added as a starter. The manufacturing procedure through milling resembles that for cheddar cheese, including packing, turning, piling, and repiling blocks of curd. The amounts of acidity developed and moisture removed depend upon which variety is being made. However, pH should be between 5.1 and 5.3 for favorable stretching. After milling, curd particles are immersed in hot water (181°F [83°C]) and worked (stretched) until they become smooth, elastic, and free of lumps.[9] This is followed by molding, often a manual process that requires dexterity and practice in handling hot plastic curd. Provolones are made in all sorts of forms—large pears, flasks, blunt-ended footballs, truncated cones, melons, cylinders, and sausages. Salting is accomplished by storing formed cheeses in chilled brine.

Varieties most responsible for recent significant increases in consumption of Italian cheeses are mozzarella and pizza. Mozzarella (as well as scamorza) contains 52 to 60% moisture and a minimum of 45% FS. Low moisture mozzarella or pizza cheese contains more than 45% and up to 52% moisture. Part-skim types are also recognized for both mozzarella and low moisture mozzarella (less than 45 to 30% FS). Because of its high moisture content, mozzarella cheese has a short shelf life (Ganesan et al., 2012). Commercial *fresh* mozzarella cheese is made by direct acidification and is stored dry or in water without added salt. This cheese has a shelf life of up to six weeks but often develops an off flavor and loses textural integrity by the fourth week, potentially due to the lack of salt and high moisture, which allows spoilage bacteria to multiply. Therefore, low moisture types are more often made when long shipments are expected (Reinbold, 1963). Limited ripening occurs in this low moisture cheese.

Provolone (*prova* = ball-shaped) cheeses are often looped with vegetable fibers for suspension in rooms in which they may be smoked as they ripen. (Liquid smoke may be added in lieu of smokehouse treatment.) If not smoked, cheeses are so labeled. The unique flavor of provolone results from a combination of penetration of smoke flavor, fatty acids released by added lipases, and an unusually high salt concentration.

Romano cheese is slightly less hard than parmesan. Traditionally it has been made from sheep milk curdled with lamb's rennet. In the United States, however, it is made from cows' milk. Italian researchers reported that pecorino romano cheese contained three to five times as much conjugated linoleic acid (CLA) as did cheeses made from cow's milk. An important factor: these sheep had received fresh grass for about 80% of their feed intake. Health benefits from an intake of 90 g/d (but not of 45 g) of this cheese included reduced LDL cholesterol and body mass index in humans who were slightly hypercholesterolemic (*Cheese Reporter*, 2009).

Lipases are often added to milk for manufacture of Italian cheeses, particularly provolone and romano, to promote development of the sharp, *piccante* flavor. The added lipases release fatty acids, especially butyric and caproic. However, they do not release components responsible for bitterness and soapiness, as does the lipase of milk.

30.5.6 Hispanic Cheeses

Hispanic-style cheeses are increasing in popularity outside their countries of origin. Variations among them arise from differences in their processing parameters, which allow their characteristics to range from hard and strongly flavored (Cotija), through semihard and meltable (asadero, Oaxaca, and menonita), to soft and crumbly (panela, queso blanco, queso fresco, and requesón).

Queso fresco (QF) represents the most popular of numerous Hispanic-type cheeses. It usually is consumed within one to two months of production. The cheese is characteristically high in pH, nonmelting, salty (about 2.5% added salt), and has a mild fresh milk flavor. Guo et al. (2011) measured the effects of salt on the chemical, functional, and rheological properties of QF during storage over an eight-week period. Cheese was produced in six separate lots with added salt ranging from 0.0 to 2.5%. QF samples with 1.5% added salt remained stable in pH and in moisture, fat, and protein content, whereas growth of contaminating bacteria in lots containing 1.0% salt or less caused pH to decrease significantly. QF with lower pH is less desirable to the consumer. Cheese with higher salt levels had a more stable color over the eight weeks. Changing salt content did not alter the nonmelting properties of QF.

Because of the high moisture and high pH of some Hispanic cheeses they can be of high risk for survival and growth of the pathogen *L. monocytogenes*. This bacterium is particularly dangerous to persons with compromised immune systems and to pregnant mothers and infants. Infective doses can be as low as 10^2. Therefore, it is imperative that manufacturing and handling procedures protect against contamination with this organism, which is found widely in the environment, especially in floor drains. Although generally *L. monocytogenes* is unable to grow in fermented dairy products having a pH less than 5.5, it can survive in many cheeses for months.

30.5.7 Brick Cheese

This semisoft cheese (44% moisture) is known for its mildly pungent yet sweet and nutty flavor. Methyl mercaptan and hydrogen sulfide contribute to the flavor. Ripening results from activities of microorganisms, especially *B. linens*, on the surface and inside the cheese. The relatively high temperature (61°F [16°C]) and relative humidity (90%) of early ripening promote surface growth of microorganisms. Brick cheese has an open texture with numerous round and irregularly shaped eyes. This is truly an American cheese, having been developed in Wisconsin.

30.5.8 Muenster (Munster) Cheese

The word Muenster is a contraction of monastere and correctly suggests that the cheese was developed in an Alsatian monastery. The best of it still comes from farms in the Munster valley not far from Colmar.

Ewen Jones

In Europe, Munster cheese is commonly surface-ripened and is strong in flavor. This cheese is protected by an Appellation d'Origine Contrôlée (AOC).

In the United States, muenster cheese is not surface-ripened but may be colored orange with a vegetable coloring. It can be made by the same procedures used for brick cheese with adjustments to allow a higher moisture content (43 to 46%). Muenster usually has a very mild flavor, a smooth and soft texture, an orange rind, and a white interior.

30.5.9 Brie and Camembert Cheeses

These cheeses are surface-ripened by molds, bacteria, and yeasts. They are characterized by a thin, feltlike layer or rind on the outside and a creamy, waxy to almost fluid consistency on the inside. Usually the curd is placed in hoops without cutting and the temperature is kept at about 68°F (20°C). Thin slices of the curd are drained in hoops with frequent turning for one to two days. Primary surface growth of grayish-white mold (*Penicillium camemberti*) is followed by a secondary bacterial fermentation characterized by sliminess and spots of yellow that change to a russet color (produced by *B. linens*).

30.5.10 Edam and Gouda Cheeses

These mild-flavored, nutty cheeses, which originated in the Netherlands, are quite similar; however, Edam contains less fat (40% FS versus 46% FS in gouda) and is, therefore, more firm than gouda. The curd is prepared much like that of cheddar cheese but is cooked in a mixture of whey and warm water of approximately equal parts. Consequently, the pH of its curd is higher (5.3 to

5.4). The curd is placed in molds that form the characteristic ball-shaped Edam or flattened spherical shape of gouda. They are surface-salted while being turned daily or by immersion in saturated brine for two to three days followed by surface salting. During ripening at 61°F (16°C) and 90% relative humidity (RH), wheels are turned and salted lightly. Excessive drying may be prevented by applying an emulsion of polyvinyl acetate or by wrapping in plastic film. The cheeses are usually covered with red-colored paraffin after being ripened. These cheeses may or may not have eyes. Their quality is often judged by "the ear," a person who thumps the cheese with his/her knuckles hoping to hear the dull thump that indicates a good cheese.

30.5.11 Monterey Cheese

This relatively high moisture (up to 44%), semisoft, mild-flavored cheese is manufactured much the same way as is Colby. Addition of cold water to the partially drained curd after cooking slows the removal of moisture and the curd is allowed to drain for less time than in the manufacture of Colby. Ripening is limited to three to six weeks.

30.5.12 Blue Mold Cheeses

The several cheeses ripened by the blue-green mold *Penicillium roqueforti* and related species are similar in appearance and taste. Important varieties are gorgonzola (from the Po Valley, Italy, AD 876), Roquefort (an industry in France for over 200 years), stilton (originating in Leicestershire, England, about 1750), and blue (United States and Canada). These cheeses vary in moisture content from 34% in stilton to a 46% maximum in blue. The mold is lipolytic and produces a distinctly rancid flavor, but one that is highly desirable to many persons and is unlike that of rancid milk. Mold spores are introduced into the curd just before it is packed into hoops to be pressed. The cheese must be open (contain holes) to allow penetration of air because the mold is highly aerobic.

Specific methyl ketones contribute to the characteristic flavor of blue cheese. Various metabolic and enzymatic pathways contribute to the generation of these methyl ketones, including the β-oxidation of fatty acids by the mold *P. roqueforti*.

Roquefort cheese is made from ewes' milk only in the Roquefort area of France where it was originally ripened in caves. A similar cheese made in France from cows' milk is labeled *bleu* cheese.

30.5.13 Feta Cheese

Feta cheese, Greek in origin and one of the oldest cheeses in the world, dates back to the Byzantine Empire. An aged, brined-curd, ripened cheese, feta is traditionally made from ewes' or goats' milk but can be made from cows' milk. Usually formed in square-shaped blocks, feta cheese has a somewhat grainy consistency (and is therefore crumbly), usually is white, and has a tangy and salty flavor. Compared with the harder form, soft feta is richer in aroma, sweeter, less spicy (piquant), and less salty. It is a table cheese, a favored ingredient in salads and pastries. Much of the feta cheese made in North America is from cows' milk and has a more creamy color than that made from sheep's milk.

30.5.14 Chèvre or Goat's Milk Cheese

Goat cheese, or chèvre, is made using the milk of goats. Although cow's milk and goat's milk contain similar amounts of fat, the higher proportion of medium-chain fatty acids (caproic, caprylic, and capric) in goat's milk contributes to the characteristic piquant flavor of the cheese. These fatty acids take their name from the Latin for goat, *capra*. France produces a great number of goat's milk cheeses, especially in the Loire Valley and Poitou, where goats are said to have been brought by the Moors in the eighth century.

Because goat cheese is often made in areas where refrigeration is limited, aged goat cheeses are often heavily rubbed with salt or placed in salt brine to prevent decay. Consequently, salt has become associated with the flavor of goat cheese, especially feta, which is heavily brined. Soft goat cheeses are made in homes around the world where cooks hang cheesecloth filled with curds in warm rooms to drain and cure for several days. If the cheese is to be aged, it is often brined so it will form a rind, and then stored in a cool cheese cave for several months to cure. Goat's milk cheese softens when exposed to heat, although, because of its smaller fat globules, it does not "oil-off" when melted, as do many cow's milk cheeses. When baked, firm goat cheeses with rinds form a gooey, warm mass that is ideal as a spread for bread.

Although in France goat's milk cheeses were made on an industrial scale as early as the 1930s, most goat milk cheeses being sold currently in North America are produced by artisan cheese makers.

Whereas the phosphatase test is applied to confirm pasteurization of cow's milk, the test is not reliable for goat's milk since the relatively low concentration of the enzyme in goat's milk (25% of that in cow's milk) permits its early denaturation during pasteurization. All manufacturers should maintain verified records of time and temperature treatments used in destroying pathogenic microorganisms in milk used in making cheeses. To fail to protect consumers with such practices is to risk loss of permits to sell finished products.

30.5.15 Flavored Cheeses

A favorite way for artisan cheese makers to attract buyers is to add flavorings to cheeses. Some examples include chives, dill, garlic, pepper, paprika, basil, oregano, red wine, and cocoa. Flavors should be balanced to assure that no single component is dominant. Concentrations of added flavoring include: 0.3% white or black pepper, 0.25% red pepper, and 0.1% herbs. Added flavorings containing fermentable carbohydrates provide substrate for microorganisms so overall cheese flavor is likely to change over time and changes will be magnified if the moisture content is increased with the added flavoring.

30.5.16 Lower Fat Cheeses

Much research has been done to develop lower fat varieties of cheeses, especially American types. Many consumers prefer reduced fat or lowfat cheddar cheeses over the full fat type, which contains about 32% fat. The general effect of decreasing fat content is slower ripening, firmer texture, and inferior flavor. Lowfat cheddar cheese differs in flavor from full fat cheese because of (1) less milkfat and (2) differences in concentrations of some key flavor compounds. Two off flavors, rosy and burnt, frequently occur in lowfat cheese (Drake et al., 2010). The lower milkfat content of lowfat cheese also changes its chemical and physical properties. Reducing fat content increases firmness/hardness (at the "first bite") of cheese while lowering cohesiveness and smoothness as perceived during chewing (Gwartney et al., 2002). Removal of fat is associated with a lower fat to protein ratio, resulting in a more dense protein structure, thus making the cheese firmer (Johnson and Chen, 1995).

Ripening greatly improves the texture of cheddar cheese while causing much less improvement in the sensory creaminess of the reduced or lowfat varieties (Hort and Le Grys, 2001). Manufacture of lowfat cheese using a starter culture producing exoploysaccharide resulted in an 8% increase in yield compared with the yield of cheese made with a nonencapsulated strain of the same *Lactococcus lactis* ssp. *cremoris* (DPC6532 and DPC6533). The researchers (Costa et al., 2010) observed a significant improvement in both textural and cooking properties without negatively affecting the flavor profiles of the cheeses.

Addition of 5 g/L of modified waxy corn starch, waxy rice starch, and instant tapioca starch has potential for improving the texture of lowfat cheese because starches are retained well in the protein during coagulation, thereby producing interruptions in the curd network and limiting protein-protein interactions (Brown et al., 2012).

30.5.17 Raw Milk Cheeses

About 30 of the 72 cheeses defined by the US Food and Drug Administration can be made legally from raw milk and sold in much of the United States. These cheeses must be ripened at temperatures above 35°F (1.6°C) for at least 60 days before sale. However, some pathogens may survive even this treatment. This behooves producers of raw milk cheeses to take special care to produce milk of the highest quality; for example, standard plate counts of < 10,000/mL and somatic cell counts of < 200,000/mL.

Four major incentives exist for making cheeses from raw milk, viz., (1) increased rate of ripening, (2) difference in flavor profile, (3) lowered processing cost with some varieties, and (4) opinion of some consumers that the product is "more natural." Whereas these factors can provide advantages, there are risks that the product may not provide the expected outcome.

The major reason for heating milk before making cheese is to destroy both pathogenic bacteria and undesirable hydrolytic enzymes that may be present in the milk and cause defects in cheese. Of course, some of milk's enzymes that can play a favorable role in cheese ripening may be partially or totally inactivated by pasteurization. However, a microbial product of the growth of *S. aureus* is a highly heat stable enterotoxin.

An analysis of 31 staphylococcal food-borne disease outbreaks in France showed that nine were either confirmed or suspected as involving raw milk cheeses. For such incidents to occur there must be a source of enterotoxin-producing *S. aureus* to contaminate the milk. The most likely source is infected mammary glands. Because numbers of *S. aureus* entering milk from an infected gland are far too small to have produced a toxic concentration of enterotoxin in fresh milk, there must be an opportunity for the bacterium to grow and produce enterotoxin. Such an opportunity is provided when the temperature of the contaminated milk is raised above about 68°F (20°C), pH of the milk remains above about 5.0, and there is limited competition from other fast-growing bacteria, especially the cheese starter culture. Time for these conditions to exist is another variable.

A significant outbreak of staphylococcal food poisoning occurred when milk contaminated with enterotoxigenic *S. aureus* was set with a phage-containing starter culture. Failure of the starter to grow adequately permitted production of toxic amounts of enterotoxin in the cheese. Whereas high salt concentrations inhibit growth of some bacteria, *S. aureus* can grow in the presence of as much as 10% salt, making this an ineffective inhibitor of the organism.

30.6 Mechanization in Cheesemaking

The cheese maker's job has traditionally involved considerable physical labor; much of it could be classified as toil. Processors and equipment manufacturers often modify, redesign, and invent devices to reduce labor requirements, to increase efficiencies, and to improve cheese quality. For manufacturers, the drain table is an excellent example. Rather than tie up cheese vats during draining and cheddaring when they could be used to set more milk, curds and whey are pumped to a drain table. Whey drains through perforations in a stainless steel plate set in the bottom of the table, leaving the curd in a thick mat that can be conveniently cheddared. Furthermore, the table is at waist height, thus eliminating the chore of reaching deep into a vat to cut and turn the cheese.

For processors, the entire cheddar cheesemaking process is a good example of mechanization and automation. Vertical, cylindrical, cone-bottom tanks are used to set milk. A pair of variable speed, counterrotating, cutter-stirrers is mounted inside the tank. From here, mechanization processes can vary. For example, in one type of system, cooked curd is pumped to a cheddaring box where it is distributed into six compartments, the whey draining away. Once filled, the box is programmed to rotate in 90° steps at preset intervals to produce the cheddaring effect. The cycle from filling to emptying requires about 3 hr.

In a second system, curd is pumped, in part of the whey, to the cheddaring machine where curd is deposited in a round, doughnut-shaped basket. The basket tips as it rotates, forcing the curd to stretch and the weight of the mass to be distributed throughout, thus accomplishing cheddaring. With the revolving basket in the vertical position, curd is dropped by gravity across a knife that cuts a 4 × 16 in (10 × 40 cm) swatch of curd. The ribbon is conveyed to a curd mill. Milled curd drops to a scaled belt where salt is added based on weight. Curd then drops into a screw conveyor that carries it to the hoop filler.

In a third system, whey and curd are separated by a prescreen ahead of a draining conveyor. The curd is then blown into a cheddaring tower 38 to 48 ft (11 to 15 m) tall in which it is cheddared by its own weight. As curd is milled from the bottom of the tower it is salted on another conveyor that drops it into hoops of 2,000-lb capacity. Here the curd is pressed and vacuum treated before being cut into blocks, usually weighing 40 lb (18 kg).

30.7 Direct Acidification

Cheese curd can be made without the aid of starter by mixing cold milk and acid. No coagulum forms at low temperatures even if the pH is lowered to the isoelectric point of casein (4.6 to 4.7). On warming, however, curd is formed. Cottage and part-skim mozzarella cheeses have been made by direct acidification. For mozzarella, after acidification to pH 5.4 to 5.6, rennet is used to complete the coagulation of milk at about 96.8°F (36°C). Phosphoric, lactic, acetic, and hydrochloric acids may be used. Moisture is removed by cooking the cubed curd.

Queso blanco, important in Latin America, is made by direct acidification, usually with glacial acetic acid. However, acid is added to hot milk (176°F [80°C]). Queso blanco resembles mozzarella but does not string or melt well when heated.

Unless starter cultures that can survive processing are added, cheeses made by direct acidification should not be expected to ripen in a predicted manner. In making such cheeses it is possible to add certain enzymes, such as lipase, to cause the development of specific flavors and aromas.

The US Food and Drug Administration has approved manufacture of cottage cheese by use of the acidulant glucono-delta-lactone (cf. section 29.10), with or without milk-clotting enzyme. Labels for such cheese must read "Directly set" or "Curd set by direct acidification" (21 CFR 133.128).

30.8 Process Cheeses

Process cheeses are made by grinding, heating, and mixing hard types of cheese followed by emulsifying, which is aided by the addition of not more than 3% inorganic salts (such as sodium citrate and sodium phosphate). Emulsifiers remove some of the calcium from the casein increasing its solubility, thus improving the emulsifying capacity of the cheese proteins. Sodium hexametaphosphate (SHMP) is commonly used as an emulsifying salt—although rarely alone. A recent study (Shirashoji et al., 2010) suggested that SHMP chelated residual colloidal Ca phosphate and dispersed casein; the newly formed Ca-phosphate complex remained trapped within the process cheese matrix, probably by cross-linking casein. Increasing the concentration of SHMP from 0.25 to 2.75% improved fat emulsification and casein dispersion during cooking, reinforcing the structure of process cheese, making it more firm.

Benefits derived from production of processed cheeses are uniformity in flavor, texture, and appearance; improved keeping quality and safety; opportunity to salvage some cheeses that are defective; and opportunities to modify properties (e.g., melting characteristics) to enhance marketability and to expand uses (e.g., addition of spices). Process cheeses do not oil off at temperatures that melt the fat.

American cheeses are most frequently used in processing. Commonly fresh (current or green) cheese is mixed with aged portions to obtain a desirable blend that particularly affects flavor and aroma. Cheese too low in fat, too high in moisture, or with defects in body and texture (such as gas holes) may be used provided they do not have excessive flavor defects. The blend may consist of one, two, or more varieties of natural cheese and may contain pimentos, fruits, vegetables, or meats. In the production of slices, three parts of cheese up to three months old can be blended with one part of cheese aged six months or more. A younger cheese, with its high amount of unhydrolyzed casein, provides the firmness needed for slice integrity.

Processed cheese is usually labeled "pasteurized" because heating at 150°F (65°C) for 30 sec is sufficient to kill pathogens. Melted cheese initially separates into fat and aqueous (serum) phases. Upon introduction of emulsifiers with agitation, pH increases, some protein is solubilized, and fat globules form. The mass becomes homogeneous and plastic. Usually only spores survive, but spores of anaerobic Clostridia may germinate, grow, and produce gas. The antibacterial agent nisin may be added to inhibit germination of these spores. In making slices, some manufacturers pump the hot emulsion onto stainless steel belts in ribbon form where it is cooled on the belts, cut into proper lengths, and stacked and packaged mechanically (figure 30.11). Minimum fat in solids is the same as required for the cheese used. However, moisture may be 1% higher than the maximum for that cheese.

Pasteurized *process cheese food* is prepared as is pasteurized process cheese except that certain dairy products (cream, milk, skim milk, cheese whey, and whey albumin) may be added. A higher temperature of pasteurization (~171°F [77°C]) is used in cooking and acid is added after emulsification to lower the pH to about 5.4. Total fat content is lower (23%) and moisture content is higher (44%) than in process cheese. Pasteurized *process cheese spreads* may contain 40 to 60% moisture and must contain 20% fat. Their higher moisture induces spreadability. They are cooked at a higher temperature (~190°F [88°C]) and have a lower pH (5.2 or lower) than process cheese foods (to promote keeping quality). Also, stabilizers, such as vegetable gums, may be added (maximum of 0.8%).

Pasteurized *blended* cheeses are similar to pasteurized process cheeses but contain no emulsifying salts. The moisture content of the blend may not exceed the arithmetic average of the maximum moisture contents as defined in the standards of identity for the varieties of cheeses blended.

Food scientists at University College Dublin developed a processed cheese product that contained 60% less sodium than full-salt versions and differed little in taste and structure (El-Bakry et al., 2011). The product is made from dry protein ingredients such as casein powder rather than through the traditional cheesemaking process. The postmanufacture hardness of the cheese was decreased due to slight changes in the way fat and moisture were distributed within the product.

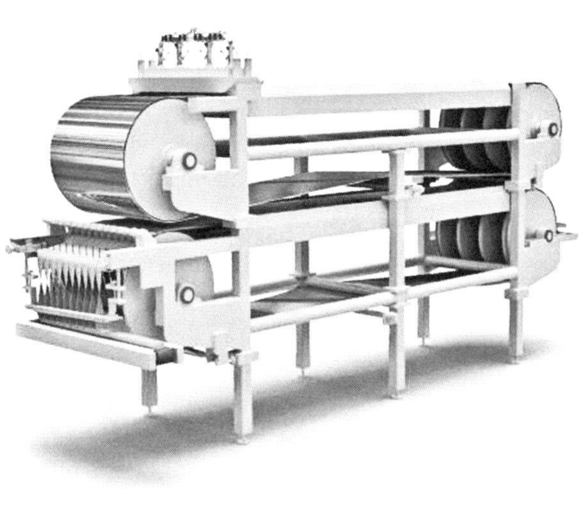

Figure 30.11 Depiction of a process cheese slice former (from Green Bay Machinery Company).

30.9 Enzyme-Modified Cheeses

Enzymatically modified cheeses (EMC) can be used in foods as the sole source of flavor or to impart selected flavor notes. The main uses of EMCs are in flavoring of processed cheese, analogue cheese, imitation cheese, cheese spreads, snack foods, soups, sauces, biscuits, dips, and pet foods. Lipases are the major types of enzymes used in preparation of EMCs. Flavors produced include blue, parmesan, romano, provolone, gouda, cheddar, and Swiss. Typically addition of only 0.1% EMC will achieve the desired flavor impact.

These products are produced by adding selected enzymes and/or microorganisms to cheese curd and incubating under controlled conditions, usually 86 to 113°F (30 to 45°C) for 24 to 72 hr, then stopping the reactions by heating to 158 to 185°F (70 to 85°C). Rennet paste, pregasteric esterase, microbial lipase, proteinases, and peptidases can be used to intensify cheese flavors.

30.10 Grading Cheeses

The US Department of Agriculture (USDA) makes available a grading service and standards. The USDA grade shield means the cheese that bears it has been inspected by a trained government grader and was produced in a USDA inspected and approved plant. Depending on the variety of cheese, grades "AA," "A," "B," and "C" are applied; "AA" is the highest, with increasing defects being allowed in each lower grade.[10] Cheeses failing to meet minimum specifications for the lowest grade are termed *undergrade.*

Bulk American cheeses for manufacturing use the following grades: (1) US extra, (2) US standard, and (3) US commercial. Grading is based on flavor, body, and texture of cheese used in manufacturing. Additionally, color, finish, and appearance are considered for cheeses that are to be marketed directly (natural cheeses). **Triers** are used to remove plugs of cheese for examination (figure 30.12).

Defects in ripened cheeses are more often associated with conditions of manufacture and ripening than with initial flavor of milk. The most frequently occurring flavor defect is high acid (sour). However, that flavor is seldom highly objectionable. Off flavors and odors associated with the growth of undesirable microorganisms or activity of enzymes include bitter, fermented, fruity, rancid, sulfide, unclean, and yeasty.

Body and texture defects usually associated with acid flavor are short (flaky or brittle), crumbly, mealy, weak, and pasty. Gassiness in cheeses without eyes (figure 30.13) suggests growth of yeasts or coliform bacteria

Figure 30.12 Triers are used to remove plugs of cheese for examination (ResolutionDigital/Shutterstock.com).

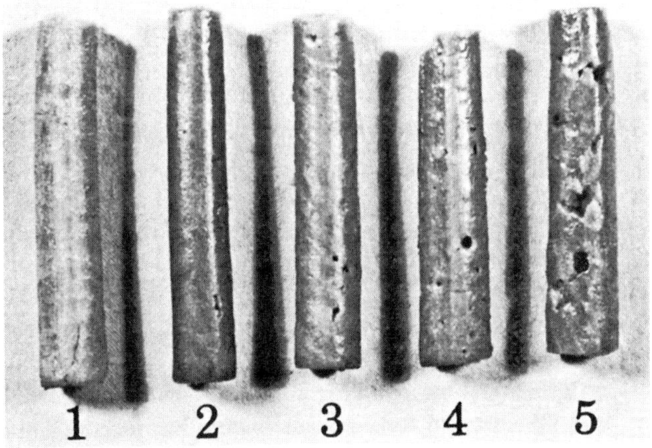

Figure 30.13 Plugs of cheddar cheese showing increasing openness and gassiness from left to right (courtesy of Missouri Agricultural Experiment Station).

and the associated flavor defects—unclean, yeasty, fermented, and fruity. Fresh cheese may have a curdy body and cheese that is too dry may be corky.

30.11 Whey

About half the original milk solids are left in whey from the manufacture of most cheeses. Average composition of whey from American cheeses is approximately 4.8% lactose, 0.9% protein (includes 0.1% casein), 0.6% ash, 0.3% fat, and 0.2% lactic acid. Whey from rennet-set cheeses, such as cheddar and Swiss, is sweet whey because the pH approximates 6.0, whereas acid whey from cottage cheese has a pH of about 4.6. The latter is therefore more difficult to dehydrate.

Acid whey can be used as a medium for growth of *Saccharomyces fragilis* (bakers' or brewers' yeast). Yeasts utilize most of the lactose and are then harvested and dried for animal feed, which contains excellent protein. Other processes (cf. chapter 33) useful for separating components of whey include electrodialysis (removes minerals), reverse osmosis (removes water), microfiltration (separates water-soluble proteins), ultrafiltration (removes protein and lactose), and combined heating and acidification (removes proteins).

Whey proteins, β-lactalbumin (~65%), α-lactalbumin (~25%), and serum albumin (~8%), are excellent as nutritional supplements. They are digested rapidly and, during exercise, open blood flow by inhibiting an angiotension-converting enzyme that constricts blood vessels, thus allowing blood flow to help repair and rebuild muscle tissues. These proteins are good sources of branched-chain amino acids, especially leucine, which plays a role in initiating protein synthesis (Fujita et al., 2007). Therefore, they are used widely in human food, especially by bodybuilders (Balch and James, 2000).[11] There also is a high demand for whey to be further processed into nutritional sports drinks and baby formula (cf. chapter 33). As of 2012, around 80% of the whey produced in the United States and the European Union was being further processed (*Cheese Reporter*, 2012).

Whey protein concentrate (29 to 89% protein) and whey protein isolate (90% protein) are being marketed. However, for every pound of this high-quality protein, uses must be found for 10 lb of lactose.

Dried whey contains approximately 70% lactose, 12% protein, and 8 to 10% ash. It is used in the dairy, baking, soup, and candy industries. Products in which considerable lactose is used include pharmaceuticals and infant formulas.

Undenatured lactalbumin has physical properties that are highly desirable in many foods. Importantly, when heated it can be coagulated and caused to form complexes with other proteins. The interaction is believed due to interchanges among disulfide pairs that result in rearrangement of and cross-linking between molecules through sulfur to sulfur (S-S) bonds. One result is increased capacity to bind water. For example, a coprecipitate with skim milk proteins will hold seven to eight times its weight of water without becoming fluid. Whey protein concentrates are particularly valuable as binders in textured vegetable proteins (Wingerd, 1971).

In the past, whey was disposed of primarily as animal feed, especially for use in milk replacers for calves and pigs. Some was distributed on land for fertilization (as such it is worth ~$1.00/ton) and some was disposed of through sewage treatment facilities. However, because of an especially high demand for oxygen when microorganisms in sewage break down whey solids, many sewage treatment plants become overloaded. Therefore, whey must not be discharged to sewers.

30.12 Summary

When the list of manufacturing variables applicable in cheesemaking is considered, it is not surprising that hundreds of varieties of cheese exist. The following is a summary of major variables by category:

- *Input variables*: (a) milk composition and quality, (b) types and number of microorganisms, and (c) enzymes contained and added.
- *Treatment variables for milk*: (a) standardization, (b) heat, and (c) homogenization.
- *Manufacturing variables*: (a) incubation time and temperature, (b) size and uniformity of curd particles, (c) cooking practices, (d) draining practices, (e) washing practices (includes pasta filata principle), (f) quantity and method of application of salt, and (g) hooping and pressing practices.
- *Ripening variables*: (a) time, (b) temperature, (c) relative humidity, (d) treatment of surfaces, and (e) package.

The ability of cheeses to be stored is closely related to moisture content and acidity, although several other factors influence keeping quality. Generally, the higher the moisture, lower the acidity (higher the pH), and lower the salt content the shorter the shelf life.

The USDA Agricultural Marketing Service (2012) provides grading services, which are outlined in *General Specifications for Dairy Plants Approved for USDA Inspection and Grading Service.* Plants that meet the requirements can label their cheddar, Colby, monterey, or Swiss cheese according to grade. Furthermore, buyers of these cheeses can select them based on grade. Other cheeses and cheese products, though not assigned grades, may carry the USDA quality approved shield if they meet quality specifications and if the manufacturing plant meets applicable sanitary requirements.

Study Questions

1. Name two primary agents that cause milk to coagulate. What protein component of milk stabilizes casein, making it insensitive to precipitation by calcium?
2. Describe the classification system for cheeses according to their relative hardness and give two examples for each class.
3. What major factors determine the hardness of cheese?
4. In what ways can the quality of raw milk affect the quality of finished cheese?
5. What are the purposes of pasteurizing milk for cheesemaking? How may pasteurization affect ripening of cheese?
6. Why would treatment with hydrogen peroxide be used in lieu of pasteurization in treating milk for some cheeses?
7. What does it mean to *set* milk? What agents are added in the setting process?
8. What sources of enzymes that coagulate milk are currently used?
9. Describe the A-C test. What is the major consideration in applying the titratable acidity test for the same purpose?
10. Why would cooking temperatures vary from 91 to 176°F (33 to 80°C) for various cheeses?
11. How can case hardening be prevented in cooking cheese curd?
12. What is the purpose of cheddaring?
13. Cite three purposes for pressing cheese.
14. What are the sources of enzymes that ripen cheeses? How may these differ between surface-ripened soft cheeses and hard cheeses that are ripened from the interior?
15. What causes the formation of eyes in cheeses such as Swiss?
16. Are defects more likely to occur in cheeses ripened at high or low temperatures? Why?
17. List some characteristics of cottage cheese that limit its shelf life.
18. What is meant by coprecipitate?
19. Why do washed curd and Colby cheeses have shorter shelf lives than cheddar and stirred curd cheeses?
20. What does pasta filata mean? Which cheeses fit this category?
21. Why are blue mold cheeses made to contain holes?
22. What acids may be used for direct acidification in cheesemaking?
23. What are the benefits of processing cheeses? Describe the differences between process cheese, process cheese food, and process cheese spread.
24. What agency offers grading service to the cheese industry? Who pays for this service?
25. Which defects of body and texture are often associated with acid flavor?
26. What is the major constituent of whey? How does whey of rennet-set cheeses differ from whey from cottage cheese?
27. Do whey proteins have unique and desirable nutritional and functional properties? Explain.
28. Identify several outlets/uses for whey and/or whey solids.

Notes

[1] κ-casein stabilizes α- and β-caseins, which in isolated form, precipitate in the presence of calcium ions.

[2] Discovery of lactic acid fermentation by Pasteur in 1857 and development of pure starter cultures later that century opened the door to industrialized fermentations of milk.

[3] Because less acid is formed in rennet-set cheeses, less calcium is solubilized to be lost in the whey than in acid-set cheeses such as cottage. On a dry weight basis, there is about three times as much calcium in American as in creamed cottage cheese (12 versus 4.3 mg/g dry matter).

[4] Pepsin is a proteolytic enzyme similar to rennin but is extracted from pig stomachs.

[5] Whereas *body* refers to resistance and elasticity, *texture* refers to degree of openness or gas space within cheese. Cheeses with considerable gas space have open or gassy texture depending on cause of the openings.

[6] For thousands of years, caves served as the main place for storing cheese under controlled conditions for ripening.

[7] In the manufacture of casein from skim milk, it is possible to coprecipitate most whey proteins. About 96% of the proteins in milk can be recovered. Combinations of heat and calcium chloride, or of heat and acid, are used to induce precipitation.

[8] The maximum moisture in cheddar cheese is 39%, but cheeses ripened for a year or more should contain approximately 36% moisture if they are to develop the best flavor, aroma, body, and texture.

[9] Cheeses so treated are referred to as the **pasta filata** (plastic curd) group.

[10] The US standards for grades of various types of cheeses were published by the USDA on the following dates: cheddar—December 5, 1958; bulk American—August 2, 1991; Colby—May 31, 1995; monterey—May 10, 1996; Swiss—February 22, 2001.

[11] For example, one company produced demineralized whey protein concentrates, composed largely of lactalbumin and high in the amino acid lysine. Such concentrates are supplements for cereal proteins that are low in lysine. These preparations exhibit minimal water absorption to the extent that solutions to which they are added show little change in viscosity in concentrations as high as 25%. However, they can be made to absorb water readily by placing them in an alkaline solution or by a specific heat treatment.

References

Balch, P. A., and F. James. 2000. *Prescription for Nutritional Healing* (3rd ed.). New York: Avery Publishing.

Broadbent, J. R., H. Cai, R. L. Larsen, J. E. Hughes, D. L. Welker, V. G. De Carvalho, T. A. Tompkins, Y. Ardö, F. Vogensen, A. De Lorentiis, M. Gatti, E. Neviani, and J. L. Steele. 2011. Genetic diversity in proteolytic enzymes and amino acid metabolism among *Lactobacillus helveticus* strains. *Journal of Dairy Science* 94(9):4313–4328.

Brown, K. W., W. R. McManus, and D. J. McMahon. 2012. Starch addition in renneted milk gels: Partitioning between curd and whey and effect on curd syneresis and gel microstructure. *Journal of Dairy Science* 95:6871–6881.

Cheese Reporter. December 4, 2009. Vol. 134, No. 23.

Cheese Reporter. February 3, 2012. Vol. 136, No. 32.

Choi, J., D. S. Horne, M. E. Johnson, and J. A. Lucey. 2004. Effects of insoluble calcium phosphate on cheese functionality. *Journal of Dairy Science* 87 (Suppl. 1):231 (Abstr).

Costa, N. E., J. A. Hannon, T. P. Guinee, M. A. E. Audy, P. L. H. McSweeney, and T. P. Beresford. 2010. Effect of exopolysaccharide produced by isogenic strains of *Lactococcus lactis* on half-fat cheddar cheese. *Journal of Dairy Science* 93:3469–3486.

Dejmek, P., and P. Walstra. 2004. "The Syneresis of Rennet-Coagulated Curd." In P. F. Fox, P. L. H. McSweeney, T. M. Cogan, and T. P. Guinee (eds.), *Cheese: Chemistry, Physics, and Microbiology,* Vol. 1, General Aspects (3rd ed., pp. 71–103). London, UK: Elsevier.

Drake, M. A., R. E. Miracle, and D. J. McMahon. 2010. Impact of fat reduction on flavor and flavor chemistry of cheddar cheeses. *Journal of Dairy Science* 93:5069–5081.

El-Bakry, M., F. Beninati, E. Duggan, E. D. O'Riordan, and M. O'Sullivan. 2011. Reducing salt in imitation cheese: Effects on manufacture and functional properties. *Food Research International* 44(2):589–596.

El-Hofi, M., E. El-Tanboly, and N. Abd-Rabou. 2011. Industrial application of lipases in cheesemaking: A review. *International Journal of Food Safety* 14:293–302.

Emmons, D. B., and S. L. Tuckey. 1967. *Cottage Cheese and Other Cultured Milk Products.* New York: Pfizer, Inc.

Everard, C. D., D. J. O'Callaghan, M. J. Mateo, M. Castillo, F. A. Payne, and C. P. O'Donnell. 2011. Effects of milk composition, stir-out time, and pressing duration on curd moisture and yield. *Journal of Dairy Science* 96:2673–2679.

Fujita, S., H. C. Dreyer, M. J. Drummond, E. L. Glynn, J. G. Cadenas, F. Yoshizawa, E. Volpi, and B. B. Rasmussen. 2007. Nutrient signaling in the regulation of human muscle protein synthesis. *Journal of Physiology* 582 (Pt 2):813–823.

Ganesan, B., D. A. Irish, C. Brothersen, and D. J. McMahon. 2012. Evaluation of microbial survival post-incidence on fresh mozzarella cheese. *Journal of Dairy Science* 95:6891–6896.

Guinee, T. P., E. O. Mulholland, J. Kelly, and D. J. O'Callaghan. 2007. Effect of protein-to-fat ratio of milk on the composition, manufacturing efficiency, and yield of cheddar cheese. *Journal of Dairy Science* 90:110–123.

Guo, L., D. L. Van Hekken, P. M. Tomasula, J. J. Shieh, and M. H. Tunick. 2011. Effect of salt on the chemical, functional, and rheological properties of queso fresco during storage. *International Dairy Journal* 21:352–357.

Gwartney, E. A., E. A. Foegeding, and D. K. Larick. 2002. The texture of commercial full-fat and reduced-fat cheese. *Journal of Food Science* 67:812–816.

Hannon, J. A., M. G. Wilkinson, C. M. Delahunty, J. M. Wallace, P. A. Morrissey, and T. P. Beresford. 2003. Use of autolytic starter systems to accelerate the ripening of cheddar cheese. *International Dairy Journal* 13:313–323.

Hort, J., and G. Le Grys. 2001. Developments in the textural and rheological properties of UK cheddar cheese during ripening. *International Dairy Journal* 11:475–481.

Ibrahim, F., S. Ruvio, L. Granlund, S. Salminen, M. Viitanen, and A. C. Ouwehand. 2010. Probiotics and immunosenescence: Cheese as a carrier. *FEMS Immunology & Medical Microbiology* 59(1):53–59.

Johnson, M. E., and C. M. Chen. 1995. "Technology of Manufacturing Reduced-Fat Cheddar Cheese." In E. L. Malin and M. H. Tunick (eds.), *Chemistry of Structure-Function Relationships in Cheese* (pp. 331–337). New York: Plenum Press.

Kenny, O. M., R. J. Fitzgerald, G. O'Cuinn, T. P. Beresford, and K. N. Jordan. 2006. Autolysis of selected *Lactobacillus helveticus* adjunct strains during cheddar cheese ripening. *International Dairy Journal* 6:797–804.

Kindstedt, P. 2012. *Cheese and Culture: A History of Cheese and Its Place in Western Civilization.* White River Junction, VT: Chelsea Green Publishing.

Kon, S. K. 1972. *Milk and Milk Products in Human Nutrition* (FAO Nutritional Studies, No. 27). Rome: Food and Agriculture Organization of the United Nations.

Lopez, C., B. Camier, and J. Y. Gassi. 2007. Development of the milk fat microstructure during the manufacture and ripening of Emmental cheese observed by confocal laser scanning microscopy. *International Dairy Journal* 17:235–247.

Lucey, J. A., and P. F. Fox. 1993. Importance of calcium and phosphate in cheese manufacture: A review. *Journal of Dairy Science* 76:1714–1724.

McMahon, D. J., H. Du, W. R. McManus, and K. M. Larsen. 2009. Microstructural changes in casein supramolecules during acidification of skim milk. *Journal of Dairy Science* 92:5854–5867.

Milesi, M. M., I. V. Wolf, C. V. Bergamini, and E. R. Hynes. 2010. Two strains of nonstarter lactobacilli increased the production of flavor compounds in soft cheeses. *Journal of Dairy Science* 93:5020–5031.

Molloy, E. M., G. Hill, P. D. Cotter, and R. P. Ross. 2011. "Bacteriocins." In J. W. Fuquay, P. F. Fox, and P. L. H. McSweeney (eds.), *Encyclopedia of Dairy Sciences*, Vol. 1 (2nd ed., pp. 420–429). San Diego, CA: Academic Press.

Olson, N. F. 1969. *Ripened Semisoft Cheeses.* New York: Pfizer, Inc.

O'Mahony, J. A., P. L. H. McSweeney, and J. A. Lucey. 2009. Rheological properties of rennet-induced skim milk gels made from milk protein concentrate solutions with different ratios of α_{s1}- : β-casein. *Milchwissenschaft* 64:135–138.

Price, W. V. 1971. Fifty years of progress in the cheese industry, a review. *Journal of Milk and Food Technology* 34:329–346.

Reinbold, G. W. 1963. *Italian Cheese Varieties.* New York: Pfizer, Inc.

Reinbold, G. W. 1972. *Swiss Cheese Varieties.* New York: Pfizer, Inc.

Sharma, S. K., L. K. Ferrier, and L. A. R. Hill. 1989. Effect of modified manufacturing parameters on the quality of cheddar cheese made from ultrafiltered (UF) milk. *Journal of Food Science* 54:573–577.

Shirashoji, N., J. J. Jaeqqi, and J. A. Lucey. 2010. Effect of sodium hexametaphosphate concentration and cooking time on the physicochemical properties of pasteurized process cheese. *Journal of Dairy Science* 93:2827–2837.

USDA Agricultural Marketing Service. 2012, June. *General Specifications for Dairy Plants Approved for USDA Inspection and Grading Service.* Washington, DC: Author.

Vallejo, J. A., J. M. Ageitos, M. Poza, and T. G. Villa. 2012. Short communication: A comparative analysis of recombinant chymosins. *Journal of Dairy Science* 95:609–613.

Wendorff, B., and K. Paulus. 2011. Impact of breed on the cheesemaking potential of milk; volume vs. content. *Dairy Pipeline* 23(1):1, 4–7.

Wilson, H. L., and G. W. Reinbold. 1965. *American Cheese Varieties.* New York: Pfizer, Inc.

Wingerd, W. H. 1971. Lactalbumin as a food ingredient. *Journal of Dairy Science* 54:1234–1236.

Websites

Center for Dairy Research (https://www.cdr.wisc.edu)

European Union, Agriculture and Rural Development (http://ec.europa.eu/agriculture/quality/index_en.htm)

Food Research International (http://www.journals.elsevier.com/food-research-international)

USDA Agricultural Marketing Service, Grading, Certification, and Verification (http://www.ams.usda.gov/AMSv1.0/DairyProductGradesandQualityApproval)

USDA Economics, Statistics, and Market Information System (http://usda.mannlib.cornell.edu)

31

Frozen Desserts from Milk

Age does not diminish the extreme disappointment
of having a scoop of ice cream fall from the cone.
Jim Fiebig

31.1 Introduction

During the first century a substance resembling ice cream was prepared for Nero from snow that runners brought from the mountains and was mixed with honey, juices, and fruit pulp. In the thirteenth century, Marco Polo returned from the Far East with a recipe for a frozen sherbet-like dish. By 1560 an Italian had written of a food "made of milk sweetened with honey and frozen. . . ." An anonymous 84-page manuscript entitled *L'art de Faire des Glaces* (The Art of Preparing Ice Cream) was written about 1700 (Dickson, 1972).

In America, Governor William Blades of Maryland served a dessert containing ice cream and strawberries in about 1744, and George Washington spent approximately $200 for ice cream in New York during the summer of 1790. Though at first a confection and a luxury food available only to the rich, primarily because refrigeration was expensive, consumption by the masses had to wait for the development of ice harvesting, insulated ice houses, and the hand-cranked ice cream freezer (invented in 1846 by Nancy Johnson and patented by W. G. Young in 1848). More important was the perfection of mechanical refrigeration in 1878 and invention of the **direct expansion** continuous ice cream freezer in 1913.

The first wholesale ice cream business in the United States was started in Baltimore by Jacob Fussell in 1851. What started as an outlet for his surplus cream became a profitable business. But for many years the small plant, the street vendor, and the milk producer selling homemade ice cream were the main sources of frozen desserts. However, introduction of the ice cream cone at the 1904 World's Fair in St. Louis gave increased impetus to the demand for ice cream.

Today's ice cream is a frozen foam consisting of components of milk, sweeteners, stabilizers, emulsifiers, and flavorings. Other ingredients often added are egg products, colorings, and hydrolysates of starch. These compose a mix that is pasteurized, homogenized, and cooled before being frozen in a swept surface heat exchanger. In the United States in 2013, the average per capita consumption of regular ice cream was 12.8 lb, while reduced fat ice cream was 5.9 lb. The total regular (10% fat or more) ice cream production in 2014 was 872 million gal/yr while production of lowfat and nonfat ice creams was 411.6 and 18.5 million gal/yr, respectively. In addition, production of frozen yogurt and sherbets was 66.8 and 45.7 million gal/yr, respectively. About 68.9 million gal/yr of water and juice ices were produced.

Because frozen desserts are at their peak of popularity during summer months, the US industry produces 30% of its ice cream during May through July but only 19% during November through January.

31.2 Types of Frozen Desserts and Selected Characteristics

The following are desirable characteristics of high-quality frozen desserts: fine flavor, smooth texture, moderately resistant body, optimal **overrun**, resistance to heat shock, attractive appearance, readily dippable, homogeneous appearance on meltdown, optimal freezing (melting) point, nutritious, economical, and refreshing.

By varying quantities of major characterizing ingredients, numerous types of frozen dairy desserts can be produced. Permissible compositional variations are outlined in US Code of Federal Regulations, Title 21, Part 135, and are also dictated by economic and quality factors as well as by federal and state standards.

31.2.1 Ice Cream

The major components of ice cream are milkfat (MF), nonfat milk solids (NMS), sweetener, stabilizer/emulsifier, and flavoring. Of all frozen dairy desserts, ice cream contains the most milk solids, at least 20% for plain and 18% for bulky-flavored products. One gallon of the finished product must weigh at least 4.5 lb and contain at least 1.6 lb of food solids. These limits establish the maximal overrun (amount of air that can be incorporated in the frozen mix). The following are minimal percentages of milkfat required in ice cream in some major ice cream producing nations: Australia, 10%; Brazil, 3% in a total fat of 8%; Canada, 10%; Greece, 8%; Germany, 10%; India, 10%; Japan, 8%; Korea, 8%; and New Zealand, 10%. Of these only Japan and Korea have minimums of total milk solids (TMS), 15 and 16% respectively.

There is considerable latitude for processors to vary content of both MF and NMS. Generally, when the quantity of either milk component is increased, quantity of the other is decreased. In the United States, ice cream contains not less than 10% MF and 10% NMS. However, when it contains additional MF at 1% increments up to 14%, NMS may be reduced by an equal amount. Packages of lower fat versions of ice cream are required to be labeled with one of the following modifiers: *reduced fat*, *light*, *lowfat*, or *nonfat*. Compared with the reference ice cream, fat content must be reduced by 25 and 50% for reduced fat and light versions, respectively. Based on 4 fl oz servings, lowfat ice cream must contain not more than 3 g of fat and nonfat ice creams not more than 0.5 g of fat. Other food fats are excluded, except as components of flavoring ingredients or in incidental amounts added for functional purposes. 21 CFR 135.110 lists each form of milk solids acceptable for use in ice cream.

Whey solids, including modified whey products, may be added to ice cream to replace up to 25% of NMS. Approved additives to mixes containing at least 20% TMS are five forms of casein and hydrolyzed milk proteins.

Frozen custards, French ice cream, and French custard ice cream conform to standards of composition for ice cream except that in finished form they must contain at least 1.4% egg yolk solids.

Plain ice creams contain no more coloring and flavoring ingredients (such as vanilla, coffee, and maple) than 5% of their unfrozen volume. Bulky-flavored ice creams may contain lower concentrations of MF and TMS than plain ice cream to the extent that flavoring and additional sweeteners displace MF and TMS (up to 2% less fat and 4% less TMS).[1]

Products containing significant amounts of flavoring and coloring ingredients (i.e., bulky-flavored ice creams) are permitted reductions in fat as follows: 2.4 times the weight of cocoa and 1.4 times the weight of fruits, fruit juices, and nuts.

Data from nutritional labels on commercial ice creams (table 31.1) show that content of calories goes up and protein goes down as fat content increases.

Over 60% of lowfat ice cream made in the United States is sold in the soft-frozen form. Likewise frozen yogurt is often sold soft frozen. Mixes for soft serve require emulsifiers and stabilizers that produce a dry, stiff, slow-melting product whereas in hard-frozen products resistance to **heat shock**[2] is more important. Also, mixes for soft serve require less sugar to avoid a low freezing point and less fat to minimize churning in the batch freezer. Their overrun is lower, usually 30 to 60% compared with 80 to 100% in hardened products, and product is drawn from the freezer at a lower temperature than hard-frozen products, 18°F (–7.5°C) versus 22°F (–5.5°C).

Formulating "better for you" frozen desserts is an objective of some manufacturers. Among the successes has been production of lactose-free ice cream for consumers whose small intestines absorb lactose poorly. Other objectives include reduction of sugars and additions of antioxidants, probiotics, dietary fibers, or omega-3 oils. However, making health claims based on modifications of formulas is quite difficult. True health claims associate intake of a nutrient(s) with mitigation of one or more disease conditions, claims that are highly regulated by the US Food and Drug Administration and must be supported with scientific evidence.

31.2.2 Mellorine and "No Sugar Added" Frozen Desserts

Mellorine is composed essentially of the same ingredients as ice cream except that MF is replaced with vegetable fat and the minimal fat content is 6%. Mellorine contains not less than 2.7% protein by weight and the minimal TMS factor is not regulated.

Dietetic and diabetic frozen desserts are artificially sweetened. They usually are labeled *fat free*; *sugar free*; or *light, no sugar added*. These products must have food solids added to replace sugar to control the freezing point and concentration of water in the mix (affecting smoothness). These replacers include polydextrose, maltodextrin, and whey proteins. Additional water-binding ingredients include mono/diglycerides, microcrystalline cellulose, gums, and carrageenan. Sugar substitutes include sorbitol and mannitol, which are assimilated slowly in the body and have a low caloric value (2 to 2.8 cal/g compared with 4 cal/g for sugar). Artificial sweeteners include acesulfame potassium, aspartame, and saccharin.

31.2.3 Frozen Yogurt

Frozen yogurt is prepared by freezing while stirring a pasteurized mix containing MF, NMS, sweetener, stabilizer, yogurt, and water. It usually is flavored with fruits. There presently is no US federal standard for frozen yogurt. Where there are state or provincial regulations, it is required commonly that the yogurt ingredient be cultured with a mixture of *Lactobacillus bulgaricus* and *Streptococcus thermophilus* bacteria after the milk has been pasteurized. Usually a very high heat treatment (e.g., 185°F [85°C] for 15 min) is given the milk before it is inoculated with the yogurt culture. The minimal acidity of 0.30% required by some regulatory agencies is used to set a theoretical minimum amount of yogurt to be added to the mix. The amount added usually ranges from 10 to 20% of the total weight of the mix.

Manufacturers limit the amount of the acetaldehyde in frozen yogurt since most consumers dislike the flavor it imparts to plain yogurt. Yogurt has a mild acidic flavor as compared with lowfat ice cream containing the same amount of fat.

Table 31.1 Average Nutritional Values per 4 fl oz (118 mL) Serving for Vanilla Ice Cream Products.

Category	Calories (kcal)	Fat (g)	Protein (g)	CHO (g)	Sugars (g)	Calcium (%DV)	Number of Samples
Nonfat	85	0.0	3.5	19	12	10	5
Lowfat	100	2.3	3.0	18	15	9	10
Light	105	3.3	3.0	17	14	9	6
Reduced fat[a]	100	5.0	3.0	13	3.5	8	4
Regular	135	7.2	2.4	15	13	9	20
Premium	165	9.4	2.3	17	14	7	7
Superpremium	250	17.0	5.0	20	19	10	1

[a] These samples were "no sugar added."

Frozen yogurt has been preferred over similar ice cream products because yogurt bacteria are thought to assist in the digestion of lactose and to have other health-promoting properties (probiotic effects). Attempts to provide high numbers of probiotic (cf. chapter 29) bacteria, including the genus *Bifidobacterium,* are hampered by susceptibility of the organisms to low pH and destruction during freezing. Addition of prebiotic components, including inulin, to frozen yogurt provides nutrients preferred by probiotic bacteria that survive in human intestinal tracts, thus improving chances the culture will colonize the host's small intestine.

The label of this class of products varies with fat content, with the same names applying as in milk products of the same fat content: < 0.5% is nonfat frozen yogurt; 0.5 to 2.0% is lowfat frozen yogurt; and at least 3.25% is frozen yogurt. Products commonly contain 3.5 to 3.8% protein and 29 to 34% total solids and have a pH ranging from 4.3 to 5.3.

Inulin, polydextrose, and sugar alcohols, such as isomalt, can be used in combination to produce a low calorie, no sugar added frozen yogurt. Inulin, a soluble dietary fiber, can be used to improve the quality of the product with nutritional and technological advantages (Isik et al., 2011). Inulin, being highly hygroscopic, binds water and forms a gel-like network. It and isomalt replace the solids sugar provided in the normal mix. Sweetness is provided by a blend of nonnutritive sweeteners.

31.2.4 Sherbet

Sherbets are characterized by a sweet but tart flavor and a low content of TMS (usually 3 to 5%). They can be made by mixing ice cream mix with water ice, or they can be compounded directly from raw ingredients such as water, nutritive sweeteners, fruit or fruit juice/flavoring, fruit acid, milk solids, stabilizer, and coloring. Minimum acid content is 0.35% calculated as lactic acid. Citric acid is more often added, but tartaric, malic, ascorbic, lactic, and phosphoric acids are permissible. Overrun in sherbets is low (about 25 to 40%); consequently, weight per gal is high (minimum is 6 lb). Because of the high sugar content, the melting point of sherbets is low. Therefore, they are softer than ice cream at the same temperature. Nearly all sherbets are sold in the hard-frozen rather than soft-frozen form. *Sherbine* is the fanciful name for the product comparable to sherbet that contains vegetable fat instead of MF.

31.2.5 Ices and Sorbets

Ices, sometimes called Italian ices or water ices, contain no dairy or egg product other than egg white (improves whippability). Usually they are composed of sugar (30%), fruit juice or fruit flavoring (20%), color, stabilizer (0.2 to 0.6%), citric acid (to adjust acidity to about 0.4% expressed as lactic acid), egg white, and water. Overrun in ices should be approximately 30%, except for quiescently frozen bars, which contain little or no air. Ices may be frozen in the same manner as ice cream but they are often dispensed as liquids into molds (forms) in which they are frozen on sticks to make Popsicles.[3]

Sorbets are an upscale version of ices in that they contain fruit, fruit juices, and/or fruit extracts rather than imitation flavorings. Sherbets and water ices are defined foods (21 CFR 135.140 and 21 CFR 135.160, respectively), but sorbets are not a defined food in the United States.

31.3 Components of Frozen Desserts

Manufacturers have several alternative sources of ingredients for frozen desserts and their choices depend upon availability, costs, functional properties, and the quality of product they wish to produce. Highest quality is usually obtained with fresh milk products but equally superior frozen desserts can be made from concentrated and preserved milk products that have been properly prepared from high-quality raw materials. Let us examine the components, their contributions to frozen desserts, and sources and rules to follow in their selection.

31.3.1 Fat

Richness of flavor, smoothness of texture, resistance of body, color (in vanilla-flavored products), and nutritive value are associated with lipids in frozen desserts. Fat imparts smoothness because it physically obstructs growth of ice crystals and concentrates at surfaces of air cells. Differences in quantity of fat are largely responsible for differences in flavor between ice cream and its lower fat versions. Fat carries many aromatic flavor compounds and controls the rate of their release in the mouth. In contrast, it forces vanilla into the aqueous phase so that, given the same concentration of vanilla in a product, the flavor intensity increases with increases in fat content. On the other hand, fat tends to mask **stale** flavors by dissolving aldehydes, making them less volatile.

Fat is in the globular state; therefore, its concentration in itself does not affect freezing point. However, when fat is replaced by soluble substances, such as corn syrup solids (to build body), the freezing point is depressed. Excessively clumped fat globules[4] tend to weaken **lamellae** of air cells, thus limiting whippability (consequently overrun). However, partial churning of fat globules is necessary to enhance foam structure, which is similar to that of whipped cream.

Sources of MF are fresh cream, frozen cream, plastic cream (80% MF), condensed milk, butter, butter oil, anhydrous milkfat (80% MF), and certain mixtures or blends of MF and sugar. Fats used in mellorine and imitation frozen desserts include coconut oil and blended fractions of partially **hydrogenated** soy and cottonseed oils. Because refinement of these oils tends to remove fat-soluble vitamins, vitamin A is added to mellorine (40

IU/g of fat). Furthermore, vitamin A is added to lower fat ice cream products.

Fats deteriorate in two major reactions, namely hydrolysis (free fatty acids cause rancidity) and oxidation (unsaturated fatty acids become oxidized to aldehydes and ketones). Rancidity usually develops prior to manufacture of frozen products, making selection of raw materials of great importance. Oxidation may occur at any stage of production or distribution. Therefore, elimination of contamination from copper, protection from light, and rapid movement of frozen product through distribution channels to reduce time of exposure to oxygen are necessary preventative procedures.

31.3.2 Nonfat Milk Solids

The major components of NMS are lactose, proteins, and minerals. Their concentrations relative to each other in frozen desserts are usually the same as in milk. However, some manufacturers may add components with different ratios, including whey solids,[5] delactosed whey solids, or modified milk proteins.

Lactose contributes to sweetness and lowers the freezing point. Too high a concentration of lactose may lead to its crystallization, causing the textural defect known as sandiness; lactose crystals dissolve slowly, leaving the impression that one has sand in the mouth. Factors that enhance lactose crystallization other than its high concentration are excessive corn solids, large air cell structure, fluctuating storage temperatures, and extended time of storage. Under usual conditions of storage there should be no more than 1 lb of NMS to each 6 lb of water in ice cream mix. However, by using selected stabilizers plus low temperature handling and storage, this limit may be exceeded.

Proteins add nutritive value, provide a background for flavor,[6] assist in foaming (incorporation of air), and absorb (bind) water. It is this latter function that produces smoothness and chewiness in the product.

Primary sources of NMS are fresh condensed milk, condensed skim milk, and nonfat dry milk. Of course, cream, milk, and other milk products contain NMS. Therefore, their composition must be known when compounding mixes. Dried buttermilk may be used and is particularly beneficial when butter or butter oil is a source of fat. Buttermilk (from the churning of cream) contains large amounts of lecithin, a natural emulsifier that aids in the emulsification of fat from butter.

31.3.3 Sweeteners

Sucrose (beet or cane sugar) traditionally has been the basic sweetener for ice cream. However, products of the hydrolysis of corn starch (corn syrup solids) are now used widely because comparatively large amounts can be used to partially replace sucrose, thus improving body and texture and lowering costs without increasing sweetness. Starch hydrolysate products having 20 to 70% of their glycosidic linkages broken are known as corn syrups. They are classified based on degree of hydrolysis as low, regular, intermediate, and high conversion with dextrose equivalent (DE) values of 28 to 38, 39 to 48, 49 to 58, and 59 to 68, respectively. As DE increases so does sweetness, but freezing point decreases. Limitations on their use primarily involve excessive lowering of the freezing point, development of syrup flavor,[7] and a masking effect on delicate flavors such as vanilla. Maximum quantities of corn syrup solids (usually 28 to 42 DE in syrup or dry form) recommended for addition to ice cream range from 25 to 35% of total sweetener, depending on quality, type, and standards established for the finished product. Use of corn syrup solids in sherbets and ices is recommended to prevent crystallization of sucrose at the surface.

Data in table 31.2 provide a basis for comparing representative sweeteners. Dextrose equivalent is a measure of the reducing sugar content; thus it is a measure of the extent of starch hydrolysis. However, it does not indicate the distribution of sizes of molecules in the sweetener.

Table 31.2 Composition of Sweeteners Frequently Used in Frozen Dairy Desserts.

Sweetener	Dextrose Equivalent	Relative Sweetness	Relative Depression of Freezing Point	Approximate Composition of Saccharides (%)				
				Mono-	Di-	Tri-	Tetra-	Higher
Sucrose		100	1.0	—	100	—	—	—
Dextrose[a]	100	80	2.0	100	—	—	—	—
Corn syrup[b]	62	88	1.14	39	28	14	4	15
Corn syrup[c]	52	56	0.96	29	18	13	10	30
Corn syrup[d]	42	50	0.78	18	14	12	10	46
Corn syrup[e]	32	42	0.61	11	10	10	9	60

[a] Also known as glucose or refined corn sugar.
[b] High-conversion corn syrup.
[c] Intermediate-conversion corn syrup.
[d] Regular-conversion corn syrup.
[e] Low-conversion corn syrup.

These may vary from the 6-carbon monosaccharides (dextrose, also known as glucose, and fructose, highly refined corn sugars) to the 12-carbon disaccharides (maltose) and on through tri-, tetra-, and larger polysaccharides (dextrine or prosugar). Relative sweetness values are important to manufacturers. For example, dextrose is only 80% as sweet as sucrose; therefore, for equivalent sweetening effect, 1.25 lb is needed to replace 1 lb of sucrose. High fructose corn sweeteners (HFCS) are made by converting dextrose to fructose enzymatically, producing syrups that are 1.8 to 1.9 times sweeter than sucrose. Only the highly hydrolyzed forms of corn starch lower the freezing point more than does sucrose. For example, nearly 1.7 lb of solids in 32 DE corn syrup are required to depress freezing point as much as 1 lb of sucrose. A molecule of sucrose (disaccharide) weighs about twice as much as one of dextrose or fructose (monosaccharides). Therefore, these sweeteners will lower freezing point twice as much as sucrose when the same weight of each is added. When used as a sweetener, HFCS must be combined with high molecular weight substances to achieve the desired freezing point depression.

Sucrose and corn syrup solids can be obtained in either liquid or crystalline forms and as blends, typically of 70% sucrose and 30% corn syrup. Liquid sugar blends are especially adaptable to use in automated mix preparation because they can be pumped and metered, thus eliminating batch weighing. In making calculations it is vital that moisture concentration in syrups be considered. The percentage by weight of sucrose in syrups is measured by the Brix scale, while the Baumé scale is used to measure total solids calculated from specific gravity. Sweetness depends on the kind and concentration of sugars and is not reflected by the Baumé scale alone.

Nonnutritive sweeteners provide sweetness needed in frozen desserts but do not provide the solids necessary to displace water. For example, only 0.05 lb of saccharine, which is 300 times as sweet as sucrose, would be needed to replace 15 lb of sugar in a regular ice cream mix. This calculation is based on relative sweetness (RS), which is a comparison of how sweet a product is compared to sucrose. Aspartame, the methyl ester of aspartic acid and phenylalanine, and asulfame potassium are approximately 200 × RS. Sucrolose, derived from chlorination of sucrose, is 600 × RS. Stevia™, 300 × RS, is **generally recognized as safe**. It is extracted from the South American plant *Stevia rebaudiana* and is therefore known as rebaudioside-A. Brazzein™, 500 × RS, is a small protein from the African sweet potato plant *Pentadiplandra brazzeana*. It is heat stable and water soluble.

Whereas nonnutritive sweeteners are readily available and of comparative cost to sucrose on a per unit of sweetening basis, choices must be made of ingredients that affect freezing point depression and build total solids (Tharp and Young, 2010). Polydextrose and maltodextrins are low-calorie bulking agents that contribute little to freezing point depression. By partial replacement of sugar (e.g., 25%) with a high intensity sweetener and an equivalent amount of bulking agent, freezing point will increase with a consequent decrease in the amount of ice that will melt in any heat shock event. As small ice crystals melt with increasing temperature they form larger ice crystals as they refreeze with decreasing temperature, making the texture coarse.

31.3.4 Stabilizers and Emulsifiers

Stabilizers are very large hydrophilic colloids[8] that lower the concentration of free water by absorbing it. They reduce the amount of water available for crystallization as ice, hence smaller crystals form and smoothness of product is enhanced. Also, stabilizers minimize crystal growth during temperature fluctuations in frozen storage. Commonly used stabilizers are extracted from kelp, moss, and algae (algin, carrageenan, agar agar); from locust bean, legumes, trees, and other plants (carob, guar, tragacanth, arabic gum, and karaya gum); from corn stalks (carboxymethyl cellulose); from calf and pork skins and bones (gelatin); and from citrus fruits (pectin). Each differs in functional properties, temperature at which it dissolves, and reaction to components of the mix, including other stabilizers. For example, gelatin may create excessive foaming and high overrun in sherbets and ices, whereas pectins are excellent stabilizers of ices and sherbets but are inadequate in ice cream; alginates hydrate well only in hot (158°F [70°C]) mixes, but guar gum becomes hydrated in cold mixes.[9] Up to 0.5% of a single or combination of stabilizers may be added to ice cream legally.

Microcrystalline cellulose (cellulose specially processed to form a colloid) produces a gel-like structure in frozen desserts. It imparts smoothness, protects against heat shock, and adds body, especially to lower-fat products. In soft-serve products it increases dryness, stiffness, and resistance to melt.[10] Up to 0.5 and 1.5% may be added to sherbets and to ice cream, respectively. Cellulose cannot be included as weight of food solids since it is practically indigestible by humans.

Labeling restrictions may affect stabilizer choice. For example, plant-based gums should be chosen as stabilizers of "all-natural" ice creams. However, reduced-fat ice cream with high overrun cannot be labeled "all-natural" because it requires blended stabilizers that may include propylene glycol esters, calcium sulfate, cellulose gum, guar gum, maltodextrin, modified food starch, and carrageenan to enhance texture and control ice crystal growth. Aerated products, such as mousse and toppings, benefit from strong surfactants like sucrose esters, which improve whippability by creating small, stable oil droplets that enhance creaminess of mouthfeel.

Emulsifiers are molecules that have both water-loving (hydrophilic) and fat-loving (lipophilic) groups. Consequently, they concentrate at interfaces of fat and aqueous phases in mixes and thereby lower surface tension, stabilizing the emulsion. They coat fat, making it less antagonistic to protein films that hold air in whipped mix. The results are improved whippability, drier appearance, smoother body and texture, and slower **meltdown**.

Two general types of emulsifiers employed are (1) mono- and diglycerides (cf. chapter 21) and (2) sorbitan esters.[11] Sorbitan esters are similar to monoglycerides in that they have a fatty acid molecule, such as stearate or oleate, attached to a sorbitol (glucose alcohol) molecule, whereas the monoglycerides have a fatty acid attached to a glycerol molecule. Of course, milk contains natural emulsifiers, including lecithin, proteins, phosphates, and citrates. Egg yolks contain large amounts of lecithin and long have been used in ice cream to improve whippability. Maximum allowable concentrations of mono- and diglycerides is 0.2% and for poly- types is 0.1%. Excessive emulsification may cause delayed melting, curdiness of meltdown, shortness or crumbliness of body when ice cream is dipped, and deemulsification, which results in churning. However, slight deemulsification is desirable to furnish "structure."

Emulsifiers are classified according to their hydrophile-lipophile balance (HLB). The number indicates essentially the percentage weight of the hydrophilic (water-loving) portion of the molecule divided by 5 (the range is 0 to 20). The more lipophilic (oil-loving) the emulsifier, the lower the HLB. Mono- and diglyceride emulsifiers have low HLB values (about 5), whereas those of polys are high (about 15). For soft-serve mixes subject to considerable abuse, an emulsifier with a low HLB is recommended.

Stabilizer/emulsifier technology has provided blended (integrated) products that improve dispersibility in frozen dessert mixes that are processed in **high-temperature short-time (HTST)** equipment. Colloidal stabilizer is mixed with melted emulsifier and the slurry is spray chilled. The mixed product hardens, forming small particles. On addition of the slurry to cold mix, gums exposed at the surface of the particles begin to hydrate through capillary action. The resultant swelling cracks the surface of the particle, exposing additional stabilizer and dispersing emulsifier to adsorb to fat. The main advantage of this product is its ready dispersibility, which allows it to hydrate and adsorb rapidly. Therefore, in mixes pasteurized by HTST, less stabilizer/emulsifier is required and fewer problems of lumping and balling are encountered in its addition to mixes.

The amount, usually 0.2 to 0.5%, and kind of stabilizer/emulsifier blend needed in a frozen dessert varies with mix composition; ingredients used; processing times, temperatures, and pressures; storage conditions; and more.

31.3.5 Flavoring Materials

Vanilla is the most popular flavor of ice cream, constituting about 30% of supermarket sales. Vanilla is added also as supplementary flavoring in frozen desserts of many flavors. Other popular flavors and their percentage of sales are: chocolate (10%); chocolate chip with and without mint (6.2%); butter pecan (4%); strawberry (3.7%); neopolitan (3%); and cookies and cream (2.6%). *Dairy Facts*, published by the International Dairy Foods Association (2015), lists 45 individual flavors.

Natural flavorings include vanilla extract, cocoa and chocolates, coffee, spices, fruits, nuts and nut extracts, and candies and confections. Artificial flavorings simulate vanilla, chocolate, many fruits, and some nuts. Three categories of flavoring in frozen desserts are recognized in federal standards. Products in category I contain only natural flavoring; those in category II contain predominantly natural flavoring; and category III products contain predominantly artificial flavoring.

Use of artificial flavorings must be indicated on the package label (21 CFR 101.22) using the name of the characterizing flavor and the words "flavored," "artificial," or "artificially flavored." Examples of labeling requirements follow:

- Category I: vanilla ice cream
- Category II: vanilla flavored ice cream
- Category III: vanilla ice cream artificially flavored

Overrun in ice creams to which insoluble bulky flavorings are added should be lower than in products containing soluble bulky flavorings, because only the soluble portion carries the overrun. Data in table 31.3 demonstrate that, at the same weight per gallon, products containing insoluble bulky flavors have more overrun in the portion that can hold air; consequently, they could have inferior body and texture if drawn at the same overrun as mix with soluble bulky flavoring. (For more detail on this subject see Marshall et al., 2003.)

31.3.5.1 Vanilla

Vanilla is extracted with alcohol from fermented vanilla beans. Single-fold vanilla contains the extractive matter from 13.35 oz of finely chopped vanilla beans per gal of 35% ethanol (equaling 10 g vanilla beans per 100 mL of alcohol). By distilling off alcohol, concentrated vanillas are prepared. Vanilla powders are available also,

Table 31.3 Comparison of Overrun in Ice Creams When Soluble and Insoluble Bulky Flavors Are Added.

	Overrun (%)	
Weight/ Gal (lb)	Soluble Bulky Flavor[a]	Insoluble Bulky Flavor (Unflavored Portion)[b]
4.5	107	128
5.0	87	103
5.5	70	83
6.0	56	66
6.5	44	52

[a] Weight of flavored mix equals 9.33 lb/gal.
[b] Weight of flavored mix equals 9.1 lb/gal. Flavoring is 18% of the mix and has no overrun.

and in some markets specks of vanilla bean are desired in ice cream. The major flavoring component of vanilla is *vanillin* (3-methoxy-4-hydroxybenzaldehyde). Since it can be synthesized, it often is used to reinforce vanilla and as the main component in imitation vanillas. The usage rate of pure vanilla extract can be reduced by up to 50% by addition of 1 oz of vanillin per fold of the pure extract. When used as the major source of flavoring, imitation vanillas often impart harsh or unnatural flavors. Vanilla flavor is more pronounced with higher serving temperatures and high sugar (up to 16%) and fat concentrations. Vanilla flavor tends to be masked by constituents of NMS.

Vanilla beans are grown in tropical regions, especially Madagascar, Réunion Island, and other areas along the Indian Ocean. *Vanilla planifolia* (syn. *V. fragrans*) is the major cultivar. Soil, environment, and conditions of fermentation affect the flavorful nature of the extracted product. Bourbon vanilla is characterized as having soft, buttery-creamy notes with only weak phenolic and floral notes, whereas Indonesian vanilla yields strong phenolic-smoky prune notes and mild buttery-creamy notes. Vanilla grown in the islands of Tahiti and Moorea, *V. Tahitensis*, is perceived quickly with anisic, floral, coumarinic, and sweet notes dominating.

Although usage rates of vanilla vary with composition of the mix and the flavor target of the producer, an approximation is as follows: 5 fl oz of single-fold vanilla in 10 gal of mix (165 mL/100 lb or 365 mL/100 kg). The usage level for multi-fold vanillas is slightly higher than that of single-fold vanilla divided by the number of folds.

31.3.5.2 Chocolate

Chocolate and cocoa are obtained from nibs of ripened and roasted **cocoa** beans. Nibs are separated from crushed beans by sieves and air currents.[12] Grinding the nibs frees fat, which amounts to about 50%, and the heat produced by grinding makes it liquid. The combined liquid fat and flavoring material are known as chocolate liquor or, when solidified, as bitter cooking chocolate. Sweet milk chocolate is prepared by adding sugar, milk solids (3.39% MF and 12% TMS), and cocoa butter to chocolate liquor. Hydraulic pressing of cocoa beans removes the practically flavorless cocoa butter (fat), leaving a cocoa cake that contains about 22% fat and nearly all the flavoring of the bean. Cocoa is produced by grinding and sifting cocoa cake. Darker and less bitter cocoa is made by the Dutch process, which involves treatment of cocoa beans with alkalis at the time of roasting. About 3% cocoa (10 to 12% fat) or about 5% chocolate liquor (50 to 60% fat) is required to flavor mix for chocolate frozen desserts, the difference being due to the higher fat content of chocolate liquor. The cocoa butter in chocolate liquor provides extra smoothness of mouthfeel over that of cocoa, but the cost is higher and consumers may object to the higher fat content that must be shown on the nutrition label of the product.

Producers of nonfat ice creams can blend defatted cocoa, containing about 2.5% fat, with regular cocoa (10 to 12% fat) to produce lowfat cocoa containing about 5% fat. Use of this blend at 2.5 to 3% of the mix makes possible production of nonfat chocolate ice cream with less than 0.5% fat, thus meeting the labeling requirement.

To allow for the additional sweetener needed in chocolate ice creams, the US Code of Federal Regulations permits reductions in the weight of MF and TMS by a factor of 2.5 times the weight of the cocoa solids. For example, when 4% cocoa is added to a mix containing 10% MF, the final MF content must be at least 9%. The formula for the calculation is as follows:

$$\left[100 - (\%\ \text{cocoa} \times \text{reduction factor})\right] \times \%\ \text{MF}$$
$$\left[100 - (4.0 \times 2.5)\right]0.1 = 9.00\%$$

In both the United States and Canada, regular chocolate ice cream must contain at least 8% MF, regardless of the amount of chocolate flavoring added.

Problems of dispersion and wettability (cf. section 33.8.5) are often encountered in adding cocoa and certain stabilizers to mixes for frozen desserts. The dispersing agent dioctyl sodium sulfosuccinate may be added to dry products to remedy such problems. This usually is done only by manufacturers of dry products.

31.3.5.3 Fruits

Fresh, frozen, preserved, canned, and dried fruits may be used as flavorings. Strawberries, peaches, and other fruits with heat-labile flavors (destroyed by heat) have been preferred in fresh or frozen form. Some flavors, such as cherries and pineapple, are stable or are improved by heat. Aseptically processed fruits and flavorings keep for months at room temperature. They are processed in swept surface heat exchangers that quickly raise the temperature of the fruit/sugar/stabilizer mixture to 190 to 250°F (88 to 121°C). Three minutes holding time is applied before the mixture passes through a series of swept surface heat exchangers to cool it to about 80°F (27°C). The product is then metered into sterile containers inside a sterile filling chamber.

Fresh fruits should be mixed with sugar and held 12 to 24 hr before use to allow equilibration of soluble components between syrup and fruit. Thus, flavorful components move into the syrup to be available to flavor the entire product, and sugar moves into the fruit to lower its freezing point, preventing it from becoming icy and hard when frozen. Fruit to sugar ratios vary from 2:1 to 9:1. Quantities of fruit required to impart desired flavor intensity vary from 10 to 25% of the weight of finished product. For example, about 15% of a 2.5:1 pack of strawberries is near optimum.

Some fruits are pureed, especially seedy types such as raspberries. Purees, concentrated fruit juices, and essences may be added to the mix just prior to freezing, but whole, sliced, diced, or crushed fruits must be added

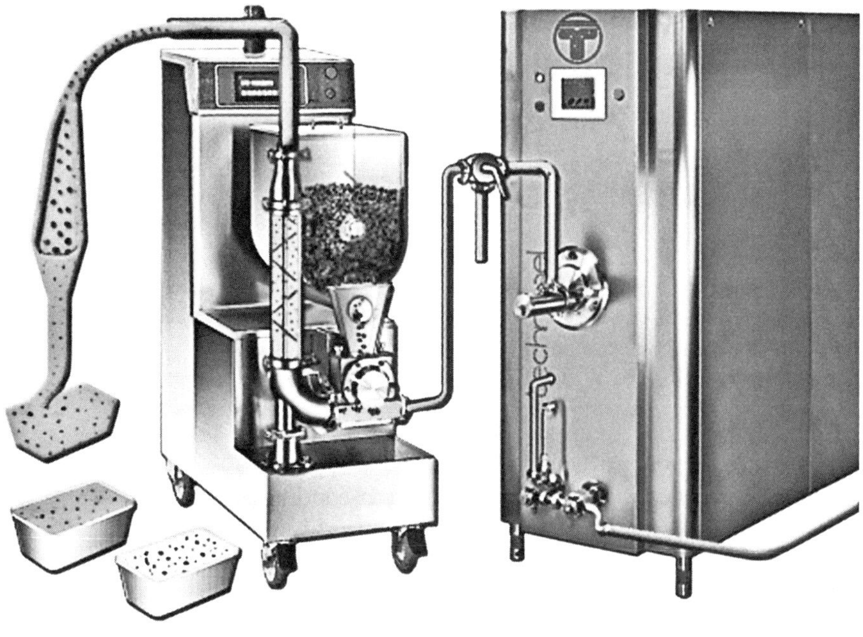

Figure 31.1 Fruit feeder positioned downstream from the freezer (courtesy of Technogel America).

to soft-frozen products via a fruit (ingredient) feeder (figure 31.1). Citrus fruits are most often used to flavor sherbets and ices. Preferred flavors of sherbets are orange, raspberry, lime, pineapple, and lemon.

31.3.5.4 Nuts

Nuts commonly added to ice cream products are pecans, black walnuts, almonds, pistachios, and peanuts. They are prepared by blanching and roasting. Most are added in the chopped form via an ingredient feeder after freezing. Peanuts may be added as peanut butter. Major product defects resulting from additions of nuts are rancid flavor and presence of pieces of shell or other extraneous materials, which are permitted up to 0.05% by US Department of Agriculture (USDA) grade standards.

Quantities of nutmeats needed in plain ice cream mix range from 3 to 6%. Butter and caramel flavorings are excellent companions with nuts. Warning labels that products contain nuts are mandatory in favor of persons who are allergic to them. Furthermore, practices to avoid cross contamination of other products with nuts are required.

31.3.5.5 Spices and Candies

Cinnamon and ginger are sometimes used as single flavorings but more often they are added with nutmeg and cloves to supplement other flavors, especially eggnog, pudding, and chocolate. Candies including peppermint, butter crunch, toffees, brittles, and chocolate chips are added using ingredient feeders. Some should be reinforced by appropriate liquid flavorings.

31.3.6 Coloring Materials

Consumer acceptance of frozen desserts is influenced by degree to which flavor is believed to be natural, and color has a psychological effect on how persons perceive flavor. Therefore, processors strive to obtain authenticity and uniformity of color. Even vanilla ice cream should contain sufficient yellow color to give it a creamy appearance. Seldom do flavoring materials, except chocolate, impart sufficient color of themselves.

According to the Federal Food, Drug, and Cosmetic Act of 1938, chemically synthesized colorings must be certified by the batch (certified colors). Natural colors, including annatto and beta carotene, which are of vegetable origin, are exempt from certification. Colorings may be purchased in dry, paste, and liquid forms. Colors purchased in powder form should be dissolved in boiling water, normally as a 3% solution.

As preservatives sodium benzoate may be added as a mold inhibitor and citric or tartaric acids as bacterial inhibitors (with pH reduced to 3.5). Some colors are soluble in oil only. Colors are usually added to the mix just prior to freezing. Normal use concentration in mixes varies from 1:500 to 1:1,000.

When a coloring has been added to butter, cheese, or ice cream, it need not be declared in the ingredient list unless such declaration is required by a regulation related to health claims. However, voluntary declaration of all colorings is recommended (21 CFR 101.22). Colors may be listed on labels by group as "color added" or "artificial color." Yellow No. 5 has been associated with certain allergic reactions, including asthma and uticaria (hives). A final rule effective in 2011 requires that labels of all foods containing the red colors cochineal extract or carmine (derived from cochineal beetles) must refer to the substances by name.

31.4 Formulas and Calculations of Mixes

A processor decides on the composition for each mix based on characteristics desired for the finished product (including meeting legal standards). For example, let's say a processor wishes to sell both ice cream and light ice cream with the following compositional formulas:

Constituent	Ice Cream	Light Ice Cream
Milkfat	11.0%	3.5%
Nonfat milk solids	11.0%	12.0%
Sucrose	15.0%	15.0%
Stabilizer-emulsifier	0.3%	0.4%
Total solids	37.3%	30.9%

The processor will use the following ingredients (based on availability, quality, and price) to comprise the formula for ice cream:

- Cream, which contains 40% fat and 5.4% NMS
- Condensed skim milk, which contains 30% NMS
- Cane sugar = 100% sugar
- Stabilizer-emulsifier = 95% of total solids (TS)
- Water

Now it is possible for the processor to calculate the quantities needed for each ingredient for the ice cream mix, which will weigh 1,000 lb. A mix is simple if each constituent is to be obtained from one source, whereas the mix is complex if constituents are obtained from several ingredients. Using the selected percentages in the processor's compositional formula, the following amounts will be needed of each constituent to achieve the formula:

- Milkfat: 11% of 1,000 lb = 110 lb
- Nonfat milk solids: 11% of 1,000 lb = 110 lb
- Cane sugar: 15% of 1,000 lb = 150 lb
- Stabilizer-emulsifier: 0.3% of 1,000 lb = 3 lb

Next the processor should calculate the quantity needed for each ingredient (see table 31.4).

Step 1. Compute the amount of each ingredient that is the only source of a particular constituent. In this problem, these are cane sugar and stabilizer-emulsifier. Cane sugar is 100% sugar, therefore 150 lb would supply the quantity needed. TS content of stabilizer-emulsifier is 95%, therefore pounds needed is divided by TS, giving the quantity of solids plus moisture (3.0 ÷ 0.95 = 3.16 lb).

Step 2. Compute the amount of each ingredient that supplies more than one constituent. Since cream is the only source of milkfat (MF), its quantity should be calculated. Then condensed skim milk can be used to furnish remaining NMS.

$$\text{Cream (lb)} = \frac{\text{MF (lb)}}{\text{\% MF in cream}} = \frac{110}{0.4} = 275 \text{ lb}$$

and,

$$\text{NMS in cream (lb)} = \text{cream (lb)} \times \text{MF (\%)} = 275 \times 0.054 = 14.85$$

Quantity of NMS to be furnished by condensed skim milk is determined by subtracting the quantity of NMS supplied by cream from the quantity needed: 110 – 14.85 = 95.15 lb. Condensed skim milk contains 30% NMS; therefore: 95.15 ÷ 0.30 = 317.17 lb (quantity of condensed skim milk required).

To obtain the quantity of water to be added, the sum of the weights of the other ingredients (275.00 + 317.17 + 150.00 + 3.16 = 745.33) is subtracted from 1,000 and the remainder, 254.67 lb, is the total water needed to complete the mix. Each column is summed, and the percentage of each constituent is calculated by dividing its weight by total weight of mix. Each calculated percentage should then be checked against that which was originally planned; an acceptable tolerance is 0.1%, except for stabilizer-emulsifier, which should vary no more than 0.01%.

Calculations of complex mixes, those to which constituents are furnished from several sources, are considerably more complicated. A similar procedure is followed, but the formulas become more involved (Goff and Hartel, 2013). In this case, computers are usually employed in calculating complex mixes and to determine the most economical combination of ingredients.

Laboratory tests should be made on finished mix to verify that contents of fat and total solids are correct. If either fails to comply with legal and/or company stan-

Table 31.4 Quantities of Constituents and Ingredients Used to Supply Each for a Defined Ice Cream Mix.

Ingredients	Weight (lb)	Milkfat (lb)	NMS (lb)	Sugar (lb)	Stabilizer-Emulsifier (lb)	TS (lb)
Cream	275.00	110.00	14.85	—	—	124.85
Condensed skim milk	317.17	—	95.15	—	—	95.15
Cane sugar	150.00	—	—	150.00	—	150.00
Stabilizer-emulsifier	3.16	—	—	—	3.00	3.00
Water	254.67	—	—	—	—	—
Total (lb)	1,000.00	110.00	110.00	150.00	3.00	373.00
Calculated (%)		11.00	11.00	15.00	0.30	37.30

dards, the mix should be standardized by modifying one or more ingredients.

31.5 Compounding and Processing Mixes

Methods of compounding, or putting the mix together, vary according to size of batch, ingredients used, and equipment available. In batch operations, each ingredient is weighed or measured into a pasteurizing vat, beginning with liquid components. Mixes can be assembled continuously from liquid ingredients that are carefully standardized before they are metered with positive metering pumps into blending equipment.

Figure 31.2 shows how ingredients can flow through an ice cream plant. Three methods can be used for continuously quantifying ingredients: (1) storage tanks can be set on load cells that allow monitoring of changes in weight, (2) meters can be placed in pipelines to monitor passage of ingredients, or (3) a batching tank, suspended on load cells, can be placed ahead of the pasteurizer. Alternatively, ingredients can be blended and the mix standardized with digital technology (figure 31.3).

Pasteurization has the following benefits: (1) renders the mix free of pathogenic microorganisms, (2) helps solubilize dry ingredients, (3) melts fats and decreases viscosity, (4) improves flavor of most mixes, and (5) extends keeping quality of the mix to a few weeks.

Mixes may be pasteurized by vat, HTST, or UHT methods. However, protection of microorganisms from heat by sugar and fat requires use of higher temperatures and/or longer times than for fluid milk and cream. Stan-

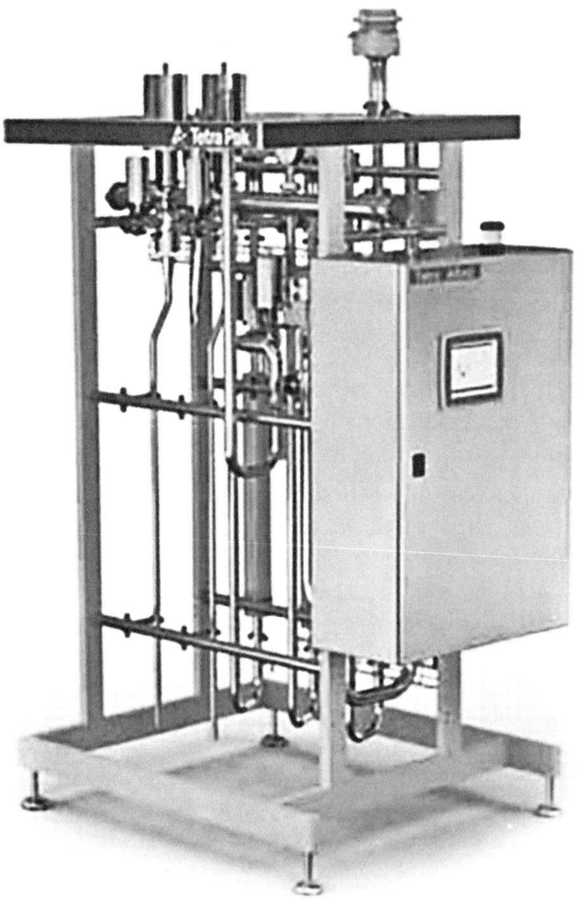

Figure 31.3 A high performance unit designed for automatic in-line blending of mix ingredients with in-line standardization of the fat, SNF, TS, and protein content in the finished product (© Tetra Pak, Inc., all rights reserved).

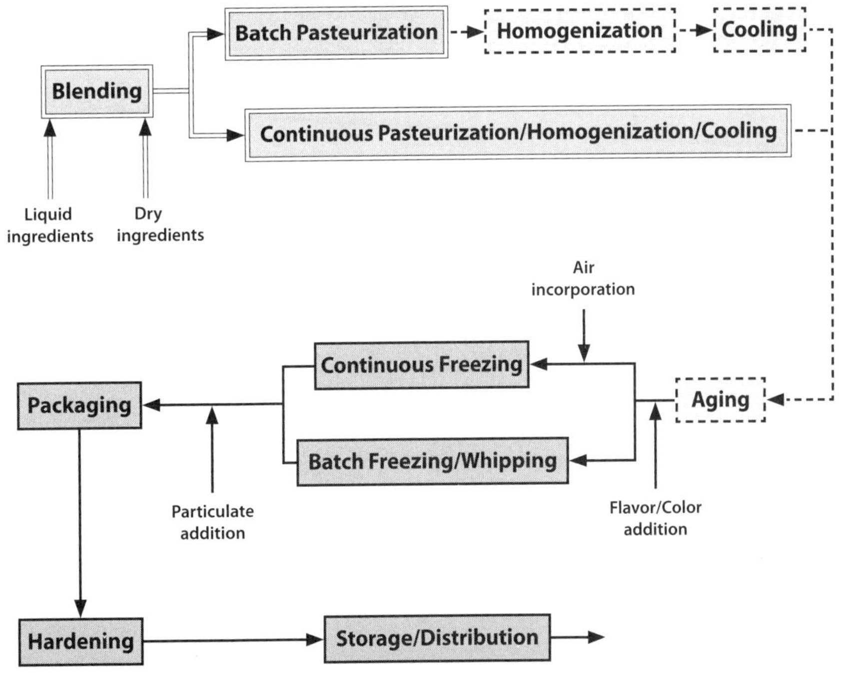

Figure 31.2 Flow diagram of the pathway of ingredients, mix, and soft-frozen product through a frozen desserts manufacturing plant (Department of Food Science, University of Guelph).

dards approved by the USDA are: (1) 155°F (69°C) for 30 min (vat method), (2) 175°F (80°C) for 25 sec (HTST method), and (3) at least 212°F (100°C) momentarily (UHT method).

Homogenization breaks up and disperses fat globules[13] and aids in mixing ingredients. It is important that fat globules do not coalesce after homogenization because clumping of fat globules causes excessive viscosity and decreased whippability. Prevention of coalescence involves homogenization at high temperatures (145 to 176°F [63 to 80°C]) and passage of mix through a second homogenizer valve operating at a lower pressure than the first (e.g., 500 vs. 1,500 psi [3.45 vs. 6.89 mPa]). If HTST pasteurization is used, the homogenizer should be located after the holding tube. This position provides the greatest length of time for dry ingredients to become hydrated and dissolved before reaching the homogenizer valves (dry ingredients are highly abrasive to these valves).

Mix should be cooled to 40°F (4°C) or below immediately after pasteurization and homogenization. Low temperatures prevent microbial growth and retard rates of chemical reactions. It is desirable to age mixes made with some stabilizers, especially gelatin, to allow their complete hydration. Additionally, fat crystallizes in cooled globules and proteins adsorb at the fat/mix interface. Thus, smoothness and whippability are improved. Usually 3 to 4 hr of aging is sufficient.

Developments in high pressure (HP) processing have shown promise for improving textural properties of ice cream, especially reduced fat types, while permitting reductions or elimination of stabilizers. According to Huppertz et al. (2010),

> Transmission electron micrographs showed a network of micellar fragments, arising from HP-induced disruption, in the HP-treated mix and ice cream prepared from it. The network of micellar fragments was believed responsible for the increased viscosity and reduced melting and was hypothesized to occur as a result of calcium-induced aggregation of caseins on decompression.

31.6 Refrigeration and Freezing

The way ice cream is frozen determines its palatability, keeping quality, and yield of finished product. Several variables must be controlled in the process to meet processing goals. Frozen dairy desserts usually can be stored several weeks with little deterioration in quality.

Freezing dairy desserts (except soft-frozen ones) is typically done in two steps: (1) dynamic freezing by which temperature is lowered to about 20°F (–7°C) as the mix is being agitated to incorporate air and produce tiny ice crystals, and (2) static freezing by which the product is hardened rapidly in a very low-temperature environment of about –21°F (–30°C). A newer process, especially designed for premium light products, involves passing the partially frozen ice cream from the conventional continuous freezer into a refrigerated twin-screw extruder, where temperature is further lowered as the product is agitated. Temperatures of the product exiting the extruder may reach as low as 5°F (–15°C). This treatment further reduces the sizes of ice crystals and air cells. Thus, products are perceived as more smooth and creamy than those frozen conventionally. Achievement of the small ice crystals can be enhanced also by pre-aeration (whipping) of the mix, producing a very fine foam as mix enters the freezer barrel.

Ammonia is the refrigerant used in most ice cream freezers (cf. 26.11). However, Freons® (halocarbons) are used in smaller installations. Their use is being restricted because their escape into the atmosphere is thought to deplete the ozone layer and contribute to global warming (Goff and Hartel, 2013).

An ice cream freezer lowers the temperature of the mix to its freezing point. This produces ice crystals and traps air that has been whipped in, producing overrun. Agitation and scraping of the heat exchange surface are essential to achieve desirable physical qualities. Ice crystals and air cells must be kept minute to obtain smoothness.[14] Air cells range in diameter from 5 to 300 μm, and lamellae between them measure from 5 to 300 μm in thickness.[15] Figure 31.4 shows an electron photomicrograph of an air cell in frozen ice cream. Overrun is the amount of increase in volume that results from incorporation of air. If it were not for overrun in frozen dairy desserts, they would be hard, icy, and unpalatable. *Package* ice cream products usually are drawn at 80 to 85% over-

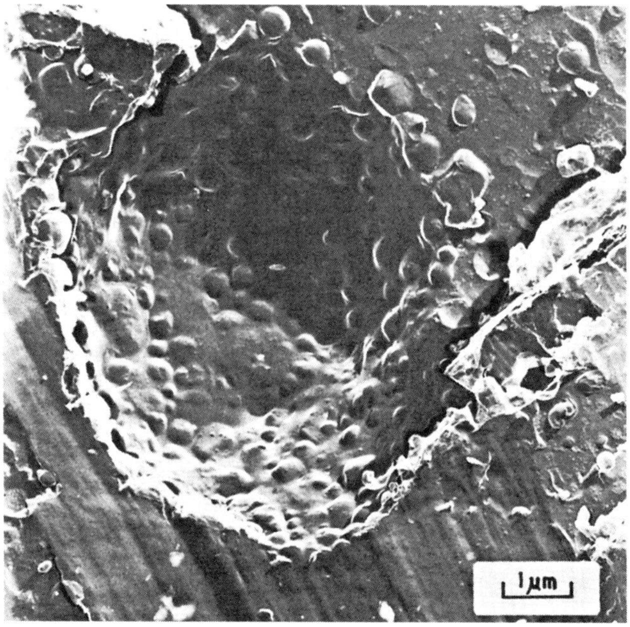

Figure 31.4 Electron photomicrograph of an air cell in frozen ice cream. The small bubble-like projections are mostly fat globules (courtesy of K. G. Berger, Lyons Central Laboratories, London. Originally published in Berger et al., 1972).

run, whereas *bulk* products are drawn at 90 to 100% overrun to allow for loss of volume in dipping. Because of compression in dipping and packing, hand-packed ice cream has a high weight per unit volume.

Overrun may be calculated using either of two formulas:

a.

$$\% \text{ overrun} = \frac{\text{volume of frozen dessert} - \text{volume of mix}}{\text{volume of mix}} \times 100$$

b.

$$\% \text{ overrun} = \frac{\text{weight of mix} - \text{weight of frozen dessert}}{\text{weight of frozen dessert}} \times 100$$

Let us consider an example using formula b. Assume 1 gal of mix weighs 9.0 lb and 1 gal of frozen dessert weighs 4.5 lb. Therefore:

$$\% \text{ overrun} = \frac{9.0 - 4.5}{4.5} \times 100 = 100\%$$

Principles of operation of a continuous ice cream freezer are illustrated in figure 31.5. A constant quantity of mix is metered by a positive-type pump to a second pump that displaces about twice the volume as does the first. Thus, sufficient suction is produced to incorporate air into the mix from an air inlet. Air and mix move into the cylindrical refrigeration chamber where whipping is accomplished by an agitator-dasher (mutator) that rotates at speeds of 150 to 200 rpm. Sharp blades scrape frozen mix from the very smooth wall on which it freezes. As liquid mix enters one end of the chamber, it displaces frozen mix through the other end. Liquid refrigerant is pumped to an insulated space surrounding the freezing chamber where it receives heat from the mix and becomes gaseous. Ice cream mix serves as the contributor of heat and is cooled.

More advanced freezers supply filtered air to the freezing chamber through an air mass controller. Mix is supplied to the freezing chamber via a pump and frozen product is pumped from the freezing chamber to filling equipment. This provides more accurate control of overrun than does the two-pump system discussed above. Automatic process control of mix flow, air flow (cylinder pressure), ice cream pump speed, and cylinder temperature exists within the industry. Some programmable freezers can be linked to the ingredient feeder and container filler to mesh speed of freezing with these operations. The computer uses inputs from sensors to calculate needed inputs of refrigeration, air pressure, and mix flow. Control signals are delivered based on preselected values for ice cream viscosity or temperature, overrun, and freezer capacity. Controls of the dasher drive and the refrigeration system are networked to control viscosity and temperature of the frozen product. Figure 31.6 shows a typical commercial freezer.

Figure 31.6 Continuous ice cream freezer equipped with a mix flow meter coupled with an air mass controller that provides accurate control of the overrun. Product viscosity is maintained: stiff for packaging and extrusion or soft for filling of molds. A product discharge pump isolates the freezing cylinder from downstream pressure variations to enhance control of freezer parameters (from WCB Ice Cream USA).

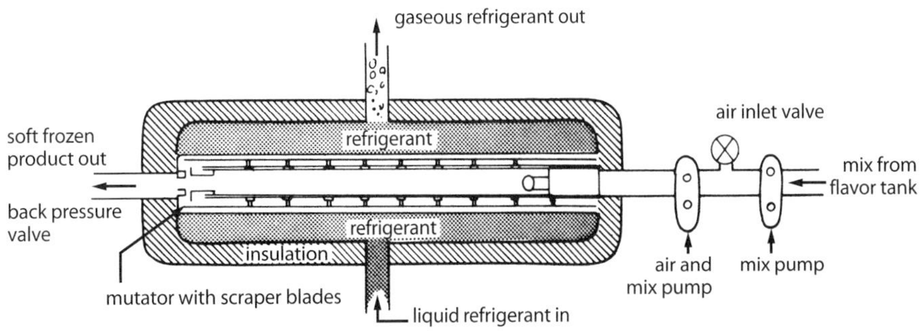

Figure 31.5 Schematic diagram of a continuous ice cream freezer showing the mutator/dasher and chambers for refrigerant and frozen dessert mix.

Air may be incorporated into the mix prior to entering the freezing cylinder, thus enabling the freezer dasher to incorporate air more uniformly and resulting in smaller air cells and less variation in overrun and package weight. Another approach is recirculation of a portion of the partially frozen mix. Recirculated mix cools the inflowing mix permitting more of the freezing cylinder to function as a freezer than as a chiller.

31.6.1 Extrusion Process

Low temperature extrusion of conventionally frozen ice cream (the slow-churned process) results in reducing the temperature of product leaving the swept surface freezer by about 16°F (9°C) as it passes through an extruder with slowly rotating screws. Shear within the extruder prevents enlargement of existing ice crystals as more water is frozen into small ice crystals, more fat globules are destabilized, and minute air cell size is maintained. The result is a more stable structure while using lower concentrations of emulsifier and with less need for static hardening (Goff and Hartel, 2013). Special filling equipment is needed for the highly viscous product exiting the extruder.

Factors that influence rate of freezing are (1) design of freezer, (2) condition of heat transfer surface and blades, (3) speed of mutator (dasher), (4) temperature and rate of flow of refrigerant, and (5) temperature and overrun at which frozen product exits the freezer. The operator has greatest control over conditions of the heat transfer surface and of overrun and temperature of the product as it exits the freezer. Scraper blades must be kept sharp and the freezer's interior wall smooth and cylindrical. Ice buildup inside the freezing chamber serves as an insulator. Additionally, oil must be kept off the opposite side (outside) of the freezer cylinder. Oil from compressors may move with refrigerant and form a layer of highly effective insulation. Oil should congeal at about –22°F (–30°C) or lower so it will drain readily from surfaces into collection sumps.

Desired overrun and drawing temperature are usually dictated by considerations of quality and economy. Therefore, it is undesirable to allow inefficiencies in freezer operation to compromise these attributes. With a properly functioning modern freezer, overrun should vary no more than 3%. Refrigeration control is the most important step in overrun control. Drawing temperature should be as low as practicable commensurate with capacity. The principle holds that the lower the temperature at the freezer outlet, the smaller the volume that can be produced by a freezer because of the additional refrigeration load. Rate of freezing largely determines the size of ice crystals for any given mix. The faster, and more completely, freezing can be accomplished in the freezer the better because completion of the freezing process in the hardening operation is certain to progress more slowly. About 50% of the water in an ice cream mix is frozen at 22°F (–5°C), but more than 60% is frozen at 19°F (–8°C). Therefore, drawing at the latter temperature produces a smoother product.

31.6.2 Soft-Frozen Products

Manufacturers provide a wide variety of freezers for making soft ice cream products, sorbets, and shakes. Mix reservoirs range in capacity from 8 to 38 L while freezer cylinder capacities range from 1.4 to 4 L. Some freezers have single barrels, others have two, and the latter permit delivery of a "twist" type product (two flavors are dispensed simultaneously from a single dispenser forming a swirl). Freezers for soft-serve products must operate intermittently. As frozen product is withdrawn, mix is added to the freezing chamber by pump or by gravity flow from a reservoir above the freezing cylinder. Overrun and freezer load may vary widely in gravity-fed systems changing with intensity of use. Overrun also depends on drawing temperature for any specific mix. For example, overrun will average 35 to 50% with a drawing temperature of 20°F (–6.7°C) but will decrease about 10% with each 1.8°F (1°C) decrease in drawing temperature. Mix pumps that incorporate air provide conditions for production of product with high overrun, up to 80%. Such a high overrun is undesirable in lower fat soft-serve products. However, 60% overrun is desirable for products containing 10% fat. Frozen custard mixes are comparatively high in solids and are frozen at a lower overrun (15 to 40%) than are typical light, lowfat, and nonfat mixes (40 to 60%). Production capacity of soft-serve freezers commonly ranges from 15 to 50 gal/hr (50 to 190 L/hr). Freezers may be either air- or water-cooled. Since heat must be carried away quickly from the mix for efficient freezing, air flow to air-cooled machines must not be restricted. All machines should be located away from direct sunlight and heating vents. When dispensing is intermittent and infrequent, mixes may churn (form butter granules) due to intermittent agitation and temperature fluctuations. Special mix formulation and well-insulated freezer barrels minimize churning.

31.7 Packaging

A major consideration in choosing the appropriate packaging for a specific ice cream product is the best shape, size, and appearance of the packaging that provides the following: consumer information and appeal, efficient handling, rapid hardening, convenience, safety, and economy (International Dairy Federation, 1995). In addition, there are federal requirements as to the inclusion of product information on the packaging. Ingredients must be listed on the label in the order of highest to lowest concentrations and nutrient content must be shown in the standard format (21 CFR 101.9). Nutrient and food component quantities are to be declared in relation to serving size, which for ice cream is 4 fl oz (one-half cup [120 mL]). This regulation, combined with the US federal standard minimum weight per gal for ice cream (4.5 lb), establishes the minimum weight per serving at 63.8 g.

A large percentage of ice cream is packaged in containers ranging in size from 48 to 64 fl oz (one-half gallon). Of approximately 200 million packages sold at retail in the United States in 2009, about half were of the 48 fl oz size and one-fourth each of the 56 and 64 fl oz size. Whereas the half-gallon rectangular paper container with a zip-open top had been favored, round containers and those shaped to nest together while empty are widely used (figure 31.7). Most packages are made of plastic-coated paper or paperboard; some are made of plastic. In choosing package composition, manufacturers must assess whether components can migrate into the food and whether the package can affect flavor or color of the food. Manufacturers of packaging materials should complete a "life cycle analysis" for each type of package (Kool and van den Berg, 1995).

Frozen desserts are packaged mechanically in most plants (figure 31.8). However, bulk containers (2 gal and larger) are sometimes filled by hand. It is important that each package be filled completely. Even more important is that the target weight be achieved. This is mainly controlled by setting and maintaining the desired overrun and, for products containing particulates or variegates, that the ingredient feeder or ripple pump be properly adjusted. A program of statistical weight control is recommended. Range charts or computerized controls of upper and lower weight limits are essential for each filler.

To facilitate handling and stacking, cartons are frequently overwrapped, two or more per package. Clear plastic film that shrinks on heating is a favorite wrapper. Cartons pass through the wrapping machine and briefly through a heating tunnel. It is essential that minimal heat pass to containers to avoid melting ice crystals. Furthermore, containers should be conveyed immediately to the hardening facility.

The ecologically perfect package is the ice cream cone, which came into its own in 1904 at the St. Louis World's Fair. When the ice cream concessionaire in a neighboring booth sold out of dishes, cones were rolled from dough of a *zalabia*, a crisp, waferlike Persian pastry, by E. A. Hamwi. Named the World's Fair Cornucopia, the idea was an immediate hit. Earlier in 1904, but unknown to Hamwi, Italo Marchiony had been granted a patent on a device to produce "small round pastry cups with sloping sides." St. Louis foundries began immediately to create baking molds for cones and the popularity of cones blossomed.

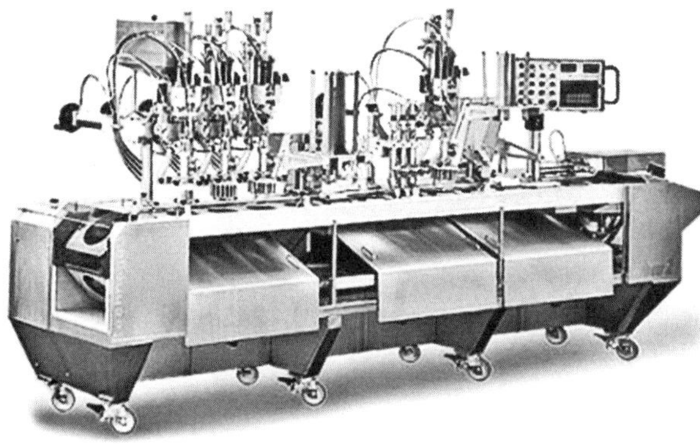

Figure 31.8 An in-line (linear) filler available in the cone/cup version with 2, 3, and 4 rows for a production rate ranging from 6,000 to 12,000 pieces per hour. A one-row version is also available for large-size containers (large packs, cakes, etc.) with a production rate of up to 3,000 pieces per hour (courtesy of Technogel America).

Figure 31.7 Typical round and nested containers used in packaging frozen dairy desserts (courtesy of Huhtamaki, De Soto, Kansas).

31.8 Hardening and Storage

Just as fast freezing in the swept surface heat exchanger favors small ice crystals and smooth ice cream, so does fast hardening. As ice crystals form, the concentration of sugar and other soluble materials increases in the unfrozen portion of the product. Eventually the concentration of these materials in the remaining water becomes so great that ice crystals cease to form. Therefore, even at a temperature of –4°F (–20°C) some water is unfrozen.[16] Fat globules are forced into smaller and smaller areas as ice crystals form.

Hardening can be considered complete when the temperature at the center of the package reaches 0°F (–18°C). Time to achieve hardening depends primarily on (1) size and shape of package (a 5-gal package requires about twice the time of the 2.5-gal), (2) amount of surface area of package exposed, (3) temperature of cooling medium (air or metal surface), (4) rate of air movement (not as important when product touches plate coolers), and (5) temperature of the product as it enters the hardening facility.

Two basic types of hardening facilities are used. In one, the blast-freeze concept is employed whereby air at –20 to –40°F (–29 to –40°C) is blown around the containers. In the other, cartons are placed in direct contact on one or two sides with plates refrigerated to about –31°F (–30°C). Systems are usually designed to move product automatically through a hardening tunnel or chamber in 24 or fewer hours (figure 31.9), then to a cold storage area in which packages are stacked on pallets. Temperature in the storage area approximates –20°F (–29°C).

Because working conditions are unfavorable to humans at such low temperatures, mechanization of operations is desirable. Some systems are mechanized and automated to the extent that packages can be placed in and removed from storage by remote control.

Two practices are vital in maintenance of product quality in cold storage: namely fluctuations in temperature must be minimal and the oldest product must be removed from storage first. Both factors affect smoothness of the product. As temperatures rise, ice crystals melt. If temperatures then decrease, new crystals form and they will be larger than those that melted. As time of storage increases so does average size of ice crystals. In addition, lactose is more likely to crystallize in products stored at fluctuating temperatures and for extended times. Finally, though most chemical reactions are slowed by cold, oxidation continues in stored frozen desserts. Therefore, it is not uncommon to detect staleness, especially at the surfaces of products stored for several months.

Ice cream probably receives greatest destructive treatment in transit to and in storage at retail stores. Temperatures in retail cabinets are less stable than those of the manufacturer. Of course, heat shock can be excessive during the process of consumer purchase and transportation home. Retailers should provide insulated bags and purchasers should not delay in moving frozen desserts to home freezers.

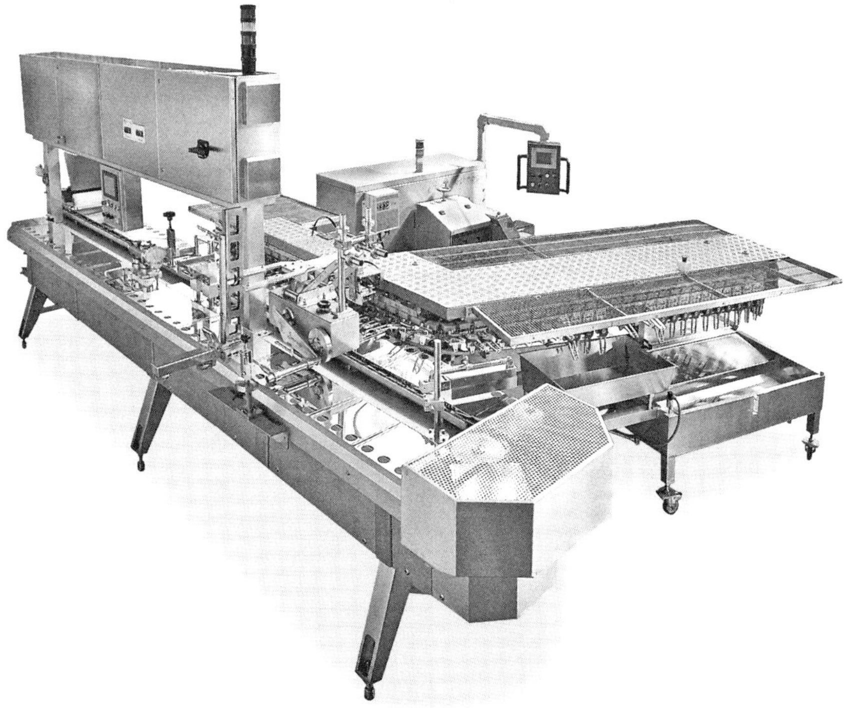

Figure 31.9 Device for moving packaged frozen desserts through a hardening tunnel (courtesy of Technogel America).

31.9 Novelties and Specials

The I-Scream Bar created by Christian Nelson in 1919 was the first of a multitude of delicious novelties and specialties that constitute about 25% of the retail value of frozen desserts produced in the United States. Nelson took Russell Stover as his partner, and Stover gave the chocolate-coated ice cream bar the new name Eskimo Pie. Together, Nelson and Stover began a nationwide franchise business. The original Eskimo Pie bars were stickless. The ingenuity of Harry Burt and the prompting of his daughter led to the development of the Good Humor Ice Cream Sucker, a chocolate-coated ice cream bar on a stick. In contrast with early days, which saw novelties sold largely by vendors and in special stands, supermarkets and food stores sell much of today's production. The favorite package size of buyers is the six-pack.

Ice cream bars are produced by filling molds (forms) with soft-frozen mix from conventional ice cream freezers and refrigerating the molds. In modern equipment (figure 31.10) the molds move to an area in which they are exposed to ammonia or Freon® (direct expansion refrigeration). These refrigerants take on heat, as in an ice cream freezer.[17] Subsequently, the molds are momentarily exposed to a heating medium that melts a thin film of ice cream at the mold's surface, freeing the bar, which falls or is lifted out. Frozen bars may then be coated with chocolate or other types of coatings. Chocolate for coating bars is specially formulated with 48 to 55% cocoa fat, which helps it cling to the frozen surface.

Popsicles are made in the same equipment except the water ice mix may not be prefrozen before being pumped directly to molds. Rated capacity of a 10-wide stick novelty machine is about 15,000/hr.

Some ice cream novelties are prepared by extruding stiffly frozen mix in an attractive form from the freezer. Portions are cut off onto a moving belt that carries them through a hardening tunnel. Ice cream sandwiches are made by extruding frozen mix between two wafers held side by side in a rotating wheel-like receptacle.

Specialties such as the sundae,[18] soda, parfait, and banana split are not technically novelties but command a portion of the same market. The smoothie is a combination of fruit and/or fruit juice blended with nonfat ice cream, frozen yogurt or sherbet, and served immediately.

31.10 Quality Assurance

Repeat sales of frozen desserts depend largely on quality. Therefore, it is vital that manufacturers systematically evaluate their products and those of competitors. Considerable expertise is required to identify specific defects, but even more valuable is the person who can correctly identify defects then relate them to cause and provide instructions for remedy.

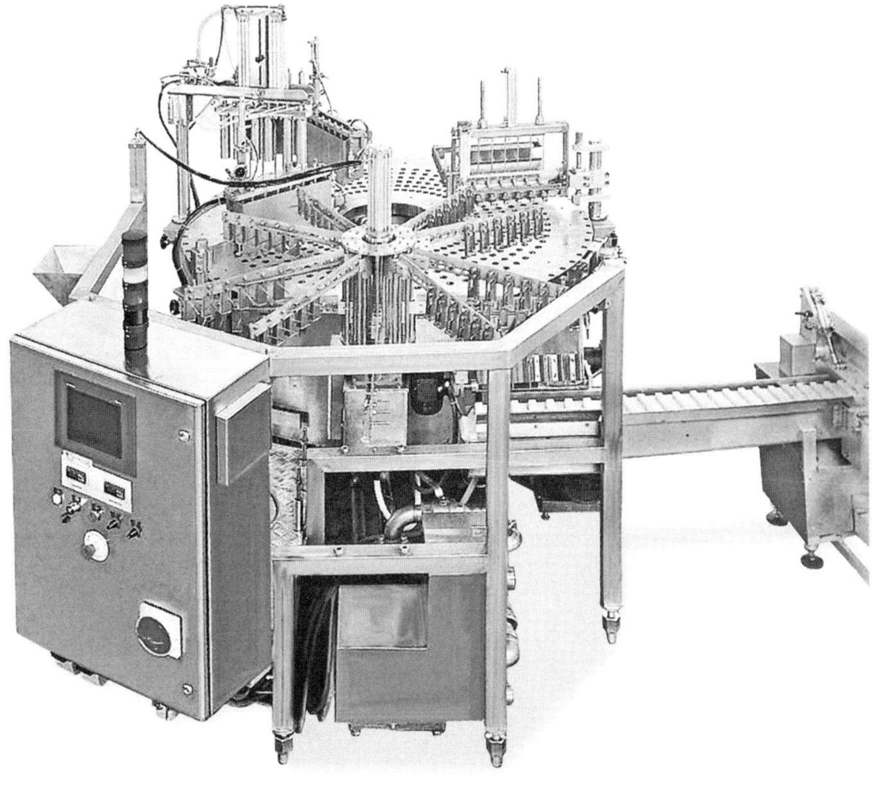

Figure 31.10 An automatic rotary machine for the production of ice cream and popsicles on sticks. It is completely mechanical and the movements of each individual station are automatically synchronized (courtesy of Technogel America).

Defects commonly occur in (1) flavor, (2) body and texture, (3) color, and (4) meltdown. In fruit or nut ice creams, distribution and appearance may be defective as well.

31.10.1 Defects in Flavor

These can be subdivided into four major categories—flavoring, sweetener, dairy products, and processing—though other types of defects may occur. Flavoring may be too intensive, lacking, or unnatural. The latter usually is most undesirable and is associated with use of artificial flavors. Sweetener may be excessive, lacking, or syrup-like. The latter is frequently observed and is associated with use of excessive or low-quality corn syrup solids. Flavors associated with quality of dairy products are acid, lacking freshness, metallic, old ingredient, oxidized, rancid, and unclean (cf. chapter 25). The major flavor caused by processing is cooked, which, unless excessive, is actually desirable to most consumers. In fact, it is the "flavor of assurance" to those who professionally evaluate ice cream and butter. This is true because cooked flavor dissipates with time (free sulfhydryl groups that impart the cooked flavor become oxidized to disulfide groups), and its presence at a mild intensity strongly suggests the absence of flavors associated with oxidation (metallic, oxidized, stale, and storage).

31.10.2 Defects of Body and Texture

We have discussed how composition, processing, and storage affect smoothness. The coarse defect is most common of all body and texture defects. Large ice crystals and air cells are responsible. At least four teams of researchers have measured ice crystals and found them to range in size from 20 to 75 μm (figure 31.11). Smoothness is associated with crystal sizes of 35 to 55 μm. Below 35 μm extreme smoothness is encountered and above 55 μm coarseness is observed. Sandiness, though similar in mouthfeel, is observed when lactose crystallizes sufficiently to be detected. This is a serious defect that renders the product unsalable.

Wide differences in resistance (body) are observed. A weak-bodied product lacks resistance and is associated with high overrun and insufficient total solids or stabilizer in the mix. The opposite defect is a heavy or soggy body. It may result from low overrun and high total solids (especially sugar) and from too much or improper stabilization. Crumbly ice cream breaks apart easily on dipping. The most frequent cause is improper emulsification or stabilization.

As frozen desserts are hardened to –22°F (–30°C) most of the water turns to ice. Expansion accompanying this physical change sets up internal stress, which may fracture weak lamellae between air cells and create channels. On warming, surface tension tends to close these channels, causing losses in overrun and **shrinkage** in volume. Products with low concentrations of NMS and those improperly stabilized are most subject to this defect.

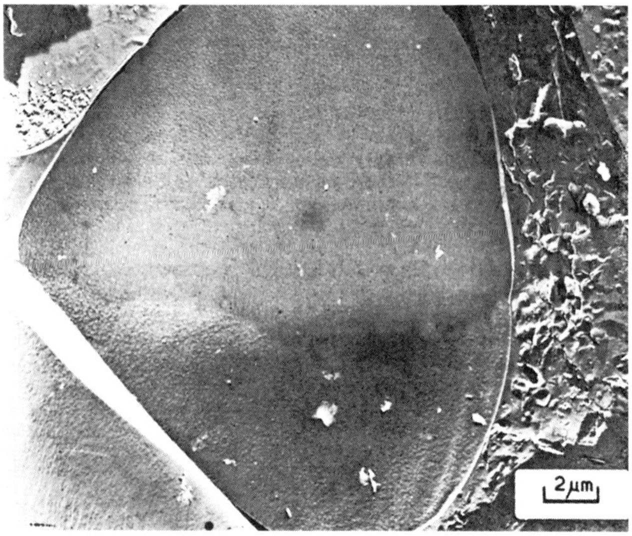

Figure 31.11 Electron photomicrograph of an ice crystal in frozen ice cream. The dimpled appearance of the crystal's surface probably was caused by expulsion of minute air bubbles on freezing (courtesy of K. G. Berger, Lyons Central Laboratory, London. Originally published in Berger et al., 1972).

31.10.3 Color

Defects usually associated with color relate to intensity and authenticity (hue and shade). Wide variations occur with some colors being unnatural. Variations in distribution of color may occur if two or more freezers are feeding frozen product to the same packaging machinery, unless the mix comes from the same flavor tank or blending of frozen product is accomplished prior to packaging.

31.10.4 Melting Quality

Commonly known as meltdown, this characteristic is affected by stabilizers, emulsifiers, salt balance in dairy ingredients, and processing or storage conditions that might destabilize proteins. A desirable product melts at a moderate rate leaving liquid with the appearance of the original mix. Common defects are foaminess, curdiness, and slow melt (figure 31.12).

31.10.5 Compositional Control

Major controls on composition are exercised in compounding and in freezing mixes. It is vital that quantities of major components of raw materials be known precisely. Therefore, tests for fat and total solids in milk products are routinely performed (cf. chapter 22). For fat in milk, cream, condensed milk, and butter, tests may be performed by the Babcock,[19] Gerber, and Roese–Gottlieb or Mojonnier methods (instrumental analyses may be

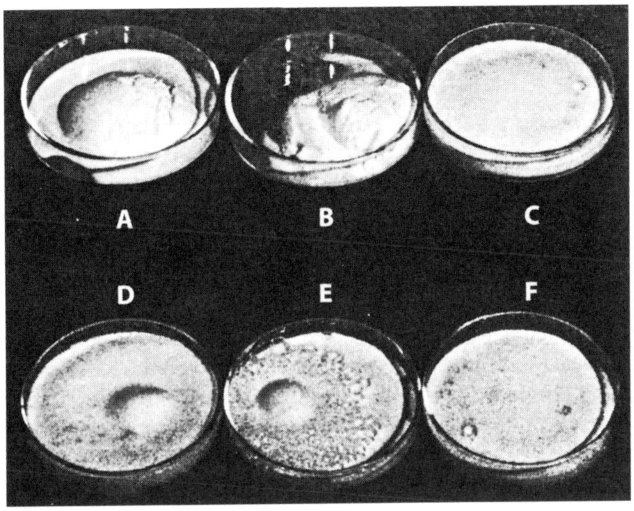

Figure 31.12 Melting quality of ice cream. Samples were dipped and held at room temperature for 10 min. They would be criticized as follows: (A) and (B) slow melting and curdy appearance; (C), (D), and (E) foamy; (F) no criticism.

used for milk). The primary test for total solids employs evaporation in a hot air oven.

Finished mix should be tested for fat and total solids before freezing. Standardization is vital if concentrations of either are outside the target range. Finally, frozen product should be tested for fat, total solids, and overrun. The latter includes periodic weight checks on filled containers as well as direct comparisons of quantities produced from original amounts of mix.

31.10.6 Microbiological Control

Standard plate counts and coliform counts are basic tools for evaluation of microbiological quality in frozen desserts. By regulation raw milk and cream should contain no more microorganisms than 3×10^5/mL and 5×10^5/mL, respectively. Standard plate counts on pasteurized milk, cream, and other fluid dairy ingredients should not exceed 10^3/mL, and those for pasteurized mix and finished frozen desserts should not exceed 10^3/g. Coliform counts for all pasteurized products should not be more than 10/g or mL. However, these numbers are considerably higher than should be accepted by manufacturers since much lower counts are achievable when recommended methods of production are used.

Yeast and mold counts are often made of fruits and nuts, whereas thermoduric bacterial counts are important with products using nonfat dry milk, cocoa, and cinnamon.

31.11 Cost Control

Other than fixed costs, such as interest on investment, there are five major sources of costs of producing frozen desserts: (1) basic ingredients, (2) flavoring, (3) containers, (4) labor and production, and (5) delivery. Costs of ingredients constitute from about 40 to 65% of the wholesale price. Therefore, it is imperative to adjust mix formulas to optimize costs as practicable within standards of quality. Flavor costs commonly vary from 7 to 11% of the wholesale price and container costs approximate 6 to 9% of the sales dollar. Significant and controllable variables are labor and production costs (5 to 27% of wholesale price), and it is in this area that costs are most loosely controlled. Included in this item are preparing mixes, freezing, packaging, hardening, and storing. Of prime consideration is output/employee/hr. Costs of delivery to stores vary from a low of 9% to as much as 21% of the sales dollar. Underloaded route trucks and high-mileage routes result in excessive delivery costs. Other less variable cost factors and approximate percentages of wholesale price are depreciation (3%); administration (3%); and selling (1%).

31.12 Summary

Ice cream and its lower fat versions are made using milkfat, nonfat milk solids, sweetener, stabilizer, flavoring, and water. Mellorine is a similar competitive product containing vegetable fats instead of milkfat. Water ices contain sweetener, flavoring (including fruit juices), citric acid, and stabilizer. Sherbets can be made using a mixture of ice cream mix and water ice mix.

Fresh or preserved dairy products may be used as ingredients and standards allow considerable flexibility in formulating frozen desserts. US standards for ice cream products specify minimum and maximum percentages of fat, minimum weight/gal of food solids, and minimum total weight/gal. Balance and quality of ingredients are major determinants of quality of finished frozen desserts.

Ingredients for mixes may all be obtained in liquid form and, as such, are readily assembled in automated operations. Pasteurization of compounded mix is followed by homogenization, cooling, flavor addition, freezing, and hardening (for hardened products).

Quality assurance includes examination of ingredients, mix, and finished products. Tests are recommended for microbial counts, percentages of fat and total solids, overrun, flavor, body, texture, color, and meltdown. When standards of high quality are met, frozen desserts become an excellent means of providing milk for humans.

Study Questions

1. Using yearly US production statistics, place in descending order the categories of frozen desserts.
2. What are the major components of ice cream?
3. Define ice cream, its lower fat versions, mellorine, sherbet, and water ice in terms of standards for composition, weight per gal, and weight of food solids per gal (except ices).

4. What is the characterizing ingredient of French ice cream?
5. How much is the maximum reduction in fat and total milk solids allowed in bulky-flavored ice cream compared with plain ice cream?
6. Explain heat shock and the reason it is objectionable.
7. What ingredients may be used to replace sugar and corn syrup solids in ice cream with no sugar added?
8. From which dairy products do manufacturers obtain milkfat for frozen desserts? Nonfat milk solids?
9. What is sandiness? Discuss its causes.
10. What three factors limit amounts of corn syrup solids used in ice cream mixes?
11. Explain dextrose equivalent. Why would dextrose lower freezing point about twice as much as an equal weight of sucrose?
12. Contrast the functions of stabilizers and emulsifiers.
13. Define *single-fold* vanilla.
14. Differentiate chocolate liquor from cocoa.
15. In what forms are fruits sometimes added to frozen desserts?
16. Compute a simple ice cream mix to be prepared from 30% cream (6.3% NMS), condensed skim milk (30% NMS), cane sugar, stabilizer-emulsifier (95% of solids), and water. The 1,000-lb mix should contain 10% fat, 12% NMS, 15% sugar, and 0.5% stabilizer-emulsifier.
17. Why should ice cream mixes be homogenized? Where in an HTST system should the homogenizer be located?
18. Why are temperatures and times of pasteurization higher for ice cream than for milk?
19. What is overrun? Why is it necessary in most frozen desserts? Why does it vary in different frozen desserts? What are the effects of too much? Too little?
20. List five factors that determine rates of hardening of frozen desserts.
21. Discuss size of ice crystals in ice cream as it relates to quality and conditions of production, hardening, storage, and distribution.
22. List four major categories of flavor defects in ice cream and give two examples of each.
23. What should be the maximum standard plate count and coliform count per gram of finished ice cream?

NOTES

[1] The amount of reduction in fat content allowable (up to 2%) is calculated as follows: weight of mix – (weight of fruit excluding sweetener × 1.4) × minimum fat content of plain mix. For example, let us add 8 lb of fruit to 100 lb of mix. Then: (100 – [8 × 1.4]) × 0.1 = 8.88; the frozen product must test at least 8.88% MF. In the above formula, the value 0.1 represents the 10% minimum fat of a plain mix.

[2] Heat shock refers to the growth of ice crystals when melting and refreezing occurs. Even small changes in temperature of frozen mix allow changes in structure.

[3] The Popsicle was invented in 1923 by Frank Epperson who, on a cold winter evening, left on a windowsill a glass of lemonade containing a spoon. By morning the lemonade had frozen. He immersed the glass in water and removed the frozen mass. He patented the invention in 1924.

[4] Salts of calcium and magnesium increase and those of citrates and phosphates decrease tendencies of fat to clump in mixes of frozen desserts.

[5] Dried whey solids contain approximately 79% lactose, 12% protein, 11% minerals, 4% moisture, and 1% milkfat. They may replace, according to federal standards, up to 25% of the NMS.

[6] Too high a concentration of NMS may cause an intense cooked flavor in the finished product, especially if high-heat nonfat dry milk is added.

[7] Syrups contain flavorful aldehydes in concentrations that depend on degree of refinement.

[8] Hydrophilic colloids are minute particles (< 1 μm in diameter) that remain suspended in solutions, though they are not dissolved, and they bind water.

[9] Hydrophilic colloids are solids and consist of billions of molecules coiled back and forth on one another with bonding forces acting to keep the molecules tightly coiled. With some types of colloids, these intermolecular bonds are sufficiently weak that stronger bonds with water will easily replace them. Such colloids are therefore easily suspended in cold mix.

[10] Substances that contribute to surface area of air cells contribute to dryness. When a sphere is divided into two equal spheres, surface area becomes four times that of the large sphere. Let us assume two separate portions of a mix contain the same amount of unfrozen water as they leave the freezer. Mix 1 contains a small number of large air cells and mix 2 contains twice as many air cells, each with half the volume of those in mix 1. Total surface area of air cells in mix 2 will be four times that of mix 1, so water is spread over four times the area of water in mix 1. Mix 2 will appear much drier.

[11] Referred to as polys, these include polyoxyethylene sorbitan monooleate and polyoxyethylene tristearate. The former may cause churning during freezing of soft-serve products because of its destabilizing effect on emulsions of fat.

[12] Sediment in chocolate-flavored drinks is caused by incomplete separation of bean shells and germ, both of which are difficult to grind and are insoluble.

[13] Ice cream containing 10% fat in the form of globules 0.5 μm in diameter would contain ~1.7×10^{12} globules/mL, and their total surface area would be ~1.3 m^2. Fat globules in ice cream are covered with an artificial membrane (compared with a natural one in unhomogenized cream or milk). Its outer layer consists of casein subunits (undenatured macromolecules, smaller than unicellular casein).

[14] In ice cream containing 100% overrun in the form of 60 μm (diameter) air cells, the number of air cells per gram is ~8.3×10^6, and their total surface area ~0.1 m^2. The number of ice crystals, assumed to be in the form of a 40 μm cube, would be ~7.8×10^6/g with a total surface area of ~0.08 m^2.

[15] Breakdown of emulsified fat during freezing frees liquid fat that spreads at the air/mix interface, displacing protein and limiting foam formation. Coalesced fat in walls of air cell lamellae limits their thinness.

[16] Approximate percentages of water frozen at 23°F, 12°F, and –22°F (–5°C, –11°C, and –30°C) are 50, 72, and 95, respectively.

[17] Some plants still employ brine tanks as the heat transfer medium. Brine is water containing calcium chloride to lower its freezing point. It is cooled by coils containing a gaseous refrigerant.

[18] The ice cream sundae, we are told, was born in Evanston, Illinois. When a law was passed forbidding the sale of ice cream sodas on Sunday, local merchants got around it by serving ice cream with just syrup. The sodaless soda came to be known as a Sunday, but objections arose about naming a dessert after the "Lord's Day," so it became a sundae.

[19] Only approximate fat tests can be made by the Babcock method and its modifications. Results normally vary ±0.3% compared with ether extraction methods. Thus, if a standard of 10.2% is established, the actual percentage might vary from 9.9% to 10.5%.

References

Berger, K. G., B. K. Bullimore, G. W. White, and W. B. Wright. 1972. The structure of ice cream. Parts 1 and 2. *Dairy Industries* 37(8):419–425, 493–497.

Dickson, P. 1972. *The Great American Ice Cream Book*. New York: Atheneum.

Goff, H. D., and R. W. Hartel. 2013. *Ice Cream* (7th ed.). New York: Springer.

Huppertz, T., M. A. Smiddy, H. D. Goff, and A. L. Kelly. 2010. Effects of high pressure treatment of mix on ice cream manufacture. *International Dairy Journal* 10:1016.

International Dairy Federation. 1995. *Technical Guide for the Packaging of Milk and Milk Products* (3rd ed.). Brussels, Belgium: Author.

International Dairy Foods Association. 2015. *Dairy Facts: 2014 Edition*. Washington, DC: Author.

Isik, U., D. Boyscioglu, E. Capanoglu, and D. Nilufer Erdil. 2011. Frozen yogurt with added inulin and isomalt. *Journal of Dairy Science* 94:1647–1656.

Kool, J., and M. J. van den Berg. 1995. "Environmental Constraints on Packaging of Dairy Products." In *Technical Guide for the Packaging of Dairy Products* (3rd ed.), Bulletin of the International Dairy Federation, No. 300. Brussels, Belgium: International Dairy Federation.

Marshall, R. T., H. D. Goff, and R. W. Hartel. 2003. *Ice Cream* (6th ed.). New York: Kluwer Academic/Plenum.

Tharp, B., and S. Young. 2010. Blending sweeteners can improve product and bottom line. *Dairy Foods Magazine* 111(2):37–39.

Websites

International Dairy Federation (http://www.fil-idf.org)

International Dairy Foods Association (http://www.idfa.org)

University of Guelph, Department of Food Science (https://www.uoguelph.ca/foodscience/book-page/ice-cream-ebook)

USDA National Agricultural Statistics Service, Dairy Products Annual Summary (http://usda.mannlib.cornell.edu/MannUsda/viewDocumentInfo.do?documentID=1054)

32

Butter and Related Products

When the making of butter originated is in the night of time.
The Greeks learned it from the Scythians, and the Romans from the Germans.
Whatever the origin, butter had nothing of its perfection before the application of modern science.
Anonymous

32.1 Introduction

The balanced array of volatile substances present in low concentrations makes butter a particularly appealing spread and a major ingredient of sauces, dressings, pastries, and confections.[1] Butter has been used as a food, and sometimes as a medicine and cosmetic, for more than 5,000 years. Ancient Hindus valued cows according to the amount of butter that could be churned from their milk. As late as 1860 farmers in the United States made butter and marketed it locally. It was not until the last half of the nineteenth century that buttermaking in the United States became a factory operation. In the earliest butter plants milk was received from farms and placed in large shallow pans for cream to rise. By about 1875 cream separation was begun on farms and cream was transported by animal power to creameries by haulers with regular routes. The remaining skim milk was used as animal feed on farms. Centrifugal separators were introduced in the 1880s.

In 1900 about 65% of US butter was made on farms.[2] Steady development of the factory system reduced farm-produced butter to 43% in 1920, less than 25% in 1940, and 12% in 1950. Today, little butter is made on US farms for resale. As the nutritional merits of skim milk for human food were realized and rapid transportation became available, creameries began purchasing milk rather than cream. This led to the current situation of the joint operation of butter and nonfat dry milk processing facilities. Skim milk obtained in separating cream from milk plus buttermilk obtained from the churning of cream are usually dried for use in numerous foods (cf. chapter 33). Presentations of the various sciences and practices applied to butter and butter products can be found in the *Encyclopedia of Dairy Sciences* (Fuquay et al., 2011).

Annual butter production in the United States averaged about 1.25 billion lb between 1999 and 2004 and reached 1.9 billion lb in 2014. Of the 86 plants producing butter in the United States, 14 are located in California. Annual per capita butter consumption in the United States is at a 40-yr all-time high of 5.6 lb.

Butter commonly is churned from cream, although it may be churned from milk. Strong agitation breaks the emulsion of fat in serum[3] and causes fat globules/particles to stick together. King (1953) offered the following hypothesis of the mechanism of churning:

> The normal churning process is confined to a comparatively narrow temperature range with an optimum value for the ratio of crystalline to solid to liquid fat. When cream foams, fat globules come to the air-serum interface of air bubbles. Liquid fat from the globules spreads at the interface along with material of fat globule membranes. The film of liquid fat cements globules into clumps. On repeated formation and destruction of foam bubbles, clumps grow to butter granules which contain modified serum (buttermilk) in the interstices among the fat globules. As the mass is "worked" (mixed vigorously), some fat globules are crushed, and their contents are added to liquid fat. Additionally, moisture droplets are subdivided and air is entrapped.

32.2 Attributes and Types of Butter

Most butter is manufactured from fresh sweet cream. Such cream is differentiated by insignificant **developed acidity**. Sour cream butter is churned from cream in which microorganisms have produced considerable acid. However, the acidity of sour cream commonly is neutralized by adding alkaline substances including calcium and magnesium hydroxides or sodium carbonate.

Cultured (ripened) cream butter is made from cream that has been inoculated with butter culture and incubated to produce acids and flavorful substances similar to those in cultured buttermilk. The cultures are *Lactococcus lactis* ssp. *cremoris* (acid producer) and *Leuconostoc mesenteroides* ssp. *cremoris* (flavor producer) in combination (cf. chapter 29). Similar qualities can be obtained by adding distillate from cultures to butter while it is in the granular stage of churning. An alternate method, often used in Europe where the cultured flavor is often preferred, involves addition of a directly acidified (with lactic acid) diacetyl-producing culture to butter granules following churning of sweet cream and draining of buttermilk. Additionally, an acid-producing culture (*Lc. cremoris*) containing salt may be added at this point. This procedure provides the desirable flavor in the butter while allowing production of sweet buttermilk that is useful in other products. When unsalted butter (called **sweet butter**) is made from ripened cream, it should be stored frozen to preserve its delicate flavor.

Cream separated from whey is often added to other cream before churning or is, itself, churned to produce butter. Quantities that may be added to other cream without seriously modifying the flavor depend on quality of the whey. Butter made from whey cream, especially from manufacture of Swiss cheese in copper kettles, is vulnerable to rapid **oxidation** and development of a tallowy flavor.

A unique quality of butter is its ability to be stored for a long time compared with many other forms of milkfat. Therefore, butter tends to be produced abundantly in spring months and the surplus is placed in cold storage for use when milk supplies are comparatively small.

Butter products include organic butter (Pasture Butter™) made from milk obtained from cows during pasture season, spreadable butters made with added vegetable oils, flavored butters, whipped butter, and European-style lower moisture butter. Whipping of butter increases spreadability at cold temperatures and increases its volume by 25 to 30%. Inert gas (nitrogen) injected during whipping limits oxidation. Spreadability is improved also by blending canola oil with butter.

32.3 Processing Milk for Buttermaking

Usual procedures for processing milk for buttermaking are clarification, separation, pasteurization, and cooling (cf. chapter 26). Undissolved extraneous matter and somatic cells are removed in clarification. Cream is separated to contain 30 to 35% fat for conventional churning or 40 to 45% fat for continuous churning. For economic reasons it is vital that less than 0.01% fat remain in skim milk.

Pasteurization temperatures must be at least 165°F (74°C) for vat pasteurization (30 min) and 185°F (85°C) for the **high-temperature short-time** method (15 sec). Because of the protective effect of fat on bacteria, these temperatures are higher than those required for pasteurization of milk by the same methods[4] but the flavor of cream is not impaired. In fact, most consumers prefer a mildly cooked flavor in butter. Pasteurization destroys most enzymes in cream, especially milk lipase, the enzyme causing rancid flavor in butter churned from unpasteurized cream. Homogenization is not performed because the small globules formed on homogenization are extremely difficult to churn. The same is true of any milk that contains minute fat globules, including goat milk and milk from certain individual cows in late lactation.

32.4 Methods and Procedures in Churning

Regardless of method of churning, butter composition averages 80.2% fat, 16.5% moisture,[5] 2% salt, and 1 to 1.5% curd. Compositional control in the laboratory is commonly based on the Kohman test for moisture in which 10 g of sample are weighed to the nearest milligram, heated to drive off moisture, cooled, and reweighed. Variations in moisture content can be caused by churns being operated under constant adjustment, therefore, duplicate moisture tests should be made and machine adjustments should be based on the average.[6] Periods when composition varies most occur following shutdown and changes in the supply of cream. The following factors interact to determine yield of butter: efficiencies of separation and churning, composition, overrun, and extent of package fill. Yield of butter from milk can be estimated using the following formulas:

$$\text{cream (lb)} = \frac{\text{milk (lb)}(\text{\% fat in milk} - \text{\% fat in skim milk})}{(\text{\% fat in cream} - \text{\% fat in skim milk})}$$

$$\text{butter (lb)} = \frac{\text{cream (lb)}(\text{\% fat in cream} - \text{\% fat in buttermilk})}{(\text{\% fat in butter} - \text{\% fat in buttermilk})}$$

For economic reasons, moisture must be incorporated into butter in the highest permissible concentrations but fat content must not be diluted below 80% because of the legal standard of identity. Moisture is the major contributor to **overrun.** Although air is sometimes present, it does not comprise overrun in butter as it does in frozen desserts. Overrun in butter is calculated from weight and not from weight per unit volume as in frozen desserts (cf. chapter 31). For example, if cream containing 400 lb of fat yields 500 lb of butter, overrun is 25%. That is,

$$\text{\% overrun} = [(500 - 400) / 400] \times 100$$

A practical goal for overrun is 23%. In addition to moisture, components contributing to overrun are salt and curd. The latter is the residue of protein and other nonfat components that originate in cream and are not removed in buttermilk or wash water.

Major concerns of the butter maker are churning within a reasonable time with minimal loss of fat in buttermilk, attainment of proper fat content, and obtaining, by working the mass, optimal distribution of water and salt without developing a sticky body. Optimal churning rate by agitation is obtained when fat content of the cream is 30 to 33% and when part of the fat is in the solid state. Solidification occurs slowly, therefore cream is usually held cold, before churning, for at least 2 hr after pasteurization.

Butter can be made also by blending anhydrous milkfat (AMF) or butter oil with water, salt, and milk solids using an emulsion pump with passage through a scraped surface heat exchanger for cooling, crystallization, and texturization. Butter produced by this process contains few phospholipids. A summary of processes and physical characteristics of butter produced by each process is presented in table 32.1.

Table 32.1 Summary of Churning Processes and Resulting Physical Characteristics.

Name of Process	Description of Process	Fat Globules in Butter (%)	Air in Butter (%)
Conventional	Slowly break emulsion of fat by agitation, then emulsify serum by working	18–28	2.5
Fritz	Break emulsion by accelerated agitation and emulsify plasma by accelerated working	not reported	5–6
Alpha	Reseparate 40% cream to 80% cream, then invert emulsion phase by working	21–33	nil

32.4.1 Conventional Churning

Conventional churning is done by the batch method in which cream is placed in a vessel (figure 32.1) that is moved in a way that will produce vigorous agitation. Churn design and materials for manufacture have changed markedly from hand-operated wood to mechanically driven stainless steel and aluminum alloyed with magnesium. Metals for churns are carefully selected and tested because butter adheres to many metals.

The usual procedure in conventional churning is as follows: (1) prepare the churn by cleaning and sanitizing; (2) pump into churn cream containing 30 to 33% fat at about 48°F (9°C) in summer or 55°F (13°C) in winter; (3) add coloring; (4) rotate churn until butter granules are formed (breaking point) and become the size of peas or popcorn; (5) drain buttermilk; (6) rinse buttermilk from interior surfaces of churn with clean, cold water; (7) wash butter with sufficient water to bring total volume to that of original cream (water that is colder than butter firms whereas water that is warmer than butter softens); (8) drain wash water; (9) add salt; (10) work butter sufficiently to bring granules and water into a compact mass; (11) sample and test for moisture; (12) add water if insufficient (below about 16%) or permit it to escape from the churn if test indicates high moisture; (13) work butter until it has a firm, waxy body; (14) sample and test for moisture, salt, and curd; and (15) remove butter from churn.

Rate of churning in conventional churns is determined by size of churn, speed at which it is turned, quantity of cream therein, and fat content, temperature, and acidity of the cream.[7] Given the same design, greater agitation occurs in larger churns. Speed must be high enough to maximize agitation but not so high it holds cream on the surface of the churn by centrifugal force. Normal load of cream is 30 to 50% of the churn volume. Overloading usually results in delayed churning and excessive losses of fat in buttermilk. Underloading decreases churning rate (less agitation) and reduces plant capacity. Reductions in churning rate accompany both low and high concentrations of fat in cream, that is, outside the range of 30 to 35%. Fat globules collide and reach surfaces of foam less frequently in light cream whereas high viscosity reduces agitation in heavy cream. Cream churns faster at higher temperatures but excessive losses of fat in buttermilk are experienced. Sour cream churns more rapidly than sweet cream if there is sufficient acidity to destabilize the casein so the serum is less viscous (allowing more frequent collisions of fat globules).

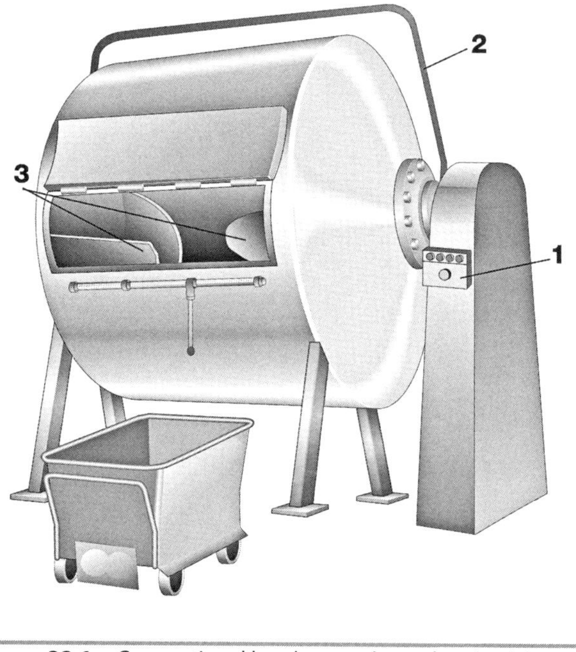

Figure 32.1 Conventional batch-type churn showing (1) control panel, (2) emergency stop, and (3) angled baffles (courtesy of Tetra Pak, *Dairy Processing Handbook*).

32.4.2 Continuous Process

The continuous buttermaking process involves two principles, namely (1) accelerated churning and working (foaming methods) and (2) reseparation with **phase inversion**.

In accelerated churning and working, also called the Fritz process or method (so named for the German scientist who pioneered its development), cream containing 40 to 42% fat is agitated vigorously in a churning cylinder by beaters driven by a variable speed motor. Rapid conversion from a true emulsion to clumped particles of fat occurs in the cylinder and, when finished, the butter granules and buttermilk pass to a draining section. A first washing of the granules sometimes occurs en route—either with water or recirculated chilled buttermilk. On leaving the working section the butter passes through a conical channel to remove most of the remaining buttermilk. The butter may then be given a second washing with water through high-pressure nozzles, thus dispersing the granules so most residual milk solids are removed. Following this stage salt may be added through a high-pressure injector.

Granules are worked into a homogeneous mass by auger movement inside refrigerated cylinders, some of which have perforated plates and working wings. Moisture is finely dispersed, air content is high (5 to 6%), and considerable fat remains in the globular form. A vacuum pump may be attached to reduce the air content of the butter equal to that of conventionally churned butter, less than 0.5%. A cutaway view of a typical continuous churn is shown in figure 32.2.

In the final or mixing section butter passes a series of perforated disks and star wheels. There is also an injector for final adjustment of water content. Moisture is controlled to within 0.2% based on measurements of **dielectric** properties (conductivity) of butter. If butter granules are not washed content of nonfat solids (curd) is about 1.5%, twice as much as in conventional butter. Capacities

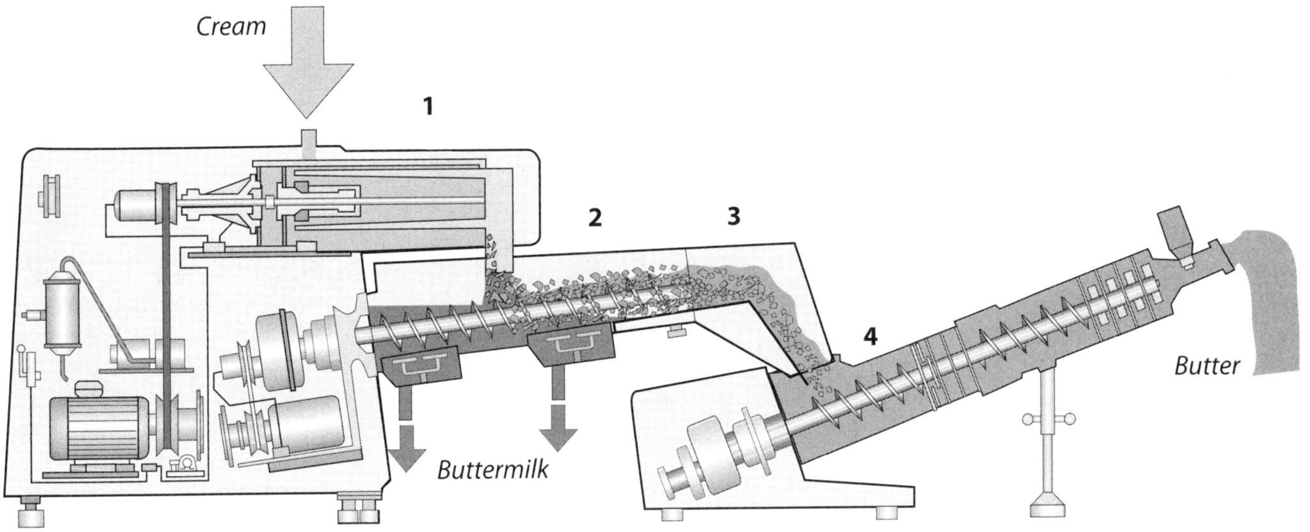

Figure 32.2 Diagram of continuous churn employing the Fritz process. (1) Churning cylinder, (2) separation section, (3) squeeze-drying section, and (4) second working section (courtesy of Tetra Pak, *Dairy Processing Handbook*).

vary from 50 kg/hr (110 lb) to 12,000 kg/hr (26,000 lb). These machines are commonly **cleaned-in-place**. Processing conditions that must be controlled to provide uniformity of butter include (1) standardization and uniformity of fat content of cream, (2) accurate temperature control of cream, and (3) uniformity of salt slurry (usually 50:50 salt to water) with maximum particle size of 50 μm diameter.

32.4.3 The Alpha Method

In the reseparation and phase inversion method (alpha process), primary cream (about 40% fat) is reseparated to 80% cream, but no butter granules are formed. Reseparated cream is cooled and worked in a **transmutator**, a device with rotors and spirally arranged ribs inside a jacket-cooled cylinder or cylinders. The 80% cream is a fat-in-serum emulsion in which fat globules are tightly packed, hence deformed. Working disrupts many fat globules and disperses plasma droplets and remaining fat globules in free fat; this is the *continuous* phase. Texture is very close because little air is incorporated. In alpha butter, and butters produced in similar systems that employ phase inversion, there is a dual aqueous emulsion. It consists of serum and cream droplets. The serum droplets are finely dispersed (1 to 7 μm in diameter) but cream droplets are larger and contain fat globules. These result from incomplete phase inversion. Butter from this method contains all membrane material and is comparably high in phospholipids.

32.4.4 Seasonality of Milk

Butter made from cream produced in winter months is often hard, crumbly, and spreads poorly. The problem arises when cows are fed rations that reduce the quantity of unsaturated fatty acids (hence the low iodine number) of milkfat. The melting point of such fat is comparatively high. Therefore, the following procedural modification is important in churning: cream is held overnight at the comparatively high temperature of 50°F (10°C) to avoid excessive crystallization of fat. After churning, wash water at the low temperature of 39°F (4°C) is added and left on the butter about 15 min to solidify fat sufficiently to firm the granules. Prolonged working completely disperses water and salt and breaks fat globules, freeing fat, thereby improving textural qualities. Finally, the butter should be chilled rapidly to prevent the slow growth of crystals because slow crystallization causes formation of large crystals that are detrimental to the body of butter.

32.4.5 Extrusion and Packaging

Once butter has been made, it is extruded and conveyed to packaging. Depending on the type of final product desired, numerous devices have been invented for handling churned butter and transporting it to the packaging area. Here we provide some examples.

1. A handling system that accepts large blocks of cold butter at around 37°F (3°C). The twin auger system processes the blocks and feeds a positive displacement pump, which discharges the butter on site or remotely via pipes (figure 32.3).
2. Blocks of butter are fed into a semiautomatic machine designed for the bulk filling of product in cartons. Digital scales are used for weight control and product is extruded with pneumatic cutters (figure 32.4).
3. Butter is discharged into a silo with a screw conveyor at the bottom. The conveyor then feeds the butter to the packaging machine. Butter silos may be used as a buffer and butter distribution vessel between the churn and packing machines (figure 32.5).

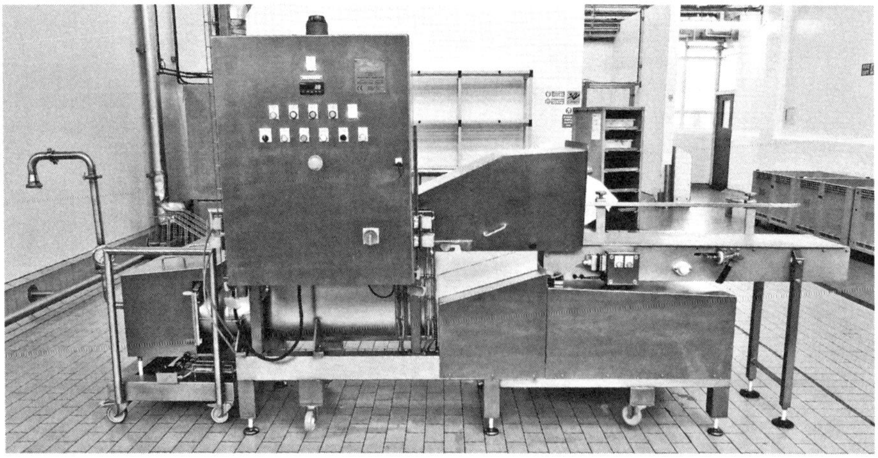

Figure 32.3 Butter pump capable of transforming large blocks of cold butter into a flowing stream and delivering it for further processing or for packaging (courtesy of Best Pump Ltd., Scotland).

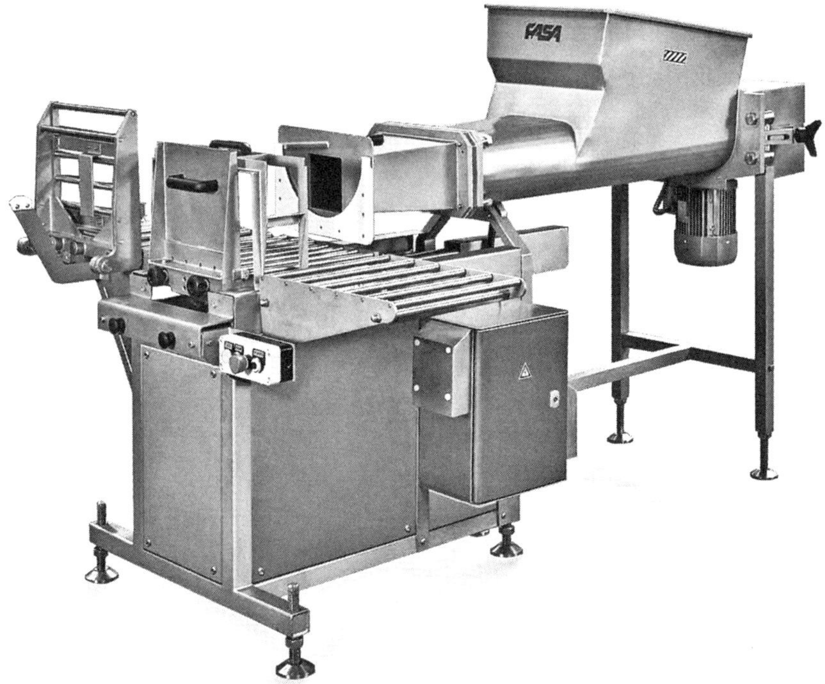

Figure 32.4 A semiautomatic filling machine designed for the bulk filling of butter/margarine in cartons (courtesy of AB FASA, Lithuania).

Figure 32.5 Butter silos are typically used as a buffer and distribution vessel between churns and packing machines. Recommended volume is 50% of churn capacity. Output can be adjusted by a variable speed drive as required by downstream packing machines (from SPX Flow Technology Copenhagen A/S).

As butter is pumped, temperature increases along piping walls producing a thin layer of liquid fat. This liquefied fat prevents interlocking of surfaces when butter is packed. Therefore, it is necessary to work pumped butter intensively in the final stages of packing to disperse liquefied fat. Minimal melting occurs if the temperature of butter being pumped is kept at 59°F (15°C) or below.

Machines for packaging butter are designed to fill various sizes and shapes of prints (figure 32.6), form and seal cups, and packages of modular design.

32.5 Microbiology of Butter

We previously discussed ripened cream butter and the role of *Lc. cremoris* and *Le. cremoris* in producing acid and flavor, respectively, in cream before churning. These organisms are added following pasteurization; therefore, many of them remain in the processed product. In such butter the total plate count of microorganisms will be high, perhaps 10^7/g.

However, in sweet cream butter the total plate count should be low because most of the microorganisms are

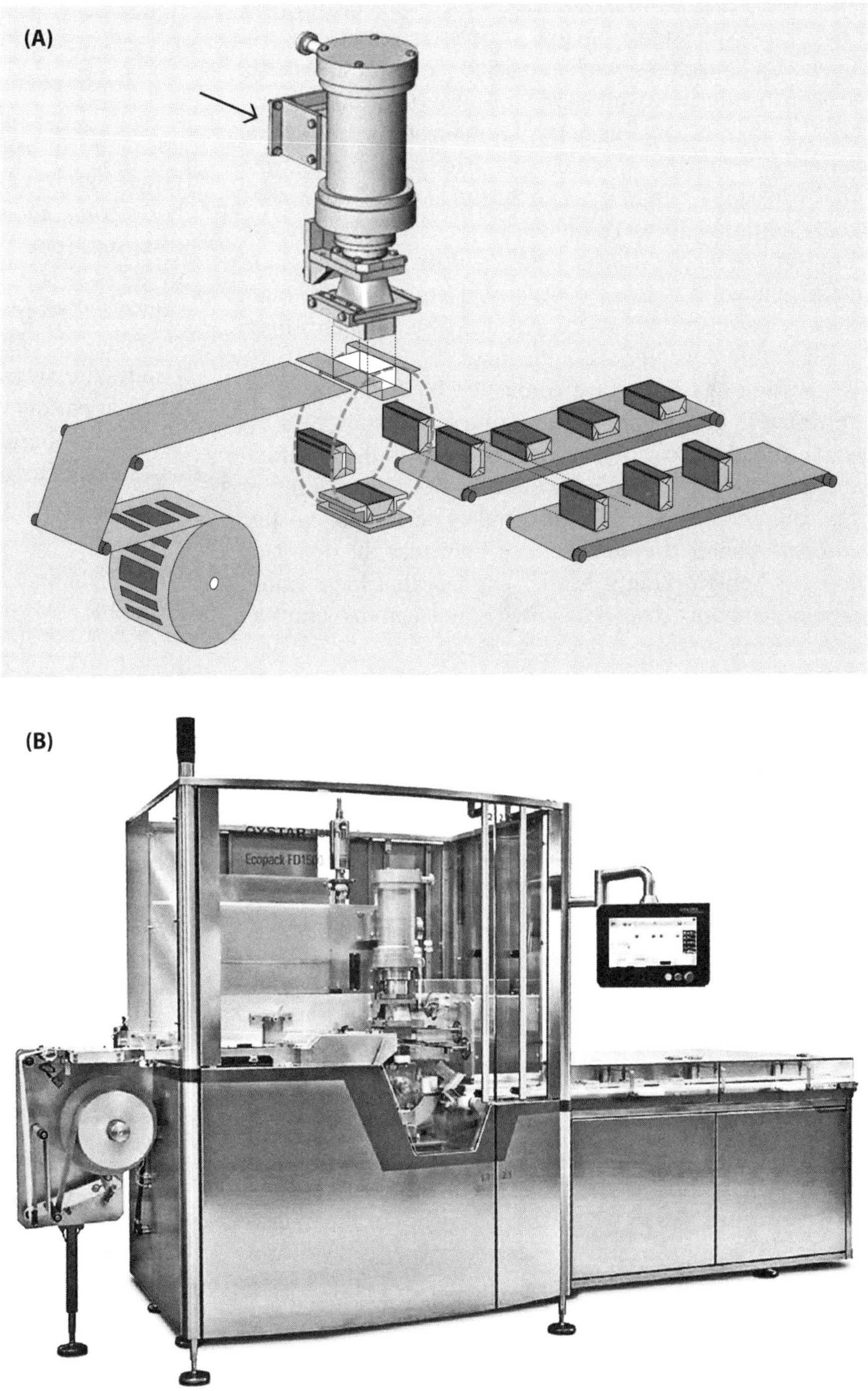

Figure 32.6 Butter wrapping machine: (A) schematic, (B) view of a model in which the dispenser, former, and wrapper are enclosed to protect from contamination (courtesy of IMA Dairy & Food Holding GmbH, Stutensee, Germany).

destroyed by pasteurization. Contamination may come from churns, wash water, personnel, equipment used in handling butter, and packaging materials. The most prominent types of microorganisms causing spoilage are proteolytic and lipolytic bacteria, yeasts, and molds (cf. chapter 23).

To grow in butter, microorganisms must be associated with the serum phase, which comprises less than 20% of the weight. Since this aqueous portion is emulsified in fat, microorganisms are isolated in individual droplets. Each cell is wholly dependent for sustenance on the droplet in which it resides, however, this relationship degrades the fat surrounding the droplet. This means that once the food supply in that droplet is exhausted, or when waste products reach a toxic concentration, growth ceases. Furthermore, all sterile droplets (i.e., those uninhabited by microorganisms) will remain sterile unless the butter is "worked," in which case bacteria are redistributed (this may enhance degradation). Since butter normally contains 10 to 18 billion droplets per gram, most droplets are sterile. Butter tends to flow in an irregular manner when it is worked. At any given moment a high velocity gradient (difference in speed of movement) may occur within a small area of the mass.

Particles of butter are connected by structures of a **thixotropic** character, by interlacing fat crystals, and by water adsorbed to surface layers of fat globules. Mulder and Den Braver (1956) concluded that when working is vigorous the number of large droplets decreases and the number of small droplets increases, whereas the opposite is true when working is gentle. The fact that large water droplets enhance bacterial growth and activity emphasizes the importance of proper working of butter.

Finally, salt is dissolved in the aqueous phase of salted butter. Assuming 2% salt is added to butter with 16% moisture, and that salt is homogeneously distributed, the concentration in the aqueous phase would be in excess of 11%:

$$\text{\% salt in aqueous phase} = [\text{\% salt} \div (\text{\% moisture} + \text{salt})] \times 100$$

Such a concentration of salt is inhibitory to most bacteria. Molds, however, are often salt-tolerant and grow well on surfaces of butter.

It is erroneous to assume uniform distribution of salt and moisture in conventionally churned butter. In fact, two different dispersed aqueous phases have been shown to exist, namely droplets of water and droplets of buttermilk. Hence, we also conclude that there is not a uniform distribution of nonfat milk solids or salt in each droplet of the aqueous phase. These characteristics of the microenvironment are certain to affect microbial growth and keeping quality of butter.

Cold-tolerant (psychotrophic) bacteria in milk and cream may be responsible for fat degradation with consequent appearance of large quantities of free fatty acids in butter. These may be detected by a test for water-insoluble acids.

32.6 Grading and Quality Aspects

Grades of butter are established and awarded by the US Department of Agriculture (USDA). Firms may contract with USDA for grading services that include plant inspection and testing of raw milk and finished products. A survey by USDA shall be made at least twice annually to determine whether facilities, equipment, method of operation, and raw materials being received are adequate and suitable for the grading service (USDA Agricultural Marketing Service, 2012).

To bear the USDA grade shield, butter must be churned from pasteurized cream (from the milk of healthy cows) in an approved plant under sanitary conditions. Grades are established based on characteristics of flavor, body, color, and salt. Scores are determined by subtracting disratings (point deductions) from the total possible score of 93. US grades "AA," "A," and "B" correspond to numerical scores of 93, 92, and 90, respectively. Samples scoring less than 90 points cannot be given a US grade. Grades are awarded at the plant where butter is manufactured or packaged; therefore, consumers depend on others in the distribution chain to ensure rapid movement and assurance of quality. Unfortunately, some wholesalers and retailers fail to appreciate the importance of proper handling, stock rotation, and inventory control. Too many consumers have been disappointed in the quality of butter purchased even though it bears the US grade "AA" label. However, consumers may be responsible for deterioration in quality when they store butter unprotected from volatile foods or leave it at room temperature for an extended time.

The typical flavor of sweet cream butter arises from many different flavorful substances, especially short-chain fatty acids, aldehydes, ketones, and lactones. US grade "AA" butter possesses a fine and highly pleasing flavor, a smooth, creamy texture with good spreadability, and well-blended, completely dissolved salt in a desirable amount (1 to 2%). Unsalted butter may be graded also. US grade "A" butter may possess definite feed and cooked flavors and a slight intensity of acid, aged, bitter, **coarse**, flat, smothered, and storage flavors. No more than 1 point may be deducted on combined scores of body, color, and salt, and then only when flavor is of grade "AA" quality. US grade "B" butter is acceptable to many consumers for table use. It is generally made from sour, rather than sweet, cream. The following off flavors are allowed: (1) in slight intensity—malty, musty, neutralizer, scorched, untensil, weed, and whey; (2) in definite intensity—acid, aged, bitter, smothered, storage, and old cream; and (3) in pronounced intensity—feed. Up to 1.5 points may be deducted for defects of body, color, and salt depending on the extent of flavor disratings.

The USDA Agricultural Marketing Service (1989) makes available the *United States Standards for Grades of Butter*; in addition, the Codex Alimentarius Commission (2010a) has a standard for butter that describes it as containing at least 80% milkfat, no more than 16% moisture, and no more than 2% nonfat milk solids. It also provides a standard for fat and blended spreads (Codex Alimentarius Commission, 2009). European Union standards (EU Council Regulation No. 2991/94) for percentage of milkfat are: 80 to 90% fat for butter, 60 to 62% fat for three-quarter fat butter, and > 29% fat for half-fat butter. Dairy spreads with the same fat contents fall into the same requirement categories.

32.6.1 Types of Defects

Common flavor defects of butter, with descriptions and causes, include:

1. *Acid*—tastes somewhat like cultured buttermilk; results from souring of cream.
2. *Bitter*—astringent to puckery sensation detectable by taste but not smell; associated with protein breakdown.
3. *Coarse*—lacking in delicate flavor or aroma; associated with acid condition but without sourness.
4. *Feed*—aromatic flavor; characteristic of feeds eaten by cows.
5. *Flat*—lacking in typical flavor.
6. *Garlic or onion*—characteristic flavor of wild garlic or onion; caused by consumption of these weeds by cows.
7. *Metallic*—a puckery sensation; caused by metal-catalyzed oxidation.
8. *Musty*—suggestive of the aroma of a damp cellar; odor may be absorbed but may result from consumption of certain feeds by cows.
9. *Neutralizer*—suggestive of bicarbonate of soda; caused by adding neutralizers to sour cream.
10. *Old cream*—lacks freshness with an unpleasant aftertaste; results from using cream in which proteins and other components have been degraded.
11. *Rancid*—high content of short-chain fatty acids from hydrolysis of fat.
12. *Scorched*—excessive concentrations of sulfides and other substances that override desirable, smooth, nutlike flavor; caused by overheating cream in pasteurization.
13. *Unclean*—similar to old cream but more intense; results from degradation of proteins, phospholipids, and other components of cream and butter.
14. *Yeasty*—flavor similar to bakers' or brewers' yeast; results from growth of yeasts in cream or butter.

Important body and texture defects include crumbly, leaky, mealy (grainy), and sticky. When particles lack cohesion, body is said to be crumbly. A leaky body is characterized by visible droplets of moisture on surfaces of sliced or plugged butter.[8] Mealy butter has a granular consistency, which results from abnormal distribution of solid and crystalline fat. When cold butter adheres to instruments used in cutting, sampling, or handling, it is said to be sticky. Ratios of crystalline to solid (amorphous) to liquid fat determine body and texture characteristics. These ratios vary with chemical composition of milkfat, temperature history of cream, churning method, method of packaging, and storage conditions. Generally, less than 20% of the fat is in crystalline form. Composition of crystals is significantly affected by the rate of cooling. In rapid cooling, only two or three types of crystals form. But in slow cooling, several types form because more time is allowed for molecules of triglycerides to become oriented in the **crystal lattice**.

32.6.2 Improving Spreadability

Spreadability of butter has long been a concern of the industry. Efforts to improve spreadability have centered on conditions of storing pasteurized cream prior to churning to provide optimal solidification of fat. However, a more suitable method would be to fractionate the milkfat and add back to cream for buttermaking the fraction with the lower melting point (below 104°F [40°C]). The authors endorse the principle of modifying milkfat in ways that will make the finished product more nearly meet consumer preferences. This includes direct addition of optimal quantities of polyunsaturated fatty acids along with appropriate antioxidants.

It is important to consider that fatty acids with 4 to 14 carbons in their chains are synthesized in the mammary gland from acetate and β-hydroxybutyrate, which are building blocks that are synthesized by bacteria in the rumen. Those with more than 16 carbons are transferred from dietary fat via circulating blood lipids, while those with 16 carbons are partly synthesized in the udder and partly transferred through the bloodstream. This helps to explain why cream from cows fed grass rather than dry forages has a lower melting point and, consequently, butter made from it tends to spread more easily. Furthermore, feeds containing soybeans, corn, linseed, or rapeseed, which are high in unsaturated fatty acids, provide softer milkfat than do feeds containing the more saturated fatty acids, such as coconut or palm oil. Flavors of butter are affected as well by the feeds cows consume and some artisanal buttermakers take advantage of flavor differences in promoting their products.

Recent introductions of butter with added canola or olive oil have provided consumers with butter spreadable from the refrigerator as well as higher ratios of unsaturated to saturated fatty acids.

32.6.3 Antioxidants and Oxidation

Antioxidants would be particularly valuable in increasing the shelf life of salted, cultured cream butter, which readily oxidizes because of interactions between salt and products of culture growth. Ascorbyl palmitate in the concentration of 10 mg/kg (10 ppm) is effective. Glucose oxidase, which is effective also, catalyzes the oxidation of glucose (which is also added) to gluconic acid and in so doing consumes oxygen from butter. Presently, it is unlawful in the United States to add antioxidants to butter or cream for buttermaking. It would be lawful to modify the cream from which butter is made by feeding cows polyunsaturated oils protected from hydrogenation by formaldehyde-treated casein. However, fat in such butter is highly susceptible to oxidation.

Whipped butter has improved spreadability but is susceptible to oxidation because of increased content of air (up to 50%). It must be manufactured from the highest quality butter if it is to retain a pleasing flavor during distribution. To reduce the tendency for oxidation, nitrogen can be introduced through a gas-injecting system during whipping with packaging done in sealed containers. Temperatures of butter at whipping should not exceed 50°F (15°C).

32.6.4 Testing for Contents

The iodine value (IV) test can be useful in estimating amounts of unsaturated fatty acids in milkfat. The higher the iodine number, the more C=C bonds present in the fat. Unsaturated fatty acids (the most important being oleic acid) have a lower melting point than saturated ones. With hard milkfat IV ranges from 26 to 31, while soft milkfat produces an IV of 37 to 42.

Testing for salt in butter can be done by titration of a melted 10 g sample using 0.1N silver nitrate ($AgNO_3$) with potassium chromate as the indicator. Color changes from yellow to brownish at the end point, and percentage of salt = milliliters $AgNO_3 \times 0.1$.

32.7 Color of Butter

The color of butter reflects the concentration of carotene (a yellow, fat-soluble pigment, also a precursor of vitamin A) present in the cream from which butter is made. The color of cream varies with seasonal changes in the carotene content of feeds; it is deep yellow when cows graze or are fed green forages and is pale yellow when dry feeds are fed.

Early buttermakers added carrot juice to feed in winter months to increase the intensity of the color in butter. Today, manufacturers add food coloring throughout the year to assure a more consistent color. Coloring may be an extract of annatto seed or synthetic β-carotene. The latter has vitamin A activity (about 500,000 IU/g in a 30% suspension). Both coloring materials are oil-soluble, therefore, when added to cream before churning, they are not lost in the buttermilk.

32.8 Anhydrous Milkfat, Butter Oil, and Ghee

AMF contains not less than 99.8% fat and no more than 0.1% moisture. It is made from fresh milk in a process involving separation of 40% cream from milk, reseparation of 40% cream to 80% cream, phase inversion of 80% cream at about 160°F (71°C),[9] and concentration of the crude oil to about 98% fat in a centrifugal separator. Serum from the last separator is recycled to recover small quantities of residual oil. Finally, oil from the last separator is homogenized to disrupt fat globule membranes, then injected into a vacuum chamber at 158 to 185°F (70 to 85°C) with about 730 mm Hg for further removal of moisture. This process can be expanded to include a fractionation step to produce more easily spreadable butter. The oil is then standardized with skim milk to the needed fat content before it is emulsified in a high pressure pump and crystallized in a swept surface heat exchanger.

Anhydrous milkfat is particularly suited for recombination in water with nonfat dry milk to make a fluid milk product for locations where sufficient fresh milk production is impractical. AMF has many other applications in both dairy and baked foods and can be fractionated into high- and low-melting components for special uses.

The Codex Alimentarius Commission (2010b) provides the *Codex Standard for Milkfat Products*, which sets a limit of 0.3% free fatty acids, calculated as oleic (C18:1), and maximum amounts of copper and iron—catalysts of oxidation. Intended for further processing or culinary use, it applies to anhydrous milkfat, ghee, butter oil, and similar products. These products are particularly suited for export/import trade.

Anhydrous butter oil (99.8% fat) is produced by melting butter and separating much of the serum by gravity. The crude oil is then separated and vacuum-treated by the same process used for AMF.

Ghee (99.6% fat) is similar to anhydrous butter oil and is characterized by a heated flavor that develops when concentrated fat is heated to drive off moisture, thus "burning" the small amount of nonfat milk solids remaining. In Asia and Africa ghee is made in the home by melting butter, **decanting** the fat following gravity separation from serum, and driving off most residual moisture by heating to near boiling. Processes in manufacturing plants are similar to processes for butter oil except that separated oil is heated in a closed system for 6 to 8 min at 221 to 230°F (105 to 110°C). Similar products made from the milk of goats, sheep, or camels in the Middle East and Africa are known mostly as *maslee* or a variant of the Arabic word *samn*.

32.9 Margarine, Lowfat Dairy Spreads, and Blends with Vegetable Oil

The market for fat- and oil-based food products is highly significant. Milkfat in the form of butter has long been the premier spread in developed countries. However, technological developments in production and processing of vegetable fats and oils have resulted in numerous products similar to butter. We discuss them here.

32.9.1 Margarine

Until recent years margarine commonly has had the same composition as butter, 80% fat and 20% serum.[10] Like butter, hard margarines consist of an emulsion of tiny droplets of serum dispersed in a fat phase, which is in a stable semicrystalline form. Whereas butter is derived from milkfat, margarine is derived primarily from plant oils and fats, however, some animal fats may be added. The serum phase often contains nonfat milk solids. In some locales it is colloquially referred to as *oleo*, short for *oleomargarine.*

Three types of margarine are common: (1) hard margarine that is used for cooking and baking; (2) soft spreads that are high in mono- or polyunsaturated fatty acids and made from safflower, sunflower, soybean, cottonseed, rapeseed, and/or olive oil; and (3) very soft margarine packaged in bottles that is used in cooking or for toppings. Fat content in margarines and vegetable fat spreads can range from 10 to 90% depending on their purpose (spreading, cooking, or baking). Softer tub margarines are made with less hydrogenated, i.e., more liquid, oils than block margarines.

In manufacture, oil is pressed from seeds and refined then blended with solid fat. If no solid fats are added to the vegetable oils, the latter undergo hydrogenation to partially solidify them. This is done by passing hydrogen through oil in the presence of a catalyst under controlled conditions. The addition of hydrogen to the unsaturated bonds (alkenic double C=C bonds) results in saturated C-C bonds. This treatment raises the melting point of the oil, thus "hardening" it. Because there are possible health benefits in limiting the amount of saturated fats in human diets, hydrogenation is limited to that sufficient to provide desired product texture. If hydrogenation is incomplete (partial hardening), the relatively high temperatures used in the hydrogenation process tend to invert some of the C=C bonds into the "trans" form. If these trans fats are present in the final margarine, consumption may pose a risk for cardiovascular disease. For this reason, partially hardened fats have limited use in today's margarine industry. Some tropical oils, including palm and coconut oils, are naturally semisolid and do not require hydrogenation.

In margarine manufacture, the fat and aqueous phases are prepared separately then mixed, emulsified, crystallized, and subjected to extensive mechanical working. In production of blends and blended fat spreads, a mixture of anhydrous milkfat and vegetable oil is normally prepared. The resulting blend is mixed with water, citric acid, carotenoids, vitamins, and milk powder. In all cases a mixture of emulsifiers, such as distilled monoacylglycerides or lecithin, is added to the lipid phase. Stabilizers including starch, gelatin, and alginates are added to the aqueous phase. Additionally, flavors, carotenoids, antioxidants, vitamins, and preservatives, especially sorbates, are added.

An emulsion of the lipid and aqueous phases is created by passage through a swept surface heat exchanger where cooling causes lipids to crystallize. After emulsification and cooling, the mixture is exposed to mechanical working. This is followed by passage through *resting cylinders* in which heat is removed from supercooled fat, permitting crystallization to be completed.

32.9.2 Blends

Many popular table spreads are blends of margarine and butter or other milk products. Blending, which improves the flavor of margarine, has long been illegal in some countries, including the United States and Australia. Under European Union directives a margarine product cannot be called "butter," even if most of it consists of natural butter. In some European countries butter-based table spreads and margarine products are marketed as "butter mixtures."

Butter mixtures now comprise a significant portion of the table spread market. The brand I Can't Believe It's Not Butter! spawned a variety of similarly named spreads marketed internationally with names that include Utterly Butterly, You'd Butter Believe it, and Butterlicious. Marketing such mixtures may avoid labeling restrictions, while strongly implying close similarity to butter. The product marketed as a spread and labeled butter with canola oil is popular in the United States.

Several blends of butter with vegetable fats are being marketed. Nutrition facts labels provide selected compositional data for a one tablespoon serving of butter and related spreads (table 32.2 on the following page). Since the overrun in whipped butter approximates 100%, the analysis is one-half that of butter.

32.9.3 Lowfat Dairy Spreads

Lowfat dairy spreads (serum in fat emulsions) vary in composition but commonly contain about 40% milkfat. When fat content is somewhat less than 40%, keeping quality is impaired because droplets of serum become too large, allowing significant microbial growth. Quantities of NMS or calcium-reduced skim milk powder included in lowfat spread formulas vary from 8 to 17%.

Table 32.2 Nutrient Facts of Butter and Related Butter-Containing Spreads per 1 tbsp Serving (14 g).

Description	Calories	Type of Fat (g)				Cholesterol (mg)
		Total	Saturated	Polyunsaturated	Monounsaturated	
Butter	100	11	7.0	0.5	3.0	30
50/50 blend[a]	100	11	5.0	1.5	3.5	15
Buttery stick[b]	100	11	5.0	1.5	3.5	15
Butter + canola I[c]	100	11	4.0	1.5	5.0	15
Butter + canola II[d]	80	9	3.5	1.5	4.0	15
Honey butter	90	8	4.0	1.5	2.0	15
Garlic butter	90	10	5.0	2.0	2.5	20
Butter + canola (light)[e]	50	5	2.0	1.0	2.5	5

[a] Butter plus palm, soy, flaxseed, and olive oils.
[b] Butter plus palm, soy, flaxseed, and fish oils.
[c] Butter plus canola oil; 14 g/serving.
[d] Butter plus canola oil; 11 g/serving.
[e] Butter plus canola oil plus water, emulsifiers, and other ingredients to reduce fat per serving.

Because of the relatively high water content (up to 45%), there is a tendency for liquid to exude from lowfat spreads. Addition of stabilizers, such as a mixture of carboxymethylcellulose, carrageenan, and locust bean gum, help retain whey. The relatively high water content also means there will be a large amount of surface area at the interface between water and fat. This condition favors oxidation, especially when salt content is high and pH is low. There are three reasons for the potentially high marketability of lowfat dairy spreads, namely (1) their relatively low calorie content, (2) a flavor much like butter, and (3) their competitive price with that of margarine.

32.10 Summary

Butter is composed mostly of fat (at least 80% by law) and moisture (about 16%). Water is emulsified in free fat but some fat is globular, that is, in minute droplets surrounded by membranes of phospholipid and protein. Fat exists in liquid, amorphous solid, and crystalline phases.

Butter normally is churned from cream, which is vigorously agitated to cause fat globules to aggregate. Conventional churning is done by the batch method or by continuous churning in larger plants. Most continuous churns employ the Fritz process, which depends on agitation, rather than the alpha or similar processes, which depend on separation of cream to about 80% fat followed by phase inversion.

Butter keeps well, especially when frozen. Microorganisms are associated with primarily the aqueous phase. If droplets of serum or water are limited in size to a few μm diameter, they provide little volume or food material for microbial growth. Additionally, most droplets are sterile in butter that contains tiny droplets. Frequently off flavors develop in butter because of oxidation, but it is unlawful to add antioxidants.

In the United States grades of butter are determined by inspectors of the USDA. The US grade "AA" label symbolizes supreme quality when awarded to butter in possession of the packer; however, quality can deteriorate during passage of butter through distribution channels and in the user's possession.

Study Questions

1. Explain the theory of churning.
2. Discuss the physical state of fat in cream and of serum in butter.
3. What is phase inversion? How is it accomplished when making butter processed with the alpha method?
4. What is overrun in butter? How much overrun should processors expect?
5. Explain how number and size of droplets in the aqueous phase affect keeping quality of butter.
6. How does the temperature history of cream influence the rate of churning and the body and texture of finished butter?
7. Compare the composition of butter, margarine, and lowfat dairy spreads.
8. What role does the USDA play in grading butter?
9. What is cultured cream butter? Whey cream butter? Sweet butter?
10. Why might butter oil and anhydrous milkfat be manufactured for export rather than butter?

Notes

1 Flavorful components include free fatty acids, methyl ketones, lactones, diacetyl, organic sulfides, aldehydes, and ketones.
2 Butter and eggs were used by farmers as barter for groceries.
3 Serum is the water and nonfat solids portion; it is sometimes referred to as plasma.

[4] Higher temperatures are required because of the insulating effect of fat that protects microorganisms from destruction by heat.

[5] The standard of identity for butter in most important dairy countries limits water content to 16.0%. This allows a single test—the moisture test—to suffice. However, the economic value of fat has caused American manufacturers to retain the standard based on fat content of 80%.

[6] Moisture is usually determined by heating a 10 g sample in a weighed pan to drive off moisture then reweighing the sample. Fat may then be removed with petroleum ether leaving salt and curd, which are weighed. Salt is dissolved and titrated. The percentage fat is calculated by subtracting the percentage of moisture, salt, and curd from 100.

[7] In continuous churns employing the Fritz process, these same factors play similar roles.

[8] The cylindrical sample withdrawn from butter with a trier is called a plug. This method is used in sampling butter for tests and for sensory evaluation.

[9] Phase inversion may be accomplished in a clarifixator, a modified homogenizer, or by reverse flow through a centrifugal pump.

[10] The technique for making margarine was developed in 1869 by French chemist Mège Mouriés. He used the semisoft fraction of beef tallow obtained by slow crystallization. A 30% emulsion was made in skim milk (including some macerated fresh udder tissues) then churned, as cream would be churned to butter. For many years animal fats, especially lard, were used in margarine. Now principally vegetable oils are used.

References

Codex Alimentarius Commission. 2009. *Standard for Fat Spreads and Blended Spreads* (256-2007). Rome, Italy: Author.

Codex Alimentarius Commission. 2010a. *Codex Standard for Butter* (279-1971). Rome, Italy: Author.

Codex Alimentarius Commission. 2010b. *Codex Standard for Milkfat Products* (280-1973). Rome, Italy: Author.

Fuquay, J. W., P. F. Fox, and P. L. H. McSweeney (eds.). 2011. *Encyclopedia of Dairy Sciences* (2nd ed.). San Diego, CA: Academic Press.

- Frede, E. "Properties and Analysis" (pp. 506–514).
- Mortensen, B. K. "The Product and Its Manufacture" (pp. 492–499).
- Mortensen, B. K. "Modified Butters" (pp. 500–505).
- Mortensen, B. K. "Anhydrous Milk Fat/Butter Oil and Ghee" (pp. 515–521).
- Mortensen, B. K. "Milk Fat-Based Spreads" (pp. 522–527).

King, N. 1953. The theory of churning. *Dairy Science Abstracts* 15:591.

Mulder, H., and F. C. A. Den Braver. 1956. The working of butter IV: Application of the theory on the working of butter to workers of different types and to printing machines. *The Netherlands Milk and Dairy Journal* 10:230.

USDA Agricultural Marketing Service. 1989, August 31. *United States Standards for Grades of Butter.* Washington, DC: Author.

USDA Agricultural Marketing Service. 2012, June 29. *General Specifications for Dairy Plants Approved for USDA Inspection and Grading Service.* Washington, DC: Author.

Websites

American Butter Institute (http://nmpf.org/ABI)

Codex Alimentarius Commission (http://www.codexalimentarius.org)

Council of the European Union, Council Regulations (http://eur-lex.europa.eu/homepage.html)

USDA Agricultural Research Service, National Nutrient Database for Standard Reference (http://ndb.nal.usda.gov/ndb/search/list)

33

Concentrated and Dried Milk Products

Concentrated and desiccated milks provide
both assurance and economy of the consumer's milk supply.
Paul G. Heineman

33.1 Introduction

As in much of the food industry, economics largely determine the use of milk in the fresh fluid form or in some concentrated and preserved form. Concentrated and **dried milk** products occupy less space, weigh less, and remain edible longer than fresh milk. Hence, they save storage and packaging space, cost less to transport, and serve as a reserve in times of short supply. Additionally, they have functional properties that may make them uniquely useful, as in the use of nonfat dry milk in dry cake mixes. Concentrated sources of milk solids are required for formulating numerous foods that would be diluted by a less concentrated form of milk.

Manufacture of many of these products primarily involves removal of water (moisture), a seemingly simple operation. However, the complex nature of raw materials used, their sensitivity to heat, and their tendencies to participate in chemical and physical reactions provide significant areas for development by dairy and food scientists and engineers.

A comprehensive review of the history and development of the condensed and evaporated milk industry can be found in *Condensed Milk and Milk Powder* by O. F. Hunziker (1946).

33.2 Definitions and Terminology

An understanding of terminology used in the industry is vital. Whereas products discussed in this chapter are concentrated, that is, they contain more solids than milk or other raw material, the generic name **concentrated milk** (alternatively *condensed milk*) refers to those products from which sufficient water is removed to concentrate the milkfat (MF) to at least 7.5% and total milk solids (TMS) to at least 25.5% (21 CFR 131.115). Concentrated milk is pasteurized to prevent spoilage but is not sterilized. It is preserved by refrigeration as is pasteurized milk, but requires less space and costs less to distribute. Most concentrated milk is referred to in the industry as *plain condensed milk* and is used as an ingredient in manufacturing other products (table 33.1).

Condensation is the process of changing a material to a denser form; however, the generic name *condensed* applied to milk refers specifically to sweetened concentrated milks. Sugar is the ingredient that preserves **sweetened condensed milk** because of the high osmotic pressure it produces when added in high concentrations. This facilitates storage of the product at room temperature.

Evaporation is the process whereby water is vaporized and removed from milk and milk products. Evaporation results in condensation or concentration of milk products. However, the generic name **evaporated milk** is reserved for the product made from homogenized milk by concentrating with heat and vacuum to obtain at least 23% TMS and 6.5% MF and by sterilizing in the can or before canning it aseptically. Vitamin D is added in such quantity that each ounce of product contains 25 IU/fl oz (within limits of good manufacturing practice). Addition of vitamin A (125 IU/fl oz) is optional, as is addition of emulsifier and stabilizer. The high-heat treatment given evaporated milk sterilized in the can produces a cooked caramelized flavor, a brownish tinge of color, and soft, fragile coagulum in the stomach upon ingestion. **Ultra-high temperature (UHT)** sterilized and aseptically canned evaporated milk has much less caramelized flavor and is lighter in color. High-quality evaporated milk can be stored several months at room temperature without significant deterioration.

Dry milk products are generally less perishable than **concentrated milks**. However, those containing considerable milkfat tend to oxidize readily unless specially protected. In the United States they may be made from grade "A" or manufacturing grade milk, but grade "A" raw materials are required if dry products are to be added in the manufacture of products that bear the grade "A" label. Commonly produced dry milks are **nonfat dry milk (NDM)** (also called skim milk powder), dry milk, dry buttermilk, dry whey, and dry casein. The following safe and suitable optional ingredients may be used in dry whole

Table 33.1 Generic Names, Composition, and Storability of Selected Concentrated and Dry Milk Products.

			Typical			
Generic Name of Milk Product	**Fat (%)**	**Total Solids (%)**	**Protein (%)**	**Lactose (%)**	**Moisture (%)**	**Storage Time**
Concentrated	7.5[a]	25.5[a]	6.0	9.0	74.5	15 days (refrigerated)
Evaporated	6.5[a]	23[a]	5.0	8.0	77.0	several months
Sweetened condensed	8.0[a]	28[a]	6.0	20	28	several months (canned)
Nonfat dry	< 1.5[a]	> 95[a]	31	48	< 5.0[a]	several months
Dry whole	26–40[a]	> 95	22[b]	35	< 5.0[a]	few months (gas packed)
Dry whey	0.2–2.0[a]	> 92[a]	12	61–75[a]	1–8[a]	several months
Dry casein	1.5	89.5	86.5	Trace	< 9	several months

[a] As required by 21 CFR 131 and 21 CFR 184.
[b] Based on 30% milkfat content and 3.1% protein in base milk.

milk and dry cream: carriers for vitamins A and D, emulsifiers, stabilizers, anticaking agents, antioxidants, and characterizing flavoring ingredients (21 CFR 131.149). Dry cream contains not less than 40% but less than 75% by weight of milkfat and may contain nutritive sweeteners. Water content of each of these dry products is limited to 5%.

Nonfat dry milk and skimmed milk powder (SMP) are very similar. Both contain, by weight, 5% or less moisture and 1.5% or less milkfat. The difference is that SMP has a minimum milk protein content of 34%, whereas NDM has no protein content standard. These products are classified for use as ingredients according to the heat treatment used in their manufacture. There are three main classifications: high heat (least soluble), medium heat, and low heat (most soluble). They are available in roller-dried and spray-dried forms, the latter being most common. Spray-dried products are available in two forms: ordinary (nonagglomerated) and **agglomerated** (instant).

33.3 Concentrating and Separating Milk and Its Derivatives

New technologies have been applied to the separation, processing, and use of **whey**, creating a high demand for its products. Two of the major processes are discussed in this section: filtration and electrodialysis.

33.3.1 The Filtration Process

Filtration is accomplished using semipermeable membranes, which allow various constituents of milk to pass through the membrane pores as pressure is applied to one side of the membrane. Components retained on the high pressure side of the membrane are called **retentate**; those passing through are called **permeate**. Figure 33.1 illustrates the general concept of filtration with four types of membranes: microfiltration, ultrafiltration, nanofiltration, and reverse osmosis. When describing the type of particulates that have passed through, it is customary to refer to the molecular weight cut off (MWCO) rather than particle size.

Performing membrane filtration at cold temperatures reduces microbial growth. Furthermore, producing casein concentrates (as retentates) from milk at different separation temperatures alters the protein profile. At cold temperatures β-casein dissociates from casein micelles so some of it passes into the serum (permeate) phase. Casein concentrates separated at various temperatures have different functional properties useful in specific food applications. Milk serum protein concentrates have favorable properties (especially color and flavor) compared with whey protein concentrates.

Filtration systems are designed to produce flow across membranes (cross-flow filtration), which creates turbulent flow to minimize concentration of retained particles at the membrane surface, thus keeping pores open for permeation.

Microfiltration is a low pressure (10 to 100 psig [6.9 to 69 kP]) process for separating larger size solutes from aqueous solutions by use of a semipermeable membrane with pores of 0.1 to 3 μm diameter. This process involves solution flow across a membrane surface under pressure. Retained solutes (such as particulate matter) leave with the flowing process stream without accumulating on the membrane surface. This process can be used to separate serum proteins and micellar caseins of skim milk (Hurt et al., 2010). Some κ-casein is found in MF permeates, likely due to heat-induced complexes of κ-casein with β-lactoglobulin as milk is heated before the MF treatment.

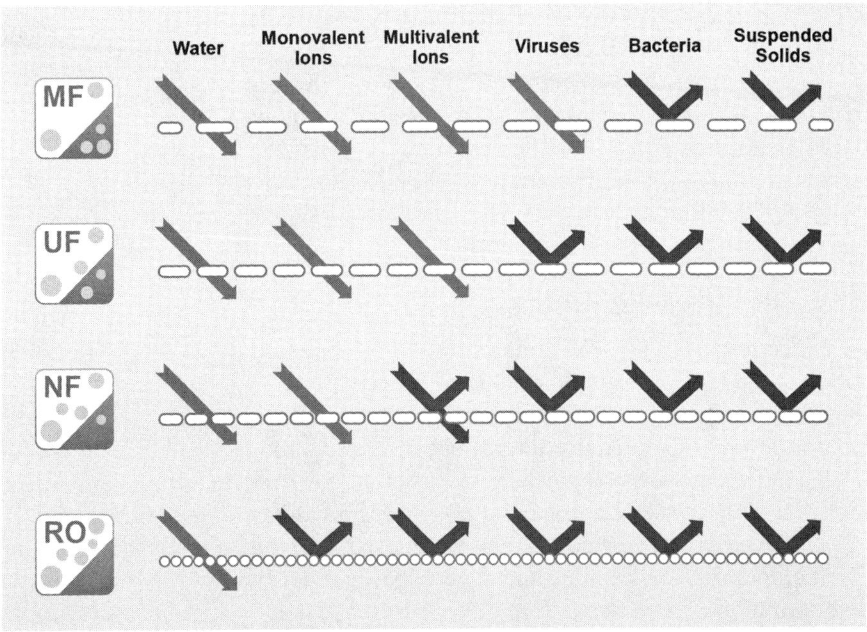

Figure 33.1 Types of semipermeable membranes used in processing dairy liquids. As pore sizes of membranes decrease, from top to bottom in this illustration, successively fewer particles are able to pass into permeate (figure courtesy of Koch Membrane Systems, Inc. [www.kochmembrane.com] and used by the publisher with permission).

Ultrafiltration (UF) is a low pressure (5 to 150 psig [3.5 to 105 kP]) process for separating larger size (0.5 to 5 µm) solutes from aqueous solutions by means of a semipermeable membrane. The membranes with pore sizes ranging from 0.005 to 0.1 µm retain lipids, proteins, bacteria, and suspended solids while allowing lactose, soluble salts, and acids to pass through. Ultrafilters retain particles and molecules that range from about 10^3 to 10^6 Da. They are used to preconcentrate milk for cheesemaking and to separate proteins of cheese whey (cf. chapter 30). Milk protein concentrate (MPC) is produced by removing some of the lactose from skim milk by UF before drying.

UF is widely used to concentrate proteins, but fractionation of one protein from another is much less common. Bhushan and Etzel (2009) examined the use of positively charged membranes to increase the selectivity of UF and allow fractionation of proteins from cheese whey. By adding a positive charge to UF membranes and adjusting the solution pH, it was possible to permeate proteins having little or no charge, such as glycomacropeptide, while retaining proteins having a positive charge. Placing a charge on the membrane markedly increased its selectivity.

Wide-pore UF membranes can be used to produce whey protein concentrate enriched with α-lactalbumin. Casein fines, lipids, and aggregated proteins are first removed by a larger-pore microfilter. The resulting prepurified whey is then processed using a polyvinylidene fluoride membrane (MWCO 50 kDa) operated at transmembrane pressure of 207 kPa.

Reverse osmosis (RO) is used in the dairy industry for production of whey protein powders, for purifying waste water and evaporator condensate, and for concentrating milk to reduce shipping costs. Pores in these membranes are very small (3 to 15 angstroms) and pressures are high (600 to 1,000 psi [4.1 to 6.9 mP]). With RO, whey is concentrated from its usual 6% total solids to 10 to 20% total solids before UF processing. UF retentate can then be used to make various whey powders, including whey protein isolate (which is used in bodybuilding formulations). Additionally, the UF permeate, which contains lactose, is concentrated by RO from 5% to 18 to 22% total solids, thus reducing costs of lactose crystallization and drying to a powder.

33.3.2 Electrodialysis Process

Electrodialysis is used to transport salt from one solution, the diluate, to another solution, concentrate, by applying an electric current. Inside an electrodialysis cell solutions are separated by alternately arranged **anion** exchange membranes, permeable only for anions, and **cation** exchange membranes, permeable only for cations. Electric current is applied as salts are moved over the membranes. The concentrate and diluate are separated by the membranes into the two respective process streams.

33.3.3 Whey and Its Products

The ability to concentrate and fractionate whey proteins into their valuable components, combined with research showing specific functionalities of these components, has provided new opportunities to use whey for humans. In earlier years whey was often fed to animals or spread on fields as a fertilizer.

Mechanical separation of whey from cheese results in high fat content in whey protein concentrate (WPC), thus limiting the maximum protein content in the final WPC powder to about 80%. Whey protein isolates (WPI), with protein concentrations above 90%, require reduction of fat content in the final product to < 0.5%. This can be accomplished by microfiltration (MF) with either ceramic or spiral wound membranes (figure 33.2).

Whey protein typically comes in three forms: concentrate, isolate, and hydrolysate. *Concentrates* (29 to 89% protein) contain small amounts of fat but higher concentrations of bioactive compounds and lactose than isolates. *Isolates* (90%+ protein) are processed to remove the fat and lactose. *Hydrolysates* are partially hydrolyzed whey proteins that, as a consequence, are easily absorbed, but their cost is comparatively high. Highly hydrolyzed whey may be less allergenic than other forms of whey. Hydrolysates may be very bitter in taste. Development of astringency in whey proteins is a complex process determined by the extent of aggregation occurring in the mouth, which depends on the whey protein beverage pH and buffering capacity in addition to saliva flow rate (Kelly et al., 2010). Vardhanabhuti et al. (2010) concluded that interactions between positively charged whey proteins and salivary proteins play a role in astringency of whey proteins at low pH.

Whey permeate powder ranges in composition as follows: 65 to 85% lactose, 8 to 20% ash/minerals, 3 to 8% protein, and a maximum of 1.5% fat. This product is used in infant formulas, bakery products, confectionary bars, milk drinks, and dessert products. It functions as a replacer of sodium and as a flavor enhancer in certain applications.

33.4 Common Manufacturing Processes

In many respects basic treatments in processing concentrated and dried milks are alike. Therefore, we discuss the common ones here.

33.4.1 Standardization

Each product has its own unique composition, making standardization of major components necessary. Standardization usually involves only addition or removal of fat. In whey products, however, lactose, minerals, or protein may be selectively removed prior to or during concentration. The ratio of fat to total solids is particularly well controlled in production of evaporated milk. It should be 1.0:3.4. Special care is taken to remove practically all fat in skim milk or whey to be dried (figure 33.3 on the following page).

Since milk proteins have exhibited increasingly higher values, it may be desirable to standardize protein content of fluid milk products. This can be done by producing protein-rich and protein-free streams by ultrafiltration then recombining those streams in desired ratios.

33.4.2 Pasteurization and Forewarming

The degree of heating prior to concentration varies with product and process. Minimal treatment is pasteurization, the purpose being to kill pathogenic bacteria and to deactivate enzymes. However, additional heat treatment is justified to activate bacterial spores, to assist in stabilizing proteins on sterilization, and to denature whey proteins. Bacterial spores are stimulated to **germinate** by heating at about 176°F (80°C). Spores that are germinating, in contrast to resting ones, are as susceptible to destructive effects of heat as are **vegetative** cells. This lessens the heat treatment required in sterilization, hence it is vital in production of evaporated milk.

Preheating and *forewarming* are synonymous terms used to describe heating prior to another treatment, such as before evaporation or pasteurization. In producing concentrated milk products, forewarming temperatures and times vary widely from 149 to 284°F (65 to 140°C) for a few seconds to 158 to 194°F (70 to 90°C) for 10 to 30 min for UHT-sterilized milk and sweetened condensed milk. Too high a forewarming treatment results in thickening of sweetened condensed milk on prolonged storage. Forewarming to stabilize proteins is of little significance in concentrated milks.

Forewarming in evaporated milk requires temperatures of about 284°F (140°C) for 25 sec, 248°F (120°C) for 3 to 4 min, or 194 to 212°F (90 to 100°C) for 10 to 25 min to impart stability in storage and desirable viscosity: Viscosity may be excessive if forewarming time or temperature is too low. Forewarming must precede evaporation because concentrating the milk lowers its stability to heat. Normally **tubular heaters** are used. After heating,

Figure 33.2 (A) Ceramic (inorganic materials) and (B) spiral wound (organic materials) membranes are applied in filtration processes (figure courtesy of Koch Membrane Systems, Inc. [www.kochmembrane.com] and used by the publisher with permission).

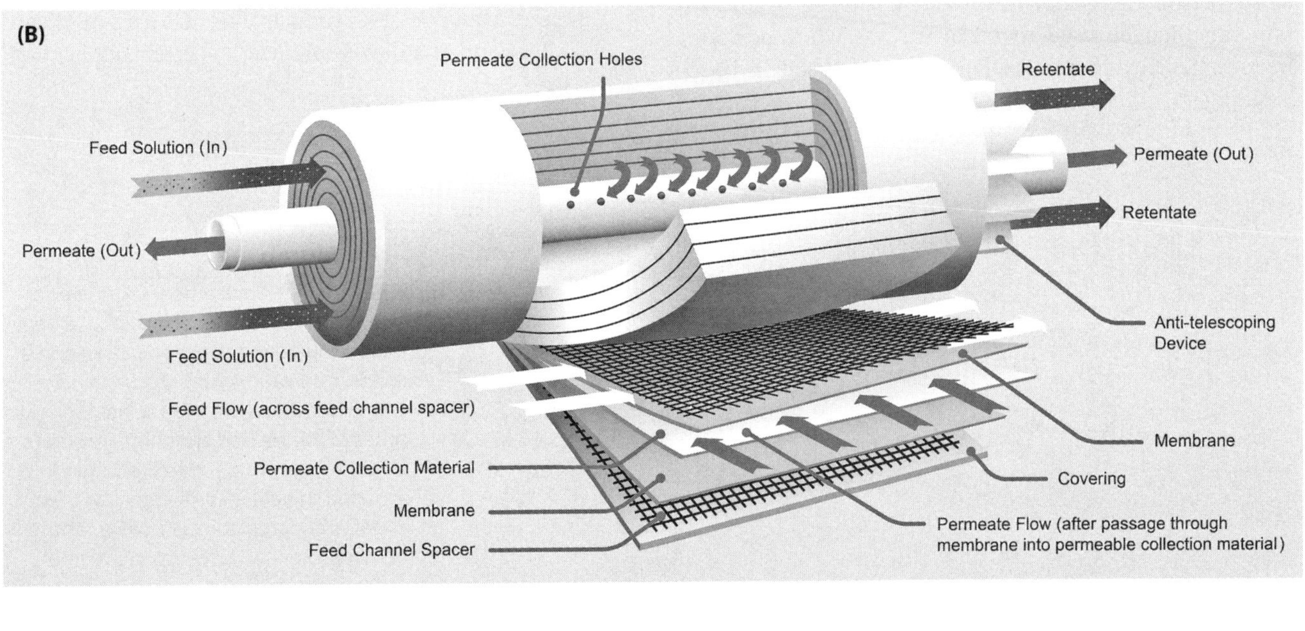

Figure 33.3 A series of separators used to produce skim milk for drying (from GEA Westfalia Separator, Ireland Limited).

the milk may be held in hot wells to provide the required holding time before evaporation.

Usually milk coagulates slowly during heating at high temperatures. The gel structure that may form if milk is not agitated during heating results from aggregation and hydration of caseinate micelles. Furthermore, whey proteins are denatured and β-lactoglobulin becomes associated (complexed) with casein. Soluble phosphates and citrates are converted to colloidal forms. Resistance to heat coagulation is termed *heat stability*, which is expressed as time required for milk to coagulate when heated quiescently at a given temperature. Heat stability is influenced dramatically by seasonal variations in the citrate and phosphate content of milk. Small amounts of disodium phosphate or sodium citrate may be added to compensate for seasonal variations, thus improving heat stability. Appropriate amounts are determined by pilot sterilization.

33.4.3 Condensing, Evaporating, or Concentrating

The fundamental objective in concentrating milk is removal of sufficient moisture while maintaining sensory attributes similar to those of pasteurized milk. Boiling milk at atmospheric pressure results in the vaporization of moisture, but flavor and color are seriously damaged. Lowering pressure on milk reduces its boiling point and allows concentration at a temperature that provides suitable quality. Moisture can be removed from the frozen product (freeze-drying), but the cost is much greater than in conventional high-temperature operations. Therefore, the temperature employed is largely determined in a compromise between cost of moisture removal and degree of heat damage.

Rapid conversion of a liquid to vapor throughout the fluid can be achieved by depressing pressure of the atmosphere above the liquid slightly below its vapor pressure. The temperature at which boiling occurs can be controlled easily by regulating air pressure above the liquid. Boiling will, of course, cease when vapor pressure of water in the air equals that of the bulk phase. Therefore, water vapor must be removed for boiling to continue.

Evaporator functions are to provide efficient heat transfer, vapor removal, and separation of solids entrained in vapor. Evaporation occurs when molecules of water obtain sufficient energy to escape as a vapor from the solution. Rate of vaporization depends on temperature of the liquid, pressure upon it, and rate of escape of moisture from the headspace. Commercial evaporators operate at temperature and vacuum levels from 100°F (38°C) and 730 mm Hg to 180°F (82°C) and 380 mm Hg. In multiple-effect evaporators a series of temperature and vacuum levels is used. Evaporator design varies widely but several components are basic to each. These are (1) heating surface, (2) vapor space, (3) vapor separator, (4) vacuum pump, and (5) condenser (figure 33.4).

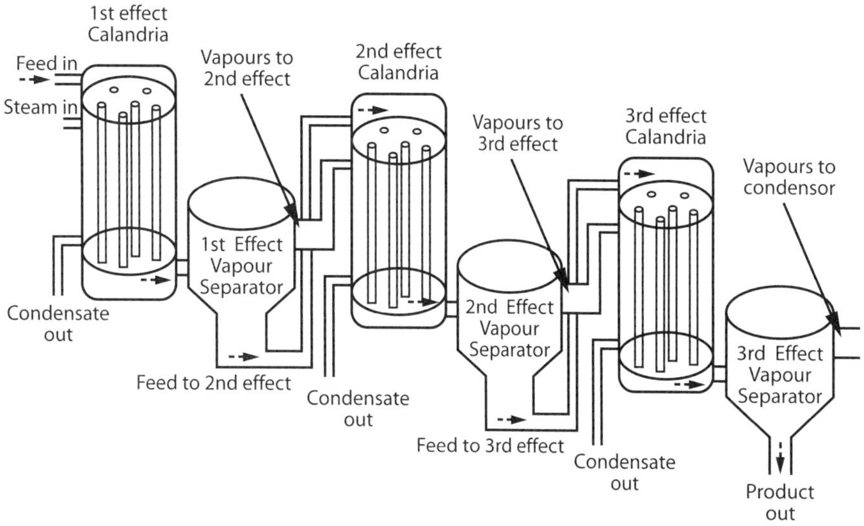

Figure 33.4 Triple effect falling film evaporator. Each effect consists of a heat transfer surface, a vapor separator and a condenser. Vapors from the first and second evaporators are used to heat milk in tubes of the second and third evaporators, respectively. Vapors from the third effect are used to heat incoming milk in the vapor preheater (courtesy of Department of Food Science, University of Guelph).

Heating surfaces may consist of tubes inside steam chests, called calandria, or a series of heat exchanger corrugated plates. Surface area inside an evaporator is large; for example, tubular steam chests may have as many as 500 heating tubes as long as 45 ft (8 m) and 3 in (7.6 mm) diameter. Less space per unit volume is required by plate-type than by tube-type evaporators. Vapor space should be large enough to allow separation of most milk solids from the vapors. Further separation is achieved with the vapor separator in which entrained solids are caused to impinge on a surface from which they drain. Impingement is usually achieved by centrifugal force. More solids are entrained as the rate of evaporation increases. Recovery of these solids is necessary to achieve economy and to avoid contamination of condensed vapors.

The condenser converts vapors back to liquid form and cools noncondensable gases so they occupy minimal volume. Water sprays are employed in most condensers for cooling. Spray (jet) condensers use cooling water equivalent to 20 to 50 times the weight of the vapor. Condensation makes possible quick removal of vapors, thus preventing buildup of pressure in the vapor space. It also reduces the load on the vacuum pump. Where used, vapor preheaters condense part of the vapors, thus heat is reused and quantity of cooling water required is reduced. Cooling water, condensed vapors, and noncondensable gases may be removed by a single pump (wet condenser) or water may be removed by one pump and noncondensables by another (dry condenser). Water may also be removed by a barometric leg.[1]

Vacuum may be produced by reciprocating positive-type pumps or by steam ejectors. The latter are used when high vacuum (above 610 mm Hg [24 in]) is required. Vacuum up to about 757 mm Hg (29.8 in) may be produced using steam ejectors in three stages.

Vapor from a vacuum pan (calandria) contains considerable latent heat.[2] Therefore vapors from one vacuum pan can be passed to the exterior of the heating surfaces of another to heat incoming product (see figure 33.4). Of course, temperature of product in the second effect (evaporator) is lower than that in the first. Therefore, a higher vacuum must be employed to vaporize moisture from it. A temperature drop of 45 to 46°F (7 to 8°C) is necessary to justify the second effect. When four evaporators are placed in a series, operating temperatures in first, second, and third effects approximate 140 to 171°F (60 to 77°C), 126 to 145°F (52 to 63°C), and 100 to 122°F (38 to 50°C), respectively.

The major advantage of multiple-effect evaporators is that less steam is required for each unit of water evaporated. Average quantities of steam required per pound of water evaporated in single-, double-, triple-, and quadruple-effect evaporators are 1.2, 0.6, 0.4, and 0.3 lb, respectively. Also, since vapors from the first effect are condensed by milk in the second (in a double-effect evaporator), much less water is used for condensing compared with a single-effect evaporator. Investment costs for multiple-effect evaporators preclude their use except for high-volume installations. Steam may be saved by compression of vapors from an evaporator. These vapors are used in heating milk inside the evaporator. Energy is added by steam jet (thermal vapor recompression) or a mechanical compressor (mechanical vapor recompression). The objective is to increase temperature of the vapor to a point above the temperature of condensation so that latent heat can be removed. The choice of type depends on comparative prices of electricity for running the mechanical compressor versus gas for making steam.

33.5 Concentrated Milks

The idea for a portable canned milk product that would keep indefinitely came to Gail Borden during a transatlantic trip on board a ship in 1852. Cows in the hold became too seasick to produce milk and an immigrant infant died from lack of milk. Borden's first condensed milk product (1854) spoiled after three days. He first thought the condensing process made milk more stable, but later realized heat killed the spoilage bacteria. Borden then added sugar to inhibit bacterial growth and was granted a patent for sweetened condensed milk in 1856. He began commercial production in 1857. His product was not initially well-received. However, the US government ordered huge amounts of it for Union soldiers during the American Civil War.

It was John Baptist Meyenberg who first suggested canned evaporated milk to his employers at the Anglo-Swiss Condensed Milk Company in Switzerland in 1866. Since the company already successfully produced sweetened condensed milk, it rejected the proposal. Meyenberg immigrated to the United States and began his own company, Helvetia Milk Condensing Co. (Pet Milk), eventually marketing unsweetened condensed milk in 1890.

Borden added evaporated milk to his product line in 1892. In 1899 Elbridge Amos Stuart developed a new process for canned, sterilized, evaporated milk and, with help from evaporated milk pioneer Meyenberg, he began its successful mass production. Evaporated milk manufacturers pioneered the use of homogenization (cf. chapter 26), but dairies producing fresh milk were slow to adopt the process.

33.5.1 Manufacturing Concentrated Milks

Manufacture of concentrated forms of milk, skim milk, buttermilk, whey, and their sour counterparts is relatively simple, involving the three basic steps described above: standardization, forewarming, and concentration. If concentrated 2:1, addition of an equal part of water restores the concentrate to the composition of fresh product.

Either refrigeration or high acidity can serve to preserve concentrated milks. Concentrated (condensed) sour skim milk, buttermilk, and whey contain up to 5% lactic acid, which prevents bacterial growth. Packaging in airtight containers inhibits growth of molds and yeasts, which are aerobic. Because of the tendency for lactose to crystallize in ice cream, partially delactosed condensed skim milk is sometimes produced for use in frozen desserts. Lactose content can be reduced by (1) crystallizing and separating the crystals, (2) enzymatically hydrolyzing lactose to glucose and galactose, or (3) by ultrafiltration.

Milk and milk components can be concentrated by membrane processes such as reverse osmosis and ultrafiltration (cf. section 33.3). For example, serum protein concentrate is composed of whey proteins concentrated directly from skim milk by UF. It has the basic composition of whey protein concentrate, which is made from cheese whey by UF. Each of these concentrates can be spray dried and are commonly sold with 80% protein content.

33.5.2 Evaporated Milk

Because of its concentrated form, evaporated milk is a multipurpose, convenient dairy product ready for a variety of uses, including creaming coffee or tea, pouring on cereals and fruits, providing consistency in meat patties and loaves, coating baked or fried meats, or in place of milk in the manufacture of candies, frostings, and pies.

Traditionally, the term *evaporated milk* has been reserved for concentrated milk sealed in a can then sterilized by heat. However, standards of identity allow sterilization prior to canning, which must be done aseptically. It is permissible to add emulsifiers and stabilizers, including disodium phosphate, sodium citrate (up to 0.1% of either or both), and **carrageenan** (up to 0.01%) (21 CFR 131.130).

Stabilizers help prevent precipitation of proteins in cans of evaporated milk during storage. Imbalance between the cations (calcium and magnesium) and the anions (phosphate and citrate) contributes to instability. Therefore, 1 to 6 oz (28 to 170 g) of phosphate or citrate per 1,000 gal (3,785 L) of milk are added. Lactic acid in raw milk destabilizes proteins on heating, therefore **titratable acidity** should not exceed 0.18 to 0.22% depending on the content of nonfat solids, which buffer milk, thus producing apparent acidity (cf. section 25.4.1). The alcohol stability test, as employed to determine protein stability, requires the addition of 70% alcohol to an equal volume of milk with vigorous mixing. Coagulation indicates instability to heat of sterilization. Both high concentrations of milk solids and high pressures of homogenization lead to instability, but neither is of practical importance to evaporated milk. Casein contributes to increased viscosity as the coagulation point of milk is approached. The rate of thickening is greatest during the last few minutes of heating. However, agitation counteracts thickening if done as coagulation commences. Viscosity of evaporated milk sometimes decreases during storage. Addition of carrageenan greatly retards separation of fat, protein, and insoluble salts and helps maintain uniform viscosity during storage. It and most stabilizing salts (if expected to be needed) are added before forewarming.

Procedures for processing evaporated milk are outlined in figure 33.5. Milk is commonly concentrated slightly more than is required in the finished product. Concentration can be determined by a hydrometer that reads degrees Baumé. The concentrate is homogenized to reduce the fat globule diameter to 1 μm or less. This prevents fat separation during storage and increases in viscosity.

Restandardization to 6.52% milkfat and 23.1% total solids is accomplished in large holding tanks. Here, also, vitamin concentrates and any additional salts needed (as determined by pilot test) are added. Cans may be filled through a small hole in the center of the lid, which is then sealed with solder, or they may be filled before the lids are affixed. Sealed cans are checked by machine for leaks and for loose pellets of solder.

Sterilization by the in-can process is accomplished as cans pass continuously through a **retort**. Spinning motion of the cans improves heat transfer and reduces burn-on inside the cans. Heat treatment necessary for sterilization depends on type and number of spore-forming bacteria and varies from 241 to 244°F (116 to 118°C) for 15 min. The objective is to obtain practical or commercial sterility, which means the occasional spore that survives will not reproduce and cause defects within the product. When the concentration of bacterial spores is high, heat treatment must be above normal. UHT sterilization involves heating at 284 to 302°F (140 to 150°C) for up to 9 sec. After UHT sterilization the product is homogenized and canned aseptically.

Browning (discoloration) results primarily from caramelization of lactose and the Maillard reaction in which carbonyl groups (–C=O) of lactose combine with amino groups ($-NH_2$) of casein. The rate of deterioration proceeds more rapidly as storage temperature and amount of stabilizing salts are increased. Little browning occurs in UHT-processed, aseptically canned evaporated milk. However, fat separation and sedimentation are more prevalent in UHT-heated products and these defects are associated with lower viscosities.

Evaporated milk keeps well when stored at 41 to 59°F (5 to 15°C). Humidity should be below 50% to prevent corrosion of cans and deterioration of labels and boxes. Periodic turning of cases helps redistribute minerals (especially tricalcium phosphate) and fat, provided the separated layer is soft and creamy.

Evaporated skim milk is concentrated to contain about 20% nonfat milk solids (NMS) (there are no federal standards). Procedures, processes, and defects are essentially the same as those for evaporated milk.

Whereas evaporated milk was used widely for home-produced infant formulas in the early to mid-twentieth century, in recent times commercially produced products more closely mimic the composition of human milk. Dur-

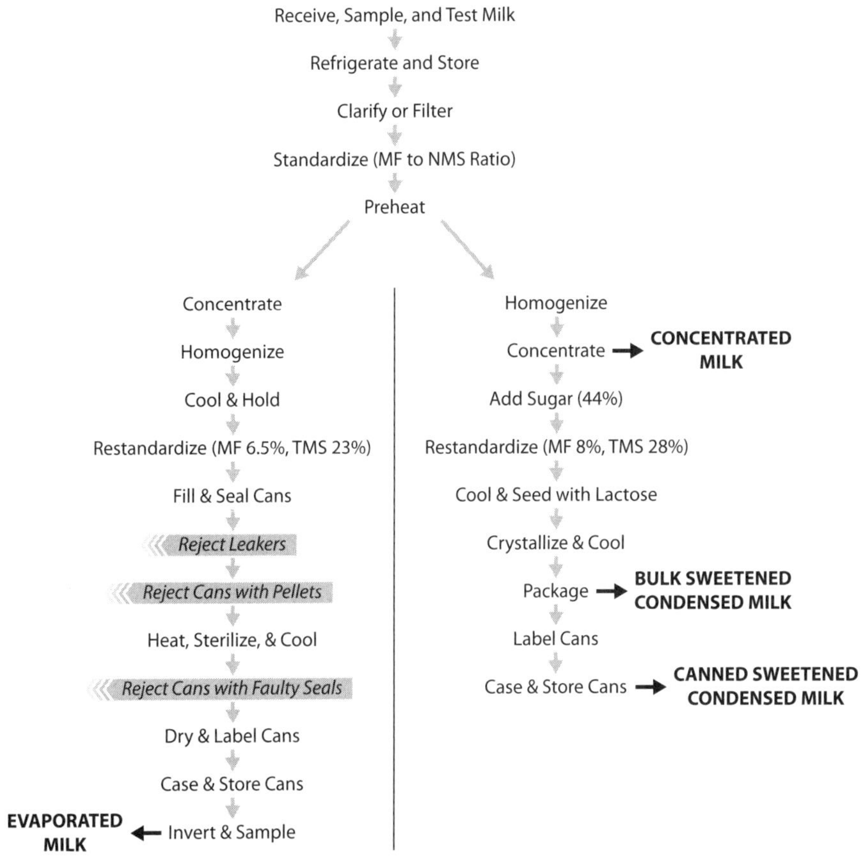

Figure 33.5 Flow diagram for the manufacture of evaporated, concentrated, and sweetened condensed milks. Tests for milk include flavor, odor, bacterial estimate, sediment, acidity, alcohol stability, and abnormal milk.

ing the First and Second World Wars, evaporated milk was a staple in the diets of military personnel. In 2014, 315.3 million lb of whole evaporated and condensed milk and 27.9 million lb of evaporated skim milk were produced.

33.5.3 Sweetened Condensed Milk

This product of milk is obtained by partial removal of water from a mixture of milk and suitable nutritive sweetener. It is pasteurized and may be homogenized. In finished form it contains, by weight, not less than 8% MF and 28% TMS. The quantity of sweetener used is sufficient to prevent spoilage. This ranges from 43.5 to 44.8% with sucrose. Sweetened condensed skim milk normally contains at least 20% milk solids (21 CFR 131.120).

Approximately 88.6 and 43.6 million lb of whole and skim sweetened condensed milks, respectively, were produced in the United States in 2014. Sweetened condensed milks are still the most favored of dairy products in many tropical countries in which refrigeration is available in few homes.

Major uses for sweetened condensed milk are in bakery products, frozen desserts, and candies. In several countries milk and sugar are concentrated by boiling them together in open kettles, as with cajeta of Mexico and keer of India. These products possess a distinct caramel flavor and are dark in color. Dulce de leche is the most common name for milk caramel in Latin America. Made as both a thick jam and a caramel candy, it is prepared by slowly heating sweetened milk to create a product that derives its taste from caramelized sugar.

When intended for industrial use, sweetened condensed milk is placed in barrels, drums, or tanks; when packed for retail it is canned. Deaeration and vacuum packaging result in improved keeping quality by limiting oxidation. Only **osmophilic** yeasts and molds grow readily in sweetened condensed milk. These microorganisms require oxygen, therefore sealing cans under vacuum to remove headspace for oxygen is recommended. Special care to prevent contamination with mold spores after forewarming is also necessary.

The flow diagram in figure 33.5 depicts the steps in the production of sweetened condensed milk. The primary differences from the method of producing concentrated milk are (1) the addition of sugar and (2) treatment to crystallize lactose. Sugar may be drawn into the batch-type evaporator with the condensed milk. However, in modern practice sugar is added and the fat/nonfat milk solids to sugar ratio is adjusted before concentration in a continuous process. Sugar in the aqueous phase inhibits microorganisms. Concentration is stated in terms of the sugar ratio and is calculated as follows:

$$\text{Sugar ratio} = \frac{\text{weight of sugar}}{\text{weight of sugar} + \text{weight of water}} \times 100$$

A ratio as low as 60% may be acceptable in bulk sweetened condensed milk that is to be used within a few weeks. However, case goods, which are sold at retail and therefore must keep several months, may require a ratio as high as 63% to maintain inhibitory conditions. Sucrose, as 65% syrup, is usually the sugar used. Up to 25% of it may be replaced by dextrose; however, this increases tendencies for age-thickening. Both sugars remain completely dissolved at the concentrations used. Other dissolved substances also increase osmotic pressure and assist in inhibiting microbial growth. These substances include lactose (approximately 10%) and milk salts (about 1.6%). However, less than half the lactose is soluble in the relatively small amount of water available. Therefore, it crystallizes as the finished product is cooled. Uncontrolled crystallization usually results in the growth of large crystals, which cause a grainy, sandy, coarse body. Crystallization may be forced by cooling the concentrate rapidly to produce a supersaturated lactose solution, then seeding it with minute lactose crystals during vigorous agitation. Crystals that form are comparatively numerous and small, yielding a smooth product. Crystallization rates are increased by increasing agitation, numbers of nuclei, and total solids of the concentrate. In the batch method of forced crystallization, agitation is continued for about an hour before completion of cooling and packaging. Continuous **seeding** may also be accomplished in a swept surface heat exchanger.

The length of the longest edge of lactose crystals associated with smooth body is approximately 10 μm, and the number of crystals/mm^3 is about 300,000. Coarse body of sweetened condensed milk is characterized by crystals of more than 15 μm on the longest edge and less than 100,000 crystals/mm^3. Usually viscosity is sufficient to prevent sedimentation of lactose crystals.

Thickening of sweetened condensed milk is sometimes caused by microorganisms producing acid or rennet-like enzymes. This differs from age-thickening, which is a gradual increase in viscosity attributed to physicochemical changes associated with casein micelles. After weeks to months of storage there is an increase in viscosity accompanied by visible gelation and irreversible aggregation of the micelles into long chains forming a three-dimensional network. Theories of causes include polymerization of casein with whey proteins (e.g., formation of complexes of κ-casein with β-lactoglobulin).

33.6 Dry Milk Products

Dry milk products can be manufactured from numerous components of milk by the removal of moisture, as illustrated in figure 33.6. Typical compositions of dry milk products are given in table 33.2. Major users of dry milk products include dairy, bakery, meat processing, confectionary, and prepared dry mix manufacturers, as well as individual consumers. The latter commonly use instantized nonfat dry milk.

Manufacturers have opportunities to fabricate dried milk products for special food applications. Among the products offered are low-, medium-, and high-heat nonfat dry milk. When milk proteins are heat modified, some of their molecular salt linkages and hydrogen bonds are broken. Proteins unfold or uncoil from their specific molecular structures into more random configurations. They tend to refold and recoil but do not revert to the original structure. Such molecular rearrangement results in changes in solubility and absorptivity. Low-heat milks (145°F [63°C] with short holding time) are preferred for table use, cakes, custards, and similar products. High-heat powders (185°F [85°C]+ with long holding time) are preferred by bakers because destabilized casein binds more water and denatured whey proteins improve loaf volume (high heat inactivates loaf volume depressant). In

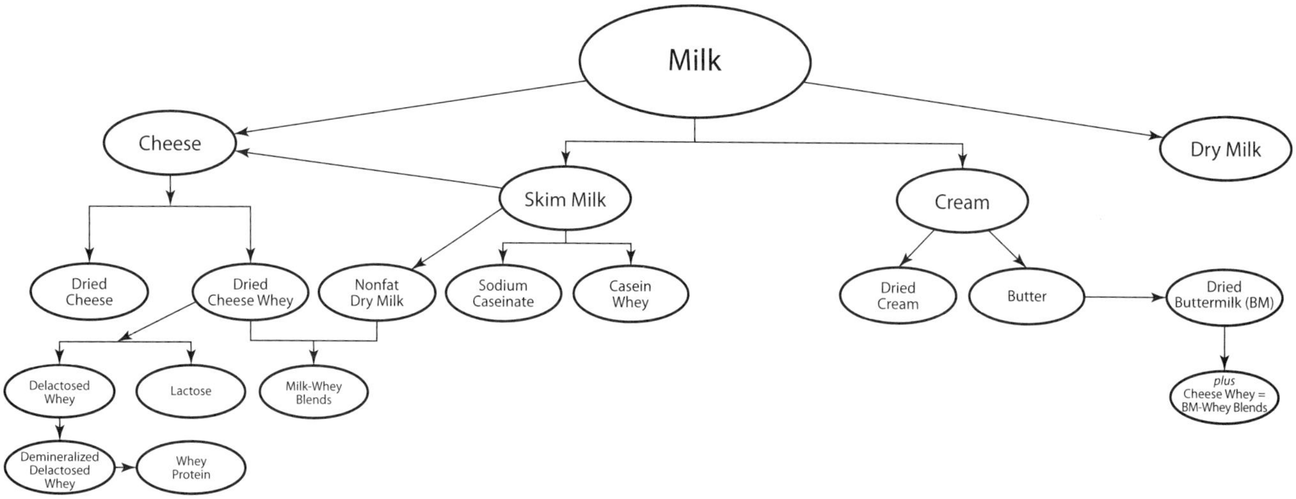

Figure 33.6 Derivation of dry milk products.

Table 33.2 Typical Composition of Dry Dairy Products in the United States.

Product Name	Typical Composition (%) Fat	Protein	Lactose	Moisture
Nonfat dry milk	0.6–1.25	34.0–37.0	49.5–52.0	3.0–4.0
Dry whole milk	26.0–28.5	24.0–27.5	36.0–38.5	3.0–5.0
Dry buttermilk	4.5–7.0	30.0–33.0	46.5–49.0	3.0–4.0
Dry buttermilk product	4.5–10.0	< 30.0	48.0–55.0	3.0–4.0
Dry whey (sweet)	1.0–1.5	11.0–14.5	63.0–75.0	3.5–5.0
Dry whey (acid)	0.5–1.5	11.0–13.5	61.0–70.0	3.5–5.0
Reduced lactose whey	1.0–4.0	18.0–24.0	52.0–58.0	3.0–4.0
Reduced minerals whey	0.5–1.8	11.0–15.0	70.0–80.0	3.0–4.0
Whey protein concentrate	1.0–10.0	34.0–80.0	10.0–55.0	3.0–4.0
Whey protein isolate	1.0	92.0	0.5	4.0–5.0
Lactose (milk sugar)	0.0	0.1	98.0	2.0–5.0
Dairy products solids	0.0–1.0	3.0–8.0	65.0–85.0	3.0–5.0

Source: American Dairy Products Institute, 2009.

white bread 1 kg of NDM can be expected to absorb 1 kg of water and to retain most of it during baking.

A simple test for heat treatment of dry milk is the undenatured whey protein nitrogen (WPN) test. Low-heat powders have at least 6.0 mg/g of WPN, whereas high-heat powders have less than 1.5 mg/g. Those powders testing 1.5 to 6.0 mg/g are defined as intermediate-heat dry milks.

Testing involves heating the rehydrated dry milk in a saturated solution of sodium chloride then filtering it to remove the denatured whey protein. The filtrate is diluted in saturated salt solution, acidified, and tested spectrophotometrically. Transmittance values are compared to a **standard curve** that plots Kjeldahl nitrogen values of standards against transmittance values for those standards (see Wehr and Frank, 2004). European standards call for use of the nitrogen solubility index (NSI). The American Dairy Products Institute (ADPI) makes available reference standards (low- and high-heat NDM samples) to be used for plotting standard curves in determining undenatured WPN by the modified Harland–Ashworth method (American Dairy Products Institute, n.d.).

Seasonality of production in the dry milk industry has been a distinct economic handicap. In the United States during the period 1965–1975, more than twice as much NDM was produced in June as in September or October. Therefore, production capacity was underutilized by 50% during certain months. More recently, in 2009 production data for 51 plants producing NDM show that only in September (86 million lb) and December (155 million lb) did the amounts produced vary by more than 10% from the overall monthly average of 126.6 million lb.

33.6.1 Nonfat Dry Milk

Nonfat dry milk far outsells all other dried milk products in the United States as revealed in production data in millions of pounds for 2011: nonfat dry milk and dry skim milk, 1,817; dry whole milk, 35; dry cream, 66. Nonfat dry milk is white to light cream in color with a clean dairy flavor. It is manufactured by removing only water from pasteurized skim milk. It contains 5% or less moisture and 1.5% or less milkfat (21 CFR 131.125). Nonfat dry milk covered by these standards shall not contain nor be derived from dry buttermilk, dry whey, or products other than skim milk, and shall not contain any added preservative, neutralizing agent, or other chemical. Product not meeting these specifications is called skim milk powder (SMP). Whereas NDM is standardized to a minimum moisture level, SMP is standardized, often with dry whey, to a minimum protein content of 34%. NDM normally contains 34 to 37% protein. The major product traded internationally is SMP whereas NDM is the form traded within the United States.

33.6.2 Dry Whole Milk

Typically dry whole milk is obtained by removing water from pasteurized, homogenized, or nonhomogenized milk. It must contain between 26 and 40% milkfat on an "as is" basis and not more than 5% moisture on a milk **solids-not-fat** basis. Vitamin A and/or D fortification of dry whole milk is an acceptable option (21 CFR 131.147).

The largest single market for dry whole milk in the United States is the candy industry, which uses products with 28 to 28.5% fat. Chocolate manufacturers prefer drum-dried milk, which contains a large quantity of free (nonglobular) fat. Foam-spray-dried milks cause foam to develop upon reconstitution, therefore they are preferred for addition in cocoa and other dry mixes. Separated dried milk proteins are useful both nutritionally and functionally. For example, sodium caseinate, made by neutralizing acid casein, is the major ingredient of coffee whiteners.

33.6.3 Dry Buttermilk

Dry buttermilk (DBM) is light cream in color with a clean, pleasing, sweet dairy taste and contains not less than 30% protein. Typically dry buttermilk is obtained by drying liquid buttermilk derived from the churning of sweet cream into butter. It is pasteurized prior to condensing. It contains no added preservative, neutralizing agent, or other chemical. The demand for dried buttermilk is based on its content of about 5% phospholipid, a natural hydrophilic emulsifier. Furthermore, it imparts a rich flavor. More dry buttermilk is used for bakery goods, including prepared dry mixes, than for any other use. Production in the United States in 2014 was 111.1 million lb.

33.6.4 Whey

Whey has traditionally been fed to animals or wasted. However, technical progress, human food needs, and problems of disposal have caused the industry to dry increasing amounts of whey for human uses. Sweet liquid whey contains about 5% lactose (dry whey 72%), which can be removed by crystallization. Delactosed whey is high in minerals, but these can be removed by reverse osmosis and ultrafiltration. Dry products of delactosed and demineralized whey can vary in composition in four different ways:

1. 29% protein, 55% lactose, and 8% minerals
2. 35% protein, 45% lactose, and 3% minerals
3. 50% protein, 33% lactose, and 3% minerals
4. 75% protein, 9% lactose, and 2% minerals

Whey may be blended with cereal products, especially soy flour, then dried. The baking industry uses these blends as substitutes for NDM. Lactose in the whey produces a desirable browning reaction (Maillard reaction) in crust and in toast, and soy proteins of the blend absorb water readily. However, the nutritive properties, buffering capacity, and body-building properties are inferior to those of NDM. The American Dairy Products Institute (2014) publishes *Whey & Whey Products; Definitions, Compositions, Standard Methods of Analysis* as well as color standards for dry whey.

Dry sweet whey is made by removing moisture from fresh sweet whey separated in the production of cheeses such as cheddar and Swiss. It is pasteurized, contains no preservatives, and has a titratable acidity of not over 0.16% on a reconstituted basis. Dry acid whey is derived from such cheeses as cottage and ricotta and has a titratable acidity of 0.35% or higher as reconstituted.

Lactose-reduced dry whey is treated by precipitation, filtration, and/or dialysis to remove a portion of the lactose before drying so that the lactose content is less than 60%, the normal amount in dry sweet whey being about 75%. Similarly, dry reduced-minerals whey is treated to reduce the ash content to 7% or less. To produce WPC, nonprotein constituents are removed from liquid whey so the dried product contains 25% or more protein. When the protein content is 90% or more the product is known as WPI. Safe and suitable ingredients may be added to adjust the acidity of these products.

Approximately 869.7 million lb of dry whey and 1.1 billion lb of lactose were produced in the United States in 2014, the latter being more than double the amount produced in 2000. At the same time about 538.2 and 81.3 million lb of WPC and WPI were produced, respectively. For details regarding products of cheese wheys, see Burrington et al. (2014).

33.6.5 Dried Cream and Butter Products

Dried high-fat products have been made sparingly because of tendencies to oxidize rapidly. Dried cream and butter powder are examples of products particularly valued for baked goods, dry mixes, sauces, icings, fillings, and frozen desserts. Butter powder, an Australian development, is prepared by drying a mixture of 100 lb cream (50% MF), 6 lb sucrose, 0.75 lb glyceryl lacto stearate (an emulsifier), 0.16 lb sodium citrate, and 0.04 lb citric acid. Products high in fat must be cooled rapidly after drying, to solidify the fat, and handled carefully to minimize shattering of the thin films of dry milk solids surrounding the fat globules. Keeping quality of butter powder is excellent when antioxidants are added and it is packaged in hermetically sealed containers.

33.6.6 Other Dried Milk Products

Cheese may be dried by tray, belt, or spray processes. Parmesan, a hard cheese, must be tray or belt dried, but cheddar and blue types are readily spray dried. Cheese is **comminuted** and mixed with a warm solution of sodium citrate to make a slurry containing 40% solids. The mixture is pasteurized, homogenized, and then atomized into the dryer in relatively large droplets to minimize loss of volatiles.

Other milk products that may be dried include ice cream mix, cultured buttermilk, malted milk, and flavored milks. Not only is there a large number of dry products of milk, but methods of drying and equipment also vary. Our approach here will be to explain and illustrate briefly the principles of drying and to highlight recent developments.

33.7 The Drying Processes

33.7.1 Drum Drying

Milk and milk products can be dried by distributing them in a thin film on the surface of a heated drum that revolves at 6 to 24 rpm toward a doctor (scraper) blade. Two drums are normally placed side by side (figure 33.7). One drum is movable so film thickness may be controlled. Unconcentrated milk or milk product may be delivered to the reservoir between the drums formed by end boards fitted against the end faces of the drums. Heat from the drums concentrates the milk, and steam is removed through a vapor hood. To minimize heat treatment it is advantageous to concentrate milk in a low-temperature evaporator, then it is sprayed on the rollers. Also, to remove vapors at a lower temperature drums may be enclosed in an airtight evacuated chamber. Usual vacuum levels are 27 to 29 in (685 to 735 mm) Hg. Steam is fed into drums through hubs at the end of each shaft. Temperatures of the saturated dry steam range up to 302°F (150°C). The temperature of dry product leaving the drum is normally within 3 to 5°F (2 to 3°C) of the temperature outside the drum. Scorching and destabilization occur under the following conditions: (1) milk supply is uneven, (2) film is incompletely removed, (3) drums are misaligned, (4) rollers are rough, (5) temperature is too high, or (6) drum speed is too low. Atmospheric drum dryers require 1.2 to 1.3 lb steam for each pound of water evaporated.

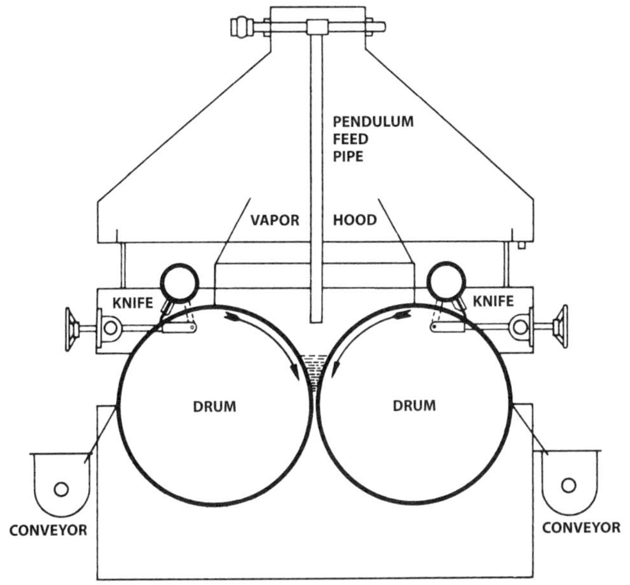

Figure 33.7 Schematic diagram of an atmospheric double-drum dryer.

Under proper drum drying conditions, a thin, light-colored film of dry product is scraped from the drum into a trough. A screw conveyor breaks up the film as it is augured to a flaker or hammer mill. Some whey is dried by the drum process in the United States, but little milk or skim milk is dried using this process.

33.7.2 Spray Drying

In this drying process, water is removed from droplets of concentrated milk (40 to 48% solids) by large volumes of heated air. A milk concentration plant is an integral part of a milk drying plant (figure 33.8).

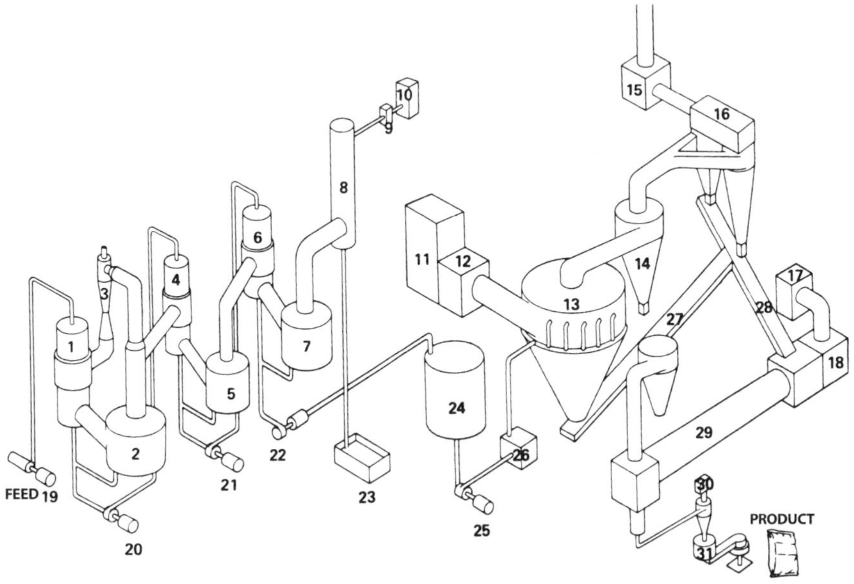

1. 1st Effect Steamchest
2. 1st Effect Separator
3. Recompressure Booster
4. 2nd Effect Steamchest
5. 2nd Effect Separator
6. 3rd Effect Steamchest
7. 3rd Effect Separator
8. Main Condenser
9. Air Ejector
10. Inter Condenser
11. Inlet Fan
12. Burner
13. Main Vertical Dryer
14. Single Pri. Separator
15. Exhaust Fan
16. Two Sec. Separators
17. Inlet Fan
18. Heater
19. 1st Effect Feed Pump
20. 2nd Effect Feed Pump
21. 3rd Effect Feed Pump
22. Discharge Pump
23. Sump
24. Crystalization Tank
25. Dryer Feed Pump
26. Hi Pressure Pump
27. Timing Belt Conv.
28. Belt Conv.
29. After Rotary Dryer
30. Fan
31. Collector

Figure 33.8 Complete spray-drying plant including milk concentration and concentrate drying sections (courtesy of Anderson IBEC, Strongsville, Ohio).

There are seven main functional parts of the spray dryer: product feed system, air supply filter, air heater, atomizing device, drying chamber, product collector or air/powder separator, and air movement device (figure 33.9). The air filter removes dust and debris from incoming air to prevent contamination of product. Direct or indirect air heaters are used. Direct heating involves drawing air across a natural gas flame. Although natural gas is the most efficient source for heating, noncombusted gas and breakdown products of it may contaminate the air.[3] Air may be heated indirectly by passing it over electrically heated elements or coils containing steam or heated materials such as therminol.[4] Air temperatures normally vary from 122 to 500°F (50 to 260°C) and relative humidity is low. Relative humidity of 100% at 98.6°F (37°C) will decrease to 3 to 4% at 302°F (150°C). Heat from exhausted air can be reused to indirectly preheat incoming cold milk.

High-pressure spray nozzles or spinning (centrifugal) discs (figure 33.10) are used to atomize concentrates. Nozzles operate under pressures of 1,500 to 5,000 psi (10.3 to 34.5 mP) and centrifugal discs rotate at speeds up to 50,000 rpm for 2-in discs and as low as 3,500 rpm for 30-in discs. The objective is to obtain uniform particles of 50 to 150 μm diameter. Such particles have adequate surface/volume ratio for efficient drying. If 1 gal of concentrate is atomized into particles 50 μm diameter, 1,390 ft^2 (129 m^2) of surface area will be formed. Nonuniform particles result in overheating of small particles and/or incomplete drying of larger ones, causing them to adhere to walls of drying equipment. Atomization also largely determines air content, bulk density, and reconstitutability. Droplet size (pressure nozzles) is controlled by size of orifice, pressure, and the spray angle (narrow angles produce larger droplets).

Air gets into milk particles in two ways: (1) primary air is mixed in during atomization and (2) secondary air gradually fills cavities after water diffuses out during drying. Little primary air is entrained in milk atomized by a pressure nozzle. However, air may be incorporated in centrifugal atomization and powder produced by this process tends to be low in density. Bowl-type centrifugal discs deliver droplets low in air content because centrifugal force separates air from concentrate on the disc. Also, flushing the interior of the atomizer disc with steam reduces the volume of the vacuoles. Steam largely replaces air in droplets. It then condenses, diffuses, and is evaporated from the droplets. Content of secondary air is determined by the drying process. Rapid mixing of droplets in hot air leads to rapid formation of a hard outer shell. Water vapor diffuses through this shell, leaving cavities within the particles.[5] Thus, it follows that the heaviest particles can be formed by slow drying, which allows particles to shrink as moisture is removed.

The drying chamber is designed to provide optimal mixing of particles of concentrate and air without excessive particle coalescence. This prevents impaction of wet particles on walls and restricts eddy currents that tend to expose some particles to overheating. The drying path must be long enough to effect complete drying of the largest particles unless there is a secondary dryer. Air may flow cocurrently (in the same direction), countercurrently (in the opposite direction), or in a combination of directions. Best thermal efficiency is associated with cocurrent flow, but larger vacuoles and less dense powder may result. Moreover, cocurrent flow affects evaporation within a relatively short distance from the atomizer, provides for minimal chamber wall temperature, and keeps product temperature relatively low. Spiral flow may be used in vertical dryers to provide a longer residence time. Cocurrent flow is necessary in box-type dryers (figure 33.11).

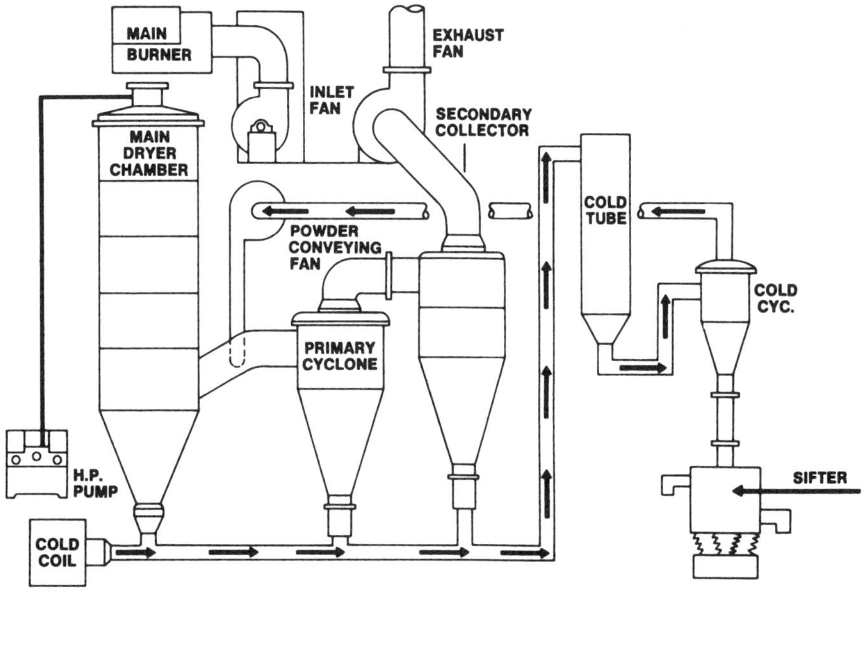

Figure 33.9 Typical flow diagram of cocurrent flow-type spray drier (courtesy of De Laval Separator Company, Poughkeepsie, New York).

Chamber design affects the direction and degree of atomization, pattern of air flow, method of discharging product, product characteristics, time required to dry product, and air flow rate. When separation of large particles from fine particles is desired, as in dryers designed to produce agglomerated particles directly, a collecting duct is placed in the mid-lower section. Fine particles (fines) are withdrawn through it with drying air, and coarse particles are removed through an air lock at the apex of the cone.

33.7.3 Drying Rate

Characteristics of product that affect drying rate include (1) moisture content, (2) amount of bound water, (3) heat capacity, (4) surface tension, (5) density, and (6) viscosity.

As concentrate changes from a liquid to a moist solid to a dry solid, the rate of evaporation decreases. Rate is constant until free water is evaporated from droplet surfaces. The solid phase then becomes a barrier to passage of moisture and thereby decreases rate of drying. Two

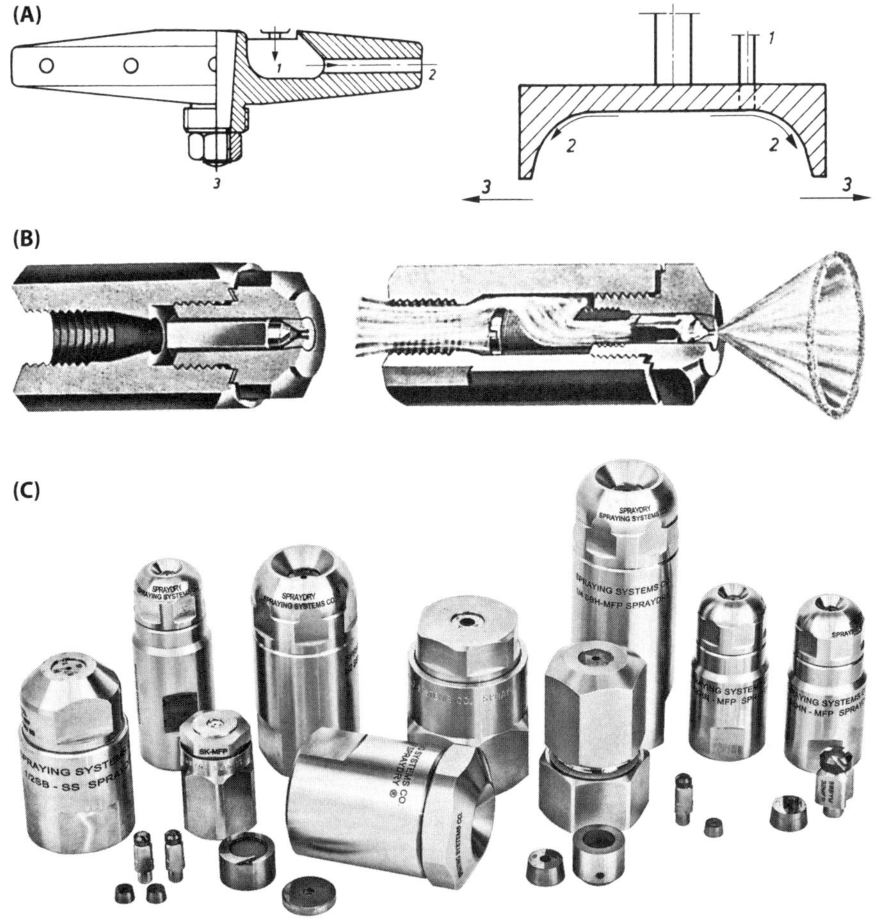

Figure 33.10 (A) Drawings of a centrifugal disc (left) and a centrifugal bowl-type disc (right). (B) Type ST pressure spray nozzle (left) and type SL pressure spray nozzle (right). Note flow pattern of concentrated product through type SL nozzle (courtesy of Spraying Systems Company, Wheaton, Illinois). (C) Examples of types of nozzles (courtesy of Spraying Systems Company, Wheaton, Illinois).

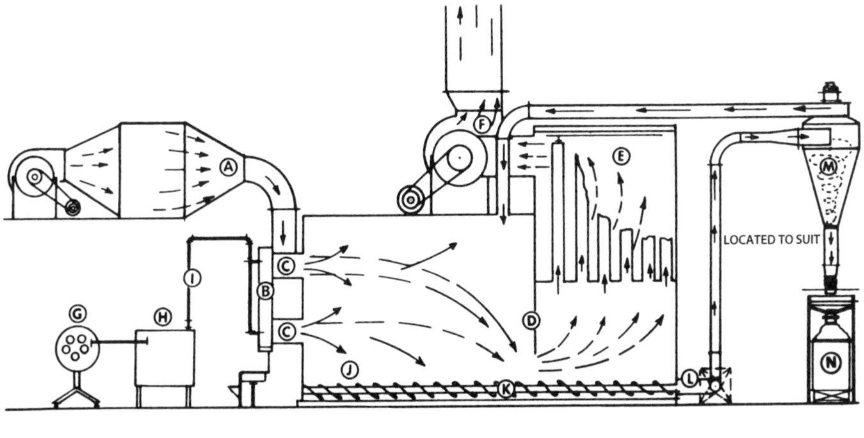

Figure 33.11 Box-type spray drier with (A) air supply, (B) concentrate distributor, (C) air inlets and spray nozzles, (D) air baffle, (E) cloth filter bags that are intermittently shaken, (F) clean air exhaust, (G) concentrate preheater, (H) high-pressure pump, (I) high-pressure line, (J) reciprocating conveyor, (K) screw conveyor, (L) chamber outlet, (M) cooling cyclone, and (N) the sifter and bagger (courtesy of C. E. Rogers Company, Mora, Minnesota).

processes occur simultaneously in drying, namely heat is transferred from air to droplets and moisture diffuses to the particle surface from which it evaporates. The particles are thus cooled. However, when the rate of evaporation slows, evaporative cooling also slows and heat damage may occur. Normally, this is controlled by monitoring the exhaust air temperature and terminating heating at the point at which temperature begins to increase. Exposure of dried product to air at 302°F (150°C) must be less than 2 sec. Theoretically, only enough heat to vaporize the moisture (latent heat of vaporization) is needed, and droplets themselves need not increase in temperature. Thus, it is possible to dry milk under low temperatures, such as 86°F (30°C). However, retention time must be long and air exceptionally dry. For example, in one process (Waite, 1964), droplets of large diameter (300 to 1,000 μm) are sprayed into dried air with 0.5 to 3% relative humidity. Air enters the bottom of a cylindrical tower that is 235 ft high and 50 to 120 ft in diameter, and ascends slowly at a speed of 0.3 to 1.6 ft/sec. The countercurrent flow of air allows the driest air to meet nearly dry particles of milk, providing maximal drying efficiency. Air leaves the tower about 95% saturated with moisture. The system is self-filtering; fine, dry particles carried upward by air are trapped by falling spray, yielding up to 99.5% recovery of solids.

33.7.4 Powder Collectors

The most commonly used product collectors are single or multiple cyclones (essentially hollow inverted metal cones) that operate on the same principle as vapor separators of evaporators, namely particles of product, because of their inertia in spiraling air, move toward the outside of the collector and gravity pulls them toward the collecting receptacle. Air, usually containing minute particles of powder, is vented through the top. Exhaust air may be channeled through a spray of milk that "scrubs" it. The mixture of milk and collected fines is returned to the balance tank for evaporation. Filter-type collectors are highly efficient but are bulky, high in initial cost, and cumbersome to clean.

Air is usually moved through the dryer with a fan or fans, which may push or draw air through the system. Some dryers employ a secondary dryer in which powder is recombined with fresh heated air, passed through a vertical drying cylinder, and separated in a cyclone. Secondary drying can be accomplished also in a fluidized bed in which the semidry particles of powder are caused to behave as a fluid when air at the desired temperature and humidity is forced through it vertically or tangentially. Thus, the particles can be dried and cooled as they pass through the bed to a collector.

Cooling of powder must be accomplished soon after drying, especially with high-fat products. In fact, it may be necessary to introduce cool air across the bottom of the main dryer to cool and solidify fat, thus preventing sticking. It is important that cooling air be practically free of microorganisms, especially pathogenic types such as salmonellae. Air may be cooled by refrigeration coils but cooling must not result in condensation of moisture in powder. Nonfat dry milk is cooled to 95 to 104°F (35 to 40°C) upon leaving the dryer. If not cooled, it becomes lumpy, turns brown, and develops off flavors rapidly. Dry milks have low capacities to conduct heat, therefore temperature falls slowly inside packages.

Conditions for producing low-heat NDM and high-heat NDM of beverage quality are presented in figure 33.12.

33.7.5 Packaging

Powder commonly is sifted before being placed in bags, drums, or bulk containers. Special precautions are taken in packaging dry products with large quantities of fat. Packages should be evacuated to reduce oxygen concentrations. If powder has been exposed to air for sufficient time to absorb oxygen, time must be permitted for that oxygen to diffuse out, then each package is evacuated again. An in-package scavenger, consisting of a palladium catalyst in an atmosphere of 90% nitrogen (N_2) and 10% hydrogen (H_2), greatly prolongs shelf life. Oxygen that may diffuse into a pouch will react with hydrogen to form water that cannot pass back out. This rapidly reduces oxygen content to about 0.001% or less.[6] Adding

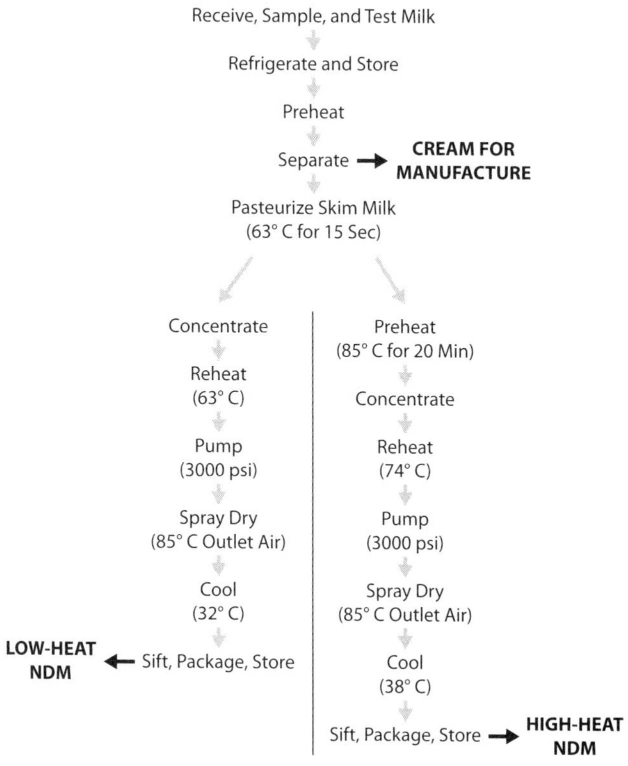

Figure 33.12 Flow diagram for the manufacture of low-heat and high-heat NDM by spray process. Tests for milk include flavor, odor, bacterial estimate, sediment, and abnormal milk.

0.01% of the antioxidants lauryl gallate, propyl gallate, and nordihydroguaiaretic acid will lengthen shelf life.[7] A desirable package for dry milk products is impervious to moisture, light, gasses, and insects; durable and resistant to corrosion; low in cost; easily filled, sealed, handled, and emptied; and reclosable if sold at retail.

33.8 Physical Properties of Dry Milks

A close relationship exists between the drying process employed and the physical structure of dry milk products. For example, in the roller drying process, a thin crinkled sheet of product is scraped from the drum, broken by a screw conveyor, and ground to a fine powder. The powder consists of platelets of irregular, angular shape with wrinkled surfaces and rough edges. No air cells within the particles are perceivable.

The fine granular product of spray drying requires no grinding. Generally the particles are spherical and contain air cells (figure 33.13). Sizes of air cells depend on amount of air introduced during atomization (primary air) and amount of vacuoles formed during drying. Air diffuses into these vacuoles (secondary air).

33.8.1 State of Lactose

Lactose is the major component of dried milk. Milk powder contains approximately 38% and nonfat dry milk about 50%. Lactose is present as concentrated (supersaturated) syrup and forms the continuous phase of the particle. However, it cannot crystallize because high viscosity limits mobility of lactose molecules. Because of high **hygroscopicity** of lactose glass, dried particles easily absorb moisture from air. Thereby the solution is diluted to such an extent that its molecules acquire sufficient mobility (at above 7% moisture) to arrange themselves into a crystal lattice and crystallization occurs. As soon as crystallization begins, the solution becomes less saturated, thus increasing its vapor tension. This allows some absorbed moisture to be given up, and the powder begins to cake. Also, the permeability of powder particles changes because of the formation of a network of crystal boundaries along which cracks may occur.

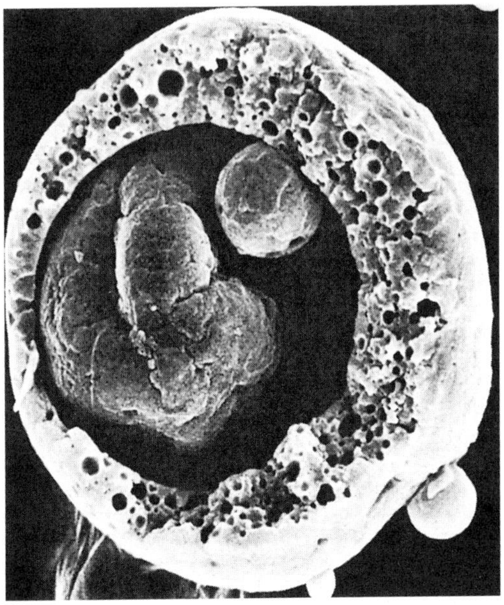

Figure 33.13 A broken particle of spray-dried milk. The large vacuole formed by primary air contains smaller particles. The small holes probably contained fat and some secondary air (courtesy of T. J. Buma, Cooperative Condenfabriek Friesland, Leeuwarden, the Netherlands and S. Henstra, Technical and Physical Engineering Society, Wageningen, the Netherlands; originally published in Buma, 1971).

Dried whey contains about 75% lactose, which must be primarily in the crystalline (alpha monohydrate) form if the powder is to be nonhygroscopic and free flowing. Furthermore, crystals must be small and numerous. Lactose in concentrated whey should be caused to crystallize as much as practicable (usually up to 85%) before drying because conditions for crystallization are easier to maintain in liquid. Temperature is controlled, the concentrate is seeded with lactose crystals, and surface layers on crystals are renewed by agitation (this cannot be accomplished in dried whey).

Further crystallization should occur as the product dries. Two processes have proven satisfactory. Precrystallized whey concentrate (approximately 50% solids) is dried in a spray dryer operating with a low outlet temperature of about 131°F (55°C). Powder leaves the drying chamber with a moisture content of 10 to 14% and drops onto a moving belt, where after-crystallization occurs. A secondary dryer then removes excess moisture. The powder consists of large agglomerates of low-bulk density. The second process has no after-crystallization stage. Therefore, nearly 85% of the lactose must be precrystallized. Spray drying proceeds at an outlet temperature that retains 6 to 7% moisture in the product, ensuring good agglomeration. Product passes to a fluidized bed for agglomeration and completion of drying. Small particles (dust) are blown away and recycled with fines from the main cyclone back to the atomizer. Agglomerates produced are relatively small and of high-bulk density. It is important to keep the crystallization process continuing. If during spray drying too much water is removed, viscosity will increase sufficiently to inhibit crystallization. If too high a temperature is used at the beginning of the fluidized bed, crystals already present can dissolve rather than act as seed for new crystals.

33.8.2 State of Proteins

Solubility of dry milk products is determined primarily by ability of casein proteins to be redispersed as stable colloidal particles. Heat, of course, is responsible for the major part of destabilization of proteins. During transition of lactose from the glassy to the crystalline state, caused by uptake of moisture, it is likely that proteins are expelled to crystal boundaries and are crowded together with each other and with milk salts. This may contribute to a decrease in the redispersibility. Also, amino-sugar complexes may be formed. This reaction accompanies browning and is enhanced by unfavorable storage conditions such as high humidity of air, high original moisture content, and high storage temperatures. At low relative vapor pressures, casein preferentially binds water, but it is after lactose and salts have absorbed moisture at relative vapor pressures above 0.5 that proteins are destabilized.

Spray drying does not materially reduce solubility of whey proteins; therefore the extent of denaturation is near that of the condensed milk from which it was produced. Furthermore, the solubility of spray-dried NDM is unaffected by preheat treatments of the fluid milk that cause extensive serum protein denaturation. However, powders produced from milk given a high-heat treatment tend to become less soluble than those from milk given low-heat treatment when they are stored at relatively high temperatures.

33.8.3 State of Fat

On reconstitution, dry milk products containing fat in an inadequately emulsified state leave greasy films on walls of containers. The degree of emulsification and dispersion of fat can be increased by homogenization before drying. However, only a thin membrane protects homogenized fat globules and rupture may occur during manufacture, thus freeing the fat. It is especially important to lower the temperature of powder quickly after drying to minimize fat migration. Dry milk particles have free fat on their surfaces, making them repel water. The amount of free fat at the surface varies in good powder from 0.5 to 3%. To overcome repellency, a surface active agent is needed at the particle surface. A method was devised whereby well-agglomerated particles of dry milk (agglomerates ranging from 100 to 250 μm) were coated with lecithin dissolved in butter oil (Pisecky and Westergaard, 1972). The resultant product was highly soluble in cold water.

Buma (1971) developed a model of extractable (free) fat in spray-dried milk. Extractable fat (f_e) consists of the following four components (figure 33.14):

- f_s = surface fat present as pools or patches of fat on the particle, particularly in surface folds and at points of contact between particles;
- f_l = outer fat layer, consisting of fat globules in the surface layer, which can be contacted directly by fat solvents;
- f_c = capillary fat, consisting of fat globules within the particles that can be contacted by fat solvents via capillary pores or cracks; and
- f_d = dissolution fat, consisting of fat globules inside particles that can be contacted by fat solvents via holes left by previously dissolved fat.

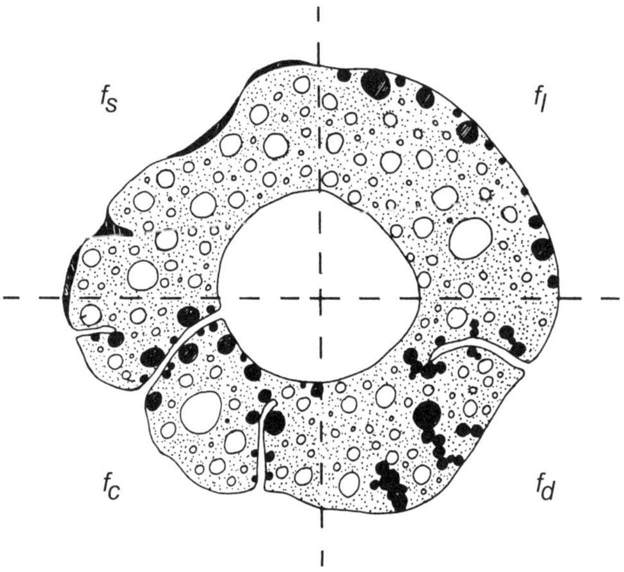

Figure 33.14 Schematic model of four forms of extractable fat in a milk powder particle with a central vacuole. Dark areas represent extractable fat (Buma, 1971).

Therefore, extractable fat can be calculated as follows:

$$f_e = f_s + f_l + f_c + f_d.$$

In experiments performed in derivation of this model, Buma concluded the following: (1) no correlation existed between quantities of extractable fat and development of oxidized flavor; (2) though surface fat affected dispersibility and wettability, no correlation was observed between quantity of extractable fat and these properties unless there was less than 20% fat in the free form; (3) no correlation existed between quantity of extractable fat and rising of cream, foaming, and churning in **reconstituted milk**; and (4) stickiness of spray-dried milk powders is influenced only by surface fat, and not by total extractable fat.

33.8.4 Surface of Spray-Dried Particles

Excellent photographs of surfaces of dry milk, nonfat dry milk, and dry whey were prepared by Buma and Henstra (figure 33.15). Apple-like structures suggest an implosion in the last stage of drying. Differences in appearance of NDM and dry whey suggest that deep surface folds are formed in the presence of casein. Variations in surface within the same preparation suggest differences in drying conditions.

33.8.5 Solubility of Dried Milks

When milk powder is added to water it passes through three stages before a reconstituted product is obtained: (1) surfaces must be intimately wetted; (2) truly soluble constituents must dissolve, namely lactose, minerals, and whey proteins; and (3) dissolved solids must spread

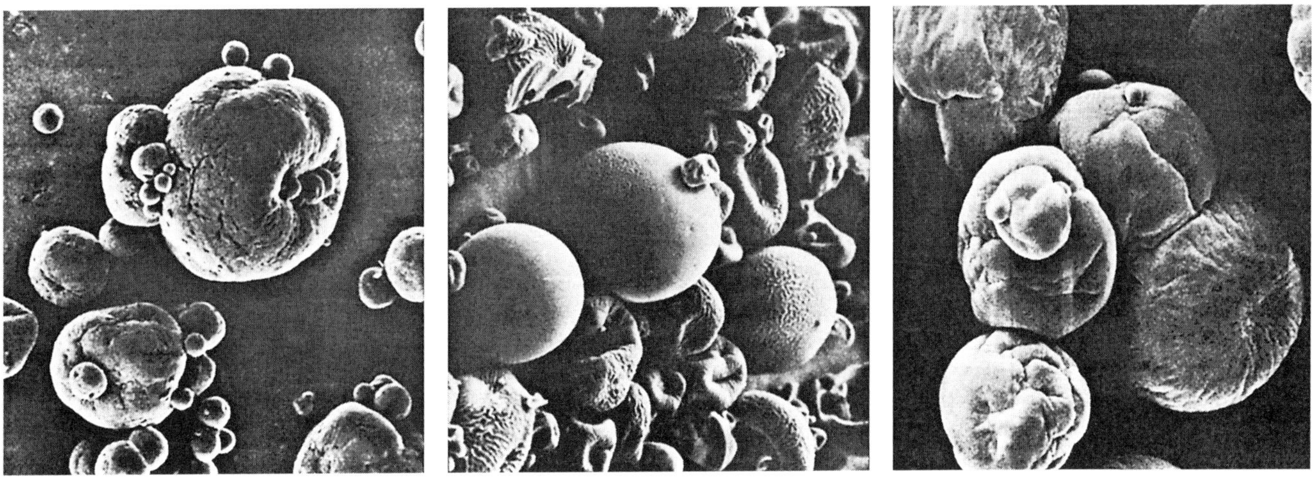

Figure 33.15 Left—Spray-dried milk (950x); center—spray-dried skim milk (290x); and right—spray-dried whey (950x) (courtesy of T. J. Buma and S. Henstra; originally published in Buma, 1971).

throughout the water phase carrying casein, fat globules, and other constituents into a uniform suspension.

Wettability is the ease with which the gaseous phase at the surface of the solid phase is replaced with a liquid phase. All three phases coexist as intermixtures as wetting occurs. Interfacial tension determines the rate at which a solid surface becomes wet.

As individual components begin to disperse and dissolve a concentrated solution of milk forms around each particle. If particles have a density higher than water they begin to sink. This highly desirable property is called *sinkability*. Density of particles depends on their composition and amount of occluded air. In early stages of reconstitution[8] density of particles decreases, mainly because lactose and minerals, the heaviest components, dissolve first. Simultaneously, density of surrounding **menstrum** increases because of its gain of solute. Therefore, differences in density between particles and their immediate environment decrease. The best way to achieve good sinkability, therefore, is to minimize quantities of occluded air.

Reconstitution of a mass of powder is more complicated. Capillary attraction for water causes particles above as well as on and below the water surface to be wetted. However, limited quantities of water usually penetrate the mass, leaving numerous air bubbles in interstitial spaces. Soon spaces between particles become filled with dissolution products, including solutions of high concentrations. This results in a mixture of sticky concentrate, unwetted powder, and residual air. Lumps with wet surfaces and dry interiors form. To obtain fully reconstituted milk in a reasonably short time with minimal effort, capillary penetrations of water into powder must be avoided. Penetration is minimal when powder consists of large agglomerates. However, there is the disadvantage that large agglomerates have a relatively small surface area exposed to water. Such agglomerates will be fully wetted and dispersed but not fully reconstituted immediately.

Usage of high protein powders, such as MPC-80 (milk protein concentrate containing 80% protein), is limited by low solubility. Commercial samples of MPC collected from 10 countries with protein contents ranging from 41 to 90% ranged in solubility from 20 to 90% (Sikand et al., 2011). Although conditions of production are responsible for differences in solubility, mineral balance is a key factor also. Solubility of MPC-80 was enhanced by addition of 50 to 150 mM NaCl during diafiltration.

33.8.6 Density of Dry Milk Products

Several aspects of density are of practical significance in dry milk products. Density influences package size, transportation costs, ability to recover fines in a cyclone, and sinkability in water. *True density* is weight per unit volume when **voids** are not present. True density of dry milk is 1.31 to 1.32 g/mL and that of NDM is 1.44 to 1.46 g/mL. However, 50 to 75% of the space occupied by spray-dried powder is air. *Apparent density* is the weight per unit volume when voids are present. It may be expressed as *bulk* (loose) or *packed* density. To determine bulk density, allow dry product to flow freely into a preweighed or tared 100 mL graduated cylinder, weigh, and calculate the net weight. Packed density is determined by tapping the same cylinder on a rubber mat until the volume of powder remains constant. Determine volume and divide net weight by volume. For example, if 100 mL of loose NDM weighs 60 g, the bulk density is 0.6. And, if the packed volume is 80 mL, the packed density is 0.75 ($60 \div 80 = 0.75$).

Advantages of low air content in powder include lower packaging costs, more rapid dispersion in water, and more rapid gas packing. The latter is possible because there is less air that has to diffuse from the powder.

Drum-dried NDM normally has an apparent density of 0.3 to 0.5 g/mL. Rugged contours and angular shapes prevent tight packing. The spherical shape of spray-dried powders permits closer packing but the amount of air space inside the particles is highly important. Apparent density of NDM usually ranges from 0.5 to 0.6 g/mL. In NDM with little occluded air, apparent density may approach 0.8 g/mL. Instantized powders have apparent densities about 50% lower than the same product that was unagglomerated because they pack less closely.

33.9 Instant Dry Milk Products

Consumers are highly conscious of convenience aspects of purchased commodities, especially foods. **Fluid milk** and most milk products are available in ready-to-eat form. However, for many uses dry milk products must be rehydrated. Rehydration, though simple in procedure, is as complex a physical process as the milk product is complex in composition. Individual particles of conventionally spray-dried products normally have good solubility but they are deficient in abilities to be wetted, to disperse, and to sink. Moreover, upon being added to water, portions tend to partially solubilize and to form a sticky substance that holds unwetted powder together in lumps. Since these lumps are virtually impervious to water, their complete reconstitution is extremely difficult, even with strong agitation. Equipment is available to food processors to disperse and reconstitute these dry products. However, consumers do not have means available to quickly rehydrate milk powder. Instead, processors have modified dry products to improve wettability, dispersibility, sinkability, and, therefore, solubility. However, the process, *agglomeration*, does not improve total solubility.

Peebles and Clary (1955) introduced the concept of instantizing product that is already dry. It involves the following: (1) uniform wetting of surfaces of dried particles in a turbulent environment of steam and/or minute water droplets, (2) agglomeration of particles—achieved by causing them to collide and adhere together in clusters, (3) equilibration of moisture within the wetted particles, (4) drying of clusters in hot air or vacuum, and (5) cooling and sizing. The latter may include breaking large clusters and screening out fine particles. Wetting of particles is sufficient to increase moisture content to 12 to 14% so lactose may change from the glassy to the crystalline form; the amount of crystallization is directly related to the time of holding rewetted powder before drying. The resulting product is granular and free flowing. It disperses and sinks readily but if agglomerates are large, the solution rate may be slow. Particle sizes for agglomeration should range from 25 to 50 μm. Low- or medium-heat powders are preferred as clusters of high-heat powders shatter excessively.

Some dryers are capable of producing agglomerated dry milks in a straight-through process (figure 33.16). Concentrate is atomized by a rotary disc to particles of about 100 μm diameter and less. Having particles varying in size gives falling droplets an uneven velocity, causing collisions between larger and smaller particles. This makes *primary* agglomeration possible. Smallest particles that pass off with drying air are returned from cyclones to the wet zone of the drying chamber where they agglomerate with wet droplets. This is *secondary* agglomeration. *Tertiary* agglomeration may occur as particles containing 5 to 9% moisture roll down the conical part of the dryer. In the dryer's horizontal section (figure 33.16), air of varying temperatures is blown upward through per-

Figure 33.16 Picture of a Vibro-Fluidizer installed beneath the conical lowest portion of a dryer. Dry instantized agglomerates exit on the left for packaging (courtesy of GEA; © GEA, GEA Vibro-Fluidizer™).

forated plates (fluidized beds) on which the particles move. The agglomerates are stabilized, dried to final moisture content, and cooled. Any larger particles at the outlet of the dry bed are screened and recirculated to the inlet. The screened and instantized particles are conveyed by the cooling air to a battery of cyclones, where they are separated from the air for packaging. A Vibro-Fluidizer may be used to coat particles of dry milk with lecithin to improve wettability.

Temperature in final drying must be carefully controlled because the Maillard reaction has its maximum velocity at about 15% moisture. At this moisture concentration only about 0.1 sec is required to denature 50% of the casein at a temperature of 212°F (100°C) (Waite, 1964).

33.10 Reconstituted and Recombined Milk Products

Most dehydrated products are rehydrated before use. The process is called *reconstitution*. When more than one milk component is mixed, the process is called *recombination*. Milk products may be recombined and reconstituted simultaneously. For example, the US armed forces have long been supplied **recombined milk** in areas of the world distant from milk supplies. Some nations in Africa are supplied a large percentage of their fluid milk as recombined product. These products are largely made from NDM (or skim milk powder) and anhydrous milkfat (or butter oil).

Mobile units are available for recombining milk in remote areas. In India and other Asian countries reconstituted skim milk powder is used to tone native buffalo milk. Thus, the high fat content of buffalo milk (up to 12%) is diluted to a minimum of 5% in India. This "toned milk" extends the local milk supply and retains most properties of the fresh buffalo milk.

Certain milk products may be manufactured from reconstituted dry milks. In particular, cottage cheese may be manufactured where the supply of fresh milk is limited (e.g., in Japan). In some countries cheeses are made from recombined milk.

33.11 Grading and Quality Assurance

Standards for grades of concentrated milks in the United States exist only for those products to be labeled grade "A"; they must be manufactured according to the US Food and Drug Administration's *Grade "A" Pasteurized Milk Ordinance* (*PMO*). These are usually bulk concentrated (condensed) whole milk or skim milk. The US Department of Agriculture's (USDA) grading program for dry milk products is voluntary. User firms pay costs of grading services. Grade "A" dry milks must be manufactured from grade "A" milk and are subject to rules published in the *PMO.* In these grading programs, not only are characteristics of the final product controlled, but any product that bears a grade on the label shall have been produced under sanitary conditions. Only fresh milk may be used in production and it shall be pasteurized before drying. Although most dry milk products are made from grade "A" raw milk, manufacturing grade milk may be used, the bacterial count of which may not exceed 500,000/mL. Under 7 CFR 58, the USDA offers the following grading services: Subpart A—Regulations Governing the Inspection and Grading Services of Manufactured or Processed Dairy Products and Subpart B—General Specifications for Dairy Plants Approved for USDA Inspection and Grading Service. It also provides *Instructions for Dairy Inspection and Grading Service* (DA Instruction Number 918-I). Section 11, "Evaluation of Product Characteristics," applies to grading of butter, cheese, dry milks, and evaporated and sweetened condensed milks and contains sections entitled prerequisites, documents and forms, monitoring, coding and marking, net weight, condition of container, sampling, and product evaluation.

Standards for finished dry products are published by USDA. Those for NDM are given in table 33.3. These same specifications apply to instant nonfat dry milk and dry whole milk, except that moisture may be as high as 4.5% and fat content of whole dry milk ranges from 26 to 40%. Limits for roller dried NDM include (1) bacterial count of 50,000/g, (2) scorched particle content of 22.5 mg, and (3) solubility index of 15 mL. Edible dry casein may have a bacterial count up to 100,000/g and must contain at least 90% protein.

Standards for US extra grade dry whey set maximum limits for each of the following: (1) standard plate count, 30,000/g; (2) coliform count, 10/g; (3) milkfat content, 1.5%; (4) moisture content, 5%; and (5) scorched particle content, 15 mg.

Adulteration of milk powder with whey powder may be detected by determining protein content or the protein to lactose ratio. Owing to natural variation in composition of milk, however, these methods give only presumptive evidence of adulteration unless deviation from

Table 33.3 Selected Maximum Specifications for US Grades of Nonfat Dry Milk.

Laboratory Tests	US Extra Grade	US Standard Grade
Standard plate count	10,000/g	75,000/g
Milkfat content	1.25%	1.50%
Moisture content	4.0%	5.0%
Scorched particle content	15.0 mg	22.5 mg
Solubility index	1.2 mL	2.0 mL
US high heat	2.0 mL	2.5 mL
Titratable acidity (lactic acid)	0.15%	0.17%

Source: USDA Agricultural Marketing Service, 2001.

normal is unusually large. A more sensitive method is based on tests for neuraminic acid containing glycomacropeptide released by the action of rennin on κ-casein. This peptide, solubilized in trichloracetic acid, can be precipitated quantitatively and specifically by phosphotungstic acid. The method can detect as little as 2% whey powder in spray-processed NDM. However, the heat of roller drying liberates some glycomacropeptides, thus limiting value of the test in that application.

33.12 Summary

To preserve milk and reduce its weight and volume, it may be concentrated and/or dried. Concentration is normally achieved by boiling at reduced pressure, with heat being added to product held in a vacuum. Concentrate may be treated to preserve it at this point; for example, sterilization produces evaporated milk and addition of sweetener produces sweetened condensed milk. Dried milk products are usually produced from concentrates. Thin films may be spread on hot drums with the dried product being scraped off. Or small droplets may be atomized into an atmosphere of hot (usually), dry air to which they release their moisture. The latter spray-dried products may be instantized to improve reconstitutability. The process may involve rewetting, agglomeration, and redrying or use of a fluid bed connected to the main dryer.

Numerous fractions of milk can be dried alone and in combination. They include skim milk, cream, cultured buttermilk, buttermilk (from churning of cream), cheese, casein, and isolated whey proteins.

The USDA and the ADPI make important contributions to quality assurance and the grading of dry milk products.

Study Questions

1. Define the following generic names of concentrated milk products: concentrated milk, evaporated milk, and sweetened condensed milk.
2. Describe how evaporated milk sterilized in the final container will differ in characteristics from that sterilized before being aseptically canned.
3. Cite three purposes of forewarming milks intended for concentrated and dried products.
4. Why is vacuum applied in the process of concentrating milk? What range of vacuum pressures is usually employed?
5. Of what major advantage is the use of multiple-effect evaporators? Cite normal minimum and maximum temperatures.
6. What are five major components of a milk evaporator? Explain the function of each.
7. The alcohol stability test would be used in conjunction with the production of what concentrated product?
8. Draw a flow diagram for production of evaporated milk, concentrated milk, sweetened condensed milk, and low- and high-heat NDM.
9. Which components of milk combine in the Maillard reaction?
10. What concentration of sugar is necessary to preserve sweetened condensed milk? How does this differ from sugar concentration expressed as the sugar ratio?
11. Discuss the purposes and methods of forced crystallization. Describe the sizes and number of crystals that can result.
12. What type of NDM is preferred by bakers? Why?
13. What are the two most important constituents of dried whey? How may they be separated and purified?
14. Name six main functional parts of a spray drier.
15. Discuss primary and secondary air in dry milk products.
16. Describe three methods of separating fine particles of dried milk from air as it passes through a spray drier
17. Why should oxygen be removed from dry milk?
18. What is the normal physical state of lactose in dry milk? What may happen to its physical state if sufficient moisture is absorbed to increase its concentration above 7%?
19. Does free fat on surfaces of particles of dried milk attract or repel water? How can wettability of dry milk be significantly improved?
20. Describe the stages in reconstituting dried milks. Why is sinkability a desirable attribute?
21. Differentiate between true and apparent densities. Describe a method for determining bulk density.
22. What are the normal steps in producing instantized NDM?
23. Name the grades of NDM recognized by the USDA. On what criteria are samples graded?
24. What are the four types of membranes that can be used in the filtration process? How does pore size and pressure applied affect the outcome?

Notes

[1] A barometric leg is placed at the bottom of the condenser. It consists of a column of water in a pipe sufficiently tall enough to provide an adequate weight of water to force removal of condensate by gravity as it is formed.

[2] Latent heat is heat required to convert a product at its boiling point into vapor. This increases as pressure decreases so that at a vacuum of 595 mm Hg (23.8 in) evaporation requires 50 kJ/lb. Heat required to raise the temperature of a liquid to its boiling point is called *sensible heat*, and this for milk is equal to 1.77 kJ/lb/°C.

[3] Dry milk from direct-type heaters typically contains higher nitrate content than that from indirect heaters.

[4] Therminol is the brand name by Solutia for heat transfer fluids intended for indirect heating purposes only.

[5] Water changes from the continuous to the discontinuous phase at about 15% moisture.

[6] One procedure involves evacuation (750 mm Hg), restoration of pressure with N_2 and CO_2 (4:1), sealing of the package, equilibration for one week, and repetition of the evacuation and gassing steps.

[7] Copper and iron act as catalysts of oxidation. Therefore, their content in dry milk is limited by US standards. Safe and suitable antioxidants are permitted in dry whole milk but not in nonfat dry milk.

[8] Reconstitutability is the ease with which dry milk is recombined with water and is dependent on dispersibility, wettability, sinkability, and solubility.

References

American Dairy Products Institute. n.d. *Standard Reference Non-Fat Dry Milk Samples.* Elmhurst, IL: Author.

American Dairy Products Institute. 2009. *Standards for Grades of Dry Milks Including Methods of Analysis* (Bulletin 916). Elmhurst, IL: Author.

American Dairy Products Institute. 2014. *Whey & Whey Products; Definitions, Compositions, Standard Methods of Analysis* (Bulletin W-16). Elmhurst, IL: Author.

Buma, T. J. 1971. Free fat in spray-dried whole milk X. A final report with a physical model for free fat in spray-dried milk. *Netherlands Milk and Dairy Journal* 25:159–174.

Bhushan, S., and M. R. Etzel. 2009. Charged ultrafiltration membranes increase the selectivity of whey protein separations. *Journal of Food Science* 74:E131–E139.

Burrington, K. J., T. Schoenfuss, and S. Patel. 2014. Technical Report: Coproducts of Milk and Whey Processing (http://www.usdairy.com/~/media/usd/public/coproducts_milk_report.pdf). Dairy Research Institute and US Dairy Export Council.

Hunziker, O. F. 1946. *Condensed Milk and Milk Powder* (6th ed.). LaGrange, IL: Author.

Hurt, E., J. Zulewska, N. Newbold, and D. M. Barbano. 2010. Micellar casein concentrate production with a 3X, 3-stage, uniform transmembrane pressure ceramic membrane process at 50°C. *Journal of Dairy Science* 93:5588–5600.

Kelly, M., B. Vardhanabhuti, P. Luck, M. A. Drake, J. Osborne, and E. A. Foegeding. 2010. Role of protein concentration and protein-saliva interactions in the astringency of whey proteins at low pH. *Journal of Dairy Science* 93(5):1900.

Peebles, D. D., and P. D. Clary. June 14, 1955. Milk Treatment Process. US Patent 2,710,808.

Pisecky, J., and V. Westergaard. 1972. Manufacture of instant whole milk powder soluble in water. *Dairy Industry* 37:144–147.

Sikand, V., P. S. Tong, S. Roy, L. E. Rodriguez-Sona, and B. A. Murray. 2011. Solubility of commercial milk protein concentrates and milk protein isolates. *Journal of Dairy Science* 94:6194–6202.

USDA Agricultural Marketing Service. 2001, February. *United States Standards for Grades of Nonfat Dry Milk (Spray Process).* Washington, DC: Author.

Vardhanabhuti, B., M. A. Kelly, P. J. Luck, M. A. Drake, and E. A. Foegeding. 2010. Roles of charge interactions on astringency of whey proteins at low pH. *Journal of Dairy Science* 93:1890.

Waite, R. 1964. Some recent developments in the manufacture of milk powders. *International Journal of Dairy Technology* 17:122–127.

Wehr, H. M., and J. F. Frank (eds.). 2004. *Standard Methods for the Examination of Dairy Products* (17th ed.). Washington, DC: American Public Health Association.

Websites

American Dairy Products Institute (https://www.adpi.org)

Dairy Research Institute (http://www.usdairy.com/about-us/about-our-research-leadership)

Eastman Chemical Company (http://www.therminol.com)

Food Safety and Standards Authority of India (http://fssai.gov.in)

Koch Membrane Systems (http://kochmembrane.com)

US Dairy Export Council (http://www.usdec.org)

Glossary

Abomasum The fourth stomach compartment (true stomach) of a **ruminant**.

Accretion The process of growth or enlargement by a gradual buildup.

Acetonemia (ketosis) A condition characterized by an elevated concentration of ketone bodies in body tissues and fluids. It is more common among high-producing cows in a negative energy balance.

Acid detergent fiber (ADF) Fiber extracted with acidic detergent in a technique employed to help appraise the quality of **forages**.

Ad libitum (ad lib.) At pleasure. Commonly used to express the availability of feed on a free-choice basis.

Adenosine triphosphate (ATP) An energy-rich substance. It carries energy from biochemical reactions that yield energy to those reactions that use energy.

Aerobic Requiring oxygen. For example, many microorganisms require oxygen for the **oxidation** of food materials, as in the Krebs cycle.

Afterbirth The placenta and allied membranes with which the fetus is connected. It is expelled from the uterus following **parturition**.

Agar An extract of seaweed that forms a firm gel at temperatures of about 40°C and lower. It is used at a concentration of about 1.5% in microbiological plating media.

Agglomeration In dry milks, the process that causes rewetted particles to collide and stick together into clusters, which are then redried. This process improves the dispersibility of **dried milks** in water.

Agglutination A reaction in which particles suspended in a liquid collect into clumps, such as occurs in an **antigen-antibody reaction**.

Allele One of a pair of genes that appears at a particular location on a particular chromosome and controls the same characteristic as the other gene.

Alopecia Baldness; loss or deficiency of hair, natural or abnormal.

Alpha-amino (α-amino) Designates the location of an $-NH_2$ group on the alpha carbon, as in amino acids.

Ambient That which surrounds. For example, the temperature about us is referred to as ambient temperature.

Anaerobic Living, active, occurring, or existing in the absence of free oxygen.

Anaphylactic shock (anaphylaxis) A state of immediate hypersensitivity following sensitization to a foreign protein or drug, usually through the **parenteral** route.

Anion An ion carrying a negative charge of electricity. Anions include the nonmetals, the acid radicals, and the hydroxyl ion.

Anionic The electrostatic status of a negatively charged ion (as one attracted to the anode in electrolysis) or any negatively charged atom or group of atoms (as opposed to **cationic**).

Anorexia Lack or loss of appetite for food.

Anthelmintic A remedy for destruction or elimination of intestinal worms.

Antibiotic A metabolic product of one microorganism that in low concentrations is detrimental to the life activities of other microorganisms. Penicillin, tetracycline, and streptomycin are antibiotics.

Antibody A protein substance (modified type of blood-serum globulin) developed or synthesized by lymphoid tissue of the body in response to an antigenic stimulus. Each **antigen** elicits production of a specific antibody. In disease defense, the animal must have had an encounter with the **pathogen** (antigen) before a specific antibody is found in its blood.

Antigen A high molecular weight substance (usually protein) that, when foreign to the bloodstream of an animal, stimulates formation of a specific **antibody** and reacts specifically **in vivo** or **in vitro** with its homologous antibody.

Antigen-antibody reaction Combination of **antigen** with **antibody** to inactivate the antigen. Commonly, the reaction results in **agglutination** and precipitation of the complex.

Apocrine A type of glandular secretion in which the apical portion of the secreting cell is released along with the secretory products.

Apoptosis A normal physiological process eliminating DNA-damaged, superfluous, or unwanted cells.

Apparent acidity The **titer** that arises from the presence of substances in fresh milk, other than acids, which react with sodium hydroxide in the **titratable acidity** test. These acidic substances are phosphates, citrates, **casein**, **whey** proteins, and dissolved carbon dioxide.

Archimedes spiral A plane curve generated by a point moving away from or toward a fixed point at a constant rate while the radius vector from the fixed point rotates at a constant rate.

Artificial insemination (AI) The introduction of male reproductive cells into the female reproductive tract by an artificial (mechanical) means.

Aseptic Preventing or free from contamination by microorganisms.

Aseptic packaging Method of filling packages without allowing the entry of microorganisms. Commonly used for sterile milk products to provide improved keeping quality.

Assay Determination of the purity or potency of a substance or the amount of any particular constituent of a mixture.

Asymptomatic Showing no evidence of disease.

Autocrine The cell signals itself through a chemical that it synthesizes and then responds to.

Autogenous vaccine Custom vaccines that consist of herd specific (homologous) **antigens**.

Bacterial count The number of bacteria that are **viable**, or both viable and nonviable, depending on method of analysis. For example, plate counts determine numbers of viable bacteria capable of growth under conditions imposed, whereas direct microscopic counts enumerate all stainable cells both living and dead.

Bactericidal An agent or substance capable of destroying bacteria.

Bacterin A suspension of killed or attenuated bacteria (vaccine) used to increase disease resistance.

Bacteriophage An ultramicroscopic, filterable virus composed of **deoxyribonucleic acid (DNA)** and a protein coat. It infects particular **strains** or species of bacteria and usually causes **lysis**, or explosive dissolution, of the infected bacterium. Bacteriophages multiply at the expense of bacterial cells. They have no enzymes of their own.

Bacteriostatic A substance that prevents the growth of bacteria but does not kill them.

Bag See **udder**.

Balanced ration The daily food allowance of an animal, mixed to include suitable proportions of nutrients required for normal health, growth, production, and well-being.

Basal metabolic rate (BMR) See **basal metabolism (BM)**.

Basal metabolism (BM) The chemical changes that occur in the cells of an animal in the **fasting** or resting state when it uses just enough energy to maintain vital cellular activity, respiration, and circulation as measured by the **basal metabolic rate (BMR)**. Basal conditions include thermoneutral environment, resting, postabsorptive state (digestive processes are quiescent), consciousness, quiescence, and sexual repose. BMR is determined in humans 14 to 18 hours after eating and when at absolute rest. It is measured by means of a calorimeter and is expressed in calories per m^2 of body surface.

Beefy (beefiness) A term used to designate the desirable physical **conformation** of a beef animal, as contrasted with a dairy animal which is lean (not beefy) and more angular.

Bioactive peptides Specific protein fragments that have a positive impact on body functions or conditions and may ultimately influence health.

Biological value The ratio of nitrogen retained in the body to nitrogen consumed under a given set of conditions. It refers to proteins.

Bioreactor A system that supports a biologically active environment.

Bleat The vocal sound made by a goat.

Blend price The price paid producers for **market milk** when **classified pricing** is used in an **individual handler pool**. The blend price is an average of class prices weighted by the quantity of milk utilized in each class.

Blind quarter A quarter of an **udder** that does not secrete milk or one that has an obstruction in the teat that prevents the removal of milk. A nonfunctional mammary gland.

Bloat A disorder of **ruminants** usually characterized by an accumulation of gas in the **rumen** (cf. chapter 13).

Blood agar A highly nutritive medium containing red blood cells. It promotes growth and production of **hemolysins** by most microorganisms, especially pathogenic bacteria. Types of hemolysins are distinguishable on blood agar.

Blood plasma The liquid portion of blood in which the corpuscles of blood cells are suspended.

Bloom A term commonly used to describe the beauty and freshness of a cow in early lactation. A dairy cow in bloom has a smooth hair coat and presents evidence of milking ability (**dairy character**). Also, a term expressing strength of gel in gelatin.

Bolus **Regurgitated** food that has been chewed and is ready to be swallowed; a large pill for dosing animals.

Bomb calorimeter An apparatus for measuring the heat of combustion, as of human foods, animal feeds, and fuel.

Bone meal Animal bones that were steamed under pressure and then ground. It contains 1.5 to 2.5% nitrogen, 12 to 15% phosphorus, and 20 to 34% calcium. It is used as a fertilizer and as a mineral **supplement** for feeding farm animals.

Bovine Pertaining to the ox or cow.

Bran The seed coat of wheat and other cereal grains that is separated from flour and used as animal food.

Breed Animals having a common origin and characteristics that distinguish them from other groups within the same species.

Breed average The average milk production of cows for a given dairy breed. (Usually, computations are based on DHIA records of the past five years.)

Breed character The physical characteristics that distinguish one breed of animals from another within the same species.

Breed type Physical characteristics peculiar to a certain breed, including special features of head, ears, and coat color.

Breeding value The genetic ability of an animal to secrete milk. One-half of this genetic ability is transmitted to offspring. See **genetic value**.

Broad-spectrum antibiotic An **antibiotic** that is active against a large number of microbial species.

Buck A sexually mature male goat.

Bulk milk Milk stored on the farm in a tank (or tanks), as compared with that stored in cans.

Bulk milk tank A refrigerated stainless steel vessel in which milk is cooled quickly to 2 to 4°C (35 to 39°F) and stored until picked up in a **bulk tank truck**.

Bulk tank truck A vehicle bearing a large stainless steel tank in which milk from farm **bulk milk tanks** is transported. Capacities range from 18,000 to 45,000 lb.

Bull A sexually mature uncastrated **bovine** male.

Buttermilk There are two kinds of buttermilk. *Natural* buttermilk is the by-product of churning cream into butter. *Cultured* buttermilk is usually made from fresh **skim milk**

(cultured buttermilk may be made from reconstituted dry skim milk or reconstituted **plain** condensed skim milk). It is prepared by heating fresh skim milk to 85°C for 30 minutes (or equivalent), cooling to 21°C, adding a lactic **starter culture**, and incubating at 21°C until the acidity develops to about 0.8%. Salt is usually added to accentuate the flavor.

Calving interval The average number of days between the latest two calving dates for all cows in the herd.

Cannula In animal physiology, a device used to connect the **rumen** with the outside environment. The opening leading from the rumen to the outside is called a **fistula**. A metal, rubber, or glass tube cannula may be inserted into a body cavity to allow the escape of fluids or gas. Liquids and other materials may be introduced into the body through a cannula.

Carcinogenic Cancer-producing tendency.

Cardiac arrhythmia An alteration or abnormality of normal cardiac rhythm.

Carrageenan A **hydrophilic** colloid extracted from the seaweed known as Irish moss. It is used as a stabilizer in frozen desserts and in **evaporated milk**. The functional group, sulfate, reacts with proteins to form a gel.

Casein The major protein of milk.

Cation An ion carrying a positive charge of electricity. Cations include the metals and hydrogen.

Cationic Refers to action of a **cation**. For example, a detergent, such as a quaternary ammonium salt, in which the cleansing action is inherent in the cation.

Cellulose The principal carbohydrate constituent of plant cell membranes. It is made available to **ruminants** through the action of cellulolytic bacteria that inhabit the **rumen**.

Centrifugal separation Removal of one substance from another mechanically by centrifugal force. Materials to be removed must exist in a separate phase, and one phase must have a higher density than the other, as with cream, which is less dense than **skim milk**.

Certified milk Milk produced and distributed under conditions that conform with high standards for cleanliness and quality set forth by the American Association of Medical Milk Commissions.

Challenge feeding Also called **lead feeding**. This concept is aimed at finding the potential of cows to secrete milk. Hence, an abundance of high-energy feed, balanced for the needs of high milk production, is provided cows above what might apparently be needed.

Checkoffs The practice of deducting a portion of the payment to a farmer for his product, the deducted portion being used by a specific agency to promote sales of the specific product. For example, checkoffs for milk may be transferred to the United Dairy Industry Association for promotional uses of milk and milk products.

Cholesterol A white, fat-soluble substance found in animal fats and oils, in bile, blood, brain tissue, nervous tissue, the liver, kidneys, and adrenal glands. It is important in metabolism and is a precursor of certain hormones.

Chymosin (rennin) A protease found in **rennet**. It is produced by newborn **ruminant** animals in the lining of the fourth stomach to curdle the milk they ingest, allowing a longer residence in the bowels and better absorption.

Cis fatty acids A natural fatty acid in which the carbon moieties lie on the same side of the double bond.

Clarification The process of removing suspended material; in milk processing, clarification is achieved by centrifugal treatment, which removes sediment (extraneous matter) and somatic cells.

Class I milk Milk used in **fluid milk products**. The highest price is received by dairypersons for this class of milk.

Class I price The minimum price that **handlers** must pay for milk used in **fluid milk products** in a federal order market.

Class II milk Milk in excess of Class I needs, consequently used in a lower price category. This includes milk used for manufacturing purposes. In some markets, **surplus milk** is divided into two classes, Class II and Class III, and milk is allocated to the respective classes based on use.

Class II price The minimum price **handlers** must pay for milk used in the Class II category.

Classified pricing A pricing plan by which **handlers** are charged for **market milk** according to the products in which it is used, hence the term *use classification.* **Class I milk** goes into fluid products and demands the highest price. Prices of **surplus milk**. Class II or III, depending on the **federal milk marketing order**, approximate the price of **manufacturing milk**.

Cleaned-in-place (CIP) Most stainless steel pipelines are cleaned by circulating washing solutions through them.

Clinical Refers to direct observation.

Close Descriptive of body and texture. In frozen desserts, it is the quality imparted by having minute ice crystals and air cells so the body is smooth (the opposite of **coarse**). In cheese, it is the absence of mechanical openings or gas holes (the opposite of **open**).

Close breeding A form of inbreeding, such as mating brothers to sisters, **sire** to daughter, and son to **dam**.

Coarse A term used in evaluating dairy products. In frozen desserts, it denotes lack of smoothness; in cottage cheese and **buttermilk**, it means an overabundance of diacetyl; in butter, a lack of sweet, pleasant, delicate flavor.

Cocoa The product derived from fermented beans of the cacao tree. Fermented beans are crushed and the nibs are separated to be hydraulically pressed. Pressing removes fats and forms a cake, which is ground to produce the cocoa.

Colligative properties Characteristics that vary with the number of **molecules** but not their nature. Freezing point, boiling point, osmotic pressure, and vapor pressure are colligative properties of solutions.

Colloidal In a state of fine subdivision or dispersion with particles ranging between 10^{-5} and 10^{-7} cm in diameter, and having no effect on the **colligative properties** of solutions. The **casein** of milk is in collodial suspension.

Colostrum The first milk secreted **pre-** and **postpartum**.

Comminuted The process in which solid materials are reduced in size by crushing, grinding, and other processes.

Compaction Closely packed feed in the stomach and intestines of an animal causing constipation and/or digestive disturbances.

Complete ration A blend of all feedstuffs (**forages** and grains) in one feed. A complete ration fits well into mechanized feeding and the use of computers to formulate least-cost rations.

Concentrate A feed high in **nitrogen-free-extract** and **total digestible nutrients** and low in **crude fiber** (less than 18%). Included are cereal grains, soybean oil meal, cottonseed meal, and by-products of the milling industry such as corn gluten and wheat **bran**. A concentrate may be low or rich in protein.

Concentrated milk The liquid food obtained by partial removal of water from milk. The **milkfat** and **total milk solids** are not less than 7.5 and 25.5%, respectively. It is pasteurized but not processed by heat so as to prevent spoilage. It may be homogenized. Also called *condensed milk*. See **concentrated milks**.

Concentrated milks Milk, **skim milk**, and components of these that have been concentrated by removing varying amounts of water under carefully controlled conditions of heat and/or vacuum. Concentrated milks may be reconstituted by adding appropriate quantities of water. See **concentrated milk**.

Condition Refers to the amount of flesh (body weight), quality of hair coat, and general health of animals.

Conformation The body form or physical traits of an animal; its shape and arrangement of parts.

Conjugated A substance formed by the union of two compounds.

Conjunctiva The mucous membrane covering the anterior portion of an eyeball.

Constitution The general strength and vigor of an animal.

Cooperative pool Each producer who belongs to a cooperative receives a common **blend price** calculated on the basis of use of all milk sold through the cooperative.

Correlation A measure of how two traits or parameters vary together. A correlation of +1.00 means that as one trait increases the other also increases proportionately—a perfect *positive* relationship. A correlation of –1.00 means that as one trait increases the other decreases proportionately—a perfect *negative,* or *inverse,* relationship. A correlation of 0.00 means that as one trait increases, the other may increase or decrease—no relationship. Thus a correlation coefficient may lie between +1.00 and 1.00.

Cow A mature female **bovine**.

Crampiness A condition among cattle, especially **bulls** in confinement, in which involuntary contractions of muscles of the hind legs result in pain and discomfort. Affected animals shift from foot to foot.

Crossbred Offspring of **purebred** parents of different breeds.

Crossbreeding Mating animals of different breeds.

Crude fiber (CF) That portion of feedstuffs composed of **cellulose**, hemicellulose, **lignin**, and other polysaccharides that serve as structural and protective parts of plants (high in **forages** and low in grains).

Crude protein (CP) Total protein in a feed. To calculate the protein percentage, a feed is first chemically analyzed for nitrogen content. Since proteins average about 16% (100 ÷ 6.25) nitrogen, the percentage of nitrogen in the analysis is multiplied by 6.25 to give the CP percentage.

Crystal lattice The organized structure of a solid chemical element or compound, the components of which are tightly packed together, such as **lactose** crystals.

Cud A **bolus** of **regurgitated** food (common to **ruminants**). See **rumination**.

Culling The process of eliminating nonproductive or undesirable animals.

Culture In microbiology, a population of microorganisms in a medium. A pure culture contains only organisms that initially arose from a single cell. Cultures are used in manufacturing cultured dairy products and most cheeses.

Curd The coagulated or thickened part of milk. Curd from whole milk consists of **casein**, fat, and **whey**, whereas curd from **skim milk** contains casein and whey but only traces of fat.

Cytokine A broad and loose category of small proteins (~ 5 to 20 kDa) that are important in cell signaling. They are released by cells and affect the behavior of other cells, and sometimes the releasing cell itself.

Dairy character Physical traits that suggest high milking ability. Important indications of this include an alert, feminine head; long thin neck; **openness** and sharpness throughout; prominent hips and pins; sharp withers; thin thighs; flat flinty bone; and thin, pliable hide with a soft silky hair coat. See **bloom**.

Dairy Herd Improvement Registry (DHIR) A modification of the DHIA program to make milk production records acceptable by dairy breed associations. An **official production record** program.

Dam The female parent.

Days dry As used in DHIA records, this is the number of days **dry** prior to calving at the start of the production record listed.

Deaeration Removal of air. In the processing of milk, it refers to **vacuumization** for removal of undesirable volatile substances that impart odors. No moisture is removed.

Decant To pour off the top layer of a multilayered solution or the liquid from a precipitate.

Dehorn To remove horns from animals or to treat young animals so horns will not develop.

Deoxyribonucleic acid (DNA) The chemical substance that is the principal nuclear material of cells. The structure of DNA determines the structure of ribonucleic acid which, in turn, determines the structure of proteins of the cell.

Developed acidity The quantity of acids produced as microorganisms ferment **lactose** (primarily) in milk, usually expressed as a percentage of lactic acid in the United States. Also called *real acidity* in contrast with **apparent acidity**.

Dicoumarol A chemical compound found in spoiled sweet clover and lespedeza hays. It is an anticoagulant and can cause internal hemorrhages when ingested by cattle.

Dielectric Conducting electricity poorly. Fat is such a substance. The moisture content in butter may be controlled through continuously monitoring the resistance to flow of electric current, because the higher the moisture content, the less the resistivity.

Digestibility The percentage of food ingested that is absorbed into the body as opposed to that which is excreted as feces.

Digestible energy (DE) That portion of energy of a feed that can be digested or absorbed into the body by an animal.

Digestible protein (DP) The amount of protein of feed that is absorbed by the digestive tract; it may be computed using the formula: DP (%) = **crude protein** of feed (%) × **digestion coefficient** for protein in the feed.

Digestion coefficient The digestion coefficient of feed ingredients (DCFI) may be calculated using the formula:

$$\text{DCFI} = \frac{\text{Wt. of ingred consumed} - \text{Wt. of undigested ingreds. in feces}}{\text{Wt. of ingred. consumed}} \times 100$$

Direct expansion A refrigeration system in which liquid refrigerant is forced through a valve into a chamber of lower pressure. As it expands and becomes a gas, it takes on heat.

Doe A sexually mature female goat.

Doeling A female goat from its first to its second birthday, or from the first birthday until first **parturition**.

Dominant Describes a gene that, when paired with its **allele**, covers up the phenotypic expression of that allele gene. For example, black and white in Holsteins (B) is dominant over red and white (b). Thus, BB and Bb Holsteins are black and white, whereas bb individuals are red and white.

Dried milk Dried milk is made from whole milk; **nonfat dry milk** is manufactured from **skim milk**. In drying milks, about 95 to 98% of the water is removed by concentrating the solids in vacuum pans and then dried by spraying the milk into a chamber of hot air or onto a slowly revolving heated drum under vacuum. Dried milks may be stored for extended periods of time in moisture-proof containers at ordinary room temperatures.

Dry Describes a nonlactating female. The dry period of cows is the time between lactations (when a female is not secreting milk).

Dry matter (DM) The moisture-free content of feeds.

Dry off To change a lactating animal to a nonlactating one. See **involution**.

Dry period Nonlactating days between lactations.

Drylot A relatively small area in which cattle are confined indefinitely as opposed to being allowed to have free access to **pasture**.

Dystocia Abnormal or difficult labor (**parturition**), causing difficulty in delivering the fetus and placenta.

Edema The presence of abnormally large amounts of fluid in the intercellular tissue spaces of the body, as in swelling of mammary glands commonly accompanying **parturition** in many farm mammals.

Efficacy Effectiveness. For example, the effectiveness of an **antibiotic** in the therapy of **mastitis**.

Electroporation A technique in which an electrical field is applied to cells in order to increase the permeability of the cell membrane, allowing chemicals, drugs, or DNA to be introduced into the cell.

Emaciation A wasted condition of the body; great losses of body weight.

Emulsion A suspension of fine particles or globules of a liquid in a liquid. It has a dispersed phase and a continuous phase.

Endemically Prevalent in or peculiar to a particular locality, region, or population.

Endocrine Chemical signals are secreted into the blood and carried by blood and tissue fluids to the cells they act upon.

Endocrine system The collection of glands that produce hormones that regulate metabolism, growth and development, tissue function, sexual function, reproduction, sleep, and mood, among other things.

Ensilage A **green chop** (**forage**) preserved by fermentation in a **silo**, pit, or stack, usually in chopped form. Also called **silage**.

Ensiling The process of making **ensilage** (silage).

Enucleated oocyte An egg cell from which the nucleus has been removed.

Ergosterol A plant sterol that, when activated by ultraviolet rays, becomes vitamin D_2; also called *provitamin* D_2 and *ergosterin*.

Eructation The act of belching or casting up gas from the stomach.

Erythema Redness of the skin that results from capillary congestion.

Essential amino acid An amino acid that cannot be synthesized in the body from other amino acids or substances, or that cannot be made in sufficient quantities for the body's needs. Eight amino acids are believed essential to humans.

Essential fatty acid A fatty acid that cannot be synthesized in the body, or that cannot be made in sufficient quantities for the body's needs. Linoleic and linolenic acids are essential for humans.

Esterify The combination of a carboxylic acid with an alcohol to form an ester plus water.

Estimated average transmitting ability (EATA) The best estimate of a cow's ability to transmit to her offspring; based on production of her paternal sisters, **dam**, maternal sisters, and daughters, in addition to the cow's own milk production. EATA is, therefore, a genetic evaluation of the cow since it evaluates both the cow's production and **pedigree**. Information from each pedigree source is weighted in accordance with how much the information contributes to the EATA. For example, production data of a cow's daughters are given greater weight than those of **half-sibs**.

Estrus (oestrus) The recurrent, restricted period of sexual receptivity (**heat**) in female mammals, commonly marked by intense sexual urge.

Ether extract (EE) Fatty substances of foods and feeds that are soluble in ether.

Etiologic Relating to **etiology**.

Etiology The science or study of the causes of disease, both direct and predisposing, and the mode of their operation. Pathogenesis.

Euglobulins One of a class of globulins characterized as being insoluble in water but soluble in saline solutions.

Evaporated milk The liquid food obtained by partial removal of water from milk. The ME and TMS contents are not less than 6.5 and 23%, respectively. It contains 25 IU vitamin D per ounce and is homogenized. The food is sealed in a container and is so processed by heat, either before or after sealing, as to prevent spoilage.

Expectoration The act of spitting out materials, as in evaluating dairy products when it is unwise to swallow samples because of accumulative effects on the ability to evaluate product quality.

Extroversion The process of exposure of tissues inside the streak canal of cows' teats. It is usually caused by prolonged milking and excessive milking vacuum.

F_1 generation The first filial generation, or the first-generation progeny, following the parental, or **P_1**, generation.

Facultative Having the power to live under different conditions, as with bacterial growth in both high and low concentrations of oxygen.

False heat The display of **estrus** by a female animal when she is pregnant.

Fastidious Literally, difficult to please. In microbiology, it means an unusual requirement for preformed nutrients or specific environmental conditions.

Fasting Abstaining from food. Certain types of experiments involve withholding food from the test animal for varying periods of time.

Fat-corrected milk (FCM) The estimated quantity of milk calculated on an equivalent energy basis. It is a means of evaluating milk production records of different dairy animals and breeds on a common basis energy-wise. The following formula is used: FCM = (0.4 × milk production) + (15 × **milkfat** production).

Federal milk market administrator An appointee of the US secretary of agriculture who supervises a **federal milk marketing order**.

Federal milk marketing order (FMMO) A milk marketing regulation issued by the US secretary of agriculture. It establishes minimum prices paid by processors to producers for milk.

Feed efficiency The units of feed consumed per unit of weight increase or unit of production (milk, meat, eggs). Also called *feed conversion*.

Feed grains Any *grain* used for livestock *feed*, including corn, grain sorghum, oats, rye, and barley.

First-calf A term commonly used to indicate the first calf born to **bovine** females.

Fistula See **cannula**.

Flaming In microbiological practice, the sterilization of surfaces such as transfer loops by holding in a flame. The Bunsen burner is commonly used to produce a flame.

Flora General plant life. See **microflora**.

Fluid milk Milk commonly marketed as fresh liquid milks and creams. It is the most perishable form of milk and commands the highest price per unit volume. Also called **market milk**.

Fluid milk products Milk, flavored milk, **concentrated milk**, filled milk, **skim milk**, fortified skim milk, **lowfat milk**, **buttermilk**, milk drinks, and cream products.

Flush season Time during the year, usually spring and early summer, of maximum milk production.

Fodder Coarse food for cattle or horses, such as corn stalks.

Forage **Roughage** of high feeding value. Grasses and **legumes** cut at the proper stage of maturity and stored to preserve quality are forage.

Fore quarter The two front quarters of a cow. Also called *fore udder*.

Free radical An atom or group of atoms having at least one unpaired electron. They participate in oxidative reactions in some dairy foods.

Free stalls Cubicles and similar housing arrangements in which dairy animals are free to enter and leave, as opposed to being confined in stanchions.

Freemartin Female born twin to a **bull** calf (about 9 out of 10 of these will not conceive).

Freshen Commonly used to designate the act of calving (**parturition**); to give birth to a calf and concurrently initiate lactation.

Full-sibs Animals having the same **sire** and **dam**, such as full brothers, full sisters, and/or a full brother-sister pair.

Gametogenesis A biological process by which diploid or haploid precursor cells undergo cell division and differentiation to form mature haploid gametes.

Gaseous products of digestion (GPD) These include the combustible gases produced in the digestive tract incident to fermentation of the **ration**. Methane constitutes the major proportion of the combustible gases produced by **ruminants**; however, nonruminants also produce methane. Trace amounts of hydrogen, carbon monoxide, acetone, ethane, and hydrogen sulfide are also produced.

Gastroenteritis Inflammation of the mucosa of the stomach and intestines.

Generally recognized as safe (GRAS) Describes food additives that have been judged as safe for human consumption by a panel of expert pharmacologists and toxicologists who use data available, including experience of common use in food. The common use factor is often the major criterion on which judgment is based.

Generation interval The time from the birth of one generation to birth of the next. In dairy cows, it is the average length of time between the birth of a cow and the birth of her replacement **heifer** (commonly five to six years).

Genetic value Total value of all genes of an animal for secreting milk or for some other trait. See **breeding value**.

Genotype The actual genetic **constitution** (makeup) of an individual as determined by its germ plasm. For example, there are two genotypes for black-and-white Holsteins—BB and Bb. See **dominant**.

Germinate To begin to grow; to sprout or evolve.

Gestation Pregnancy (**gravidity**). The period from conception to birth.

Ghee Concentrated **milkfat** prepared by melting butter, **decanting** the fat after gravity separation from the serum, and driving off most of the remaining moisture by heating. About 1% moisture remains. It is used mostly in Asia and Africa.

Glycolysis The conversion of carbohydrates to **lactate** by a series of catalysts, the breaking down of sugars into simpler compounds.

Goitrogenic Producing or tending to produce goiter.

Gossypol A toxic yellow pigment found in cottonseed. Heat and pressure tend to bind it with protein and thereby render it safe for animal consumption.

Grade Animals showing the predominant characteristics of a given breed. They usually have at least one **purebred** parent, ordinarily the **bull**.

Grade "A" milk Milk produced and processed under rigid sanitary regulations. Milking and processing facilities are approved and inspected by public health officials. **Fluid milk products** must meet this inspection standard. Ordinances controlling production and processing are usually patterned after the *Pasteurized Milk Ordinance* recommended by the US Food and Drug Administration.

Grading up The continued use of **purebred** sires on **grade** dams.

Gram-positive Cells that retain the crystal violet stain when alcohol is applied in the procedure developed by Hans Christian Gram. These cells have a thick layer of peptidoglycan that retains the stain. Cells that lose the stain are called Gram-negative.

Grass tetany A magnesium-deficiency disease of cattle characterized by hyperirritability, muscular spasms of legs, and convulsions.

Gravidity See **gestation**.

Graze To consume standing vegetation, as by livestock or wild animals.

Green cheese A term applied to the fresh uncured product. It refers to flavor, odor, body, and texture, not color.

Green chop **Forages** harvested (cut and chopped) in the field and hauled to livestock. This minimizes the loss of moisture, color, nutrients, and wastage. Also called *zero grazing* or *fresh forage*.

Gross energy (GE) The amount of heat, measured in calories, released when a substance is completely oxidized in a **bomb calorimeter**.

Growthy Describes an animal that is large and well-developed for its age.

Half-sib In genetics, a half brother or half sister.

Handlers Processors or dealers of milk who commonly purchase **raw milk** and sell pasteurized milk and milk products.

Haplotypes A combination of DNA sequences (alleles) at adjacent locations (loci) on a chromosome that are inherited together.

Hay Dried **forage** (e.g., grasses, alfalfa, clovers) used for feeding farm animals.

Haylage **Low-moisture silage** (35 to 55% moisture). Grass and **legume** crops are cut and wilted in the field to a lower moisture level than normal for grass silage, but the crop is not sufficiently dry for baling. It is commonly stored in a sealed, or airtight, storage system.

Heart girth The circumference of the body just back of the shoulders of an animal. It is used to estimate body weight.

Heat See **estrus**.

Heat increment (HI) The increase in heat produced following consumption of food when an animal is in a thermoneutral environment. It consists of calories released in fermentation and nutrient metabolism. When environmental temperature is below critical temperature, this heat may be used to keep the body warm; therefore, it is not wasted. Also called *work of digestion*.

Heat shock In frozen desserts, the melting and refreezing of ice crystals resulting in a **coarse**, icy body and texture. Water from small ice crystals tends to combine and to form larger crystals on refreezing. Some ice melts with even a small change in temperature below 0°C.

Heifer A **bovine** female less than three years of age who has not borne a calf. Young cows with their first calves are sometimes called *first-calf heifers*.

Hemolysin A substance that causes the dissolution of red blood cells.

Hemolysis The liberation of hemoglobin. Hemolysis consists of the separation of hemoglobin from the red blood corpuscles and its appearance in the fluid in which corpuscles are suspended. Freezing blood, for example, causes hemolysis. Many microorganisms are able to hemolyze red blood cells by production of **hemolysins**.

Herdmate comparison Removes, from the evaluation of **breeding value**, complications arising from herd, year, and season of calving production variations, and assumes that measured production differences are due to inheritance (cf. chapter 6).

Herdmates A term used when comparing records of a **sire's** daughters with those of their nonpaternal herdmates, both being milked during the same season of the year under the same conditions. Each daughter is compared with her herdmates who calved over a five-month period (two months prior and two months succeeding, plus the month the daughter calves). This is done to adjust for the effect of seasonal differences. Also called *stablemates* or *contemporaries*.

Heritability A term used by animal breeders to indicate what proportion (fraction) of differences in a trait, such as milk production, is due to differences in **genetic value** rather than environmental factors. It is the variation due to genetic effects divided by the total variation (genetic plus environmental).

Heritability estimate An estimate of the proportion of total phenotypic variation between individuals for a certain trait that is due to heredity. It is the hereditary variation due to additive gene action.

Heterosis The increased growth or production exhibited by the **F_1 generation** over the average of its two parent breeds. It is usually expressed as a percentage. Also called **nicking** or *hybrid vigor*.

High-moisture silage Silage usually containing 70% or more moisture.

High-temperature short-time (HTST) The temperature and time combination for **pasteurization** of milk and milk products. The minimum for milk is 161°F (72°C) for at least 15 sec. Higher temperatures are used in pasteurizing products of high solids because of the protection afforded bacteria by solids.

Holding area An area commonly large enough to accommodate one to four groups of cows. A group means the number of cows required to fill the milking parlor stalls.

Homogenization See **homogenized milk**.

Homogenized milk Milk that has been treated to ensure breakup of fat globules to such an extent that, after 48 hr of quiescent storage at 7°C, no visible cream separation occurs on the milk, and the fat percentage of the top 100 mL of milk in 1 qt, or of proportionate volumes in containers of other sizes, does not differ by more than 10% from the fat percentage of the remaining milk as determined after thorough mixing. The reduced size of fat particles results in formation of a softer **curd** in the stomach.

Homozygous Having identical pairs of genes for any given pair of hereditary characteristics.

Hot-pack A process of filling containers with hot product, closing the package, and cooling.

Hydrogenated Having hydrogen added. For example, hydrogenated fats have hydrogen atoms added at double bonds in the carbon chain of unsaturated fatty acids.

Hydrophilic Water loving.

Hygroscopicity The degree of tendency to absorb and retain moisture. For example, **nonfat dry milk** is hygroscopic in that it tends to absorb moisture.

Hyperemia Panting (as in dogs); deep breathing.

Hypertonic A solution that contains a higher concentration of solutes outside the cell than inside the cell.

Hypocalcemia Below normal concentration of ionic calcium in blood, resulting in convulsions, as in **tetany** or **parturient paresis**.

Hypoglycemia A reduction in concentration of blood glucose below normal.

Ice milk A frozen product resembling ice cream, except that it contains less fat (2 to 5% versus 10%) and more **nonfat milk solids** (12% versus 10%) than ice cream. Both ice milk and ice cream contain stabilizers and emulsifiers and about 15% sugar.

Immunity The power an animal has to resist and/or overcome an infection to which most of its specie is susceptible. *Active* immunity is attributable to the presence of antibodies formed by an animal in response to an antigenic stimulus. *Passive* immunity is produced by the administration of preformed **antibodies**.

Impaction Constipation. See **compaction**.

In milk Designating a lactating female.

In utero In the uterus.

In vitro Within an artificial environment, as within a test tube.

In vivo Within the living body.

Individual handler pool A pool in which each producer selling **raw milk** to a single **handler** (dealer) receives the same

blend price for milk regardless of the use made of any individual producer's milk.

Integron A genetic element that possesses a site, *attI*, at which additional DNA, in the form of gene cassettes, can be integrated by site-specific recombination, and which encodes an enzyme, integrase, that mediates these site-specific recombination events.

Interface The boundary between two phases in a heterogeneous system, such as the boundary between fat globules and **skim milk** (serum) in milk or between free fat and **buttermilk** droplets in butter.

Intermediate form In microbiology, a microorganism that is taxonomically between bacteria and fungi.

Intravenously The infusion of liquid substances directly into a vein.

Involute See **involution**.

Involution A decline in size or activity of tissues and/or organs. For example, the mammary gland tissues normally **involute** with advancing lactation. It is a part of the process of lactating cows wherein they **dry off**.

Isoelectric point (pI) In **colloidal** chemistry, the pH at which a colloidal particle has a net charge of zero. For example, **casein** has an isoelectric point of 4.6, and it precipitates from suspension at this pH. This is the principle on which much cheesemaking is based.

Keto analogue A **molecule** having the same composition as an amino acid except that there is an oxygen on the alpha-carbon instead of a hydrogen and an amino group.

Ketosis See **acetonemia**.

Kid A young goat up to its first birthday.

Kidding **Parturition** in the goat; the act of giving birth to young.

Kilocalorie (kcal) Equivalent to 1000 calories.

Kwashiorkor A syndrome produced by a severe protein deficiency; characterized by changes in pigmentation of skin and hair, **edema**, anemia, and apathy.

Lachrymation The act of tearing; secreting and conveying tears.

Lactate To secrete or produce milk; also, salt of beta-hydroxy propionic acid.

Lactation period The number of days an animal secretes milk following each **parturition**.

Lactometer An instrument used to determine the specific gravity of milk. The lactometer reading is a factor in formulas for calculating the percentage of solids in milk.

Lactose The 12-carbon sugar unique to milk; a disaccharide composed of glucose and galactose.

Lactose intolerance A condition in which there are **clinical** signs (abdominal pain, diarrhea, bloating, flatulence) consistently following the ingestion of milk. **Lactose** is not adequately hydrolyzed in the intestine because of insufficient quantities of the enzyme lactase, and the high osmotic pressure that results causes infiltration of fluids into the intestines. Gas-producing microorganisms of the lower intestine may utilize lactose to produce gas.

Lamella A thin layer of substances surrounding an air cell, such as the air cell membrane in frozen desserts.

Laminar In layers. Often used to designate a type of flow of fluids that is the opposite of turbulent flow. For example, in water flowing slowly through a pipe, the portion at the center flows faster than that at the surface of the pipe because of the differences in friction. Turbulent flow is needed for efficient cleaning in **CIP** pipelines. Laminar air flow (air moving in one direction) is desirable in clean rooms, microbiological transfer hoods, and around food-packaging equipment to aid in prevention of contamination by microorganisms.

Lead feeding This is the practice of increasing grain fed cows to a level equal to 1.0 to 1.5% of body weight beginning about three weeks prior to the predicted calving date. Following **parturition**, grain is increased until a cow reaches maximum production or maximum feed intake, whichever comes first. Also see **challenge feeding**.

Legume Clovers, alfalfa, and similar crops that can absorb nitrogen directly from the atmosphere and use it through action of bacteria that live in their roots.

Leukopenia (leucopenia) A *decrease* below the normal number of leukocytes in peripheral blood. Also known as *leukocytopenia*.

Libido Sexual desire or instinct.

Lignin A compound which, with **cellulose**, forms the cell walls of plants. It is practically indigestible.

Line breeding A form of inbreeding in which an attempt is made to concentrate the inheritance of some ancestor in the **pedigree**.

Lipid Any one of a group of organic substances that are insoluble in water though soluble in alcohol, ether, chloroform, and other fat solvents, and have a greasy feel. They include fatty acids and soaps, neutral fats, waxes, steroids, and phosphatides (by United States terminology).

Lowfat milk Milk containing at least 8.25% **nonfat milk solids** and from which sufficient **milkfat** has been removed to produce, within limits of good manufacturing practice, a milkfat content of 0.5, 1.0, 1.5, or 2.0%. It is **pasteurized** or **ultrapasteurized**, contains 2,000 IU vitamin A per quart, and may or may not be **homogenized** and protein-fortified.

Low-moisture silage Silage that contains 35 to 55% moisture.

Lyophilization The evaporation of water from a frozen product with the aid of high vacuum. Also called *freeze-drying*.

Lysis The process of disintegration or solution; cell destruction. For example, the breakdown of red blood cells of **blood agar** by many pathogenic bacteria is lysis.

Maillard reaction The reaction of **lactose** with milk protein, notably with the epsilon-amino group of lysine. Amino compounds catalyze sugar dehydration, fragmentation, and condensation. Also known as the *browning reaction* because milk turns brown as high heat induces the reaction. Concomitantly, a caramelized flavor develops.

Manufactured milk products These include cheeses, butter, evaporated whole milk, condensed whole and **skim milks**, whole milk powder, **nonfat dry milk**, ice cream, frozen desserts, aerated cream, frozen and plastic creams, and milk used in candy, soup, bakery products, and animal feeds.

Manufacturing grade milk Milk produced under regulations less strict than those for **grade "A" milk** production. The USDA *Milk for Manufacturing Purposes and its Production and Processing: Recommended Requirements* are guidelines for state milk control agencies. Acceptable milk contains not more than 3,000,000 bacteria per milliliter.

Manufacturing milk Milk for the manufacture of dairy products, such as cheese, butter, powdered milk, ice cream, and condensed milk.

Market milk Milk produced and handled under conditions that qualify it for use as fluid milk in an organized marketing area. **Grade "A" milk** meets this requirement. See **fluid milk**.

Marketwide pool A pool in which each producer supplying a market receives the same **uniform price** for milk calculated on the basis of use of all milk received by **handlers** in the market.
Mastitis An inflammation of the mammary gland (or glands).
Mature equivalency (ME) Age-conversion formulas (provided by the USDA and dairy breed associations) applied to milk production records of young cows to predict their expected milk production potential as mature cows. Within-breed ME factors are used for comparative purposes in selecting and mating dairy animals.
Medium-moisture silage Silage that commonly contains 55 to 70% moisture.
Megacalorie (Mcal) Equivalent to 1000 kcal or 1,000,000 cal. A megacalorie is equivalent to a **therm**.
Meltdown In frozen desserts, the rate of melting as well as the attributes of the melted product. Normally, the melted frozen dessert should have the appearance of the mix before freezing.
Menstrum A solvent, commonly one that extracts certain principles from entire plant or animal tissues.
Merocrine A type of glandular secretion in which cell contents are excreted via exocytosis from secretory cells into an epithelial-walled duct or ducts and thence into the lumen.
Metabolic weight The weight of an animal raised to the three-quarter power ($W^{0.75}$).
Metabolizable energy (ME) Food-intake **gross energy** minus fecal energy, energy in the **gaseous products of digestion** (mostly methane), and urinary energy.
Metritis An inflammation of the uterus.
Micelles In aqueous solution, **molecules** having both polar or charged groups and nonpolar regions (amphiphilic molecules) form aggregates called micelles.
Microflora Microbial life characteristic of a region, such as the bacteria and protozoa populating the **rumen**.
Milk equivalent The quantity of milk, as produced, required to furnish the milk solids in manufactured dairy products. For example, approximately 10 and 20 lb of milk are required to manufacture 1 lb of cheddar cheese and butter, respectively.
Milk serum The nonfat components of milk.
Milk soil Residue on equipment, utensils, or containers after emptying.
Milk well The opening in the abdominal wall through which milk veins (**subcutaneous** abdominal veins) enter to join the vena cava and return blood from **udder** to heart.
Milkfat (MF) The **lipid** components of milk.
Milkfat differential The amount added to or subtracted from the **blend price** for each 0.1% that the **milkfat** is above or below 3.5%. For example, if the blend price is $9.00/cwt and the fat differential is 10 cents, a dairyperson with milk of 4.0% milkfat would receive $9.50/cwt.
Milkshed An area in which milk is produced for a given milk marketing area.
Milkstone Residue on equipment, utensils, or containers used for handling milk. It is characterized by a high mineral content, especially calcium. It is normally soluble in acid.
Miscible Describing two or more substances that, when mixed together, form a uniformly homogeneous solution.
Mitosis A type of cell division in which cells with the 2n (diploid) number of chromosomes produce daughter cells that also possess the 2n (diploid) number of chromosomes. Thus, the daughter cells receive the full complement of chromosomes existing in the original cell before division.
Moiety One of two approximately equal parts. A part or portion, especially a **molecule**, generally complex, having a characteristic chemical or pharmacological property.
Molecule A group of atoms held together by chemical forces. A molecule is the smallest unit of matter that can exist by itself and retain all its chemical properties.
Mount To climb onto, as demonstrated by females in **heat** and by males in **natural service**.
Mycobacterium tuberculosis The species of bacteria that causes tuberculosis. Transmissible from cattle to humans via milk.
Natural immunity Immunity to a disease or infestation that results from qualities inherent in an animal. See **immunity**.
Natural service In farm animals, it means to allow natural mating, as opposed to **artificial insemination**.
Neonatal Pertaining to a newborn animal.
Neonate A newborn infant.
Net energy (NE) The difference between **metabolizable energy** and **heat increment**. It includes the amount of energy used either for maintenance only or for maintenance plus production.
Net merit (NM$) The relative economic value of a dairy cow based on milk produced, fat and protein content, health traits, productive life, and fertility.
Neutral detergent fiber (NDF) Technique used to measure most of the structural components in plant cells (e.g., **lignin**, hemicelluloses, and **cellulose**), but not pectin.
Nicking See **heterosis**.
Nicotinamide adenine dinucleotide (NAD) A coenzyme in metabolism that serves as a hydrogen and electron transfer agent by virtue of reversible **oxidation** and reduction. Formerly called **diphosphopyridine nucleotide (DPN)**.
Nicotinamide adenine dinucleotide phosphate (NADP) A coenzyme in metabolism that serves as a hydrogen and electron transfer agent by virtue of reversible **oxidation** and reduction.
Nitrogen balance Nitrogen in the food consumed minus nitrogen in feces and nitrogen in urine (nitrogen retention).
Nitrogen-free extract (NFE) Consisting of carbohydrates, sugars, starches, and a major portion of materials classed as hemicellulose in feeds. When **crude protein**, fat, water, ash, and fiber are added and the sum is subtracted from 100, the difference is NFE.
Nonfat dry milk (NDM) The product obtained by removing water from pasteurized **skim milk**. It contains not more than 5% moisture and not more than 1.5% **milkfat** unless otherwise indicated.
Nonfat milk solids (NMS) Protein, **lactose**, and minerals of milk. Also called *nonfat solids of milk* (*NFS*) or **solids-not-fat (SNF)**.
Nonprotein nitrogen (NPN) For example, **urea**.
Normality Concentration or strength of a chemical on the normal scale. A normal solution contains one gram atom of replaceable hydrogen or its equivalent per liter.
Nucleotide Building block of nucleic acids, composed of a nitrogenous base, a five-carbon sugar (ribose or deoxyribose), and at least one phosphate group.
Nutrient broth A liquid nutritional medium formulated to support microbial growth.
Obligate parasite A parasite incapable of living without a host.
Off feed Having ceased eating with a healthy and normal appetite.

Official production record Standard **DHIA** and DHIR records pertaining to milk production made under the supervision of an unbiased individual. Such records are used for management purposes (e.g., feeding, breeding, and **culling**), genetic evaluation (**sire summaries**), publicity, and sales. See **DHIR**.
Oligosaccharides A polymer of simple sugars (monosaccharides), commonly 3 to 9.
Omasum The third division of the stomach of **ruminants**. Also called *manifold*, *many-plies*, and *psalterium*.
Open A term commonly used of female farm mammals to indicate a nonpregnant status. Also, a characteristic of cheeses that have an excessive amount of mechanical openings and/or gas holes.
Openness A quality in dairy cattle associated with length of body and width and length of rib.
Opsonization The process by which a **pathogen** is marked for ingestion and destruction by a **phagocyte**.
Osmophilic Loving a high osmotic pressure. For example, osmophilic yeasts may grow in **sweetened condensed milk**.
Osteopontin (OPN) OPN belongs to a family of secreted acidic proteins whose members have an abundance of negatively charged amino acids such as Asp and Glu. It plays an important role in bone remodeling.
Overrun The increase in quantity of product output over raw material input. In frozen desserts, overrun results from the incorporation of air, which increases the volume of product. In butter, overrun is calculated based on the input of fat and the output of butter. Components of butter other than fat are moisture, salt, and **curd**, and these may legally comprise no more than 20% of butter.
Owner-sampler (OS) An unofficial milk production record system, the records of which originate with the breeder (owner) and are summarized and supervised by the USDA and land-grant colleges of agriculture throughout the United States.
Oxidation Chemically, the increase of positive charges on an atom or the loss of negative charges. There may be a loss of one electron (univalent 0) or two electrons (divalent 0). It is also the combining of oxygen with another element to form one or more new substances. Burning is one kind of oxidation.
P_1 generation The parental generation.
Paracrine Chemical signals that diffuse into the area and interact with receptors on nearby cells.
Parakeratosis Any abnormality of the stratum corneum (horny layer of epidermis) of skin, especially a condition caused by **edema** between the cells, which prevents the formation of keratin.
Parenteral Administration by injection, not through the digestive tract (e.g., **subcutaneous**, intramuscular, intrasternal, intravenous).
Parity price A guaranteed level of farm prices intended to provide the same approximate purchasing power for farmers as enjoyed during a preceding base period. For example, the support price for **manufacturing grade milk** might be 90% of the price during a five-year period.
Parturient paresis A condition (observed especially in high-producing dairy cows) characterized by a low blood calcium concentration that results in partial to complete paralysis soon after **parturition**.
Parturition The act or process of giving birth to young.
Pasta filata Plastic curd. This pertains to Italian cheeses in which the warmed **curd** is stretched.
Pasteurization The process of heating milk to at least 62.8°C (145°F) and holding it at that temperature for not less than 30 min (holding method); see **high-temperature short-time pasteurization** and **ultrapasteurization**.
Pasture Plants, as grass, grown for feeding or grazing animals; also to feed cattle and other livestock on pasture.
Patent Open, unobstructed, or not closed. In milking, it is used to indicate weak, relaxed, or incompetent **sphincter** muscles.
Pathogen Any microorganism that produces disease (bacteria, viruses, yeasts, molds, and helminths).
Paunch The large, first stomach compartment of a **ruminant**. Also called **rumen**.
Pediculicide An agent that destroys lice.
Pedigree A record of an animal's ancestors, usually only those of the five closest generations. A complete pedigree may include milk production, **type classification**, and other information pertaining to dairy cattle.
Pellicle A thin membrane, film, or skin; a film on the surface of a liquid, such as the film that often appears on hot chocolate, which is coagulated **whey** protein of milk.
Pendant Something suspended. See **pendulous**.
Pendulous Suspended so as to swing freely, such as a loosely attached **udder**.
Peptide bond A bond joining two amino acids, in which the amino group of one acid is condensed with the carboxyl group of another, resulting in the elimination of water and formation of a –CO • NH-linkage. Amino acids of proteins are joined in this manner.
Permeate (filtrate) Fraction of milk or other liquid that passes through a semipermeable membrane.
Pernicious anemia A deficiency in the production of red blood cells resulting from a lack of vitamin B_{12}.
Persistent The ability of lactating animals to maintain milk production over a period of time.
Phagocytes From the Greek *phago*, meaning "eat," and *kylos*, meaning "cell." Defensive cells of the body (leukocytes, or white blood cells) that ingest and destroy bacteria and other infectious agents.
Phagocytosis The engulfing of microorganisms, cells, or foreign particles by **phagocytes**.
Phase inversion In butter manufacture, the reversal of the **emulsion** phase from fat-in-serum to a serum-in-fat. In the former, serum (**buttermilk**) is the continuous phase, whereas in the latter, free fat is the continuous phase.
Phenotype The expression of genes that can be measured by our senses—what we physically see of some trait in an animal. For inherited traits, such as color in Holsteins, we see either black and white or red and white.
Pica A craving for unnatural articles of food such as is observed in potassium-deficient animals; a depraved appetite.
Plain A term suggesting general inferiority; **coarse**; lacking the desired quality or **breed character**. Also refers to unflavored frozen dessert mixes.
Plasma Milk minus its fat; also called **skim milk** or *nonfat milk*.
Plasmid A genetic structure in a cell that can replicate independently of the chromosomes; typically a small circular DNA strand in the cytoplasm of a bacterium or protozoan.
Polled Describing a naturally hornless animal.
Polymerase chain reaction (PCR) A technology in molecular biology used to amplify a single copy or a few copies of a

piece of DNA across several orders of magnitude, generating thousands to millions of copies of a particular DNA sequence.

Polymorphic Having or occurring in several forms, as a substance crystallizing in different forms.

Polymorphonuclear leukocyte The mature neutrophil leukocyte, so-called because of its segmented and irregularly shaped nucleus.

Polyunsaturated A compound, often a fatty acid, that contains more than one double bond. Linoleic and linolenic fatty acids contain two and three double bonds, respectively, and are essential in the diet of humans.

Positive-type pump A mechanical device for moving a fluid at a constant rate under varying head pressures. It may be of piston, gear, or lobe design.

Post-leggedness A condition in which the hind legs are too straight, so that the springy quality of the hock and pastern is lost.

Postpartum Occurring after birth of the offspring.

Prebiotics Nondigestible carbohydrates that act as food for **probiotics**.

Predicted difference (PD) A measure of a **bull's** ability to transmit milk-producing capacity to his daughters. The PD may be positive or negative, depending on whether the bull's daughters yield more or less milk than daughters of other bulls (**herdmates**) under the same conditions.

Predicted transmitting ability (PTA) The average **genetic value** for a certain trait that an animal transmits to an offspring.

Prepartum Occurring before birth of the offspring.

Prepuce The fold of skin that covers the head of the penis.

Probiotics Live microorganisms that confer a beneficial health effect on the host when administered in sufficient amounts.

Producer settlement fund A fund maintained by the market administrator into which some processors pay money, because they use above average proportions of milk as Class I, and from which other processors receive money, because they use below average proportions of milk as Class I. This fund is used in conjunction with a **marketwide pool**. Also called *equalization fund*.

Profile A curve, graph, or other schema presenting quantitatively or descriptively the chief characteristics of something, as of an organ, process, or person.

Progeny testing Evaluating the **genotype** of an individual by a study of its offspring.

Prolactin A hormone released by the pituitary gland that stimulates mammary gland development and milk production.

Prophylactic A preventive, preservative, or precautionary measure that tends to ward off disease or an undesirable condition.

Prophylaxis The prevention of disease.

Protected designation of origin (PDO) A status granted a product if there is a link between its characteristics and the geographical *origin*. As used in Europe foods may qualify for either a protected geographical indication (PGI) or a PDO.

Protein equivalent A term indicating the total nitrogenous contribution of a substance in comparison with the nitrogen content of protein (usually plant). For example, the **nonprotein nitrogen** compound **urea** contains approximately 45% nitrogen and has a protein equivalent of 281% (6.25 × 45%).

Protein supplements Feed products that contain 20% or more of protein.

Proven sire A **bull** whose genetic transmitting ability has been measured by comparing the milk production performance of his daughters with that of the daughter's **dam** and/or **herdmates** under similar conditions. *Plus-proven* sires are those whose daughters exceeded their dams and/or herdmates in milk production whereas daughters of *minus-proven* sires have milk production below that of their dams and/or herdmates.

Psychrophilic The ability to grow best at refrigerated temperatures, that is, 4 to 15°C.

Psychrotrophic Cold-tolerant. Also called *psychrotolerant*.

Psychrotrophs Microorganisms that grow at low temperatures (39 to 59°F [4 to 15°C]), but may have an optimum temperature above this range. These organisms especially affect the **shelf life** of refrigerated dairy products such as cottage cheese.

Purebred The offspring of purebred parents of the same breed. Ancestors of purebreds can be traced to foundation stock in the original herdbook. Purebreds are not all **registered**.

Purulent Containing, consisting of, or being **pus**.

Pus A liquid product of inflammation consisting of leukocytes, lymph, bacteria, dead tissue cells, and fluid derived from their decomposition.

Putrefaction The microbial decomposition of proteins.

Pyometra An accumulation of **pus** in the uterus.

Rate of passage The time taken by undigested residues from a given meal to reach the feces. (A stained undigestible material is commonly used to estimate rate of passage.)

Ration The food allowed an animal for 24 hours. See **balanced ration**.

Raw milk Fresh, untreated milk as it comes from a cow, goat, or other mammal.

Receiving station The facility at which milk is collected from farm **bulk milk tanks**, stored, then shipped, usually in large semitrailer trucks, to another destination. Only cooling, **clarification**, and **standardization** of milk are likely to be performed here.

Recombined milk The product resulting from the mixing together and rehydration of multiple individual components of milk such as **milkfat**, **nonfat dry milk**, and water in proportions to yield the constituent percentage occurring in normal milk.

Reconstituted milk The product resulting from the mixing together and rehydration of a dried product of milk with water. For example, **nonfat dry milk** and water yield reconstituted **skim milk**.

Refractometer A laboratory or field device for the measurement of an index of refraction of light waves.

Registered **Purebred** animal registered in the herdbook of the proper breed association. Certain associations (for example, the Red & White Association) will register nonpurebred cattle providing they satisfy certain other criteria.

Regurgitate To cast up undigested food from the stomach to the mouth, as done by **ruminants**.

Rennet An extract of the stomach of certain mammals containing **rennin**. It is used in the coagulation of milk proteins in the manufacture of most cheeses.

Rennin The milk-coagulating enzyme found in the gastric juice of the fourth stomach compartment of certain mammals. Also called **chymosin**.

Repeatability The term used in dairy cattle breeding to indicate the precision with which **predicted differences** for milk production will estimate a **bull's** transmitting ability for milk production. Repeatability, then, is a confidence factor that increases with increases in the (1) number of daughters, (2) number of herds in which daughters are located, and (3) number of lactations per daughter.

Retained placenta Placental membranes not expelled at parturition.

Retentate (concentrate) The substance retained by a semipermeable membrane in the course of ultrafiltration.

Reticulum The second division or stomach compartment of a **ruminant**. Also called *honeycomb*.

Retort A vessel in which canned foods are exposed to heat in the form of steam under pressure for sterilization.

Reynolds number The value representing the degree of turbulence of flowing liquid. Reynolds numbers below 2,000 indicate **laminar** flow, whereas those above 3,000 indicate turbulent flow, numbers between being in the transitional flow range. Larger values mean more turbulent flow.

Ride To be transported in a **mounted** position, as one cow mounted on another during **estrus**.

Ring test A test for brucellosis performed by mixing stained *Brucella* bacteria with **raw milk**. If **antibodies** to *Brucella* are present, the stained cells agglutinate (clump) and rise to the surface with the cream to form a blue ring.

Roan A close mixture of red and white colors such as is often found in Milking Shorthorn cattle.

Rolling herd average The average milk production per herd per year based on the 12 months just past. Upon completion of a new test record, the record for the same period of the previous year is deducted and the new record is added, then a new rolling 365-day average is calculated. New herds have a rolling average computed after they have been on test at least 350 days.

Roughage Consists of **pasture**, **silage**, **hay**, or other dry **fodder**. It may be of high or low quality. Roughages are usually high in **crude fiber** (more than 18%) and relatively low in **nitrogen-free extract** (approximately 40%).

Rugged When referring to an animal, it means large and strong.

Rumen The first and largest stomach compartment of a **ruminant**; also called **paunch**.

Rumen microflora Microorganisms of the **rumen**.

Ruminant One of the order of animals having a stomach with four complete cavities (**rumen**, **reticulum**, **omasum**, and **abomasum**) through which food passes in digestion. These animals chew their **cud**. They include cattle, sheep, goats, deer, and camels.

Rumination The casting up or regurgitation of food (**cud**) to be chewed a second time, as cattle do; a chewing of the cud by **ruminants**.

Sanitize To kill or remove injurious microorganisms but not necessarily to **sterilize**. Dairy equipment is commonly sanitized with heat or chemicals.

Saponify To hydrolyze a fat with alkali to form a soap and glycerol.

Saturated fat A completely **hydrogenated** fat, that is, each carbon is associated with the maximum number of hydrogens.

Scours A persistent diarrhea in animals.

Scrub An animal from nonpurebred parents not showing the predominating characteristics of any breed.

Seeding Placing seed in an environment in which it can grow. In microbiology, placing microorganisms in fresh cultural media. In production of **lactose** or **sweetened condensed milk**, the addition of finely ground lactose to induce growth of lactose crystals.

Selection The causing or allowing of certain individuals in a population to produce the next generation. *Artificial selection* is that practiced by man; *natural selection* is that practiced by nature.

Septicemia The presence in blood of microorganisms and their associated poisons (commonly called *blood poisoning*). If the microorganisms are bacteria, the condition is *bacteremia*.

Sequester To combine with and hold. For example, phosphate salts combine with calcium and prevent its entry into certain reactions.

Serotype The type of microorganism as determined by the kind and combination of constituent **antigens** associated with the cell.

Serovars A group of closely related microorganisms distinguished by a characteristic set of **antigens**.

Service A term commonly used in animal breeding, denoting the mating of male to female. Also called *serving* or *covering*.

Set In cheese and **yogurt** making, the process of achieving coagulation of milk protein.

Setting temperature Temperature (°F or °C) at which **starter culture** is added to milk in a cheese vat to begin fermentation with acid production and **curd** formation.

Shelf life The time after processing during which a product remains suitable for human consumption, especially the time a food remains palatable.

Shrinkage A term used to indicate the amount of loss in body weight, as in dairy **steers**, when exposed to various conditions and/or slaughter. Also, the decrease in volume of ice cream and other dairy products during storage, and the loss of milk or milk solids in processing.

Sickle-hocked Describes an animal having a crooked hock, which causes the lower part of the leg to be bent forward out of a normal perpendicular straight line.

Silage Green **forage**, such as grass or clover, or **fodder**, such as field corn or sorghum, that is chopped and blown into a chamber or **silo** (preferably airtight), where it is packed or compressed to exclude air and undergo an acid fermentation (lactic and acetic acids) that retards spoiling. It usually contains 65 to 70% moisture. Also called **ensilage**.

Silo A vertical cylindrical structure, pit, trench, or other relatively airtight chamber in which chopped green crops, such as corn, grass, **legumes**, or small grain and other livestock feeds are fermented and stored.

Sire The male parent. The verb means to father or to beget.

Sire summary A summary of milk production records of daughters of **sires** to aid in the selection of the best genetic material available for breeding dairy cattle. Published by the USDA, NDHIA, and herd associations.

Skim milk Milk from which sufficient cream has been removed to reduce its **milkfat** content to less than 0.5% (usually less than 0.1%). Skim milk contains as much protein, **lactose**, minerals, and water-soluble vitamins and only half as many calories as whole milk. In the final beverage form, it has been **pasteurized** or **ultrapasteurized** and contains added vitamin A. Since the fat is separated off into cream, skim milk is practically **cholesterol** free.

Solids-not-fat (SNF) Proteins, **lactose**, minerals, and other water-soluble constituents of milk. This is the same as **nonfat solids** and **nonfat milk solids**.

Sparging A technique that involves bubbling a chemically inert gas, such as nitrogen, through a liquid.

Sphincter A ring-shaped muscle that closes an opening, such as the sphincter muscles in the lower end of a cow's teat.

Stale A period when an animal does not work or **lactate** at the normal standards, as opposed to **bloom**. Also refers to milk products that have deteriorated in storage.

Standard curve A line constructed between points that represent values determined in analyzing a reference substance. For example, in the dye-binding test for milk protein, values are often based on a standard curve constructed from data obtained in performing Kjeldahl tests on milks of varying protein content.

Standardization The process of adjusting the composition of a dairy product or ingredient thereof to a selected level.

Standby pool An arrangement between cooperatives in markets that have periods of milk shortages and cooperatives in areas of **surplus milk** production. Cooperatives in the deficit market make monthly payments to certain cooperatives in the surplus regions (particularly Wisconsin and Minnesota) to assure that they have sources of milk available at reasonable prices when local milk does not supply their Class I needs.

Starter culture See **culture**.

Statistically significant Of importance in a statistical sense. In research, it usually refers to tests for differences resulting from treatments. The reliability of such differences is expressed as degree of probability or the percentage of time an observation from a normal population would be expected to fall outside a certain range of variation from the mean.

Steer A male **bovine** castrated before development of secondary sex characteristics.

Sterilize To remove or kill all living organisms. Also, to make barren or unproductive, as a vasectomy in **bulls**.

Stillbirth Born lifeless; dead at birth.

Stover **Fodder**; mature cured stalks of grain from which seeds have been removed, such as stalks of corn without ears.

Strains Microorganisms within a species that differ slightly from others of the same species but not sufficiently to be classified separately.

Streptococcus cremoris A species of **Gram-positive** cocci that ferments **lactose**. They divide in one plane only, like rods. They are used extensively in manufacturing cultured dairy products and cheeses. *Streptococcus lactis* is a closely related species used for the same purpose.

Stromal cells Connective tissue cells of any organ (e.g., in the uterine mucosa).

Subclinical A disease condition without **clinical** manifestations.

Subcutaneous Situated or occurring beneath the skin.

Sublimation The direct transition from solid to vapor (bypassing the liquid form). Also, it is the process of vaporizing and condensing a solid substance without melting it.

Substrate A substance upon which cells or organisms may live and be nourished.

Supplement To add minerals, vitamins, or other minor ingredients (volumewise) to a **ration**.

Surplus milk The quantity of **grade "A" milk** in excess of that needed for Class I purposes.

Surprise tests These are tests for verification performed on cows whose projected 2X, 305D, ME records exceed certain levels of milk and/or fat as established by breed associations. Surprise tests include the entire herd. When possible, surprise tests are made by a supervisor other than the one who normally tests the herd.

Suspensoid An apparent solution that is seen, by a microscope, to consist of small particles of solid dispersed material in active Brownian movement.

Sweet butter Unsalted butter.

Sweetened condensed milk The food obtained by partially removing water from a mixture of milk and a safe and suitable nutritive sweetener. The finished food contains not less than 8 and 28% of **milkfat** and **total milk solids**, respectively. The quantity of nutritive sweetener used is sufficient to prevent spoilage. The food is **pasteurized** and may be **homogenized**.

Switch The brush of hair on the end of a **bovine's** tail.

Symbiosis A relationship between two living things that depend on each other when living together.

Synthesizing Utilizing various processes and/or operations necessary to build up a compound. Also, it is processing to build complex compounds from elements or simple compounds.

Systemic Of, pertaining to, or involving the body considered as a functional whole, as in systemic circulation.

Tanker load In milk collection and distribution, large quantities of milk in a large stainless steel tank borne on a transport truck—the method of transporting milk.

Teat meatus Small canal located in the end of each teat; also called streak canal. The teat meatus is approximately 1 cm long and is kept closed by **sphincter** muscles. It prevents the escape of milk between milkings and serves as a barrier to the entrance of microorganisms and foreign material; its size influences rate of milk withdrawal (fast-milking cows usually have a teat meatus of larger diameter than slow-milking ones).

Term The **gestation** period.

Tester One who tests milk.

Tetany A condition in an animal in which there are localized, spasmodic muscular contractions.

Therm The amount of heat required to raise the temperature of 1,000 kg of water 1°C, or 1,000 lb of water 4°F. One therm is 1,000 cal, or 1 Mcal.

Thermistor A small solid-state device that changes resistance markedly with changes in temperature, hence it is used to convert changes in temperature to changes in voltage or current.

Thermoneutral zone A state of thermal balance between an organism and its environment so that thermoregulatory mechanisms of the body are inactive. Also referred to as the *comfort zone.*

Thixotropic The property of certain gels, when mechanically agitated, to become fluid without a change in temperature. They reconvert to the gel state when allowed to stand.

Titer The quantity of a substance required to produce a reaction with a given volume of another substance, or the amount of one substance required to correspond with a given amount of another substance. Agglutination titer is the highest dilution of a serum that causes clumping of bacteria. See **titratable acidity**.

Titratable acidity The quantity of components of milk and milk products that neutralize a base expressed as percentage of lactic acid. The test for titratable acidity is used in determining milk quality and in monitoring the progress of fermentations of cultured milk products and cheeses. See **apparent acidity** and **developed acidity**.

Tocopherol A class of organic methylated phenols of which many have vitamin E activity.

Toe out To walk with the feet pointed outward. Also called *splay-footed.*

Total digestible nutrients (TDN) A standard evaluation of the nutritional merit of a particular feed for farm animals that includes all the digestible organic nutrients—protein, fiber, **nitrogen-free extract**, and **lipids**.

Total milk solids (TMS) Primarily **milkfat**, proteins, **lactose**, and minerals.

Toxins The poisons produced by certain microorganisms. They are products of cell metabolism. The symptoms of bacterial diseases, such as diphtheria, tetanus, botulism, and staphylococcal food poisoning, are caused by toxins.

Toxoid A detoxified **toxin**. It retains the ability to stimulate the formation of antitoxin in an animal's body. The discovery that toxin treated with formalin loses its toxicity is the basis for preventive immunization against such diseases as diphtheria and tetanus.

Trade barriers Rules and regulations that hamper the trade of commodities. The courts have ruled that some public health regulations (e.g., that milk must be pasteurized within the city it is purchased in) are unlawful barriers to trade.

Transmutator A device used to aid in changing the physical or chemical condition of a substance. In freezing ice cream and similar desserts, it is the device inside the freezing chamber that stirs and scrapes frozen product from the walls of the chamber. It is also called a *mutator.*

Transposon A chromosomal segment that can undergo transposition, especially a segment of bacterial DNA that can be translocated as a whole between chromosomal, phage, and **plasmid** DNA in the absence of a complementary sequence in the host DNA.

Trier An instrument for sampling butter and cheese. It is normally the shape of half a tube, sawed through the longitudinal plane, with its edges sharpened and being tapered toward one end, and a handle on the opposite end.

Tubular heaters Tubes through which milk is pumped to pick up heat from steam, hot water, or on other heating medium that surrounds the tubes. Designs are numerous.

Type The physical **conformation** of an animal.

Type classification A program sponsored by breed associations whereby a **registered** animal's **conformation** may be compared with the "ideal" or "true" type animal of that breed by an official inspector (classifier).

Udder The encased group of mammary glands provided with teats or nipples, as in a cow, ewe, mare, or sow. Also called **bag**.

Ultrahigh temperature (UHT) Temperatures of 185°F (85°C) and higher as applied to **pasteurization** or **sterilization** of milk in a relatively short time.

Ultrapasteurization The process of heating milk at **ultrahigh temperatures** for a sufficient time to kill all pathogenic microorganisms; e.g., 190°F (88°C) for 1 sec.

Uniform price The price paid producers for **market milk** when **classified pricing** is used in a **marketwide pool**. The uniform price is an average of class prices weighted by the quantity of milk in each class. See **blend price**.

Unthriftiness The quality or state of being **unthrifty** in animals.

Unthrifty Lack of vigor; poor growth or development.

Urea A nonprotein organic nitrogenous compound (NH_2CONH_2). It is made synthetically by combining ammonia and carbon dioxide.

Vacuumization In milk processing, the subjection of heated milk to vacuum with the purpose of removing volatile off flavors and odors and/or moisture. Sometimes called *vacreation*. See **deaeration**.

Vascular Consisting of, pertaining to, or provided with vessels (such as blood vessels).

Vasoconstrictor A compound or substance that causes constriction of blood vessels.

Vat pasteurization The heating of milk, while stirring in a vessel, to a temperature of at least 62.8°C for 30 minutes.

Vealer A calf fed for early slaughter (usually at less than 3 months of age in the United States). Most vealers are males of dairy breeds.

Vegetative In microbiology, the growing form as opposed to the resting form, the spore.

Viable Living; capable of living; likely to live.

Virulence The degree of pathogenicity (ability to produce disease) of a microorganism as indicated by case fatality rates and/or its ability to invade the tissues of a host.

Virulent Poisonous or harmful; deadly; of a microorganism, able to cause a disease by breaking down the protective mechanisms of a host. Also, it refers to fully active organisms.

Voids Empty or unfilled spaces; in spray-dried milks, the open space or spaces within the particles.

Volatile fatty acids (VFA) Commonly used in reference to acetic, propionic, and butyric acids produced in the **rumen** of cattle, and in goats and sheep. They are also produced by **starter culture** bacteria in cultured milk products.

Whey The water and solids of milk that remain after the **curd** is removed. It contains about 93.5% water and 6.5% **lactose**, protein, minerals, enzymes, water-soluble vitamins, and traces of fat.

Xerophthalmia A dry and thickened condition of the **conjunctiva**, sometimes following chronic conjunctivitis, a disease of the lacrimal apparatus, or vitamin A deficiency.

Yearling A male or female farm animal (especially cattle and horses) during the first year of life.

Yogurt Fermented milk, **lowfat milk**, or **skim milk**, sometimes protein-fortified. *Lactobacillus bulgaricus* and *Streptococcus thermophilus* are the fermenting bacteria. Fruit, flavors, and sugar may be added. Milk solids content is commonly 15%. Most yogurt is high in protein and low in calories.

Zoonotic A disease communicable from animals to humans under natural conditions.

Index